m.F.

3275952/2510A4

RMA/xMR

The Biochemistry of Plants

A COMPREHENSIVE TREATISE

Volume 15

P. K. Stumpf and E. E. Conn
EDITORS-IN-CHIEF
*Department of Biochemistry
and Biophysics
University of California
Davis, California*

Volume 1 **The Plant Cell** *N. E. Tolbert, Editor*
Volume 2 **Metabolism and Respiration** *David D. Davies, Editor*
Volume 3 **Carbohydrates: Structure and Function** *Jack Preiss, Editor*
Volume 4 **Lipids: Structure and Function** *P. K. Stumpf, Editor*
Volume 5 **Amino Acids and Derivatives** *B. J. Miflin, Editor*
Volume 6 **Proteins and Nucleic Acids** *Abraham Marcus, Editor*
Volume 7 **Secondary Plant Products** *E. E. Conn, Editor*
Volume 8 **Photosynthesis** *M. D. Hatch and N. K. Boardman, Editors*
Volume 9 **Lipids: Structure and Function** *P. K. Stumpf, Editor*
Volume 10 **Photosynthesis** *M. D. Hatch and N. K. Boardman, Editors*
Volume 11 **Biochemistry of Metabolism** *David D. Davies, Editor*
Volume 12 **Physiology of Metabolism** *David D. Davies, Editor*
Volume 13 **Methodology** *David D. Davies, Editor*
Volume 14 **Carbohydrates** *Jack Preiss, Editor*
Volume 15 **Molecular Biology** *Abraham Marcus, Editor*
Volume 16 **Intermediary Nitrogen Metabolism** *B. J. Miflin, Editor*

THE BIOCHEMISTRY OF PLANTS

A COMPREHENSIVE TREATISE

Volume 15
Molecular Biology

Abraham Marcus, editor
*Institute for Cancer Research
Fox Chase Cancer Center
Philadelphia, Pennsylvania*

ACADEMIC PRESS, INC.
Harcourt Brace Jovanovich, Publishers
San Diego   New York   Berkeley   Boston
London   Sydney   Tokyo   Toronto

COPYRIGHT © 1989 BY ACADEMIC PRESS, INC.
ALL RIGHTS RESERVED.
NO PART OF THIS PUBLICATION MAY BE REPRODUCED OR
TRANSMITTED IN ANY FORM OR BY ANY MEANS, ELECTRONIC
OR MECHANICAL, INCLUDING PHOTOCOPY, RECORDING, OR
ANY INFORMATION STORAGE AND RETRIEVAL SYSTEM, WITHOUT
PERMISSION IN WRITING FROM THE PUBLISHER.

ACADEMIC PRESS, INC.
San Diego, California 92101

United Kingdom Edition published by
ACADEMIC PRESS LIMITED
24-28 Oval Road, London NW1 7DX

Library of Congress Cataloging-in-Publication Data

(Revised for vol. 15)

The Biochemistry of plants.

  Vol. 15 edited by Abraham Marcus.
  Includes bibliographies and indexes.
  Contents: v. 1. The plant cell.—v. 2. Metabolism
and respiration.—[etc.]—v. 15. Molecular biology
  I. Botanical chemistry. I. Stumpf, Paul K.
(Paul Karl), Date. II. Conn, Eric E.
QK861.B48   581.19′2   88-13168
ISBN   0-12-675402-0 (v. 2)
ISBN   0-12-675415-2 (v. 15, alk. paper)

PRINTED IN THE UNITED STATES OF AMERICA
89 90 91 92   9 8 7 6 5 4 3 2 1

Contents

List of Contributors	xi
General Preface	xiii
Preface to Volume 15	xv

1 Regulation of Plant Gene Expression: General Principles
JACK K. OKAMURO AND ROBERT B. GOLDBERG

I.	Introduction	2
II.	The Plant Genome	4
III.	Measuring Gene Activity during Plant Development	16
IV.	Differential Gene Activity during Plant Development	28
V.	Regulation of Plant Gene Expression	39
VI.	The Role of cis-Acting Elements and trans-Acting Factors in Plant Gene Regulation	51
	References	72

2 Transposable Element Influence on Plant Gene Expression and Variation
LILA O. VODKIN

I.	History and Perspectives	83
II.	Molecular Isolation and Characterization of Transposable Element Families	90

III.	Transposable Element Action	112
IV.	Gene Tagging with Transposable Elements	123
	References	128

3 The Chloroplast Genome
MASAHIRO SUGIURA

I.	Introduction	133
II.	Chloroplast DNA	134
III.	Genes for the Genetic Apparatus	140
IV.	Genes for the Photosynthetic Apparatus	143
V.	Conclusions	146
	References	147

4 Chloroplast RNA: Transcription and Processing
WILHELM GRUISSEM

I.	Introduction	151
II.	The Chloroplast Transcription Apparatus	152
III.	RNA Processing	169
IV.	Transcriptional and Posttranscriptional Regulation of Plastid Gene Expression	179
	References	185

5 Protein Synthesis in Chloroplasts
ANDRÉ STEINMETZ AND JACQUES-HENRY WEIL

I.	Introduction	193
II.	*In Organello* and *in Vitro* Synthesis of Chloroplast Proteins	194
III.	Structure of Chloroplast Messenger RNAs	195
IV.	Translation	199
V.	Maturation of Proteins	217
VI.	Posttranscriptional Regulation of Chloroplast Gene Expression	218
VII.	Concluding Remarks	220
	References	221

6 The Plant Mitochondrial Genome
DAVID M. LONSDALE

I.	Introduction	230
II.	Physical Parameters of Mitochondrial DNA	230

III.	Mitochondrial Genome: Composition and Organization	233
IV.	Cells—Mitochondria—Mitochondrial DNA	250
V.	Promiscuous DNA	251
VI.	Genetic Complexity of Plant Mitochondrial Genomes	258
VII.	Mitochondrial Coding Sequences	260
VIII.	Gene Copy Number and Mapping	274
IX.	Transcription	277
X.	Ribosome Binding and Translation Initiation	281
XI.	Genomic Reorganization	283
XII.	Conclusions	286
	References	287

7 The Biochemistry and Molecular Biology of Seed Storage Proteins
MARK A. SHOTWELL AND BRIAN A. LARKINS

I.	Globulin Storage Proteins	298
II.	Synthesis and Deposition of Storage Globulins	304
III.	Organization and Structure of Storage Globulin Genes	311
IV.	Regulation of Globulin Gene Expression	313
V.	Prolamine Storage Proteins	316
VI.	Synthesis and Deposition of Cereal Prolamines	327
VII.	Organization and Structure of Prolamine Genes	330
VIII.	Regulation of Prolamine Gene Expression	334
IX.	Summary	337
	References	338

8 Stress-Induced Proteins: Characterization and the Regulation of Their Synthesis
TUAN-HUA DAVID HO AND MARTIN M. SACHS

I.	Introduction	347
II.	Temperature Stress	348
III.	Drought and Salt Stress-Induced Proteins	359
IV.	Anaerobic Stress	362
V.	Response to Ultraviolet Light Exposure	367
VI.	Heavy Metal—Induced Proteins and Peptides	368
VII.	Biological Stress	369
VIII.	Summary and Perspective	372
	References	374

9 The Thaumatins
H. VAN DER WEL AND A. M. LEDEBOER

I.	Introduction	379
II.	Isolation and Characterization of Thaumatins	380
III.	Biochemistry and Physiology	382
IV.	Molecular Genetics of the Thaumatins	383
V.	Study of the Natural Genes Encoding Thaumatin	384
VI.	Production of Thaumatin by Microorganisms	385
VII.	Expression of Thaumatin in Plants Other Than *Thaumatococcus daniellii*	389
VIII.	Conclusions	389
	References	390

10 Cytoskeletal Proteins and Their Genes in Higher Plants
DONALD E. FOSKET

I.	The Cytoskeleton—A Definition	394
II.	The Structure of Cytoskeletal Elements	395
III.	Dynamics of the Cytoskeleton in Plant Cells	396
IV.	Microtubule Proteins and Their Genes	399
V.	Actin and Other Microfilament Proteins	434
VI.	Intermediate Filament Proteins	441
	References	446

11 Calmodulin and Calcium-Binding Proteins
ELIZABETH ALLAN AND PETER K. HEPLER

I.	Introduction	455
II.	Calmodulin: Structure	456
III.	The Function of Calmodulin in Plant Cells	458
IV.	Concluding Remarks	477
	References	477

12 Plant Hydroxyproline-Rich Glycoproteins
ALLAN M. SHOWALTER AND JOSEPH E. VARNER

I.	Introduction	485
II.	Cell Wall Hydroxyproline-Rich Glycoproteins or "Extensins"	486
III.	Arabinogalactan Proteins	506
IV.	Solanaceous Lectins	513

Contents

    V. Summary and Insights into Future Plant Hydroxyproline-Rich Glycoprotein Research   516
         References   517

13 Protein Degradation
RICHARD D. VIERSTRA

    I. Introduction   521
    II. Functions of Protein Degradation   522
    III. Mechanisms for Degrading Proteins   525
    IV. Conclusions   533
      References   533

14 Viroids
T. O. DIENER AND R. A. OWENS

    I. Introduction   537
    II. The Biochemical Uniqueness of Viroids   538
    III. The Biochemical Significance of Viroids   539
    IV. Molecular Structure   539
    V. Viroid Function   544
    VI. Analysis of Structure/Function Relationships   552
    VII. Mechanisms of Pathogenicity   557
    VIII. Possible Viroid Origins   559
      References   559

15 Biochemistry of DNA Plant Viruses
ROBERT J. SHEPHERD

    I. Introduction   563
    II. Caulimoviruses (Double-Stranded DNA Viruses)   566
    III. Geminiviruses (Single-Stranded DNA Viruses)   593
    IV. Prospects for Using DNA Viruses as Gene Vectors   606
      References   610

16 Tumor Formation in Plants
A. POWELL AND M. P. GORDON

    I. Introduction   617
    II. Crown Gall Tumors   618
    III. Virus-Induced Tumors of Plants   636
    IV. Habituated Plant Tissues and Genetic Tumors   638
    V. Transfer of Genetic Information in the Biosphere   641
      References   643

17 Genetic Manipulation of Plant Cells
ANTHONY J. CONNER AND CAROLE P. MEREDITH

I.	Introduction	653
II.	Cell Selection	654
III.	Protoplast Fusion	664
IV.	Transformation	671
V.	Concluding Discussion	680
	References	682

Index   689

Contents of Other Volumes   699

List of Contributors

Numbers in parentheses indicate the pages on which the authors' contributions begin.

Elizabeth Allan (455), Department of Cell Biology, John Innes Institute, Norwich NR4 7UH, England

Anthony J. Conner (653), Crop Research Division, Department of Scientific and Industrial Research, Christchurch, New Zealand

T. O. Diener[1] (537), Microbiology and Plant Pathology Laboratory, Beltsville Agricultural Research Center, Beltsville, Maryland 20705

Donald E. Fosket (393), Department of Developmental and Cell Biology, University of California, Irvine, Irvine, California 92717

Robert B. Goldberg (1), Department of Biology, University of California, Los Angeles, Los Angeles, California 90024

M. P. Gordon (617), Department of Biochemistry, University of Washington, Seattle, Washington 98195

Wilhelm Gruissem (151), Department of Botany, University of California, Berkeley, Berkeley, California 94720

Peter K. Hepler (455), Department of Botany, University of Massachusetts, Amherst, Massachusetts 01003

Tuan-Hua David Ho (347), Department of Biology, Washington University, St. Louis, Missouri 63130

[1] Current address: Center for Agricultural Biotechnology and Department of Botany, University of Maryland, College Park, Maryland, and Agricultural Research Service, U.S. Department of Agriculture, Beltsville, Maryland 20705

Brian A. Larkins (297), Department of Botany and Plant Pathology, Purdue University, West Lafayette, Indiana 47907

A. M. Ledeboer (379), Unilever Research Laboratorium, Vlaardingen, The Netherlands

David M. Lonsdale (229), Department of Molecular Genetics, Institute of Plant Science Research, Cambridge Laboratory, Trumpington, Cambridge CB2 2JB, England

Carole P. Meredith (653), Department of Viticulture and Enology, University of California, Davis, Davis, California 95616

Jack K. Okamuro (1), Laboratorium Genetika, Rijksuniversiteit Gent, B-9000 Gent, Belgium

R. A. Owens (537), Microbiology and Plant Pathology Laboratory, Beltsville Agricultural Research Center, Beltsville, Maryland 20705

A. Powell (617), Hirzhrunnenschanze 37, 4058 Basle, Switzerland

Martin M. Sachs (347), Department of Biology, Washington University, St. Louis, Missouri 63130

Robert J. Shepherd (563), Department of Plant Pathology, University of Kentucky, Lexington, Kentucky 40506

Mark A. Shotwell (297), Department of Botany and Plant Pathology, Purdue University, West Lafayette, Indiana 47907

Allan M. Showalter (485), Department of Botany, Ohio University, Athens, Ohio 45701

André Steinmetz (193), Institut de Biologie Moléculaire et Cellulaire du C.N.R.S., Université Louis Pasteur, Strasbourg, France

Masahiro Sugiura (133), Center for Gene Research, Nagoya University, Chikusa, Nagoya 464, Japan

Joseph E. Varner (485), Department of Biology, Washington University, St. Louis, Missouri 63130

Richard D. Vierstra (521), Department of Horticulture, University of Wisconsin–Madison, Madison, Wisconsin 53706

Lila O. Vodkin (83), Department of Agronomy, University of Illinois, Urbana, Illinois 61801

Jacques-Henry Weil (193), Institut de Biologie Moléculaire et Cellulaire du C.N.R.S., Université Louis Pasteur, Strasbourg, France

H. van der Wel (379), Unilever Research Laboratorium, Vlaardingen, The Netherlands

---

[2] Current address: Friedrich Miescher Institute, 4002 Basle, Switzerland

General Preface

In 1950, a new book entitled "Plant Biochemistry" was authored by James Bonner and published by Academic Press. It contained 490 pages, and much of the information described therein referred to animal or bacterial systems. This book had two subsequent editions, in 1965 and 1976.

In 1980, our eight-volume series entitled "The Biochemistry of Plants: A Comprehensive Treatise" was published by Academic Press; this multivolume, multiauthored treatise contained 4670 pages.

Since 1980, the subject of plant biochemistry has expanded into a vigorous discipline that penetrates all aspects of agricultural research. Recently a large number of research-oriented companies have been formed to explore and exploit the discipline of plant biochemistry, and older established chemical companies have also become heavily involved in plant-oriented research. With this in mind, Academic Press and the editors-in-chief of the treatise felt it imperative to update these volumes. Rather than have each chapter completely rewritten, it was decided to employ the approach used so successfully by the editors of *Methods in Enzymology*, in which contributors are invited to update those areas of research that are most rapidly expanding. In this way, the 1980 treatise constitutes a set of eight volumes with much background information, while the new volumes both update subjects that are rapidly developing and discuss some wholly new areas. The editors-in-chief have therefore invited the editors of the 1980 volumes to proceed on the basis of this concept. As a result, new volumes are forthcoming on lipids; general metabolism, including respiration; carbohydrates; amino acids; molecular biology; and photosynthesis. Additional volumes will be added as the need arises.

Once again we thank our editorial colleagues for accepting the important task of selecting authors to update chapters for their volumes and bringing their volumes promptly to completion. And once again we thank Mrs. Billie Gabriel and Academic Press for their assistance in this project.

P. K. Stumpf
E. E. Conn

Preface to Volume 15

Volume 6 of this treatise, published in 1981, presented information pertinent to gene expression available at the time. While it was already clear that much regulation of gene expression occurs at the nucleic acid level, we wished to emphasize that a molecular understanding of biological systems requires knowledge of both the structure and function of specific proteins and titled the volume "Proteins and Nucleic Acids." In the interim, much progress has been made in describing specific gene systems and defining regulatory regions within the genes. Recognizing these developments, we have titled the current volume "Molecular Biology." Still, wherever significant information on proteins is available, we have given this special emphasis—two such notable areas being the cytoskeletal proteins and the hydroxyproline-rich glycoproteins. Knowledge of organelle genes and gene products has advanced rapidly in the past decade, and major aspects of these areas are reviewed in four chapters.

Another important mechanism for regulating gene expression that has recently become the focus of attention is the selective turnover of gene products, and a chapter is included describing current insights into protein degradation. The extracellular carriers of genetic information remain the best analyzed biological entities, and three chapters reviewing viral and plasmid systems relevant to plants are included. Finally, with the advent of transformation as a viable and practical technology, we include a chapter describing the scope of its current application and some thoughts of future potential.

I want to express personal gratitude to the authors of the different chapters for their cooperation and patience. Reading their manuscripts has been an illuminating and insightful experience.

Abraham Marcus

Regulation of Plant Gene Expression: General Principles

JACK K. OKAMURO
ROBERT B. GOLDBERG

I. Introduction
    A. Unique Processes of Plant Development
    B. Central Questions of Plant Gene Expression
II. The Plant Genome
    A. DNA Reassociation Kinetics
    B. Diversity of Genome Size and Genome Complexity in Higher Plants
    C. Representation of DNA Sequences in a Plant mRNA Population
    D. Constancy of the Plant Genome during Development
    E. Summary
III. Measuring Gene Activity during Plant Development
    A. RNA/DNA Hybridization Kinetics
    B. Plant mRNA Diversity versus Abundance: mRNA/cDNA Hybridization Kinetics
    C. RNA/Single-Copy DNA Saturation Hybridization
    D. Summary
IV. Differential Gene Activity during Plant Development
    A. Gene Activity in the Sporophyte
    B. Gene Activity in the Male Gametophyte
    C. Total Number of Genes Expressed during Plant Development
    D. Summary
V. Regulation of Plant Gene Expression
    A. Nuclear versus Cytoplasmic RNA Complexity
    B. Transcriptional and Posttranscriptional Regulation of Plant Gene Expression
    C. Transcriptional and Posttranscriptional Regulation of Specific Plant Genes
    D. Cytoplasmic Regulation of Plant Gene Expression: mRNA Stability and Turnover
    E. Summary
VI. The Role of cis-Acting Elements and trans-Acting Factors in Plant Gene Regulation
    A. Light-Regulated Gene Expression
    B. Plant Heat Shock Gene Expression
    C. Seed Protein Gene Regulation
    D. Combinatorial Model for Seed Protein Gene Regulation

E. Networks of cis-Elements and trans-Factors Coordinate the Expression of Structural Gene Sets in Eukaryotic Organisms
F. Identifying Plant trans-Acting Regulatory Factors
G. Molecular Analysis of Plant trans-Acting Regulatory Factors
H. Summary
References

I. INTRODUCTION

A. Unique Processes of Plant Development

Higher plants undergo many unique developmental processes. First, as shown in Fig. 1, plants have an alternation of diploid spore-producing and haploid gamete-producing generations. The sporophytic generation, or dominant phase of the plant life cycle, begins at fertilization and ends with the production of haploid microspore and megaspore cells which are formed in distinct floral organs. The spores divide mitotically and differentiate into the three-celled male gametophyte or pollen grain, and the seven-celled female gametophyte or embryo sac. These give rise to the sperm and egg, respectively. The cells of the male and female gametophyte are functionally and morphologically distinct from one another and arise as a consequence of differential expression of the gametophytic genome.

Second, plant development occurs in the absence of cell movement due to the presence of a semirigid cell wall. Despite this limitation, plants generate

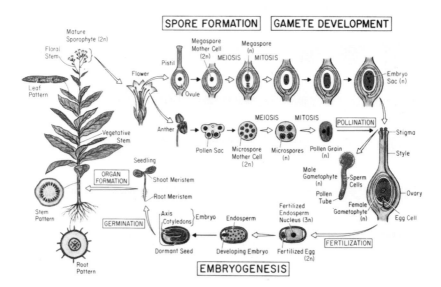

Fig. 1. Life cycle of the flowering plant.

complex organ systems containing more than 50 distinguishable cell and tissue types by the processes of asymmetric cell division and cell division in different planes. By contrast, differentiation of many animal cell and tissue types such as germ cells, the nervous system, white blood cells, and the muscle cells of the limb, does not occur without cell migration.

Third, plants undergo a continuous program of differentiation and development. Unlike animals, that complete their major morphogenetic events during embryogeny, plants contain meristematic cells that are established during embryogenesis and are capable of producing the organ systems of the sporophyte throughout their life cycle.

Fourth, environmental factors have a major influence on plant development and can induce dormancy, stimulate periods of rapid growth, or trigger a switch from vegetative to reproductive states. Light, temperature, and water conditions influence plant ontogeny by interacting with cellular factors that control differentiation and development. How environmental cues are translated into specific regulatory signals that affect plant gene expression is largely unknown.

Finally, unlike animal cells, many differentiated plant cells remain totipotent and can be developmentally reprogrammed to divide and regenerate into a mature, fertile plant (Conger *et al.*, 1981). The cellular and molecular processes required to induce a differentiated plant cell to enter regenerative development are not known. However, the totipotency of plant cells provides the best evidence that the plant genome remains informationally constant during development and that differential gene activity is the fundamental process responsible for plant cell differentiation.

B. Central Questions of Plant Gene Expression

No plant system develops in precisely the same way. Nevertheless, to illustrate basic principles of differential gene expression and gene regulation in higher plants, we review in this chapter studies from a diverse collection of plant systems that examine the changing patterns of gene expression and different levels of gene regulation used by the plant cell. To understand how the expression of the genetic program stored within each plant cell is regulated we focus on four questions central to understanding plant gene expression and gene regulation at the molecular level.

1. What Is the Genetic Potential of the Plant Genome?

Higher plant genomes vary enormously in size and complexity and contain a much higher proportion of repeated DNA sequences than most animal genomes. Because of these unique characteristics, it is important to ask what fraction of the plant genome is expressed during the life cycle and what role, if any, do repeated DNA sequences play in the regulation of structural gene sets during plant development?

2. What Role Does Differential Gene Expression Play in Plant Development?

In addition to reviewing the expression and regulation of many cloned plant genes, we summarize studies that characterize plant gene expression at the RNA population level using RNA/DNA hybridization techniques. These studies illustrate the complexity of the plant developmental program and provide an important overview of the dramatically changing patterns of gene activity and gene regulation during sporophytic development.

3. How Does the Plant Cell Regulate Gene Expression?

Genetic and molecular studies in both animal and plant systems suggest that gene expression in the eukaryotic cell is governed by a multilayered network of *cis*-acting control elements and *trans*-acting regulatory factors that coordinate gene expression at both the transcriptional and posttranscriptional levels. The development of plant cell transformation technology has given plant biologists the ability to identify *in vivo* the *cis*-elements that regulate plant gene expression. We review in this chapter progress made in characterizing control elements that determine developmental-specific patterns of plant gene transcription.

4. How Are the Processes of Cell Differentiation and Development Regulated in Plants?

Conceptually, there is a quantum jump from understanding how a single gene is regulated in a plant cell to understanding how complex developmental processes such as embryogenesis and organ formation are programmed and executed during the plant life cycle. To dissect these complex processes at the molecular level requires identifying genes that control the expression of ontological pathways. We conclude this chapter by discussing genetic and biochemical strategies for identifying and cloning plant regulatory genes.

II. THE PLANT GENOME

What is the genetic potential of the plant genome and how much of that potential is utilized during plant development? Some of the most important advances in our understanding of plant development and gene regulation at the molecular level were derived from genome studies using DNA/DNA reassociation analyses. The kinetic analysis of repetitive and nonrepetitive DNA sequences as well as transcribed and nontranscribed regions of the plant genome provided important clues to the molecular complexity of plant gene expression and development. Moreover, these investigations were essential for the development of molecular tools for analyzing differential gene activity during plant development. Therefore, for historical as well as conceptual reasons, we review the principles of DNA reassociation analysis, define terms used throughout this chapter to discuss the plant genome and gene regulation, and highlight the contributions of DNA/DNA reassociation analyses to our understanding of the plant genome. For a more comprehensive treatment of DNA reassociation theory see Britten *et al.* (1974) and Davidson (1976).

A. DNA Reassociation Kinetics

DNA reassociation is a second-order reaction whereby complementary single-stranded DNA molecules reassociate to form double-stranded duplexes according to Eq. (1), the fundamental equation of DNA reassociation kinetics.

$$\frac{C}{C_0} = \frac{1}{1 + kC_0 t} \quad (1)$$

In this expression, t is the time of reaction in seconds, C_0 represents the initial concentration of single-stranded DNA (moles of nucleotide per liter), C is equal to the concentration of single-stranded DNA sequences at t_c, and k is the second-order rate constant. The fraction of single-stranded DNA (C/C_0) at time (t) can be calculated using Eq. (1), if the second-order rate constant and the initial concentration of single-stranded DNA are known. Hence a "Cot curve," as defined by this expression, describes the reassociation of single-stranded DNA into double-stranded structures over time.

Using Eq. (1), the second-order rate constant k can be determined experimentally from the $C_0 t$ value at which 50% of the single-stranded DNA has reassociated or the $C_0 t_{1/2}$ as shown in Eq. (2).

$$0.5 = \frac{C}{C_0} = \frac{1}{1 + kC_0 t_{1/2}} \quad (2)$$

Solving for k illustrates how these terms are inversely related [Eq. (3)].

$$k = 1/C_0 t_{1/2} \quad (3)$$

Because a reassociation reaction is a collision-dependent reaction, the rate constant k is a function of DNA complexity or sequence diversity. *Genome complexity* is defined as the total number of nucleotide pairs in a genome expressed as nonrepeating sequences. As genome complexity increases, the concentration of each sequence in the reaction decreases, reducing the probability that complementary strands of DNA will find each other and reassociate. Thus, k is inversely proportional to DNA complexity as shown in Eq. (4). This is the fundamental rule of DNA reassociation analysis.

$$k \propto \frac{1}{\text{Complexity}} \quad (4)$$

1. Genome Size versus Genome Complexity

Because the rate constant of a reassociation reaction is inversely proportional to DNA complexity [Eq. (4)], the product of $k \times C$ must be constant under standard reassociation conditions of salt, temperature, and DNA fragment length. Therefore, the complexity of an unknown genome can be determined experimentally from its reassociation rate constant and the complexity and rate constant of a DNA standard as shown in Eq. (5).

$$k^{\text{unknown}} \times C^{\text{unknown}} = k^{\text{standard}} \times C^{\text{standard}} \tag{5}$$

The second-order rate constant can also be used to determine genome size. *Genome size* is defined as the total amount of DNA per haploid nucleus. In cases where DNA sequences are represented equally in a genome, as in *Escherichia coli*, genome complexity C is equal to genome size G. However, because higher plant and animal genomes contain a diverse spectrum of repetitive DNA sequences that contribute disproportionately to genome size with respect to their complexity (reviewed by Walbot and Goldberg, 1979; Murray and Thompson, 1981), genome complexity is always less than genome size. For example, a hypothetical organism with the genome dAGAT:dTCTA has a genome size and genome complexity of 4 nucleotide pairs (NTP). However, an organism with the genome $d(GC)_n : d(CG)_n$ has a genome complexity of 2 NTP but a genome size of $2n$ NTP. Thus, genome complexity reflects sequence diversity rather than size.

2. Complex Reassociation Kinetics

Most plant and animal genomes contain a significant fraction of repetitive DNA sequences. In fact, repetitive DNA sequences constitute 35–75% of most higher plant genomes (reviewed by Walbot and Goldberg, 1979; Murray and Thompson, 1981). The reassociation kinetics of genomic DNA containing repetitive DNA sequences can be expressed as the sum of a series of second-order reaction components each behaving according to Eq. (1). The reassociation of genomic DNA is represented by Eq. (6), where n is the total number of DNA sequence components, α_j is the fraction of the genome each component represents, k_j is the observed rate constant, and β is the fraction of DNA remaining unreassociated at the end of the reaction.

$$\frac{C}{C_0} = \beta + \frac{\alpha_1}{1 + k_1 C_0 t} + \frac{\alpha_2}{1 + k_2 C_0 t} + \frac{\alpha_3}{1 + k_3 C_0 t} + \cdots \frac{\alpha_n}{1 + k_n C_0 t} \tag{6}$$

In practice, however, it is seldom possible to resolve a DNA reassociation reaction into more than three or four kinetic components. Therefore, each component is a numerical average and each rate constant k_j reflects the average reiteration frequency of the DNA sequences contained in that component.

Because the mathematical solution for a plant DNA reassociation reaction contains two or more second-order components, it is necessary to calculate

Regulation of Plant Gene Expression

two forms of the rate constant, k^{whole} and k^{pure}, to determine genome complexity and genome size kinetically. k^{whole} is the rate constant for a reaction component in the presence of total nuclear or whole DNA, that is, with all other sequence components present. k^{pure} is the rate constant of a purified DNA component isolated from whole DNA by reassociation and hydroxyapatite fractionation (Galau et al., 1974). Because the purified component is free of other DNA sequences, it is enriched for its own sequences. Therefore, k^{pure} is always greater than k^{whole}. The relationship between k^{pure} and k^{whole} is described in Eq. (7) where F is the fraction of whole DNA represented by a specific kinetic component.

$$k^{pure} = k^{whole}/F \tag{7}$$

The complexity of a repetitive or nonrepetitive DNA component can be obtained from k^{pure} using a known standard as described in Eq. (5). In theory, each repetitive sequence in the genome contributes to the total genome complexity. In practice, however, the complexity of the nonrepetitive or single-copy DNA component determines genome complexity.

3. Determining Genome Size

The genome size G of any eukaryotic organism can be calculated from the single-copy DNA complexity C^{SC} and the fraction of single-copy DNA F^{SC} in whole DNA [Eq. (8)].

$$G = C^{SC}/F^{SC} \tag{8}$$

C^{SC} is obtained from $k^{pure\ SC}$ or $C_0 t_{1/2}^{pure\ SC}$ relative to a DNA standard according to Eq. (5).

Alternatively, genome size can be obtained directly from the $k^{whole\ SC}$ as shown in Eq. (9) because $k^{whole\ SC} \times F^{SC} = k^{pure\ SC}$.

$$G = \frac{C^{standard}}{k^{whole\ SC}} = k^{standard} \tag{9}$$

Therefore, $k^{whole\ SC}$ is used to measure genome size [Eq. (9)] and $k^{pure\ SC}$ is used to determine genome complexity [Eq. (5)].

4. Determining Repeated Sequence Copy Number

The average reiteration frequency R of each component in genomic DNA can be obtained using Eq. (10), where G represents genome size, and C is the complexity of the specific reaction component.

$$R = GF/C \tag{10}$$

Alternatively, this relationship can be stated in terms of the rate constants [Eq. (11)].

$$R = k^{whole\ repetitive\ component}/k^{whole\ SC} \tag{11}$$

This simply states that the rate of repetitive sequence reassociation is directly proportional to repeated sequence copy number.

B. Diversity of Genome Size and Genome Complexity in Higher Plants

Given this theoretical background, we can examine the relationship between the nuclear genome, gene expression, and gene regulation during plant ontogeny. Is there a typical plant genome?

As shown in Fig. 2, plant genomes range in size from 7×10^4 kilobase pairs (kb) in the crucifer *Arabidopsis thaliana* (Leutwiler *et al.*, 1984) to 4×10^8 kb in the lily *Fritillaria assyriaca* (Bennett and Smith, 1976). Most plant genomes investigated, however, range from 10^5 to 10^7 kb in size (Table I; Bennett and Smith, 1976). By comparison, the genome size of *Drosophila melanogaster* is 1.5×10^5 kb (Manning *et al.*, 1975) and that of man is 2.0×10^6 kb (Schmid and Deininger, 1975). Thus, despite the fact that plants have fewer organs and specialized cell types than most animals, their genomes are as large or larger than the genomes of many complex animals, including man. Is this observation significant in terms of structural gene expression?

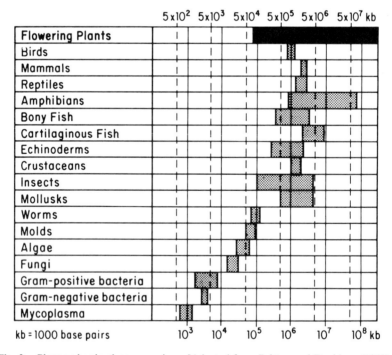

Fig. 2. Plant and animal genome sizes. [Adapted from Britten and Davidson (1969).]

TABLE I

Plant and Animal Genome Size and Genome Complexity

Species	Genome size[a] (kb)	Genome complexity[b] (kb)	Genome complexity relative to Arabidopsis[c]	Reference
Arabidopsis thaliana	7.0×10^4	5.5×10^4	—	Leutwiler et al. (1984)
Cotton (Gossypium hirsutum)	7.2×10^5	5.1×10^5	9.3	Walbot and Dure (1976)
Flax (Linum usitatissimum)	1.5×10^5	6.8×10^4	1.2	Cullis (1981)
Maize (Zea mays)	5.7×10^6	2.3×10^6	42	Hake and Walbot (1980)
Mung bean (Vigna radiata)	4.7×10^5	2.6×10^5	4.6	Murray et al. (1979)
Parsley (Petroselinum sativum)	3.8×10^6	1.3×10^6	23	Kiper and Herzfeld (1978)
Pea (Pisum sativum)	4.5×10^6	1.3×10^6	25	Murray et al. (1978)
Pearl millet (Pennisetum americanum)	3.8×10^5	1.0×10^5	1.8	Wimpee and Rawson (1979)
Soybean (Glycine max)	1.3×10^6	6.9×10^5	13	Gurley et al. (1979)
	1.8×10^6	7.3×10^5	13	Goldberg (1978)
Tobacco (Nicotiana tabacum)	1.5×10^6	6.4×10^5	12	Zimmerman and Goldberg (1977)
	2.4×10^6	1.0×10^6	18	Okamuro and Goldberg (1985)
Wheat (Triticum aestivum)	5.2×10^6	6.2×10^5	11	Smith and Flavell (1975)
Man (Homo sapiens)	2.0×10^6	1.0×10^6	18	Schmid and Deininger (1975)
Mouse (Mus musculus)	1.6×10^6	9.1×10^5	16	Cech and Hearst (1976)
Fruit fly (Drosophila melanogaster)	1.5×10^5	1.1×10^5	2	Manning et al. (1975)
Nematode worm (Caenorhabditis elegans)	8.0×10^4	7.0×10^4	1.2	Sulston and Brenner (1974)
Water mold (Achyla bisexualis)	4.2×10^4	3.4×10^4	0.62	Hudspeth et al. (1977)
Escherichia coli	4.2×10^3	4.2×10^3	0.076	Britten and Kohne (1968)

[a] Determined by DNA reassociation kinetics using $k^{\text{whole sc}}$ according to Eq. (9).
[b] Determined by DNA reassociation kinetics using $k^{\text{pure sc}}$ according to Eq. (5).
[c] The ratio of the genome complexity of each species to the genome complexity of Arabidopsis thaliana.

The genome of *Arabidopsis thaliana* (7×10^4 kb), which is only 17-fold larger than the *E. coli* genome, suggests that a complete set of structural genes sufficient for carrying out plant development can be packaged into a genome no larger than that of the nematode *Caenorhabditis elegans* (see Table I). By contrast, the tobacco genome (1.5×10^6 kb) is 20-fold larger than that of *Arabidopsis*. One explanation for the compact size of the *Arabidopsis* genome is that unlike most higher plants, the *Arabidopsis* genome does not contain major repeated sequence components other than rDNA sequences (Leutwiler *et al.*, 1984; Pruitt and Meyerowitz, 1986). Evidence for dramatic differences in repetitive DNA content has been noted for diverse plant species (Flavell *et al.*, 1974; Thompson *et al.*, 1979; Flavell *et al.*, 1979; Bedbrook *et al.*, 1980; Walbot and Cullis, 1985), suggesting that repetitive DNA sequences that do not encode structural genes are not essential to all plant cells. Changes in repetitive DNA content alone, however, cannot account for the extreme range of genome sizes found in higher plants.

Table I, which also compares plant genome complexities, shows that even when repeated DNA sequences are "subtracted" from the soybean, pea, tobacco, cotton, and maize genomes, there remains a 10- to 40-fold difference in genome *complexity* between these plant species and that of *Arabidopsis*. In some cases, the differences in genome complexity can be attributable to ancient or "fossilized" repetitive sequences as seen in pea (Thompson *et al.*, 1980; Murray *et al.*, 1981). Allopolyploidization can also precipitate an evolutionarily "sudden" increase in genome complexity as seen in tobacco (Okamuro and Goldberg, 1985). In addition, variable intron size can cause differences in genomic complexity. For example, the *Arabidopsis* EPSP gene (5-enolpyruvateshikimate-3-phosphate synthase) is approximately one third as long as the corresponding petunia gene due to differences in intron sizes (Klee *et al.*, 1987). Still the underlying basis and functional significance, if any, of plant complexity differences are not known.

Bennett (1972) suggested that genome size may influence plant development directly by affecting the duration of mitosis and meiosis and therefore a species generation time. Grime *et al.* (1985) extended this proposal by suggesting that nuclear DNA content also influences the timing of cell division and cell growth by seasons. Clearly, from both biochemical and molecular analyses there is no such thing as a typical plant genome. Rather, plant genomes are unusually heterogeneous in size and evolutionarily dynamic. There is a need to determine whether the dramatic differences in plant genome size and genome complexity are also reflected in the number and diversity of structural genes expressed during development in plants like tobacco and *Arabidopsis thaliana*. Later in this chapter we compare structural gene numbers from several plant species.

C. Representation of DNA Sequences in a Plant mRNA Population

Do most plant genes belong to the repetitive or the single-copy fraction of the genome? The reiteration frequency of genes represented in a population of diverse mRNAs from a plant cell or organ can be determined kinetically using DNA-excess DNA/cDNA hybridization analysis (Bishop *et al.*, 1974). In this approach, a trace amount of labeled DNA is hybridized with total genomic DNA. Because the amount of labeled DNA is negligible (>20,000:1 mass excess), there is no discernible effect of the cDNA on the kinetics of genomic DNA reassociation. According to Eq. (11), the rate of labeled cDNA hybridization is proportional to the number of genes encoding the mRNA.

Using this kinetic approach, a number of studies showed that genes that encode most of the mRNA mass as well as the mRNA diversity are represented once per haploid genome, or belong to low-copy-number gene families (Goldberg *et al.*, 1978; Cashmore, 1979; Siflow *et al.*, 1979; Galau and Dure, 1981; Goldberg *et al.*, 1981a; Murray *et al.*, 1981). For example, Fig. 3

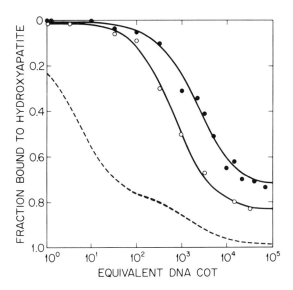

Fig. 3. The majority of the plant mRNA mass is transcribed from unique genes or low-copy-number gene families. Trace amounts of tobacco leaf [^{3}H]cDNA (○) or tobacco ^{3}H-single-copy DNA (●) were hybridized with greater than 20,000:1 mass excess of leaf nuclear DNA (Goldberg *et al.*, 1978). The second-order rate constants for the [^{3}H]cDNA and ^{3}H-single-copy DNA reactions were 1.5×10^{-3} and 4.8×10^{-4} M^{-1} sec^{-1}, respectively, indicating that the average reiteration frequency of genes that code for the majority of the leaf mRNA mass is roughly 3. The dashed curve represents the reassociation of unlabeled leaf nuclear DNA (Zimmerman and Goldberg, 1977).

represents the hybridization of tobacco leaf cDNA with an excess of nuclear DNA. By comparing the hybridization rate of labeled leaf cDNA to the reassociation rate of labeled single-copy DNA, it is apparent that the genes that encode greater than 95% of the leaf mRNA mass are unique or present only a few times (<5 copies) per genome.

To determine whether these conclusions would hold for individual cloned genes, Bernatzky and Tanksley (1986) analyzed genetically the number of independent loci in the tomato genome that corresponded to 34 randomly chosen leaf cDNA clones. They found that 53% of the clones mapped to a single locus containing 1–2 copies of each structural gene. An additional 32% of the clones mapped to only 2 loci per haploid genome. The remaining 15% of the clones corresponded to small multigene families containing 3–5 loci per haploid genome. Together, these analyses indicate that the majority of plant structural genes are present in 1–2 copies per haploid genome.

Not all plant genes are unique or belong to low-copy-number gene families. Genes that encode highly abundant mRNA species are often represented many times in the genome. Table II shows that seed protein genes, genes for the small subunit of ribulose-bisphosphate carboxylase, and genes that encode the light harvesting chlorophyll *a/b* binding protein belong to multigene families that range from 3–50 copies per haploid genome. Interestingly, in *Arabidopsis thaliana* the size of a multigene family, the number of introns in a gene, and the size of the introns are generally smaller than in plants with much larger genomes (Table II), suggesting that selective pressures have forced this plant to economize genetically (Chang and Meyerowitz, 1986).

Gene redundancy also provides the cell with the opportunity to evolve new patterns of gene regulation and new physiological functions (Perutz *et al.*, 1960). Qualitative as well as quantitative differences in gene expression have been described for members of several plant multigene families, including glutamine synthetase (Gebhardt *et al.*, 1986; Tingey *et al.*, 1987), the small subunit of ribulose-bisphosphate carboxylase (Coruzzi *et al.*, 1984; Dean *et al.*, 1985b; Fluhr *et al.*, 1986a), sucrose synthase (Choury and Nelson, 1976), and actin (Hightower and Meagher, 1985). The physiological significance of these intrafamily differences in gene expression, if any, is not known.

D. Constancy of the Plant Genome during Development

Plant cell totipotency provides direct evidence that cells with distinct differentiated states have identical DNA sequence contents; that is, all cells of a plant contain the same genome. As shown in Fig. 4, DNA reassociation analysis of tobacco floral and stem DNAs confirms this principle, although much less rigorously because small differences in DNA sequence content would not be revealed. To a first approximation, the genomes of stem cells

TABLE II
Plant Gene Copy Number

Gene/plant species	Gene copy number	Gene size (kb)	mRNA size (kb)	Ratio of gene size to mRNA size	Reference
Light-regulated genes					
LHCP					
Arabidopsis thaliana	3–4	0.9	0.9^a	1	Leutwiler et al. (1986)
Petunia	16	1.0	1.0^a	1	Dunsmuir et al. (1983)
Pea	8	1.0	1.0^a	1	Coruzzi et al. (1983); Cashmore (1984)
Duckweed	10–12	$1.0–1.1^b$	1.0^a	1.0–1.1	Tobin et al. (1984); Karlin-Neumann et al. (1985)
Wheat	≥7	1.1	1.1	1	Lamppa et al. (1985a)
RbcS					
Arabidopsis thaliana	4	1.0	0.8	1.3	E. Krebbers (unpublished)
Petunia	8	1.2–2.0	0.9	1.3–2.2	Dean et al. (1985a)
Pea	≥5	0.9	0.7^a	1.2	Coruzzi et al. (1984)
Duckweed	13	1.0	0.9	1.1	Wimpee et al. (1983)
Wheat	≥10	1.1	0.8	1.3	Broglie et al. (1983)
Soybean	≥10	1.4	0.9^a	1.5	Berry-Lowe et al. (1982)
Seed storage protein genes					
Arabidopsis thaliana					
12S globulin	3	2.2–2.6	1.7	1.3–1.5	P. Pang and E. Meyerowitz (unpublished)
2S arabin	4	0.7	0.7^a	1	E. Krebbers (unpublished)
Rape					
12S cruciferin	≥5	ND^c	1.7	—	M. L. Crouch (unpublished)
2S napin	10–16	0.7	0.7^a	1	Scofield and Crouch (1987); Josefsson et al. (1987)
Pea					
11S legumin	8	2.0	1.7^a	1.2	Lycett et al. (1984); Domoney and Casey (1985)
7S vicilin	11	ND^c	1.7	—	Domoney and Casey (1985)

(continued)

TABLE II (*continued*)

Gene/plant species	Gene copy number	Gene size (kb)	mRNA size (kb)	Ratio of gene size to mRNA size	Reference
Soybean					
11S glycinin	≥6	2.7	1.7[a]	1.6	Fischer and Goldberg (1982); R. L. Fischer, G. N. Drews, and R. B. Goldberg (unpublished)
7S conglycinin	≥15	2.2,2.7	1.5[a],2.1[a]	1.5,1.3	Doyle et al. (1986); Barker et al. (1988); J. J. Harada and R. B. Goldberg (unpublished)
Maize					
22-kDa zein	≥24	0.9	0.9	1	Hu et al. (1982); Wilson and Larkins (1984)
19-kDa zein	≥50	0.9	0.9	1	Pedersen et al. (1982)
15-kDa zein	2–3	0.7	0.7	1	Pedersen et al. (1986)
Alcohol dehydrogenase					
Arabidopsis thaliana	1	2.0	1.4[a]	1.4	Change and Meyerowitz (1986)
Maize	2	3.4	1.6[a]	2.1	Dennis et al. (1984)
Pea	≥3	2.3	1.4[a]	1.7	Llewellyn et al. (1987)
Histone H3					
Arabidopsis thaliana	5–7	0.5	0.5	1	Chaboute et al. (1987)
Maize	60–80	0.5	0.5	1	Chaubet et al. (1986)
Histone H4					
Arabidopsis thaliana	5–7	0.4	0.4	1	Chaboute et al. (1987)
Maize	100–120	ND[c]	ND[c]	—	Chaubet et al. (1986)
Wheat	100–125	0.4	0.4	1	Tabata et al. (1983)

[a] Does not include mRNA poly(A) tail.
[b] Not all members of the Lemna gibba LHCP gene family contain an intron.
[c] ND, Not determined.

primer and viral reverse transcriptase. Purified cDNA is then mixed with a large excess of homologous RNA and allowed to hybridize. The formation of RNA/cDNA duplexes over time is generally monitored by S1 nuclease resistance (Young and Anderson, 1986).

The analysis of a complex RNA/cDNA hybridization reaction is similar to the resolution of a DNA Cot curve into individual kinetic components. Theoretically, a cDNA hybridization reaction can be described as the sum of n independent pseudo-first-order reactions as shown in Eq. (18).

$$R/R_0 = \beta + \alpha_1 e^{-k_1 R_0 t} + \alpha_2 e^{-k_2 R_0 t} \cdots + a_n e^{-k_n R_0 t} \tag{18}$$

In this expression, n is the number of different RNA abundance components in the population, β represents the fraction of nonreactable cDNA molecules, α_j denotes the fraction of the RNA mass represented by each transcript, and k_j represents the observed rate constant for each component in the hybridization reaction. In practice, however, it is rarely possible to resolve a cDNA hybridization reaction into more than four pseudo-first-order reaction components. Therefore, each component represents a set of transcripts found at a similar abundance level or concentration in the cell. Generally, the first sequences to hybridize in a RNA/cDNA hybridization represent the most abundant but least complex set of transcripts in an RNA population. The last sequences to hybridize represent the least abundant but most diverse set of transcripts.

1. Analyzing Plant Gene Expression Using mRNA/cDNA Hybridization Kinetics

A typical plant RNA/cDNA hybridization reaction can be resolved into several pseudo-first-order hybridization components by computer analysis (Pearson *et al.*, 1977). The term k^{pure} represents the rate constant of a pure hybridization component, and is used to calculate sequence complexity. The value of k^{pure} is always greater than k^{whole} because it represents the rate at which a set of RNAs would hybridize with their complementary cDNA sequences if they were separated from other RNA sequences. This relationship is described in Eq. (19), where F_R is the fraction of the cDNA mass represented by the hybridizing RNA.

$$k^{\text{pure}} = k^{\text{whole}}/F_R \tag{19}$$

The complexity of a specific RNA component C_R is derived from k^{pure} using Eq. (16). The number of diverse RNAs or structural gene transcripts represented in a kinetic component N is computed using C_R according to Eq. (17).

2. Messenger RNA Abundance Classes

Table III illustrates the complex nature of plant gene expression throughout development as analyzed by RNA/cDNA hybridization kinetics. In gen-

TABLE III

Abundance Distribution of Plant mRNA Sequence Sets

Organ/tissue	Organism	Prevalence class[a]	Percent mRNA mass	Number of diverse mRNAs	Percent mRNA mass per sequence[b]	Number of molecules per cell per sequence	Reference
Leaf	Soybean	1	8	2	4	27,000	Goldberg et al. (1981)
		2	17	45	0.40	2,000	
		3	35	900	0.04	200	
		4	40	35,000	0.0011	5	
	Tobacco	1	9	10	0.9	4,500	Goldberg et al. (1978)
		2	52	770	0.07	340	
		3	39	11,300	0.0035	17	
Shoot	Barley	1	15	12	1.2	—	Heinze et al. (1980)
		2	28	2,000	0.014	—	
		3	57	31,300	0.0018	—	
	Pea	1	42	210	0.20	1,300	de Vries et al. (1983)
		2	38	3,150	0.012	79	
		3	19	12,500	0.0015	9	
Root	Soybean	1	19	50	0.40	1,200	Auger et al. (1979)
		2	36	2,000	0.02	60	
		3	45	18,000	0.0025	8	
	Pea	1	35	120	0.29	1,170	de Vries et al. (1983)
		2	50	4,150	0.012	48	
		3	15	11,900	0.0012	5	
Root tip	Vicia faba	1	37	26	1.4	2,500	Buffard et al. (1982)
		2	34	610	0.56	100	
		3	29	11,700	0.0025	4	

Tissue	Species	Component					Reference
Root nodule	Soybean	1	24	1	24	110,000	Auger et al. (1979)
		2	25	510	0.05	230	
		3	51	22,000	0.0022	10	
Embryo 30 DAF	Soybean	1	38	180	0.20	800	Goldberg et al. (1981)
		2	62	14,000	0.0044	17	
Embryo 75 DAF	Soybean	1[c]	20	<1	20	150,000	Goldberg et al. (1981)
		2	31	6	5.2	19,000	
		3	27	180	0.15	550	
		4	22	32,000	6.9×10^{-4}	3	
Embryo cotyledons 50 mg	Cotton	1[c]	29	<1	29	35,000	Galau and Dure (1981)
		2	18	4	4.5	5,400	
		3	11	28	0.40	470	
		4	15	700	0.02	26	
		5	27	13,000	0.002	3	
Pollen	*Tradescantia paludosa*	1	15	44	0.34	26,000	Willing and Mascarenhas (1984)
		2	61	1,400	0.043	3,400	
		3	24	18,000	0.0013	100	
Suspension culture	Soybean	1	18	60	0.30	—	Silflow et al. (1979)
		2	44	1,900	0.02	—	
		3	38	30,000	0.0013	—	

[a] Minimum number of reaction components required to describe the RNA/cDNA hybridization reaction based on computer analysis of the reaction kinetics.

[b] The total number of mRNA molecules per cell ranges from approximately 120,000 in 50 mg cotton embryo cotyledons (Galau and Dure, 1981) to 610,000 in soybean root nodules (Auger et al., 1979).

[c] Probably not a prevalence class but due to homologies between related mRNAs (for example, storage protein mRNAs; Goldberg et al., 1981; Galau et al., 1981).

TABLE IV

Messenger RNA Abundance Classes

I. Superabundant
  A. May represent 15–90% of the mRNA mass
  B. Negligible contribution to mRNA diversity (<10 structural gene transcripts)
  C. >5000 molecules per cell per sequence
  D. Present in HIGHLY SPECIALIZED CELL TYPES
  E. Polypeptide products CAN be visualized by gel electrophoresis
II. Abundant
  A. Usually 50–75% of the mRNA mass
  B. <5% of the mRNA diversity (~200–1000 structural gene transcripts)
  C. 500–2500 molecules per cell per sequence
  D. Present in most cell types
  E. Polypeptide products can be visualized by gel electrophoresis
III. Rare or Complex
  A. Usually <25% of the total mRNA mass
  B. ≥95% of the mRNA diversity (i.e., most structural genes code for rare class mRNAs)
  C. 1–10 molecules per cell per sequence
  D. Individual mRNAs comprise <0.01% of the mRNA mass
  E. Products CANNOT be visualized by gel electrophoresis
  F. Present in most cell types

eral, a plant mRNA/cDNA hybridization reaction can be resolved into three to four kinetic components representing as many as 30,000 diverse mRNAs. By convention, these mRNA sequences are grouped into three abundance classifications according to their cytoplasmic concentration and sequence diversity: superabundant, abundant, and rare or complex class mRNAs. These terms conceptually simplify the complex architecture of a mRNA population. In reality, a typical eukaryotic mRNA population contains an almost continuous spectrum of mRNA species with unique cellular concentrations (Flytzanis *et al.*, 1982). The general characteristics of each mRNA abundance classification are outlined in Table IV.

a. Superabundant Class Transcripts. Superabundant mRNAs are unique to highly specialized cell types that must produce large amounts of specific polypeptides. Genes encoding superabundant mRNAs in plants include the small subunit of ribulose-bisphosphate carboxylase (Cashmore, 1979), seed storage proteins (Goldberg *et al.*, 1981b), and leghemoglobin (Auger *et al.*, 1979). The corresponding mRNAs are represented by 5000 molecules per cell or more, and are so prevalent they can be visualized as distinct bands on an ethidium bromide (EtBr)-stained agarose gel. Figure 6 shows, however, that superabundant mRNAs represent less than 1% of the structural gene set expressed in a plant cell.

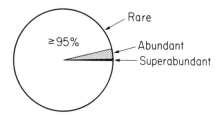

Fig. 6. The majority of genes expressed in a plant cell produce rare class mRNAs.

b. Abundant Class Transcripts. The second most prevalent class of mRNA sequences in the cell are called abundant class mRNAs, and are found in virtually all cell types investigated. They constitute 50–75% of the mRNA mass, but together with superabundant transcripts still represent less than 5% of the total mRNA sequence diversity. On the average, there are 200–2000 different abundant class gene sequences in the cell represented by approximately 500–2500 mRNA molecules per cell (Table III). Generally, the translation products of abundant and superabundant class transcripts can be seen on a two-dimensional protein gel.

c. Rare Class Transcripts. The last sequences to hybridize in an RNA/cDNA hybridization reaction represent rare class transcripts or the lowest detectable level of gene expression in the cell. Unlike superabundant and abundant class mRNAs, rare class transcripts cannot be visualized as bands on an EtBr-stained agarose gel or be detected by radioactively labeling *in vitro* translation products on two-dimensional acrylamide gels (Dure *et al.*, 1981). A rare class mRNA sequence is represented by only 1–10 molecules per cell or approximately 0.001–0.010% of the mRNA mass. Collectively, however, these transcripts constitute 20–60% of the mRNA mass and greater than 95% of the mRNA diversity in the plant cell, as shown in Fig. 6. Clearly, one of the important conclusions illustrated by the RNA/cDNA hybridization studies summarized in Table III is that the majority of structural genes expressed during plant development encode rare class mRNA transcripts.

Theoretically, there are two classes of rare class transcripts that cannot be distinguished in an RNA/DNA hybridization reaction: (1) transcripts represented in all cells at a low concentration and (2) transcripts expressed in a restricted population of specialized cells at a much higher level. Genes encoding rare class mRNAs demonstrating this second pattern of regulation have been identified by *in situ* hybridization in several plant organs (K. Cox, G. N. Drews, and R. B. Goldberg, unpublished).

What cellular functions do rare class mRNA products fulfill? In animal cells, a small number of genes encoding rare class mRNAs have been identified, including α-interferon, dihydrofolate reductase, and thymidine kinase (Mantei and Weissman, 1982; Ragg and Weissman, 1983; Leys and Kellems, 1981; Leys et al., 1984; Merrill et al., 1984). At present, few plant genes encoding rare class mRNA transcripts have been studied (Fischer and Goldberg, 1982; Okamuro et al., 1986; R. B. Goldberg et al., unpublished). RNA/DNA hybridization experiments described in the next section of this chapter reveal that approximately 70% of the plant genes encoding rare class mRNAs are developmentally regulated during sporophyte ontogeny, suggesting that despite their low levels of expression, rare class transcripts may play an important role in plant cell differentiation and organogenesis.

3. Limitations of cDNA Hybridization Analysis

The determination of mRNA sequence complexity by cDNA hybridization kinetics is subject to several possible sources of error. First, the kinetic determination of cDNA hybridization rate under the best conditions is generally subject to a twofold margin of error (Pearson et al., 1977; Axel et al., 1976; Hereford and Rosbash, 1977; Galau et al., 1977b; Goldberg and Timberlake, 1980). Factors that retard hybridization rate such as RNA degradation, changes in solution viscosity, and disparity in RNA and DNA lengths (Van Ness and Hahn, 1982; Chamberlin et al., 1978) can contribute to errors in RNA complexity, particularly at the end of a reaction when rare class sequences are hybridizing.

Second, not all eukaryotic structural gene transcripts have poly(A) tails. For example, in *Drosophila melanogaster* approximately 70% of the mRNA diversity is poly(A)$^-$ (Zimmerman et al., 1980). In mouse brain, 1.1×10^5 kb of the mRNA sequence set expressed after birth is poly(A)$^-$ (Chaudhari and Hahn, 1983). In plants, only 35–50% of the cytoplasmic mRNA is polyadenylated (Key and Silflow, 1975; Gray and Cashmore, 1976; Ragg et al., 1977; Goldberg et al., 1978; Galau et al., 1981). In cotton embryo, Galau et al. (1981) showed that poly(A)$^+$ and poly(A)$^-$ mRNA populations contain the same RNA sequences as determined by cDNA hybridization analysis; that is, each mRNA species is represented by polyadenylated and nonadenylated transcripts. A similar conclusion was reached for tobacco leaf mRNA (Goldberg et al., 1978). By contrast, the data of Silflow et al. (1979) suggest that only 70% of the total cell RNA complexity in soybean suspension culture cells is represented in the total cell poly(A)$^+$ RNA population. Therefore, it is possible that not all plant cytoplasmic polysomal poly(A)$^+$ and poly(A)$^-$ mRNA populations are identical with respect to sequence representation.

C. RNA/Single-Copy DNA Saturation Hybridization

The most reliable method for determining RNA sequence diversity is with single-copy DNA saturation hybridization (Davidson and Hough, 1971; Galau et al., 1974). This approach is preferred to cDNA hybridization kinetics because the measurement of RNA complexity is *rate independent* and therefore not affected by factors that influence hybridization rate. Because single-copy DNA sequences are equimolar, RNA complexity is determined by the fraction of labeled single-copy DNA hybridized with RNA (Fig. 5B). RNA complexity is calculated from the single-copy DNA saturation value according to Eq. (20).

$$C_R = C^{SC} \times \text{fraction single-copy DNA hybridized} \times 2 \quad (20)$$

The factor of 2 corrects for asymmetric transcription. The value of C^{SC} is obtained from the reassociation kinetics of purified single-copy DNA as described in Eq. (5).

A second advantage of this approach over the kinetic method for estimating RNA complexity is that the fraction of single-copy DNA hybridized can be determined very precisely. In tobacco, for example, the margin of error can be as little as ±0.03% of the single-copy DNA complexity or approximately 300 genes (Kamalay and Goldberg, 1980). Moreover, since this approach does not rely on the presence of a mRNA poly(A) tail, the RNA transcripts in total polysomal mRNA, polysomal (A)$^+$ mRNA, and nuclear RNA can be compared.

A hypothetical RNA/single-copy DNA hybridization reaction is shown in Fig. 5B. Because rare class RNAs constitute greater than 95% of the cytoplasmic RNA diversity (Table III), most of the hybridized single-copy DNA is complementary to this RNA abundance class. In fact, if the reaction kinetics are plotted on a log scale as shown for the cDNA reaction in Fig. 5A, only one kinetic component is apparent, the hybridization of rare class transcripts. Of course, superabundant and abundant class RNA sequences may be present in the reaction but, because of their low complexity, they represent only a small fraction of the single-copy DNA hybridized and cannot be observed kinetically.

Analyzing Plant Gene Expression Using RNA/Single-Copy DNA Hybridization

Table V shows that single-copy DNA hybridization experiments demonstrate that plant gene activity during development and in each plant organ is strikingly complex. For example, the leaf, stem, and root express approximately 10,000–30,000 structural genes in all plant species examined. Similar studies showed that plant gene activity during embryogenesis and in the reproductive organs of the flower is equally complex (Table V). The different organs and tissues of a plant express similar numbers of structural genes, regardless of their specialized structure or function. In tobacco, for example, leaf, stem, root, ovary, anther, and petal express 25,000–28,000 diverse structural genes (Kamalay and Goldberg, 1980). This high level of gene activity over the course of sporophyte development contrasts sharply with

TABLE V

Sequence Complexity of Plant mRNA Populations

Organ system and/or developmental state	Organism	Type of RNA[a]	Complexity[b] (kb)	Average mRNA size[c] (kb)	Number of average-sized mRNAs[d]	Percent single-copy DNA[e]	Method[f]	Reference
Leaf	Tobacco	A	3.3×10^4	1.2	28,000	5.2	scDNA	Goldberg et al. (1978)
	Soybean	A	4.0×10^4	1.1	36,000	10.0	cDNA	Goldberg et al. (1981)
	Parsley	B	1.4×10^4	1.4	10,000	6.2	scDNA	Kiper et al. (1979)
Shoot	Pea	C	2.5×10^4	1.2	19,000	7.8	scDNA	de Vries et al. (1983)
	Pea	C	2.0×10^4	1.2	15,000	6.3	cDNA	de Vries et al. (1983)
	Tradescantia paludosa	D	3.4×10^4	1.2	29,000	—	cDNA	Willing and Mascarenhas (1984)
Stem	Tobacco	A	3.2×10^4	1.2	27,000	5.0	scDNA	Kamalay and Goldberg (1980)
Root	Tobacco	A	3.0×10^4	1.2	25,000	4.7	scDNA	Kamalay and Goldberg (1980)
	Soybean	C	2.4×10^4	1.2	20,000	6.2	cDNA	Auger et al. (1979)
	Pea	C	2.4×10^4	1.2	20,000	7.6	scDNA	de Vries et al. (1983)
	Pea	C	2.0×10^4	1.2	16,000	6.4	cDNA	de Vries et al. (1983)
Root tip	Broad bean	C	1.7×10^4	1.3	13,000	3.7	scDNA	Buffard et al. (1982)
	Broad bean	C	1.6×10^4	1.3	12,000	3.2	cDNA	Buffard et al. (1982)
Root nodule	Soybean	C	2.8×10^4	1.2	23,000	7.2	cDNA	Auger et al. (1979)
Root callus	Parsley	B	2.0×10^4	1.4	14,000	8.6	scDNA	Kiper et al. (1979)
	Parsley	C	2.0×10^4	1.4	14,000	8.6	scDNA	Kiper et al. (1979)

Organ	Species	Type of RNA[a]	Complexity[b]	Avg mRNA size (kb)[d]	Number of mRNAs[d]	Method[f]	Reference
Anther	Tobacco	A	3.2×10^4	1.2	27,000	scDNA	Kamalay and Goldberg (1980)
Ovary	Tobacco	A	3.1×10^4	1.2	26,000	scDNA	Kamalay and Goldberg (1980)
Petal	Tobacco	A	3.3×10^4	1.2	28,000	scDNA	Kamalay and Goldberg (1980)
Embryo—75 DAF[g]	Soybean	A	2.0×10^4	1.4	14,000	scDNA	Goldberg et al. (1981)
Embryo axis—75 DAF	Soybean	A	2.6×10^4	1.4	18,000	scDNA	Goldberg et al. (1981)
Embryo—75 DAF cotyledon	Soybean	A	2.1×10^4	1.4	15,000	scDNA	Goldberg et al. (1981)
Embryo—50 mg	Cotton	D	2.1×10^4	1.5	14,000	cDNA	Galau and Dure (1981)
Endosperm	Wheat	C	1.1×10^4	0.8	13,000	—	Pernollet and Vaillant (1984)
Pollen grain	*Tradescantia*	D	2.3×10^4	1.2	19,000	cDNA	Willing and Mascarenhas (1984)

[a] Type of RNA: A, EDTA-released polysomal poly(A); B, total polysomal; C, polysomal poly(A); D, total cell poly(A).

[b] For experiments using single-copy DNA saturation hybridization, RNA complexity was calculated directly from the saturation hybridization plateau. The complexity of RNA analyzed by the cDNA hybridization kinetics method represents the sum of the complexity of each kinetic component. Recently, the sizes of four rare leaf mRNAs complementary to cloned genes were measured (Fischer and Goldberg, 1982; Okamuro et al., 1986). These mRNAs averaged 1.1 kb in length, a value similar to that of the total leaf mRNA population (Goldberg et al., 1978).

[c] Since average mRNA lengths were calculated from RNA mass measurements, they may be biased toward prevalent mRNA classes.

[d] Quotient of mRNA complexity and average mRNA size.

[e] Actual saturation hybridization plateau value obtained with the single-copy DNA method. Calculated for the cDNA method by the relationship: (mRNA complexity/single-copy DNA complexity) × 100. The single-copy DNA complexities used were barley, 1.5×10^6 kb (Flavell et al., 1974); cotton, 4.4×10^5 kb (Walbot and Dure, 1976); parsley, 2.3×10^5 kb (Kiper et al., 1979); soybean, 3.9×10^5 kb (Goldberg, 1978); tobacco, 6.4×10^5 kb (Zimmerman and Goldberg, 1977; Goldberg et al., 1978); pea, 4.5×10^6 kb (Sivolap and Bonner, 1971; Murray et al., 1978).

[f] cDNA refers to the use of RNA/cDNA hybridization kinetics to analyze gene expression. scDNA refers to the single-copy DNA saturation hybridization method of analysis.

[g] DAF, Days after flowering.

the progressive decline in structural gene activity observed during sea urchin development (Galau *et al.*, 1976), but resembles the complex programs of gene expression found in the organs of higher animals. In fact, with the exception of brain (Chaudhari and Hahn, 1983), the number of genes expressed in each tobacco organ is comparable to the number of genes expressed in animal organs such as the liver (24,000; Savage *et al.*, 1978), kidney (12,000; Hastie and Bishop, 1976), and oviduct (16,000; Axel *et al.*, 1976). Single-copy DNA hybridization experiments alone, however, do not reveal what fraction of the structural gene set expressed in each plant organ is developmentally regulated. That is, does each organ system express a developmental-specific set of abundant and rare class mRNAs? This question is addressed by comparing the sequences in each RNA population.

D. Summary

Gene activity in plant cells is represented by a spectrum of mRNA transcripts that can be classified into three general classes according to their cellular prevalence. Greater than 95% of the structural genes expressed during plant development encode rare class transcripts. The function of these transcripts remains largely unknown. By contrast, a small proportion of the mRNA diversity but a large proportion of the mRNA mass is represented by abundant and superabundant mRNA sequences.

RNA/cDNA hybridization kinetics and RNA/single-copy DNA saturation hybridization are complementary tools for characterizing gene activity at the RNA population level. RNA/cDNA hybridization kinetics is used to analyze the abundance distribution and sequence complexity of an RNA population. Because RNA/single-copy DNA hybridization is rate independent, however, RNA complexity is most accurately determined by single-copy DNA hybridization.

IV. DIFFERENTIAL GENE ACTIVITY DURING PLANT DEVELOPMENT

What role does differential gene expression play in establishing and maintaining the differentiated state of plant cells? Studies of plant gene expression at the protein level (Dure *et al.*, 1981; Sung and Okimoto, 1981), and studies using cloned gene hybridization probes, demonstrate that the expression of unique sets of developmentally regulated genes is associated with each stage of plant development. However, we know from RNA/DNA hybridization experiments that these studies explore only a small fraction of the structural gene set expressed in plant cells. To obtain a complete overview of the changing pattern of gene activity during plant development, it is necessary to analyze the developmental regulation of rare class mRNA transcripts as well as abundant and superabundant mRNAs using RNA/single-copy DNA hybridization. The RNA/DNA hybridization studies summarized below provide insight into the developmental regulation of structural gene sets in plants. Together with the analysis of cloned plant genes, these studies reveal a level of informational and regulatory complexity in plants comparable to that of higher animals.

A. Gene Activity in the Sporophyte

As pointed out earlier, plants are relatively simple organisms morphologically. The mature sporophyte has just three vegetative organs (leaf, stem, root) and three reproductive organ systems (petal, pistil, stamen). The RNA/DNA hybridization studies summarized in Tables III and V clearly show that most cloned plant genes investigated to date do not represent the majority of structural genes expressed in a plant cell; that is, genes encoding rare class transcripts. To obtain an overview of the changes in gene expression that occur during sporophyte development, Kamalay and Goldberg (1980) compared the total spectrum of genes expressed in tobacco leaf, stem, root, ovary, anther, and petal using mRNA/single-copy DNA hybridization. To distinguish mRNA sequence sets present on the polysomes of different organs, two nonoverlapping single-copy DNA fractions, referred to as leaf mDNA and null-mDNA, were isolated by hybridization with a vast excess of leaf polysomal mRNA. Figure 7 shows schematically how these sequences were isolated.

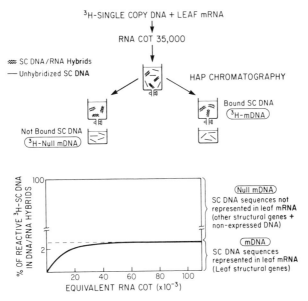

Fig. 7. Isolation of mDNA and null-mDNA hybridization probes from tobacco nuclear DNA by hydroxyapatite fractionation.

Leaf mDNA contains single-copy DNA sequences that are represented on leaf polysomes. Hybridization of leaf mDNA with mRNA from a heterologous organ measures the number of structural genes expressed in leaf that are also expressed in the heterologous organ. By contrast, leaf null-mDNA represents single-copy DNA that is depleted of sequences complementary to leaf mRNA, that is, the single-copy DNA sequence set that is not represented on leaf polysomes. Hybridization of leaf null-mDNA with mRNA from a heterologous organ measures the complexity of the structural gene set expressed in the heterologous organ that is not expressed in leaf. The use of these purified single-copy DNA fractions to analyze plant mRNA populations provides a sensitive assay for detecting differences in gene activity during plant development.

1. Gene Activity in Sporophytic Organ Systems Is Overlapping

Results of the hybridization reaction between tobacco stem, root, anther, ovary, and petal mRNA with leaf mDNA and null-mDNA probes are summarized in Table VI (Kamalay and Goldberg, 1980). A major conclusion of this study is that each organ shares a diverse set of mRNAs with leaf. The extent of leaf mDNA hybridization with the mRNA from different organs ranged from 57% in ovary to 98% in petal, which is equivalent to 15,000 and 27,000 structural genes, respectively. A second major observation is that the mRNA from four of the five tobacco organ systems also hybridized to leaf null-mDNA, indicating that these organs also express a large set of nonleaf structural genes equivalent to 6,000 to 10,000 average-sized mRNAs (Table VI). By contrast, the petal does not detectably express a set of nonleaf mRNAs. Because the leaf and petal are morphologically very similar (Esau, 1977), these observations suggest that rare class mRNA sequence sets help establish and maintain the differentiated state of these organ systems.

2. Each Sporophytic Organ Expresses a Unique Set of Structural Genes

Are the mRNA sequences from each organ that hybridized with leaf null-mDNA developmentally regulated? That is, does each organ system have a developmental-specific mRNA set not expressed in other organs? To address this question, Kamalay and Goldberg (1980) hybridized leaf null-mDNA with an equal mass of mRNA from two or more organs. If the nonleaf mRNA sequences in each organ are developmental-specific, then the fraction of the leaf null-mDNA hybridized by a mRNA mixture should equal the sum of the individual mRNA/null-mDNA reactions. Conversely, if a fraction of the nonleaf mRNA sequences in one organ is a subset of the nonleaf mRNA sequences in a second organ, then the fraction of leaf null-mDNA hybridized by the mRNA mixture should be less than the sum of the individual mRNA reactions.

TABLE VI
Summary of Tobacco Leaf mDNA and Null-mDNA Hybridization Reactions

mRNA	mDNA reactions[a]			Null-mDNA reactions[b]				
	Percent of leaf mRNA sequence set expressed	Complexity (nucleotides)	Number of mRNAs shared with leaf[c]	Percent null-mDNA hybridized	Complexity (nucleotides)	Number of mRNAs not shared with leaf[c]	Total mRNA complexity (nucleotides)	Number of diverse mRNAs
Leaf	100	3.33×10^7	27,000	$<0.03^d$	—	—	3.33×10^7	27,000
Stem	75 ± 3	2.48×10^7	20,000	0.63 ± 0.06	7.6×10^6	6,000	3.24×10^7	26,000
Root	65 ± 4	2.18×10^7	17,500	0.68 ± 0.07	8.2×10^6	6,500	3.00×10^7	24,000
Petal	99 ± 1	3.30×10^7	27,000	$<0.03^d$	$<3.7 \times 10^5$	<300	3.30×10^7	27,000
Anther	58 ± 4	1.94×10^7	15,500	1.06 ± 0.06	1.3×10^7	10,500	3.23×10^7	26,000
Ovary	57 ± 4	1.90×10^7	15,000	1.00 ± 0.10	1.2×10^7	10,000	3.11×10^7	25,000
Anther + ovary				2.11 ± 0.09	2.56×10^7	20,700		
Root + stem				1.30 ± 0.13	1.58×10^7	12,700		
Anther + ovary + root + stem				3.84 ± 0.52	4.66×10^7	37,600		

[a] The leaf mRNA sequence set has a complexity of 3.33×10^7 nucleotides (Goldberg, 1978).
[b] Tobacco single-copy DNA has a complexity of 6.4×10^8 nucleotide pairs (Zimmerman and Goldberg, 1977). The null-mDNA complexity equals that of total single-copy DNA (6.4×10^8) less the complexity of leaf mDNA (3.33×10^7; Goldberg et al., 1978), or 6.07×10^8 nucleotide pairs. Hence C_{NL} = (% of reactive null-mDNA hybridized) $\times (6.07 \times 10^8) \times 2$.
[c] The number average size of poly(A) mRNA in leaf polysomes is 1340 nucleotides, which includes a 100-nucleotide stretch of poly(A) (Goldberg et al., 1978). Poly(A) mRNAs of other organ systems were found to be similar in size. Thus the number of different mRNAs in a population = $C/1240$. This calculation assumes that abundant and rare class messages have the same size distribution.
[d] Represents the upper limit of diverse mRNA sequences that could not be reliably detected by these methods.

The results obtained from the mRNA mixing experiments are summarized in Table VI. The most striking result obtained from these experiments is that the hybridization of leaf null-mDNA with the mRNA sequences from each tobacco organ is completely additive, indicating that with the exception of leaf and petal, each organ system expresses a unique and complex set of structural genes. As shown in Fig. 8, the organ-specific mRNA sequences represent 23 to 40% of the structural gene set expressed in each organ. Therefore, to a first approximation, each set of rare class organ-specific mRNAs, and the corresponding structural genes, are required to both establish and maintain the differentiated state of the mature sporophyte.

Abundant and superabundant class mRNAs constitute only a small fraction of the mRNA complexity in each organ (Table IV). Therefore, develop-

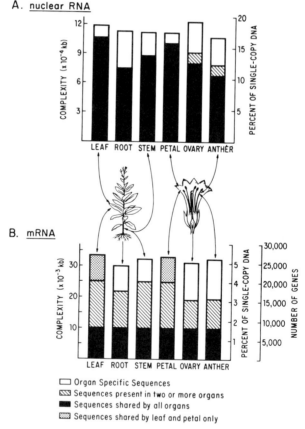

Fig. 8. (A) Nuclear RNA and (B) mRNA sequence sets are developmentally regulated in plants. Columns represent the RNA complexity of each organ system expressed in kilobases and percent single-copy DNA.

mental changes in the expression of genes encoding prevalent class mRNAs, as well as quantitative changes in gene expression, are not distinguishable by mDNA/null-mDNA saturation hybridization analysis. Both of these processes, however, also play an important role in plant cell differentiation.

3. Housekeeping Genes

Do plant cells express a constitutive set of genes that encode products required by all cells? Kamalay and Goldberg (1980) hybridized leaf mDNA with different pairs of organ mRNAs to determine the extent of sequence overlap between the shared mRNA populations of tobacco leaf, stem, root, ovary, anther, and petal. If two organs express the same leaf structural gene set, or if the leaf mRNAs of one organ are a subset of those in another organ, then the extent of leaf mDNA hybridization should equal the maximum value obtained with one of the two mRNA populations alone. On the other hand, if two organs express different fractions of the leaf structural gene set, then the extent of mDNA hybridization with a RNA mixture should be greater than either mRNA alone. The results showed that each organ system shares a significant fraction of the leaf mRNA set with all other organs. The largest amount of sequence overlap observed (59% of the leaf mRNA set) was between stem and anther while the smallest amount of sequence overlap (28% of the leaf mRNA set) was between stem and ovary. Thus, the maximum number of mRNA sequences shared by all organ systems is 28% of the leaf structural gene set or approximately 8000 genes, as shown in Fig. 8. These ubiquitous mRNA sequences probably encode proteins responsible for the maintenance of all plant cells, for example, respiration, growth, replication, and gene expression. Similar sets of housekeeping genes have been noted in animal cells as well (10,000, Hastie and Bishop, 1976; 10,000, Axel *et al.*, 1976; 1,000–1,500, Galau *et al.*, 1976).

4. Tissue-Specific and Cell-Specific Patterns of Plant Gene Expression

What is the biological significance of organ-specific mRNA sets? There are at least 50 different cell and tissue types in a flowering plant (Esau, 1977) and many more that cannot be distinguished morphologically. Each cell or tissue has a distinct structural or physiological function that requires a unique set of structural genes. In the leaf, for example, mesophyll cells harvest light energy, guard cells regulate gas exchange, and xylem and phloem cells conduct water, minerals, and photosynthate to other plant parts. The petal, stem, root, anther, and ovary have other unique, organ-specific cell types (Esau, 1977). Obviously, the analysis of gene expression by RNA/DNA hybridization does not distinguish gene expression according to cell types. Rather, the specialized cell types in each organ contribute collectively to the organ-specific mRNA sequence set described by RNA/single-copy DNA hybridization.

In situ hybridization is a useful tool for distinguishing cell and tissue-

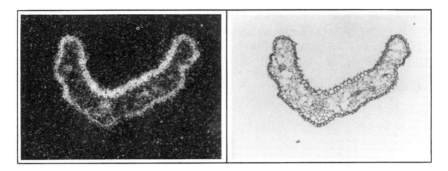

Fig. 9. Chalcone synthase gene expression is spatially restricted to epidermal cells in the tobacco petal. (From G. N. Drews and R. B. Goldberg, unpublished.) A cross-section from tobacco petal was hybridized to labeled tobacco chalcone synthase transcripts synthesized *in vitro* (Cox *et al.*, 1984; K. Cox and R. B. Goldberg, unpublished). The distribution of silver grains indicates that chalcone synthase gene activity is restricted to pigmented regions of the petal epidermis.

specific patterns of gene regulation within a plant organ. For example, *in situ* hybridization showed that the expression of the S_2-incompatibility glycoprotein gene in *Nicotiana alata* is restricted to the transmitting tissue of the style (Anderson *et al.*, 1986; Cornish *et al.*, 1987). As shown in Fig. 9, tobacco chalcone synthase mRNA is highly prevalent within the pigmented petal epidermis but not detectable in petal mesophyll (G. N. Drews and R. B. Goldberg, unpublished results). In fact, most plant organ-specific genes analyzed by *in situ* hybridization display tissue or cell-type specific patterns of gene regulation, and in some cases have revealed new, previously unrecognized cell types (R. B. Goldberg *et al.*, unpublished).

In situ hybridization is also useful for looking at the pattern of gene induction during development. For example, Fig. 10 shows that at the peak of seed protein gene expression during soybean embryogenesis (60 DAF) virtually every parenchymal cell in the cotyledon contains seed protein gene transcripts (L. Perez-Grau and R. B. Goldberg, unpublished). In young embryos (15 DAF), however, seed protein transcripts do not accumulate in all parenchymal cells simultaneously but first accumulate in a thin layer of cells at the outer perimeter of the cotyledon. As the embryo matures, seed protein gene expression then spreads inward in a wave of gene activity. The physiological significance of this observation is not known, but may reflect gene induction by a diffusible signal molecule or differences in cell ages.

5. *Intracellular Localization of mRNA Sequences*

How newly synthesized proteins find their appropriate sites within a cell is a central question of cell biology. Little is known about the distribution of

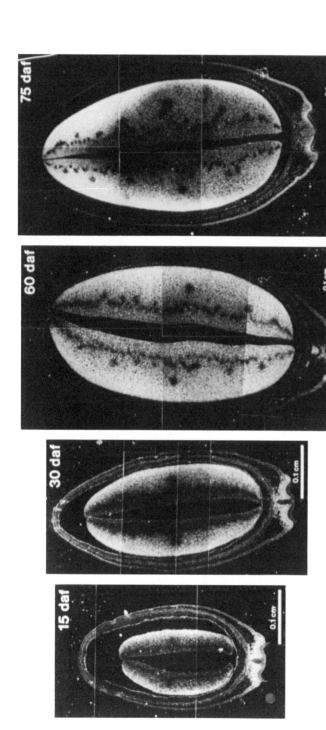

Fig. 10. Seed protein gene expression is temporally and spatially regulated during soybean embryogenesis. (L. Perez-Grau and R. B. Goldberg, unpublished.) Cross-sections from embryo cotyledons taken at different times during embryogenesis were hybridized to labeled seed lectin gene transcripts synthesized *in vitro*. The temporal and spatial pattern of seed protein gene activity is revealed by the distribution of silver grains. Lectin gene activity is first detectable in a thin layer of parenchyma/mesophyll cells at the cotyledon perimeter (15 DAF). At the mid-maturation stage of embryogenesis all parenchymal cells, but no vacular elements, contain lectin transcripts (60 DAF). During late maturation (75 DAF), as the embryo prepares for dormancy, lectin transcripts disappear from the cotyledon stochastically. A similar pattern of expression was seen for Kunitz trypsin inhibitor gene expression (L. Perez-Grau and R. B. Goldberg, unpublished).

specific mRNA sequences in the plant cell cytoplasm. In animal cells, the intracellular localization of mRNA sequences can play an important role in determining where cytoplasmic proteins are synthesized and deposited. For example, actin, vimentin, and tubulin mRNAs are distributed asymmetrically in different regions of the chicken myoblast according to the cellular distribution of their protein products (Lawrence and Singer, 1986). By contrast, *in situ* hybridization with labeled poly(U) showed that polyadenylated RNA was distributed uniformly throughout the cell. In animal eggs, the nonrandom distribution of mRNA molecules may play an important role in cell determination during early embryogenesis (Rodgers and Gross, 1978; Weeks and Melton, 1987; reviewed by Davidson, 1986). Together, these studies suggest that mRNA sequences may contain sequence information which determines their cytoplasmic distribution and site of expression in the cell. In plants, the localization of mRNAs to specific cell regions has not yet been demonstrated. The polar nature of some plant cell processes such as egg cell cleavage, megasporocyte development, and pollen tube growth, are possible subjects for the investigation of intracellular localization of mRNA by *in situ* hybridization.

6. Quantitative Regulation of Plant Gene Expression

During the course of plant development quantitative as well as qualitative differences in gene expression play an important role in cell differentiation and cell function. Individual mRNA species may differ more than 100-fold in concentration within a typical mRNA population (Table III). Specific superabundant mRNA transcripts can vary more than 1000-fold in concentration during development. For example, during embryogenesis seed storage protein mRNAs constitute as much as 50% of the embryo mRNA mass (Goldberg *et al.*, 1981; Dure *et al.*, 1983). However, during embryogenesis the concentration of storage protein mRNAs and other quantitatively regulated mRNA sequence sets fluctuates dramatically as the cotyledon cells pass through different periods of development such as cell division, cell expansion and differentiation, maturation, dehydration, and dormancy (Galau and Dure, 1981; Goldberg *et al.*, 1981a,b). Figure 11 shows that the changing prevalence of storage protein mRNAs can be visualized as changes in the intensity of mRNA bands on an EtBr-stained gel. Both transcriptional and posttranscriptional regulatory processes contribute to these quantitative changes in gene expression (Walling *et al.*, 1986).

By contrast, there is little qualitative change in plant gene activity during embryogenesis. Greater than 95% of the mRNA mass (Galau and Dure, 1981) and greater than 90% of the embryonic mRNA diversity are represented continuously during most of embryogenesis (Goldberg *et al.*, 1981a). Therefore, an important conclusion from these studies is that the quantitative control of cytoplasmic mRNA levels also plays a critical role in plant ontogeny.

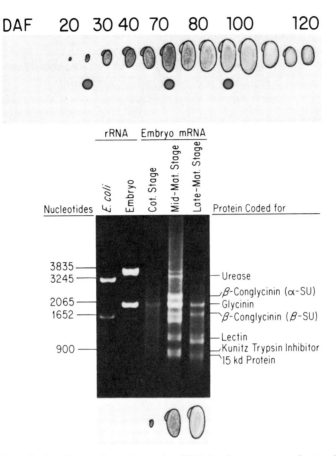

Fig. 11. Quantitative changes in seed protein mRNA levels accompany plant embryogenesis. Five micrograms of mRNA from 20-, 70-, and 100-day-old embryos was electrophoresed on a denaturing agarose gel and stained with ethidium bromide. Superabundant seed protein mRNAs bands are labeled.

B. Gene Activity in the Male Gametophyte

Unlike animals, plants do not have a germline. Instead, the sperm and egg are produced mitotically by the haploid two-celled male gametophyte (pollen grain) and seven-celled female gametophyte (embryo sac) that develop from haploid spores in the floral stamen and pistil (Fig. 1). Little is known about gene expression in the female gametophyte because this phase of plant development is buried deep within the ovule. However, gene expression in the male gametophyte has been analyzed by RNA/DNA hybridization kinetics.

Willing and Mascarenhas (1984) investigated the diversity and abundance distribution of mRNA sequences in *Tradescantia paludosa* pollen using

RNA/cDNA hybridization kinetics. They showed that approximately 2.2×10^4 kb of poly(A) RNA, or approximately 19,000 mRNA species, are stored in the mature pollen grain for expression during germination and pollen tube growth. By comparison, they found that the vegetative shoot (the terminal 5 cm of the shoot minus the leaf blade) of *Tradescantia* expresses roughly 29,000 different genes. Similar to sporophytic organ systems and tissues, the male gametophyte mRNA population contains a diverse spectrum of RNA sequences with different cellular abundancies. The complexity and prevalence of these RNAs is summarized in Table IV.

Do genes expressed in the male gametophyte overlap with genes active in the sporophyte? When *Tradescantia* pollen cDNA was hybridized with shoot mRNA, only 64% of the reactable cDNA hybridized. Therefore, 36% of the pollen mRNA mass is not represented in the shoot and appears to be unique to gametophytic development. The number of genes uniquely expressed in pollen, however, could not be determined from this experiment because the authors did not determine whether the pollen-specific RNAs represented rare or abundant class transcripts. If the pollen cDNA sequences that did not hybridize with shoot RNA represent rare class transcripts, then the difference in hybridization reactions is equivalent to approximately 18,000 different mRNAs (see Table IV). On the other hand, if the difference in cDNA hybridization is due to abundant class pollen mRNAs, then the difference in mRNA between the mature male gametophyte and the vegetative shoot of *Tradescantia* is 300 mRNAs or less. Nevertheless, these findings clearly demonstrate that there are qualitative differences in gene expression between the gametophytic and sporophytic generations in plants. Additional experiments are required to determine how many structural genes are uniquely expressed during male gametophyte development and whether a unique set of genes is also expressed in the female gametophyte.

C. Total Number of Genes Expressed during Plant Development

The study of structural gene expression in tobacco by Kamalay and Goldberg (1980) provides a limit estimate of the total number of genes required to program sporophytic development. If the organ-specific structural gene sets shown in Table VI are added to the leaf structural gene set, then at least 7×10^4 kb of mRNA or approximately 60,000 structural genes are expressed in the mature sporophyte. Together, these genes represent 11% of the tobacco single-copy DNA or only 4.6% of the tobacco genome. It is important to point out that these measurements do not distinguish between *different* mRNA sequences that share greater than 80% sequence homology. Similarly, mRNAs that are less than 80% homologous but members of the same gene family were scored as different. Nor do these measurements include male and female gametophyte-specific mRNAs that may represent as many as 18,000 additional structural genes as discussed above (Willing and Mascarenhas, 1984). Thus, they provide only a limit estimate of the total number of structural genes required to program plant development.

The estimate of the number of structural genes expressed during tobacco development is also complicated by the fact that tobacco is an allotetraploid (Goodspeed and Clausen, 1927, 1928) formed by the hybridization of two highly divergent genomes of *Nicotiana sylvestris* and *Nicotiana tomentosiformis* (Okamuro and Goldberg, 1985). It could be argued that the expression of two heterologous genomes in tobacco cells could cause an overestimate of the number of genes necessary to program tobacco development. That is, mRNAs from homologous genes within each genome having less than 80% sequence identity will be scored as unique. However, this is probably only a minor factor because gene expression in the organs of other diploid plants is equally complex (Table V), and because most tobacco structural genes investigated to date are sufficiently conserved between the progenitor genomes to form stable duplexes on DNA gel blots (R. B. Goldberg, K. Cox, G. N. Drews, and J. Truettner, unpublished); that is, their mRNAs would be scored as identical and therefore would not add to the gene number estimate. The total complexity of structural genes expressed in tobacco (7×10^4 kb) is equal to or greater than the total complexity of the *Arabidopsis* genome (4.2×10^4 to 7×10^4 kb; Leutwiler *et al.*, 1984). Therefore, it is important that similar DNA/RNA hybridization experiments be used to measure gene number in *Arabidopsis thaliana* to estimate the minimum number of genes necessary to program plant development.

D. Summary

Gene activity during plant development is complex and highly regulated. Over the course of sporophytic and gametophytic development a plant expresses a minimum of 60,000 structural genes or 7×10^4 kb of DNA. Each organ of the sporophyte is programmed by a unique set of structural genes that encode primarily rare class mRNAs. With the exception of leaf and petal, whose mRNA sequence sets overlap almost completely, 25–40% of the structural genes expressed in each organ are organ-specific. Approximately 30% of the genes expressed in each organ encode a set of ubiquitous or "housekeeping" mRNAs. Furthermore, quantitative as well as qualitative changes in gene expression play an important role in the establishment and maintenance of the unique structure and function of each plant organ and tissue types.

V. REGULATION OF PLANT GENE EXPRESSION

How does the plant cell regulate the expression of 60,000–80,000 different structural genes during the course of development? A simplistic scheme outlining the processes that regulate gene expression in a plant cell is shown in Fig. 12. These include (1) differential gene transcription; (2) nuclear RNA modification, splicing, and turnover; (3) selective RNA transport from the nucleus to the cytoplasm; (4) cytoplasmic mRNA turnover; (5) translation, posttranslational processing, compartmentalization, and protein turnover. Each of these processes plays an important role in establishing the expressed state of a gene, that is, the functional protein product. We focus our discussion, however, on the transcriptional and posttranscriptional control processes that regulate the composition of developmentally specific mRNA sets.

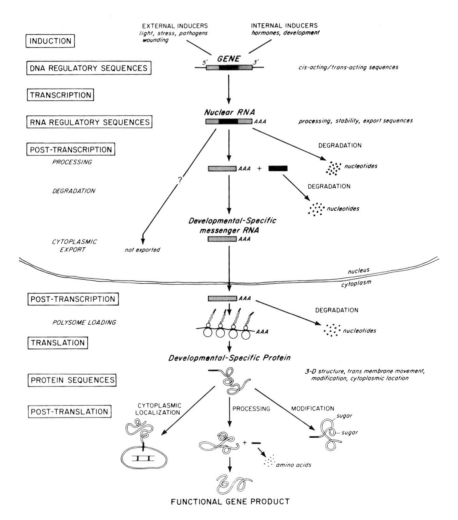

Fig. 12. Regulation of gene expression in a eukaryotic cell.

A. Nuclear versus Cytoplasmic RNA Complexity

To what extent do transcriptional and posttranscriptional regulatory processes control the flow of sequence information from the nucleus to the cytoplasm? Several laboratories have determined the complexity of nuclear and total cell RNA during sporophytic development and in plant tissue culture. These measurements, summarized in Table VII, show that plant nuclear RNA complexity ranges from 5.7×10^4 kb in parsley root callus (Kiper *et al.*, 1979) to 1.2×10^5 kb in tobacco organ systems (Goldberg *et al.*, 1978; Kamalay and Goldberg, 1984). These values are 3–4 times greater than that of the corresponding mRNA population (Table V), indicating that 60–70% of the nuclear RNA sequence information is confined to the nucleus, and that posttranscriptional selection processes play a major role in establishing the sequence content of plant mRNA sets.

The ratio of nuclear to cytoplasmic mRNA complexity in plants is similar to that found in animal cells (reviewed by Minty and Birnie, 1981; Davidson, 1986) but is much higher than that of simple eukaryotes such as yeast and water molds where the ratio is approximately 1 (Timberlake *et al.*, 1977; Hereford and Rosbach, 1977). One obvious explanation for the difference between nuclear and cytoplasmic RNA complexity is that plant genes contain introns that increase their size by a factor of three or four compared to the corresponding processed mRNA. Alternatively, the plant cell may transcribe a set of genes from which only a small subset of transcripts is selectively exported to the cytoplasm. In fact, both situations occur in plant cells.

Are Introns Responsible for the Difference between Nuclear and Cytoplasmic mRNA Complexity?

To the best of our knowledge, there have been no general studies of plant nuclear RNA structure and stability. In tobacco, *in vivo* labeled leaf nuclear poly(A) RNA has a mass average size of 2.6 kb (Kamalay, 1981). By contrast, leaf cytoplasmic poly(A) mRNA has a mass average size of 1.2 kb. If every leaf nuclear transcript reached the cytoplasm, then the size difference between nuclear RNA and cytoplasmic mRNA molecules could be explained by posttranscriptional processing reactions such as the removal of introns. However, this is not a valid assumption. In mouse, *Drosophila melanogaster, Xenopus laevis,* and sea urchin, only 5–10% of the newly synthesized nuclear RNA mass reaches the cytoplasm as mRNA, indicating that the majority of the nuclear transcripts in these organisms turn over in the nucleus (reviewed by Davidson, 1986). In plants, it is experimentally difficult to determine what fraction of the nuclear RNA molecules reach the cytoplasm as mRNA. However, we can estimate the contribution of introns to nuclear RNA complexity by examining the structure of cloned plant genes.

The presence of introns in a small number of plant genes like alfalfa

TABLE VII

Sequence Complexity of Plant Nuclear RNA Populations

Organism	Organ system and/or developmental state	Percent single-copy DNA	Complexity[a] (nucleotides)	$\frac{C^{mRNA[b]}}{C^{HnRNA}}$	Reference
Parsley	Root callus	24.8	5.7×10^7	0.34	Kiper et al. (1979)
Soybean	Suspension culture	12.4[c]	6.4×10^7	0.53	Silflow et al. (1979)
Tobacco	Leaf	18.7	1.2×10^8	0.28	Goldberg et al. (1978)
	Leaf	18.8	1.2×10^8	0.28	Kamalay and Goldberg (1984)
	Stem	17.8	1.1×10^8	0.29	Kamalay and Goldberg (1984)
	Root	17.6	1.1×10^8	0.27	Kamalay and Goldberg (1984)
	Anther	15.6	1.0×10^8	0.32	Kamalay and Goldberg (1984)
	Ovary	18.4	1.2×10^8	0.26	Kamalay and Goldberg (1984)
	Petal	18.2	1.2×10^8	0.28	Kamalay and Goldberg (1984)

[a] Calculated from the relationship: complexity = (% single-copy DNA hybridized × single-copy DNA complexity)/100. Single-copy DNA complexities are 2.3×10^5 kb, 5.1×10^5 kb, and 6.4×10^5 kb for parsley (Kiper et al., 1979), soybean (Silflow et al., 1979), and tobacco (Zimmerman and Goldberg, 1977; Goldberg et al., 1978), respectively.
[b] Ratio of mRNA to nuclear RNA complexity. mRNA complexities were taken from the data in Table V.
[c] Based on total cell RNA complexity.

glutamine synthetase (Tischer *et al.*, 1986) and maize sucrose synthase (Werr *et al.*, 1985) results in a primary transcript that is 2–3 times greater than the size of the mature mRNA. A far greater number of cloned plant genes, however, lack or have only small introns. From the small sample of plant genes in Table II, it does not appear that introns alone account for the total difference in complexity between plant nuclear and cytoplasmic mRNA populations. To a first approximation, posttranscriptional selection events must also regulate the flow of sequence information from the nucleus to the cytoplasm. Thus, it is important to address the issue of whether each plant nuclear RNA population contains the same sequence information or whether there are also changes in the patterns of plant gene transcription during development that result in developmentally specific nuclear RNA sequence sets.

B. Transcriptional and Posttranscriptional Regulation of Plant Gene Expression

One strategy for determining the contribution of transcriptional and posttranscriptional control processes to the regulation of plant gene expression is to compare nuclear RNA sequence sets during development. If we assume that the complexity of steady-state nuclear RNA is a measure of the diverse sequences transcribed in the cell, then changes in the composition of a nuclear RNA population as measured by RNA/single-copy DNA hybridization should reflect the contribution of differential gene transcription to the general pattern of plant gene expression.

The absence of a sequence from a steady-state nuclear RNA population, however, is not definitive proof that a gene is transcriptionally inactive. It is always arguable that some transcripts are rapidly degraded after transcription and therefore not detectable in the steady-state nuclear RNA population. However, unstable transcripts would require a half-life less than one-hundreth that of most steady-state nuclear RNA molecules in order to go undetected in an RNA/DNA hybridization reaction (Kamalay and Goldberg, 1984). There is no evidence for a complex highly unstable population of nuclear transcripts in either plants or animals. In sea urchin, total nuclear RNA undergoes the same synthesis and turnover kinetics, as does steady-state nuclear RNA analyzed by single-copy DNA hybridization experiments (Hough *et al.*, 1975). Therefore, it is assumed in plants that the steady-state nuclear RNA represents the vast majority of sequences transcribed by the cell.

1. *Nuclear RNA Sequence Sets in Sporophytic Organ Systems Are Overlapping*

Is the sequence composition of plant nuclear RNA developmentally regulated? Kamalay and Goldberg (1984) compared tobacco organ nuclear RNA sequences using RNA/single-copy DNA hybridization. To carry out this analysis, they isolated two single-copy DNA probes, referred to as leaf HnDNA and leaf null-HnDNA, using a single-copy DNA fractionation approach analogous to that shown in Fig. 7 for preparing leaf mDNA and null-mDNA. Leaf HnDNA represents the set of single-copy DNA sequences complementary to leaf nuclear RNA. Hybridization between leaf HnDNA and the nuclear RNA from another organ is a direct measure of the nuclear RNA sequence overlap between these organs. On the other hand, leaf null-HnDNA contains single-copy DNA sequences not detectably represented in leaf nuclei. Therefore, hybridization between leaf null-HnDNA and the nuclear RNA from another organ measures the set of nuclear transcripts not detectable in leaf nuclei but present in the heterologous organ.

Hybridization of leaf HnDNA and null-HnDNA with nuclear RNA from each tobacco organ system revealed that each organ possesses a unique set of nuclear transcripts (Kamalay and Goldberg, 1984). Figure 8 shows that 60 (anther) to 90% (petal) of the leaf nuclear RNA sequence set is represented in the nuclei of heterologous organs. Nonleaf nuclear transcripts represent from 10 (petal) to 40% (anther) of the total nuclear RNA sequence complexity of each organ. Together, these results indicate that although the nuclear RNA sequence complexity of each organ is similar, no tobacco organ contains the same set of nuclear transcripts.

2. *Each Sporophytic Organ Has a Developmentally Specific Set of Nuclear Transcripts*

Kamalay and Goldberg (1984) hybridized leaf null-HnDNA with nuclear RNA mixtures to determine whether the nonleaf nuclear RNA transcripts of each organ are developmentally specific. If all or part of the nonleaf nuclear RNA sequence set in each organ is organ-specific, then the extent of null-HnDNA hybridization obtained with a nuclear RNA mixture from two or more organs should be greater than the individual hybridization reactions alone. The experiments summarized in Fig. 8 demonstrate that the nonleaf nuclear RNA sequence sets in root, stem, and petal are completely additive and therefore developmentally specific. Moreover, the anther and ovary nonleaf nuclear RNA sequence sets overlap by only 10% or 1.0×10^4 kb. Together, these findings indicate that transcription during plant development is highly regulated and that each organ has a unique nuclear RNA sequence set derived from the transcription of a restricted genomic single-copy DNA fraction.

3. *A Large Genomic Fraction Is Transcribed during Plant Development*

The data summarized in Fig. 8 permit an estimate to be made of the extent to which single-copy DNA sequences are transcribed and represented collectively in the nuclear RNA populations of a mature flowering plant. Adding the organ-specific nuclear RNA complexity of each organ system to the complexity of leaf nuclear RNA (Fig. 8) yields an aggregate complexity of almost 3×10^5 kb. Thus, at least 45% ($3 \times 10^5/6.4 \times 10^5$ kb) of the single-copy DNA is expressed at the nuclear RNA level in an entire plant. This is clearly a minimal estimate but indicates that a large amount of genetic information is used by a plant to achieve its developmental potential.

4. Regulation of Developmental-Specific mRNA Transcripts in the Nucleus

To determine whether organ-specific mRNA transcripts are represented in the set of organ-specific nuclear RNA transcripts, Kamalay and Goldberg (1984) hybridized tobacco leaf null-HnDNA with stem, root, and anther polysomal mRNAs. The results, summarized in Table VIII, showed that stem mRNA *did not* react detectably with the leaf null-HnDNA probe. This finding indicates that, within the limits of detection, the majority of the 6000 stem-specific genes are also transcribed in leaf nuclei, even though these transcripts are not detectable on leaf polysomes (Fig. 8). Using an analogous experimental approach, leaf mDNA hybridized to the same extent with both leaf and stem nuclear RNA, indicating that the genes that encode the leaf mRNA transcripts not detectable in the stem mRNA population are also represented in stem nuclei (Kamalay and Goldberg, 1980). Similar post-transcriptional processes play a major role in regulating developmental-specific mRNA sets in sea urchin (Wold *et al.*, 1978), suggesting that this is a commonly used regulatory strategy in eukaryotic organisms.

By contrast, both stem and anther mRNAs hybridized with leaf null-HnDNA. This result indicates that the stem and anther developmental-specific nuclear RNAs contain structural gene transcripts. Results summarized in Table VIII indicate that 100% of the anther-specific mRNA set and 70% of the root-specific mRNA set are represented in the corresponding developmental-specific nuclear RNA sequence set. Therefore, transcriptional processes regulate the developmental expression of these gene sets.

Figure 13 summarizes the transcriptional and posttranscriptional regulatory processes in leaf that lead to the production of developmentally regulated nuclear and cytoplasmic RNA transcripts detected by RNA/DNA hybridization. Clearly, no single mechanism regulates rare class gene expression in the cell. Instead, transcriptional and posttranscriptional-level selection mechanisms together control the flow of genetic information from the nucleus to the cytoplasm.

C. Transcriptional and Posttranscriptional Regulation of Specific Plant Genes

A central issue of this chapter is whether the concepts inferred from population hybridization experiments apply to the regulation of individual

TABLE VIII

Null-HnDNA Hybridization with Polysomal mRNAs

mRNA	Percent hybridization[a]	Complexity[b] (kb × 10⁻⁴)	Number of diverse organ-specific mRNAs[c]	
			Observed	Expected[d]
Stem	<0.02	<0.02	<200[e]	6,000
Root	0.54 ± 0.20	0.56	4,500	6,500
Anther	1.24 ± 0.13	1.3	11,000	10,000

[a] Average of three independent reactions corrected for null-HnDNA reactivity.

[b] The mRNA complexity was calculated from the relationship: C_m = (percent hybridization) × (5.2 × 10⁵ kb) × 2 × 0.01, where 5.2 × 10⁵ kb is the null-HnDNA complexity (Kamalay and Goldberg, 1984).

[c] Number of mRNAs = C_m/1.25 kb, where 1.25 kb is the number average mRNA size (Kamalay and Goldberg, 1980; Goldberg et al., 1978).

[d] The number of diverse mRNAs undetectable in the polysomes of other organ systems (Kamalay and Goldberg, 1980). This represents the maximum number of mRNA species that could be absent from the nuclear and cytoplasmic RNAs of other organs.

[e] Value represents the maximum number of stem-specific mRNAs that could be represented in the null-HnDNA population and would not be reliably detected by hybridization.

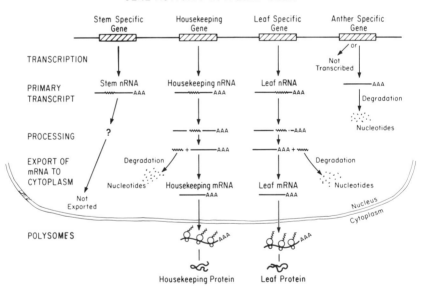

Fig. 13. Transcriptional and posttranscriptional regulation of gene activity in leaf. Steady-state nuclear RNA and cytoplasmic mRNA population analyses indicate that transcriptional and posttranscriptional regulatory processes determine the pattern of gene expression in leaf. The set of anther-specific and root-specific genes are not detectably transcribed in leaf. Stem-specific structural genes are transcribed but are not detectable on leaf polysomes.

cloned plant genes. Because most genes investigated by recombinant DNA technology produce abundant or superabundant class mRNAs, regulatory trends obtained by the study of cloned genes may not reflect those obtained with genes encoding rare class mRNA sets measured by RNA-excess RNA/single-copy DNA hybridization procedures. Therefore, it is important to compare the cellular processes that regulate the expression of specific genes during plant development to establish unifying principles of plant gene regulation and to formulate molecular models for the global regulation of plant differential gene activity and plant development.

In Vitro *Analysis of Plant Gene Transcription*

In vitro nuclear run-off transcription studies with isolated nuclei are used for determining whether specific genes are regulated at the transcriptional or posttranscriptional levels or both (Hofer and Darnell, 1981; Hofer *et al.*, 1982). In such experiments, isolated nuclei are used to incorporate labeled uridine triphosphate into RNA transcripts initiated *in vivo* (Groudine *et al.*, 1981). This procedure was adapted for studies in plants by Luthe and Quatrano (1980a,b) and applied to studies with specific plant genes by Willmitzer and Wagner (1981) and Walling *et al.* (1986) among others (see below). The transcriptional activity of a gene is proportional to the fraction of labeled nuclear RNA that hybridizes with a specific gene clone by filter or gel blot hybridization procedures. In animal nuclei, the initiation of new transcripts does not occur *in vitro* (Weber *et al.*, 1977; Evans *et al.*, 1977; Tsai *et al.*, 1978; McKnight and Palmiter, 1979), selective turnover is not detectable (Blanchard *et al.*, 1978), and results obtained with *in vitro* synthesized transcripts agree well with those obtained with *in vivo* labeled nuclear RNAs (Hofer and Darnell, 1981; Derman *et al.*, 1981; Hofer *et al.*, 1982). Therefore, unlike steady-state nuclear RNA comparisons, studies with *in vitro* labeled nuclear RNA directly reflect transcriptional activity in the cell. In plants, it is assumed but has not been demonstrated that nuclear RNA sequences synthesized *in vitro* also represent gene activity *in vivo*.

a. **Seed Protein Genes.** Recently, Evans *et al.* (1984), Beach *et al.* (1985), and Walling *et al.* (1986) used *in vitro* nuclear run-off transcription experiments to examine the regulation of seed protein gene expression during embryogenesis and in organ systems of the mature plant. These studies demonstrated that transcriptional control mechanisms affect seed protein gene expression both qualitatively and quantitatively during embryogenesis. For example, Walling *et al.* (1986) showed that soybean seed protein genes are transcriptionally activated and then repressed during embryogenesis and are inactive or transcribed at very low levels in the mature plant. Moreover, embryonic seed protein gene transcription rates are modulated during embryogenesis and changes in transcription closely parallel the increases and decreases in seed protein mRNA levels. In addition, these experiments demonstrated that seed protein and nonseed protein mRNAs that differ by 10,000-fold in concentration have the same relative transcription rates. This observation indicates that posttranscriptional control mechanisms (e.g., nuclear RNA processing, turnover, nuclear transport, or mRNA stability) also play an important role in establishing seed protein mRNA levels during development.

b. Light-Regulated Genes. The effect of light on plant gene transcription has also been analyzed by nuclear run-off transcription experiments. LHCP (light harvesting chlorophyll *a/b*-binding protein) and rbcS (small subunit of ribulose-bisphosphate carboxylase) gene transcription is induced 10- to 30-fold by light (Gallagher and Ellis, 1982; Silverthorne and Tobin, 1984; Berry-Lowe and Meagher, 1985). This induction is mediated by the photoreceptor phytochrome (Silverthorne and Tobin, 1984) and a blue-light receptor that has not yet been identified (Fluhr and Chua, 1986).

Phytochrome also mediates gene repression by light. The NADPH-protochlorophyllide oxidoreductase gene (Apel, 1981; Batschauer and Apel, 1984; Mösinger *et al.*, 1985) and phytochrome itself (Colbert *et al.*, 1983, 1985; Otto *et al.*, 1983) are negatively affected by red light. Because phytochrome is located in the cytoplasm (McCurdy and Pratt, 1986; Speth *et al.*, 1986), the effects of light on gene transcription must be transmitted to the nucleus by one or more secondary messengers (Silverthorne and Tobin, 1987). The nature of this putative phytochrome-induced signal in the plant cell has not yet been determined.

c. Hormonally Regulated Gene Expression. Phytohormones are pleiotrophic regulators of higher plant development. We know little at the molecular level about how these general signal molecules control the program of plant development and the regulation of developmental-specific mRNA sets. We do know, however, that these hormones regulate gene expression, in part, at the transcriptional level. Auxin- (Hagen *et al.*, 1984), ethylene- (Nichols and Laties, 1984), and gibberellin- (Jacobsen and Beach, 1985) induced gene transcription has been reported. However, the cellular circuitry of *cis*-acting elements and *trans*-acting factors that coordinate the expression of genes encoding both rare and abundant class mRNA sets that respond to these hormones at the transcriptional and posttranscriptional level has not yet been described.

The demonstration that specific plant genes are transcriptionally regulated by environmental and physiological regulators of development supports the general conclusions of the steady-state nuclear RNA and mRNA population comparisons. That is, differential gene transcription plays an important role in the regulation of structural gene sets in plants and contributes significantly to the establishment of plant developmental states.

D. Cytoplasmic Regulation of Plant Gene Expression: mRNA Stability and Turnover

Messenger RNA stability is an important but often overlooked level of gene regulation in plants. The relative stability of a mRNA sequence determines in part its abundance in the cell, particularly for highly prevalent mRNAs. Globin mRNA, for example, constitutes 90% of the erythrocyte mRNA mass, partly because its half-life is 3.5 times greater than the rest of the erythrocyte mRNA sequences (Lodish and Small, 1976). To determine whether plant mRNA populations contain sequence sets with different decay rates, Silflow and Key (1979) analyzed the turnover rate of mRNA in soybean suspension culture cells by pulse–chase labeling. They detected two mRNA populations with half-lives of 0.6 and 30 hr. After 1 hr of labeling, the short-lived component represented approximately 90% of the total labeled mRNA and the stable component represented 10% of the labeled mRNA. Under steady-state conditions, however, the stable RNA component constituted the majority of the polysomal poly(A) RNA. Similar populations of mRNA with different half-lives have been described in mammalian and insect cell cultures (Lengyel and Penman, 1977; Spradling *et al.,* 1975), indicating that the regulation of mRNA stability is a universal mechanism for the regulation of gene expression in eukaryotes. To date, quantitative measurements of specific plant mRNA half-lives have not been carried out. Nor have definitive studies been carried out on the synthesis, decay, and cytoplasmic export of any known plant transcript. Clearly, we are left with animal models, or with indirect references from "run-off" transcription studies with isolated nuclei, in order to evaluate the contribution of posttranscriptional RNA metabolism to the establishment of plant differentiated states.

In animal cells, differential mRNA stability plays an important role in the developmental regulation of large sets of structural genes (Kafatos and Gelinas, 1977). In *Dictyostelium discoideum,* the half-life of approximately 2500 different "aggregation-specific" mRNAs was reduced to approximately 30 min in response to disaggregation (Mangiarotti *et al.,* 1982). By contrast, the half-life of the unaffected "conserved" mRNA population was greater than 3.5 hr under these conditions. Similar processes may regulate the pattern of plant gene expression. Beach *et al.* (1985) demonstrated that sulfur conditions affect seed protein mRNA prevalence during embryogenesis. The level of legumin mRNA increased 20-fold during a 48-hr period of recovery from sulfur starvation. However, legumin gene transcription increased only twofold during this same period of time. They concluded that legumin mRNA was differentially stabilized with respect to the rest of the mRNA mass. An alternative explanation is that legumin transcripts may be processed more efficiently in the nucleus or preferentially exported to the cytoplasm. It is clear, however, that posttranscriptional processes play a role in establishing the level of the mRNA. Similar processes may play a fundamental role in the developmentally programmed accumulation of seed protein mRNAs during embryogenesis (Gatehouse et al., 1982; Walling et al., 1986).

How does the plant cell control cytoplasmic mRNA abundance levels? Cytoplasmic mRNA abundance can be regulated at several processing steps (Fig. 12). Of these processes, mRNA turnover or stability in the cytoplasm has been analyzed most thoroughly in animal and prokaryotic systems. Messenger RNA stability and instability in both prokaryotic (Belasco et al., 1986; Wong and Chang, 1986) and eukaryotic cells are specified by discrete sequence elements located at both the 5' and 3' ends of a mRNA transcript. For example, sequences in the 3' half of the hsp70 mRNA in *Drosophila melanogaster* are responsible for the rapid destabilization of hsp mRNA after release from heat shock (Simcox et al., 1985). In contrast, the first 20 nucleotides at the 5' end of the human histone H3 mRNA are responsible for the selective destabilization of this mRNA when DNA synthesis is terminated (Morris et al., 1986). Shaw and Kamen (1986) identified a highly conserved 51-nucleotide AUUUA sequence from the 3' untranslated region of the lymphokine granulocyte–monocyte colony stimulating factor mRNA that specifically destabilizes the normally very stable rabbit β-globin mRNA. The half-life of the globin-AUUUA fusion mRNA was reduced from greater than 2 hr to less than 30 min. Nucleotide sequence analysis suggests that this AU element may regulate the mRNA levels of 15 other transiently expressed genes in animal cells (Shaw and Kamen, 1986). Specific sequences affecting mRNA stability in plant cells have not yet been identified but could be an important tool for manipulating properties of crop plants, including the synthesis of specific polypeptides.

E. Summary

Both transcriptional and posttranscriptional regulatory processes play a major role in regulating differential gene activity in plants. The comparison of steady-state nuclear and cytoplasmic RNA sequence sets indicates that as much as 45% of the plant single-copy genome or 3×10^5 kb of DNA is transcribed during sporophyte development. However, only 25% of that sequence information reaches the cytoplasm. The analysis of cloned plant gene structure indicates that introns alone do not account for the total difference between nuclear and cytoplasmic RNA complexity. Therefore, selective RNA transport from the nucleus to the cytoplasm and RNA turnover must also play a major role in regulating differential gene activity in the plant cell.

The comparison of steady-state nuclear RNA sequence sets from vegetative and reproductive plant organ systems indicates that plant nuclear RNA transcripts can be divided into developmentally specific and shared sets of nuclear transcripts. Depending on the plant organ, both sets of nuclear transcripts may contain organ-specific mRNA sequences. Therefore, both transcriptional and posttranscriptional control mechanisms contribute to the developmental regulation of plant mRNA sequence sets.

VI. THE ROLE OF *CIS*-ACTING ELEMENTS AND *TRANS*-ACTING FACTORS IN PLANT GENE REGULATION

How does the plant cell coordinate the temporal and spatial regulation of differentially expressed structural gene sets? One hypothesis, first proposed by Britten and Davidson (1969), is that genes expressed under the same developmental circumstances share a set of repetitive *cis*-acting control elements that together form a regulatory network or circuit. *Cis*-control elements are in turn recognized by sequence-specific *trans*-acting factors that regulate gene expression at the transcriptional or posttranscriptional level. Recent studies of coordinately regulated fungal and animal gene systems indicate that regulatory networks play a central role in the developmental programming of gene expression in eukaryotic organisms in general. Little is known, however, about the DNA elements and *trans*-acting factors that coordinate plant gene expression.

The identification and functional analysis of *cis*-control elements that regulate differential gene transcription in plants is a central objective of plant molecular biology. With the development of methods for introducing DNA into the plant cell genome (Zambryski *et al.*, 1983; Caplan *et al.*, 1983; Fraley *et al.*, 1985) a rapidly growing catalogue of developmental-specific, organ-specific, and environmentally induced *cis*-control regions has been described in plants (Table IX). Although the nucleotide sequences of most plant *cis*-control elements have not been precisely determined, many putative control sequences have been suggested. More importantly, however, the characterization of each control region and its effect on the activity of heterologous gene promoters has provided fresh insight into how the plant cell may regulate structural gene sets during development.

This section focuses on several coordinately regulated gene systems that illustrate our current understanding of plant gene regulation. These include the light-regulated and organ-specific genes of the small subunit of ribulose-bisphosphate carboxylase and the light-harvesting chlorophyll *a/b* binding protein, plant heat-shock protein genes, and embryo-specific seed storage protein genes.

A. Light-Regulated Gene Expression

Many different plant genes are regulated by light, including those encoding ribulose-bisphosphate carboxylase, the light-harvesting chlorophyll *a/b*

TABLE IX
Location and Regulatory Properties of Transcriptional Control Elements in Plants

Gene/species		Cis-control region borders[a]		Regulatory properties[b]							Reference
		5'	to 3'	A	B	C	D	E	F		
Light-regulated genes											
Ribulose-bisphosphate carboxylase											
Pea	rbcS ss3.6	−973	to −4	Pos	ND	ND	ND	ND	I,II,III		Herrera-Estrella et al. (1984)
		−973	to −90	Pos	+[c]	+	ND	W	I,II,III		Timko et al. (1985)
Pea	rbcS 3A	−327	to −48	Pos	+[d]	+	+	ND	I,II,III, II*,III*		Fluhr et al. (1986b)
		−330	to −112	Pos	+[d]	ND	ND	ND	I,II,III, II*,III*		Kuhlemeier et al. (1987b)
		−410	to −170	Pos	+[d]	+[h]	ND	ND	II,*III*		Kuhlemeier et al. (1988)
		−170	to −50	Pos	ND	ND	ND	ND	I,II,III		Kuhlemeier et al. (1988)
		−169	to −112	Neg	+[e]	ND	ND	ND	I,II,III		Kuhlemeier et al. (1987b)
		−152	to −137	Neg[g]	+[e]	ND	ND	ND	II		
		−136	to −50	Neg	+[e]	ND	ND	ND	III		
Pea	rbcS 3C	−2000	to +1800	Pos	ND	ND	ND	ND	I,II,III		Fluhr and Chua (1986); Nagy et al. (1986b)
Pea	rbcS E9	−317	to −82	Pos	+[c]	+	+	ND	I,II,III		Fluhr et al. (1986b)
		−1052	to −352	Pos	ND	ND	ND	ND	—		Morelli et al. (1985)
		−35	to +2	Pos	ND	ND	ND	ND	—		Nagy et al. (1986b)
Light-harvesting chlorophyll a/b binding protein											
Pea	LHCP AB80	−347	to −100	Pos/Neg	+[c]	+	ND	ND	—		Simpson et al. (1986a)
Wheat	cab1	−357	to −124	Pos	+[d]	+	ND	ND	—		Nagy et al. (1987)
Chalcone synthase											
Antirrhinum majus	chs	−1200	to −357	Pos	ND	ND	ND	ND	—		Kaulen et al. (1986)
		−357	to −39	Pos	ND	ND	ND	ND	—		

	Clone	Region							Reference
Seed protein genes									
Conglycinin									
Soybean	Gmg 17.1	−257 to −143	Pos	ND	ND	ND	ND	IV	Chen et al. (1986)
		−257 to −78	Pos	+^{d}	+	ND	−	IV	Chen et al. (1988)
Phaseolin									
Phaseolus vulgaris	λ177.4	−863 to +3789	ND	ND	ND	ND	ND	—	Sengupta-Gopalan et al. (1985)
Lectin									
Soybean	Le1	−522 to +5000	Pos	ND	ND	ND	ND	VI,VII,VIII	J. K. Okamuro and R. B. Goldberg (unpublished)
Phaseolus vulgaris	Pdlec2	−499 to −399	Pos	ND	ND	ND	ND	—	Voelker et al. (1986)
Glycinin									
Soybean	G1	−970 to −65	Pos	ND	ND	ND	ND	VI,VII,VIII	T. L. Sims, J. T. Truettner, and R. B. Goldberg (unpublished)
Kunitz trypsin inhibitor									
Soybean	KTi1&2	−1527 to +2492	ND	ND	ND	ND	ND	VI,VIII	K. D. Jofuku and R. B. Goldberg (unpublished)
	KTi2	−366 to +1020	ND	ND	ND	ND	ND	VI,VIII	
Glutenin									
Wheat	LMW Glutenin	−326 to +30	Pos	ND	ND	ND	ND	V	Colot et al. (1987)
		−938 to −140	Pos	+^{d}	+	+	ND	V	
		−438 to −5	Pos	ND	ND	ND	ND	V	
Wheat	Glu-D1 (HMW Glutenin)								
Other plant genes									
Alcohol dehydrogenase									
Maize	Adh1	−1094 to −81	Pos	+^{d}	ND	ND	ND	XI,XII	Walker et al. (1987)
		−140 to −99	Pos	ND	ND	ND	ND	XI,XII	
Leghemoglobin									
Soybean	lbc$_3$	−1100 to −950	Pos	ND	ND	ND	ND	—	Stougaard et al. (1987)
		−230 to −170	Pos	ND	ND	ND	ND	—	
		−139 to −102	Pos	ND	ND	ND	ND	X	
		−102 to −49	Neg	ND	ND	ND	ND	—	

(*continued*)

TABLE IX (*continued*)

Gene/species		Cis-control region borders[a]		Regulatory properties[b]						Reference
		5' to	3'	A	B	C	D	E	F	
Tubulin										
Soybean	Sb1	−2000 to	+2300	ND	ND	ND	ND	ND	—	Guiltinan et al. (1987)
Proteinase inhibitor II										
Potato	J32	−3000 to	+2200	ND	ND	ND	ND	ND	—	Sanchez-Serrano et al. (1987)
Heat shock										
Soybean	Gmhsp 17.5E	−1175 to	−95	Pos	ND	ND	ND	ND	IX	Gurley et al. (1986)
		−95 to	+1250	Pos	ND	ND	ND	ND	IX	
Constitutively expressed genes										
Cauliflower mosaic virus 35S		−343 to	−46	Pos	+	+	+	+	—	Odell et al. (1985); Nagy et al. (1987)
Nopaline synthase		−292 to	−116	Pos	+[f]	+	ND	ND	—	Koncz et al. (1983)
Octopine synthase		−193 to	−178	Pos	+[f]	+	ND	W	XIII	Ellis et al. (1987a,b)

[a] Control regions tested *in vivo*. These regions do not define the functional limits of the control elements tested, only the physical boundaries of the DNA sequences tested.
[b] Regulatory properties of plant *cis*-control regions. A, the region described contains positive (P), negative (N), or positive and negative (P/N) control elements. B, *cis*-dominant regulation of the following heterologous promoters: [c]nopaline synthase; [d]cauliflower mosaic virus 35S promoter and upstream enhancer sequences; [f]maize Adh. C, orientation-independent regulation of transcription. D, regulates transcription from various positions from the transcription start site. E, functions from 3' to the transcription start site. W, weakly. F, putative regulatory sequences: I, TTTCAAA (Box I); II, GTGTGGTTAATATG (Box II); II*, GTGAGGTAATAT (Box II*); III, ATCATTTTCACT (Box III); III*, CATTTACACT (Box III*); IV, AGCCCA; V, TGTAAAGTGAATAAG; VI, ATT(T/A)AAT; VII, CATGCAT; VIII, AACACA(C/A); IX, CT--AA--TTC-A; X, AAAGAT; XI, CTGCAGCCCGGTTC; XII, CCGTGGTTTGCTTGCC; XIII, ACGTAAGCGCTTACGT. ND, Not determined.
[g] One copy only is weakly regulatory. Three copies are needed to repress the CaMV 35S promoter.
[h] Functions weakly in reverse orientation.

binding protein, a 32-kDa thylakoid membrane protein, NADPH-protochlorophyllide reductase, and phytochrome (see Lamb and Lawton, 1983). The small subunit of ribulose-bisphosphate carboxylase (rbcS) and the light-harvesting chlorophyll a/b binding protein (LHCP) are two of the best characterized gene systems for analyzing both light-regulated and organ-specific gene expression in plants (reviewed by Tobin and Silverthorne, 1985; Fluhr *et al.*, 1986b; Nagy *et al.*, 1986b; Silverthorne and Tobin, 1987; Kuhlemeier *et al.*, 1987a). Each step in rbcS and LHCP gene regulation can be reproduced in transgenic plants. RbcS and LHCP gene transcription is light-inducible, organ-specific, and phytochrome-mediated (Broglie *et al.*, 1984; Herrera-Estrella *et al.*, 1984; Lamppa *et al.*, 1985b; Nagy *et al.*, 1986a; Fluhr and Chua, 1986; Simpson *et al.*, 1986b). Nascent polypeptides are correctly processed and exported to the chloroplast (Broglie *et al.*, 1984; Schreier *et al.*, 1985) where they are assembled in the stroma (Broglie *et al.*, 1984) and thylakoid membrane, respectively.

1. Are the Light-Induced Cis-Control Elements of RbcS and LHCP Genes Transcriptional Enhancers?

The rbcS and LHCP gene *cis*-acting transcriptional control elements display organizational features and complex interactions that are typical of eukaryotic regulatory sequences in general. First, the light-responsive regulatory elements of rbcS and LHCP genes are *multipartite* and *redundant*. They are located 5' to the transcription start site and can be separated into at least two autonomous control regions (Morelli *et al.*, 1985; Timko *et al.*, 1985; Simpson *et al.*, 1985; Kuhlemeier *et al.*, 1987b, 1988; Castresana *et al.*, 1988). Like the classical transcriptional enhancers of SV40 and the immunoglobulin genes (Banerji *et al.*, 1981, 1983; Gillies *et al.*, 1983; Serfling *et al.*, 1985), these control regions can be separated from the rbcS and the LHCP promoters and placed 5' to a constitutive test promoter where they function in a *cis*-dominant manner to direct light-regulated and phytochrome-mediated gene transcription in transgenic plants as well as in transformed calli (Timko *et al.*, 1985; Fluhr *et al.*, 1986b; Simpson *et al.*, 1986b; Nagy *et al.*, 1987; Kuhlemeier *et al.*, 1988). Furthermore, the rbcS and LHCP transcriptional control regions function in either orientation (Timko *et al.*, 1985; Fluhr *et al.*, 1986b; Simpson *et al.*, 1986a; Kuhlemeier *et al.*, 1988; Nagy *et al.*, 1987), and from at least two different positions from the transcription start site (Fluhr *et al.*, 1986b). Therefore, they are referred to as "enhancer-like" control elements (Timko *et al.*, 1985; Fluhr *et al.*, 1986b; Simpson *et al.*, 1986a). Unlike true enhancers, however, the rbcS and LHCP gene regulatory regions function poorly or not at all from a position 3' to the coding sequence (Timko *et al.*, 1985; Simpson *et al.*, 1986a). For this reason, the rbcS and LHCP control regions are not strict enhancer elements but functionally resemble the enhancer-like UAS control elements of yeast

(Guarente and Hoar, 1984; Struhl, 1984, 1986). The significance of this distinction is not yet known.

Three highly conserved repetitive elements, Box I, II, and III (Table IX) are found in the upstream control region of each rbcS gene in pea (Coruzzi *et al.*, 1984; Fluhr *et al.*, 1986a,b; Kuhlemeier *et al.*, 1987a). The function of Box I is not known. Box II (GTGTGGTTAATATG) is homologous to the mammalian core enhancer concensus sequence $\text{GTGG}^{TTT}_{AAA}\text{G}$ (Weiher *et al.*, 1983; Zenke *et al.*, 1986; Herr and Clarke, 1986) and is required for light-induced gene transcription together with Box III (Kuhlemeier *et al.*, 1988). A GG → CC transversion of the Box II element rendered it inactive, confirming the regulatory function of this element (Kuhlemeier *et al.*, 1988). Box III is also homologous to several mammalian enhancer elements (Kuhlemeier *et al.*, 1987b, 1988). Similar control sequences were found in the upstream control region of the *Cab-E* gene from *Nicotiana plumbaginifolia* (Castresana *et al.*, 1988) suggesting that the regulation of the rbcS and LHCP gene families may be linked in the cell by common *cis*-elements and *trans*-factors.

Green *et al.* (1987) used *in vitro* protein binding experiments and DNaseI footprint analysis to identify a leaf nuclear factor, GT-1, that binds to both distal and proximal copies of Box II and III. This factor is present in both light-grown and dark-grown plants, suggesting that if GT-1 binds differentially to this control region *in vivo* to regulate rbcS transcription, its activity must be blocked in the dark.

No transcriptional control functions have been identified in the rbcS and LHCP gene coding sequences, introns, or 3' flanking DNA (Nagy *et al.*, 1986a,b; Fluhr and Chua, 1986) as is sometimes found for developmentally regulated animal genes (Mercola *et al.*, 1983; Wright *et al.*, 1984; Hultmark *et al.*, 1986).

2. *Quantitative Regulation of RbcS and LHCP Gene Expression*

RbcS and LHCP transcriptional control sequences also regulate light-induced gene expression quantitatively. A series of 5' deletions showed that the truncation of each control region reduced the level of light-induced gene expression by increments (Morelli *et al.*, 1985; Simpson *et al.*, 1985; Kuhlemeier *et al.*, 1988). Moreover, a chimeric gene containing two copies of a LHCP control region produced twice the level of light-induced mRNA of a gene containing just one copy (Simpson *et al.*, 1986a). In addition, the quantitative effect of these control elements is influenced by the stage of plant development; young leaves being sensitive to control element copy number whereas mature leaves are not (Kuhlemeier *et al.*, 1988).

3. *Negative Control Elements*

Simpson *et al.* (1986a) observed that the LHCP control region repressed nopaline synthetase promoter activity in root cells despite being exposed to light. Since the nopaline synthetase promoter is normally constitutive (Simpson et al., 1985), the LHCP upstream control region therefore contains a developmental-specific *negative* control element or silencer in addition to the positive light-induced enhancer-like control elements. Kuhlemeier et al. (1987b) demonstrated that the *rbcS-EA* gene from pea is also under negative regulation; the negative control region is 58 bp in length and overlaps with the positive control region containing the repetitive elements Box II and III described above (Table IX; Kuhlemeier et al., 1988). A chimeric gene containing three copies of Box II inserted between the CaMV 35S enhancer and promoter was downregulated in the dark, indicating (1) that the positive and negative regulatory elements of the rbcS gene family are overlapping or may be one and the same; and (2) that the Box II element can function independently of Box III as a negative element but not as a positive regulator. Thus, the positive and negative regulatory functions of Boxes II and III may represent two distinct regulatory pathways that operate on similar *cis*-elements.

How complex is the network of controls that regulates light-regulated gene expression in plants? Light-regulated gene expression in plants is mediated by at least three spectrally distinct photoreceptors, phytochrome, a UV/blue-light receptor, and protochlorophyllide (reviewed by Kuhlemeier et al., 1987a). In addition, LHCP and rbcS gene activity is influenced by chloroplast development. If proplastid differentiation and carotenoid biosynthesis are blocked in etiolated leaves and shoots, then LHCP and rbcS mRNA does not accumulate in the light (Mayfield and Taylor, 1984; Eckes et al., 1985; Batschauer et al., 1986; Simpson et al., 1986b). Run-on transcription experiments (Batschauer et al., 1986), and the analysis of LHCP and rbcS chimeric gene constructs (Simpson et al., 1986b), indicate that gene activity in these plants is blocked at transcription. Thus, proplastid differentiation also influences factors that regulate LHCP and rbcS gene transcription. Identifying the control elements that regulate the expression of light-induced genes under different spectral and developmental conditions at the transcriptional and posttranscriptional levels is important for further defining the regulatory pathways that control light-regulated gene expression in the plant cell.

B. Plant Heat Shock Gene Expression

Plants, like all eukaryotic and prokaryotic organisms investigated, have a set of thermally induced heat-shock protein (*hsp*) genes whose products are correlated with increased heat tolerance in cells (Sachs and Ho, 1986 and this volume; Lindquist, 1986; Pelham, 1986). In plants, the *hsp* genes encode a complex array of 30–50 polypeptides that are grouped into three general size classes of 69–110, 20–27, and 15–18 kDa (Sachs and Ho, 1986). At least one plant gene, the maize 70-kDa *hsp* gene, is homologous to the *Drosophila hsp70* gene by nucleotide sequence comparison (Rochester et al., 1986).

The Coordinate Regulation of hsp Gene Expression in Plants

The regulation of heat-shock gene expression in plant cells is similar to that observed in *Drosophila*. A shift in temperature from 28° to 40°C causes a rapid reduction in the synthesis of normal proteins and the induced synthesis of heat-shock proteins (Key *et al.*, 1981). This striking change in gene expression is regulated at the level of mRNA accumulation (Czarnecka *et al.*, 1984) and translation (Key *et al.*, 1981).

The 5' transcriptional control region of all plant *hsp* genes examined (Schöffl *et al.*, 1984; Nagao *et al.*, 1985; Czarnecka *et al.*, 1985) contain multiple sequence elements 80–90% homologous to the 10-bp "enhancer-like" heat shock transcriptional control element, CT-GAA--TTC-AG (Pelham, 1982; Pelham and Bienz, 1982). This sequence is found in the 5' flanking DNA of all *Drosophila hsp* genes investigated, and under thermally induced conditions binds the transcription factor HSTF required for heat-shock gene transcription in *Drosophila* (Wu, 1984; Parker and Topol, 1984a,b; Topol *et al.*, 1985; Shuey and Parker, 1986). The Drosophila *hsp70* gene has three HSTF binding domains (Topol *et al.*, 1985) with two contiguous heat-shock elements in the promoter–proximal HSTF binding domain (sites 1 and 2). Only when both sites are occupied can transcription be initiated (Topol *et al.*, 1985). The binding of HSTF to site 1 alters HSTF contact with site 2, suggesting that HSTF molecules interact to facilitate transcription (Shuey and Parker, 1986). Maximum transcription was obtained *in vitro* only when all four copies of the heat-shock control element were present.

The analysis of soybean *hsp* gene regulation in transgenic sunflower calli demonstrated that a 95-bp control region of *Gmhsp17.5E*, containing two partially overlapping copies of the heat-shock control element allowed a low level of thermally induced *hsp* gene expression (Gurley *et al.*, 1986). The addition of adjacent upstream sequences containing two *hsp* elements increased expression 20-fold, suggesting that this repetitive control element family may regulate plant heat shock gene expression in a manner similar to animal cells.

C. Seed Protein Gene Regulation

As pointed out earlier, there are approximately 20,000 genes expressed at any given stage of plant embryogenesis (Goldberg *et al.*, 1981a; Galau and Dure, 1981). Seed protein genes represent one striking example of a highly regulated embryonic gene set. The *cis*-acting DNA sequences that control embryo-specific seed protein gene transcription have been localized for several different legume storage protein genes (Table IX). Preliminary studies indicate that, like the light-induced control elements of rbcS and LHCP, seed protein gene regulatory elements are multipartite (Chen *et al.*, 1986; R. B. Goldberg, J. K. Okamuro, T. L. Sims, and B. B. Mathews, unpublished), can confer embryo-specific expression on a heterologous promoter

(Chen et al., 1988; E. Krebbers, unpublished), and function from either orientation (Chen et al., 1988). The analysis of legume seed protein gene expression in transgenic petunia and tobacco seed by *in situ* hybridization (Barker et al., 1988; L. Perez-Grau, and R. B. Goldberg, unpublished) and developmental protein and RNA gel blots (Sengupta-Gopalan et al., 1985; Beachy et al., 1985; Okamuro et al., 1986) demonstrated that the spatial as well as the temporal patterns of seed protein gene expression are conserved in transgenic tobacco and petunia plants. Moreover, the nascent polypeptide chains are glycosylated, the hydrophobic signal sequence is correctly processed (Sengupta-Gopalan et al., 1985; Beachy et al., 1985), and the subunits are assembled into multimeric proteins (Beachy et al., 1985) and transported to protein bodies where they are stored until germination (Greenwood and Chrispeels, 1985). Together these studies demonstrate that the regulatory circuitry of *cis*-elements and *trans*-factors responsible for the temporal, spatial, and subcellular regulation of seed protein gene expression is conserved between the Leguminosae and Solanaceae. Thus, these plants provide an excellent system for the molecular dissection of the mechanisms that regulate seed protein gene expression.

Putative Embryo-Specific Cis-*Control Elements that May Regulate Seed Protein Gene Expression*

Is there a developmental-specific seed protein gene transcriptional control element(s)? Several 5' regulatory regions necessary for seed protein gene expression have been identified by deletion/transformation experiments (Table IX). The search for conserved oligonucleotide sequences within these control regions has identified four possible embryo-specific seed protein gene regulatory elements.

Chen et al. (1986) noted that a 98-bp 5' control region from the α'-subunit of β-conglycinin contains two 28-bp imperfect repeats and five copies of a 6-bp repeated sequence AGCCCA$\genfrac{}{}{0pt}{}{A}{C}$. This sequence is also found in the 5' flanking control region of the β-subunit of β-conglycinin (S. J. Barker and R. B. Goldberg, unpublished) and the β-subunit of phaseolin in French bean (Sengupta-Gopalan et al., 1985). When a 170-bp fragment containing this region was placed 5' to a chimeric gene containing the CaMV 35S promoter and the bacterial chloramphenicol acetyltransferase (CAT) gene coding sequence, CAT activity was enhanced at least 25-fold in transgenic tobacco seed (Chen et al., 1988). There was no positive or negative effect of this control region on CAT activity in leaf, stem, or root, nor did orientation affect the level of CAT activity, indicating that this enhancer-like control region is developmental-specific and does not have a detectable negative regulatory activity.

The activity of the β-conglycinin control region was dramatically affected by position. If the 170-bp fragment was moved to the CAT gene 3' noncoding region, CAT activity dropped more than 15-fold in tobacco seeds but was still reproducibly enhanced 2- to 5-fold (Chen *et al.*, 1988). Moving the control region further downstream 3' to the polyadenylation signal sequence abolished its effect. The β-conglycinin seed protein gene control region is thus subject to dramatic position effects over relatively small distances, much like the rbcS and LHCP gene enhancer-like control regions (Timko *et al.*, 1985; Fluhr *et al.*, 1986a; Simpson *et al.*, 1986a).

The probability that the unique cluster of repeats in the β-conglycinin 5' control region occurred by chance alone is less than 2% (Davidson *et al.*, 1983). However, this cluster is not found in the 5' control regions of other soybean seed protein genes such as glycinin, seed lectin, or Kunitz trypsin inhibitor (R. B. Goldberg *et al.*, unpublished). If these repetitive elements are regulatory in function, as suggested by Chen *et al.* (1986, 1988), they would appear to control only a subset of the seed protein genes in soybean.

Jofuku *et al.* (1987) identified a second putative seed protein gene control element by mapping the binding sites of an embryo nuclear DNA-binding protein to the soybean seed lectin gene 5' control region. DNaseI footprint experiments showed that this nuclear protein recognizes two regions 12 and 40 bp long containing a 7-bp core element ATT^A_TAAT found in the 5' control region of the soybean lectin, Kunitz trypsin inhibitor, glycinin, and β-conglycinin seed protein genes. The activity of this DNA-binding protein is temporally regulated during embryogenesis and correlates well with developmental changes in lectin gene transcription. Embryo nuclear protein extracts protect the same sequence motif from the Kunitz trypsin inhibitor gene 5' region from DNaseI digestion, and competition binding experiments confirmed that the Kunitz trypsin inhibitor gene 5' region and the lectin gene 5' region are recognized by the same DNA-binding protein *in vitro* (C. Reeves and R. B. Goldberg, unpublished). Although suggestive of a role in seed protein gene expression, the function of this putative control element network and interacting DNA-binding protein remain to be established.

Two other repetitive elements, CATGCAT (N. Nielsen, unpublished) and AACACA^A_C (T. L. Sims and R. B. Goldberg, unpublished; Goldberg, 1986), are found in the 5' flanking sequences of the soybean glycinin, β-conglycinin, and seed lectin genes as well as the seed protein genes from French bean and cotton. Clearly, there is no shortage of possible seed protein gene control elements in plants. Obviously, the next step is to test the function of each sequence *in vivo*.

D. Combinatorial Model for Seed Protein Gene Regulation

Not all seed protein genes are expressed solely in the cotyledons of the embryo. For example, the soybean lectin gene *Le1* is expressed at low levels in the ground meristem of postgerminated seedlings (L. Perez-Grau and

R. B. Goldberg, unpublished) and at rare class mRNA levels in the roots and leaves of the mature plant (Okamuro et al., 1986; B. Mathews and R. B. Goldberg, unpublished). By contrast, soybean Kunitz trypsin inhibitor gene transcripts are highly abundant in both the axis and cotyledons of the embryo (L. Perez-Grau and R. B. Goldberg, unpublished), and are also found at rare class mRNA levels in the leaf, stem, and root of the mature plant (K. D. Jofuku and R. B. Goldberg, unpublished). These observations indicate that, while all seed protein genes are expressed at high levels during embryogenesis, some genes also display unique patterns of gene expression in the organs of the mature sporophyte. How does the plant cell generate such diverse patterns of seed protein gene regulation?

Theoretically, only a small number of positive and/or negative *cis*-control elements are necessary to generate a diversity of developmental, tissue, and cell-specific patterns of gene transcription (E. Davidson, unpublished; Dickinson, 1988). The number of possible combinations of n *cis*-elements, taken r elements at a time, can be calculated using Eq. (21).

$$C_r^n = \frac{n!}{r!(n-r)!} \quad (21)$$

Figure 14 illustrates how the number of possible combinations increases exponentially with the number of *cis*-elements. A set of 20 control elements

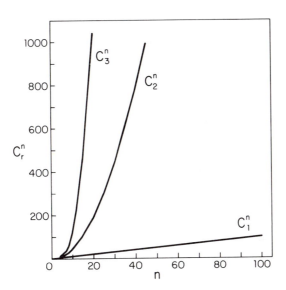

Fig. 14. A combinatorial model illustrates the power of combining different *cis*-control elements to generate an array of developmentally complex or developmentally specific patterns of gene expression. Each curve was generated according to Eq. (21) where n represents the number of types of control elements in the cell and r represents the number of different elements used to regulate the expression of each gene. C_r^n is the number of different combinations of *cis*-elements possible under these conditions.

(n) taken 3 elements at a time (r) can generate more than 1000 developmentally unique combinations of control elements. Therefore, the potential for generating different developmental, tissue, and cell-specific patterns of gene transcription, given a simple repertoire of control elements, is immense. In reality, the situation may be more complex.

Recently, Strittmatter and Chua (1987) showed that by combining a 36-bp heat-shock control region from the soybean *hsp17.3B* gene with the light-inducible 5' control region from pea *rbcS-3A*, a chimeric reporter gene was rendered light-inducible and organ-specific under heat-shock conditions, thus demonstrating that unique patterns of gene regulation are possible by generating novel combinations of *cis*-control elements. In animals, developmentally complex patterns of gene transcription are determined by different combinations of physically separable and functionally autonomous *cis*-control elements; for example, the *Drosophila hsp26* and *yellow* genes (Cohen and Meselson, 1985; Geyer and Corces, 1987), human metallothionein II$_A$ (Karin *et al.*, 1984), yeast *HO* (Breeden and Nasmyth, 1987), mouse α-fetoprotein (Hammer *et al.*, 1987), and the SV40 enhancer (Schirm *et al.*, 1987).

Different combinations of developmental-specific control elements may explain the contrasting pattern of soybean seed protein gene expression during development. For example, all seed protein genes, including lectin and Kunitz trypsin inhibitor, may have one or more copies of a *cis*-acting control element (E_1) responsible for regulating cotyledon-specific seed protein gene transcription during embryogenesis. The embryo-specific control element (E_1) would interact with a sequence-specific nuclear DNA-binding factor ($SETF_{E1}$), such as that described by Jofuku *et al.* (1987), to activate transcription at the appropriate time during embryogenesis. Seed protein genes expressed in the organs of the mature plant, like lectin and Kunitz trypsin inhibitor, may also contain organ-specific or tissue-specific control elements (L, S, R_{GM}) that interact with other developmentally regulated sequence-specific transcription factors to trigger transcription in the relevant organ system. For example, soybean lectin gene expression in root ground meristem would be induced by the interaction of a developmental-specific transcription factor $SRTF_{GM1}$ with the *cis*-control element R_{GM1}.

E. Networks of *cis*-Elements and *trans*-Factors Coordinate the Expression of Structural Gene Sets in Eukaryotic Organisms

Evidence for a network of *cis*-elements and *trans*-factors that coordinate the expression of structural gene sets in eukaryotic cells is accumulating rapidly. Genetic and molecular studies in yeast and animal cells have provided many elegant examples (Table X). Regulatory networks control the expression of genes involved in such diverse cell processes as cell-type determination (yeast mating type control), hormone-induced gene expression (glucocorticoid regulated genes), environmentally induced gene expression (heat shock), cell-type specific gene expression (immunoglobulin heavy and light chain synthesis), chromatin packaging (histone synthesis), and respiration (cytochrome synthesis). Perhaps the most elegant example is the regulation of amino acid biosynthesis in *Saccharomyces cerevisiae*.

In yeast, at least 24 and perhaps as many as 50 unlinked genes involved in a variety of amino acid biosynthetic pathways are under common control. This horizontal form of regulatory circuitry is known as the general amino acid control system or GCN (reviewed by Jones and Fink, 1982; Fink, 1986; Hinnebusch, 1988). In response to starvation for a single amino acid the set of genes under GCN control is coordinately derepressed. For example, starvation for tryptophan results in the derepression of genes encoding enzymes in the tryptophan, arginine, histidine, isoleucine, and lysine biosynthetic pathways.

Genetic analysis by Hinnebusch and Fink (1983) demonstrated that genes under general amino acid control are regulated by several *trans*-acting loci, including the "master-control" gene *GCN4*. The *GCN4* gene product is a DNA-binding factor (Hope and Struhl, 1985) that recognizes a short repeated *cis*-acting control element TGACTC required for the expression of virtually all genes under general amino acid control (Struhl, 1982; Hinnebusch and Fink, 1983; Donahue *et al.*, 1983; Lucchini *et al.*, 1984; Hinnebusch *et al.*, 1985). The construction of a chimeric *GCN4* regulatory protein demonstrated that the GCN4 protein has a regulatory domain that is physically separable from its DNA-binding function, suggesting that it must interact with other transcription factors to regulate genes under general control. This elegant example of coordinate gene regulation, together with other examples from yeast and animal gene systems (Table X) illustrate how the eukaryotic cell utilizes a network of repetitive DNA sequences and sequence-specific nuclear DNA-binding proteins to coordinate the transcriptional activity of genes involved in diverse cellular activities. Clearly, the principle of coordinate gene regulation through repetitive *cis*-control elements and sequence-specific *trans*-acting factors, first proposed by Britten and Davidson (1969) has emerged as one of the fundamental tenets of gene regulation in eukaryotic cells.

F. Identifying Plant *trans*-Acting Regulatory Factors

To understand how the temporal and spatial pattern of plant gene expression is programmed during development, it is obvious that we must also study the expression and regulation of plant *trans*-acting regulatory genes. In the final section of this chapter we discuss recent attempts to identify and clone plant *trans*-acting regulatory genes using a powerful combination of genetic and molecular approaches.

TABLE X

Regulation of Gene Networks in the Eukaryotic Cell

Organism/gene network	Regulation	Consensus sequence
Drosophila melanogaster		
hsp70, hsp83, hsp26, hsp27, hsp68, hsp23, hsp22, hsp82	Induced by heat shock	CTNGAATNTTCTAGA
Man		
Immunoglobulin μ	B cell	ATTTGCAT
Immunoglobulin k	specific	
Mouse		
Tyr-aminotransfersae	Glucocorticoid	GGTACANNNTGTTCT
Tryp-oxygenase	controlled	
Man		
Metallothionein II$_A$	Glucocorticoid	GGTACANNNTGTTCT
Growth hormone	controlled	
Placental lactogen	Glucocorticoid controlled	GGTACANNNTGTTCT
Chicken		
Lysozyme	Glucocorticoid controlled	GGTACANNNTGTTCT
Man		
c-fos	Induced by serum factors	GATGTCCATATTAGGACATC
Xenopus		
γ-actin	Induced by serum factors	GATGTCCATATTAGGACATC
Man		
Metallothionein I$_A$	Induced by	CTNTGCRCNCGGCCC
Metallothionein II$_A$	heavy metals	
Yeast		
HIS3, HIS4, ARG4, TRP5 ILV1, ILV2	Regulated by GCN4	GTGACTC
GAL1, GAL7, GAL10	Regulated by GAL4	CGGA^G_CGACACTC^C_GTCCG
CYC1, CYC7, Cytochrome b_2	Regulated by heme and HAP1	TGGCCGGGGTTTACGGACGATGA CCCTCGCTATTATCGCTATTAGC

Function tested	Copy number	Nuclear protein binding site	Reference
Yes	4(hsp70)	Yes/HSTF	Pelham (1982); Wu (1984); Topol et al. (1985)
Yes	1	Yes/NF-A	Mason et al. (1985)
Yes	1	Yes/NF-A	Sen and Baltimore (1986)
Yes	3	Gluc-receptor	Jantzen et al. (1987)
			Danesch et al. (1988)
Yes	2	Gluc-receptor	Karin et al. (1984)
	2	Gluc-receptor	Slater et al. (1985)
			Moore et al. (1985)
	2	Gluc-receptor	Eliard et al. (1985)
Yes	2	Gluc-receptor	Renkawitz et al. (1984)
Yes	1	Yes/SRF	Treisman (1986)
	1	Yes/SRF	Treisman (1986)
			R. Mohum and R. Treisman (unpublished)
Yes	8		Karin et al. (1984)
	7		Stuart et al. (1985)
HIS3	6(HIS3)	Yes/GCN4	Hope and Struhl (1985)
HIS4	5(HIS4)	Yes	Donahue et al. (1983)
	3(ILV1)	Yes	Arndt and Fink (1986)
	1(ILV2)	Yes	
Yes	4(GAL1/GAL10)	Yes/GAL4	Giniger et al. (1985)
CYC1	2	Yes/HAP1	Pfeifer et al. (1987a)
CYC7	1	Yes/HAP1	Pfeifer et al. (1987b)
			Lalonde et al. (1986)

(continued)

TABLE X (*continued*)

Organism/gene network	Regulation	Consensus sequence
STE3, MFα1, MFα2	Regulated by MATα1	TTTCCTAATTAGGAAA
STE6, BAR1, STE2, MFa1, MFa2	Repressed by MATα2	5'CATGTAATTACC-AATA AGGAAATTTACATG-T3'
HO, MATα1, STE5	Regulated by MATa1 + MATα2	TC^A_GTGTNN^A_TNANNTACATCA
H2A-H2B (TRT1) H3-H4 (CI) H3-H4 (CII)	Cell cycle regulated	GCGAAAAANTNRGAAC

1. Genetic Analysis of Plant Gene Regulation

Some of the most compelling evidence for a network of *trans*-acting regulatory genes controlling plant gene expression comes from genetic studies in maize. For example, the maize zein gene superfamily consists of four sets of zein genes that encode polypeptides of 22, 19, 15, and 10 kDa. Together these polypeptides represent greater than 50% of the endosperm protein (Mosse, 1966). The 22-, 19-, and 15-kDa zein gene families contain 24, 54, and 3 genes per haploid genome, respectively (Wilson and Larkins, 1984), and are dispersed on several chromosomes (Viotti *et al.*, 1979; Soave *et al.*, 1978; Valentini *et al.*, 1979). At least six *trans*-acting regulatory gene loci coordinate zein synthesis during endosperm development. Mutations at *floury-2* (*fl2*), *Mucronate* (*Mc*), and *opaque-6* (*o6*) suppress the expression of the 22- and 19-kDa zein gene families equally (Di Fonzo *et al.*, 1979; Salamini *et al.*, 1983; Soave *et al.*, 1976). *Opaque-2* (*o2*) and *De*-B30* mutants affect the expression of the 22-kDa zein gene family preferentially over the 19-kDa gene set (Soave *et al.*, 1976; Salamini *et al.*, 1979), while *opaque-7* mutants (*o7*) preferentially suppress 19-kDa zein gene family expression (Di Fonzo *et al.*, 1979). In the double mutant *o2o7*, the effect of the two mutations on 22- and 19-kDa, zein gene expression is additive (Ma and Nelson, 1974; Di Fonzo *et al.*, 1980). Thus, *opaque-2* and *opaque-7* appear to represent independent regulatory pathways. On the other hand, *Mucronate* and *opaque-2* work synergistically to suppress zein synthesis (Salamini *et al.*, 1983). In the double mutant *o2Mc*, zein levels are reduced by 90%, greater than the simple sum of the two mutations alone. Hence, zein gene superfamily expression in maize endosperm is controlled by a minimum of two regulatory pathways under the combined influence of at least six *trans*-acting regulatory gene products.

Function tested	Copy number	Nuclear protein binding site	Reference
STE3	1(STE3) 3(MFα1) 1(MFα2)	Yes/α1	Bender and Sprague (1987)
STE6	1× all genes	Yes/α2	Johnson and Herskowitz (1985)
Yes	10(HO) 1(MATα1) 2(STE5)		Miller et al. (1985a)
Yes	3 1 3		Osley et al. (1986)

The molecular details of zein gene regulation are still sketchy. Zein mRNA levels in homozygous *o2, o7,* and *fl2* plants suggest that the products of these gene loci are necessary for 22- and 19-kDa zein mRNA accumulation specifically (Pedersen *et al.,* 1980; Landridge *et al.,* 1982; Burr and Burr, 1982; Marks *et al.,* 1985). The expression of nonzein endosperm-specific mRNAs is not affected by these mutations (Salamini *et al.,* 1983). One hypothesis is that these genes encode a set of zein gene-specific transcription factors that recognize a *cis*-control element common to one or more classes of zein genes. Brown *et al.* (1986) found one highly conserved 15-bp element in the 5' flanking region of zein and several other endosperm-specific monocot seed protein genes. However, *o2, o7,* and *fl2* may just as easily encode posttranscriptional regulatory factors that regulate zein nuclear RNA turnover, nuclear transport, or zein mRNA stability.

2. Cloning of Plant Regulatory Gene Loci

To begin analyzing a genetically identified *trans*-acting regulatory gene at the molecular level, it is necessary to clone the mutant gene from a mutant genome library. Transposon mutagenesis (Bingham *et al.,* 1981) is a powerful strategy for tagging and cloning mutant gene loci. However, a very limited number of plant species contain active transposons and an even smaller number have been cloned (see Nevers *et al.,* 1986). Thus, the opportunities for cloning plant regulatory genes by transposon tagging are restricted by the distribution of cloned transposons. Table XI lists some of the regulatory and structural genes cloned from maize and snapdragon using transposable elements. Recently, Baker *et al.* (1986) and VanSluys *et al.* (1987) demonstrated that the maize transposable element *Ac* transposes at a high frequency in transgenic tobacco, carrot, and *Arabidopsis* cells, suggesting that

TABLE XI
Plant Regulatory and Structural Genes Cloned by Transposon Tagging

Plant species	Gene locus cloned	Gene function	Transposable element system	Reference
Maize	*Bz1*	UDP glycosyltransferase	Ac	Federoff *et al.* (1984)
	A1	NADPH-dependent reductase	Spm(En); Mu	O'Reilly *et al.* (1985)
	C2	Chalcone synthase	Spm(En)	Wienand *et al.* (1986)
	C	Regulates *C2* and *Bz1*	Spm(En); Ds	Paz-Ares *et al.* (1986)
				Cone *et al.* (1986)
	Bz2	Anthocyanin synthesis	Ds	Theres *et al.* (1987)
	R	Regulates *C2* and *Bz1*	Ac	Dellaporta *et al.* (1988)
	O2	Regulates zein synthesis	Spm	Schmidt *et al.* (1987)
			Ac	Motto *et al.* (1988)
Snapdragon	*Pal*	Anthocyanin synthesis; homologous to the *A1* gene from maize	Tam3	Martin *et al.* (1985)

it may be possible to use transposons from a heterologous system as an insertional mutagen in a variety of plant species.

The maize regulatory gene locus *o2* was cloned using the transposon *Spm* (Schmidt *et al.*, 1987). RNA gel blots indicate that the *O2* gene is expressed in wild-type endosperm but not in homozygous *o2* endosperm or leaf. Homozygous *o7* and *fl2* plants produce normal levels of *O2* mRNA, suggesting that *o2* gene expression is not regulated by *o7* or *fl2*. Whether *O2* encodes a sequence-specific DNA-binding transcription factor or another regulatory function must await further analysis of this gene. The molecular analysis of *o2*, *o7*, *fl2*, and other *trans*-acting loci may soon provide fresh insight into the regulation of the zein gene "superfamily" during kernel development.

Another system having at least five *trans*-acting regulatory loci (*C*, *R*, *Vp*, *Clf*, *Pl*) regulates flavonoid synthesis in maize. One of these loci *R*, coordinates the expression of chalcone synthase (C2), UDP glucosyltransferase (Bz), and other enzymes involved in flavonoid synthesis in the aleurone (Dooner, 1983). Complex alleles of this locus display altered patterns of anthocyanin tissue-specificity. For example, one allele *R-r:standard*, is composed of two homologous elements, seed (S) and plant (P), that regulate anthocyanin synthesis in kernel aleurone and in various parts of the seedling and mature plant, respectively (Dooner and Kermicle, 1971). P and S can be separated by meiotic recombination into seed only (*R-g*) and plant only (*r-r*) alleles (Dooner and Kermicle, 1971, 1974, 1976). The molecular basis of P and S developmental specificity is not known, but it appears that after duplication of the original gene there evolved a new pattern of developmental regulation.

Several regulatory and structural genes in the maize anthocyanin biosynthetic pathway were cloned by transposon mutagenesis (Table XI; Dellaporta *et al.*, 1988). Molecular analysis of *R-r:standard* and other *R* alleles by DNA gel blot experiments confirmed the genetic model of *R-r* organization, indicating that P and S are homologous at the DNA level. A search for *R* encoded transcripts identified a 2.5-kb mRNA in strongly anthocyanin-pigmented seedling tissue (S. Wessler and S. Dellaporta, unpublished). The function of this transcript is unknown. The cloning and analysis of *R* locus developmental-specific *trans*-acting regulatory alleles should provide new insight into the differential regulation of the anthocyanin biosynthetic pathway.

Unlike *R*, the regulatory effect of *C* on chalcone synthase and flavonoid 3-O-glucosyltransferase (UFGT) gene expression is restricted to the aleurone and scutellum. Using *Spm* and *Ds* tagging, Paz-Ares *et al.* (1986) and Cone *et al.* (1986) cloned the maize *C1* locus and found four endosperm mRNA transcripts complementary to *C1* DNA, several of which were cloned and sequenced. Two transcripts appear to originate from the *C* locus (Paz-Ares *et al.*, 1987; Cone *et al.*, 1986). One clone, encoding a 1.4-kb mRNA transcript, was approximately 40% homologous to the *myb* protooncogenes from man, chicken, mouse, and *Drosophila*. These genes encode nuclear DNA-binding proteins of unknown function. The region of homology spanned the basic DNA-binding domain of these proteins, which prompted Paz-Ares *et al.* (1987) to propose that *C* might encode a sequence-specific DNA-binding protein that regulates the transcription of structural genes in the anthocyanin biosynthetic pathway such as *A1*, *C2*, and *Bz1*.

Determining the cellular function of *trans*-acting regulatory genes is a difficult challenge. In *Drosophila melanogaster*, many gene loci affecting segmentation, embryonic polarity, and cell fate during embryogenesis have been thoroughly analyzed at the molecular level (Lehmann and Nüsslein-Volhard, 1986; Edgar *et al.*, 1986; Mlodzik and Gehring, 1987). Despite a wealth of genetic and molecular information on these genes, their function in most cases is speculative, often inferred from sequence homology to animal and fungal genes (Rosenberg *et al.*, 1986; Villares and Cabrera, 1987). Thus, both genetic and molecular approaches are critical to developing a conceptual understanding of the cellular processes that control plant differentiation.

G. Molecular Analysis of Plant *trans*-Acting Regulatory Factors

Unlike prokaryotic RNA polymerase holoenzyme, purified RNA polymerase II does not have a specific affinity for promoter DNA *in vitro* but binds to DNA indiscriminately (Roeder, 1976). Studies of both viral and cellular gene transcription show that eukaryotic nuclei contain additional transcription factors that regulate promoter recognition by RNA polymerase II, thereby controlling transcription initiation (Jones *et al.*, 1985; Carthew *et al.*, 1985). Some transcription factors are nonspecific (Parker and Topol, 1984a; Hawley and Roeder, 1985; Sawadogo and Roeder, 1985a), while others are promoter-specific (Parker and Topol, 1984b; Sawadogo and Roeder, 1985b). Promoter-specific transcription factors are generally DNA-binding proteins that recognize specific *cis*-acting transcriptional control elements (Table X). This sequence-specific DNA-binding activity makes the characterization, purification, and cloning of these *trans*-acting regulatory factors possible by techniques such as DNA-binding gel electrophoresis (Fried and Crothers, 1981; Garner and Revzin, 1981; Strauss and Varshavsky, 1984), footprint analysis (Galas and Schmitz, 1978), and affinity chromatography (Kadonaga and Tjian, 1986; Briggs *et al.*, 1986; Kasher *et al.*, 1986). Unlike the genetic approach, cloning specific DNA-binding regulatory factors can theoretically be applied to almost any gene system.

The characterization of sequence-specific DNA-binding factors in plants is just beginning. In maize, Maier *et al.* (1987) showed that endosperm nuclei contain a DNA-binding protein that recognizes a 22-bp region 5' to a maize 19-kDa zein gene. This region contains 14 nucleotides of the 15-bp repetitive element CACATGTGTAAAGGT which is located 300 bp from the transcription start site and is found 5' to all maize 19- and 22-kDa zein genes investigated (Brown et al., 1986). A similar sequence was found 5' to the endosperm-specific maize sucrose synthase gene (Werr et al., 1985) and the barley β-hordein and wheat α-gliadin genes (Forde et al., 1985), suggesting that this DNA-binding activity and its corresponding DNA-binding site may coordinate the tissue-specific transcription of these genes during seed development.

It is not known whether the DNA-binding activity found in maize endosperm is genetically linked to any of the four trans-acting regulatory loci (o2, c, fl2, Mc) that affect zein gene expression in maize described above. However, three of these loci affect zein mRNA levels in endosperm (Pedersen et al., 1980; Landridge et al., 1982; Burr and Burr, 1982; Marks et al., 1985). Thus, one or more of these genes may be involved in regulating zein gene transcription.

In soybean, Jofuku et al. (1987) showed that the developing embryo contains a nuclear DNA-binding protein that interacts with two regions in the lectin gene 5' flanking DNA. The two regions share a 7-nucleotide core sequence ATT$_A^T$AAT that is strongly protected from DNaseI digestion when treated with embryo nuclear protein extract. Competition binding experiments demonstrated that this DNA-binding protein recognizes a similar sequence 5' to other soybean seed protein genes including Kunitz trypsin inhibitor, glycinin, and β-conglycinin, but not DNA sequences from the lectin gene coding region, soybean leghemoglobin, or a Drosophila blastoderm-specific gene (Jofuku et al., 1987; C. Reeves and R. B. Goldberg, unpublished). Furthermore, the DNA-binding activity is not detectable in nuclear protein extracts from leaf, stem, or roots and is temporally regulated during embryogenesis, fluctuating quantitatively with the increases and decreases in soybean lectingene transcription. Together, these data suggest that this DNA-binding activity and its corresponding binding site may represent elements of the regulatory circuitry responsible for the developmental regulation of soybean seed protein gene transcription during embryogenesis.

Although it is still too early to generalize about trans-acting regulatory factors in plants, there are some general features of sequence-specific DNA-binding transcription factors emerging in yeast and animal cells. As in prokaryotes, there are positive or gene "activating" proteins (Bram and Kornberg, 1985; Breeden and Nasmyth, 1987) and negative or "repressor-like" proteins (Johnson and Herskowitz, 1985; Sternberg et al., 1987; Shore and Nasmyth, 1987). Eukaryotic transcription factors and their corresponding DNA sequences, however, unlike prokaryotic regulatory factors, can regulate transcription over large and variable distances from the gene promoter. DNA-binding proteins have, in general, two distinct structural features. Prokaryotic activators and repressors contain a "helix-turn-helix" structure that is intimately involved in DNA attachment (see Pabo and Sauer, 1984).

Eukaryotic DNA-binding proteins contain a series of "finger-like" metal-binding domains that are necessary for sequence specific DNA binding (Miller *et al.*, 1985b; Hartshorne *et al.*, 1986; Rosenberg *et al.*, 1986; Hollenberg *et al.*, 1987). However, these features are not typical of all DNA-binding regulatory factors (Hope and Struhl, 1986).

The DNA-binding and regulatory functions of most but not all transcription factors analyzed to date represent functionally separable protein domains (Brent and Ptashne, 1985; Hope and Struhl, 1986). Therefore, the regulation of gene activity is not simply a function of DNA binding but involves cooperative interaction with other transcription factors.

H. Summary

Plant gene expression is governed by an elaborate system of *cis*-acting control elements and *trans*-acting gene products affecting gene expression at the transcriptional and posttranscriptional level. These processes are responsible for establishing the cell, tissue, and organ-specific patterns of gene expression as well as the abundance of mRNA transcripts. The mechanisms of plant gene regulation appear to be fundamentally similar to those found in all eukaryotic cells despite the unique reproductive, morphogenetic, and physiological processes of plants.

Plant transcriptional control elements are generally found in the 5' flanking region of a gene within 1–2 kb of the transcription start site. Plant transcriptional control regions analyzed to date appear to be redundant, multipartite, function from either orientation, and contain short repetitive oligonucleotide sequences that act as recognition sites for nuclear DNA-binding proteins. The regulatory function of these short repeats has as yet not been thoroughly tested *in vivo*.

REFERENCES

Anderson, M. A., Cornish, E. C., Mau, S.-L., Williams, E. G., Hoggart, R., Atkinson, A., Bonig, I., Grego, B., Simpson, R., Roche, P. J., Haley, J. D., Penschow, J. D., Niall, H. D., Tregear, G. W., Coghlan, J. P., Crawford, R. J. and Clarke, A. E. (1986). *Nature* **321**, 38–44.
Apel, K. (1981). *Eur. J. Biochem.* **120**, 89–93.
Arndt, K. and Fink, G. R. (1986). *Proc. Natl. Acad. Sci. U.S.A.* **83**, 8516–8520.
Auger, S., Baulcombe, D., and Verma, D. P. S. (1979). *Biochim. Biophys. Acta* **563**, 496–507.
Axel, R., Feigelson, P., and Schutz, G. (1976). *Cell* **7**, 247–254.
Baker, B., Schell, J., Lorz, H., and Federoff, N. (1986). *Proc. Natl. Acad. Sci. U.S.A.* **83**, 4844–4848.
Banerji, J., Rusconi, S., and Schaffner, W. (1981). *Cell* **27**, 299–308.
Banerji, J., Olson, L., and Schaffner, W. (1983). *Cell* **33**, 729–740.
Barker, S. J., Harada, J. J., and Goldberg, R. B. (1988). *Proc. Natl. Acad. Sci. U.S.A.* **85**, 458–462.

Batschauer, A., and Apel, K. (1984). *Eur. J. Biochem.* **143,** 593–597.
Batschauer, A., Mösinger, E., Kreuz, K., Dörr, K., and Apel, K. (1986). *Eur. J. Biochem.* **154,** 625–634.
Beach, L. R., Spencer, D., Randall, P. J., and Higgins, T. J. V. (1985). *Nucl. Acids Res.* **13,** 999–1013.
Beachy, R. N., Chen, Z.-L., Horsch, R. B., Rogers, S. G., Hoffman, N. J., and Fraley, R. T. (1985). *EMBO J.* **4,** 3047–3053.
Bedbrook, J. R., Jones, J., O'Dell, M., Thompson, R., and Flavell, R. B. (1980a). *Cell* **19,** 545–560.
Bedbrook, J. R., O'Dell, M., and Flavell, R. B. (1980b). *Nature* **208,** 133–137.
Belasco, J. G., Nilsson, G., von Gabain, A., and Cohen, S. N. (1986). *Cell* **46,** 245–251.
Bender, A., and Sprague Jr., G. F. (1987). *Cell* **50,** 681–691.
Bennett, M. D. (1972). *Proc. R. Soc. Lond. B.* **181,** 109–135.
Bennett, M. D., and Smith, J. B. (1976). *Proc. R. Soc. Lond. B.* **274,** 227–274.
Bernatzky, R., and Tanksley, S. D. (1986). *Mol. Gen. Genet.* **203,** 8–14.
Berry, J. O., Nikolau, B. J., Carr, J. P., and Klessig, D. F. (1986). *Mol. Cell. Biol.* **6,** 2347–2353.
Berry-Lowe, S. L., and Meagher, R. B. (1985). *Mol. Cell. Biol.* **5,** 1910–1917.
Berry-Lowe, S. L., McKnight, T. D., Shah, D. M., and Meagher, R. B. (1982). *J. Mol. Appl. Gen.* **1,** 483–498.
Bingham, P. M., Lewis, R., and Rubin, G. M. (1981). *Cell* **25,** 693–704.
Bishop, J. O., Morton, J. G., Rosbach, M., and Richardson, M. (1974). *Nature* **250,** 199–204.
Blanchard, J.-M., Weber, J., Jelinek, W., and Darnell, J. E. (1978). *Proc. Natl. Acad. Sci. U.S.A.* **75,** 5344–5348.
Blumberg, D. D., and Lodish, H. F. (1980). *Dev. Biol.* **78,** 285–300.
Bram, R. J., and Kornberg, R. D. (1985). *Proc. Natl. Acad. Sci. U.S.A.* **82,** 43–47.
Breeden, L., and Nasmyth, K. (1987). *Cell* **48,** 389–397.
Brent, R., and Ptashne, M. (1985). *Cell* **43,** 729–736.
Briggs, W. R., and Iiono, M. (1983). *Philos. Trans. R. Soc. London Ser. B.* **303,** 347–359.
Briggs, M. R., Kadonaga, J. T., Bell, S. P., and Tjian, R. (1986). *Science* **234,** 47–52.
Britten, R. J., and Davidson, E. H. (1969). *Science* **165,** 349–357.
Britten, R. J., and Davidson, E. H. (1986). *In* "Nucleic Acid Hybridisation" (B. D. Hames and S. J. Higgins, eds.), pp. 3–15. IRL Press, Washington, D.C.
Britten, R. J., and Kohne, D. E. (1968). *Science* **161,** 529–540.
Britten, R. J., Graham, D. E., and Neufeld, B. R. (1974). *In* "Methods in Enzymology" (L. Grossman and K. Moldave, eds.), Volume XXIX Part E, pp. 363–405. Academic Press, London.
Broglie, R., Coruzzi, G., Lamppa, G., Keith, B., and Chua, N.-H. (1983). *Bio/Technology* **1,** 55–61.
Broglie, R., Coruzzi, G., Fraley, R. T., Rogers, S. G., Horsch, R. B., Niedermeyer, J. G., Fink, C. L., Flick, J. S., and Chua, N.-H. (1984). *Science* **224,** 838–843.
Brown, J. W. S., Wandelt, U., and Feix, G. (1986). *Eur. J. Biochem.* **42,** 161–170.
Buffard, D., Vaillant, V., and Esnault, R. (1982). *Eur. J. Biochem.* **126,** 129–134.
Burr, F. A., and Burr, B. (1982). *J. Cell Biol.* **94,** 201–206.
Caplan, A., Herrera-Estrella, L., Inze, D., Van Haute, E., Van Montagu, M., Schell, J., and Zambryski, P. (1983). *Science* **222,** 815–821.
Carthew, P. W., Chodosh, L. A., and Sharp, P. A. (1985). *Cell* **43,** 439–448.
Cashmore, A. R. (1979). *Cell* **17,** 383–388.
Cashmore, A. R. (1984). *Proc. Natl. Acad. Sci. U.S.A.* **81,** 2960–2964.
Castresana, C., Garcia-Laque, I., Alonso, E., Malik, V. S., and Cashmore, A. R. (1988). *EMBO J.* **7,** 1929–1936.
Cech, T. R., and Hearst, J. E. (1976). *J. Mol. Biol.* **100,** 227–256.
Chamberlin, M. E., Galau, G. A., Britten, R. J., and Davidson, E. H. (1978). *Nuc. Acids Res.* **5,** 2073–2094.

Chang, C., and Meyerowitz, E. (1986). *Proc. Natl. Acad. Sci. U.S.A.* **83,** 1408–1412.
Chaubet, N., Philipps, G., Chaboute, M.-E., Ehling, M., and Gigot, C. (1986). *Plant Mol. Biol.* **6,** 253–263.
Chauboute, M.-E., Chaubet, N., Philipps, G., Ehling, M., and Gigot, C. (1987). *Plant Mol. Biol.* **8,** 179–191.
Chaudhari, N., and Hahn, W. E. (1983). *Science* **220,** 924–928.
Chen, Z.-L., Schuler, M. A., and Beachy, R. N. (1986). *Proc. Natl. Acad. Sci. U.S.A.* **83,** 8560–8564.
Chen, Z.-L., Pan, N., and Beachy, R. N. (1988). *EMBO J.* **7,** 297–302.
Chourey, P. S., and Nelson, O. E. (1976). *Biochem. Genet.* **14,** 1041–1055.
Coen, E. S., Carpenter, R., and Martin, C. (1986). *Cell* **47,** 285–296.
Cohen, R. S., and Meselson, M. (1985). *Cell* **43,** 737–746.
Colbert, J. T., Hershey, H. P., and Quail, P. H. (1983). *Proc. Natl. Acad. Sci. U.S.A.* **80,** 2248–2252.
Colbert, J. T., Hershey, H. P., and Quail, P. H. (1985). *Plant Mol. Biol.* **5,** 91–101.
Colot, V., Robert, L. S., Kavanaugh, T. A., Goldsbrough, A. P., Bevan, M. W., and Thompson, R. D. (1987). *EMBO J.* **6,** 3559–3564.
Cone, K. C., Burr, F. A., and Burr, B. (1986). *Proc. Natl. Acad. Sci. U.S.A.* **83,** 9631–9635.
Conger, B. V., Hughes, K. W., Skirvin, R. M., Bottino, P. J., and Mott, R. L. (1981). In "Cloning Agricultural Plants Via In Vitro Techniques" (B. V. Conger, ed.), CRC Press, Boca Raton, Florida.
Cornish, E. C., Pettitt, J. M., Bonig, I., and Clarke, A. E. (1987). *Nature* **326,** 99–102.
Coruzzi, G., Broglie, R., Cashmore, A., and Chua, N.-H. (1983). *J. Biol. Chem.* **258,** 1399–1402.
Coruzzi, G., Broglie, R., Edwards, C., and Chua, N.-H. (1984). *EMBO J.* **3,** 1671–1679.
Cox, K. H., DeLeon, D. V., Angerer, L. M., and Angerer, R. C. (1984). *Dev. Biol.* **101,** 485–502.
Cullis, C. A. (1976). *Heredity* **36,** 73–79.
Cullis, C. A. (1981). *Biochim. Biophys. Acta* **652,** 1–15.
Cuming, A. C., and Bennett, J. (1981). *Eur. J. Biochem.* **118,** 71–80.
Czarnecka, E., Edelman, L., Schöffl, F., and Key, J. L. (1984). *Plant Mol. Biol.* **3,** 45–58.
Czarnecka, E., Gurley, W. B., Nagao, R. T., Mosquera, L. A., and Key, J. L. (1985). *Proc. Natl. Acad. Sci. U.S.A.* **82,** 3726–3730.
Davidson, E. H. (1976). "Gene Activity in Early Development," 2nd edition, Academic Press, Orlando, Florida.
Davidson, E. H. (1986). "Gene Activity in Early Development," 3rd edition, Academic Press, Orlando, Florida.
Davidson, E. H., and Britten, R. J. (1969). *Science* **204,** 1052–1059.
Davidson, E. H., and Hough, B. R. (1971). *J. Mol. Biol.* **56,** 491–506.
Davidson, E. H., Jacobs, H. T., and Britten, R. J. (1983). *Nature* **301,** 468–470.
Dean, C., van den Elzen, P., Tamaki, S., Dunsmuir, P., and Bedbrook, J. (1985a). *Proc. Natl. Acad. Sci. U.S.A.* **82,** 4964–4968.
Dean, C., van den Elzen, P., Tamaki, S., Dunsmuir, P., and Bedbrook, J. (1985b). *EMBO J.* **4,** 3055–3061.
Dellaporta, S. L., Greenblatt, I., Kermicle, J., Hicks, J. B., and Wessler, S. (1988). Stadler Genetics Symposia (in press).
Dennis, E. S., Gerlach, W. L., Pryor, A. J., Bennetzen, J. L., Inglis, A., Llewellyn, D., Sachs, M. M., Ferl, R. J., and Peacock, W. J. (1984). *Nuc. Acids Res.* **12,** 3983–4000.
Derman, E., Krauter, K., Walling, L., Weinberger, C., Ray, M., and Darnell, J. E. (1981). *Science* **23,** 731–739.
de Vries, S. C., Springer, J., and Wessels, J. G. H. (1983). *Planta* **158,** 42–50.
Dickinson, W. J. (1988). *Bioessays* **8,** 204–208.
Di Fonzo, N., Gentinetta, E., Salamini, F., and Soave, C. (1979). *Plant Sci. Lett.* **14,** 345–354.

Di Fonzo, N., Fornasari, E., Salamini, F., Reggiani, R., and Soave, C. (1980). *J. Hered.* **71,** 397–402.
Domoney, C., and Casey, R. (1985). *Nuc. Acids Res.* **13,** 687–699.
Donahue, T. F., Daves, R. S., Lucchini, G., and Fink, G. R. (1983). *Cell* **32,** 89–98.
Dooner, H. (1983). *Mol. Gen. Genet.* **189,** 136–141.
Dooner, H., and Kermicle, J. L. (1971). *Genetics* **67,** 427–436.
Dooner, H. K., and Kermicle, J. L. (1974). *Genetics* **78,** 691–701.
Dooner, H. K., and Kermicle, J. L. (1976). *Genetics* **82,** 309–322.
Doyle, J. J., Schuler, M. A., Godette, W. D., Zenger, V., Beachy, R. N., and Slightom, J. L. (1986). *J. Biol. Chem.* **261,** 9228–9238.
Dunsmuir, P., Smith, S. M., and Bedbrook, J. (1983). *J. Mol. Appl. Genet.* **2,** 285–300.
Dure, L., III, Greenway, S. C., and Galau, G. A. (1981). *Biochemistry* **20,** 4162–4168.
Dure, L., III, Pyle, J. B., Chlan, C. A., Baker, J. C., and Galau, G. A. (1983). *Plant Mol. Biol.* **2,** 199–206.
Eckes, P., Schell, J., and Willmitzer, L. (1985). *Mol. Gen. Genet.* **199,** 216–224.
Eckes, P., Rosahl, S., Schell, J., and Willmitzer, L. (1986). *Mol. Gen. Genet.* **205,** 14–22.
Edgar, B. A., Weir, M. P., Schubiger, G., and Kornberg, T. (1986). *Cell* **47,** 747–754.
Eliard, P. H., Marchand, M. J., Rousseau, G. G., Formstecher, P., Mathy-Hartert, M., Belayew, A., and Martial, J. A. (1985). *DNA* **4,** 409–417.
Ellis, J. G., Llewellyn, D. J., Dennis, E. S., and Peacock, W. J. (1987a). *EMBO J.* **6,** 11–16.
Ellis, J. G., Llewellyn, D. J., Walker, J. C., Dennis, E. S., and Peacock, W. J. (1987b). *EMBO J.* **6,** 3203–3208.
Esau, K. (1977). "Anatomy of Plants," 2nd edition, Wiley and Sons, New York.
Evans, R. M., Fraser, N., Ziff, E., Weber, J., Wilson, J., and Darnell, J. E. (1977). *Cell* **12,** 733–739.
Evans, I. M., Gatehouse, J. A., Croy, R. R. D., and Boulter, D. (1984). *Planta* **160,** 559–568.
Federoff, N. V., Furtek, D. B., and Nelson, O. E. (1984). *Proc. Natl. Acad. Sci. U.S.A.* **81,** 3825–3829.
Fink, G. R. (1986). *Cell* **45,** 155–156.
Fischer, R. L., and Goldberg, R. B. (1982). *Cell* **29,** 651–660.
Flavell, R. B., Bennett, M. D., Smith, J. B., and Smith, D. B. (1974). *Biochem. Genet.* **12,** 257–269.
Flavell, R. B., Bedbrook, J., Jones, J., O'Dell, M., Gerlach, W. L., Dyer, T. A., and Thompson, R. D. (1979). *Proc. John Innes Symp.,* Fourth, 1979, 15–29.
Fluhr, R., and Chua, N.-H. (1986). *Proc. Natl. Acad. Sci. U.S.A.* **83,** 2358–2362.
Fluhr, R., Moses, P., Morelli, G., Coruzzi, G., and Chua, N.-H. (1986a). *EMBO J.* **5,** 2063–2071.
Fluhr, R., Kuhlemeier, C., Nagy, F., and Chua, N.-H. (1986b). *Science* **232,** 1106–1112.
Flytzanis, C. N., Brandhorst, B. P., Britten, R. J., and Davidson, E. H. (1982). *Dev. Biol.* **91,** 27–35.
Forde, B. G., Heyworth, A., Pywell, J., and Kreis, M. (1985). *Nucl. Acids. Res.* **13,** 7327–7339.
Fraley, R. T., Rogers, S. G., Horsch, R. B., Eichholtz, D. A., Flick, J. S., Fink, C. L., Hoffman, N. L., and Sanders, P. R. (1985). *Bio/Technology* **3,** 629–635.
Fried, M., and Crothers, D. M. (1981). *Nucleic Acids Res.* **9,** 3047–3060.
Galas, D. J., and Schmitz, A. (1978). *Nucl. Acids Res.* **5,** 3157–3170.
Galau, G. A., and Dure, L. S., III (1981). *Biochem.* **20,** 4169–4178.
Galau, G. A., Britten, R. J., and Davidson, E. H. (1974). *Cell* **2,** 9–20.
Galau, G. A., Klein, W. H., Davis, M. M., Wold, B. J., Britten, R. J., and Davidson, E. H. (1976). *Cell* **7,** 487–505.
Galau, G. A., Klein, W. H., Britten, R. J., and Davidson, E. H. (1977b). *Arch. Biochim. Biophys.* **179,** 584–599.
Galau, G. A., Legocki, A. B., Greenwood, S. C., and Dure, L. S., III (1981). *J. Biol. Chem.* **256,** 2551–2560.

Gallagher, T. F., and Ellis, R. J. (1982). *EMBO J.* **1,** 1493–1498.
Garner, M. M., and Revzin, A. (1981). *Nucl. Acids. Res.* **9,** 3047–3060.
Gatehouse, J. A., Evans, I. M., Bown, D., Croy, R. R. D., and Boulter, D. (1982). *Biochem. J.* **208,** 119–127.
Gebhardt, C., Oliver, J. E., Forde, B. G., Saarelainen, R., and Miflin, B. (1986). *EMBO J.* **5,** 1429–1435.
Geyer, P. K., and Corces, V. G. (1987). *Genes Dev.* **1,** 996–1004.
Gillies, S. D., Morrison, S. L., Oi, V. T., and Tonegawa, S. (1983). *Cell* **33,** 717–728.
Giniger, E., Varnum, S. M., and Ptashne, M. (1985). *Cell* **40,** 767–774.
Goldberg, R. B. (1978). *Biochem. Genet.* **16,** 45–68.
Goldberg, R. B. (1986). *Phil. Trans. R. Soc. Lond. B* **314,** 343–353.
Goldberg, R. B., and Timberlake, W. E. (1980). *Nature* **283,** 601–603.
Goldberg, R. B., Hoschek, G., and Kamalay, J. C. (1978). *Cell* **14,** 123–131.
Goldberg, R. B., Hoschek, G., Tam, S. H., Ditta, G. S., and Breidenbach, R. W. (1981a). *Dev. Biol.* **83,** 201–217.
Goldberg, R. B., Hoschek, G., Ditta, G. S., and Breidenbach, R. W. (1981b). *Dev. Biol.* **83,** 218–231.
Goodspeed, T. H., and Clausen, R. E. (1927). *Univ. Calif. Publ. Bot.* **11,** 127–140.
Goodspeed, T. H., and Clausen, R. E. (1928). *Univ. Calif. Publ. Bot.* **11,** 245–256.
Gray, R. E., and Cashmore, A. R. (1976). *J. Mol. Biol.* **108,** 595–608.
Green, P. J., Kay, S. A., and Chua, N.-H. (1987). *EMBO J.* **6,** 2543–2549.
Greenwood, J. S., and Chrispeels, M. J. (1985). *Plant Physiol.* **79,** 65–71.
Grime, J. P., Shacklock, J. M. L., and Band, S. R. (1985). *New Phytol.* **100,** 435–445.
Groudine, M., Peretz, M., and Weintraub, H. (1981). *Mol. Cell. Biol.* **1,** 281–288.
Guarante, L., and Hoar, E. (1984). *Proc. Natl. Acad. Sci. U.S.A.* **81,** 7860–7864.
Guiltinan, M. J., Velten, J., Bustos, M. M., Cyr, R. J., Schell, J., and Fosket, D. E. (1987). *Mol. Gen. Genet.* **207,** 328–334.
Gurley, W. B., Hepburn, A. G., and Key, J. L. (1979). *Biochim. Biophys. Acta* **561,** 167–183.
Gurley, W. B., Czarnecka, E., Nagao, R. T., and Key, J. L. (1986). *Mol. Cell. Biol.* **6,** 559–565.
Hagen, G., Kleinschmidt, A., and Guilfoyle, T. (1984). *Planta* **162,** 147–153.
Hake, S., and Walbot, V. (1980). *Chromosoma* **79,** 251–278.
Hammer, R. E., Krumlauf, R., Camper, S. A., Brinster, R. L., and Tilghman, S. M. (1987). *Science* **235,** 53–58.
Hartshorne, T. A., Blumberg, H., and Young, E. (1986). *Nature* **320,** 283–287.
Hastie, N. D., and Bishop, J. O. (1976). *Cell* **9,** 761–774.
Hawley, D. K., and Roeder, R. G. (1985). *J. Biol. Chem.* **260,** 8163–8172.
Heinze, H., Herzfeld, F., and Kiper, M. (1980). *Eur. J. Biochem.* **111,** 137–144.
Hereford, L. M., and Rosbach, M. (1977). *Cell* **10,** 453–462.
Herr, W., and Clarke, J. (1986). *Cell* **45,** 461–470.
Herrera-Estrella, L., Van den Broeck, G., Maenhaut, R., Van Montagu, M., Schell, J., Timko, M., and Cashmore, A. (1984). *Nature* **310,** 115–120.
Hightower, R. C., and Meagher, R. B. (1985). *EMBO J.* **4,** 1–8.
Hinnebusch, A. G. (1985). *Mol. Cell. Biol.* **5,** 2349–2360.
Hinnebusch, A. G. (1988). *Microbiol. Rev.* **52,** 248–273.
Hinnebusch, A. G., and Fink, G. R. (1983). *J. Biol. Chem.* **258,** 5238–5247.
Hinnebusch, A. G., Lucchini, G., and Fink, G. R. (1985). *Proc. Natl. Acad. Sci. U.S.A.* **82,** 498–502.
Hofer, E., and Darnell, Jr., J. E. (1981). *Cell* **23,** 585–593.
Hofer, E., Hofer-Warbinek, R., and Darnell, J. E. (1982). *Cell* **29,** 887–893.
Hollenberg, S. M., Giguere, V., Segui, P., and Evans, R. M. (1987). *Cell* **49,** 39–46.
Hope, I. A., and Struhl, K. (1985). *Cell* **43,** 177–188.
Hope, I. A., and Struhl, K. (1986). *Cell* **46,** 885–894.

Hough, B. R., Smith, M. J., Britten, R. J., and Davidson, E. H. (1975). *Cell* **5**, 291–299.
Hu, N.-T., Peifer, M. A., Heidecker, G., Messing, J., and Rubenstein, I. (1982). *EMBO J.* **1**, 1337–1342.
Hudspeth, M. E. S., Timberlake, W. E., and Goldberg, R. B. (1977). *Proc. Natl. Acad. Sci. U.S.A.* **74**, 4332–4336.
Hultmark, D., Kiemenz, R., and Gehring, W. J. (1986). *Cell* **44**, 429–438.
Ingham, P. W. (1988). *Nature* **335**, 25–34.
Jacobsen, J. V., and Beach, L. R. (1985). *Nature* **316**, 275–277.
Jantzen, H.-M., Strahle, U., Gloss, B., Stewart, F., Schmid, W., Boshart, M., Miksicek, R., and Schutz, G. (1987). *Cell* **49**, 29–38.
Jofuku, K. D., Okamuro, J. K., and Goldberg, R. B. (1987). *Nature* **328**, 734–737.
Johnson, A. D., and Herskowitz, I. (1985). *Cell* **42**, 237–247.
Jones, E. W., and Fink, G. R. (1982). *In* "The Molecular Biology of the Yeast Saccharomyces: Metabolism and Gene Expression," (J. N. Strathern, E. W. Jones, and J. R. Broach, eds.), pp. 181–300. Cold Spring Harbor Laboratory, Cold Spring Harbor, NY.
Jones, J. D. G., Dunsmuir, P., and Bedbrook, J. (1985). *EMBO J.* **4**, 2411–2418.
Jones, K. A., Yamamoto, K. R., and Tjian, R. (1985). *Cell* **42**, 559–572.
Josefsson, L.-G., Lenman, M., Ericson, M. L., and Rask, L. (1987). *J. Biol. Chem.* **262**, 12196–12201.
Kadonaga, J. T., and Tjian, R. (1986). *Proc. Natl. Acad. Sci. U.S.A.* **83**, 5889–5893.
Kafatos, F. C., and Gelinas, R. (1974). *In* "Biochemistry of Cell Differentiation," (J. Paul, ed.), pp. 223–264. University Park Press, Baltimore.
Kamalay, J. C. (1981). Ph.D. dissertation. University of California, Los Angeles.
Kamalay, J. C., and Goldberg, R. B. (1980). *Cell* **19**, 935–946.
Kamalay, J. C., and Goldberg, R. B. (1984). *Proc. Natl. Acad. Sci. U.S.A.* **81**, 2801–2805.
Karin, M., Haslinger, A., Holtgreve, H., Richards, R. I., Krauter, P., Westphal, H. M., and Beato, M. (1984). *Nature* **308**, 513–519.
Karlin-Neumann, G. A., Kohorn, B. D., Thornber, J. P., and Tobin, E. M. (1985). *J. Mol. Appl. Genet.* **3**, 45–61.
Kasher, M. S., Pintel, D., and Ward, D. C. (1986). *Mol. Cell. Biol.* **6**, 3117–3127.
Kaulen, H., Schell, J., and Kreuzaler, F. (1986). *EMBO J.* **5**, 1–8.
Key, J. L., and Silflow, C. (1975). *Plant Phys.* **56**, 364–369.
Key, J. L., Lin, C. Y., and Chen, Y. M. (1981). *Proc. Natl. Acad. Sci.* **78**, 3526–3530.
Kiper, M., and Herzfeld, F. (1978). *Chromosoma* **65**, 335–351.
Kiper, M., Bartels, D., Hertzfeld, F., and Richter, G. (1979). *Nucl. Acids Res.* **6**, 1961–1978.
Klee, H. J., Muskopf, Y. M., and Gasser, C. S. (1987). *Mol. Gen. Genet.* **210**, 437–442.
Koncz, C., Kreuzaler, F., Kalman, Z., and Schell, J. (1983). *EMBO J.* **2**, 1597–1603.
Koornneef, M., van Eden, J., Hanhart, C. J., Stam, P., Braaksma, F. J., and Feenstra, W. J. (1983). *J. Heredity* **74**, 265–272.
Kuhlemeier, C., Green, P. J., and Chua, N.-H. (1987a). *Ann. Rev. Plant Physiol.* **38**, 221–257.
Kuhlemeier, C., Fluhr, R., Green, P., and Chua, N.-H. (1987b). *Genes Dev.* **1**, 247–255.
Kuhlemeier, C., Cuozzo, M., Green, P., Goyvaerts, E., Ward, K., and Chua, N.-H. (1988). *Proc. Natl. Acad. Sci. U.S.A.* **85**, 4662–4666.
Lalonde, B., Arcangioli, B., and Guarente, L. (1986). *Mol. Cell. Biol.* **6**, 4640–4696.
Lamb, C. J., and Lawton, M. A. (1983). *In* "Encyclopedia of Plant Physiology (NS): Photomorphogenesis," (W. Shropshire and H. Mohr, eds.), Vol. 16, pp. 213–257. Springer-Verlag, Berlin.
Lamppa, G. K., Morelli, G., and Chua, N.-H. (1985a). *Mol. Cell. Biol.* **5**, 1370–1378.
Lamppa, G., Nagy, F., and Chua, N.-H. (1985b). *Nature* **316**, 750–752.
Landridge, P., Pintor-Toro, J. A., and Feix, G. (1982). *Planta* **156**, 166–170.
Lawrence, J. B., and Singer, R. H. (1985). *Nucl. Acids Res.* **13**, 1777–1799.
Lawrence, J. B., and Singer, R. H. (1986). *Cell* **45**, 407–415.
Lehmann, R., and Nüsslein-Volhard (1986). *Cell* **47**, 141–152.

Lengyel, J. A., and Penman, S. (1977). *Dev. Biol.* **57**, 243–253.
Leutwiler, L. S., Hough-Evans, B. R., and Meyerowitz, E. M. (1984). *Mol. Gen. Genet.* **194**, 15–23.
Leutwiler, L. S., Meyerowitz, E. M., and Tobin, E. M. (1986). *Nuc. Acids Res.* **14**, 4051–4064.
Leys, E. J., and Kellems, R. E. (1981). *Mol. Cell Biol.* **1**, 961.
Leys, E. J., Crouse, G. F., and Kellems, R. E. (1984). *J. Cell Biol.* **99**, 180.
Lindquist, S. (1986). *Ann. Rev. Biochem.* **55**, 1091–1117.
Llewellyn, D. J., Finnegan, E. J., Ellis, J. G., Dennis, E. S., and Peacock, W. J. (1987). *J. Mol. Biol.* **195**, 115–123.
Lloyd, A. M., Barnason, A. R., Rogers, S. G., Byrne, M. C., Fraley, R. T., and Horsch, R. B. (1986). *Science* **234**, 464–466.
Lodish, H. F., and Small, B. (1976). *Cell* **7**, 59–65.
Lucchini, G., Hinnebusch, A. G., Chen, C., and Fink, G. R. (1984). *Mol. Cell. Biol.* **4**, 1326–1333.
Luthe, D. S., and Quatrano, R. S. (1980a). *Plant Physiol.* **65**, 305–308.
Luthe, D. S., and Quatrano, R. S. (1980b). *Plant Physiol.* **65**, 309–313.
Lycett, G. W., Croy, R. R. D., Shirsat, A. H., and Boulter, D. (1984). *Nuc. Acids Res.* **12**, 4493–4506.
McCurdy, D. W., and Pratt, L. H. (1986). *Planta* **167**, 330–336.
McKnight, G. S., and Palmiter, R. D. (1979). *J. Biol. Chem.* **254**, 9050–9058.
Ma, Y., and Nelson, O. (1974). *Maize Genet. Coop. News.* **48**, 16–19.
Maier, U.-G., Brown, J. W. S., Toloczyki, C., and Feix, G. (1987). *EMBO J.* **6**, 17–22.
Mangiarotti, G., Lefebvre, P., and Lodish, H. F. (1982). *Dev. Biol.* **89**, 82–91.
Manning, J. E., Schmid, C. W., and Davidson, N. (1975). *Cell* **4**, 141–155.
Mantei, N., and Weissman, C. (1982). *Nature (London)* **297**, 128.
Marks, D. M., Lindell, J. S., and Larkins, B. A. (1985). *J. Biol. Chem.* **260**, 16445–16450.
Martin, C., Carpenter, R., Sommer, H., Saedler, H., and Coen, E. (1985). *EMBO J.* **4**, 1625–1630.
Mason, J. O., Williams, G. T., and Neuberger, M. S. (1985). *Cell* **41**, 479–487.
Mayfield, S. P., and Taylor, W. C. (1984). *Eur. J. Biochem.* **44**, 79–84.
Meinke, D. W. (1985). *Theor. Appl. Genet.* **69**, 543–552.
Mercola, M., Wang, X.-F., Olsen, J., and Calame, K. (1983). *Science* **22**, 663–665.
Merrill, G. F., Harland, R. M., Groudine, M., and McKnight, S. L. (1984). *Mol. Cell Biol.* **4**, 1769.
Miller, A. M., MacKay, V. L., and Nasmyth, K. (1985a). *Nature* **314**, 598–603.
Miller, J., McLachlan, A. D., and Klug, A. (1985b). *EMBO J.* **4**, 1609–1614.
Minty, A. J., and Birnie, G. D. (1981). In "Biochemistry of Cellular Recognition." (M. J. Clemens, ed.), pp. 43–82. CRC Press, Boca Raton, Florida.
Mlodzik, M., and Gehring, W. (1987). *Cell* **48**, 465–478.
Moore, D. D., Marks, A. R., Buckley, D. I., Kapler, G., Payvar, F., and Goodman, H. (1985). *Proc. Natl. Acad. Sci. U.S.A.* **82**, 699–702.
Morelli, G., Nagy, F., Fraley, R. T., Rogers, S. G., and Chua, N.-H. (1985). *Nature* **315**, 200–204.
Morris, T., Marashi, F., Weber, L., Hickey, E., Greenspan, D., Bonner, J., Stein, J., and Stein, G. (1986). *Proc. Natl. Acad. Sci. U.S.A.* **83**, 981–985.
Mösinger, E., Batschauer, A., Schäfer, E., and Apel, K. (1985). *Eur. J. Biochem.* **147**, 137–142.
Mosse, J. (1966). *Fed. Proc.* **25**, 1663–1669.
Motto, M., Maddaloni, M., Ponziani, G., Brembilla, M., Marotta, R., DiFonzo, N., Soave, C., Thompson, R., and Salamini, F. (1988). *Mol. Gen. Genet.* **212**, 488–494.
Murray, M. G., and Thompson, W. F. (1981). In "Biochemistry of Plants," (A. Marcus, ed.) pp. 1–81. Academic Press, San Diego.
Murray, M. G., Cuellar, R. E., and Thompson, W. F. (1978). *Biochem.* **17**, 5781–5790.

Murray, M. G., Palmer, J. D., Cuellar, R. E., and Thompson, W. F. (1979). *Biochem.* **18,** 5259–5266.
Murray, M. G., Peters, D. L., and Thompson, W. F. (1981). *J. Mol. Evol.* **17,** 31–42.
Nagao, R. T., Czarnecka, E., Gurley, W. B., Schoffl, F., and Key, J. L. (1985). *Mol. Cell. Biol.* **5,** 3417–3428.
Nagy, F., Morelli, G., Fraley, R. T., Rogers, S. G., and Chua, N.-H. (1985). *EMBO J.* **4,** 3063–3068.
Nagy, F., Kay, S. A., Boutry, M., Hsu, M.-Y., and Chua, N.-H. (1986a). *EMBO J.* **5,** 1119–1124.
Nagy, F., Fluhr, R., Kuhlemeier, C., Kay, S., Boutry, M., Green, P., Poulsen, C., and Chua, N.-H. (1986b). *Phil. Trans. R. Soc. Lond. B* **314,** 493–500.
Nagy, F., Boutry, M., Hsu, M.-Y., Wong, M., and Chua, N.-H. (1987). *EMBO J.* **6,** 2537–2542.
Nelson, T., Harpster, M. H., Mayfield, S. P., and Taylor, W. C. (1984). *J. Cell Biol.* **98,** 558–564.
Nevers, P., Shepherd, N. S., and Saedler, H. (1986). *Adv. Bot. Res.* **12,** 104–203.
Nicols, S. E., and Laties, G. G. (1984). *Plant Mol. Biol.* **3,** 393–401.
Odell, J. T., Nagy, F., and Chua, N.-H. (1985). *Nature* **313,** 810–812.
Okamuro, J. K., and Goldberg, R. B. (1985). *Mol. Gen. Genet.* **198,** 290–298.
Okamuro, J. K., Jofuku, K. D., and Goldberg, R. B. (1986). *Proc. Natl. Acad. Sci. U.S.A.* **83,** 8240–8244.
O'Reilly, C., Shepherd, N. S., Pereira, A., Schwarz-Sommer, Z., Bertram, I., Robertson, D. S., Peterson, P. A., and Saedler, H. (1985). *EMBO J.* **4,** 877–882.
Osley, M. A., Gould, J., Kim, S., Kane, M., and Hereford, L. (1986). *Cell* **45,** 537–544.
Otto, V., Mösinger, E., Sauter, M., and Schäfer, E. (1983). *Photochem. Photobiol.* **38,** 693–700.
Pabo, C. O., and Sauer, R. T. (1984). *Annu. Rev. Biochem.* **53,** 293–321.
Parker, C. S., and Topol, J. (1984a). *Cell* **36,** 357–369.
Parker, C. S., and Topol, J. (1984b). *Cell* **37,** 273–283.
Paz-Ares, J., Wienand, U., Peterson, P. A., and Saedler, H. (1986). *EMBO J.* **5,** 829–833.
Paz-Ares, J., Ghosal, D., Wienand, U., Peterson, P. A., and Saedler, H. (1987). *EMBO J.* **6,** 3553–3558.
Pearson, W. R., Davidson, E. H., and Britten, R. J. (1977). *Nucl. Acids Res.* **4,** 1727–1737.
Pedersen, K., Bloom, K. S., Anderson, J. N., Glover, D. V., and Larkins, B. A. (1980). *Biochemistry* **19,** 1644–1650.
Pederson, K., Devereux, J., Wilson, D. R., Sheldon, E., and Larkins, B. A. (1982). *Cell* **29,** 1015–1026.
Pedersen, K., Argos, P., Naravana, S. V. L., and Larkins, B. A. (1986). *J. Biol. Chem.* **261,** 6279–6284.
Pelham, H. R. B. (1982). *Cell* **30,** 517–528.
Pelham, H. R. B. (1986). *Cell* **46,** 959–961.
Pelham, H. R. B., and Bienz, M. (1982). *EMBO J.* **1,** 1473–1477.
Pernollet, J.-C., and Vaillant, V. (1984). *Plant Physiol.* **76,** 187–190.
Perutz, M. F., Rossman, M. G., Cullis, A. F., Muirhead, H., Will, G., and North, A. T. C. (1960). *Nature* **185,** 416–422.
Pfeifer, K., Arcangioli, B., and Guarente, L. (1987a). *Cell* **49,** 9–18.
Pfeifer, K., Prezant, T., and Guarente, L. (1987b). *Cell* **49,** 19–27.
Pruitt, R. E., and Meyerowitz, E. M. (1986). *J. Mol. Biol.* **187,** 169–183.
Ragg, H., and Weissman, C. (1983). *Nature (London)* **303,** 439.
Ragg, H., Schröder, J., and Hahlbrock, K. (1977). *Biochim. Biophys. Acta* **474,** 226–233.
Renkawitz, R., Schutz, G., von der Ahe, D., and Beato, M. (1984). *Cell* **37,** 503–510.
Rochester, D. E., Winter, J. A., and Shah, D. M. (1986). *EMBO J.* **5,** 451–458.
Rodgers, W. H., and Gross, P. R. (1978). *Cell* **14,** 279–288.

Roeder, R. G. (1976). *In* "RNA Polymerase," (R. Losick and M. Chamberlin, eds.), pp. 285–330. Cold Springer Harbor Laboratory, Cold Spring Harbor, New York.
Rosahl, S., Eckes, P., Schell, J., and Willmitzer, L. (1986). *Mol. Gen. Genet.* **202**, 368–373.
Rosbash, M., Ford, P. J., and Bishop, J. O. (1974). *Proc. Natl. Acad. Sci. U.S.A.* **71**, 3746–3750.
Rosenberg, U. B., Schröder, C., Preiss, A., Kienlin, A., Côté, S., Riede, I., and Jäckle, H. (1986). *Nature* **319**, 336–339.
Sachs, M. M., and Ho, T.-H. D. (1986). *Ann. Rev. Plant Physiol.* **37**, 363–376.
Salamini, F., Di Fonzo, N., Gentinetta, E., and Soave, C. (1979). *In* "Seed Protein Improvement in Cereals and Grain Legumes," Vol. I, pp. 97–108. International Atomic Energy Agency, Vienna.
Salamini, F., Di Fonzo, N., Fornasari, E., and Gentinetta, E. (1983). *Theor. Appl. Genet.* **65**, 123–128.
Sanchez-Serrano, J. J., Keil, M., O'Connor, A., Schell, J., and Willmitzer, L. (1987). *EMBO J.* **6**, 303–306.
Savage, M. J., Sala-Trepat, J. M., and Bonner, J. (1978). *Biochemistry* **17**, 462–467.
Sawadogo, M., and Roeder, R. B. (1985a). *Proc. Natl. Acad. Sci. U.S.A.* **82**, 4394–4398.
Sawadogo, M., and Roeder, R. G. (1985b). *Cell* **43**, 165–175.
Schirm, S., Jiricny, J., and Schaffner, W. (1987). *Genes Devel.* **1**, 65–74.
Schmid, C. W., and Deininger, P. L. (1975). *Cell* **6**, 345–358.
Schmidt, R. J., Burr, F. A., and Burr, B. (1987). *Science* **238**, 960–963.
Schöffl, F., Raschke, E., and Nagao, R. T. (1984). *EMBO J.* **3**, 2491–2497.
Schreier, P. H., Seftor, E. A., Schell, J., and Bohnert, H. J. (1985). *EMBO J.* **4**, 25–32.
Scofield, S. R., and Crouch, M. L. (1987). *J. Biol. Chem.* **262**, 12202–12208.
Sen, R., and Baltimore, D. (1986). *Cell* **46**, 705–716.
Sengupta-Gopalan, C., Reichert, N. A., Barker, R. F., Hall, T. C., and Kemp, J. D. (1985). *Proc. Natl. Acad. Sci. U.S.A.* **82**, 3320–3324.
Serfling, E., Jasin, M., and Schaffner, W. (1985). *TIG* **August 1985**, 224–230.
Shah, D., Horsch, R., Klee, H., Kishore, G., Winter, J., Tumer, N., Hironaka, C., Sanders, P., Gasser, C., Aykent, S., Siegel, N., Rogers, S., and Fraley, R. (1986). *Science* **233**, 478–481.
Shaw, G., and Kamen, R. (1986). *Cell* **46**, 659–667.
Shore, O., and Nasmyth, K. (1987). *Cell* **51**, 721–732.
Shuey, D. J., and Parker, C. S. (1986). *J. Biol. Chem.* **261**, 7934–7940.
Silflow, C. D., and Key, J. L. (1979). *Biochemistry* **18**, 1013–1018.
Silflow, C. D., Hammett, J. R., and Key, J. L. (1979). *Biochemistry* **18**, 2725–2731.
Silverthorne, J., and Tobin, E. M. (1984). *Proc. Natl. Acad. Sci. U.S.A.* **81**, 1112–1116.
Silverthorne, J., and Tobin, E. (1987). *Bioessays* **7**, 18–23.
Simcox, A. A., Cheney, C. M., Hoffman, E. P., and Shearn, A. (1985). *Mol. Cell. Biol.* **5**, 3397–3402.
Simpson, J., Timko, M. P., Cashmore, A. R., Schell, J., Van Montagu, M., and Herrera-Estrella, L. (1985). *EMBO J.* **4**, 2723–2729.
Simpson, J., Schell, J., Van Montagu, M., and Herrera-Estrella, L. (1986a). *Nature* **323**, 551–554.
Simpson, J., Van Montagu, M., and Herrera-Estrella, L. (1986b). *Science* **233**, 34–38.
Sivolap, Y. M., and Bonner, J. (1971). *Proc. Natl. Acad. Sci. U.S.A.* **68**, 387.
Slater, E. P., Rabenau, O., Karin, M., Baxter, J. D., and Beato, M. (1985). *Mol. Cell. Biol.* **5**, 2984–2992.
Smith, D. B., and Flavell, R. B. (1975). *Chromosoma* **50**, 223–242.
Soave, C., Righetti, P. G., Lorenzoni, C., Gentinetta, E., and Salamini, F. (1976). *Maydica* **21**, 61–75.
Soave, C., Suman, N., Viotti, A., and Salamini, F. (1978). *Theor. Appl. Genet.* **52**, 263–267.

Speth, V., Otto, V., and Schöfer, E. (1986). *Planta* **168**, 299–304.
Spradling, A., Hui, A., and Penman, S. (1975). *Cell* **4**, 131–137.
Sternberg, P. W., Stern, M. J., Clark, I., and Herskowitz, I. (1987). *Cell* **48**, 567–577.
Stougaard, J., Sandal, N. N., Gron, A., Kuhle, A., and Marcker, K. A. (1987). *EMBO J.* **6**, 3565–3569.
Strauss, F., and Varshavsky, A. (1984). *Cell* **37**, 889–901.
Strittmatter, G., and Chua, N.-H. (1987). *Proc. Natl. Acad. Sci. U.S.A.* **84**, 8986–8990.
Struhl, K. (1982). *Nature* **300**, 284–287.
Struhl, K. (1984). *Proc. Natl. Acad. Sci. U.S.A.* **81**, 7865–7869.
Struhl, K. (1986). *In* "Maximizing gene expression." (W. Reznikoff and L. Gold, eds.), Butterworths, Boston.
Stuart, G. W., Searle, P. F., and Palmiter, R. D. (1985). *Nature* **317**, 828–831.
Sulston, J. E., and Brenner, S. (1974). *Genetics* **77**, 95–104.
Sung, Z. R., and Okimoto, R. (1981). *Proc. Natl. Acad. Sci. U.S.A.* **78**, 3683–3687.
Tabata, T., Sasaki, K., and Iwabuchi, M. (1983). *Nuc. Acids Res.* **11**, 5865–5875.
Theres, N., Scheele, T., and Starlinger, P. (1987). *Mol. Gen. Genet.* **209**, 193–197.
Thompson, W. F., Murray, M. G., and Cuellar, R. E. (1979). *In* "Genome Organization in Plants." (C. J. Leaver, ed.), pp. 1–15. Plenum Press, New York.
Timberlake, W. E., Shumard, D. S., and Goldberg, R. B. (1977). *Cell* **10**, 623–632.
Timko, M. P., Kausch, A. P., Castresana, C., Fassler, J., Herrera-Estrella, L., Van den Broeck, G., Van Montagu, M., Schell, J., and Cashmore, A. R. (1985). *Nature* **318**, 579–582.
Tingey, S. V., Walker, E. L., and Coruzzi, G. M. (1987). *EMBO J.* **6**, 1–9.
Tischer, E., DasSarma, S., and Goodman, H. (1986). *Mol. Gen. Genet.* **203**, 221–229.
Tobin, E. M., and Silverthorne, J. (1985). *Ann. Rev. Plant Physiol.* **36**, 569–593.
Tobin, E. M., and Slovin, J. P. (1982). *Planta* **154**, 465–472.
Tobin, E. M., Wimpee, C. F., Silverthorne, J., Stiekema, W. J., Neumann, G. A., and Thornber, J. P. (1984). *In* "Biosynthesis of the Photosynthetic Apparatus: Molecular Biology, Development and Regulation." pp. 325–334. Alan R. Liss, New York.
Topol, J., Ruden, D. M., and Parker, C. S. (1985). *Cell* **42**, 527–537.
Treisman, R. (1986). *Cell* **46**, 567–574.
Tsai, S. Y., Roop, D. R., Tsai, M.-J., Stein, J. P., Means, A. R., and O'Malley, B. W. (1978). *Biochem.* **17**, 5773–5780.
Valentini, G., Soave, C., and Ottaviano, E. (1979). *Heredity* **42**, 33–40.
VanNess, J., and Hahn, W. E. (1982). *Nucl. Acid Res.* **10**, 8061–8077.
VanSluys, M. A., Tempe, J., and Federoff, N. (1987). *EMBO J.* **6**, 3881–3889.
Villares, R., and Cabrera, C. V. (1987). *Cell* **50**, 415–424.
Viotti, A., Sala, E., Marotta, R., Alberi, P., Balducci, C., and Soave, C. (1979). *Eur. J. Biochem.* **102**, 211–222.
Voelker, T. A., Staswick, P., and Chrispeels, M. J. (1986). *EMBO J.* **5**, 3075–3082.
Walbot, V., and Cullis, C. A. (1985). *Ann. Rev. Plant Physiol.* **36**, 367–396.
Walbot, V., and Dure, L. S. (1976). *J. Mol. Biol.* **101**, 503–536.
Walbot, V., and Goldberg, R. B. (1979). *In* "Nucleic Acids in Plants" (T. C. Hall and J. W. Davies, eds.), Vol. 1, pp. 3–40. CRC Press, Boca Raton, Florida.
Walker, J. C., Howard, E. A., Dennis, E. S., and Peacock, W. J. (1987). *Proc. Natl. Acad. Sci. U.S.A.* **84**, 6624–6628.
Walling, L., Drews, G. N., and Goldberg, R. B. (1986). *Proc. Natl. Acad. Sci. U.S.A.* **83**, 2123–2125.
Weber, J., Jelinek, W., and Darnell, J. E., Jr. (1977). *Cell* **10**, 611–616.
Weeks, D., and Melton, D. A. (1987). *Cell* **51**, 861–867.
Weiher, H., König, M., and Gruss, P. (1983). *Science* **219**, 626–631.
Werr, W., Frommer, W.-B., Maas, C., and Starlinger, P. (1985). *EMBO J.* **4**, 1373–1380.

Wienand, U., Weydemann, U., Niesbach-Klösgen, U., Peterson, P. A., and Saedler, H. (1986). *Mol. Gen. Genet.* **203,** 202–207.
Willing, R. P., and Mascarenhas, J. P. (1984). *Plant Physiol.* **75,** 865–868.
Willmitzer, L., and Wagner, K. G. (1981). *Exp. Cell Res.* **135,** 69–77.
Wilson, D. R., and Larkins, B. A. (1984). *J. Mol. Evol.* **20,** 330–340.
Wilson, K. L., and Herskowitz, I. (1986). *Proc. Natl. Acad. Sci. U.S.A.* **83,** 2536–2540.
Wimpee, C. F., and Rawson, J. R. Y. (1979). *Biochem. Biophys. Acta* **562,** 192–206.
Wimpee, C. F., Stiekma, W. J., and Tobin, E. M. (1983). *Plant Molecular Biology, UCLA Symposia on Molecular and Cellular Biology, New Series* (R. B. Goldberg, ed.), Vol. 12, pp. 391–401, Alan R. Liss, New York.
Wold, B. J., Klein, W. H., Hough-Evans, B. R., Britten, R. J., and Davidson, E. H. (1978). *Cell* **14,** 941–950.
Wong, H. C., and Chang, S. (1986). *Proc. Natl. Acad. Sci. U.S.A.* **83,** 3233–3237.
Wright, S., Rosenthal, A., Flavell, R., and Grosveld, F. (1984). *Cell* **38,** 265–273.
Wu, C. (1984). *Nature* **311,** 81–87.
Yamamoto, K. R. (1985). *Ann. Rev. Genet.* **19,** 209–252.
Young, B. D., and Anderson, L. M. (1986). *In* "Nucleic Acid Hybridisation." (B. D. Hames and S. J. Higgins, eds.), pp. 47–72. IRL Press, Washington, D.C.
Zambryski, P., Joos, H., Genetello, C., Leemans, J., Van Montagu, M., and Schell, J. (1983). *EMBO J.* **2,** 2143–2150.
Zimmerman, J. L., and Goldberg, R. B. (1977). *Chromosoma* **59,** 227–252.
Zimmerman, J. L., Fouts, D. L., and Manning, J. E. (1980). *Genetics* **95,** 673–691.
Zufluh, L. I., and Guilfoyle, T. J. (1982). *Plant Physiol.* **69,** 338–340.

Transposable Element Influence on Plant Gene Expression and Variation

2

LILA O. VODKIN

I. History and Perspectives
  A. Discovery of Maize Transposable Elements
  B. Overview of Mobile Elements in Bacteria
  C. Current Perspectives
II. Molecular Isolation and Characterization of Transposable Element Families
  A. The Maize Element, Activator (Ac)
  B. The Enhancer/Suppressor-Mutator Element Family
  C. Mutator (Mu) Sequences
  D. Retroviral-Like Elements
  E. Tam Elements in Snapdragon
  F. Tgm Insertions in Soybean
III. Transposable Element Action
  A. Effects on Gene Expression
  B. Models of Transposition in Plants
  C. Role in Evolution
  D. Origin and Regulation
IV. Gene Tagging with Transposable Elements
  A. Isolating Genes Whose Products Are Unknown
  B. Examples of Genes Isolated by Transposon Tagging
  C. Potential for Tagging in Heterologous Systems
    References

I. HISTORY AND PERSPECTIVES

A. Discovery of Maize Transposable Elements

Unlike most genes which occupy fixed positions within the chromosomes of the cell, transposable elements are units of DNA capable of changing their positions in the genome. To a large degree, the history of the discovery of transposable elements reflects a fascination of geneticists with variegation and mosaicism. Variegation in plants is most easily recognized as irregular spotting or striping of coloration in the seed or leaves. The seminal experiment which led to the eventual discovery of transposition was conducted by Barbara McClintock in 1944. McClintock, a cytogeneticist of exceptional ability, had been investigating the effects of X-rays on maize chromosomes, specifically chromosome breakage induced by radiation and the new mutations caused by large-scale chromosome damage. Newly ruptured chromosome ends will fuse, leading to dicentric chromosomes which subsequently rupture at various locations during anaphase as the centromeres are pulled to opposite poles of the cell. The newly broken ends then set in motion a repetition of the fusion-bridge-breakage cycle in subsequent cell divisions. In 1944, McClintock self-pollinated 450 plants which had all received a newly ruptured chromosome from each parent (reviewed by McClintock, 1984). When she analyzed the progeny from each of these plants for new mutations, she found an unexpected array that affected many different plant loci. Many of the mutations were unstable, leading to variegation in pigmentation patterns.

Particularly striking were those unstable mutations leading to chlorophyll variegation in the seedling leaves. She observed that variegated plants sometimes gave rise to twin sectors in which the frequency of green streaks on a white background were higher in one half of the twin and lower in the other half with respect to the frequency in the rest of the leaf. Since the sectors were side by side, it appeared that they were clonal lineages representing the products of a mitotic division in which each daughter cell had a modified pattern of gene expression with respect to the original cell. McClintock reasoned that the different effects on gene expression in these lineages reflected that "one daughter cell had gained something that the other daughter cell had lost" (McClintock, 1984). After 6 years of genetic and cytogenetic experimentation, she obtained formal proof that the physical entities which one cell had lost and the other cell had gained were genetic units which had transposed to new locations in the genome. (Transpositions of elements from genes affecting pericarp color and leading to twin sectors in the kernels are described in more detail in Section III,B.) Originally, McClintock termed these genetic units "controlling elements" because of the effects that they have on gene expression. Today there is increasing evidence that various stresses to the cell or "shocks to the genome" can lead to the activation of previously quiescent transposable elements. The maize lines in which McClintock originally found transposable elements had undergone extensive trauma created by repeated chromosome breakage.

While the effects produced by transposable elements were originally observed in seedling and plant characteristics such as chlorophyll expression in leaves, kernel traits were easier to analyze. One of the instabilities from McClintock's original experiment affected a gene on the short arm of chromosome 9 known as the *C* locus. It is one of several loci involved in pigmentation of the aleurone layer, the outermost cell layer of the seed endosperm. Kernels of the genetic constitution *C/C* or *C/c* have purple kernels due to the presence of the dominant *C* allele. The *C-I* allele is an inhibitor of pigmentation, and is dominant to the *C* locus. McClintock's seed should have had the genotype *C-I/C*, producing colorless kernels. Several aberrant kernels stood out because they had sectors of color on a colorless background, indicating the loss of expression of the *C-I* allele at certain times. Since maize is an excellent plant for classical genetic mapping because of the wealth of marker genes, McClintock undertook to map the factor causing the loss of expression of the *C-I* allele (reviewed in Fedoroff, 1983). In fact, expression of several other genes was concomitantly lost with the *C-I* gene, indicating that a chromosomal segment was being lost. She found that the reason was a specific site of chromosome breakage that mapped just to the centromere side of the *waxy* (*Wx*) locus, another endosperm marker gene which is closer to the centromere than the *C* locus. McClintock named the factor causing the site-specific chromosome breakage, the Dissociation (Ds) element, because of its propensity to dissociate or break chromosomes at its location. Chromosome dissociation could also be observed cytologically, but it is more easily monitored by the loss of dominant genetic markers. When chromosome 9 breaks at the site of the Ds element, the genes carried on the fragment distal to Ds will be lost in subsequent cell divisions as the fragment doesn't carry a centromere. These cell lineages develop tissue sectors with various recessive phenotypes.

Many kernels were scored for the loss of genetic markers associated with Ds-induced chromosome breakage. In some of the progeny, it was found that the Ds element which originally mapped near the *Wx* gene had disappeared from this location and now mapped at the *C* locus, creating an unstable, recessive mutation known as *c-m1* (mutable allele at the *C* locus) in which chromosome breakage now occurred at this position. Thus, the first proof of transposition came from the finding that the Ds element could change its location.

The genetic segregation ratios obtained in mapping Ds also proved to McClintock that two genetic elements were required to cause the specific chromosome breakage at the location of Ds or the transposition of Ds. She mapped the second genetic unit to a position on the long arm of chromosome 9, far away from the Ds element at the *Wx* locus. She named the second unit Activator (Ac) because this unit was necessary to obtain chromosome dissociation at the Ds location or transposition of Ds away from its original site. The fact that Ac could exert its influence on Ds from a distance implied the production of a *trans*-acting factor by Ac that could diffuse to the Ds location and promote transposition of Ds and subsequent chromosome breakage. Ac was later found inserted into genes of known phenotypes, as the *Wx* locus, where it created unstable, recessive mutations in which both somatic and germinal reversions occurred at a certain frequency to produce phenotypically wild-type or variegated kernels.

Ac was autonomous in the sense that it mediated its own instability or transposition. The Ds element, however, could move only when an Ac element was present elsewhere in the genome. Ac was also observed to convert to forms that could no longer transpose without another active Ac element in the genome. Thus, the autonomous Ac element would sometimes generate a defective element that behaved as a Ds. These observations indicated that Ac and Ds were structurally related and recent molecular data reveal this to be the case (see below). Ds elements were also found which can transpose under the influence of Ac but which do not promote chromosome breakage. In fact, it is now known that Ds elements which induce chromosome breakage have a more complicated structure than most Ds elements (see Section II,A).

McClintock outlined the conclusions about the nature and action of maize controlling elements in a series of publications and presented them at a meeting at Cold Spring Harbor in 1951. The scientific community for the most part was unreceptive or disbelieving. One reason may have been that her concepts and data could not be readily understood except by scientists specifically versed in the life cycle of maize and its genetics (Keller, 1983). A major factor was the belief by most geneticists as a basic tenet that genes occupied unchanging positions within chromosomes. All data since the rediscovery of Mendel in 1921 seemed to indicate this. The orderly behavior of chromosomes during mitosis and meiosis and the fact that genes could be consistently mapped to specific locations was evidence of stability of the genome. Fueled by the discovery of DNA as the genetic material and its detailed structure in the 1940s and early 1950s, the emphasis shifted from the study of higher organisms to simpler model systems of bacteria and phages. The significance of McClintock's findings began to be recognized when mobile elements were found in bacteria and were physically isolated and characterized in the late 1960s.

B. Overview of Mobile Elements in Bacteria

Certain mutations in bacteria were found to be very different from the point mutations, or small base changes in DNA sequence, which were well described by the late 1960s. The new classes of mutations were caused by large insertions of DNA which interrupted gene function. Insertion sequence (IS) elements were physically isolated and DNA hybridization studies showed that they occurred in multiple copies in the bacterial chromosome. ISs are also present in plasmids and bacteriophages. These elements could transpose as discrete units and integrate into new locations by mechanisms which did not require DNA sequence homology as did recombination events. It soon became clear that in prokaryotes there were many classes of

ISs and transposable elements which differed in their DNA sequences (reviewed by Calos and Miller, 1980; Kleckner, 1981; Iida et al., 1983).

ISs are generally 800 to 1400 bp in length and encode a transposase needed for their movement to other locations. Two characteristics were found to be common for different ISs: (1) the presence of inverted repeat sequences at the ends of the element and (2) the duplication of a short segment of the target site DNA upon integration of the element into a new site. The termini are perfect or nearly perfect inverted repeats of about 10–40 bp which apparently serve as recognition sequences for the transposase action (Iida et al., 1983). Alterations within the inverted repeats often affect transposition activity of the elements. Duplication of a short segment of the target DNA during integration of an IS element into a new site is shown in Fig. 1A, for the *lacI* gene of *Escherichia coli*. The 9-bp direct repeat which flanks each end of the inserted IS1 element occurs only once in the normal *lacI* gene sequence. The sequence of the terminal inverted repeats of IS elements and the number of base pairs duplicated at the target site have been used as criteria for classifying IS elements.

Transposons (Tn elements) were initially distinguished from ISs because they carry along other genes in addition to those needed for transposition. Many drug resistances found in nature are carried as part of a transposable unit. Subsequent analyses made it clear, however, that there is a close relationship between IS and Tn elements (Calos and Miller, 1980). Many Tn elements are flanked by long inverted or direct repeats of 800–1500 bp which resemble IS elements. Sometimes the IS units are found alone, while in other cases, they are found only as part of a composite transposon, i.e., two insertion element sequences flanking additional genes. The terminal IS elements are responsible for mobilization of the intervening material within the transposon. In contrast to IS elements, movement of Tn elements can be followed, if the genes carried in the transposon have selectable or scoreable phenotypes. A composite Tn element, Tn10, is shown in Fig. 1B. The middle 6500 bp of Tn10 is composed of a tetracycline resistance gene and other material of unknown function. The 1400-bp repeats found in inverted orientations at the left and right ends of Tn10 are structurally intact IS elements, each of which has 13-bp near-perfect inverted repeat termini important for transposition. IS10-Right has a long open reading frame of about 1200 nucleotides which encodes a transposase. IS10-Right has been observed to move on its own and promotes normal levels of Tn10 transposition even when IS10-Left is inactivated (Kleckner, 1983). IS10-Left, on the other hand, provides only one-tenth of the transposition activity of IS10-Right. The two elements are similar but not identical in sequence. It is speculated that Tn10 originated when a copy of an IS element similar to IS10-Right transposed to a location flanking the tetR determinant. One of the IS elements may have degenerated to give IS10-Left since only one active element is needed to

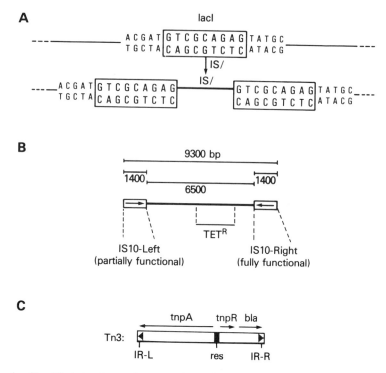

Fig. 1. Simplified structures of some prokaryotic transposable elements. (A) Illustration of a simple insertion sequence, IS1, and the consequences of its insertion into the *lacI* gene of *E. coli*. The normal *lacI* sequence is shown on the upper line and the 9-bp target sequence which is duplicated upon insertion of the IS1 element is boxed. IS1 is 768 bp in length and terminates in imperfect inverted repeats (not shown) of which 20 of the terminal 23 bases at the 5' and 3' ends are complementary. (From Calos and Miller, 1980.) (B) Structure of a composite transposon, Tn10. The 9300-bp element is composed of two 1400-bp insertion elements, IS10-Left and IS10-Right, which can mobilize a tetracycline resistance gene. (From Kleckner, 1983.) (C) Structure of Tn3. The 4957-bp element contains inverted repeats of 38 bp at each end as denoted by arrows. The element encodes three polypeptides, a transposase (tnpA) and a repressor-resolvase (tnpR) which are both needed for transposition and a β-lactamase (bla) which confers ampicillin resistance. The internal resolution site (res) is a region of site-specific recombination used by the tnpR protein to resolve cointegrate plasmids formed during the transposition process. (From Grindley, 1983.)

produce a transposase which can mobilize sequences residing between the 13-bp terminal inverted repeats.

A third major class of prokaryotic transposable element is represented by Tn3, shown in Fig. 1C. Tn3 carries a β-lactamase gene that confers resistance to ampicillin, but it is not flanked by two intact IS elements. Instead it has small, inverted repeat termini of 38 bases. Two other proteins that are both needed for transposition, a transposase and a repressor-resolvase, are encoded within the 4957-bp element. The latter is a bifunctional protein which regulates the expression of the transposase gene and is also a site-specific endonuclease. During transposition of Tn3 from a donor to a recipient plasmid, a cointegrate plasmid forms that contains the original and a replicated copy of the Tn3 element. The site-specific resolvase recognizes and cuts a specific region of each Tn3 element called the internal resolution site, leading to a recombination between DNA strands and resolution of the cointegrate plasmid into two separate units each now containing a copy of Tn3.

C. Current Perspectives

In addition to bacteria and plants, mobile elements or sequences which have structural properties characteristic of transposable elements have been found in many other organisms, including yeast, *Drosophila*, and mammals (Roeder and Fink, 1983; Rubin, 1983; Syvanen, 1984; Georgiev et al., 1986). In certain eukaryotic elements, known as retrotransposons, transposition proceeds via RNA intermediates (reviewed in Weiner et al., 1986).

McClintock's discovery of genetic elements that could change location was not fully appreciated until mobile sequences were found and characterized at the molecular level. However, analysis of maize transposable elements was pioneered by several geneticists in addition to McClintock. Discussion of the genetic properties of maize transposable elements and the attributes of the maize life cycle are included in several reviews (Peterson, 1987; Fedoroff, 1983, 1988). Mutable alleles in maize are found at 20 different genetic loci (Nevers et al., 1986). Also listed in the review by Nevers et al. (1986) are numerous examples of mutable alleles in other higher plant systems that are potentially due to transposable elements. These include 35 mutable genes in snapdragon (*Antirrhinum majus*) and more than 56 alleles in 31 other plant species. The present review focuses only on those plant species and elements for which molecular characterization has been obtained. Section II details the isolation and characterization of transposable elements families with emphasis on the initial loci from which the elements were cloned. Additional information of the molecular properties of plant transposable elements can be found in reviews by Freeling (1984), Doring (1985), and Doring and Starlinger (1984, 1986).

The effects of transposable elements on gene expression and an emerging understanding for the molecular basis of "changes in state," "changes in phase," "presetting," and "cycling" of element activities are reviewed in Section III. Current models for transposable element excision and integration and the role of transposable elements in creating genetic diversity are also discussed. There is increasing evidence that mobile elements have a positive role in evolution and may serve to promote rapid genomic change in organisms challenged by stressful environments. A contrasting view has argued that transposable elements are molecular parasites or symbionts whose only function is to survive using their ability to move as a way to avoid being eliminated from the genome (Doolittle and Sapienza, 1980). There are currently no examples in which transposable elements regulate normal plant development (Schwarz-Sommer and Saedler, 1987), but it may be too early to rule this function out. Other types of genome instability also result from specific DNA rearrangements which alter gene expression. Mating type switching in yeast, antigenic phase variation in prokaryotes and trypanosomes, and immunoglobulin rearrangements in mammals (reviewed in Borst and Greaves, 1987) are phenomena that involve programmed rearrangements but these are not mediated by transposable elements. Transposable elements, however, are certainly one cause of increased variation in the genome; genomic flexibility in response to environmental changes is now recognized as an important survival mechanism for both plant and animal life forms (Walbot and Cullis, 1985).

Much current interest in transposable elements derives from their utility to identify genes whose biochemical functions are unknown. An element clone can be utilized as a hybridization probe, or molecular tag, to isolate a genomic clone carrying a transposable element-induced mutation at a specific genetic locus. The DNA region flanking the element can then be used to isolate the normal gene. Termed "transposon tagging," this approach was first applied in prokaryotes and *Drosophila*. Section IV covers examples of higher plant genes which have been isolated using transposable elements and the potential to develop gene tagging systems by introducing transposable elements into heterologous plant species.

II. MOLECULAR ISOLATION AND CHARACTERIZATION OF TRANSPOSABLE ELEMENT FAMILIES

A. The Maize Element, Activator (Ac)

In order to physically isolate the first member of a transposable element family, it was necessary to clone the structural gene into which the element had inserted. Of the maize genes known to have transposable elements which interrupted their functions, only a few code for abundant, identifiable products and thus are amenable to cloning of the structural genes responsible for the observed phenotypes. One such gene is a classic maize marker called the *waxy* (*Wx*) locus (reviewed in Nelson, 1987), so named because it affects the starch composition of the kernel endosperm. One can easily observe the difference between the shiny, waxy appearance of the mature maize kernels containing a functional *Wx* gene and the recessive *wx* kernels which are dull and opaque. The functional basis for the observed phenotype is a defect in amylose content of endosperm tissue. By a simple test with iodine, one can distinguish *Wx* versus *wx* kernels (or sectors of kernels) since *Wx* endosperm will stain blue whereas *wx* mutant endosperm appears red. The ease of testing for phenotypic expression of this gene allowed its fine structure to be studied by classical genetic means of intragenic recombination analysis. Nelson detailed its fine structure (1968) and McClintock identified several alleles of the *Wx* locus which contained the maize controlling elements Ac and Ds.

The Ac/Ds system was one of the first controlling elements studied by McClintock (McClintock, 1950; Fedoroff, 1983; reviewed in Section I,A). The Dissociation (Ds) element was initially identified by its ability to induce chromosome breakage or dissociation at its location, hence its name. Ds, however, could not move or cause chromosome breakage on its own. Instead Ds depended on an autonomous element named Activator (Ac). Based on evidence that the *Wx* locus encoded a starch-granule-bound UDP-glucose starch transferase (Nelson and Rines, 1962), the molecular isolation and identification of the *Wx* locus was achieved by Shure *et al.* (1983) with the aim of using it, in turn, to isolate the Ac and Ds elements from the *wx* mutant alleles. To accomplish this, a cDNA clone containing a partial sequence for the 58-kDa glucose starch transferase was identified and used to isolate the wild-type *Wx* structural gene from normal maize strains. The cloned *Wx* gene was then used to isolate the mutant allele from one of McClintock's lines, designated *wx-m9 Ac* and known to contain an Ac element interrupting the *Wx* gene function. Heteroduplex analysis and restriction endonuclease mapping showed that the *wx-m9 Ac* DNA contained a 4.3-kb insertion of extra DNA in the *wx* transcription unit (Fedoroff *et al.*, 1983). Behrens *et al.* (1984) also cloned an Ac element from another waxy mutation, designated *wx-m7 Ac*, and likewise found it to be a 4.3-kb insertion.

The sequences of both independently isolated elements, Ac7 and Ac9, are identical (Pohlman *et al.*, 1984; Muller-Neumann *et al.*, 1984). Each is 4563 bp long and terminates with 11-bp imperfect inverted repeats. A duplication of 8 bp of the *wx* locus DNA occurs immediately adjacent to its inverted repeat termini. With regard to these two characteristics of inverted repeat termini and duplication of small regions of the target DNA, the maize Ac element structurally resembles transposable elements from bacteria and *Drosophila*.

Although Ac7 and Ac9 are identical in DNA sequence, they show different phenotypic effects. Unlike *wx-m9 Ac*, the *wx-m7 Ac* allele is not a complete null mutation and some amylose is produced in the endosperm. Also, somatic reversion events occur earlier in *wx-m7 Ac* than in *wx-m9 Ac*. The orientation of the Ac element is the same in both alleles, but in *wx-m7 Ac*, the Ac element is inserted about 2.5 kb 5' of its position in the *wx-m9 Ac* allele. These different positions within the *wx* gene may lead to the observed differences in their effects on gene expression and in their somatic reversion rates (Muller-Neumann *et al.*, 1984).

Fedoroff et al. (1983) also isolated a Ds element from the maize line wx-m9, which is a derivative of the line containing wx-m9 Ac. The wx-m9 allele is not capable of independent transposition but is unstable in the presence of Ac; thus it behaves as a classic Ds element derived from an Ac insertion. Molecular analysis of Ds9 showed that it was a deletion derivative of Ac9. As illustrated in Fig. 2, Ds9 is identical to Ac9 except that 194 bp of sequence in the middle of the element has been deleted within one of the open reading frames of the Ac9 element. Since Ds9 is defective in autonomous transposition, the location of the Ds9 deletion indicates that the ORF1 reading frame encodes a transposase. Another Ds element, Ds6, arose by transposition to the wx locus of the Ds element first identified by McClintock as the cause of chromosome instabilities. The 2040-bp Ds element is a deletion derivative of Ac and is identical to an element responsible for the unstable mutations studied at the *shrunken* locus as discussed below.

Before the isolation of an Ac element and Ds derivatives from the wx locus, several different Ds elements had been cloned and characterized from other maize genes. The shrunken (*Sh1*) gene of corn has been the focus of study by several laboratory groups since mutations affecting this gene have a collapsed endosperm phenotype that is easily scorable. McClintock isolated recessive *sh1* mutants caused by the presence of Ds in the vicinity of the gene (McClintock, 1953). *Sh1* encodes sucrose synthase, a major endosperm protein, that catalyzes a reaction in starch metabolism (Chourey and Nelson, 1976). Sucrose synthase cDNA clones were isolated and used to compare genomic Southern blots of DNAs from the normal locus and Ds-induced

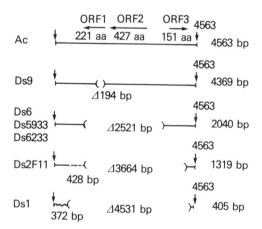

Fig. 2. Schematic relationship of several Ds deletion derivatives to the maize Ac element. The size of each Ds element is indicated to the right. The amount of Ac sequence deleted is indicated within parentheses. The solid line denotes sequences identical to Ac while the dashed and wavy lines denote sequences which are not found in Ac. (From Doring and Starlinger, 1984.)

mutants of the locus (Burr and Burr, 1981; Doring et al., 1981; Courage-Tebbe et al., 1983). These studies initially revealed a complex picture. One mutant allele, *sh-m6233* contains an insertion of a "double-Ds" element, which is composed of a 2040-bp Ds element inserted into an identical copy of itself (Courage et al., 1984; Weck et al., 1984). The 2040-bp unit copy of this 4-kb double-Ds element is identical to the 2040-bp element described by Fedoroff et al. (1983) at the *wx* locus (Fig. 1). The interior element of Ds6233 is inverted in orientation with respect to the outer copy and is also flanked by 8 bp of duplicated sequence. Apparently, an internal deletion of Ac which gave rise to the 2040-bp Ds element was followed by an insertion of the new Ds element into a copy of itself.

Both elements of a double-Ds contain the 11-bp inverted repeats characteristic of the Ac–Ds termini. Thus, there are four copies of these repeats, with two sets of adjacent repeats in direct rather than inverse orientation. Normally, a transposase acts on inverted repeats to excise the element and religate the flanking DNA in a manner that does not disrupt the chromosome. If through looping of DNA strands, however, the direct repeats within the double-Ds structure were brought into juxtaposition as inverted repeats, a breakage and ligation of the cleaved ends at these "false" inverted termini would lead to chromosome breakage. It is speculated that such double-Ds structures are the molecular explanation for the ability of Ds elements to induce chromosome breaks in the presence of Ac (Courage et al., 1984). Ac alone has not been observed to induce chromosome breaks.

Another Ds-induced mutant of the sucrose synthase gene, *sh-m5933*, is a very large insertion (Burr and Burr, 1981; Doring et al., 1981; Geiser et al., 1982). Ds5933 is 30 kb long and one end of the insertion consists of a 4-kb double-Ds as described above with the same 2040-bp unit structure (Doring et al., 1984a). At the other end of the 30-kb insertion is a 3-kb structure that represents part of the double-Ds. The 30-kb insertion is not, however, analogous to a bacterial transposon. A model of how it may have been derived, based on chromosome breakage and religation of sister chromatids which would capture a region of chromosomal DNA between the double-Ds structures, has been presented (Courage et al., 1984).

The Ds elements described above are clearly internal deletions of Ac. However, the first Ds element isolated and several others described since that time are substantially different in pattern and are sometimes referred to as "aberrant" Ds elements. A Ds-induced mutant in the alcohol dehydrogenase gene *Adh1* was identified (Osterman and Schwartz, 1981) and isolated (Sutton et al., 1984). It differs from the normal *Adh1* gene by the insertion of 402 bp of DNA which terminates with the 11-bp inverted repeats and 8-bp flanking duplications that are characteristic of the Ac–Ds family. Except for the termini, however, the internal sequence of this element, designated Ds1, does not resemble that of the Ac and Ds elements. Since Ds1 moves in response to Ac, one can deduce that the only requirement for a piece of DNA to respond to transposase function from Ac is the presence of the 11-bp inverted repeats. Another Ds element isolated from the *Adh1* gene is Ds2F11, a 1319-bp element that has homology for several hundred base pairs with the Ac element but which differs from Ac in its central core (Doring *et al.*, 1984b; Merckelbach *et al.*, 1986). The relationship between Ds elements is shown diagrammatically in Fig. 2. The origin of aberrant Ds elements and the total number of Ds elements in the maize genome are unknown. At least 40 bands are detected when genomic DNA is probed with the terminal areas of Ac, while the number of sequences hybridizing to the central region of Ac is much smaller, about 4 to 10 (Fedoroff *et al.*, 1983; Starlinger, 1985).

Recent evidence indicates that ORF2 is also involved in production of an Ac transposase. Dooner *et al.* (1986) described a Ds element from the bronze locus, *bz-m2(DI)*, which contains ORF1 but has a deletion of 1312 bp in the ORF2 region. Since the element is defective in autonomous transposition as is the Ds9 allele, which is defective only in ORF1, the implication is that both reading frames are needed to produce a transposase. Additionally, an Ac transcript of 3.5 kb is found at low abundance in all organs of maize strains carrying the *wx-m7* Ac allele but not in lines genetically free of Ac. Comparison of cDNA of this transcript with genomic sequences indicates that the Ac transcription unit is 4.2 kb, which covers most of the Ac element and includes four small introns. A 652-bp leader sequence is present and a protein of 86 kDa is potentially encoded by the open reading frame (Kunze *et al.*, 1987).

B. The Enhancer/Suppressor-Mutator Element Family

A second transposable element family in maize was identified by Peterson (1953, 1960) as the basis for a pale green mutable phenotype in leaves, and named Enhancer (En). McClintock independently isolated an element responsible for mutability at the *a1-m1* allele, a gene involved in anthocyanin biosynthesis (1954, 1961). She termed the element Spm (Suppressor-mutator) and studied its properties in detail. The Spm family has parallels to the Ac–Ds family in that it contains both autonomous elements (analogous to Ac) and defective elements (analogous to Ds elements) that respond to an active En/Spm element in the genome. McClintock did not name the defective, or receptor, Spm elements whereas Peterson referred to the defective forms of En as Inhibitor (I) elements. Genetic tests demonstrated that both En–I and Spm are functionally identical since the En element could control excision of defective Spm derivatives and Spm could act on I elements (Peterson, 1965). The En–I and Spm class of elements is currently being characterized in detail at the molecular level by several groups. Researchers who have used McClintock's stocks refer to the element as Spm while those using Peterson's stocks use the name En. Sometimes the element is dually referred to as the En/Spm family, a terminology adhered to in this article.

Unlike Ac, none of the loci containing an active En/Spm element were amenable to direct cloning, as the gene products encoded by the loci were not known. Therefore, in order to isolate the En/Spm element, En insertions into the previously cloned *waxy* locus (Schwarz-Sommer *et al.*, 1984) were selected for genetically. The *wx* locus has also been used to obtain a clone for the Ac element as discussed previously (Fedoroff *et al.*, 1983; Section II,A). From a cross in 1983, several mutant seed having the recessive *wx* phenotype were observed (Pereira *et al.*, 1985). One of these mutant kernels, designated wx-844, showed variegated sectors of normal and waxy mutant endosperm. The presence of En/Spm in the *wx-844* allele was confirmed genetically and a phage library of the DNA of the mutant plant was constructed. Comparison of the mutant clone to the normal *Wx* gene showed the presence of an 8.2-kb insertion in the *waxy* transcription unit as determined by heteroduplex analysis and restriction mapping.

Salient features of the 8287-bp element are presented in Fig. 3 (Pereira *et al.*, 1986). The ends of En/Spm are 13-bp inverted repeats which have homology to the Tam1, Tam2, and Tgm1 elements in other plant species (Schwarz-Sommer *et al.*, 1984). Its internal borders can be represented as extended, imperfectly paired termini or as a series of palindromic repeats

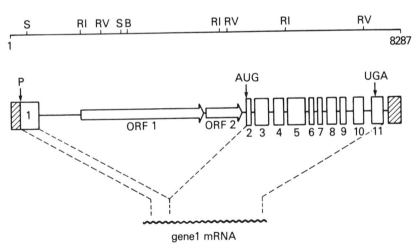

Fig. 3. Mosaic structure of the En-1 (Spm) gene in maize. The top line denotes the cleavage sites within the 8287-bp sequence for the following restriction enzymes: B, *Bam*HI; RI, *Eco*RI; RV, *Eco*RV; S, *Sal*I. The 11 exons which produce the 2.5-kb transcript are denoted by numbered boxes and the two open reading frames within the large, first intron are marked. The hatched areas are the subterminal border regions which contain repeated palindromic sequences. Positions of the promoter (P) and translation start and stop codons are indicated. (From Pereira *et al.*, 1986.)

(Section II,F). The internal area of En/Spm has a complicated structure which was pieced together by comparing the cDNA sequence of an En/Spm specific transcript to that of its genomic sequence. Messenger RNA from immature endosperms of *wx-844* :: *En-1* plants contains a 2.5-kb transcript that hybridizes to an internal 1.4-kb *Eco*RI fragment of En/Spm. A nearly full length cDNA clone containing 2361 nucleotides of this transcript was isolated and sequenced (Pereira *et al.*, 1986). The gene that produced this transcript, designated gene 1, has a complicated mosaic structure which spans nearly the full length of the 8287-bp element and contains 11 exons (Fig. 3). The transcription start site at nucleotide 209 was determined by S1 mapping and a possible promoter sequence, TATGAA, occurs 30 bp upstream of the transcription start site immediately adjacent to the extended subterminal repeats. Two possible translational start sites occur at the beginning of exon 2. If the first of these is used, then gene 1 of the En/Spm element would encode a peptide of 621 amino acids. The product of the gene 1 has been expressed in *E. coli* and it appears to have DNA binding activity which recognizes the repetitive palindromic sequences found in the element borders (Gierl *et al.*, 1987).

All of the introns are less than 200 nucleotides except for the first one which is 4434 bp. Intron 1 contains two large open reading frames of 2714 and 761 nucleotides. The only coding sequence similarities found to date between maize En/Spm, Tam1 of snapdragon, and Tgm elements in soybean is a region of approximately 1.5 kb in the ORF1 of the first large intron (Sommer *et al.*, 1987; Rhodes and Vodkin, 1988). Thus, it is possible that the transposase function is actually encoded by this region rather than by the 3.5-kb major transcript produced by gene 1. Minor transcripts of 1.7 and 1.1 kb in endosperm mRNA preparations hybridize with the internal 3.0-kb RI fragment that is located entirely within the intron 1 sequence (Pereira *et al.*, 1986).

A number of nonautonomous elements of the En/Spm family have been isolated. Similar to the situation with Ac and Ds elements, these defective elements are generally internal deletion derivatives of the autonomous transposable element. An En/Spm receptor element, Spm-I8 (or dSpm-I8), was described genetically by McClintock and was isolated by Schwarz-Sommer *et al.* (1984). It is a 2242-bp insertion into the *waxy* locus whose excision is dependent on an active En/Spm element in the genome. dSpm-I8 is apparently derived by an internal deletion of 6046 bases from the autonomous element (position 861–6906) (Gierl *et al.*, 1985; Pereira *et al.*, 1986). Otherwise, its sequence is identical to the autonomous En/Spm sequence except for 4 single base substitutions. These substitutions may reflect the different maize lineages of the En element of Peterson and the Spm element of McClintock.

From an analysis of mutable alleles of the *bronze* (*bz*) locus, a series of defective En/Spm elements (designated dSpm elements) has been characterized and correlated with their effects on gene expression (Fedoroff et al., 1984; Schiefelbein et al., 1985). A 2.2-kb insertion, dSpm-13, found in the *bz-m13* allele, appears, based on restriction mapping and hybridization studies, to be identical to dSpm-I8 isolated from the *wx* locus. In the presence of an active Spm element, a number of derivatives from the original *bz-m13* allele were generated. These derivatives are listed in Table I and are distinguished from one another on the basis of their varied effects on gene expression at the *bz* locus (Schiefelbein et al., 1985). Heritable changes in the frequency and timing of reversion events at an allele containing an insertion element were referred to by McClintock as "changes in state" of the locus. At the molecular level, some of these changes in state were found to be internal deletion derivatives of the original dSpm-13 element without a change in the position of the element within the *bz* locus. Likewise, three different insertions of defective Spm elements in the *a1* locus, which is involved in anthocyanin biosynthesis, have likewise been shown to be internal deletions of the active En/Spm element (Schwarz-Sommer et al., 1985b; Tacke et al., 1986). An active En/Spm element and a number of defective derivatives that affect gene expression at the *a1-m2* allele have also been cloned and compared (Masson et al., 1987). An autonomous element in *a1-m2*, referred to as Spm-s or standard Spm, is essentially the same element as the En element isolated from the *wx* locus (Pereira et al., 1985), proving not only the genetic functionality but also the structural near identity of the elements studied by McClintock and Peterson. Except for four fewer bases and six single base substitutions, the Spm-s-7991A element is identical to the En element. Four of the dSpm derivatives which affect expression of the *a1-m2* locus in different ways were found to have deletions of internal sequence. A fifth defective element was full length but presumably has a small lesion not detected by restriction mapping.

In summary, the En/Spm family in maize is a two-element system as is the Ac–Ds mobile element family. The nomenclature for this family can be somewhat confusing because of the independent names given to the same element. Briefly, the autonomous element is called either Spm or En, sometimes referred to jointly as En/Spm. The defective, nonautonomous elements are termed either Inhibitor elements (En-I or Spm-I) or defective Spm elements (dSpm). These are generally internal deletion derivatives of the autonomous element. Table I summarizes the molecular properties of the different En/Spm elements which have been analyzed to date. Three different transposition defective En/Spm elements from different genes apparently have the same structure. These are dSpm-I8 from the *wx-m8* allele, dSpm-13 from *bz-m13,* and dSpm-I6078 from *a1-m1*. This does not imply a site-specific deletion but rather that all of these alleles are traceable to a common progenitor stock, the nonautonomous *a2-m1* allele (Schwarz-Sommer et al., 1985b). A defective element at that locus may have transposed to the other three sites.

TABLE I
Molecular Descriptions of Higher Plant Transposable Elements[a]

Element	Host gene	Size (bp)[b]	Inverted repeats Size (bp)	Inverted repeats Sequence[c]	Size of flanking duplication (bp)	References and remarks
Zea mays						
Ac9	wx-m9 Ac	4563	11	tagggatgaaa (g)	8	Fedoroff et al. (1983); Pohlman et al. (1984)
Ac7	wx-m7 Ac	4563	11	tagggatgaaa (g)	8	Behrens et al. (1984); Muller-Neuman et al. (1984)
Ds9	wx-m9	4369	11	tagggatgaaa (g)	8	Fedoroff et al. (1983); Pohlman et al. (1984)
Ds6	wx-m6	2.0 kb				Fedoroff et al. (1983)
Ds5933	sh-m5933	2040	11	tagggatgaaa	8	Doring et al. (1984a); Courage et al. (1984) (single unit of a double Ds; similar to Ds6 and Ds6233 basic unit)
Ds6233	sh-m6233	2.0 kb	11	tagggatgaaa		Weck et al. (1984)
Ds1	Adh-Fm335	402	11	tagggatgaaa		Sutton et al. (1984) (an aberrant Ds element)
Ds2	Adh-2F11	1321	11	tagggatgaaa	8	Doring et al. (1984b); Merckelbach et al. (1986) (an aberrant Ds element)
Ac	bz-m2 Ac	4.5 kb				Fedoroff et al. (1984) (bz gene isolated by Ac tagging)
Ds	bz-m2(DII)	3.7 kb				Schiefelbein et al. (1985)
Ds	bz-wm	0.4 kb				Schiefelbein et al. (1985) (an aberrant Ds element)
Ds	bz-m2(DI)	3251				Dooner et al. (1985, 1986)
Ds	bz-m1	1.1 kb				Dooner et al. (1985)
Ds2	bz2-m	1321	11	tagggatgaaa	8	Theres et al. (1987) (bz2 gene isolated by Ds tagging)

Ds	c1-m1	2.2 kb				Cone et al. (1986)
Ds	c1-m2	2.5 kb				Paz-Ares et al. (1986); Cone et al. (1986)
Ac	P-vv	4.5 kb				J. Chen et al. (1987) (P gene isolated by Ac tagging)
En-1	wx-844 :: En-1	8287				Pereira et al. (1985, 1986)
dSpm-18	wx-m8	2242				Schwartz-Sommer et al. (1984)
dSpm-13	bz-m13	2.2 k				Schiefelbein et al. (1985) [dSpm-13 appears to be identical to dSpm-I8 and dSpm-I6078. CS refers to "changes of state" of the original bz-m13 allele induced by the presence of an active En/Spm element]
dSpm-13CS1	bz-m13	2.2 kb	13	cactacaagaaaa	3	
dSpm-13CS3	bz-m13	1.6 kb	13	cactacaagaaaa	3	
dSpm-13CS5	bz-m13	0.6 kb				
dSpm-13CS6	bz-m13	2.2 kb				
dSpm-13CS9	bz-m13	0.7 kb				
dSpm-13CS12	bz-m13	0.9 kb				
En	a1-m(papu)	8.2 kb				O'Reilly et al. (1985) (a1 gene tagged with En element)
dSpm-I6078A-1	a1-m1	2.2 kb	13	cactacaagaaaa	3	Schwartz-Sommer et al. (1985b) [changes in state of the a1-m1 allele induced by an active En/Spm element]
dSpm-15719A-1	a1-m1	789	13	cactacaagaaaa	3	
dSpm-I1112A-1	a1-m1	945	13	cactacaagaaaa	3	Tacke et al. (1986) (a change in state progeny of the a1-m1 6078 allele)
Spm-s-7991A	a1-m2	8283	13	cactacaagaaaa	3	Masson et al. (1987) (a standard Spm nearly identical in sequence to En-1)
Spm-w-8011	a1-m2	6570				(an autonomous but weakly active Spm element) (a series of "changes of state" derived from the a1-m2 allele)
dSpm-7995	a1-m2	3493				
dSpm-7977	a1-m2	1277				
dSpm-8004	a1-m2	1090				
dSpm-8167	a1-m2	8.2 kb				
dSpm-8417	a1-m2	1.8 kb				
En	c1-m668655 :: En	8.3 kb				Paz-Ares et al. (1986) [c1 gene isolated by En/Spm tagging from two independent mutations]
En	c1-m668613 :: En	8.3 kb				
Spm	c1-m5	8.3 kb				Cone et al. (1986) [c1 gene isolated by En/Spm tagging]
dSpm	c1-m858	1.1 kb				Cone et al. (1986)

(continued)

TABLE I (continued)

Element	Host gene	Size (bp)	Inverted repeats Size (bp)	Sequence	Size of flanking duplication (bp)	References and remarks
Spm	c2-m1	(Partial clone)				Weinand et al. (1986) [c2 gene isolated by En/Spm tagging]
dSpm-I2	c2-m2	3.0 kb				Weinand et al. (1986)
Mu1	Adh1-S3034	1367	213/215 (95% homology)		9	Freeling et al. (1982); Bennetzen et al. (1984)
Mu1	Adh1-S4477	1.4 kb				Freeling et al. (1982)
Mu1	Adh1-S4478	1.4 kb				
Mu1	a1-Mum2	1.4 kb			9	O'Reilly et al. (1985); Shepherd, personal communication (a1 locus tagged with Mu1)
Mu1	bz-mu1	1.4 kb				Taylor et al. (1986)
Mu2 or Mu1.7	bz-mu2	1.7 kb				Taylor et al. (1986)
Mu1	bz2-mu1	1.4 kb				McLaughlin and Walbot (1987) (bz2 gene isolated by Mu1 tagging)
Bs1	Adh1-S5446	3.3	304 perfect direct repeats		6	Mottinger et al. (1984) (induced in BSMV infected plants)
Tz86	sh-m5586	3.6 kb	No inverted or direct repeats		10	Dellaporta et al. (1984) (induced in BSMV infected plants)
Cin1-001	—	691	6	tgttgg	5	Shepherd et al. (1984); Hehl et al. (1985)

Cin1-102	—		693	tgttgg	6	5	(similar sequences have been found in teosinte, a maize relative)
Cin1-102	—		679	tgttgg	6	5	
rDt	am-1 : Cache		704	cagtgttttaaatc (t) (t)	14	8	Brown, et al. (1989) (receptor of dotted element)
rMrh	a-mrh		0.25 kb				Shepherd, personal communication (receptor of Mrh)
Antirrhinum majus (snapdragon)							
Tam1	niv-98		17 kb	cactacaacaaaa	13	3	Bonas et al. (1984)
Tam2	niv-43		5.0 kb	cactacaacaaaa	13	3	Upadhyaya et al. (1985) (inverted repeats similar to Tam1, but internal sequence is different)
Tam3	niv-98		3.5 kb	taaagatgtgaa	12	8	Sommer et al. (1985)
Tam3	pal-rec		3.5 kb	taaagatgtgaa	12	5	Martin et al. (1985); Coen et al. (1986) (unique example of an element generating different sizes of target site duplications)
Glycine max (soybean)							
Tgm1	lel		3550	cactattagaaaa (g)	13	3	Vodkin et al. (1983); Rhodes and Vodkin (1985)
Tgm2 to 7	—		1.6 to 12 kb	as above	13	3	Rhodes and Vodkin (1988) (isolated from a genomic library, some contain an ORF similar to En/Spm)

[a] In order to more effectively differentiate between the active En/Spm elements and nonautonomous derivatives, the defective derivatives have been given the prefix "d" following the terminology used by some authors (Schiefelbein et al., 1985; Masson et al., 1987).
[b] If the element has not been sequenced, the approximate size is shown in kilobase pairs (kb) rather than base pairs.
[c] The sequence of the 5' end of the element is shown. The 3' ends are inverse complements unless otherwise noted by bases underneath in parentheses. As shown, the first and last bases of some Ac and Ds elements are not inverse complements.

En/Spm elements have a wide array of effects on gene expression as well as on the frequency and developmental timing of their excision. The significance of various deletion derivatives or "changes of state" of an allele on gene expression and reversion will be discussed in more detail in Sections III,A and B.

C. Mutator (Mu) Sequences

A mutator system (Mu) which increases the frequency of mutation rates about 30-fold for a number of loci was discovered in a maize line (Robertson, 1978). Many of the induced mutations were recessive and unstable, suggesting that they could be due to insertions. To examine this, an experiment was set up to induce mutations in the alcohol dehydrogenase (*Adh1*) gene. ADH is expressed in the pollen and mutants can readily be selected with allyl alcohol. Also, the *Adh1* gene had been cloned and sequenced (Gerlach *et al.*, 1982). Comparison of the mutator-induced *Adh1-S3034* allele with the normal gene showed a 1.4-kb insertion in the first intron of the gene and a duplication of 9 nucleotides at the insertion site (Freeling *et al.*, 1982; Bennetzen *et al.*, 1984). Designated Mu1, the insertion has long inverted terminal repeats of 213 and 215 bp showing 95% homology. Within the element are four possible open reading frames and internal direct repeats of 104 bp with 95% homology (Barker *et al.*, 1984).

Although it is not known whether Mu1 is the cause of Mutator activity, correlations have been made between Mu sequences and Mutator activity in maize lines. In Mutator active lines (those which segregate new recessive mutants at high frequency) there are generally 10–50 copies of Mu1-related elements, whereas in non-Mutator lines there are no Mu sequences or only a few copies (Alleman and Freeling, 1986). Thus, it appears that the Mutator activity is due to the movement of these elements. Most Mu1-related elements are similar in size and structure, as determined by cleaving maize DNA with *Hin*fI or *Tth*111I restriction enzymes which cleave once within the terminal repeats of Mu1 but do not cleave internally. However, a slightly larger element of 1.7 kb, designated Mu2, has been found in some lines and is also the cause of a new mutation at the *bronze* locus *bz-mu2* (Taylor *et al.*, 1986, 1987). Mu1 appears to be a deletion derivative of the slightly larger Mu2 element. A third element of about 1.8 kb, Mu3, was found in another *Adh* mutation and appears to be substantially different in internal sequence from Mu1 and Mu2 (Lillis and Freeling, 1986; C.-H. Chen *et al.*, 1987). It is not yet clear whether any of these elements encode the proteins required for transposition. Mu2 is known to transpose but Mu1 sequences can also transpose in lines that do not possess Mu2 elements (Alleman and Freeling, 1986; Taylor *et al.*, 1986).

The Mu sequences have been followed in crosses of Mutator stocks to Mutator-negative lines by using a restriction enzyme that does not cut within the Mu elements. In this way, the number of copies of the element in the F1 generation was found to be approximately the same as in the Mutator parent line rather than the expected dilution by half (Alleman and Freeling, 1986; Bennetzen et al., 1987). Many of the parental fragments containing Mu elements are present in the progeny, but there are also many nonparental restriction bands (approximately 25%), demonstrating that new transposition events have occurred. The first generation from self-pollinated Mutator plants also shows an increase in the number of Mu element sequences. These molecular data suggest that Mu sequences transpose by a replicative mechanism at high frequency in the germ cells before fertilization (Alleman and Freeling, 1986; Bennetzen et al., 1987).

It is not uncommon for lines which are negative for Mutator activity to contain copies of Mu sequences. In fact, Mutator activity can be lost in about 10% of the outcross progeny between Mutator-active and Mutator-negative lines. Additionally, after four generations of inbreeding, Mutator activity is lost from lines which were originally Mutator active (Lillis and Freeling, 1986). One likely explanation is the inactivation of Mu sequences by DNA methylation. Chandler and Walbot (1986) obtained evidence of Mu methylation while observing a loss of Mutator activity in a line carrying the *bz-mu2* allele that contains an unstable 1.7-kb Mu2 element. Somatic revertants of *bz-mu2* to wild type can be readily identified as spots of dark aleurone pigment against colorless background kernels. Eight plant lines were derived from the original stock and a few of these lineages lost the somatic mutability of the *bz-mu2* allele. In Southern blots of DNA from the active Mutator lines, all of the Mu sequences fell, as expected, into two bands of 1.3 kb (Mu1) or 1.7 kb (Mu2) upon digestion with the enzyme *Hin*fI which cleaves the element twice, once in each of the termini. However, in the lines which lost somatic mutability, Mu sequences also lost the ability to be cut by *Hin*fI and were found as a series of higher-molecular-weight fragments. It appeared that the Mu sequences are methylated since *Hin*fI is sensitive to methylation of cytosine residues. An inactive element, which is methylated in non-Mutator lines, has been isolated and shown to be nearly identical in structure to Mu1 (Chandler et al., 1986). Hypermethylation of Mu1 sequences also correlates with loss of Mutator activity in maize lines which have been intercrossed (Bennetzen, 1987). Thus, methylation may be a mechanism by which Mu regulates its movement.

Recently, an extrachromosomal, closed circular form of Mu has been identified (Sundaresan and Freeling, 1987). This is surprising since Mu is not thought to resemble retroviral elements which replicate through circular intermediates. The circular forms appear to be identical to Mu1 and Mu2 but they are found at less than one copy per cell in Mutator active lines and are not found in inactive lines. Whether they are intermediates in transposition or whether they represent excision products is unknown.

In summary, the Mu element is very different from Ac–Ds and En/Spm families. Mu elements are distinguished by the long inverted repeat termini, by a very high transposition frequency, by the maintenance of copy number control, and apparently by a replicative mechanism of transposition as opposed to a loss/gain mechanism used by other elements (Section III,B).

D. Retroviral-Like Elements

An unstable mutation which inactivated the maize *Adh1* gene was selected from plants which had been infected with barley stripe mosaic virus (BSMV) (Mottinger *et al.*, 1984). Analysis of this mutation, *Adh1-S5446*, revealed the presence of a 3.3-kb insertion, designated Bs1, that has 304-bp perfect direct repeats at the ends of the element and duplicates 6 bp of the *Adh1* gene (Johns *et al.*, 1985). The long direct repeats are not characteristic of any other plant transposable elements but do have similarities to the retroviral-like transposons of *Drosophila* (copia elements) and yeast (Ty elements), including identical TGTT sequences at the 5' end of the repeats and less exact homologies further in. No short inverted repeat structure occurs at the ends of the direct repeats of Bs1, however, as found for the copia and Ty elements. There are from 1 to 5 copies of Bs1 sequences in 11 maize lines that were examined and most of the copies are similar. No free copies of the terminal direct repeats were detected (as is the case for the Ty element repeats in yeast) nor were any large deletions of the elements found. Bs1 does not hybridize to the BSMV genome and it is hypothesized that the BSMV infection induced the mobility of the Bs1 element.

Cin1 is a 700-bp insert found by comparing two unique genomic clones (of unknown identity) from two maize lines. Clone CL1 was originally isolated as a putative positive in a screen of Line C for the maize chalcone synthase gene using a heterologous parsley cDNA as a probe (Wienand *et al.*, 1982). When LC1 was used to probe genomic DNA from the maize variety Northern Flint, the corresponding *Eco*RI fragment was approximately 700 bp larger than its size in Line C. The analogous fragment from the Northern Flint line was subsequently cloned and found to harbor Cin1, one of a family of dispersed repeats estimated to be about 1000 copies per haploid genome (Shepherd *et al.*, 1982; Gupta *et al.*, 1983). The repeated elements range in size up to a maximum of 700 bp. Several of the larger elements have been sequenced and shown to have 6-bp perfect inverted repeat termini, of which the first 5 are identical to those of the copia element in *Drosophila* (Shepherd *et al.*, 1984; Hehl *et al.*, 1985). The Cin1 elements are flanked by 5-bp direct duplications. In some respects, the Cin elements are similar to a single LTR (long terminal repeat) found at the ends of certain retrotransposons such as copia in *Drosophila* and Ty elements in yeast. The various Cin elements show approximately 90% sequence homology to one another and similar sequences are found in teosinte, a maize relative.

The Cin4-1 element, found in a wild-type allele of the *A1* locus, is but one of a family of elements that resemble retrotransposons in having open frames that potentially code for proteins with sequence similarity to retroviral reverse transcriptase (Schwarz-Sommer *et al.*, 1987b). A nonmaize retroviral-like element has been found by comparing alleles of a storage protein gene in wheat (Harberd *et al.*, 1987).

E. Tam Elements in Snapdragon

To date, snapdragon (*Antirrhinum majus*) is the only plant system other than maize in which there are both genetic and molecular characterizations of transposable elements. Several stable and unstable alleles are known at the *nivea* locus which encodes chalcone synthase, an enzyme involved in early steps of anthocyanin biosynthesis (Spribille and Forkmann, 1982). Initially, a heterologous chalcone synthase cDNA clone from parsley was used to identify the *nivea* gene in *Antirrhinum* (Wienand *et al.*, 1982; Sommer and Saedler, 1986). With the *nivea* clone in hand, a series of molecular studies by researchers at the Max-Planck Institute yielded three transposable elements, Tam1, Tam2, and Tam3, from various alleles of the *nivea* gene.

Several snapdragon mutants that exhibited unusual red and white variegated flowers resulted from pollen mutagenesis. One of these, designated *nivea-recurrens* or *niv-53*, has a 17-kb insertion (Tam1, Transposon *Antirrhinum majus*) 17 bp upstream of the chalcone synthase promoter (Bonas *et al.*, 1984; Saedler *et al.*, 1984). The Tam1 insertion generates a 3-bp duplication of the chalcone synthase DNA and has 13-bp perfect inverted terminal repeats which are similar to the termini of the maize En/Spm element and the soybean Tgm1 element. The entire sequence of Tam1 shows an organization similar to En/Spm in that it is highly mosaic and contains a number of introns, but there is no sequence similarity except for a 1.5-kb region of an intron which corresponds to the part of the En/Spm ORF1 sequence (Sommer *et al.*, 1987).

Stable red revertants derived from the *niv-53* stock were found to have lost the Tam1 element but retained an additional 2 bp of the target site duplication. The Tam1 element, as well as a number of other transposable elements, frequently leave behind many small duplications or deletions at the target site (see Section III,B). Two stable white derivatives of the *niv-53* line (Tam1-46 and Tam1-49) were recovered and these retain the Tam1 element except that the five terminal bases of the inverted repeats have been deleted (Hehl *et al.*, 1987). The stable nature of these derivatives again indicates the importance of the inverted repeats as a substrate for transposase action.

The unstable *niv-53* allele due to the Tam1 insertion can be influenced by the paramutagenic allele *niv-44*. Paramutation is an unexplained phenomenon of interaction between two alleles at a locus whereby one of the alleles leads to a heritable change of the other with a high frequency. The affected allele is referred to as paramutable (in this case *niv-53* :: *Tam1*) and the one causing the change is termed paramutagenic (*niv-44*). Homozygous *niv-53* plants have highly variegated flowers with red spots or flakes on a white background while plants homozygous for the recessive *niv-44* allele produce stable white flowers. When *niv-53* and *niv-44* plants are crossed, instead of producing variegated plants with the phenotype of the *niv-53* allele, a majority of the F1 progeny are light-rose colored, often with dark flakes. The *niv-53* phenotype (red flakes on white background) also fails to be recovered in the next two generations or in backcrosses, indicating a heritable change (Harrison and Carpenter, 1973). In order to determine the molecular basis of paramutation, the *niv-44* allele was isolated and found to contain a 5-kb insertion, designated Tam2, in the first intron of the chalcone synthase gene (Saedler *et al.*, 1984; Upadhyaya *et al.*, 1985).

Tam2 has inverted repeats similar to Tam1 and also duplicates a three-base segment of target DNA. Although the first 700 bp of Tam2 are 80% similar to those of Tam1, the rest of the element shows no homology (Krebbers *et al.*, 1987). Thus, Tam2 is not an internal deletion derivative of Tam1. Although Tam2 has structural characteristics of transposable elements, the flowers of *niv-44* plants are stably white. However, crosses between two stable white flowered lines (*niv-44* :: *Tam2* and *niv-53* :: *Tam1-46*) produced F1 plants that were variegated, indicating an interaction between Tam1 and Tam2 (Hehl *et al.*, 1987). Molecular analysis showed that the Tam2 element can excise from the *niv* locus in these plants. The Tam1 element cannot excise because its inverted repeat termini have a small deletion as discussed above. When both elements are in the same plant, however, a putative transposase from the Tam1-46 element can act on the inverted repeat termini of Tam2 and cause it to excise.

The molecular basis for the paramutagenic effect of the *niv-44* :: *Tam2* allele on the *niv-53* :: *Tam1* allele is not completely understood but it does not appear to involve large molecular rearrangements of the two elements (Krebbers *et al.*, 1987). It is postulated that the genetic background of the *niv-44* :: *Tam2* line contains diffusable suppressors which can reduce or suppress the action of a transposase encoded by Tam1 elements in *niv-44* × *niv-53* crosses (Krebbers *et al.*, 1987; Hudson *et al.*, 1987). Derivatives in which Tam2 has excised from the *niv* locus do not lose paramutagenic potential; therefore the Tam2 element is not likely to be the source of a suppressor. The suppressors could be encoded by Tam2-related sequences which are present in line 44 and presumably have more coding potential than the Tam2 element itself.

Another transposable element in snapdragon, Tam3, which is quite different from Tam1 and Tam2, was also isolated from the *nivea* locus (Sommer *et al.*, 1985). Tam3 originally came from the *pallida* gene in snapdragon which is also involved in the production of anthocyanin pigment of the flowers.

Unlike the *nivea* locus, its protein product is unknown, so attempts were made to capture Tam3 in the *nivea* gene. An unstable allele, *pallida-recurrens* (*pal-rec*), like *niv-53*, produces variegated red spots on white flowers. One revertant derived from this line produces full red color background with some small pale and white spots. The *pal+* revertant allele was made homozygous and called the TR75 line. The high frequency of pale and white spots in the TR75 line is termed "resurgent instability" or "reverse variegation." It has been postulated that these white spots represent reintegration of the transposable element into genes involved in anthocyanin biosynthesis (Harrison and Carpenter, 1979). Line TR75 was crossed onto tester lines which do not produce anthocyanin in order to uncover new unstable alleles. Thirty four of 2034 flowers showed sectors of instability. Sommer *et al.* (1985) derived an unstable mutant which was shown to be an allele of the *nivea* locus. This mutation, *niv-98*, was made homozygous; after molecular isolation of the *niv-98* locus, comparison with the normal chalcone synthase gene revealed the presence of a 5-kb insertion, designated Tam3 (Sommer *et al.*, 1985). Tam3 is integrated into the promoter region and generates an 8-bp duplication of target sequence. Its 12-bp inverted repeat termini are distinctly different from Tam1 and Tam2 but are similar to those of the maize Ac–Ds elements.

Because the genetic derivation of Tam3 in the *niv-98* mutation was from a line originally carrying the *pal-rec* allele, it was assumed that the *pal-rec* mutation may also be due to an element similar to Tam3. The cloned Tam3 sequence was thus used as a transposon tag to identify and isolate the *pallida* locus (Martin *et al.*, 1985; see Section IV,B). Surprisingly, the Tam3 elements in both the *pal* and *niv* loci are inserted just upstream of the TATA promoter, 41 bp upstream in the *pal* allele and 29 bp in the *niv* locus. Also, sequences of 5 and 6 bases to the left and right of each insertion site are identical in both the *pal* and *niv* loci. This suggests that the insertions of Tam3 in these two genes were not random events. Oddly enough, Tam3 duplicates an 8-bp target sequence of the *niv* gene but only 5 bp of the *pal* locus. To date, this is a unique example in plants of variable target site duplication by a transposase during independent integration events.

The *pal-rec* allele is germinally as well as somatically unstable and gives rise to a series of *pal* alleles with new flower color phenotypes. Some of these produce full red revertant flowers and others show uniform color but of varying intensities of pigmentation from near red to very pale. Another type are stable alleles with different patterns of color on the flower parts while a third class are unstable alleles with altered frequencies and patterns of spotting. Coen *et al.* (1986) have analyzed these at the molecular level by restriction mapping with a probe of the *pal* locus and found that five of six of the stable full-colored alleles have the normal map expected for the *pallida* locus, suggesting that the Tam3 element has excised from the locus. One of these carries an extra 100 bp due to part of the element remaining. Sequence analysis of several of the revertant alleles reveals minor deletions at the target site which presumably effect the expression of the *pallida* locus. One of the near-full-color alleles has a 1-bp addition at the target site, and a *pal* allele which produces very pale flowers has a 13-bp deletion with respect to the wild-type allele. For two of three lines having an altered spatial pattern of flower pigmentation, the element also had excised while the third line gave a much smaller fragment than expected. One of the new unstable alleles had a small deletion of Tam3 near its right end and another had a more complex rearrangement of the *pal* gene sequence. Thus, large variations in gene expression can arise through the Tam3 insertion and excision process. Variation caused by transposable elements is thought to be of evolutionary significance in generating new genetic diversity (Section III,C).

The behavior of Tam3 at both the *niv* and *pal* loci is greatly influenced by temperature, with excision frequencies 1000-fold higher at 15°C as compared to 25°C (Carpenter *et al.*, 1987). Tam3 is also influenced by an unlinked gene, *Stabilizer,* which considerably reduces its excision frequency. Tam1 is not as sensitive to temperature and is not influenced by *Stabilizer*.

F. Tgm Insertions in Soybean

Aside from *Zea mays* and *Antirrhinum majus,* there are currently no other plants for which there is both molecular and genetic evidence of transposition. In soybean there are several examples of variegation produced by nuclear genes which may be due to the action of transposable elements. These include the yellow mutable (*Y18-m*) allele that conditions a variegated yellow and green leaf color (Peterson and Weber, 1969), the *W4-m* allele that produces variegated white and purple flowers (Groose *et al.*, 1988), and the *r-m* allele that has a variegated black and brown seed coat (Bernard and Weiss, 1973). Both somatic and germinal revertants are found in all three examples. Revertants of the *r-m* allele are not completely stable but can cycle through active and inactive phases both somatically and germinally, which suggests the involvement of a transposable element (Chandlee and Vodkin, 1989).

To date, there are no documented examples of movement from an element at any of the three soybean alleles that show variegation effects to other loci. However, there is molecular evidence for the presence of at least one transposable element family in soybean. A soybean insertion, Tgm1 (transposable element, *Glycine max*), arose by a transposition mechanism, although there is no genetic evidence that it is unstable. Soybean lectin is a moderately abundant protein produced in the cotyledons during seed development. Normally, lectin accounts for about 2–5% of the total seed protein, but several lines do not express lectin (Pull *et al.*, 1978), and the absence of lectin is inherited stably as a recessive single gene trait, *le1* (Orf *et al.*, 1978). In lectin negative lines, the *le1* gene coding region is interrupted by a 3.5-kb insertion, Tgm1, which blocks expression of the lectin gene (Goldberg *et al.*, 1983). The ends of the element have inverted repeats flanked by a 3-bp duplication of the lectin target sequence (Vodkin *et al.*, 1983). These structural features demonstrate that Tgm1 entered the lectin gene by a transposition mechanism. The complete sequence of the element revealed a complex series of small inverted repeat borders (Rhodes and Vodkin, 1985) but no large open reading frames. Tgm1 is a deletion derivative of much larger (>12 kb) elements some of which contain an open reading frame with homology to ORF1, the open frame found within an intron of maize En/Spm (Rhodes and Vodkin, 1988). Conservation of an open reading frame sequence between these three diverse plant species (maize, soybean, and snapdragon) suggests a functional significance for the ORF and possibly it encodes a transposase.

Comparisons of the terminal sequences among several plant elements show strong sequence similarities at the 13-bp termini between En/Spm in maize, Tam1 and Tam2 in snapdragon, and Tgm1 in soybean (Saedler *et al.*, 1984; Rhodes and Vodkin, 1985). The termini of these elements are compared in Fig. 4. All begin with the same five bases, CACTA, and all duplicate a three-base sequence of the target DNA. In addition to these features, all of these elements possess sequences just internal to the 5' and 3' termini which are composed of a series of repeated sequences, most of which occur as inverted repeats. The repeating sequences are referred to by various authors

ELEMENT SPECIES	5' TERMINUS		REPEATED PALINDROME CHARACTERISTIC OF BORDERS			3' TERMINUS	
TGM1 SOYBEAN	CACTA	TTAGAAAA -	TTAACATCGGTTTT	2-10 NT LOOP	AAAACCGATGTTAA -	TTTTGTAA	TAGTG
EN(SPM) MAIZE	CACTA	CAAGAAAA -	AAGAGTGTCGG	2-10 NT LOOP	CCGACACTCTT -	TTTTCTTG	TAGTG
TAM1&2 SNAPDRAGON	CACTA	CAACAAAA -	TCTTGGGACACT	4-23 NT LOOP	AGTGTCCCAAGA -	TTTTGTTG	TAGTG
LEGC GENE PEA		?	CACCTTAGAGGGCGCTTT	6 NT LOOP	AAAGCGCCCTCTAAAGTG -	TTTTGGCG	TAGTG

Fig. 4. Comparison of termini and border repeat palindromes among several plant elements. The five bases that are identical in each element are boxed. The inverted repeat represents one of the recurring palindromes that are characteristic of the element borders. Within each element the palindromic sequence is highly conserved; however, as shown, the sequences vary among elements of different plant species. Sequences having similarity to the lectin gene target region are underlined in Tgm1 and some homology to the chalcone synthase gene was observed in the Tam1 border region. A sequence from the pea legumin C gene (Lycett *et al.*, 1985) appears to represent the 3' end of an element similar to those of the En/Spm family. (From Vodkin and Rhodes, 1986.)

as the "common motif" region, the "extended palindromic borders," or the "subterminal repeats." The repeated sequence, however, is different for each element, as illustrated in Fig. 4.

The subterminal repeats can be represented in two different ways. In one format, they are depicted as a series of hairpins formed by pairing adjacent inverted repeat sequences as shown in Fig. 5B. Tgm1 has 13 hairpins in 726 bp of 5' border sequence and only 2 hairpins within 137 nucleotides of 3' sequence. In the left border, hairpins 9–13 represent a highly conserved

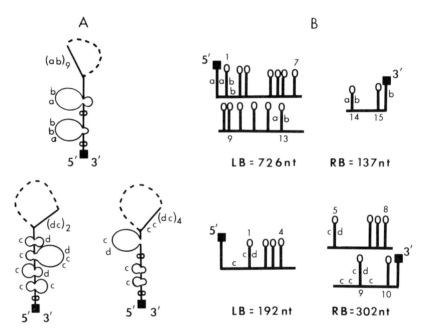

Fig. 5. Alternative representations of the subterminal repeat areas for soybean Tgm1 (top) and the maize En/Spm element (bottom). (A) Pairing of the 5' and 3' subterminal repeat areas while (B) shows pairing of adjacent palindromes to form hairpin structures. The dark box denotes the 13 nucleotides (nt) of the 5' and 3' termini which are similar in both elements. Single-stranded regions which are not part of the extended double-stranded termini are denoted as loops in A. The largest of these are approximately 50 nt. The dotted line denotes the remainder of the Tgm1 or En/Spm elements which do not include border sequence. LB, left border; RB, right border. The stem of each hairpin contains a sequence similar to the consensus sequence shown in Fig. 4. The sequence and its complement are denoted by "a" and "b" for the Tgm1 element and by "c" and "d" for En/Spm. Small circles denote the single-stranded loops of about 10 nt. One copy of a and b is also found within the terminal 30 nt of the 5' and 3' ends of Tgm1. One occurrence of b for which there is not a corresponding a is found between hairpins 1 and 2, as marked. Likewise, the conserved inverted-repeat sequences found within the En/Spm element are denoted by c and d. Note that extended pairing of the termini as depicted in A for both elements leads to many occurrences of unpaired a, b, c, or d sequences in the large single-stranded loops. (From Vodkin and Rhodes, 1986.)

tandem repeat unit of 54 bp. The other hairpins are degenerate forms of this basic repeat unit, within which the stem region remains more conserved than the interhairpin sequence or the loop of the stem. Each stem contains the common motif sequence ACATCGG and its inverse, which has been designated "a" and "b" in Fig. 5. In total there are 33 repetitions of this sequence, its inverse complement, or slightly derivative forms. When drawn in this manner, En/Spm borders have 4 hairpin repeats within 192 bases of the left border and 6 in approximately 302 bases of the right border of the element. As in Tgm1, the stems of the hairpins are highly conserved within the En/Spm element but the sequence is not the same as that in Tgm1, as shown in Fig. 4 and denoted by "c" and "d" in Fig. 5.

An alternative representation of the borders is also shown in Fig. 5. In this format, the 5' and 3' borders of the element are paired into a long double-stranded region that extends beyond the perfectly paired 13-bp termini of the element. However, the exact pairing could be done in a number of ways and many of the "ab" and "cd" palindromes are not a part of the pairing, as illustrated in Fig. 5A. They would fall into the single-stranded loop regions which can vary up to 50 bases.

The functional significance of these subterminal repeats or possible secondary structures within them is unknown. The sequence of the borders is highly conserved, however, in Tgm1-related elements found at other locations in the soybean genome (Rhodes and Vodkin, 1988). Also, it has been noted that there is some homology between the stems of Tgm1 and the lectin target gene (Vodkin *et al.*, 1983) and those of Tam1 and the chalcone synthase target gene (Bonas *et al.*, 1984), suggesting that there may be a site preference for insertion that may be modulated by the border sequences. A number of defective En/Spm elements which show a reduced excision frequency have internal deletions that omit one part of the element repeated motif region (Schiefelbein *et al.*, 1985; Schwarz-Sommer *et al.*, 1985b; Tacke *et al.*, 1986). Taken together, the evidence suggests that there is a functional role for the borders of the En/Spm-Tam-Tgm1 class of transposons. A model has been proposed in which the En/Spm-encoded transposase binds not only to the 13-bp termini but also to the extended repeating motif sequences and can lead to deletions (Schwarz-Sommer *et al.*, 1985b). The first direct evidence shows that a protein produced in *E. coli* from the En/Spm gene 1 transcript is a DNA-binding protein that interacts with the common motif sequence of the elements (Gierl *et al.*, 1987). Another model proposes that a positive regulatory gene product of the En/Spm element (perhaps a part of the transposase protein) interacts with the subterminal repeats in a way to regulate transcription of the element (Masson *et al.*, 1987). It is interesting to note that the TATA promoter region of the En/Spm element is located immediately adjacent to the subterminal repeat region, as shown in Fig. 3.

III. TRANSPOSABLE ELEMENT ACTION

A. Effects on Gene Expression

1. Site of Insertion within the Target Gene

Expression of the target gene can be profoundly affected by the insertion of a transposable element. The site of insertion (intron, exon, or promoter region) can influence whether the mutant gene has a null phenotype or an intermediate phenotype due to residual gene expression. An insertion into a promoter of the *Adh* gene has also been shown to affect its organ specificity (C.-H. Chen *et al.*, 1987). Cumulatively, there are a number of examples of transposable element insertions into all three regions of plant genes (Doring and Starlinger, 1986). Mu1 insertions into the first intron of the maize *Adh1*' gene produce a normal-sized mRNA but at reduced levels (Bennetzen *et al.*, 1984). Thus, the Mu1 insertion does not abort transcription and is removed during intron processing. Using runoff transcription assays, Rowland and Strommer (1985) found that the reduced Adh mRNA level resulted from lowered transcription rates and not from an inefficiency in processing of the nuclear Adh RNAs that contain the Mu1 transcript. The mechanism by which an insertion downstream of the target gene promoter can affect upstream transcription is unknown. Transcription of soybean seed lectin, which has no introns, is prevented by the insertion of Tgm1, which interrupts the gene in the 3' side of the coding region (Goldberg *et al.*, 1983).

Ds and dSpm insertions into exons do not always completely block synthesis of a functional protein or a normal-size transcript. In several cases, it has been shown that a splice site provided by sequences in or near the element termini can lead to removal of the element by intron splicing (Kim *et al.*, 1987; Wessler *et al.*, 1987; Simon and Starlinger, 1987). Thus, alternative splice sites within an element may have a functional role to lessen the deleterious impact of element insertions on gene activity.

2. Change of State and Spm-Suppressible Alleles

The effects of En/Spm on gene expression are often more complex than that of Ac. Spm derived its name from the two functions of the element which can be discerned genetically, the suppressor and mutator functions. Mutator function controls excision of the element or its transposition-defective derivatives from the host gene. Mutator function is thus analogous to a bacterial transposase. McClintock inferred the existence of the suppressor function from the effect of an active Spm element on dSpm insertions at the *A1* locus in maize (McClintock, 1951), a locus encoding an unknown enzyme in the anthocyanin biosynthetic pathway. Kernels with a normal *A1* gene (and all other dominant color genes) contain intense purple pigment in the aleurone layer of the endosperm. The suppressor function was discovered by the effects of an active En/Spm element on certain defective elements located in the *a1* gene, as illustrated in Fig. 6A. In the absence of an active En/Spm element elsewhere in the genome, certain alleles carrying defective En/Spm elements are partially expressed, leading to a pale aleurone color rather than a completely colorless kernel. The dSpm element is stable since there is no autonomous element elsewhere in the genome which can provide a transposase to excise the element. In the presence of an active En/Spm element, however, the insertion can excise, leading to spots of intense color where the gene is fully expressed against a completely colorless aleurone

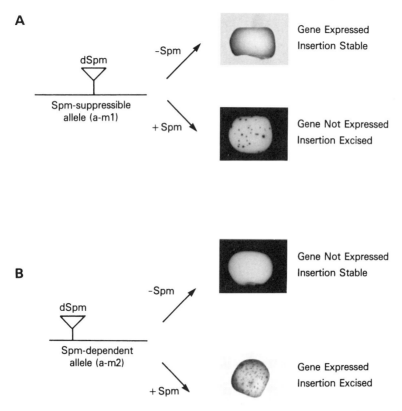

Fig. 6. Differential effects of an En/Spm element on expression of *a1* alleles containing dSpm insertions. (A) An Spm-suppressible allele. In the absence of an active element the dSpm insertion is stable and the *a1* gene is partially expressed, leading to pale pigmentation in the kernel aleurone. In the presence of *trans*-acting factors provided by an active Spm, *a1* gene expression is suppressed except in areas where the dSpm insertion has excised. As illustrated in B, spm-dependent alleles such as those in the *a1-m2* series, are not expressed in the absence of Spm. In the presence of Spm, the gene is partially expressed despite the presence of the dSpm insertion. In areas where the dSpm has excised, expression is fully restored, leading to spots of intense color. (From Masson *et al.*, 1987.)

background, instead of a background having a pale color. Somehow, an active En/Spm element has suppressed the residual expression of the *a1* allele harboring a dSpm element. Alleles which show this effect are termed Spm-suppressible alleles.

One of the objectives of recent analysis of the En/Spm element is to understand the molecular basis for the suppressor function and to explain in molecular terms the nature of "changes of state" which frequently occur at loci under the control of Spm. "Changes of state" is the classical term applied by McClintock to different heritable phenotypes induced by En/Spm at a locus. For example, McClintock obtained several stable transposition-defective derivatives of the *a1-m1* locus which originally contained an autonomous En/Spm element. The derivatives, or new states of the locus, affected the phenotypic expression of the *a1* gene, producing kernels that ranged from colorless to pale or almost fully colored. In the presence of an active En/Spm, the new derivatives also varied in the frequency and timing of excision of the defective element from the locus. Some of the new alleles were Spm-suppressible and others were not.

Change in state behavior could be due to a change in the location or orientation of the defective Spm within the *a1* gene or the element could have undergone structural alterations leading to changes in its effects on gene expression and its response to an active En/Spm element. In order to determine which of these occurred, three derivatives of the *a1-m1* locus, states 6078, 5719A-1, and 1112, have been physically isolated and characterized (Schwarz-Sommer *et al.*, 1985b; Tacke *et al.*, 1986). Phenotypically, state 6078 is colorless in the absence of an active element while in the presence of En/Spm it has large colored sectors representing excision events early in development. In contrast, state 1112, which is genetically derived from state 6078, has nearly fully colored kernels in the absence of En/Spm. In the presence of an active element, however, the *a1* gene activity is suppressed except in the areas where the Spm element has excised, leading to spots of color on a near colorless background. State 1112, and also the third state, 5719A-1, are classic examples of Spm-suppressible alleles (see Fig. 6A). It is also likely that state 6078, which is not a suppressible allele, is the parent allele of state 5719A-1. Molecular analysis of the three states showed that the dSpm insertions within the locus did not change positions and that all three elements were deletion derivatives of Spm, as shown in Fig. 7 (see also Section II,B). For the Spm-suppressible alleles of states 5719A-1 and 1112, the deletions extend into the repeating sequence motif of the Spm extended borders, indicating a possible functional significance for the repeats.

A model to explain the effects of En/Spm on gene activity postulates that the suppressor and mutator functions of the element are the same protein. This is supported by the fact that genetic data suggest a strong correlation between the suppressor and mutator functions. Schwarz-Sommer *et al.* (1985b) propose that readthrough transcription of the dSpm elements occurs

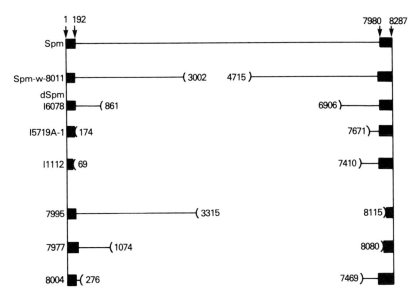

Fig. 7. Defective Spm elements are internal deletion derivatives of the active element. The first line indicates the active En/Spm element. Nucleotide numbers of the deletion endpoints are shown in parentheses and the black boxes represent the subterminal repeated regions. The I5719A-1 and I1112 insertions are Spm-suppressible alleles of the *a1-m1* series and the 7995, 7977, and 8004 are Spm-dependent alleles of the *a1-m2* series; Spm-w-8011 is a weakly active element which carries a deletion in the first intron of the element. (Data from Schwarz-Sommer et al., 1985b; Tacke et al., 1986; and Masson et al., 1987.)

in some insertions which show partial activity of the host allele. In the presence of an active Spm element, however, the Spm protein (transposase) binds to the extended terminal sequences of the dSpm insertion in such a way as to block readthrough transcription and thus suppress the host gene activity. While transcription data are not yet available to test this hypothesis for the different states of the dSpm insertions in the *a1-m1* alleles, Gierl and co-workers (1985) have analysed transcripts from the *wx-m8* allele and found that transcription from the *wx* gene promoter which terminates in the dSpm-I8 element is suppressed in plants carrying an active En/Spm element.

3. Changes of State and Spm-Dependent Alleles

In contrast to the effect of an active En/Spm element of Spm-suppressible alleles, another class of dSpm insertions within the *a1* gene, termed Spm-dependent alleles, shows exactly the opposite effect on gene expression in the presence of an active En/Spm element (Fig. 6B). Spm-dependent alleles normally have a stable null phenotype in the absence of En/Spm. In the presence of an autonomous element, however, the *a1* locus is partially expressed even in tissue areas in which the element has not excised. The *a1-m2* alleles are a series of change-of-state dSpm insertions found at another site within the *a1* locus and some of them are Spm-dependent. Masson *et al.* (1987) have isolated a series of these *a1-m2* derivatives. Similar to the case with the *a1-m1* alleles, the dSpm insertions are all in the same position and orientation as the original Spm in the *a1-m2* progenitor mutation and are all internal deletion derivatives of the element as shown in Fig. 7 for states 7995, 7977, and 8004. Additionally, the deletion breakpoints occur in the 12-bp subterminal repeat sequences or in internal regions having strong homology to the 12-bp consensus repeat sequence (Masson *et al.*, 1987). These investigators propose that the En/Spm "suppressor" is actually a positive regulatory protein that binds to the subterminal repeat sequences in such a way as to promote the expression of En/Spm encoded genes. They note that the En/Spm TATA box lies immediately adjacent to the subterminal repeat region. They suggest that probably the "suppressor," or positive regulatory function, and the "mutator," or transposition function, are properties of the same bifunctional protein.

Whereas the dSpm elements within the *a1-m1* and the *a1-m2* alleles are deletion derivatives and do not change positions within the gene, their respective sites of insertion within the *a1* gene are different. This difference could explain the difference between the Spm-suppressible versus the Spm-dependent effect on *a1* gene activity. The *a1-m1* insertions are found within the coding region at nucleotide position 480 whereas the insertions of the *a1-m2* series are found upstream of the promoter at position −99 (Schwarz-Sommer *et al.*, 1987a). The orientation of the element is the same in both locations. Interaction of a transposase/regulator protein from an active element with the dSpm termini in the *a1-m1* alleles may prevent readthrough transcription, leading to suppression of *a1* expression. On the other hand, interaction of this protein with the *a1-m2* insertions may enhance gene expression of the adjacent *a1* promoter.

4. Presetting of dSpm Containing Alleles

Certain dSpm alleles show "presetting" phenomena in which the expression of an Spm-dependent allele can be carried through a number of cell generations after the active Spm element has been lost. For example, a plant heterozygous for the active Spm element but homozygous for a dSpm insertion at the *a1-m2* locus will lose the Spm element in some kernels of the next generation through meiotic segregation. However, the Spm-dependent alleles in these kernels will still produce clonal sectors of *a1* gene expression as if they are still responding to the *trans*-acting factors produced by an active Spm element. Presetting possibly reflects molecular interactions between the Spm protein and its target dSpm sequences at a site that interferes with an inactivating mechanism. Although the mechanism is unknown, one hypothesis is that reversible DNA modifications by methylation is the cause of presetting (Masson *et al.*, 1987). It has been noted that the dSpm-8004 element (Fig. 7), which lacks a very GC rich region in the element's first exon, is not able to be preset whereas the 1.3-kb dSpm-7977 element which contains this region is able to be preset by En/Spm. This region contains a cluster of CG and CNG nucleotides which are commonly methylated in plant DNA.

5. *Changes in Phase or Cycling of Element Activity*

Some En/Spm and Ac elements can undergo frequent reversible changes from an active to an inactive state that McClintock termed "changes in phase" of the element's activity. Elements exhibiting this behavior are also referred to as cycling elements. It has been postulated that the molecular basis of these reversible changes in activity may be due to modification of element sequences (Chandler and Walbot, 1986; Fedoroff, 1988). As previously discussed for the Mu element, loss of mutator activity is correlated with modification of all genomic copies of the element (Chandler and Walbot, 1986). Chomet and co-workers (1987) examined the restriction patterns of Ac-related sequences in maize lines carrying an active Ac element (*wx-m7* :: *Ac active*) versus a change-in-phase Ac element which produced an inactive element (*wx-m7* :: *Ac inactive*) at the *waxy* locus in maize. Ac contains *Pvu*II sites which are sensitive to methylation. Genomic DNA from lines containing the active Ac in the *waxy* locus could be cleaved with *Pvu*II to yield the expected fragments whereas these sites were not restricted in lines harboring the inactive Ac element. To verify that the *Pvu*II sites are present in the inactive element, it was cloned into *E. coli*. The sites were present and unmodified in *E. coli* and the map of the inactive element was shown to be the same as that of the active Ac element. Thus, methylation at certain sites within the Ac element is associated with its conversion to an inactive phase. Schwartz and Dennis (1987) have shown that an inactive Ac element at the *wx-m9* site can revert to the active form at rates from 0 to 10%. The inactive Ac is methylated at all *Hpa*II sites in the element, while the revertant active Ac is only partially methylated.

B. Models of Transposition in Plants

Although movement from one location to another is the salient feature of transposable elements, transposition usually occurs at low frequency and is often difficult to demonstrate genetically and unequivocally. Excision events that can be readily seen as variegation in somatic tissues do not necessarily represent transposition of the excised element to another locus. Both germinal and somatic transpositions can occur but it is not known whether the mechanisms are the same in both cases. A molecular understanding of transposition in plants is far from complete.

1. Mechanism of Mp(Ac) Transposition from the P Locus

Genetic studies have indicated that transposition of Ac and En/Spm is accompanied by the loss of the element at the original locus with the gain of the element at another locus. Such a loss/gain mechanism is in contrast to a replicative mechanism of transposition in which a new copy of the element transposes to a new location while the original element remains in position. Extensive genetic experiments on the mechanism of transposition of the Mp element (genetically identical to Ac) have been conducted using the *P* locus in corn which is involved in pigmentation of the pericarp (Greenblatt, 1968, 1974; reviewed in detail by Fedoroff, 1983, and Nevers *et al.*, 1986). Mp at the *P* locus blocks its expression so that kernels are colored only when Mp has somatically excised from the gene. The amount of red striping in the resultant variegated pericarp shows a negative dosage effect with respect to the number of Mp elements in the genome. With one copy of Mp, the kernels are medium variegated, whereas with two copies, they are lightly variegated. This dosage effect of Mp on its excision from the *P* locus allows its transposition to be followed genetically.

Twin sectors of full red versus lightly variegated kernels often arise on a background of medium variegated ears in heterozygous plants which have the genotype *p/p-Mp*, where *p* is a stable recessive allele for pericarp color. The twinned sectors of kernels represent the clonal lineages from a single progenitor cell in which a transposition event has occurred during mitosis at a time just before or after the chromosome segment carrying the *P* locus replicates. One daughter cell and its lineage inherit the chromatid from which the Mp element has excised and transposed. These kernels (now *p/P* and no Mp in the genome) will have full red color restored. In contrast, the other daughter cell (still *p/p-Mp*) resulting from the mitotic division will inherit the sister chromatid now carrying two copies of the Mp element, one at the original *P* locus and another at a new location. Because of the dosage effect on Mp excision, lightly variegated kernels result from this cell lineage. The excised Mp element integrates into the newly replicated sister chromatid rather than to a site on the original chromatid from which it excised. Genetic tests on the kernels that arise in the twinned red and lightly variegated sectors were used to determine the new location of the Mp elements. The transposed element is found on the same chromosome in 61–67% of the cases. Thus, there is a preference for transposition to nearby locations; however, transposition to independently assorting chromosomes does occur.

Isolation and analysis of the *P* locus have provided molecular evidence supporting the genetic model that Ac transposition takes place after replication of the donor site but before completion of replication at the target site (J. Chen *et al.*, 1987). Whether other types of transpositions occur at lower frequencies is not known. Such events would include transpositions that occur in nondividing chromosomes. Another mechanism is replicative transposition in which the original element does not move but rather a replicated copy of the element goes to a new location. Replicative transposition is a common mechanism for a number of prokaryotic elements, e.g., Tn3 (Heffron, 1983). The maize Mu elements which have a high transposition rate also appear to move by a replicative mechanism, as discussed in Section II,C (Alleman and Freeling, 1986; Bennetzen *et al.*, 1987).

2. A Molecular Model for Element Excision and Integration

A molecular model for the mechanism of plant transposable element excision and integration has been proposed (Saedler and Nevers, 1985; Nevers *et al.*, 1986). The model deals primarily with the finding that excision events in plants do not restore the host gene sequence exactly to wild type. Instead, the imprecise excisions leave behind the duplication of the target area flanking the insertion site or remnants of this area. The model is a simple cleavage and repair mechanism, and, other than the transposase, all of the enzymes required are part of the normal DNA repair mechanism of the host cells. During the excision process, a transposase encoded by the element is presumed to create staggered nicks on each side of the duplicated target sequences that flank the element termini. The transposase then draws the termini of the element together into a DNA–protein complex which brings the duplicated target sequences and the four nicks into close proximity. It is postulated that a 5' exonuclease activity could then degrade the duplicated target sequences attached to the element termini, leaving no extra nucleotides attached to the element itself. Meanwhile, a DNA polymerase will repair the gap left in the target duplication area using the complementary DNA strand as template. The blunt-ended chromosomal segments can then be religated. The net result is the excision of a transposable element with no extra nucleotides attached whereas the host gene would be left with a target site duplication rather than a completely restored original sequence. If the 5' exonuclease were also to remove some of the flanking target site sequences on the complementary strand before the polymerase fills in the gap, the result would be deletions of part or all of the duplicated sequence on either side of the element termini. Other imprecisions in the process can lead to small additions or inversions of bases at the target sites. Supporting evidence that many of these imprecise events actually occur was obtained by examining the sequences of revertant alleles resulting from the excision of transposable elements (Saedler and Nevers, 1985).

Integration of the excised element, which presumably continues to exist as a DNA–protein complex, would also begin with the generation of staggered nicks by a transposase at the new target site DNA. These nicks are a characteristic number of base pairs apart for each element. After insertion of the element, a polymerase fills in the single-stranded gaps at the target site, resulting in the duplicated sequences that flank each element. Integration events appear to be very precise. Only one example has been found where the number of bases duplicated at the target site is not the same for each element regardless of its location. Tam3 is flanked by an 8-bp duplication in the *niv-98* gene and 5 bp in the *pal-rec* allele of snapdragon (Coen et al., 1986).

3. Analysis of Integration Sites

Why elements integrate into particular DNA regions is not known. However, preference for particular sites is inferred from analysis of the target areas of several genes. Analysis of nine different insertion sites for Ac, Ds, Mu, and En/Spm revealed small (6–19 bp) perfect or near perfect direct repeats within about 50 bp of the insertion sites (Doring and Starlinger, 1984), indicating that areas having short direct repeats may be favored. Tam3 was found to insert just upstream of the promoter in two different anthocyanin genes, the *pal* and *niv* loci. In both genes, 5 of 6 bases that occur just to the left and right of the target duplication are similar in both genes (Coen et al., 1986). The repeating hairpin units in the borders of the soybean Tgm1 insertion as well as within 30 bases of the termini of the element each contain the 7-base sequence ACATCGG and its complement (Rhodes and Vodkin, 1985). Interestingly, this sequence is found within 17-bp inverted repeats of the lectin gene that are spaced about equal distances of 65 to 80 bp to either side of the insertion site in the lectin gene (Vodkin et al., 1983). The Tam1 element which also shows a repeating border format has homology between the repeating palindromic sequences and regions of the chalcone synthase gene (Bonas et al., 1984). The significance of these host gene sequence similarities to the repeats of the extended termini is unknown but is suggestive of a site preference perhaps through interaction between proteins that bind to these sequences.

C. Role in Evolution

Plant transposable elements can generate mutations and genetic diversity by the simple act of inserting into a genetic locus, leading to a disruption of the normal control of the gene. Defective elements generated by subsequent internal deletions within the element proper can have new effects on gene expression leading to stable null, intermediate, or full expression phenotypes. Subsequent excision of active or defective elements should lead to wild-type revertants. Surprisingly, however, the host gene DNA sequence is not usually restored to wild type but maintains vestiges of the transposable element target duplications, due to imprecise excision as discussed in the previous section. This finding has further implications for the role of plant transposable elements in generating mutations and genetic diversity needed for evolutionary selection.

The vestiges of imprecise excision events are often referred to as transposon footprints. As might be expected, footprints left behind by elements which have excised from introns generally have the least impact on gene expression. But even in exons where selection pressure to maintain function should be the greatest, footprints still occur at a much higher frequency than precise excisions, as demonstrated by Schwarz-Sommer and co-workers (1985a). These researchers analyzed 2 germinal and 16 somatic reversion events of the dSpm-I8 element from an exon of the *wx-m8* gene in maize. Some of the germinal revertants, which are phenotypically wild type and occur in the presence of an active En/Spm element at a frequency of about 1%, had footprints at the initial insertion site that produced an in-frame codon for an additional amino acid in the waxy protein. Somatic reversion events were analyzed for footprints in two ways. In the first test, excision of the dSpm-I8 element from the *waxy* locus was detected by Southern blots of leaf DNA from plants with the genotype *wx-m8* + En. When probed with a *wx* gene fragment, two bands are found. One represents the gene containing the insert and the other, 2.5 kb smaller, represents the gene left behind by an excision event. The relative intensities of the bands indicated an excision frequency of 10 to 20%. Cloned fragments from the excision event band were isolated and sequenced to examine the empty target site. In another approach, normal-sized *waxy* transcripts appear in the endosperm of *wx-m8* + En plants at about 10% of the wild-type levels. These mRNAs represent excision events and were examined by cloning their cDNAs. Nine revertants from leaf tissue and seven from the endosperm tissue were analyzed but only one precisely restored the wild-type sequence. Seven events resulted in the addition of an amino acid to the coding region, one event resulted in the deletion of two amino acids, and five were frameshift mutations. Thus, in addition to chromosomal rearrangements (such as those promoted by Ds elements) that induce chromosome breakage and the effects of large insertional mutations on gene activity, many small mutations in the genome may result from transposable element action. Such changes can add or delete amino acids and possibly change protein structure and function. Additionally, imprecise excision events which occur in the promoter region of the *pal* gene of snapdragon have been demonstrated to result in altered spatial patterns of gene activity as well as different levels of gene expression (Coen *et al.*, 1986).

Differences in the sequence between two wild-type sucrose synthase alleles from different maize varieties have been attributed to former visitations by transposable elements (Zack *et al.*, 1986). Four small duplications present in the introns of one allele but not the other were interpreted to be due to previous insertion and excision of transposable elements. Three small insertions of 9 to 140 nucleotides flanked by short direct repeats were also present in one allele and not the other. An 84-nucleotide intron in the chlorophyll a/b protein of *Lemna gibba*, a small aquatic plant, has small inverted repeats flanked by imperfect direct repeats (Karlin-Newmann *et al.*, 1985). Based on these features, the authors speculate that the intron may represent the vestiges of a class of transposable elements that inserts into genes but does not inactivate them since the insertion can be eliminated by the splicing reaction.

D. Origin and Regulation

The release of transposable element activity in an organism is often related to a stress or a shock to the genome. The Ac and Ds elements were found in progeny of maize kernels that carried broken chromosomes and had undergone extensive "breakage-bridge-fusion" cycles (McClintock, 1984). Other factors that induce chromosome damage, such as X-ray and UV radiation, can also induce transposable element movement. Plants regenerated from tissue culture show new genetic and cytogenetic variation, termed somaclonal variation, which may in part be due to the activation of transposable elements during the process of tissue culture. As evidence that tissue culture generates transposable element activity, 11 occurrences of active Ac elements were found in a total of 301 plants regenerated from embryogenic cells of a maize line that previously did not show transposable element activity (Peschke et al., 1987). Active elements could be released because of chromosome breakage, which is the most frequent cytogenetic event observed in regenerated corn plants (Benzion et al., 1986). These researchers speculate that the breaks result from delayed replication of heterochromatic regions in cultured cells. Regions not replicated by anaphase could result in anaphase bridges and lead to chromosome breakage.

The induction of new mutations in maize lines exposed to infection by virus or other pathogens is another form of stress that may activate transposable element movement. For example, the Bs1 insertion in the maize *Adh1* gene was recovered from infection of maize lines with barley stripe mosaic virus (Mottinger et al., 1984).

If silent or cryptic elements exist in plants, how do they become activated and how are they regulated? It would seem that they must be regulated in some manner to prevent the mutation rate caused by element movement from becoming too high to tolerate. There is increasing evidence that methylation of element sequences may be an important mechanism of regulating element activity. Although genetic analysis usually reveals the presence of only one active Ac or En/Spm element, molecular hybridization studies support the notion that cryptic elements are often present in multiple copies in lines that do not show genetic instability. While many of these sequences may be defective elements with internal deletions, it is possible that potentially active elements are masked by methylation. Active Ac elements are undermethylated and the En/Spm active element can be distinguished from other sequences using methyl-sensitive enzymes (Cone et al., 1986). Some elements undergo frequent reversible changes from an active to an inactive state which McClintock termed "changes in phase" of the element's activity. These changes in phase have been shown to correlate with methylation of certain internal restriction sites in the Ac element (Chomet et al., 1987; Schwartz and Dennis, 1987). Demonstration of a correlation between loss of Robertson's Mutator activity in maize lines with methylation of Mu sequences is also supportive evidence that methylation is involved in regulating transposable element action (Chandler and Walbot, 1986; Chandler et al., 1986).

Thus, transposable elements may serve as a survival mechanism by which populations can respond to stressful environmental changes through increased mutation rates. For example, a traumatic event that disrupts chromosome replication and division may allow potentially active mobile elements to escape from the normal cellular functions which keep them quiescent. Delayed methylation at the affected chromosomal region could be a triggering event. But once activated, an element-encoded transposase would promote its own movement to new genomic locations as well as the movement of defective elements scattered throughout the genome. It is possible that an active element in a new genomic location may not be inactivated due to differing methylation patterns at the new sites. There, it could remain active until internal deletions that arise at high frequency convert it to a defective element. The propensity of certain elements to generate internal deletions derivatives which are no longer capable of producing a transposase may be viewed as a self-limiting mechanism to reduce the deleterious effects of an element on the host genome (Masson et al., 1987; Fedoroff, 1988). Imprecise excision events generated from the new burst of element movements will leave behind a legacy of newly induced mutations until the element is inactivated by methylation, internal deletion, or other unknown mechanisms. As with most mutagenic agents, many of the mutations generated by transposable elements will be deleterious. On the other hand, some will provide subtle changes in gene expression which in turn lead to new functions or forms in plant growth and development that will be selectively valuable over evolutionary time.

IV. GENE TAGGING WITH TRANSPOSABLE ELEMENTS

A. Isolating Genes Whose Products Are Unknown

Entire genomes can be cloned easily using recombinant DNA methods. However, the process of identifying a specific gene among thousands is still a difficult task. Generally, some knowledge of the protein product is required in order to identify cDNA clones or to make synthetic oligonucleotides which can be used to screen genomic libraries. We still know very little about the products of the vast majority of plant genes, including genes that control developmental processes, plant growth and morphological form, and disease resistances. Isolation of such genes is an initial step toward understanding their mechanisms. This basic information will also contribute to improving the productivity and quality of commercially important plants.

Because transposable elements act as mutagens when inserted into loci, they are useful in the isolation and identification of such genes. As molecular tags, transposable elements allow a direct correlation between a mutant phenotype and a DNA sequence without prior knowledge of the products or pathways in which the gene is involved. The basic process of transposon tagging begins with a genetic cross in which one of the parents contains an active element. The progeny are then screened for recessive mutations in the trait of interest that occur at frequencies of 10^{-4} to 10^{-3}, which are higher than the spontaneous mutation rate. These mutations are likely to be due to the insertion of the transposable element into or near the gene responsible for the mutant phenotype. A genomic library from the mutant plant is then created and screened with a radioactive probe of the transposable element sequence. In this way, the transposable element serves as a molecular tag to identify the mutant gene since at least part of its sequence will also be carried on a clone that contains the element sequence. The regions which flank the element can be subcloned and used, in turn, as probes to screen a genomic library from normal plants in order to isolate the wild-type gene. Transposon tagging has been used to identify many genes in prokaryotes and, more recently, in several eukaryotes, including *Drosophila* and maize.

The above is a simplified description of transposon tagging and there are experimental difficulties in using transposon tagging in plants. The procedure is currently possible only in plant systems with genetically active elements that have been identified and cloned. To date, this limits the approach to maize and snapdragon. Another problem is the large number of plants (typically 10,000 or more) that must be screened for the mutant phenotype. For genes that encode readily observable traits in the kernels or seedlings, this is a relatively easy task. However, large field plots may be required for screening other traits expressed only in mature plants. If a trait is not observable, the labor required to assay for the mutant phenotype may become prohibitive. Once a mutant plant has been generated, there are also molecular difficulties that must be overcome. Chief among these is the fact that many related elements, either defective or cryptic, are generally present in the genome. This means that a number of clones at locations other than the one of interest will be selected by the transposon probe in a screen of a genomic library. Consequently, once a putative clone has been isolated, its identity must be confirmed by analyzing DNA from normal, mutant, and revertant plants to definitively correlate presence of the element with the gene in question. However, revertants may not always be available so that even this approach may not be straightforward.

B. Examples of Genes Isolated by Transposon Tagging

The plant genes initially isolated by transposon tagging have been involved in anthocyanin biosynthesis. These genes were obvious candidates because many of them have transposable element insertions which were identified by maize geneticists over the last 40 years. At least nine loci are known to be involved in anthocyanin biosynthesis in the aleurone tissue of maize kernels. Placed in a tentative order of gene action, the loci which have been defined genetically are *vp, c, r, c2, pr, a1, a2, bz, bz2* (Paz-Ares *et al.*, 1986). To date, the *c, r, c2, a1, bz,* and *bz2* loci have been isolated by tagging with transposable elements (Fedoroff *et al.*, 1984; O'Reilly *et al.*, 1985; Cone *et al.*, 1986; Paz-Ares *et al.*, 1986; Weinand *et al.*, 1986; Doring and Starlinger, 1986; McLaughlin and Walbot, 1987; Theres *et al.*, 1987). The products for these genes are unknown except for the *c2* locus which encodes chalcone synthase and the *bz* locus which encodes UDPglucose-flavonal glucosyltransferase. The *vp, c,* and *r* loci appear to be regulatory. With appropriate probes of the loci becoming available, an understanding of the pathway and its regulation will likely follow.

The first locus isolated by tagging was the *bronze* locus (Fedoroff *et al.*, 1984). Although its product was known, attempts to clone cDNAs of the messages had not been successful. An Ac insertion had been located at the *bronze* locus in the *bz-m2* allele (McClintock, 1951). A genomic library from DNA of *bz-m2* plants was screened with a probe representing the middle of the Ac9 element since the middle of the element hybridizes with fewer related Ds elements than do the sequences near the Ac termini. Restriction mapping of 25 initial clones with homology to the Ac9 probe showed that only 2 had internal structures identical to Ac9 and were flanked by new genomic sequences. Regions flanking the Ac9 elements in these two clones were subcloned, and those not containing repetitive DNAs were used to probe genomic DNA from plants containing different mutations at the *bz* locus. With one of the regions, normal *Bz* plants and wild-type revertants of the *bz-m2* allele had a 6.3-kb *Bgl*II fragment whereas *bz-m2* mutant plants had a 10.8-kb fragment, indicating the presence of the Ac element. The *bz*-specific probe was, in turn, used to isolate the *Bz* locus from nonmutant plants.

The En/Spm element has also been used as a transposon tag to isolate the *a1* gene whose product is unknown (O'Reilly *et al.*, 1985). DNA was cloned from plants carrying the *a1-m(papu)* allele which carries an autonomous En element. Although internal regions of the En element are found at high copy number, only 6 to 50 initial λ clones hybridized with equal intensity to two probes representing 1.6- and 3.0-kb *Eco*RI internal regions of the En element. Restriction mapping identified only one clone that had the same structure as the En-1 element isolated from the *wx* locus. Confirmation that this clone represented the *a1* locus was obtained by demonstrating that the flanking gene regions were identical in a clone which contained the Mu1 element at the *a1* locus was obtained by demonstrating that the flanking gene regions were identical in a clone which contained the Mu1 element at the *a1* locus and no insertions in the wild-type allele. Using the *a1* gene specific probe, a number of different alleles of the *a1* locus have been investigated to examine the effects of defective En/Spm insertions on expression of the locus as discussed previously (Section III,A; Schwarz-Sommer et al., 1985b, 1987a; Tacke et al., 1986; Fedoroff et al., 1987). The *a1* gene has also been used to demonstrate that a new flower color is created in petunia flowers when petunia mutants blocked in the anthocyanin pathway are transformed with a clone of the maize *A1* gene (Meyer et al., 1987).

The *C1* locus is involved in regulating the anthocyanin pathway. In mutants homozygous for the recessive *c1* allele no pigment is produced and enzyme activities of the *C2* (chalcone synthase) and *Bz* loci (UDP glucose-flavonal glycosyltransferase) are not detectable. Cloning *C1* was accomplished using En/Spm-induced mutants *c1-m668655*, *c1-m668613* (Paz-Ares et al., 1986), and *c1-m5* (Cone et al., 1986). Fourteen clones with homology to left, middle, and right sections of En/Spm were identified; of these, one contained a restriction map indistinguishable from that of the authentic En element (Paz-Ares et al., 1986). Cone and co-workers (1986) used a somewhat different approach to isolate the En/Spm element residing in the *c1-m5* allele. Since active elements are usually undermethylated in genomic DNA, they reasoned that most of the defective and inactive elements would be uncut by methyl-sensitive restriction enzymes. Indeed, using *Sal*I, they identified an 8.3-kb band in genomic DNA that segregated with the *c1-m5* allele. Genomic DNA from the 8.3-kb *Sal*I region, purified on agarose gels, was cloned and screened with a dSpm-I8 probe. The 8.3-kb clone was found to have a map identical to the En/Spm element, and contained 350 bp of flanking c1 DNA. This region represented single-copy DNA and was used to isolate the full genomic clone of the *C1* locus from nonmutant plants. Several other alleles of the *C1* locus were also isolated, including some containing small Ds elements (Table I). Using a probe from the *C1* gene, transcripts from developing kernels were examined. Transcripts of approximately 1.5 and 1.2 kb are found in *C1* plants. The regulatory nature of the *C1* gene was confirmed by examining transcripts of the *C1*, *Bz*, and *A1* loci in plants carrying the *C1* versus *C1-I* alleles. *C1-I* is a dominant inhibitor of aleurone color. All three transcripts are produced in *C1*, *Bz*, *A1* plants. *C1-I* plants produce transcripts slightly larger than the wild-type gene that hybridize to *C1* but no *Bz* and *A1* transcripts are detected (Cone et al., 1986). The sequence of the *C1* gene predicts an encoded protein with a short acidic domain at the carboxy end and a basic domain that has 40% amino acid similarity to *myb* protooncogenes in animals which are DNA-binding proteins (Paz-Ares et al., 1987). These similarities indicate that the *C1* gene may regulate transcription through DNA–protein interactions.

Thus, both Ac and En/Spm have been used successfully for transposon tagging in maize. Since the structure of active elements appears to be highly conserved, molecular distinctions can be made between the active element which tags a locus and the related sequences present elsewhere in the genome. Screening of libraries with probes from different regions of the transposon also helps to narrow down the initial number of clones which must be handled. The fact that active elements are in genomic regions that are undermethylated can sometimes be used to enrich for a specific DNA fragment prior to cloning, thus removing many unwanted related sequences. The maize Mu element has not been as widely used for tagging. Most of its related sequences have a very similar structure in contrast to the Ac-Ds and En/Spm elements. However, the recent finding that many Mu sequences become methylated (Chandler and Walbot, 1986) may allow identification strategies to be developed for Mu. A novel approach was used by McLaughlin and Walbot (1987) to clone the *bz2-mu1* allele in maize which contains a 1.7-kb Mu1 insertion. A small bank of all sequences hybridizing to Mu1 was selected from a genomic library of *bz2-mu1* DNA. The Mu sequence which contained flanking *bz2* DNA was identified by using groups of the clones to probe RNA isolated from maize husks which were green (*bz2*) or purple (*Bz2*).

Transposon tagging has also been used successfully to isolate genes in snapdragon. The Tam3 element (originally isolated from the chalcone synthase gene) was used to isolate the *pallida* locus, whose role in anthocyanin biosynthesis is unknown, by first isolating a variegated mutant, the *pal-rec* allele which contains a Tam3 element. This was accomplished by comparing several *pal-rec* lines with *Pal+* revertants (Martin *et al.*, 1985). Even internal regions of Tam3 hybridized strongly with more than 15 bands on Southern blots. Fortunately, an *Eco*RI digest consistently showed a 7.3-kb band in mutant plants but not revertants. *Eco*RI genomic fragments of this approximate size were selected from agarose gels and cloned. Sixteen strongly hybridizing clones were picked from 150,000 clones screened and one of these was identified as having the expected structure for a Tam3 insertion. Flanking sequences were used to identify the single gene *pal* locus and compare genomic DNA from normal and mutant plants.

The first nonanthocyanin gene to be tagged and isolated was the *opaque* locus in maize, which acts as a positive transcription activator of zein storage protein synthesis (Schmidt *et al.*, 1987).

C. Potential for Tagging in Heterologous Systems

Since transposable elements which can be manipulated for transposon tagging have been found in only few plant species to date, there is growing interest in determining whether elements will function in heterologous systems. Toward this aim, the maize Ac element has been shown to transpose in tobacco tissue by several criteria (Baker *et al.*, 1986). Tobacco callus was initiated from protoplasts transformed with *Agrobacterium tumefaciens* vectors containing the Ac element and a small amount of the *waxy* gene flanking each side of the element. In four of nine independently transformed lines, there were a number of extra genomic bands hybridizing to the Ac probe and empty *waxy* sites (lacking Ac) were found in the tobacco DNA when the *wx* gene was used as probe. Some of the Ac-containing fragments were cloned and found in new locations in the tobacco genome. Additionally, the empty *waxy* sites were cloned and the sequence revealed that the Ac element had excised imprecisely in the tobacco tissue as it does in maize. In contrast, the Ds9 element which differs from the active Ac by a 200-bp deletion in an open reading frame does not transpose when introduced into tobacco tissue. Thus, the active Ac element appears to promote element movement in a nonrelated species as it does in maize. Ac also transposes in carrot and *Arabidopsis* cultures, and regenerated plants were produced (Van Sluys *et al.*, 1987). These results open the possibility that transposon tagging may be applicable to a wider range of plant species.

ACKNOWLEDGMENTS

I am grateful to Drs. P. R. Rhodes, J. M. Chandlee, and R. L. Frank for critical reading of the manuscript. I thank Dr. N. V. Fedoroff for photographs of seed and many researchers who supplied manuscript preprints.

REFERENCES

Alleman, M., and Freeling, M. (1986). *Genetics* **112**, 107–119.
Baker, B., Schell, J., Lorz, H., and Fedoroff, N. (1986). *Proc. Natl. Acad. Sci. U.S.A.* **83**, 4844–4848.
Baker, B., Coupland, G., Fedoroff, N., Starlinger, P., and Schell, J. (1987). *EMBO J.* **6**, 1547–1554.
Barker, R. F., Thompson, D. V., Talbot, D. R., Swanson, J., and Bennetzen, J. L. (1984). *Nucleic Acids Res.* **12**, 5955–5967.
Behrens, U., Fedoroff, N., Laird, A., Muller-Neumann, M., Starlinger, P., and Yoder, J. (1984). *Mol. Gen. Genet.* **194**, 346–347.
Bennetzen, J. L. (1987). *Mol. Gen. Genet.* **208**, 45–51.
Bennetzen, J. L., Swanson, J., Taylor, W. C., and Freeling, M. (1984). *Proc. Natl. Acad. Sci. U.S.A.* **81**, 4125–4128.
Bennetzen, J. L., Fracasso, R. P., Morris, D. W., Robertson, D. S., and Skogen-Hagenson, M. J. (1987). *Mol. Gen. Genet.*, **208**, 57–62.
Benzion, G., Phillips, R. L., and Rines, H. W. (1986). In "Cell Culture and Somatic Cell Genetics of Plants," Vol. 3 (I. K. Vasil, ed.). Academic Press, Orlando, Florida, pp. 435–448.
Bernard, R. L., and Weiss, M. G. (1973). In "Soybeans: Improvement, Production, and Uses" (B. E. Caldwell ed.), pp. 117–154. Am. Soc. Agron., Madison, Wisconsin.
Bonas, U., Sommer, H., and Saedler, H. (1984). *EMBO J.* **3**, 1015–1019.
Borst, P., and Greaves, D. R. (1987). *Science* **235**, 658–667.
Brown, J. L., Mattes, M., O'Reilly, C., and Shepherd, N. S. (1989). *Mol. Gen. Genet.*, in press.
Burr, B., and Burr, F. A. (1981). *Genetics* **98**, 143–156.
Calos, M. P., and Miller, J. H. (1980). *Cell* **20**, 579–595.
Carpenter, R., Martin, C., and Coen, E. S. (1987). *Mol. Gen. Genet.* **207**, 82–89.
Chandlee, J. M., and Vodkin, L. O. (1989). *Theor. Appl. Genet.*, in press.
Chandler, V. L., and Walbot, V. (1986). *Proc. Natl. Acad. Sci. U.S.A.* **83**, 1767–1771.

Chandler, V., Rivin, C., and Walbot, V. (1986). *Genetics* **114,** 1007–1021.
Chen, C.-H., Oishi, K., Kloeckener-Gruissem, B., and Freeling, M. (1987). *Genetics* **116,** 469–477.
Chen, J., Greenblatt, I. M., and Dellaporta, S. (1987). *Genetics* **117,** 109–116.
Chomet, P. S., Wessler, S., and Dellaporta, S. L. (1987). *EMBO J.* **6,** 295–302.
Chourey, P. S., and Nelson, O. E. (1976). *Biochem. Genet.* **14,** 1041–1055.
Coen, E. S., Carpenter, R., and Martin, C. (1986). *Cell* **47,** 285–296.
Cone, K., Burr, F. A., and Burr, B. (1986). *Proc. Natl. Acad. Sci. U.S.A.* **83,** 9631–9635.
Courage, U., Doring, H.-P., Frommer, W.-B., Kunze, R., Laird, A., Merckelbach, A., Muller-Neumann, M., Riegel, J., Starlinger, P., Tillmann, E., Weck, E., Werr, W., and Joder, J. (1984). *Cold Spring Harbor Symp. Quant. Biol.* **49,** 329–331.
Courage-Tebbe, U., Doring, H.-P., Fedoroff, N., and Starlinger, P. (1983). *Cell* **34,** 383–393.
Dellaporta, S. L., Chomet, P. S., Mottinger, J. P., Wood, J. A., and Yu, S. M. (1984). *Cold Spring Harbor Symp. Quant. Biol.* **49,** 321–328.
Doolittle, W. F., and Sapienza, C. (1980). *Nature (London)* **284,** 601–607.
Dooner, H. K., Weck, E., Adams, S., Ralston, E., Favreau, M., and English, J. (1985). *Mol. Gen. Genet.* **200,** 240–246.
Dooner, H. K., English, J., Ralston, E., and Weck, E. (1986). *Science* **234,** 210–211.
Doring, H.-P. (1985). *BioEssays* **3,** 164–171.
Doring, H.-P., and Starlinger, P. (1984). *Cell* **39,** 253–259.
Doring, H.-P., and Starlinger, P. (1986). *Annu. Rev. Genet.* **20,** 175–200.
Doring, H.-P., Geiser, M., and Starlinger, P. (1981). *Mol. Gen. Genet.* **184,** 377–380.
Doring, H.-P., Tillmann, E., and Starlinger, P. (1984a). *Nature (London)* **307,** 127–130.
Doring, H.-P., Freeling, M., Hake, S., Johns, M. A., Kunze, R., Merckelbach, A., Salamini, F., and Starlinger, P. (1984b). *Mol. Gen. Genet.* **193,** 199–204.
Fedoroff, N. V. (1983). In "Mobile Genetic Elements" (J. A. Shapiro, ed.), pp. 1–63. Academic Press, New York.
Fedoroff, N. V. (1988). In "Developmental Genetics of Higher Organisms" (G. Malacinski, ed.), pp. 97–126.
Fedoroff, N., Wessler, S., and Shure, M. (1983). *Cell* **35,** 235–242.
Fedoroff, N. V., Furtek, D. B., and Nelson, O. E. (1984). *Proc. Natl. Acad. Sci. U.S.A.* **81,** 3825–3829.
Freeling, M. (1984). *Annu. Rev. Plant Physiol.* **35,** 277–298.
Freeling, M., Cheng, D., and Alleman, M. (1982). *Dev. Genet.* **3,** 179–196.
Geiser, M., Weck, E., Doring, H.-P., Werr, W., Courage-Tebbe, U., Tillmann, E., and Starlinger, P. (1982). *EMBO J.* **1,** 1455–1460.
Georgiev, G., Hyin, R. V., Ryskov, A. P., and Gerasimova, T. I. (1986). In "DNA Systematics" (S. K. Dutta, ed.), Vol. 1, pp. 20–46. CRC Press, Boca Raton, Florida.
Gerlach, W. L., Pryor, A. J., Dennis, E. S., Ferl, R. J., Sachs, M. M., and Peacock, W. J. (1982). *Proc. Natl. Acad. Sci. U.S.A.* **79,** 2981–2985.
Gierl, A., Schwarz-Sommer, S., and Saedler, H. (1985). *EMBO J.* **4,** 579–583.
Gierl, A., Schwarz-Sommer, S., Peterson, P. A., and Saedler, H. (1987). *Proc. Int. Symp. Plant Transposable Elem.* p. 18.
Goldberg, R. B., Hoschek, G., and Vodkin, L. O. (1983). *Cell* **33,** 465–475.
Greenblatt, I. M. (1968). *Genetics* **58,** 585–597.
Greenblatt, I. M. (1974). *Genetics* **77,** 671–678.
Grindley, N. D. F. (1983). *Cell* **32,** 3–5.
Groose, R. W., Weigelt, H. D., and Palmer, R. G. (1988). *J. Hered.,* **79,** 263–267.
Gupta, M., Bertram, I., Shepherd, N. S., and Saedler, H. (1983). *Mol. Gen. Genet.* **192,** 373–377.
Harberd, N. P., Flavell, R. B., and Thompson, R. D. (1987). *Mol. Gen. Genet.* **209,** 326–332.
Harrison, B. J., and Carpenter, R. (1973). *Heredity* **31,** 309–323.
Harrison, B. J., and Carpenter, R. (1979). *Mutat. Res.* **63,** 47–66.

Heffron, F. (1983). *In* "Mobile Genetic Elements" (J. A. Shapiro, ed.), pp. 223–260. Academic Press, New York.
Hehl, R., Shepherd, N. S., and Saedler, H. (1985). *Maydica* **30,** 199–207.
Hehl, R., Sommer, H., and Saedler, H. (1987). *Mol. Gen. Genet.* **207,** 47–53.
Hudson, A., Carpenter, R., and Coen, E. S. (1987). *Mol. Gen. Genet.* **207,** 54–59.
Iida, S., Meyer, J., and Arber, A. (1983). *In* "Mobile Genetic Elements" (J. A. Shapiro, ed.), pp. 159–213. Academic Press, New York.
Johns, M. A., Mottinger, J., and Freeling, M. (1985). *EMBO J.* **4,** 1093–1102.
Karlin-Neumann, G. A., Kohorn, B. D., Thornber, J. P., and Tobin, E. M. (1985). *J. Mol. Appl. Genet.* **3,** 45–61.
Keller, E. F. (1983). "A Feeling for the Organism: The Life and Work of Barbara McClintock," pp. 139–151. Freeman, New York.
Kim, H.-Y., Schiefelbein, J. W., Raboy, V., Furtek, D. B., and Nelson, O. E. (1987). *Proc. Natl. Acad. Sci. U.S.A.* **84,** 5863–5867.
Kleckner, N. (1981). *Annu. Rev. Genet.* **15,** 341–404.
Kleckner, N. (1983). *In* "Mobil Genetic Elements" (J. A. Shapiro, ed.), pp. 261–298. Academic Press, New York.
Krebbers, E., Hehl, R., Piotrowiak, R., Lonnig, W.-E., Sommer, H., and Saedler, H. (1987). *Mol. Gen. Genet.* **209,** 499–507.
Kunze, R., Stochaj, U., Laufs, J., and Starlinger, P. (1987). *EMBO J.* **6,** 1555–1563.
Lillis, M., and Freeling, M. (1986). *Trends Genet.* July, 183–187.
Lycett, G. W., Croy, R. R. D., Shirsat, A. H., Richards, M., and Boulter, D. (1985). *Nucleic Acids Res.* **13,** 6733–6743.
McClintock, B. (1950). *Proc. Natl. Acad. Sci. U.S.A.* **36,** 344–355.
McClintock, B. (1951). *Year Book—Carnegie Inst. Washington* **61,** 448–461.
McClintock, B. (1953). *Year Book—Carnegie Inst. Washington* **52,** 227–237.
McClintock, B. (1954). *Year Book—Carnegie Inst. Washington* **53,** 254–260.
McClintock, B. (1961). *Year Book—Carnegie Inst. Washington* **60,** 469–476.
McClintock, B. (1984). *Science* **226,** 792–801.
McLaughlin, M., and Walbot, V. (1987). *Genetics* **117,** 771–776.
Martin, C., Carpenter, R., Sommer, H., Saedler, H., and Coen, E. S. (1985). *EMBO J.* **4,** 1625–1630.
Masson, P., Surosky, R., Kingsbury, J. A., and Fedoroff, N. V. (1987). *Genetics* **177,** 117–137.
Merckelback, A., Doring, H.-P., and Starlinger, P. (1986). *Maydica* **31,** 109–122.
Meyer, P., Heidmann, I., Forkmann, G., and Saedler, H. (1987). *Nature (London)* **330,** 677–678.
Mottinger, J. P., Johns, M. A., and Freeling, M. (1984). *Mol. Gen. Genet.* **195,** 367–369.
Muller-Neumann, M., Yoder, J. I., and Starlinger, P. (1984). *Mol. Gen. Genet.* **198,** 19–24.
Nelson, O. E. (1968). *Genetics* **60,** 507–524.
Nelson, O. E. (1987). *Genetics* **116,** 339–342.
Nelson, O. E., and Rines, H. W. (1962). *Biochem. Biophys. Res. Commun.* **9,** 297–300.
Nevers, P., Shepherd, N. S., and Saedler, H. (1986). *Adv. Bot. Res.* **12,** 103–203.
O'Reilly, C., Shepherd, N. S., Pereira, A., Schwarz-Sommer, S., Bertram, I., Robertson, D. S., Peterson, P. A., and Saedler, H. (1985). *EMBO J.* **4,** 877–882.
Orf, J. H., Hymowitz, T., Pull, S. P., and Pueppke, S. G. (1978). *Crop. Sci.* **18,** 899–900.
Osterman, J. C., and Schwartz, D. (1981). *Genetics* **99,** 267–273.
Paz-Ares, J., Wienand, U., Peterson, P. A., and Saedler, H. (1986). *EMBO J.* **5,** 829–833.
Paz-Arez, J., Ghosal, D., Wienand, U., Peterson, P. A., and Saedler, H. (1987). *EMBO J.* **6,** 3553–3558.
Pereira, A., Schwarz-Sommer, S., Gierl, A., Bertram, I., Peterson, P. A., and Saedler, H. (1985). *EMBO J.* **4,** 17–23.
Pereira, A., Cuypers, H., Gierl, A., Schwarz-Sommer, S., and Saedler, H. (1986). *EMBO J.* **5,** 835–841.

Peschke, V. M., Phillips, R. L., and Gengenback, B. G. (1987). *Science* **238,** 804–807.
Peterson, P. A. (1953). *Genetics* **38,** 682–683.
Peterson, P. A. (1960). *Genetics* **45,** 115–133.
Peterson, P. A. (1965). *Am. Nat.* **99,** 391–398.
Peterson, P. A. (1987). *In* "Critical Reviews in Plant Sciences," pp. 105–208. CRC Press, Boca Raton, Florida.
Peterson, P. A., and Weber, C. R. (1969). *Theor. Appl. Genet.* **39,** 156–162.
Pohlman, R. F., Fedoroff, N. V., and Messing, J. (1984). *Cell* **37,** 635–643.
Pull, S. P., Pueppke, S. G., Hymowitz, T., and Orf, J. H. (1978). *Science* **200,** 1277–1279.
Rhodes, P. R., and Vodkin, L. O. (1985). *Proc. Natl. Acad. Sci. U.S.A.* **82,** 493–497.
Rhodes, P. R., and Vodkin, L. O. (1988). *Genetics* **120,** 597–604.
Robertson, D. S. (1978). *Mutat. Res.* **58,** 21–28.
Roeder, G. S., and Fink, G. R. (1983). *In* "Mobile Genetic Elements" (J. A. Shapiro, ed.), pp. 300–328. Academic Press, New York.
Rowland, L. J., and Strommer, J. N. (1985). *Proc. Natl. Acad. Sci. U.S.A.* **82,** 2875–2879.
Rubin, G. M. (1983). *In* "Mobile Genetic Elements" (J. A. Shapiro, ed.), pp. 329–361. Academic Press, New York.
Saedler, H., and Nevers, P. (1985). *EMBO J.* **4,** 585–590.
Saedler, H., Bonas, U., Gierl, A., Harrison, B. J., Klosgen, R. B., Krebbers, E., Nevers, P., Peterson, P. A., Schwarz-Sommer, Z., Sommer, H., Upadhyaya, U., and Wienand, U. (1984). *In* "The Impact of Gene Transfer Techniques in Eukaryotic Cell Biology," (J. S. Shell and P. Starlinger, eds.) pp. 54–64. Springer-Verlag, Berlin, Federal Republic of Germany.
Schiefelbein, J. W., Raboy, V., Fedoroff, N. V., and Nelson, O. E. (1985). *Proc. Natl. Acad. Sci. U.S.A.* **82,** 4783–4787.
Schmidt, R. J., Burr, F. A., and Burr, B. (1987). *Science* **238,** 960–963.
Schwartz, D., and Dennis, E. (1987). *Mol. Gen. Genet.* **205,** 476–482.
Schwarz-Sommer, S., and Saedler, H. (1987). *Mol. Gen. Genet.* **209,** 207–209.
Schwarz-Sommer, S., Gierl, A., Klosgen, R. B., Wienand, U., Peterson, P. A., and Saedler, H. (1984). *EMBO J.* **3,** 1021–1028.
Schwarz-Sommer, S., Gierl, A., Cuypers, H., Peterson, P. A., and Saedler, H. (1985a). *EMBO J.* **4,** 591–597.
Schwarz-Sommer, S., Gierl, A., Berndtgen, R., and Saedler, H. (1985b). *EMBO J.* **4,** 2439–2443.
Schwarz-Sommer, S., Shepherd, N., Tacke, E., Gierl, A., Rohde, W., Leclercq, L., Mattes, M., Berndtgen, R., Peterson, P. A., and Saedler, H. (1987a). *EMBO J.* **6,** 287–294.
Schwarz-Sommer, S., Leclercq, L., Goebel, E., and Saedler, H. (1987b). *EMBO J.* **6,** 3873–3880.
Shepherd, N. S., Schwarz-Sommer, Z., Wienand, U., Sommer, H., Deumling, B., Peterson, P. A., and Saedler, H. (1982). *Mol. Gen. Genet.* **188,** 266–271.
Shepherd, N. S., Schwarz-Sommer, Z., Blumberg vel Spalve, J., Gupta, M., Wienand, W., and Saedler, H. (1984). *Nature (London)* **307,** 185–187.
Shure, M., Wessler, S., and Fedoroff, N. V. (1983). *Cell* **35,** 225–233.
Simon, R., and Starlinger, P. (1987). *Mol. Gen. Genet.* **209,** 198–199.
Sommer, H., and Saedler, H. (1986). *Mol. Gen. Genet.* **202,** 429–434.
Sommer, H., Carpenter, R., Harrison, B. J., and Saedler, H. (1985). *Mol. Gen. Genet.* **199,** 225–231.
Sommer, H., Hehl, R., Krebbers, E., and Saedler, H. (1987). *Proc. Int. Symp. Plant Transposable Elem.* p. 27.
Spribille, R., and Forkmann, G. (1982). *Phytochemistry* **21,** 2231–2234.
Starlinger, P. (1985). *Biol. Chem. Hoppe-Seyler* **366,** 931–937.
Sundaresan, V., and Freeling, M. (1987). *Proc. Natl. Acad. Sci. U.S.A.* **84,** 4924–4928.
Sutton, W. D., Gerlach, W. L., Schwarz, D., and Peacock, W. J. (1984). *Science* **223,** 1265–1268.

Syvanen, M. (1984). *Annu. Rev. Genet.* **18,** 271–293.
Tacke, E., Schwarz-Sommer, S., Peterson, P. A., and Saedler, H. (1986). *Maydica* **31,** 83–91.
Taylor, L. P., Chandler, V., and Walbot, V. (1986). *Maydica* **31,** 31–45.
Taylor, L. P., and Walbot, V. (1987). *Genetics* **117,** 297–307.
Theres, N., Scheele, T., and Starlinger, P. (1987). *Mol. Gen. Genet.* **209,** 193–197.
Upadhyaya, K. C., Sommer, H., Krebbers, E., and Saedler, H. (1985). *Mol. Gen. Genet.* **199,** 201–207.
Van Sluys, M. A., Tempe, J., and Fedoroff, N. (1987). *EMBO J.* **6,** 3881–3889.
Vodkin, L. O., and Rhodes, P. R. (1986). *In* "Molecular Biology of Seed Storage Proteins and Lectins" (L. M. Shannon and M. J. Chrispeels, eds.), pp. 97–105. Am. Soc. Plant Physiol., Rockville, Maryland.
Vodkin, L. O., Rhodes, P. R., and Goldberg, R. B. (1983). *Cell* **34,** 1023–1031.
Walbot, V., and Cullis, C. A. (1985). *Annu. Rev. Plant Physiol.* **36,** 367–396.
Weck, E., Courage, U., Doring, H.-P., Fedoroff, N., and Starlinger, P. (1984). *EMBO J.* **3,** 1713–1716.
Weiner, A. M. (1986). *Annu. Rev. Biochem.* **55,** 631–661.
Wessler, S. R., Baron, G., and Varagona, M. (1987). *Science* **237,** 916–918.
Wienand, U., Sommer, H., Schwarz-Sommer, Z., Shepherd, N., Saedler, H., Kreuzaler, F., Ragg, H., Fautz, E., Hahlbrock, K., Harrison, B., and Peterson, P. A. (1982). *Mol. Gen. Genet.* **187,** 195–201.
Wienand, U., Weydemann, U., Niesbach-Klosgen, U., Peterson, P. A., and Saedler, H. (1986). *Mol. Gen. Genet.* **203,** 202–207.
Zack, C. D., Ferl, R. J., and Hannah, L. C. (1986). *Maydica* **31,** 5–16.

The Chloroplast Genome 3

MASAHIRO SUGIURA

I. Introduction
II. Chloroplast DNA
III. Genes for the Genetic Apparatus
  A. Ribosomal RNA Genes
  B. Transfer RNA Genes
  C. Ribosomal Protein Genes
  D. Other Putative Genes
IV. Genes for the Photosynthetic Apparatus
  A. Gene for the Large Subunit of RuBisCO
  B. Photosystem I Genes
  C. Photosystem II Genes
  D. Cytochrome b/f Complex Genes
  E. ATP Synthase Genes
  F. Other Putative Genes
V. Conclusions
  References

I. INTRODUCTION

Chloroplasts are the intracellular organelles present in plants that contain the machinery for the process of photosynthesis. The discovery at the beginning of this century of non-Mendelian mutants of the chloroplast phenotype suggested the existence of a separate genetic system in chloroplasts. The demonstration of a unique DNA species in chloroplasts (Sager and Ishida, 1963) served as the impetus for intensive studies of both the structure of the chloroplast genome and its expression. This chapter presents a summary of structures of sequenced genes in chloroplast genomes. Other aspects of chloroplast genomes are presented in several recent reviews (Dyer, 1984; Palmer, 1985; Weil, 1987; Rochaix, 1987; Sugiura, 1987).

II. CHLOROPLAST DNA

Chloroplast DNAs are double-stranded, circular molecules ranging in size from 120 to 160 kilobase pairs (kb). One of the outstanding features of

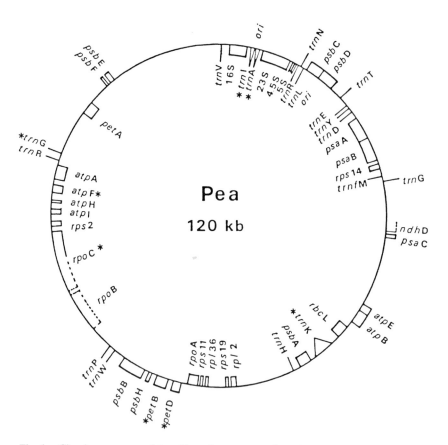

Fig. 1. Circular gene map of the chloroplast genome of pea (*Pisum sativum*). Genes shown inside the circle are transcribed clockwise and genes shown outside the circle are transcribed counterclockwise. Asterisks indicate split genes. The position of the replication origin is indicated by *ori*. Gene names are explained in Table I. This map was redrawn based mainly on an updated map of J. C. Gray and pea data found in Cozens and Walker (1986), Cozens *et al.* (1986), Meeker *et al.* (1988), Phillips and Gray (1984), Purton and Gray (1987a,b), Rasmussen *et al.* (1984, 1987), Shapiro and Tewari (1986), Willey *et al.* (1984), Zurawski *et al.* (1986a,b).

The Chloroplast Genome

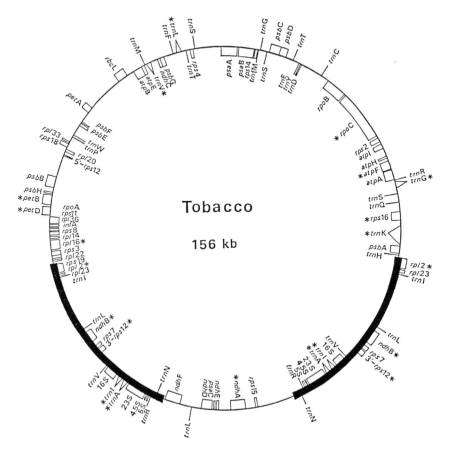

Fig. 2. Circular gene map of the chloroplast genome of tobacco (*Nicotiana tabacum*). Inverted repeats are shown by bold lines. Genes shown inside the circle are transcribed clockwise and genes shown outside the circle are transcribed counterclockwise. Asterisks indicate split genes. Gene names are explained in Table I. This map was redrawn from the map in Fig. 1 of Sugiura (1987) with extensive modifications.

chloroplast DNAs of most plants is the presence of a large inverted repeat (IR). The segments of the IR are separated by a large and a small single-copy region (LSC and SSC, respectively). Pea and broad bean chloroplast DNAs are exceptions to this pattern and lack IRs (Koller and Delius, 1980; Palmer and Thompson, 1981). The chloroplast DNA from *Euglena gracilis* contains three tandem repeats, each of which contains an rRNA gene cluster (Gray and Hallick, 1978). Therefore, chloroplast DNAs can be classified into three groups: chloroplast DNAs lacking IRs (group I), chloroplast DNAs containing IRs (group II), and chloroplast DNAs with tandem repeats (group III).

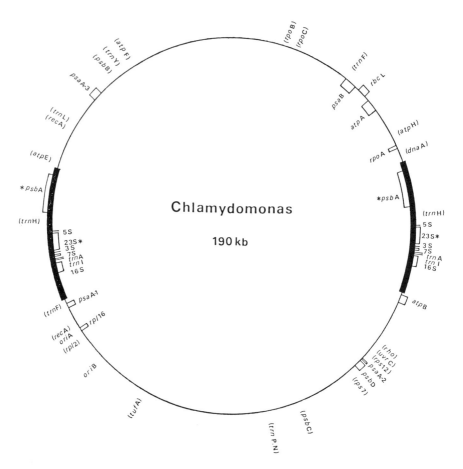

Fig. 3. Circular gene map of the chloroplast genome of *Chlamydomonas reinhardtii*. Inverted repeats are shown by bold lines. Genes shown inside the circle are transcribed clockwise and genes shown outside the circle are transcribed counterclockwise. Asterisks indicate split genes. Genes in parentheses have been assigned by heterologous hybridization and their positions are tentative. The position of the replication origin is indicated by *ori*A and *ori*B. Gene names are explained in Table I. This map was drawn based mainly on the map in Fig. 3 of Rochaix (1987) and *Chlamydomonas* data found in Bergmann *et al.* (1985), Dron *et al.* (1982), Erickson *et al.* (1984a), Hallick (1984), Kück *et al.* (1987), Lou *et al.* (1987), Rochaix *et al.* (1984), Rochaix and Malnoe (1978), Schmidt *et al.* (1985), Schneider and Rochaix (1986), Surzycki *et al.* (1986), Vallet and Rochaix (1985), Watson and Surzycki (1983), Woessnes *et al.* (1986).

Chloroplast DNAs are known to contain all the chloroplast rRNA genes (3–5 genes), tRNA genes (about 30 genes), and probably all the genes for proteins synthesized in the chloroplast (100–150 genes). Structural studies of chloroplast genomes and identification of their genes have been pursued

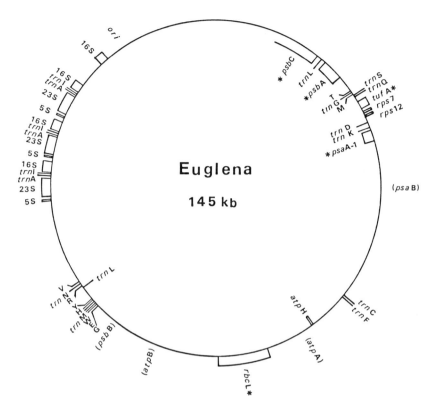

Fig. 4. Circular gene map of the chloroplast genome of *Euglena gracilis* (strain Z). Genes shown inside the circle are transcribed clockwise and genes shown outside the circle are transcribed counterclockwise. Asterisks indicate split genes. Genes in parentheses have been assigned by heterologous hybridization and their positions are tentative. The position of the replication origin is indicated by *ori*. Gene names are explained in Table I. The map was drawn using *Euglena* data from Gray and Hallick (1978), Hallick *et al.* (1984), Jenni and Stutz (1979), Koller and Delius (1982), Karabin *et al.* (1984), Keller and Stutz (1984), Koller *et al.* (1984), Manzara *et al.* (1987), Montandon and Stutz (1983, 1984), Montandon *et al.* (1986), Passavant and Hallick (1985), Ravel-Chapuis *et al.* (1982), Schantz (1985).

extensively in maize, spinach, pea, tobacco, *Euglena*, and *Chlamydomonas*. The complete nucleotide sequences of chloroplast DNAs from tobacco (155,844 bp, Shinozaki *et al.*, 1986b), liverwort (121,024 bp, Ohyama *et al.*, 1986), and rice (134,525 bp, Hiratsuka *et al.*, 1989) have been determined. Figures 1–4 show physical and gene maps of chloroplast DNA from pea, representing group I, tobacco and *Chlamydomonas*, group II, and *Euglena*, group III. Chloroplast genes that have been sequenced (including putative genes) are presented in Table I, with nomenclature following the proposal of Hallick and Bottomley (1983).

TABLE I

Chloroplast Genes[a]

Genes[b]	Products	Tobacco	Pea	*Chlamydomonas*	*Euglena*
Genes for the genetic apparatus					
23S rDNA	23S rRNA	++	+	**	+++
16S rDNA	16S rRNA	++	+	++	++++
7S rDNA	7S rRNA	−	−	++	−
5S rDNA	5S rRNA	++	+	++	+++
4.5S rDNA	4.5S rRNA	++	+	−	−
3S rDNA	3S rRNA	−	−	++	−
*trn*A-UGC	Ala-tRNA(UGC)	**	*	++	+++
*trn*R-ACG	Arg-tRNA(ACG)	++	+		+
*trn*R-UCU	Arg-tRNA(UCU)	+	+		
*trn*R-CCGΔ	Arg-tRNA(CCG)	−			
*trn*N-GUU	Asn-tRNA(GUU)	++	+	(+)	+
*trn*D-GUC	Asp-tRNA(GUC)	+	+		+
*trn*C-GCA	Cys-tRNA(GCA)	+			+
*trn*Q-UUG	Gln-tRNA(UUG)	+			+
*trn*E-UUC	Glu-tRNA(UUC)	+	+		+
*trn*G-GCC	Gly-tRNA(GCC)	+	+		+
*trn*G-UCC	Gly-tRNA(UCC)	*	*		+
*trn*H-GUG	His-tRNA(GUG)	+	+	(+)	+
*trn*I-GAU	Ile-tRNA(GAU)	**		++	+++
*trn*I-GAU	Ile-tRNA(CAU)	++			
*trn*L-UAA	Leu-tRNA(UAA)	*		(+)	+
*trn*L-CAA	Leu-tRNA(CAA)	++	+		
*trn*L-UAG	Leu-tRNA(UAG)	+			+
*trn*K-UUU	Lys-tRNA(UUU)	*	*		+
*trn*fM-CAU	fMet-tRNA(CAU)	+	+		+
*trn*M-CAU	Met-tRNA(CAU)	+			+
*trn*F-GAA	Phe-tRNA(GAA)	+		(+)	+
*trn*P-UGG	Pro-tRNA(UGG)	+	+	(+)	
*trn*S-GGA	Ser-tRNA(GGA)	+			
*trn*S-UGA	Ser-tRNA(UGA)	+			
*trn*S-GCU	Ser-tRNA(GCU)	+			+
*trn*T-GGU	Thr-tRNA(GGU)	+	+		
*trn*T-UGU	Thr-tRNA(UGU)	+			+
*trn*W-CCA	Trp-tRNA(CCA)	+	+		+
*trn*Y-GUA	Tyr-tRNA(GUA)	+	+	(+)	+
*trn*V-GAC	Val-tRNA(GAC)	++	+		
*trn*V-UAC	Val-tRNA(UAC)	*			+
*rps*2	30S ribosomal protein CS2	+	+		
*rps*3	30S ribosomal protein CS3	+			
*rps*4	30S ribosomal protein CS4	+			
*rps*7	30S ribosomal protein CS7	+		(+)	+
*rps*8	30S ribosomal protein CS8	+			
*rps*11	30S ribosomal protein CS11	+	+		
*rps*12	30S ribosomal protein CS12	**	+	(+)	+

TABLE I (*continued*)

Genes[b]	Products	Tobacco	Pea	*Chlamydomonas*	*Euglena*
*rps*14	30S ribosomal protein CS14	+			
*rps*15	30S ribosomal protein CS15	+			
*rps*16	30S ribosomal protein CS16	*			
*rps*18	30S ribosomal protein CS18	+			
*rps*19	30S ribosomal protein CS19	+	+		
*rpl*2	50S ribosomal protein CL2	**	+	(+)	
*rpl*14	50S ribosomal protein CL14	+	+		
*rpl*16	50S ribosomal protein CL16	*		+	
*rpl*20	50S ribosomal protein CL20	+			
*rpl*21Δ	50S ribosomal protein CL21	−			
*rpl*22	50S ribosomal protein CL22	+			
*rpl*23	50S ribosomal protein CL23	++			
*rpl*33	50S ribosomal protein CL33	+			
*rpl*36	50S ribosomal protein CL36	+	+		
*rpo*A	RNA polymerase, subunit α	+	+	(+)	+
*rpo*B	RNA polymerase, subunit β	+	+	(+)	
*rpo*C	RNA polymerase, subunit β'	*	*	(+)	
*tuf*A	Elongation factor Tu	−		(+)	*
*inf*A	Initiation factor 1	+			
Genes for the photosynthetic apparatus					
*rbc*L	RuBisCO, large subunit	+	+	+	*
*rbc*S	RuBisCO, small subunit	−	−	−	−
*psa*A	PSI, P700 apoprotein A1	+	+	*	*
*psa*B	PSI, P700 apoprotein A2	+	+	+	(+)
*psa*C	PSI, 9-kDa polypeptide	+	+		
*psb*A	PSII, D1 protein	+	+	**	*
*psb*B	PSII, 47-kDa polypeptide	+	+	(+)	(+)
*psb*C	PSII, 43-kDa polypeptide	+	+	(+)	*
*psb*D	PSII, D2 protein	+	+	+	
*psb*E	PSII, cytochrome b_{559} (8 kDa)	+	+		
*psb*F	PSII, cytochrome b_{559} (4 kDa)	+	+		
*psb*G	PSII, G protein	+			
*psb*H	PSII, 10-kDa phosphoprotein	+	+		
*pet*A	b/f complex, cytochrome *f*	+	+		
*pet*B	b/f complex, cytochrome b_6	*	*		
*pet*D	b/f complex, subunit IV	*	*		
*atp*A	H^+-ATPase, subunit α	+	+	+	(+)
*atp*B	H^+-ATPase, subunit β	+	+	+	(+)
*atp*E	H^+-ATPase, subunit ε	+	+	(+)	
*atp*F	H^+-ATPase, subunit I	*	*	(+)	
*atp*H	H^+-ATPase, subunit III	+	+	(+)	+
*atp*I	H^+-ATPase, subunit IV	+	+		

(*continued*)

TABLE I (*continued*)

Genes[b]	Products	Tobacco	Pea	Chlamydomonas	Euglena
ndhA	NADH dehydrogenase, ND1	*			
ndhB	NADH dehydrogenase, ND2	**			
ndhC	NADH dehydrogenase, ND3	+			
ndhD	NADH dehydrogenase, ND4	+	+		
ndhE	NADH dehydrogenase, ND4L	+			
ndhF	NADH dehydrogenase, ND5	+			

[a] References are given in the text and the legend to Figs. 1–4. +, Continuous genes present; *, split gene present; − absent; (+), genes detected by heterologous hybridization; numbers of + and * indicate copies per genome, and Δ transcripts not reported.

[b] Translation products for rps2–19, rpl2–36, rpoA–C, tufA, infA, and ndhA–F were not analyzed.

III. GENES FOR THE GENETIC APPARATUS

A. Ribosomal RNA Genes

Chloroplasts contain the 70S class of ribosomes distinct from those found in the plant cell cytoplasm. The 23S, 5S, and 4.5S rRNAs are associated with the 50S subunit and the 16S rRNA is associated with the 30S subunit. The 4.5S rRNA species has not been found in organisms other than the higher plants (Bowman and Dyer, 1979). Maize chloroplast rRNA genes (rDNAs) were the first chloroplast genes cloned (Bedbrook *et al.*, 1977) and the 16S and 23S rDNAs were later sequenced (Schwarz and Kössel, 1980; Edwards and Kössel, 1981). The first chloroplast 4.5S and 5S rDNAs sequenced were those from tobacco (Takaiwa and Sugiura, 1980). The order of chloroplast rDNAs in higher plants is 16S–23S–4.5S–5S.

The rDNAs in most higher plants and *Chlamydomonas* are located in IRs and therefore present in two copies per genome. Exceptions to this organization are broad bean and pea in which there is only one copy each of the rDNAs. In *Euglena* (strain Z), there are three copies of the rDNA cluster arranged tandemly (Gray and Hallick, 1978) and an extra copy of the 16S rDNA (Jenni and Stutz, 1979). *Euglena* strains which have one to five rDNA clusters have been reported. In *Chlamydomonas reinhardtii* the rDNA cluster consist of 16S, 7S, 3S, 23S, and 5S in this order (Rochaix and Malnoe, 1978). Unique features of this cluster include the presence of an 888-bp intron in the 23S rDNA and the two small rDNAs (3S and 7S) which precede the 23S rDNA. The *Chlamydomonas* 23S rDNA was the first split gene found in the chloroplast genome. Its intron can be folded with a secondary structure typical of group I introns of fungal mitochondrial genes, and it contains a 489-bp open reading frame (ORF) encoding a potential polypeptide related to mitochondrial maturases (Rochaix *et al.*, 1985).

The rDNAs of chloroplasts so far described are all arranged as they are in *Escherichia coli* (16S–23S–5S). However, the rDNA cluster of *Chlorella ellipsoidea* is split into two back-to-back operons: namely, operon 1, 16S rRNA–tRNA^{Ile} (GAU), and operon 2, tRNA^{Ala} (UGC)–23S rRNA–5S rRNA (Yamada and Shimaji, 1986). The tRNA^{Ile} and tRNA^{Ala} genes contain no introns, in contrast to higher plant chloroplasts (see below), while the 23S rDNA contains an intron of 243 bp (Yamada and Shimaji, 1987).

B. Transfer RNA Genes

Chloroplast genomes are believed to encode all the tRNA species used in chloroplast protein synthesis. The presence of tRNAs and their genes (*trn*) in chloroplasts was demonstrated by separating 30–50 tRNA spots by two-dimensional gel electrophoresis and hybridizing 21–34 of these to chloroplast DNA fragments (e.g., Driesel *et al.*, 1979).

The tobacco chloroplast DNA fragments that hybridized to total chloroplast tRNAs have been sequenced, and 30 different tRNA genes were found (Wakasugi *et al.*, 1986). With the determination of the complete nucleotide sequence of the tobacco and rice chloroplast genomes, more tRNA genes were sought by computer analysis, but none were found. Hence these 30 tRNA genes are probably all of the tRNA genes encoded in the tobacco and rice genomes (Shinozaki *et al.*, 1986b; Hiratsuka *et al.*, 1989). Blot hybridization analysis to total tobacco chloroplast tRNA has confirmed that all of the 30 tRNA genes are expressed in the chloroplast. From the DNA sequence of liverwort chloroplast DNA, 31 tRNA genes were deduced, a putative tRNA^{Arg} (CCG) gene constituting the additional member (Ohyama *et al.*, 1986).

All possible codons are used in the sequences coding for polypeptides in chloroplasts (e.g., Wakasugi *et al.*, 1986). The minimum number of tRNA species required for translation of all codons is thought to be 32. As shown in Table I, no tRNAs which recognize codons CUU/C(Leu), CCU/C(Pro), GCU/C(Ala), and CGC/A/G(Arg) have been found. If the "two-out-of-three" mechanism operates in the chloroplast, as has been shown for an *in vitro* protein synthesizing system from *E. coli*, the single tRNA^{Pro} (UGG), tRNA^{Ala} (UGC), and tRNA^{Arg} (ACG) species can read all four Pro, Ala, and Arg codons, respectively (GC pairs in the first and second codon–anticodon interactions). There is a gene in which the tRNA^{Leu} anticodon would be UAG and, if this tRNA has an unmodified U in the first position of the anticodon, it can read all four Leu codons (CUN) by U:N wobble. The bean, spinach, and soybean tRNAs^{Leu} (UAG) have unmodified U in their anticodons (UA^{m7}G) (Pillay *et al.*, 1984). These 30 tRNAs are therefore likely to be sufficient to read all codons in the chloroplast system using the above two mechanisms (Shinozaki *et al.*, 1986b).

General features of the chloroplast tRNAs and tRNA genes deduced from the DNA sequences are as follows. No tRNA genes code for the 3'-CCA end. Six tRNA genes in higher plants harbor long single introns (0.5–2.5 kb). The presence of introns in chloroplast tRNA genes was first demonstrated in maize *trn*I and *trn*A by Koch *et al.* (1981), genes that are located in the spacer separating the 16S and 23S rDNAs. Interestingly, *trn*G-UCC contains an intron in the D stem region (Deno and Sugiura, 1984). This seems to be a unique feature of chloroplast genomes. *trn*K has the longest intron (2.5 kb) and contains an ORF in the intron, potentially encoding a 60-kDa protein (Sugita *et al.*, 1985). All the tRNAs can form the cloverleaf structure and none has an abnormal structure as has been reported for some mammalian mitochondrial tRNAs. All the tRNA sequences show closer homology to the corresponding bacterial tRNAs than to the corresponding eukaryotic cytoplasmic tRNAs.

In higher plants, the tRNA genes are scattered over the chloroplast genome, while in *Euglena* most of the tRNA genes are clustered (Hallick *et al.*, 1984). *Euglena* tRNA genes have been mapped to at least nine different loci on the genome, and the DNA sequences of 22 different tRNA genes have been determined. None of the genes contain introns.

C. Ribosomal Protein Genes

Chloroplast ribosomes contain about 60 ribosomal proteins, one-third of which are thought to be encoded by chloroplast DNA (Eneas-Filho *et al.*, 1981). Genes for chloroplast ribosomal proteins (*rpl* for large subunit proteins and *rps* for small subunit proteins) have been deduced through their homology with *E. coli* ribosomal protein genes. The tobacco *rps*19 was the first potential ribosomal protein gene to be identified in a chloroplast genome (Sugita and Sugiura, 1983); subsequently, additional putative genes for ribosomal proteins have been reported in a number of plants (e.g., Subramanian *et al.*, 1983). Twenty different ORFs potentially coding for polypeptides homologous to *E. coli* ribosomal proteins have been found in the tobacco, liverwort and rice chloroplast genomes (Shinozaki *et al.*, 1986b; Ohyama *et al.*, 1986; Hiratsuka *et al.*, 1989). Tobacco and rice genomes lack *rpl*21 but contain *rps*16, which has an intron (Shinozaki *et al.*, 1986a). The *rpl*23, *rpl*2, *rps*19, *rpl*22, *rps*3, *rpl*16, *rpl*14, *rpl*8, *rpl*36, and *rps*11 sequences are clustered in this order and the arrangement corresponds to that of the homologous genes in the *E. coli* S10 and *spc* operons (Tanaka *et al.*, 1986).

An intron within a potential chloroplast ribosomal protein gene was first found in *Nicotiana debneyi rpl*2 (Zurawski *et al.*, 1984). The *rps*16 and *rpl*16 sequences also contain introns. The most striking feature is that *rps*12 consists of three exons with its 5' exon (5'-*rps*12) located far from the other two exons (3'-*rps*12) in tobacco, liverwort, and rice (Fromm *et al.*, 1986; Torazawa *et al.*, 1986; Fukuzawa *et al.*, 1986; Hiratsuka *et al.*, 1989). We have designated this gene structure as a "divided" gene (Shinozaki et al., 1986b). In contrast, *Euglena rps*12 is not split (Montandon and Stutz, 1984). Reverse transcription analysis indicated that *trans*-splicing between tobacco 5'-*rps*12 and 3'-*rps*12 transcripts occurs *in vivo* (Zaita et al., 1987). *Trans*-spliced *rps*12 mRNA has also been observed in the electron microscope using an artificial DNA construct linking tobacco exons 1, 2, and 3 (Koller et al., 1987).

D. Other Putative Genes

It had been suggested that higher plant chloroplast RNA polymerase might be encoded in the nuclear genome. However, DNA sequences hybridizing with the α, β, and β' subunit genes of *E. coli* RNA polymerase were reported in the *Chlamydomonas* chloroplast genome (Watson and Surzycki, 1983). Chloroplast DNA regions potentially coding for polypeptides similar to *E. coli* RNA polymerase α subunit (*rpo*A) were sequenced in spinach (Sijben-Müller et al., 1986) and sequences similar to the β subunit (*rpo*B) were sequenced in tobacco (Ohme et al., 1986). Subsequently, sequences similar to the β' subunit gene of *E. coli* RNA polymerase (*rpo*C) were sequenced in tobacco, pea, liverwort, and rice (Shinozaki et al., 1986b; Cozens and Walker, 1986; Ohyama et al., 1986; Hiratsuka et al., 1989).

A putative gene, the initiation factor IF-1 (*inf*A), was found in spinach (Sijben-Müller et al., 1986). A sequence similar to the *E. coli* EF-Tu gene (*tuf*A) has been found in *Euglena* (Montandon and Stutz, 1983, but neither has been found in the tobacco, liverwort, or rice chloroplast genomes. *Euglena tuf*A contains three short introns (Montandon et al., 1987).

The "genes" for protein components of the chloroplast transcriptional and translational apparatus have all been found through homology with the corresponding *E. coli* genes. Transcripts have been detected from most of these "genes" but few translation products have been identified. We therefore refer to these as putative genes.

IV. GENES FOR THE PHOTOSYNTHETIC APPARATUS

A. Gene for the Large Subunit of RuBisCO

Ribulose-1,5-bisphosphate carboxylase/oxygenase (RuBisCO or fraction I protein) is the major soluble protein of chloroplasts. The enzyme is composed of eight identical large subunits (LS) of 53 kDa and eight identical small subunits (SS) of 12–14 kDa. LS is encoded in the chloroplast DNA and SS is encoded in the nuclear DNA. The only exception so far is a sea alga, *Olisthodiscus lutens*, whose small subunit gene (*rbc*S) is found in the chloroplast genome (Reith and Cattalico, 1986).

The gene for LS (*rbc*L) was first cloned and sequenced in maize (McIntosh *et al.*, 1980) and then in various other plants (e.g., Zurawski *et al.*, 1981). The *rbc*L genes of higher plants and *Chlamydomonas* have been found not to contain introns (e.g., Dron *et al.*, 1982). In the *Euglena rbc*L gene, however, nine introns have been found (Koller *et al.*, 1984).

B. Photosystem I Genes

Thylakoid membranes of plants have five functionally distinct complexes (Herrmann *et al.*, 1985; Gray, 1987): photosystem I (PSI), photosystem II (PSII), the light-harvesting chlorophyll protein complex (its proteins are all nuclear-coded), the cytochrome b/f complex, and the ATP synthase complex.

At least three components of PSI are encoded in the chloroplast DNA. Genes for two subunits (A1 and A2) of $P700$ chlorophyll a apoprotein (*psa*A and *psa*B) were localized in the middle of the spinach chloroplast LSC (Westhoff *et al.*, 1983a) and were first sequenced from maize (Fish *et al.*, 1985). The *psa*A and *psa*B genes in higher plants contain no introns and are situated tandemly and bear about 45% similarity toward each other at the amino acid level. In *Chlamydomonas* the *psa*A gene is split into three exons scattered around the chloroplast genome, while the *psa*B gene is uninterrupted (Kück *et al.*, 1987). The three distantly separated exons of *psa*A constitute a functional gene which probably operates by a *trans*-splicing mechanism. *Euglena psa*A also contains intron(s) (Manzara *et al.*, 1987).

Recently, a gene for a 9-kDa polypeptide (an apoprotein for the iron–sulfur centers A and B) of PSI (*psa*C) has been located on the chloroplast genome (Hayashida *et al.*, 1987; Høj *et al.*, 1987). This gene was identified by first comparing the N-terminal amino acid sequence of the spinach 9-kDa polypeptide with the sequence of tobacco chloroplast DNA and then detecting the appropriate transcripts.

C. Photosystem II Genes

At least eight components of PSII are encoded in the chloroplast genome. The 33-, 24-, and 18-kDa polypeptides which function as the water-splitting apparatus are encoded in the nuclear DNA. The gene (*psb*A) for the 32-kDa protein Q_B or D1 protein was the first PSII component gene sequenced in spinach and *N. debneyi* (Zurawski *et al.*, 1982a). This gene has been studied intensively in a number of plants. In *Euglena* the *psb*A gene is interrupted by four introns (Karabin *et al.*, 1984; Keller and Stutz, 1984). The *Chlamydomonas psb*A gene also contains four introns and is located entirely within the IR, and is thus present as two copies per genome (Erickson *et al.*, 1984a). The 32-kDa protein has been found to bind to herbicides such as atrazine and DCMU. *psb*A genes isolated from herbicide-resistant mutants of *Amaran-*

thus hybridus (Hirschberg and McIntosh, 1983), *Solanum nigrum* (Goloubinoff *et al.*, 1984), and *Chlamydomonas* (Erickson *et al.*, 1984b) have point mutations resulting in substitution from serine to glycine or alanine at codon 264 of the 32-kDa protein.

The genes for the 47- and 43-kDa polypeptides (*psb*B and *psb*C) were localized in the spinach genome (Westhoff *et al.*, 1983b) and their nucleotide sequences were determined (Morris and Herrmann, 1984; Alt *et al.*, 1984; Holschuh *et al.*, 1984). The *Euglena psb*C gene seems to contain multiple introns, one of which is large (1.6 kb) and contain an ORF (Montandon *et al.*, 1986).

The gene for the D2 protein or "32-kDa-like" protein (*psb*D) was first mapped and sequenced in *Chlamydomonas* (Rochaix *et al.*, 1984). In higher plants, the *psb*D gene overlaps *psb*C by about 50 bp (e.g., Alt *et al.*, 1984; Holschuh *et al.*, 1984).

Genes for the 9- and 4-kDa subunits of cytochrome b_{559} (*psb*E and *psb*F, respectively) have been characterized from spinach (Herrmann *et al.*, 1984) and several other plants. An ORF on the maize chloroplast genome was identified as the gene for the 24-kDa G protein (*psb*G) by Western blotting analyses using antibodies against synthetic oligopeptides deduced from the DNA sequence (Steinmetz *et al.*, 1986). The gene for the 10-kDa phosphoprotein (*psb*H) was identified by comparing the N-terminal amino acid sequence of the spinach peptide (Farchaus and Dilley, 1986) with the sequence of tobacco chloroplast DNA (Shinozaki *et al.*, 1986b).

D. Cytochrome *b/f* Complex Genes

The cytochrome *b/f* complex consists of six components, with cytochrome *f* (*pet*A), cytochrome b_6 (*pet*B), and subunit IV (*pet*D) coded for by the chloroplast genome, and the Rieske Fe–S protein (*pet*C), plastocyanin (*pet*E), and ferredoxin-NADP oxidoreductase (*pet*F) being nuclear-encoded. *pet*A was first sequenced in pea (Willey *et al.*, 1984). *pet*B and *pet*D are located close together in higher plants and were first sequenced in pea and spinach (Phillips and Gray, 1984; Heinemeyer *et al.*, 1984). The *pet*B and *pet*D genes are clustered with *psb*B and *psb*H in higher plants. Both *pet*B and *pet*D contain single introns with short first exons (6–8 bp) in tobacco, liverwort, and rice (Tanaka *et al.*, 1987; Fukuzawa *et al.*, 1987; Hiratsuka *et al.*, 1989).

E. ATP Synthase Genes

ATP synthase consists of two parts, CF_1 and CF_0. CF_1 is located on the outer surface of the thylakoid membrane and is composed of five different subunits (α, β, γ, δ, and ε). CF_0 is located within the membrane and composed of four different subunits (I, II, III, and IV). The genes for six subunits

(α, β, ε, I, III, and IV) are present in the chloroplast genome. Genes for the β and ε subunits (*atp*B and *atp*E, respectively) are located upstream from *rbc*L on the opposite strand and were first sequenced from maize and spinach (Krebbers *et al.*, 1982; Zurawski *et al.*, 1982b). The *atp*B and *atp*E genes in most higher plants overlap by 4 bp, so that the first two bases of the TGA stop codon of *atp*B and the adenosine preceding it form the ATG initiation codon of *atp*E. In pea the two genes do not overlap and are separated by a 25-bp spacer (Zurawski *et al.*, 1986a), and in *Chlamydomonas* they are located far from each other (Rochaix, 1987). The gene for the α subunit (*atp*A) is located far from the *atp*B–*atp*E cluster (Westhoff *et al.*, 1981) and was initially sequenced in tobacco (Deno *et al.*, 1983).

The genes for the IV, III, and I subunits (*atp*I, *atp*H, and *atp*F, respectively) are clustered in this order just before *atp*A (Howe *et al.*, 1982; Bird *et al.*, 1985; Cozens *et al.*, 1986; Hennig and Herrmann, 1986). The *atp*F genes from higher plants contain single introns (e.g., Bird *et al.*, 1985). The amino acid sequences of these six subunits deduced from the DNA show homology with their counterparts in *E. coli*.

F. Other Putative Genes

Six tobacco chloroplast DNA sequences whose predicted amino acid sequences resemble those of components (ND1, 2, 3, 4, 4L, and 5) of the respiratory-chain NADH dehydrogenase from human mitochondria have been found (Shinozaki *et al.*, 1986b). As these sequences are highly expressed in tobacco chloroplasts, they are likely to be the genes for components of a chloroplast NADH dehydrogenase (*ndh*A, B, C, D, E, and F) (Matsubayashi *et al.*, 1987). The tobacco *ndh*A and *ndh*B genes contain single introns and their transcripts are spliced quickly in tobacco leaves. Similar sequences have also been found in several other plants (Meng *et al.*, 1986; Ohyama *et al.*, 1986). These observations suggest the existence of a respiratory chain in the chloroplasts of higher plants.

V. CONCLUSIONS

So far, about 35 RNA and 55 definite and putative polypeptide genes have been found in the chloroplast genome (Table I) but dozens of ORFs remain to be identified. Translation products from putative genes, namely *ndh*A–F, *rpo*A–C, and *rps*/*rpl* genes, remain to be isolated and characterized.

The sequence and expression analyses have revealed both prokaryotic and eukaryotic features of chloroplast genes. Genes coding for rRNAs, tRNAs, and some proteins (e.g., ATP synthase subunits) have substantial sequence homology with their prokaryotic counterparts. The basic regulatory sequences (promoters, terminators, and ribosomal binding sites) are also similar to those found in prokaryotic genomes. Some gene clusters resemble corresponding clusters in *E. coli* and cyanobacteria (e.g., rDNA, *rps/rpl*, and *atp* clusters). At the same time, some chloroplast genes contain introns similar to those found in eukaryotic genomes. However, introns found in the tRNA genes are very long (up to 2.5 kb) and one intron is located in an unusual position, namely the D-stem region of *trn*G-UCC. The *rps*12 genes in higher plants and the *psa*A gene in *Chlamydomonas* are divided into three parts very distant from each other and requiring *trans*-splicing (divided genes). The endosymbiotic theory, which proposes that chloroplasts derived from an ancestral photosynthetic prokaryote related to cyanobacteria, is supported in part by comparisons between chloroplast and cyanobacterial *rrn* operons. This leads to the speculation that ancestral photosynthetic prokaryotes had introns in their genomes and that existing chloroplast genomes have retained these intron sequences (Tomioka and Sugiura, 1984).

ACKNOWLEDGMENTS

I am grateful to Dr. J. C. Gray for the pea chloroplast gene map and for unpublished data and to Dr. J. D. Rochaix for the *Chlamydomonas* chloroplast gene map and other information. I also thank Dr. R. Whittier for his critical reading of this manuscript.

REFERENCES

Alt, J., Morris, J., Westhoff, P., and Herrmann, R. G. (1984). *Curr. Genet.* **8**, 597–606.
Bedbrook, J. R., Kolodner, R., and Bogorad, L. (1977). *Cell* **11**, 739–749.
Bergmann, P., Schneider, M., Burkard, G., Weil, J. H., and Rochaix, J. D. (1985). *Plant Sci.* **39**, 133–140.
Bird, C. R., Koller, B., Auffret, A. D., Huttly, A. K., Howe, C. J., Dyer, T. A., and Gray, J. C. (1985). *EMBO J.* **4**, 1381–1388.
Bowman, C. M., and Dyer, T. A. (1979). *Biochem. J.* **183**, 605–613.
Cozens, A. L., and Walker, J. E. (1986). *Biochem. J.* **236**, 453–460.
Cozens, A. L., Walker, J. E., Phillips, A. L., Huttly, A. K., and Gray, J. C. (1986). *EMBO J.* **5**, 217–222.
Deno, H., and Sugiura, M. (1984). *Proc. Natl. Acad. Sci. U.S.A.* **81**, 405–408.
Deno, H., Shinozaki, K., and Sugiura, M. (1983). *Nucleic Acids Res.* **11**, 2185–2191.
Driesel, A. J., Crouse, E. J., Gordon, K., Bohnert, H. J., Herrmann, R. G., Steinmetz, A., Mubumbila, M., Keller, M., Burkard, G., and Weil, J. H. (1979). *Gene* **6**, 285–306.
Dron, M., Rahire, M., and Rochaix, J. D. (1982). *J. Mol. Biol.* **162**, 775–793.
Dyer, T. A. (1984). In "Chloroplast Biogenesis" (N. R. Baker and J. Barber, eds.), pp. 23–69. Elsevier, Amsterdam.
Edwards, K., and Kössel, H. (1981). *Nucleic Acids Res.* **9**, 2853–2869.
Eneas-Filho, J., Hartley, M. R., and Mache, R. (1981). *Mol. Gen. Genet.* **184**, 484–488.
Erickson, J. M., Rahire, M., and Rochaix, J. D. (1984a). *EMBO J.* **3**, 2753–2762.
Erickson, J. M., Rahire, M., Bennoun, P., Delepelaire, P., Diner, B., and Rochaix, J. D. (1984b). *Proc. Natl. Acad. Sci. U.S.A.* **81**, 3617–3621.

Farchaus, J., and Dilley, R. A. (1986). *Arch. Biochem. Biophys.* **244,** 94–101.
Fish, L. E., Kuck, U., and Bogorad, L. (1985). *J. Biol. Chem.* **260,** 1413–1421.
Fromm, H., Edelman, M., Koller, B., Goloubinoff, P., and Galun, E. (1986). *Nucleic Acids Res.* **14,** 883–898.
Fukuzawa, H., Kohchi, T., Shirai, H., Ohyama, K., Umesono, K., Inokuchi, H., and Ozeki, H. (1986). *FEBS Lett.* **198,** 11–15.
Fukuzawa, H., Yoshida, T., Kohchi, T., Okumura, T., Sawano, Y., and Ohyama, K. (1987). *FEBS Lett.* **220,** 61–66.
Goloubinoff, P., Edelman, M., and Hallick, R. B. (1984). *Nucleic Acids Res.* **12,** 9489–9496.
Gray, J. C. (1987). *In* "Photosynthesis" (J. Amesz, ed.), pp. 319–342. Elsevier, Amsterdam.
Gray, P. W., and Hallick, R. B. (1978). *Biochemistry* **17,** 284–289.
Hallick, R. B. (1984). *FEBS Lett.* **177,** 274–276.
Hallick, R. B., and Bottomley, W. (1983). *Plant Mol. Biol. Rep.* **1,** 38–43.
Hallick, R. B., Hollingsworth, M. J., and Nickoloff, J. A. (1984). *Plant Mol. Biol.* **3,** 169–175.
Hayashida, N., Matsubayashi, T., Shinozaki, K., Sugiura, M., Inoue, K., and Hiyama, T. (1987). *Curr. Genet.* **12,** 247–250.
Heinemeyer, W., Alt, J., and Herrmann, R. G. (1984). *Curr. Genet.* **8,** 543–549.
Hennig, J., and Herrmann, R. G. (1986). *Mol. Gen. Genet.* **203,** 117–128.
Herrmann, R. G., Alt, J., Schiller, B., Widger, W. R., and Cramer, W. A. (1984). *FEBS Lett.* **176,** 239–244.
Herrmann, R. G., Westhoff, P., Alt, J., Tittgen, J., and Nelson, N. (1985). *In* "Molecular Form and Function of the Plant Genome" (L. van Vloten-Doting, G. S. P. Groot, and T. C. Hall, eds.), pp. 233–256. Plenum, New York.
Hiratsuka, J., Shimada, H., Whittier, R., Sakamoto, M., Ishibashi, T., Mori, M., Honji, Y., Kondo, C., Nishizawa, Y., Kanno, A., Hirai, A., Sun, C. R., Meng, B. Y., Li, Y. Q., Shinozaki, K., and Sugiura, M. (1988). *Mol. Gen. Genet.* Submitted.
Hirschberg, J., and McIntosh, L. (1983). *Science* **222,** 1346–1349.
Holschuh, K., Bottomley, W., and Whitfeld, P. R. (1984). *Nucleic Acids Res.* **12,** 8819–8834.
Høj, P. B., Svendsen, I., Scheller, H. V., and Moller, B. L. (1987). *J. Biol. Chem.* **262,** 12676–12684.
Howe, C. J., Auffret, A. D., Doherty, A., Bowman, C. M., Dyer, T. A., and Gray, J. C. (1982). *Proc. Natl. Acad. Sci. U.S.A.* **79,** 6903–6907.
Jenni, B., and Stutz, E. (1979). *FEBS Lett.* **102,** 95–99.
Karabin, G. D., Farley, M., and Hallick, R. B. (1984). *Nucleic Acids Res.* **12,** 5801–5812.
Keller, M., and Stutz, E. (1984). *FEBS Lett.* **175,** 173–177.
Koch, W., Edwards, K., and Kössel, H. (1981). *Cell* **25,** 203–213.
Koller, B., and Delius, H. (1980). *Mol. Gen. Genet.* **178,** 261–269.
Koller, B., and Delius, H. (1982). *EMBO J.* **1,** 995–998.
Koller, B., Gingrich, J. C., Stiegler, G. L., Farley, M. A., Delius, H., and Hallick, R. B. (1984). *Cell* **36,** 545–553.
Koller, B., Fromm, H., Galun, E., and Edelman, M. (1987). *Cell* **48,** 111–119.
Krebbers, E. T., Larrinua, I. M., McIntosh, L., and Bogorad, L. (1982). *Nucleic Acids Res.* **10,** 4985–5002.
Kück, U., Choquet, Y., Schneider, M., Dron, M., and Bennoun, P. (1987). *EMBO J.* **6,** 2185–2195.
Lou, J. K., Wu, M., Chang, C. H., and Cuticchia, A. J. (1987). *Curr. Genet.* **11,** 537–541.
McIntosh, L., Poulsen, C., and Bogorad, L. (1980). *Nature (London)* **288,** 556–560.
Manzara, T., Hu, J. X., Price, C. A., and Hallick, R. B. (1987). *Plant Mol. Biol.* **8,** 327–336.
Matsubayashi, T., Wakasugi, T., Shinozaki, K., Shinozaki-Yamaguchi, K., Zaita, N., Hidaka, T., Meng, B. Y., Ohto, C., Tanaka, M., Kato, A., Maruyama, T., and Sugiura, M. (1987). *Mol. Gen. Genet.* **210,** 385–393.
Meeker, R., Nielsen, B., and Tewari, K. K. (1988). *Mol. Cell. Biol.* **8,** 1216–1223.
Meng, B. Y., Matsubayashi, T., Wakasugi, T., Shinozaki, K., Sugiura, M., Hirai, A., Mikami, T., Kishima, Y., and Kinoshita, T. (1986). *Plant Sci.* **47,** 181–184.

Montandon, P. E., and Stutz, E. (1983). *Nucleic Acids Res.* **11**, 5877–5892.
Montandon, P. E., and Stutz, E. (1984). *Nucleic Acids Res.* **12**, 2851–2859.
Montandon, P. E., Vasserot, A., and Stutz, E. (1986). *Curr. Genet.* **11**, 35–39.
Montandon, P. E., Knuchel-Aegerter, C., and Stutz, E. (1987). *Nucleic Acids Res.* **15**, 7809–7822.
Morris, J., and Herrmann, R. G. (1984). *Nucleic Acids Res.* **12**, 2837–2850.
Ohme, M., Tanaka, M., Chunwongse, J., Shinozaki, K., and Sugiura, M. (1986). *FEBS Lett.* **200**, 87–90.
Ohyama, K., Fukuzawa, H., Kohchi, T., Shirai, H., Sano, T., Sano, S., Umesono, K., Shiki, Y., Takeuchi, M., Chang, Z., Aota, S., Inokuchi, H., and Ozeki, H. (1986). *Nature (London)* **322**, 572–574.
Palmer, J. D. (1985). *Annu. Rev. Genet.* **19**, 325–354.
Palmer, J. D., and Thompson, W. F. (1981). *Proc. Natl. Acad. Sci. U.S.A.* **78**, 5533–5537.
Passavant, C. W., and Hallick, R. B. (1985). *Plant Mol. Biol.* **4**, 347–354.
Phillips, A. L., and Gray, J. C. (1984). *Mol. Gen. Genet.* **194**, 477–484.
Pillay, D. T. N., Guillemaut, P., and Weil, J. H. (1984). *Nucleic Acids Res.* **12**, 2997–3001.
Purton, S., and Gray, J. C. (1987a). *Nucleic Acids Res.* **15**, 1873.
Purton, S., and Gray, J. C. (1987b). *Nucleic Acids Res.* **15**, 9080.
Rasmussen, O. F., Stummann, B. M., and Henningsen, K. W. (1984). *Nucleic Acids Res.* **12**, 9143–9153.
Rasmussen, O., Jepsen, B., Stummann, B., and Henningsen, K. W. (1987). *Nucleic Acids Res.* **15**, 854.
Ravel-Chapuis, P., Heizmann, P., and Nigon, V. (1982). *Nature (London)* **300**, 78–81.
Reith, M., and Cattalico, R. A. (1986). *Proc. Natl. Acad. Sci. U.S.A.* **83**, 8599–8603.
Rochaix, J. D. (1987). *FEMS Microbiol. Rev.* **46**, 13–34.
Rochaix, J. D., and Malnoe, P. (1978). *Cell* **15**, 661–670.
Rochaix, J. D., Dron, M., Rahire, M., and Malnoe, P. (1984). *Plant Mol. Biol.* **3**, 363–370.
Rochaix, J. D., Rahire, R., and Michel, F. (1985). *Nucleic Acids Res.* **13**, 975–983.
Sager, R., and Ishida, M. R. (1963). *Proc. Natl. Acad. Sci. U.S.A.* **50**, 725–730.
Schantz, R. (1985). *Plant Sci.* **40**, 43–49.
Schmidt, R. J., Hosler, J. P., Gillham, N. W., and Boynton, J. E. (1985). In "Molecular Biology of the Photosynthetic Apparatus" (K. Steinback, S. Bonitz, C. J. Arntzen, and L. Bogorad, eds.), pp. 417–427. Cold Spring Harbor Lab., Cold Spring Harbor, New York.
Schneider, M., and Rochaix, J. D. (1986). *Plant Mol. Biol.* **6**, 265–270.
Schwarz, Z., and Kössel, H. (1980). *Nature (London)* **283**, 739–742.
Shapiro, D. R., and Tewari, K. K. (1986). *Plant Mol. Biol.* **6**, 1–12.
Shinozaki, K., Deno, H., Sugita, M., Kuramitsu, S., and Sugiura, M. (1986a). *Mol. Gen. Genet.* **202**, 1–5.
Shinozaki, K., Ohme, M., Tanaka, M., Wakasugi, T., Hayashida, N., Matsubayashi, T., Zaita, N., Chunwongse, J., Obokata, J., Yamaguchi-Shinozaki, K., Ohto, C., Torazawa, K., Meng, B. Y., Sugita, M., Deno, H., Kamogashira, T., Yamada, K., Kusuda, J., Takaiwa, F., Kato, A., Tohdoh, N., Shimada, H., and Sugiura, M. (1986b). *EMBO J.* **5**, 2043–2049.
Sijben-Müller, G., Hallick, R., Alt, J., Westhoff, P., and Herrmann, R. G. (1986). *Nucleic Acids Res.* **14**, 1029–1044.
Steinmetz, A. A., Castroviejo, M., Sayre, R. T., and Bogorad, L. (1986). *J. Biol. Chem.* **261**, 2485–2488.
Subramanian, A. R., Steinmetz, A., and Bogorad, L. (1983). *Nucleic Acids Res.* **11**, 5277–5286.
Sugita, M., and Sugiura, M. (1983). *Nucleic Acids Res.* **11**, 1913–1918.
Sugita, M., Shinozaki, K., and Sugiura, M. (1985). *Proc. Natl. Acad. Sci. U.S.A.* **82**, 3557–3561.
Sugiura, M. (1987). *Bot. Mag.* **100**, 407–436.
Surzycki, S. J., Hong, T. H., and Surzycki, J. A. (1986). In "Regulation of Chloroplast Differentiation" (G. Akoyunoglou and H. Senger, eds.), pp. 511–516. Liss, New York.

Takaiwa, F., and Sugiura, M. (1980). *Mol. Gen. Genet.* **180**, 1–4.
Tanaka, M., Wakasugi, T., Sugita, M., Shinozaki, K., and Sugiura, M. (1986). *Proc. Natl. Acad. Sci. U.S.A.* **83**, 6030–6034.
Tanaka, M., Obokata, J., Chunwongse, J., Shinozaki, K., and Sugiura, M. (1987). *Mol. Gen. Genet.* **209**, 427–431.
Tomioka, N., and Sugiura, M. (1984). *Mol. Gen. Genet.* **193**, 427–430.
Torazawa, K., Hayashida, N., Obokata, J., Shinozaki, K., and Sugiura, M. (1986). *Nucleic Acids Res.* **14**, 3143.
Vallet, J. M., and Rochaix, J. D. (1985). *Curr. Genet.* **9**, 321–324.
Wakasugi, T., Ohme, M., Shinozaki, K., and Sugiura, M. (1986). *Plant Mol. Biol.* **7**, 385–392.
Watson, J. C., and Surzycki, S. J. (1983). *Curr. Genet.* **7**, 201–210.
Weil, J. H. (1987). *Plant Sci.* **49**, 149–157.
Westhoff, P., Nelson, N., Bunemann, H., and Herrmann, R. G. (1981). *Curr. Genet.* **4**, 109–120.
Westhoff, P., Alt, J., Nelson, N., Bottomley, W., Bunemann, H., and Herrmann, R. G. (1983a). *Plant Mol. Biol.* **2**, 95–107.
Westhoff, P., Alt, J., and Herrmann, R. G. (1983b). *EMBO J.* **2**, 2229–2237.
Willey, D. L., Auffret, A. D., and Gray, J. C. (1984). *Cell* **36**, 555–562.
Woessner, J. P., Gillham, N. W., and Boynton, J. E. (1986). *Gene* **44**, 17–28.
Yamada, T., and Shimaji, M. (1986). *Nucleic Acids Res.* **14**, 3827–3839.
Yamada, T., and Shimaji, M. (1987). *Curr. Genet.* **11**, 347–352.
Zaita, N., Torazawa, K., Shinozaki, K., and Sugiura, M. (1987). *FEBS Lett.* **210**, 153–156.
Zurawski, G., Perrot, B., Bottomley, W., and Whitfeld, P. R. (1981). *Nucleic Acids Res.* **9**, 3251–3270.
Zurawski, G., Bohnert, H. J., Whitfeld, P. R., and Bottomley, W. (1982a). *Proc. Natl. Acad. Sci. U.S.A.* **79**, 7699–7703.
Zurawski, G., Bottomley, W., and Whitfeld, P. R. (1982b). *Proc. Natl. Acad. Sci. U.S.A.* **79**, 6260–6264.
Zurawski, G., Bottomley, W., and Whitfeld, P. R. (1984). *Nucleic Acids Res.* **12**, 6547–6558.
Zurawski, G., Bottomley, W., and Whitfeld, P. R. (1986a). *Nucleic Acids Res.* **14**, 3974.
Zurawski, G., Whitfeld, P. R., and Bottomley, W. (1986b). *Nucleic Acids Res.* **14**, 3975.

Chloroplast RNA: Transcription and Processing

WILHELM GRUISSEM

I. Introduction
II. The Chloroplast Transcription Apparatus
    A. Chloroplast RNA Polymerase
    B. *In Vitro* Transcription Systems
    C. Chloroplast Promoter Regions
    D. Transcription Termination
III. RNA Processing
    A. Ribosomal RNA Processing
    B. Processing of tRNA
    C. Processing of Chloroplast mRNA
    D. Chloroplast Introns–Structure and Processing
    E. Trans-Splicing of Cholorplast mRNAs
IV. Transcriptional and Posttranscriptional Regulation of Plastid Gene Expression
    A. Transcriptional Regulation
    B. Regulation of Plastid RNA Levels during Plant Development
    C. Conclusions
        References

I. INTRODUCTION

The chloroplast DNA molecule encodes several, but not all, proteins that have important structural and functional roles in the photosynthetic enzyme complexes. A large number of genes for photosynthetic proteins have a nuclear location. The separation of the genetic function of the chloroplast on two genomic locations in the cell has been extensively reviewed and remains a subject of evolutionary speculation (Bogorad, 1975; Ellis, 1981; Wallace, 1982). The presence of protein coding genes on the chloroplast genome requires the maintenance of a complete apparatus for their expression, i.e., transcription and translation systems operational within this organelle. Several components of the chloroplast transcription and translation machinery are encoded by the chloroplast's own genome, most notably the genes for ribosomal RNAs and for a full complement of tRNAs (Wakasugi *et al.*, 1986). In addition, chloroplast genes encode ribosomal proteins, translation factors as well as putative subunits of a chloroplast RNA polymerase. The organization and molecular structure of genes on the chloroplast genome have been discussed in detail (Bohnert *et al.*, 1982; Whitfeld and Bottomley, 1983), and complete DNA sequences are now available for the chloroplast genomes from *Nicotiana tabacum* (Shinozaki *et al.*, 1986b) and *Marchantia* (Ohyama *et al.*, 1986). This has allowed the precise location of all genes and has revealed the evolutionary conservation of individual transcription units.

The organization of genes into polycistronic transcription units and their differential expression in photosynthetically active leaf chloroplasts and other specialized plastids (e.g., etioplasts, amyloplasts, chromoplasts) indicate the existence of regulatory events that control the synthesis of the individual gene products. This chapter discusses our present understanding of transcriptional and posttranscriptional regulatory mechanisms in higher plant plastids, emphasizing recent developments and current problems.

II. THE CHLOROPLAST TRANSCRIPTION APPARATUS

A. Chloroplast RNA Polymerase

1. *Chloroplast RNA Polymerase Purification*

Expression of chloroplast genes requires one or more RNA polymerases for transcription of the different genes. The application of different schemes for preparing DNA-dependent RNA polymerase from chloroplasts has led to the intriguing idea that chloroplasts of algae and higher plants may contain at least two different RNA polymerase activities distinguishable by their preference for specific genes and biochemical parameters (Greenberg *et al.*, 1984b). Extraction of the stromal fraction from lysed intact chloroplasts in the presence of moderate salt concentrations [0.5 M $(NH_4)_2SO_4$ or 0.5 M KCl] results in the isolation of a soluble enzyme dependent on a DNA template. Such soluble RNA polymerases have been isolated from maize chloroplasts (Bottomley *et al.*, 1971), wheat (Polya and Jagendorf, 1971), maize (Orozco *et al.*, 1985b), mustard (Link, 1984a), spinach (Briat and Mache 1980; Gruissem *et al.*, 1983a), pea (Joussaume, 1973; Tewari and Goel, 1983), and *Euglena gracilis* (Gruissem *et al.*, 1983b). Some of these preparations have been further purified, resulting in enzyme fractions with subunit compositions of 7 to 14 polypeptides, ranging in molecular weight from 25,000 to 180,000 (Kidd and Bogorad, 1979; Tewari and Goel, 1983; Lerbs et al., 1983).

A second chloroplast RNA polymerase activity can be isolated in the form of a DNA–protein complex. Such transcriptionally active chromosome (TAC) was first isolated from *Euglena* chloroplasts (Hallick et al., 1976), and similar DNA–RNA polymerase complexes were subsequently prepared from *Chlamydomonas* (Dron et al., 1979) and spinach (Briat et al., 1979). The RNA polymerase associated with TAC is tightly bound to the DNA and efforts to dissociate the enzyme in an active and specific form have not been successful. It has been possible, however, to obtain highly purified TAC preparations from *Euglena* by gel filtration at high ionic strength (i.e., 1 M KCl). The RNA polymerase in such complexes contains only one or two major polypeptide subunits, with molecular weights of 118,000 and 85,000, but it still retains the transcriptional properties and specificity of less purified preparations (Narita et al., 1985).

2. Properties of Chloroplast RNA Polymerase Activities

Besides the presence or absence of DNA in the different RNA polymerase preparations from chloroplasts, the two enzyme activities can also be distinguished by a number of biochemical parameters (Table I). Both the soluble and DNA-bound RNA polymerase (TAC) activities require Mg^{2+}, with the optimal ion concentration varying for different enzyme preparations. Both activities are inhibited by Mn^{2+}, which may have implications for the regulation of transcription in the chloroplast (Job et al., 1987). For example, upon illumination, the Mg^{2+} concentration in chloroplasts rises from 2 to 5 mM, which may directly affect the activity of the enzyme *in vivo*. Light induces an increase in the transcription of maize, mustard, and spinach chloroplast DNA (Apel and Bogorad, 1976; Reiss and Link, 1985; Deng and Gruissem, 1987), and at least for the maize enzyme it has been concluded that this change is not accompanied by either an increase in RNA polymerase concentration or qualitative changes in the purified enzyme.

The chloroplast TAC RNA polymerases from spinach, pea, and *Euglena* have optimal activities at salt concentrations between 100 and 300 mM, and are still active at higher concentrations. The soluble chloroplast RNA polymerases from higher plants and *Euglena* are inactive at these salt concentrations. In spinach, mustard, and *Euglena*, the soluble enzyme is also inhibited by heparin, which is known to compete with DNA for binding to RNA polymerases (Walter et al., 1967). Transcription by the TAC enzyme is not blocked with heparin, confirming that this RNA polymerase is tightly bound to chloroplast DNA. The TAC enzyme from spinach and *Euglena* can initiate transcription *in vitro* (Hallick et al., 1976; Briat et al., 1979). These properties have led to the proposal that the RNA polymerase in the TAC complex is mainly engaged in open promoters and/or elongating ternary

TABLE I

Properties of Chloroplast RNA Polymerase Activities[a]

	Soluble RNA polymerase activities[a]					Transcriptionally active chromosomes (TAC)[a]		
	Spinach[b]	Euglena[c]	Maize[d]	Mustard[e]	Pea[f]	Spinach[g]	Euglena[h]	Mustard[i]
Optimal [Mg^{2+}] (mM)	10	10	30	5–10	10	20	10	10
[Mn^{2+}][j]	Inhibited	Inhibited	Inhibited	—	—	Inhibited	Inhibited	—
Optimal [KCl] (mM)	40	50	80–100	50	—	300	100–600	50
Optimal [NH$_4^+$] (mM)	80	20	0–30	—	—	—	—	—
Optimal [NaCl] (mM)	—	—	—	—	100	—	—	—
Heparin	Inhibited	Inhibited	—	Inhibited	—	Active	Active	—
Optimal °C	25	25	48	25	30	35	30	30
Gene products	tRNA/mRNA rRNA(?)	tRNA	tRNA/mRNA	mRNA	rRNA[k]	rRNA mRNA/tRNA(?)	rRNA	rRNA mRNA/tRNA(?)

[a] See text for explanation.
[b] From Gruissem et al., 1983a; Orozco et al., 1985a.
[c] From Gruissem et al., 1983b; Greenberg et al., 1984b.
[d] From Bottomley et al., 1971; Schwarz et al., 1981.
[e] From Link, 1984a.
[f] From Tewari and Goel, 1983; Sun et al., 1986.
[g] From Briat et al., 1979, 1982.
[h] From Hallick et al., 1976; Rushlow et al., 1980; Narita et al., 1985.
[i] From Reiss and Link, 1985; conditions used in the transcription reaction.
[j] Mn^{2+} added to the transcription reaction in the presence of optimal [Mg^{2+}].
[k] Transcripts from other chloroplast DNA regions have been observed.
(—) No experimental results reported.

complexes (Briat *et al.*, 1986). Both RNA polymerase activities appear to preferentially transcribe different chloroplast genes. In its most purified form, the TAC enzyme from spinach and *Euglena* specifically transcribes the ribosomal RNA (*rrn*) operon, even though the complete chloroplast genome is available as a template (Rushlow *et al.*, 1980; Briat *et al.*, 1982). This preference for the transcription of ribosomal RNAs may be related to the structure of the *rrn* operon as an inverted or tandem repeat, because in pea chloroplasts, which lack an inverted repeat structure, the TAC enzyme seems to transcribe the entire chloroplast genome, although transcription initially starts at the *rrn* operon (Tewari and Goel, 1983). In contrast, the soluble RNA polymerase in transcription extracts from spinach (Gruissem *et al.*, 1983a; Gruissem and Zurawski, 1985a,b), mustard (Link, 1984a), maize (Orozco *et al.*, 1985), and *Euglena* (Gruissem *et al.*, 1983b; Greenberg *et al.*, 1984b) synthesizes tRNA and mRNA from supercoiled plasmid template DNAs. These crude soluble RNA polymerase preparations initiate transcription *in vitro* correctly at the *in vivo* 5' transcription start sites, and thus appear to retain faithful promoter recognition. Highly purified fractions of the soluble chloroplast RNA polymerase have not been tested with mRNA and tRNA gene templates. Taken together, the present data suggest two RNA polymerase activities in chloroplasts, but more rigorous experiments are necessary to establish the differences and specificity of these activities.

3. *Chloroplast RNA Polymerase—A Prokaryote-Type Enzyme?*

The similarity of the DNA sequences in chloroplast promoter regions with the *Escherichia coli* consensus promoter sequence, and the ability of *E. coli* RNA polymerase to transcribe chloroplast genes suggest similarity between the chloroplast RNA polymerase and the bacterial enzyme. The predominant *E. coli* RNA polymerase consists of four polypeptides, the 155-kDa β'-subunit, the 150-kDa β-subunit, the 36-kDa α-subunit, as well as the 70-kDa σ-subunit, which together comprise the holoenzyme. Gene probes for the *E. coli* DNA polymerase β-subunit hybridize to both nuclear and chloroplast DNA in *Chlamydomonas* (Watson and Surzycki, 1983). Using a solid-phase sandwich enzyme immunoassay and antibodies against the *E. coli* RNA polymerase subunits, Lerbs *et al.* (1985) found immunological cross-reactions between the β'- and β-subunits of the bacterial enzyme with a purified spinach chloroplast RNA polymerase fraction, indicating structural relatedness of these subunits of the chloroplast enzyme to the bacterial RNA polymerase subunits. Additionally, three open reading frames of the chloroplast genome in tobacco (Ohme *et al.*, 1986; Shinozaki *et al.*, 1986b), spinach (Sijben-Müller *et al.*, 1986), and *Marchantia* (Ohyama *et al.*, 1986) have significant sequence similarity with the genes for the *E. coli* RNA polymerase α-subunit (*rpo*A), and β- (*rpo*B) and β'-subunits (*rpo*C). Although the *rpo* genes are transcribed in the above plants, it is not known if the transcripts are translated into functional proteins. Chloroplast RNA polymerases are insensitive to rifampicin, a potent inhibitor of the bacterial enzyme. In addition, it has been found that the spinach chloroplast enzyme in the chloroplast extract does not efficiently recognize *E. coli* promoter regions (Gruissem and Zurawski, 1985b; W. Gruissem, unpublished results). Thus the analogy to the bacterial enzyme is at best only limited. Finally, it is also noteworthy that the chloroplast RNA polymerase is not inhibited by α-amanitin, and thus is clearly distinguishable from nuclear RNA polymerases II and III, which are, respectively, fully and partially inhibited by this drug.

For the precise recognition of promoter sequences, bacterial, mitochondrial, and nuclear RNA polymerases require additional factors that are not part of the core enzyme complexes. Although the subunit composition and the structure of the transcription initiation complex of the chloroplast RNA polymerase are still unknown, chromatographic separations result in protein fractions that, when recombined with the purified RNA polymerase, can increase the activity and specificity of the enzyme. Such a protein fraction (S-factor) has been isolated from maize chloroplasts, and its stimulatory effect for the maize chloroplast RNA polymerase is not obtained by the *E. coli* σ-subunit. Addition of the S-factor fraction to *E. coli* core RNA polymerase also does not increase the activity of the bacterial enzyme (Jolly and Bogorad, 1980). Similar experiments with supernatants from spinach leaf homogenates, however, have yielded chromatographic protein fractions which apparently contain α-like and σ-like polypeptides (Lerbs *et al.*, 1987). These protein fractions, when added back to a purified spinach chloroplast RNA polymerase, allow correct initiation and transcription of the spinach *rbc*L gene. It is important to note, however, that the stimulatory factors have not been purified, nor have other fractions been analyzed for nucleolytic activities that may obscure additional chloroplast-specific factors regulating chloroplast RNA polymerase activity. Taken together, the chloroplast and prokaryotic RNA polymerases may share structural similarities, but more work is needed to fully understand the unique properties of the chloroplast enzyme.

B. *In Vitro* Transcription Systems

Although chloroplast RNA polymerases have now been isolated from several higher plants and algae, the preparation and purification of the enzyme is rather time consuming and the resulting enzyme activity is relatively unstable. In addition, as discussed above, potential regulatory factors may have been removed from these RNA polymerases during the chromatographic purifications. To characterize the potential regulatory functions of chloroplast promoter regions and other DNA sequences involved in transcription initiation by the chloroplast RNA polymerase, *in vitro* transcription systems have been developed to study chloroplast gene expression.

1. In Organello *Transcription*

Since the first demonstration of RNA synthesis in isolated chloroplasts (Spencer and Whitfeld, 1967), considerable efforts have been made to demonstrate the transcriptional regulation of chloroplast genes, and to determine the role of light in the control of chloroplast gene expression. Incubation of isolated intact chloroplasts with radioactive uridine demonstrated incorporation into newly synthesized RNA in the light, but failed to show labeled RNA products in the dark (Hartley and Ellis, 1973). Recently, transcription experiments with isolated maize plastids from dark- and light-grown plants showed that ribonucleotide triphosphates can be incorporated into newly synthesized RNA equally well in both etioplasts and chloroplasts (Altman *et al.*, 1984). The incorporation rate over a 2-hr period was low, most likely due to the poor uptake of nucleotide triphosphates by the intact chloroplast membrane. This and the undefined roles of transcription reinitiation and RNA turnover make the intact chloroplast transcription system impractical to study changes in the relative transcriptional activities of plastid genes during plant development.

2. *Soluble* in Vitro *Transcription Systems*

The similarities between chloroplast and bacterial genes have prompted the use of *E. coli* RNA polymerase to express plastid genes *in vivo* and *in vitro* (Bottomley and Whitfeld, 1979; Gatenby *et al.*, 1981). *E. coli* RNA polymerase holoenzyme can bind to plastid DNA and initiate transcription at discrete sites (Tohdoh *et al.*, 1981; Zech *et al.*, 1981; Koller *et al.*, 1982), although the relationship between the binding sites of the prokaryotic enzyme and chloroplast transcription initiation is unknown. Identical 5' termini for *E. coli* and chloroplast transcripts have been shown for the gene of the large subunit of ribulose-1,5-bisphosphate carboxylase (Shinozaki and Sugiura, 1982; Erion *et al.*, 1983); information for other chloroplast transcription units is not available.

The utilization of prokaryotic organisms and enzymes to study chloroplast gene expression is valid only if coupled to parallel studies with plastid enzymes. The development of chloroplast *in vitro* transcription systems has allowed the identification and detailed analysis of several chloroplast promoter regions. The chloroplast transcription system, consisting of a crude stromal extract from which membranous material and DNA have been removed, has been prepared from *Euglena* (Gruissem *et al.*, 1983b; Greenberg *et al.*, 1984b), spinach (Gruissem *et al.*, 1983a, 1986a; Orozco *et al.*, 1985, 1986), mustard (Link, 1984a), pea, and maize (Orozco *et al.*, 1985). The chloroplast RNA polymerase present in such extracts retains its full activity and is presumably associated with all the factors required for specific and correct transcript initiation and elongation *in vitro*. Removal of contaminating nucleic acids is usually achieved by diethylaminoethyl (DEAE) chromatography of the extract with a step salt gradient elution, although treatment with micrococcal nuclease has also been employed successfully. Specific transcription of chloroplast tRNA genes has been assayed by direct visualization of radioactively labeled tRNA product or by RNase T1 fingerprint analysis of the tRNA product (Gruissem et al., 1983a,b), while that of protein genes has been shown by S1 nuclease or primer extension analysis of the mRNA transcripts (Orozco et al., 1985). Although the enzymatic properties of the chloroplast RNA polymerase in the crude extracts are difficult to measure, certain parameters of the reaction can be varied to optimize production of *in vitro* transcripts (Gruissem, 1984; Gruissem et al., 1986; Orozco et al., 1986).

3. Plastid Run-On Transcription

Although it has been demonstrated that chloroplast *in vitro* transcription extracts can be used to identify and analyze wild-type and mutant chloroplast promoter regions (reviewed by Hanley-Bowdoin and Chua, 1987), run-on transcription systems are more useful to study the regulation of chloroplast gene expression during developmental changes. For example, significant alterations in plastid RNA levels that occur during chloroplast development or plastid differentiation (e.g., Link, 1982; Rodermel and Bogorad, 1985; Herrmann et al., 1985; Piechulla et al., 1985, 1986; Kreuz et al., 1986) could be regulated at the transcriptional or posttranscriptional level. It has already been shown for several animal and plant cytoplasmic mRNAs that their differential accumulation or decrease during development is accompanied by changes in the activity of their nuclear genes, which can be measured by the amount of RNA synthesized in isolated nuclei (for review see Darnell, 1982; Tobin and Silverthorne, 1985). Active run-on transcription systems to distinguish transcriptional and posttranscriptional regulation of differential RNA accumulation have been developed for plastids of spinach and barley (Deng and Gruissem, 1987; Deng et al., 1987; Mullet and Klein, 1987), and for plastids of tomato fruit to analyze the differential RNA decrease during chromoplast differentiation (Gruissem et al., 1987b). Although isolated intact chloroplasts synthesize only low levels of RNA (Altman et al., 1984), they rapidly incorporate radioactive nucleotide triphosphates into high-molecular-weight RNA after mechanical lysis in a hypotonic transcription reaction mixture (Deng et al., 1987; Mullet and Klein, 1987). The addition of exogenous chloroplast DNA or heparin to the lysed chloroplasts does not increase or inhibit transcription, respectively. This indicates that RNA is strictly synthesized by preinitiated transcription complexes, and thus the activity of individual chloroplast genes is directly reflected by the amount of synthesized RNA. The chloroplast run-on transcription system in combination with Northern analysis can therefore be used to determine the relationship between changes in transcription and

RNA levels in plastids during chloroplast development or plastid differentiation, and thus provides an important complement to the *in vitro* transcription extracts.

C. Chloroplast Promoter Regions

1. Promoter Identification

The 5′ regions of many chloroplast genes contain DNA sequences similar to prokaryotic promoters (Hawley and McClure, 1983). Since *E. coli* RNA polymerase can initiate transcription of chloroplast genes *in vivo* and *in vitro*, presumably by recognizing chloroplast sequences that resemble prokaryotic promoters, it was suggested that such sequences are of functional importance for the chloroplast RNA polymerase (for review, see Whitfeld and Bottomley, 1983). This notion was reinforced by S1 nuclease analysis of chloroplast mRNAs, in which the 5′ end of several mRNAs was located immediately 3′ of DNA sequences which resemble the prokaryotic "-10" consensus promoter sequence. Based on such structural analyses and sequence comparisons, sequences of 5′ upstream regions of several chloroplast genes have been compiled, and the prokaryote-like promoter sequences present in these upstream regions have been suggested as potential chloroplast promoters (Kung and Lin, 1985).

The promoter structure of three chloroplast tRNA genes from spinach (*trn*M2, *trn*R1, and *trn*S1; Gruissem and Zurawski, 1985a; Gruissem *et al.*, 1986) and the genes for the large subunit of ribulose-1,5-bisphosphate carboxylase (*rbc*L), the 32-kDa protein of photosystem II (*psa*A), and the β/ε-subunit of the ATpase (*atp*BE) have been analyzed in chloroplast transcription systems (Link, 1984a; Gruissem and Zurawski, 1985a,b; Hanley-Bowdoin *et al.*, 1985; Bradley and Gatenby, 1985). These studies have confirmed that DNA sequences similar to the prokaryotic consensus promoter DNA sequences are required for transcription initiation by the chloroplast RNA polymerase. However, *in vitro* transcription experiments have also revealed that, unlike prokaryotic promoters, deletion of DNA sequences from chloroplast promoter regions that resemble the bacterial "-35" consensus sequence does not completely inactivate transcription initiation (Link, 1984a). More importantly, prokaryote-like promoter sequences are not present in the upstream region of some chloroplast tRNA genes, although these genes are actively transcribed *in vitro* (Gruissem *et al.*, 1986b). Thus, structural similarities to prokaryotic promoter regions are only of limited applicability in locating chloroplast promoter regions. Rather, such regions can be identified unequivocally only by a functional assay in homologous chloroplast transcription systems.

2. Chloroplast tRNA Promoter Regions

The spinach chloroplast trnM2 locus, which codes for tRNA^{Met}, was the first chloroplast gene in which the complete promoter region was identified by mutational analysis (Gruissem and Zurawski, 1985a). The tRNA^{Met} gene, located between the atpBE and trnV1 transcriptional units, is transcribed toward the 3' end of atpBE. The tRNA^{Val} gene contains a 600-bp intron. Both trnM2 and trnV1 are transcribed in the chloroplast extract, but the intron in the tRNA^{Val} transcript is not processed (Gruissem et al., 1983a). Deletion of nucleotides in the sequence TTGCTT (ctp1), which resembles the prokaryotic "-35" consensus promoter sequence TTGACA (Hawley and McClure, 1983), results in a progressive decrease in the tRNA^{Met} product. Seventeen base pairs downstream from this sequence, a DNA sequence (ctp2) occurs which is identical to the prokaryotic "-10" consensus promoter sequence TATAAT. Internal deletion mutants have demonstrated that this sequence is also required for transcription initiation. Both promoter elements are highly conserved for the trnM2 loci from spinach, tobacco, maize, and barley, but DNA sequences proximal and distal to these elements are less highly conserved. The mutational analysis has confirmed that the DNA sequence in these less conserved regions is not critical for promoter function, but the distance between the "-10" and "-35" elements can affect the transcription initiation frequency in vitro. Thus, the minimal DNA sequence requirement for transcription of the spinach chloroplast trnM2 locus by the homologous RNA polymerase resembles the prokaryotic promoter organization.

Although 5' prokaryote-like promoter elements are clearly of functional significance for trnM2 transcription, this concept may not apply to all higher plant chloroplast tRNA genes. When several spinach chloroplast tRNA genes were analyzed for their promoter function, a subpopulation of tRNA genes was identified that did not require upstream promoter elements for transcription (Gruissem et al., 1986b). The absence of functional promoter elements in the 5' region of at least two tRNA genes, trnR1 and trnS1, is supported by several observations. First, comparison of the upstream DNA sequences from these genes with the defined promoter region of trnM2 does not reveal regions of significant DNA sequence similarity. Second, deletion of 5' DNA sequences immediately upstream of the trnR1 and trnS1 coding regions does not significantly affect transcription of these genes in vitro. Third, DNA sequences from the 5' upstream region of these genes do not support transcription of a trnM2 promoter deletion mutant when fused to this gene. At present it is unknown which sequences are recognized by the chloroplast RNA polymerase for transcription of trnR1 and trnS1, but it cannot be excluded that DNA sequences internal to these genes may function as promoter regions. Such a mechanism has been well documented for the transcription of nuclear tRNA genes by RNA polymerase III (Galli et al., 1981), but the demonstration of a similar mechanism awaits the analysis of mutations in the coding region of the chloroplast tRNA genes. Based on their organization in the chloroplast genome, it is also possible that both *trn*R1 and *trn*S1 are part of polycistronic transcripts *in vivo,* in which case there is a decreased constraint on the 5' upstream promoter regions, but not on DNA sequences within the tRNA genes that are required for the structure of these molecules and that may function as a promoter *in vitro*.

3. *Promoter Region of the Chloroplast* rrn *Operon*

Promoter regions were first assigned to the *rrn* operon in maize (Schwarz *et al.,* 1981), tobacco (Tohdoh *et al.,* 1981), spinach (Briat *et al.,* 1982), and *Spirodela* (Keus *et al.,* 1983) based on sequence similarity with the prokaryotic consensus promoter sequence, S1 nuclease mapping, and binding of *E. coli* RNA polymerase to chloroplast DNA. These experiments, however, did not exclude the possibility that the *trn*V2 gene located upstream of the 16S rRNA gene is cotranscribed with the *rrn* operon. *In vitro* capping experiments with isolated RNA from maize chloroplasts have demonstrated, however, that the tRNA^{Val} gene appears to be a separate transcription unit (Strittmatter *et al.,* 1985). Analysis of the spinach 16S rRNA 5' region in *E. coli* has shown that the bacterial enzyme recognizes two promoter regions (P1 and P2) downstream of *trn*V2 (Lescure *et al.,* 1986), but only one of the two promoter regions (P1) is used *in vivo* and *in vitro* by the chloroplast RNA polymerase (Briat *et al.,* 1987). This observation demonstrates again that results obtained with *E. coli* are not directly applicable to higher plant chloroplasts. In maize, in which the DNA sequence corresponding to the P1 promoter is not present, transcription starts 3' of a DNA sequence positioned as P2 in the spinach chloroplast *rrn* 5' region (Strittmatter *et al.,* 1985). The DNA sequence in the spinach chloroplast *rrn* P2 region is also highly conserved in other plants. It should be noted, however, that, although the 5'region identified in maize (Strittmatter *et al.,* 1985), pea (Sun *et al.,* 1986), and spinach (Lescure *et al.,* 1985; Briat *et al.,* 1987) most likely constitutes the promoter of the chloroplast *rrn* operon in these plants, the functional requirement for these DNA sequences for the initiation of transcription of ribosomal RNA has not been demonstrated.

4. *Structure of Promoter Regions for Chloroplast Protein Genes*

The DNA sequences flanking the 5' ends of the primary transcripts from the genes of the ribulose-1,5-bisphosphate carboxylase large subunit (*rbc*L), the β/ε-subunits of ATPase (*atp*BE), and the 32-kDa photosystem II core protein (*psb*A) are highly conserved in monocotyledonous and dicotyledonous plants. Using a method termed "evolutionary filtering," Zurawski and Clegg (1987) have compared the *rbc*L 5' upstream sequences from several plants, from which they have derived a consensus promoter region shown in Fig. 1. The 5' regions of the three genes contain sequences that are similar to the prokaryotic "-35" and "-10" consensus promoter elements. The signifi-

```
Zm   aagattagggtTTGGGTTGCGCtATAtcTATcAAAGAGTATAcAATAATgATGgATTTggtgAATCAAAT

Hv   gggatta   atTTGGGTTGCGCtATAccTATcAAAGAGTATAcAATAATgATGgATTTggtaAATCAAAT

Ta   ggaatta   atTTGGGTTGCGCtATAtcTATcAAAGAGTATAcAATAATtATGgATTTggtaAATCAAAT

Ps   aaaaaaaacggTTGGGTTGCGCcATAcaTATgAAAGAGTATAgAATAATgATGtATTTcccaAATCAAAT

So   aacggttacggTTGGGTTGCGCcATAtaTATgAAAGAGTATAcAATAATgATGtATTTggcgAATCAAAT

Nt   aaaaagaaaaaTTGGGTTGCGCtATAtaTATgAAAGAGTATAcAATAATgATGtATTTggcaAATCAAAT

Le   aaaaatcaaaaTTGGGTTGCGCtATAtaTATgAAAGAGTATAcAATAATgATGtATTTggcaAATCAAAT


            TTGCGC              TACAAT

            ctp1                ctp2
```

Fig. 1. DNA sequence conservation in *rbc*L promoter regions from different plants. DNA sequences shown are from *Zea mays* (Zm; McIntosh *et al.*, 1980), *Hordeum vulgare* (Hv; Zurawski *et al.*, 1984), *Triticum aestivum* (Ta; Howe *et al.*, 1985), *Pisum sativum* (Ps; Zurawski *et al.*, 1985), *Spinacia oleracea* (So; Zurawski *et al.*, 1981), *Nicotiana tabacum* (Nt; Shinozaki and Sugiura, 1982), and *Lycopersicon esculentum* (Le; T. Manzara and W. Gruissem, unpublished results). Capital letters represent conserved nucleotides, lower case letters indicate sequence divergence. The conserved promoter elements ctp1 and ctp2 required for transcription *in vitro* are indicated (Gruissem and Zurawski, 1985a). (Modified from Zurawski and Clegg, 1987.)

cance of these sequences for promoter function was first demonstrated for the spinach *rbc*L, *atp*BE, and *psb*A loci in a homologous chloroplast transcription extract. It was shown that templates constructed using oligonucleotides spanning the 5' upstream regions of these genes and fused to a *trn*M2 promoter-deletion mutant are able to support the transcription of tRNA^{Met} *in vitro*. These experiments have defined the essential 5' sequences of *rbc*L, *atp*BE, and *psb*A (Gruissem and Zurawski, 1985b). The promoter for the *rbc*L gene has also been identified in maize. Multiple transcripts have been detected for the large subunit, but only the longest transcript can be labeled with [α-^{32}P]GTP and guanylyl transferase (Crossland *et al.*, 1984; Mullet *et al.*, 1985). Approximately 40 bp flanking the transcription start site of the primary transcript is sufficient for the maize RNA polymerase to transcribe the homologous *rbc*L gene *in vitro* (Hanley-Bowdoin and Chua, 1987). The maize *atp*BE promoter region has been analyzed both in *E. coli* and in a maize chloroplast transcription extract. Similar to the spinach *atp*BE gene, the essential maize promoter elements are confined to a region 35 bp proximal to the 5' end of the primary transcript (Bradley and Gatenby, 1985).

The experimentally defined promoter regions located in the upstream regions of the *rbc*L, *atp*BE, and *psb*A genes contain consensus sequences, which are likely to function as promoter elements in most plants. Nevertheless, it is possible that DNA sequences eliminated in these experiments are required in addition to the essential promoter elements for efficient transcription of these genes *in vivo*. It has been reported that the 49-bp promoter region derived from the maize *rbc*L gene is recognized less efficiently by the maize RNA polymerase in a chloroplast extract than a template which includes an additional 26 bp 5' proximal to the promoter region and 34 bp of 3' DNA (Hanley-Bowdoin and Chua, 1987). The addition of these sequences does not, however, affect the transcription initiation frequency of the heterologous pea RNA polymerase at this promoter region. In mustard, a DNA sequence has been identified between the ''-35'' and ''-10'' promoter elements of *psb*A that may allow the expression of this gene at a basal level in the absence of the -35 region (Link, 1984a). Base changes in the conserved promoter regions of chloroplast genes from different plants may be of functional importance for the chloroplast RNA polymerase. It has been found, for example, that pea *trn*M2 is a less efficient template for the spinach chloroplast RNA polymerase (Gruissem and Zurawski, 1985a). Thus, chloroplast promoter regions having DNA sequences not defined by the canonical prokaryotic promoter may be important for the species-specific modulation of transcription in chloroplasts.

5. *Regulation of Promoter Function*

The relative transcriptional activity of chloroplast genes may be regulated by differential promoter strength. This question has been investigated in spinach using chloroplast promoter regions fused to the *trn*M2 promoter deletion mutant as a reporter gene. Quantitation of the tRNA^{Met} product revealed that the promoter regions of *rbc*L, *atp*BE, and *psb*A transcribe the *trn*M2 gene with different efficiencies (Gruissem and Zurawski, 1985b). Based on these results, the promoter regions of the four genes have been grouped according to their relative strength (Fig. 2). To determine whether changes in the DNA sequence of promoter elements influence their strengths, mutations in promoter regions have been analyzed for the spinach *rbc*L, *psb*A, and *trn*M2 genes (Gruissem and Zurawski, 1985a,b; G. Zurawski and W. Gruissem, unpublished results), and the maize *rbc*L and *atp*BE genes (Hanley-Bowdoin et al., 1985; Bradley and Gatenby, 1985). The results are difficult to interpret. In general, most base changes in the ''-35'' (ctp1) and ''-10'' (ctp2) promoter elements decrease the efficiency of the promoters. On the other hand, specific mutations in the spinach *psb*A and *trn*M2 promoter regions that increase their similarity to the prokaryotic consensus promoter do not increase their efficiency. Small changes in the distance between the ctp1 and ctp2 promoter elements significantly affect transcription of the spinach *trn*M2 and the maize *rbc*L genes, but have little effect on the efficiency of the *psb*A promoter *in vitro*. Thus, although spe-

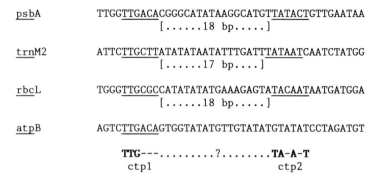

Fig. 2. Comparison of spinach chloroplast promoter regions. The promoter regions were identified and analyzed in a chloroplast transcription extract (Gruissem and Zurawski, 1985a,b). The underlined DNA sequences represent the essential promoter elements ctp1 and ctp2 that affect transcription when deleted or mutagenized. The conserved nucleotides in these promoter elements are indicated. The promoter regions are shown in the order of their relative strengths in the chloroplast extract (Gruissem and Zurawski, 1985b) and the chloroplast run-on transcription system (Deng and Gruissem, 1987).

cific DNA sequences have been identified as basic elements of chloroplast promoters, the precise requirements in the interaction of the chloroplast RNA polymerase with these sequences awaits further characterization.

It has been suggested that the promoter activity of chloroplast genes may be regulated by local changes in superhelical densities of the DNA template (Stirdivant *et al.,* 1985; Thompson and Mosig, 1987; Lam and Chua, 1987). Energetic and structural changes in the topologically constrained supercoiled DNA affect transcription of specific genes in prokaryotic cells (for review, see McClure, 1985; Wang, 1985). Thus, it is not surprising that changes in the superhelical density of plasmid DNA-containing chloroplast promoter regions also affect their relative activity *in vitro,* as has been demonstrated for the maize *rbc*L and *atp*BE promoter regions (Stirdivant *et al.,* 1985). Although this seems an attractive model for the differential regulation of these genes in mesophyll and bundle sheath cells in maize leaves (Link *et al.,* 1978; Sheen and Bogorad, 1985), it is unknown if both genes are regulated at the transcriptional level *in vivo.* The stimulation of a *Chlamydomonas* chloroplast promoter *in vivo* and the decrease of *rbc*L and *atp*BE promoter activity *in vitro* by novobiocin, a specific inhibitor of topoisomerase II, have been used as evidence that changes in template topology may be a mechanism by which chloroplast genes are differentially regulated (Thompson and Mosig, 1987; Lam and Chua, 1987). However, it can not be excluded that novobiocin interferes with the site-specific interaction of transcription factors, as has been demonstrated for various RNA polymerase III promoters (Van Dyke and Roeder, 1987), and thus such interference may explain the observed changes in promoter activity.

6. Promoter–Protein Interactions

In addition to specific DNA sequences that may determine the different strengths of chloroplast promoter regions, the relative transcriptional activity of chloroplast genes may also be determined by the interaction of *trans*-acting factors with the DNA sequences and RNA polymerase(s). In earlier experiments to examine promoter-specific DNA-binding proteins, labeled chloroplast DNA fragments were used in protein blotting and binding experiments with purified RNA polymerase from spinach (Lerbs *et al.*, 1985) and the TAC proteins from mustard (Bülow *et al.*, 1987). Although the chloroplast DNA fragments bound to some or all of the RNA polymerase subunits, this experimental approach may be subject to nonspecific binding reactions and requires rigorous controls. For example, it was found for spinach RNA polymerase that both promoter-containing and coding region DNA fragments bind the polypeptides with the same efficiency (Lerbs *et al.*, 1983). In the case of the mustard TAC proteins a DNA fragment from the *psb*A 5' upstream region was used in the binding studies, but other controls were not included (Bülow *et al.*, 1987).

A different approach to identify promoter-specific DNA-binding proteins takes advantage of the characterized chloroplast promoter regions. These DNA fragments of defined length and function bind proteins in the chloroplast *in vitro* transcription extract, and these promoter–protein complexes can be resolved in native gel systems in the presence of nonspecific competitor DNA (Gruissem *et al.*, 1987a). In the case of the spinach chloroplast *psb*A promoter, which is shown as an example in Fig. 3, several retarded complexes are observed that can be competed with unlabeled promoter fragments. The different retarded complexes may represent the binding of different proteins to the *psb*A promoter, or the binding of two or more proteins to the same DNA fragment. Although the function of the bound proteins is currently under investigation, it is reasonable to speculate that such DNA–protein complexes represent the interaction of specific proteins with the promoter regions as part of the transcription initiation complex. Further experiments will verify if these proteins have a direct function in establishing and maintaining differential promoter strength in chloroplasts, or a role in modulating overall transcription levels in developing and differentiating plastids.

D. Transcription Termination

1. Chloroplast 3' Inverted Repeat DNA Sequences

Transcription termination in both prokaryotic and eukaryotic cells has been recently reviewed (Platt, 1986). Efficient terminators in bacterial cells generally consist of a series of uridine residues at the 3' end of transcripts, preceded by a GC-rich inverted repeat (IR) sequence. Experimental evi-

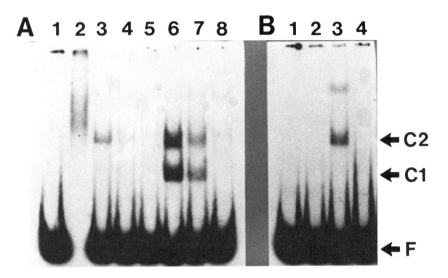

Fig. 3. Binding of chloroplast proteins to the spinach *psb*A promoter region. After incubation of the labeled 50-bp *psb*A promoter region (Gruissem and Zurawski, 1985b) in the chloroplast transcription extract, and electrophoresis in a nondenaturing polyacrylamide gel, distinct DNA–protein complexes can be visualized (H. Jones and W. Gruissem, unpublished results). The free promoter fragment is labeled F, and C1 and C2 represent DNA–protein complexes in the presence of nonspecific competitor. (A) Lane 1, free *psb*A promoter fragment; lane 2, addition of chloroplast proteins in the absence of nonspecific competitor; lanes 3–5, competition of DNA-binding proteins with unlabeled *psb*A promoter fragment in the presence of poly-(dI)(dC); lanes 6–8, competition of DNA-binding proteins with the unlabeled *psb*A promoter fragment in the presence of poly(dA)(dT). Poly(dI)(dC) appears to compete more effectively for the binding of proteins to the *psb*A promoter, indicating that only the protein binding in the presence of both nonspecific competitors may be specific for this promoter. (B) No retardation of the *psb*A promoter region is observed with bovine serum albumin in the presence of poly-(dI)(dC) in lane 1, and poly(dA)(dT) in lane 2. The binding activity is destroyed after heating of the chloroplast extract to 100°C (lane 4), but not to 65°C (lane 3).

dence confines the transcription termination signal to sequences within the RNA molecule (Farnham and Platt, 1982; Ryan and Chamberlin, 1983; Sharp and Platt, 1984), which supports the model that the IR allows the formation of intramolecular stem-loop structures in the RNA transcript, causing a pause in transcription elongation by RNA polymerase. The uridine residues facilitate dissociation of the transcript from the template, since RNA–DNA hybrids with rU-dA pairing are exceptionally unstable (Platt, 1986). It is generally assumed that no additional protein factors are required for termination at these signals. In contrast, ρ factor-dependent termination in *E. coli* is less well understood, but it appears that IR sequences contained in the termination region cannot alone be responsible for termination, and that sequences considerably upstream of the RNA 3' ends also convey the signal for termination, presumably via interactions between the ρ factor and the

RNA molecule (Platt, 1986). In addition to their function in termination, it is also possible that IR sequences located at the 3' end of mRNA molecules function as RNA processing sites or protective structures against nucleolytic degradation, as has been suggested for such sequences in photosynthetic and nonphotosynthetic bacteria and animal cells (Birchmeier et al., 1984; Belasco et al., 1985; Mott et al., 1985; Wong and Chang, 1986; Newbury et al., 1987).

Both mono- and polycistronic protein-coding transcription units and some tRNA genes found in the chloroplast genome of higher plants and algae are flanked at their 3' ends by IR sequences that permit the formation of intramolecular hairpin structures in the RNA transcript. However, not all of these IR sequences are followed by a series of uridine residues. The 3' ends of the IR sequences approximately coincide with the 3' ends of mature mRNAs. Based on sequence comparisons with known E. coli terminator regions and S1 nuclease protection experiments using chloroplast RNA, the original hypothesis was that the IR sequences in the chloroplast genome have a general function as transcription terminators (e.g., Zurawski et al., 1981; Steinmetz et al., 1983; Heinemeyer et al., 1984; Deno et al., 1984; Holschuh et al., 1984; Sugita and Sugiura, 1984; Kirsch et al., 1986).

2. Chloroplast Transcription Termination and mRNA Processing

The role of spinach chloroplast IR sequences has been examined directly in the chloroplast transcription extract (Gruissem et al., 1987a; Stern and Gruissem, 1987). IR sequences from the 3' end of psbA, petD, rbcL, and rpoA were inserted between the functionally defined psbA promoter region (Gruissem and Zurawski, 1985b) and a trnM2 promoter deletion mutant, and the ability of these sequences to terminate transcription in vitro was determined using tRNA^{Met} as an assayable marker. It was found that transcription of such constructs results in efficient read-through of the IR sequences, thus demonstrating that these IR sequences in the chloroplast genome do not have a general function as transcription terminators (Table II). These results do not reflect the inability of RNA polymerase to terminate at IR sequences, since a known bacterial terminator region (tec1) is recognized by the enzyme with high efficiency (Table II). In addition, transcription terminates with high efficiency in the 3' region of some, but not all tRNA genes tested (e.g., trnH1, trnS1; Table II), indicating that the lack of transcription termination at the 3' IR sequences is not merely due to the absence of termination factors in the chloroplast extract (Stern and Gruissem, 1987). It is conceivable that, if termination at the 3' IR sequences does not occur in vivo, RNA polymerase may proceed into other transcription units, resulting in the formation of antisense RNA. Antisense RNA has in fact been detected for the spinach chloroplast atpBE (Deng et al., 1987) and atpI-atpH-atpF-atpA (B. Koller, personal communication) transcription units, but the physiological significance and consequences of such antisense transcription is unknown.

TABLE II

Termination Efficiencies of Chloroplast 3′ Inverted Repeat Sequences

Location of 3′ IR[a]	Strand[b]	Termination frequency (%)[c]
rpoA[d]	+	0
petD[d]	+	38
rbcL	+	0
rbcL	−	29
psbA	+	0
psbA	−	17
tec1[e]	+	74
trnH[f]	+	85
trnS[f]	+	86

[a] Listed are the respective genes for which the termination efficiency of their 3′ regions was assayed in a spinach chloroplast transcription extract (Stern and Gruissem, 1987).

[b] + indicates the coding, − the noncoding DNA strand.

[c] The termination frequency was inferred from the ratio of IR-containing RNA to tRNA produced in the chloroplast transcription extract (Stern and Gruissem, 1987).

[d] rpoA and petD are located on opposite DNA strands. Their 3′ ends share a common IR sequence.

[e] tec1 is a coliphage T7 "early" terminator, and functions as an effective terminator with *E. coli* RNA polymerase *in vitro* (Neff and Chamberlin, 1980).

[f] trnH and trnS do not contain 3′ IR DNA sequences. Although their precise termination sites are unknown, efficient termination does occur within 30 bp distal to the tRNA coding region (Gruissem *et al.*, 1986).

Transcription of the chloroplast 3′ IR sequences in the chloroplast extract has provided evidence that these structures may function as RNA processing signals and stabilizing structures *in vitro* (Stern and Gruissem, 1987). When synthetic RNAs encompassing the 3′ IR sequence and sequences distal to the stem-loop structure from *rbc*L, *psb*A, and *pet*D are added to the chloroplast extract, rapid 3′ to 5′ exonucleolytic processing occurs such that the 3′ ends of these mRNAs are identical or nearly identical to the 3′ ends of the mature mRNAs *in vivo*. These processed RNA molecules possessing IR sequences at their 3′ ends are substantially more stable in the chloroplast extract than control RNAs not containing IR sequences. Kinetic measurements have also demonstrated that each IR-containing RNA has a unique decay rate (Stern and Gruissem, 1987). Consequently, the mechanism of site-specific termination in chloroplasts, if such a process occurs, is not yet understood. The above studies suggest that efficient termination can occur at certain tRNA genes (e.g., *trn*S, *trn*H). These tRNAs may be located in strategic positions on the chloroplast genome to function as efficient terminators. Other tRNA genes, which are part of polycistronic transcription units, may function as processing sites within long primary transcripts, similar to the maturation pathway for the human mitochondrial DNA primary transcripts (Ojala *et al.*, 1981). Ineffective termination at the IR sequences by the chloroplast RNA polymerase is corrected by rapid and precise RNA processing, a mechanism which may be responsible for the generation of the nearly homogeneous 3' ends found *in vivo*.

III. RNA PROCESSING

Most chloroplast genes are organized into polycistronic transcription units. Extensive processing and splicing events and stability of partially processed intermediates results in the accumulation of complicated sets of overlapping RNAs transcribed from several of these gene clusters. As with prokaryotic and eukaryotic RNAs, posttranscriptional processing of polycistronic and intron-containing RNAs most likely represents an important regulatory step in the control of chloroplast gene expression. Some processing reactions in the chloroplast have an unequivocal impact on the structure and function of the processed RNA species. For example, processing of ribosomal RNA is required for its assembly into functional ribosomes. Processing is also required to form active tRNAs from precursors. Similarly, the removal of intron sequences is necessary to form functional RNA molecules. The significance of other RNA processing events in the chloroplast, however, is less well understood, but the availability of DNA sequences and of chloroplast extracts containing processing activity should make it possible to study these reactions in more detail.

A. Ribosomal RNA Processing

1. Organization of the Ribosomal RNA Operon

The ribosomal RNA (*rrn*) operon is located in the inverted repeat on the chloroplast genome of all higher plants except legumes, in which the inverted repeat is lost (for review, see Kössel *et al.*, 1985). In *Euglena*, the *rrn* operon is organized as a tandem repeat of two to five copies (Gray and Hallick, 1978; Jenni and Stutz, 1978; Rawson *et al.*, 1978; Helling *et al.*,

1979; Koller and Delius, 1982a). All chloroplast *rrn* operons show the typical organization of the prokaryotic *rrn* operons, with the individual genes transcribed in the order 16S, 23S, 4.5S (in higher plants), and 5S rRNA. Until recently, the transcription of the 4.5S and 5S rRNAs within the same primary transcript as 16S and 23S rRNA was questionable, but it was concluded from S1 nuclease mapping and primer extension analysis of rRNA processing intermediates and mature rRNA species using total RNA from maize chloroplasts that these rRNAs are part of the polycistronic transcript and not transcribed from separate promoters (Strittmatter and Kössel, 1984). The spacer region between the 16S and 23S rRNA genes ranges from 1.6 to 2.4 kb in all higher plants, and includes the two intron-containing tRNA^{Ile} and tRNA^{Ala} genes (Koch *et al.*, 1981; for review, see Crouse *et al.*, 1984). Although the same tRNA genes are located in the 16S–23S spacer region in *Chlamydomonas* and *Euglena*, they do not contain introns (Graf *et al.*, 1980; Rochaix, 1981).

2. 5'/3' Terminal Processing and Cleavage of the Ribosomal RNA Precursor

The processing of rRNA has been studied in detail mostly in *E. coli* and other prokaryotic cells (for review, see King *et al.*, 1986), but little information is available for organelle and cytoplasmic rRNA in eukaryotic cells. Processing of rRNA involves multiple cleavage reactions at the 5' and 3' ends of each species. Some of the cleavage reactions occur on the free rRNA molecule, while other reactions are completed after assembly of the partially processed rRNA species with ribosomal proteins. The first cleavage that occurs in *E. coli* separates the 16S and 23S rRNAs from each other, the 5S rRNA and tRNAs, and is mediated by RNase III. The enzyme cleaves the rRNA precursor in the double-stranded regions formed by complementary sequences flanking both 16S and 23S rRNA (Young and Steitz, 1978; Bram *et al.*, 1980). The ensuing maturation of the 16S, 23S, and 5S rRNAs is accomplished by several different enzyme activities that have been partially purified (for review, see Deutscher, 1984).

Limited information about rRNA processing in chloroplasts was obtained primarily from analysis of radiolabeled RNA synthesized *in vivo* and in isolated chloroplasts by gel electrophoresis and saturation– and competition–hybridization experiments, or hybridization of the transcripts to filter-immobilized rDNA fragments (reviewed by Crouse *et al.*, 1984). In experiments with isolated chloroplasts, label accumulates rapidly in the 5S rRNA position, but not in 4.5S rRNA (Hartley, 1979). The model that was based on the above experiments suggests that an approximately 7.9-kb rRNA precursor is first cleaved into 6.8-kb and 5-kb intermediary products, which are subsequently processed into the 23S–4.5S rRNA precursor, the 16S rRNA, and the spacer tRNA^{Ile} and tRNA^{Ala}. The 23S–4.5S rRNA precursor is then processed into the 23S and 4.5S rRNAs (Bohnert *et al.*, 1977; Hartley *et al.*,

1977; Hartley and Head, 1979; Zenke *et al.*, 1982; reviewed by Crouse *et al.*, 1984). The above processing scheme excludes the 5S rRNA from the primary transcript and requires the separate transcription of this gene (Dyer and Bedbrook, 1980; Takaiwa and Sugiura, 1980; Keus *et al.*, 1983). However, recent experiments in maize and spinach have verified the cotranscription of this gene as part of the *rrn* operon (Strittmatter and Kössel, 1984; Audren *et al.*, 1987). A specific cleavage reaction that is less well understood in chloroplasts results in two major hidden breaks in the 23S rRNA. The position of the hidden breaks within the molecule has been determined by S1 nuclease mapping using maize rRNA (Kössel *et al.*, 1985). They coincide with two stem-loop regions from which several nucleotides have been removed at the position of the hidden breaks. The significance of the hidden breaks for ribosome function, as well as the requirements for the specific cleavage reaction, are unknown.

B. Processing of tRNA

1. Organization of Chloroplast tRNA Genes

The chloroplast genome in higher plants and algae encodes a complete set of tRNAs required for protein synthesis on chloroplast 70S ribosomes (for review, see Whitfeld and Bottomley, 1983; Sugiura *et al.*, 1985). Thirty seven tRNA genes have been identified in the complete nucleotide sequence of the tobacco chloroplast genome, seven of which are located in the inverted repeat (Wakasugi *et al.*, 1986; Shinozaki *et al.*, 1986b), and several tRNA genes have been mapped and sequenced from chloroplast genomes of other plants (for review, see Crouse *et al.*, 1984). The chloroplast tRNA genes in higher plants are organized as independent transcription units, except for the *trn*I and *trn*A genes located in the spacer between the 16S and 23S rRNA genes in the *rrn* operon (Koch *et al.*, 1981; Takaiwa and Sugiura; 1982) and the *trn*E-*trn*Y-*trn*D genes, which are organized into a polycistronic transcription unit (Ohme *et al.*, 1985). Cotranscription of tRNA genes as part of longer transcripts has as yet not been studied, and based on the finding that some chloroplast tRNA genes do not contain 5' promoter regions (Gruissem *et al.*, 1986b), it is possible that these tRNAs are cotranscribed and processed from larger primary transcripts. Cotranscription of the spinach *psb*A-*trn*H genes occurs in *E. coli* (Thomas *et al.*, 1987). Such cotranscription may also occur in the chloroplast, although it has been found that both tobacco and spinach *trn*H genes are transcribed separately from the *psb*A gene in chloroplast extracts (M. Sugita, D. Stern, and W. Gruissem, unpublished results). In contrast to higher plants, most of the tRNA genes in *Euglena* are organized in polycistronic transcription units reminiscent of prokaryotic tRNA operons (Hallick *et al.*, 1984). Six of the tRNA genes in tobacco contain long introns, but no introns have been found in tRNA genes from *Euglena*.

2. Processing of Chloroplast tRNA Precursor Molecules

The primary transcription products of tRNA genes are generally not identical to the functional tRNA molecules. The processing of tRNA involves a collection of enzymatic reactions, for which the enzymes have been best characterized in *E. coli*. At least four different enzyme activities participate in the cleavage of multicistronic tRNA transcripts and the maturation of tRNA 5′ and 3′ ends (for review, see King *et al.*, 1986). RNase P cleaves the tRNA precursor endonucleolytically to generate a mature 5′ end, and may also be responsible for the cleavage of polycistronic tRNA transcripts (for review, see Deutscher, 1984). RNase P-like activities have been identified in eukaryotic cells and yeast mitochondria (Koski *et al.*, 1976; Hollingsworth and Martin, 1986). The enzyme requires an RNA component for its full catalytic activity both in prokaryotic and eukaryotic cells (Guerrier-Takada *et al.*, 1983; Miller and Martin, 1983). Exonucleolytic removal of sequences from the tRNA precursor establishes the mature 3′ end, but the enzyme(s) involved in this reaction are less well characterized.

Correct processing of chloroplast tRNAs was first demonstrated in a spinach chloroplast extract for the spinach tRNA^{Met} and tRNA^{Ile} transcripts, and the *Euglena* tRNA^{Leu} and polycistronic tRNA^{Val}-tRNA^{Asn}-tRNA^{Arg} transcripts (Gruissem *et al.*, 1983b; Greenberg *et al.*, 1984a). The correct processing of the *Euglena* polycistronic tRNA transcript was also confirmed in a homologous chloroplast extract (Gruissem *et al.*, 1983a), and the analysis of *Euglena* tRNA processing was later extended to 15 other *Euglena* tRNAs (Greenberg and Hallick, 1986). The *in vitro* processing reactions confirmed that an RNase P-like activity produces the mature 5′ ends of the chloroplast tRNAs. This conclusion results from studies of the processing of tobacco tRNA^{Phe} in a tobacco chloroplast lysate (Yamaguchi-Shinozaki *et al.*, 1987). An RNA component seems to be required for the catalytic function of the chloroplast RNase P-like enzyme, since the activity is undetectable after treatment of the extract with micrococcal nuclease. A 3′-exonuclease is responsible for the maturation of the tRNA 3′ termini. However, an endonuclease activity that cleaves 3′ to the tRNA has also been detected in the spinach chloroplast extract, since the tRNA^{Met} transcript includes a 3′ trailer with a stem-loop structure which is removed prior to 3′ end maturation (Gruissem *et al.*, 1983a). A similar endonucleolytic processing has been observed for the tRNA^{Phe} in a tobacco chloroplast lysate (Yamaguchi-Shinozaki *et al.*, 1987). It is unclear, however, if the same or different endonucleolytic enzyme(s) participate in the processing of mono- and polycistronic tRNA and other transcripts. From the observed processing reactions *in vitro*, it is possible to propose a pathway for the processing of chloroplast tRNA molecules (Fig. 4). Although polycistronic chloroplast tRNA transcript can also be processed in eukaryotic cell extracts, the processing pathways are considerably different (Gruissem *et al.*, 1982).

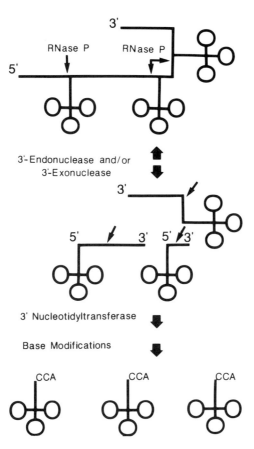

Fig. 4. Model of the chloroplast tRNA processing pathway. Shown here is a polycistronic transcript of three tRNAs which is sequentially processed into individual tRNAs. The bidirectional arrow after the first cleavage event indicates that RNase P and 3'-endo-/exonuclease cleavage may occur simultaneously, or that 3' processing occurs prior to RNase P cleavage. The model is based on results reported by Greenberg *et al.* (1984a) and Yamaguchi-Shinozaki *et al.* (1987).

3. Transfer RNA 3' End Maturation and Base Modifications

In chloroplasts, the tRNA 3'-CCA is not encoded in the tRNA genes but is required for aminoacylation (for review, see Deutscher, 1982). Consequently, the 3'-CCA must be added posttranscriptionally to the processed tRNAs by tRNA nucleotidyltransferase in the chloroplast (Hallick *et al.*, 1983; Whitfeld and Bottomley, 1983). A tRNA nucleotidyltransferase activity has been detected in a spinach chloroplast extract (Greenberg *et al.*, 1984a), and the enzyme has been partially purified from spinach and *Euglena* chloroplasts by chromatography on DE-52 cellulose (Greenberg, 1985). The spinach activity could be separated from the endogenous tRNAs and RNA polymerase activity, and is equally active with tRNAs from spinach chloroplasts, *Euglena* chloroplasts, and *E. coli*. In *Euglena,* the tRNA nucleotidyltransferase labels approximately 30 tRNAs, which is close to the expected number of chloroplast tRNA species, and thus seems to synthesize 3′-CCA for all chloroplast tRNAs. It is unknown at present if addition of the 3′-CCA occurs only after the 5′ terminus is mature.

After transcription of a tRNA sequence, specific nucleotides within the tRNA are modified by rearrangements and substitutions of additional groups (for review, see McCloskey and Nishimura, 1977). The modified nucleotides are located in specific regions of the tRNA molecule, and include the first position of the anticodon, the residue next to the 3′-side of the anticodon, the TC-loop (which contains ribothymidine and pseudouridine), and several other positions in the molecule. The modification of nucleotides is accomplished by a number of enzymes, several of which have been isolated and purified (for review, see Mazzara *et al.,* 1980). Modified bases have also been identified in chloroplast tRNAs (Pirtle *et al.*, 1981; Chang *et al.*, 1981; Francis and Dudock, 1982; Greenberg *et al.*, 1984a; Schön *et al.*, 1986), but no information is available for the enzymes catalyzing these reactions. Base modification has been detected in the spinach chloroplast transcription extract that modifies uridine nucleosides into pseudouridine by a pseudouridylate synthetase in positions of the spinach chloroplast tRNA^{Met}, which are also modified *in vivo* (Pirtle *et al.*, 1981; Greenberg *et al.*, 1984a). Although the exact sequence of processing events for chloroplast tRNAs remains to be established, base modifications is not a prerequisite for 3′-CCA addition *in vitro* (Greenberg *et al.,* 1984a).

C. Processing of Chloroplast mRNA

1. 5′ Processing of mRNA Termini

The analysis of chloroplast mRNAs has revealed that several transcripts contain multiple 5′ and 3′ ends. When the *rbc*L and *atp*BE transcription units from spinach, maize, and pea were compared, multiple 5′ ends were found for *rbc*L in all three plants (Crossland *et al.*, 1984; Mullet *et al.*, 1985). S1 nuclease and primer extension analysis of the *rbc*L transcripts showed that in spinach and pea the two transcripts differ by approximately 100 nucleotides (nt) in the length of the nontranslated leader region, and the two maize transcripts differ by approximately 240 nt at their 5′ termini. Two different *rbc*L transcripts comparable to the maize RNAs also exist in barley (Poulsen, 1984). Three mRNAs are synthesized from the *atp*BE transcription unit in spinach that differ in the length of their 5′ nontranslated leader region by approximately 100 and 80 nt (Mullet *et al.*, 1985). However, only a single 5′ end has been found in *atp*BE transcripts in tobacco (Shinozaki and

Sugiura, 1982), maize (Krebbers *et al.,* 1982; Mullet *et al.,* 1985), and pea (Mullet *et al.,* 1985). Labeling of chloroplast RNA from spinach, maize, and pea with [α-^{32}P]GTP by guanylyl transferase has confirmed that only the longest RNAs represent the primary transcript from the *rbc*L genes (Crossland *et al.,* 1984; Mullet *et al.,* 1985), and thus the shorter transcripts must result from processing events at the 5' terminus. Processing of the primary maize *rbc*L transcript into the shorter mRNA has been demonstrated in homologous chloroplast extracts (Crossland *et al.,* 1984; Hanley-Bowdoin *et al.,* 1985), thus confirming the above results. In addition, DNA fragments excluding the 5' region of the longest *rbc*L transcript do not support *in vitro* transcription of the shorter RNAs, indicating that the only promoter region is located 5' upstream of the primary transcript (Hanley-Bowdoin, 1986).

Multiple 5' ends differing by approximately 120 nt have also been detected for the *psb*B-*psb*H-*pet*B-*pet*D transcript in spinach. However, transcription of this region in the chloroplast extract produces only the longer transcript, again indicating processing of the 5' terminus (Westhoff, 1985). The significance of the 5' processing of chloroplast mRNAs is not well understood. It is possible that the heterogeneity in the 5' end of the transcripts may be related to their translational properties. In barley and maize, the longer *rbc*L transcripts accumulate after illumination (Poulsen, 1984; Crossland *et al.,* 1984), whereas in spinach there is only a small increase in the ratio of primary and processed transcript (Mullet *et al.,* 1985). Thus, it is possible that, at least in monocots, 5' processing of the *rbc*L transcripts provides a regulatory mechanism for expression of the gene during chloroplast development.

2. Processing of Polycistronic Chloroplast Transcripts

Several chloroplast polycistronic transcription units give rise to many overlapping RNAs. Some of the resulting processed RNAs are monocistronic and others are not. The removal of introns in certain chloroplast mRNAs and the stable accumulation of their unspliced precursors account for some of this RNA heterogeneity. The polycistronic transcription units in chloroplasts of higher plants for which complex overlapping transcript patterns have been observed include *psb*B-*psb*H-*pet*B-*pet*D in spinach (Heinemeyer *et al.,* 1984; Westhoff, 1985), maize (Barkan *et al.,* 1986; Rock *et al.,* 1987), and pea (Berends *et al.,* 1986), and *atp*I-*atp*H-*atp*F-*atp*A in spinach (Westhoff *et al.,* 1985), maize (Barkan *et al.,* 1986), and pea (Cozens *et al.,* 1986). Chloroplast loci organized into polycistronic transcription units, but which are transcribed into single long transcripts that may be processed at their 5' end, include *psa*A-*psa*B (Westhoff *et al.,* 1983, Berends *et al.,* 1987) and *atp*B-*atp*E in spinach (Zurawski *et al.,* 1982).

The most detailed analysis of a polycistronic chloroplast transcript has been achieved for the *psb*B-*psb*F-*pet*B-*pet*D operon in maize (Rock *et al.,*

1987) and spinach (Westhoff and Herrmann, 1988). This transcription unit is unusual since its genes encode subunits of two functionally distinct photosynthetic complexes (photosystem II and cytochrome b_6/f complex). Cytochrome b_6/f, but not photosystem II subunits, accumulate in the dark in spinach (Herrmann et al., 1985) and maize (A. Barkan, unpublished results), and the relative activities and polypeptide levels differ between the photosynthetically distinct chloroplasts of bundle sheath and mesophyll cells in maize (Ghirardi and Melis, 1983; Schuster et al., 1985). The three largest transcripts of 5.4, 4.5, and 3.8 kb from this region in maize encode all four proteins. This pattern of RNAs is highly conserved between spinach, pea, and maize, suggesting that this mode of expression is functionally significant (Rock et al., 1987). The two single introns that interrupt the N-terminus in petB and petD are removed from some of the stable RNAs, as demonstrated by S1 mapping experiments using small DNA probes that span the putative splice junctions (Rock et al., 1987). Some unspliced RNAs are quite stable, accumulating to levels comparable to those of spliced RNAs, and thus accounting in part for the RNA size heterogeneity. Differences in the relative abundance of different RNA species occur during development. For example, maize proplastids are specifically depleted of the spliced 4.5- and 3.8-kb RNAs, suggesting that splicing of these RNAs may be a regulatory event during chloroplast development. (A. Barkan, unpublished results). It is not known which of the overlapping RNAs serve as functional mRNAs *in vivo*. However, all of the *psb*B-*psb*H-*pet*B-*pet*D RNAs, including the large precursor RNAs, copurify with polysomes, suggesting that they are translationally active (Barkan, 1988).

D. Chloroplast Introns—Structure and Processing

1. *Classes of Chloroplast Introns*

Intron sequences in genes of organelle genomes have been separated into three different groups (I, II, and III) based on their conserved intron–exon junction sequences and the RNA secondary structures that bring these junction sequences into relatively close proximity (Michel and Dujon, 1983; Shinozaki et al., 1986b). Group I introns have conserved internal sequences consistent with secondary structure and can be self-excised as linear molecules (Kruger et al., 1982). Group II introns also have substantial secondary structure and conserved junction sequences (Michel and Dujon, 1983), and may be self-excised as lariat structures (Peebles et al., 1986). Group III introns have been compiled for chloroplast genes in higher plants, and contain conserved boundary sequences GTGCGNY at the 5' ends and ATCNRYY(N)YYAY (Y = C, T; R = A, G; N = A, G, C, T) at the 3' ends (Shinozaki et al., 1986a). Group II and III introns, however, appear to be closely related.

Introns are present in several tRNA genes and a few protein-coding genes in higher plant chloroplasts (Shinozaki et al., 1986b; Ohyama et al., 1986), but no introns have been found in chloroplast tRNA genes in *Euglena* (Hallick et al., 1984). *Euglena* chloroplast protein-coding genes, however, are often interrupted by multiple introns (Koller and Delius, 1984; Hallick et al., 1985). Intron sequences in chloroplast ribosomal RNA genes have only been reported for the 23S rRNA gene in *Chlamydomonas* (Allet and Rochaix, 1979). With the sequence information available from tobacco (Shinozaki et al., 1986b) and *Marchantia* (Ohyama et al., 1986), single introns have been found in *atp*F, *pet*B, *pet*D, *ndh*A, *ndh*B, *rpl*2, *rpl*16, and *rps*16. The single introns present in *trn*K(UUU), *trn*G(UCC), *trn*L(UAA), *trn*V(UAC), *trn*I(GAU), and *trn*(UGC) are all very long (up to 2526 bp). The intron of *trn*L in chloroplasts of higher plants can be folded into a secondary structure and contains the conserved T at the 3' end of the 5' exon and G at the 3' end of the intron, which is common to all group I introns (Steinmetz et al., 1982; Bonnard et al., 1984). Similarly, the introns in *psb*A and the 23S rRNA gene in *Chlamydomonas* can be included into this group (Erickson et al., 1984; Rochaix et al., 1985). The introns in *trn*A and *trn*I, which are cotranscribed with the ribosomal RNA genes (Koch et al., 1981; Takaiwa and Sugiura, 1982), can be folded into a secondary structure typical of group II introns. On the basis of the boundary sequences, all other chloroplast introns have been included into group III (Shinozaki et al., 1986a). The sequence of the intron/exon junctions are similar to those compiled for the *Euglena* intron sequences (Hallick et al., 1985), and resemble most closely the introns found in nuclear protein-coding genes (Sharp, 1985; Cech, 1986). It is interesting to note that the group III introns in chloroplasts from higher plants also include several tRNA genes.

2. Intron-Encoded Open Reading Frames

Several mitochondrial and chloroplast introns contain open reading frames (ORF). The possible functions of a few of the putative proteins have been identified in mitochondria (e.g., Lazowska et al., 1980; Bonitz et al., 1980; Burke and RajBhandary, 1982; Osiewacz and Esser, 1984; Lang et al., 1985; Michel and Lang, 1985), but their function (if any) in chloroplasts is unknown. It was first recognized that the 23S rRNA intron in *Chlamydomonas* is structurally related to mitochondrial reading frames that potentially code for maturases (Rochaix et al., 1985). The higher plant chloroplast *trn*K gene is interrupted by an approximately 2.5-kb intron in tobacco (Sugita et al., 1985), liverwort (Ohyama et al., 1986), mustard (Neuhaus and Link, 1987), and spinach (X. W. Deng and W. Gruissem, unpublished results), and contains an ORF of 509, 524, and 370 amino acids in tobacco, mustard, and liverwort, respectively. Sequence comparison of the tobacco and mustard ORF shows striking similarity, with up to 90% sequence conservation for certain regions within the ORF and an overall sequence similarity of 66%. It has been noted that the derived amino acid sequence of the mustard *trn*K ORF has stretches of residues near the C-terminal end which are structurally related to maturases (Neuhaus and Link, 1987). The 1.6-kb intron of the *Euglena psb*C gene encodes an ORF of 458 codons potentially coding for a basic protein of 54 kDa (Montandon *et al.*, 1986), with no apparent sequence homology to mitochondrial intron proteins.

3. Mechanisms of Cis-*Splicing*

Autocatalytic and enzyme-mediated splicing mechanisms have been investigated for several years (Sharp, 1985; Cech, 1986), and *in vitro* systems have been developed to study individual steps in the splicing of intron sequences in nuclear RNAs (e.g., Aebi *et al.*, 1986; Reed and Maniatis, 1986; Konarska and Sharp, 1986). Similar systems are currently not available for higher plant and algal chloroplast introns, although several attempts have been made to establish *in vitro* splicing reactions in chloroplast extracts. All chloroplast intron sequences are correctly and efficiently removed, since precursor RNA molecules from most intron-containing genes do not accumulate. Thus it is unclear why splicing reactions are not supported *in vitro*. A systematic study of splicing conditions for the different chloroplast introns is required to distinguish self-splicing from splicing reactions requiring specific enzymes.

E. *Trans*-Splicing of Chloroplast mRNAs

Trans-splicing mechanisms were first proposed for the synthesis of mRNAs in the parasite *Trypanosoma brucei* that all have a common 35-nucleotide sequence at their 5' termini (for review, see Borst, 1986). The generality of this mechanism was demonstrated when actin mRNAs in *Caenorhabditis elegans* (Kraus and Hirsch, 1987) and late mRNAs of the vaccinia virus (Bertholet *et al.*, 1987; Schwer *et al.*, 1987) were found to have similar features. In all the above cases, the apparent *trans*-splicing attaches a 5' noncoding leader region to the coding region of the mRNA (for review, see Sharp, 1987). The suggestion that coding exons for a single protein may be *trans*-spliced was first made after the sequence for the chloroplast ribosomal protein *rps*12 was obtained in liverwort (Fukuzawa *et al.*, 1986) and tobacco (Torazawa *et al.*, 1986). The gene is interrupted twice, and exons 2 and 3 are located in the inverted repeat of the chloroplast DNA. The tobacco 5' *rps*12 exon is located in the large single copy region, and is separated by 28 and 86 kb from the other two exons (Shinozaki *et al.*, 1986b). Electron microscopy and primer extension analysis have unequivocally demonstrated the integrity of the mature mRNA (Koller *et al.*, 1987; Zaita *et al.*, 1987). Although the *trans*-splicing mechanism is unknown in chloroplasts, it has already been demonstrated that mRNA, which is usually spliced in *cis*, will splice in *trans* in nuclear extracts if the RNAs are initially bound to each other by extensive secondary structure (Konarska et al., 1985). Such secondary structure could potentially be formed during the *trans*-splicing event of the tobacco *rps*12, since the 5' exon is part of a longer transcript that includes *rpl*20 and ORF130, while the 3' exons are co-transcribed with *rps*7. In these two polycistronic transcripts, homologous RNA regions, termed transons 1 and 2, could pair to form a secondary structure that may aid the *trans*-splicing of *rps*12 exons 1 and 2. A *trans*-splicing mechanism has also been proposed for the synthesis of the *psa*A mRNA in *Chlamydomonas*, which is encoded in three exons separated by approximately 50 and 90 kb (Kück et al., 1987). *psa*A and *psa*B RNA is synthesized as a polycistronic transcript in higher plants, and *psa*B is present as an uninterrupted gene in *Chlamydomonas*.

IV. TRANSCRIPTIONAL AND POSTTRANSCRIPTIONAL REGULATION OF PLASTID GENE EXPRESSION

The maintenance of the genetic apparatus in plastids raises several important questions about the regulation of gene expression in this organelle. Many of the proteins required in photosynthetically competent chloroplasts are encoded by the nuclear genome, and thus it is reasonable to speculate that the transcriptional regulation of the nuclear genes and the availability and import of their gene products are important regulatory steps in chloroplast development and plastid differentiation (see Ellis, 1984). In addition, dramatic changes in the accumulation of plastid RNAs and proteins during chloroplast development and differentiation indicate that regulatory events within the organelle also play a critical role in the control of plastid gene expression. Although this regulation can occur at several levels, we consider here only processes that control the differential accumulation of chloroplast RNA at the transcriptional and posttranscriptional level.

A. Transcriptional Regulation

1. Plastid DNA Levels and RNA Polymerase Activity

Plastids in higher plants and algae contain multiple copies of their genome, and significant changes in the genome copy number are observed in various plastid types and during chloroplast development (Lamppa and Bendich, 1979; Boffey and Leech, 1982; Scott and Possingham, 1983; Tymms et al., 1983; Scott et al., 1984; Lawrence and Possingham, 1985; Cannon et al., 1986). Explanations for the multiplicity of the chloroplast genomes are speculative, but it has been proposed that the excess of plastid genomes over nuclear genomes is due to an increased need for plastid ribosomes in chloroplasts which is satisfied by the increased rRNA gene number that results from plastid genome multiplication (Bendich, 1987). Although this hypothesis provides a convenient model for the increased requirement of ribosomal RNA synthesis, no direct correlation of overall transcriptional activity with changes in plastid DNA levels was found in tests of plastids at different stages of chloroplast development in spinach or chromoplast differentiation in tomato (Deng and Gruissem, 1987; Gruissem et al., 1987b). Most strikingly, while the chloroplast to nuclear DNA ratio increases approximately 30% during spinach leaf maturation, the transcriptional activity in the chloroplast decreases approximately 5-fold on a per DNA basis (Deng and Gruissem, 1987). Similarly, the decrease in overall transcriptional activity during chromoplast differentiation in tomato fruit is not associated with significant changes in the level of plastid DNA (Gruissem et al., 1987b). These results suggest, therefore, that the differences in overall transcriptional activity and RNA accumulation or decrease are not simple consequences of changes in plastid DNA levels.

It is possible, however, that template availability is limiting for the transcriptional activity in plastids. This model would predict that the different strengths of plastid promoter regions are the primary determinants for the relative transcriptional activities of individual genes at all developmental stages. Consequently, the lack of significant promoter competition (i.e., stronger promoters competing more effectively for limiting RNA polymerases) should result in only small changes in the relative transcription rates of plastid genes. This model is supported by the result that the different strengths of spinach chloroplast promoters *in vitro* (Gruissem and Zurawski, 1985b) are reflected by the relative transcription rates of their genes in run-on experiments (Deng and Gruissem, 1987; Deng et al., 1987). Additional studies are required to determine the fraction of the plastid genomes that is engaged in transcription and the molecular conformation of the supercoiled plastid DNA at different developmental stages. Alternatively, changes in the transcriptional activity during chloroplast development and plastid differentiation may reflect changes in the level of RNA polymerase. Thus, in the presence of a limiting level of RNA polymerase, stronger promoter regions (e.g., *psb*A), in the absence of regulatory, *trans*-acting factors, would compete more effectively for RNA polymerase. Such a model is currently not supported by the experimental data. However, information about RNA polymerase levels in plastids is as yet not available.

2. Regulation of rRNA Transcription

Conserved DNA sequences have been found in the 5' region of the chloroplast 16S rRNA gene which can be folded into three stem-loop structures, termed H1, H2, and H3 (Briat et al., 1983). Two of these structures, H2 and H3, are mutually exclusive. H1 and H2, but not H3, can form sequentially when the RNA polymerase initiates transcription. In addition, the resulting transcript could potentially be translated into a short polypeptide, suggesting the possibility that a prokaryote-type attenuation mechanism may be operating for rRNA transcription in chloroplasts cells (Platt, 1981). Based on the putative secondary structure, it has been proposed that the formation of H1 and H2 in the absence of plastid ribosomes results in transcription of the *rrn* operon. In the presence of ribosomes transcription is prevented, since translation of the peptide leader would occur, resulting in formation of H3 and premature termination of transcription (Briat *et al.*, 1986). Alternatively, transcription of the chloroplast *rrn* operon could be regulated by a mechanism reminiscent of the antitermination mechanism proposed for the *E. coli rrn*G and *rrn*C operons (Li *et al.*, 1984; Holben and Morgan, 1984; Briat *et al.*, 1986). Applicability of these mechanisms to the regulation of rRNA transcription in plastid awaits further experimental tests.

B. Regulation of Plastid RNA Levels during Plant Development

1. Differential RNA Accumulation during Chloroplast Development

Light-induced plastid RNA accumulation was first demonstrated for the *psb*A mRNA in developing maize leaves (Bedbrook *et al.*, 1978) and *Spirodela* (Reisfeld *et al.*, 1978), and a similar light-dependent accumulation of *psb*A mRNA has been observed during chloroplast development in mustard cotyledons (Link, 1982). Different results were obtained for *rbc*L, however, in that the mRNA appeared to be induced after illumination of pea seedlings (Smith and Ellis, 1981), but was present at high levels in dark-grown mustard cotyledons (Link, 1981). Subsequent hybridization studies using cloned probes for plastid DNA regions has allowed plastid RNAs to be categorized into three different classes: RNAs that accumulate after illumination and during chloroplast development (transiently or continuously), RNAs that remain approximately constant in the dark and in the light, and RNAs that decrease after illumination (e.g., Sasaki *et al.*, 1983; Link, 1984b; Nelson *et al.*, 1984; Palmer *et al.*, 1984; Rodermel and Bogorad, 1985; Berry *et al.*, 1985; Zhu *et al.*, 1985; Herrmann *et al.*, 1985; Kreuz *et al.*, 1986; Klein and Mullet, 1987). A developmental control appears to be superimposed on the light-regulated differential accumulation of *rbc*L and *psa*A/B mRNAs during chloroplast development in barley, since after an initial rise in the light, these mRNAs decline to lower levels if plastid mRNA levels of equal plastid volumes are correlated (Klein and Mullet, 1987). If, however, total RNA is used to normalize the RNA levels, no decrease in *rbc*L mRNA levels was observed in maize seedlings after 8 days in continuous light (Nelson *et al.*, 1984). Although differential RNA accumulation during chloroplast development has now been reported for several plant species, no definite pattern for specific chloroplast genes has emerged from these studies, reflecting in part the adaptation of particular regulatory mechanisms for individual RNAs in different plants. Some differences may also result from the different experimental conditions in these studies.

The accumulation of plastid RNAs during chloroplast development may be a consequence of control at the transcriptional level (Rodermel and Bogorad, 1985; Zhu et al., 1985). Recent studies, however, show posttranscriptional mechanisms as a principal mode of gene regulation during plastid development in spinach cotyledons and barley leaves, at which time major adjustments occur in mRNA levels for specific loci (Deng and Gruissem, 1987; Deng et al., 1987; Mullet and Klein, 1987). Although adjustments of the plastid transcriptional activity may also affect specific genes in barley, such changes do not alter significantly the relative transcription rate of nine spinach chloroplast genes. The adjustments at the transcriptional level in most cases, however, are not sufficient to explain the differential accumulation of specific RNAs, suggesting that changes in the stability of individual RNA species occur during chloroplast development.

2. Changes in RNA Levels during Plastid Differentiation

Plastids in higher plants differentiate into nonphotosynthetic chromoplasts in flower petals and fruits and amyloplasts in root tissue (for review, see Thomson and Whatley, 1980). For example, the differentiation of chromoplasts from photosynthetically active chloroplasts in tomato fruits is accompanied by major morphogenetic changes (Harris and Spurr, 1969a,b). Although no changes are detected in the structure or copy number of chloroplast DNA during chromoplast differentiation (Piechulla et al., 1986), significant adjustments occur in the levels of individual plastid RNAs (Piechulla et al., 1985, 1986, 1987). Run-on transcription experiments with plastids from different fruit developmental changes have shown that, although an overall decrease in transcription activity can be detected, the relative transcription rate of individual plastid genes is approximately maintained (Gruissem et al., 1987b). Thus, the differential decline of specific RNAs most likely reflect changes in RNA stability during chromoplast differentiation. In contrast, transcriptional control of plastid gene expression was proposed in amyloplasts of sycamore suspension culture cells (Macherel et al., 1986). Most of the plastid RNAs except rRNA are undetectable in these plastid types. However, analysis of plastid RNA levels from dark-grown spinach roots indicate that most (if not all) plastid RNAs are also present in amyloplasts. Transcription run-on experiments with isolated amyloplasts demonstrate that all plastid genes are actively transcribed, although at a low rate, and the relative transcription rate of individual genes in amyloplasts is similar to the relative rate of these genes in chloroplasts (Deng and Gruissem, 1988).

Adjustment of rbcL mRNA levels occurs during the light-induced differentiation of chloroplasts in mesophyll and bundle-sheath cells in maize leaves (Link *et al.*, 1978; Sheen and Bogorad, 1985). In one study, it was concluded that the suppression of *rbc*L expression in mature chloroplasts of mesophyll cells is controlled at the transcriptional level, either by synthesis of a light-induced repressor or by light-mediated elimination of a transcription stimulator (Sheen and Bogorad, 1985). In addition to chloroplast development and plastid differentiation, adjustment of plastid mRNA levels also operates during adaptation of chloroplasts to different environments. For example, the different reaction center ratios in the thylakoid membrane of plants growing in yellow or red light is accompanied by a differential accumulation of mRNAs for chloroplast reaction center proteins (Glick *et al.*, 1986). Run-on transcription experiments with chloroplasts from spinach plants growing in red or yellow light again demonstrate that the relative transcription rate of most plastid genes is maintained, suggesting that the synthesis of chloroplast membrane complexes and the assembly of photosystems are regulated by light quality at a posttranscriptional level (X. W. Deng, J. Tonkyn, and W. Gruissem, manuscript submitted).

C. Conclusions

A scheme for the control of chloroplast gene expression emerges from the available data that emphasizes the importance of posttranscriptional and translational processes in the formation of stable mRNA and protein products during chloroplast development (Fig. 5). Evidence is emerging that posttranscriptional and translational regulatory mechanisms also operate in nonphotosynthetic plastids of plant tissues other than leaves. It appears that transcriptional control is based on two principal mechanisms: The strength of promoter regions for respective plastid genes, and the modulation of overall transcriptional activity at different developmental stages. Overall, it seems clear that the role of transcription in controlling chloroplast gene expression is limited, and may be operational exclusively in the control of a few plastid genes. Transcription termination in the chloroplast is at best inefficient and does not reflect a significant regulatory step.

The discordant changes in transcriptional activity and differential mRNA accumulation or decline in developing and differentiating chloroplasts provide strong evidence that gene expression in plastids of higher plants is controlled at the posttranscriptional level. This step involves changes in mRNA stability which may be achieved by nucleolytic processing events and/or RNA–protein interactions that specifically affect the stability of 3' ends of individual mRNAs. In addition, the differential stabilities of 5' ends of chloroplast mRNAs may also contribute to this posttranscriptional control step. Superimposed on these regulatory events are changes in the translatability of mRNAs at different developmental stages, which may serve as important short-term control steps in the assembly and maintenance of functional photosynthetic complexes. Obviously, the relative importance of the

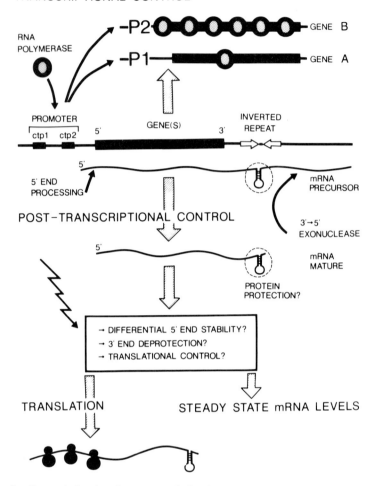

Fig. 5. Transcriptional and posttranscriptional control of chloroplast gene expression: a model. At the transcriptional level, P1 and P2 represent promoter regions of different strengths that, together with RNA polymerase and possible specific DNA-binding proteins, establish the relative transcriptional activities of genes A and B. RNA polymerase proceeds through 3' inverted repeat sequences to produce mono- and polycistronic transcripts, which undergo 3' processing to establish the mature 3' termini. The differential stability and accumulation of chloroplast transcripts may be mediated by the binding of specific proteins to the 3' inverted repeat sequences. Transcripts can be further modified by 5' processing. Polycistronic transcripts are processed into complex sets of smaller RNAs. Light may regulate the synthesis of proteins that specifically recognize mRNA 3' termini, resulting in different levels of individual RNAs during chloroplast development. Superimposed on the posttranscriptional regulation of mRNA accumulation is the translational control of individual chloroplast proteins.

different control steps must be defined more precisely for individual genes or classes of mRNAs and proteins. Yet, it seems clear that, in contrast to the importance of transcriptional control of bacterial and nuclear genes, posttranscriptional and translational control steps appear to be more significant in regulating chloroplast gene expression in higher plants.

ACKNOWLEDGMENTS

The author thanks Drs. Alice Barkan and Helen Jones for their critical comments and suggestions. Unpublished research from the author's laboratory was supported by grants from the National Institute of Health and the Department of Energy.

REFERENCES

Aebi, M., Hornig, H., Padgett, R. A., Reiser, J., and Weissmann, C. (1986). *Cell* **47**, 555–565.
Allet, B., and Rochaix, J. D. (1979). *Cell* **18**, 55–60.
Altman, A., Cohen, B., Weissbach, H., and Brot, N. (1984). *Arch. Biochem. Biophys.* **235**, 26–33.
Apel, K., and Bogorad, L. (1976). *Eur. J. Biochem.* **67**, 615–620.
Audren, H., Bisanz-Seyer, C., Briat, J. F., and Mache, R. (1987). *Curr. Genet.* **12**, 263–270.
Barkan, A. (1988). *EMBO J.* **7**, 2637–2644.
Barkan, A., Miles, D., and Taylor, W. C. (1986). *EMBO J.* **5**, 1421–1427.
Bedbrook, J. R., Link, G., Coen, D. M., Bogorad, L., and Rich, A. (1978). *Proc. Natl. Acad. Sci. U.S.A.* **75**, 3060–3064.
Belasco, J. G., Beatty, J. T., Adams, C., von Gabain, A., and Cohen, S. N. (1985). *Cell* **40**, 171–181.
Bendich, A. J. (1987). *BioEssays* **6**, 279–282.
Berends, T., Kubicek, Q., and Mullet, J. E. (1986). *Plant Mol. Biol.* **6**, 125–134.
Berends, T., Gamble, P. E., and Mullet, J. E. (1987). *Nucleic Acids Res.* **15**, 5217–5240.
Berry, J. O., Nikolau, B. J., Carr, J. P., and Klessig, D. F. (1985). *Mol. Cell. Biol.* **5**, 2238–2246.
Bertholet, C., Van Meier, E., ten Heggeler-Bordier, B., and Wittek, R. (1987). *Cell* **50**, 153–162.
Birchmeier, C., Schümperli, D., Sconzo, G., and Birnstiel, M. L. (1984). *Proc. Natl. Acad. Sci. U.S.A.* **81**, 1057–1061.
Boffey, S. A., and Leech, R. M. (1982). *Plant Physiol.* **69**, 1387–1391.
Bogorad, L. (1975). *Science* **188**, 891–898.
Bohnert, H. J., Driesel, A. J., and Herrmann, R. J. (1977). In "Nucleic Acids and Protein Synthesis in Plants" (J. H. Weil and L. Bogorad, eds.), pp. 213–218. Centre National de la Recherche Scientifique, Paris.
Bohnert, H. J., Crouse, E. J., and Schmitt, J. M. (1982). *Encycl. Plant Physiol., New Ser.* **14B**, 475–530.
Bonitz, S. G., Coruzzi, G., Thalenfeld, B. E., and Tzagoloff, A. (1980). *J. Biol. Chem.* **255**, 11927–11941.
Bonnard, G., Michel, F., Weil, J. H., and Steinmetz, A. A. (1984). *Mol. Gen. Genet.* **194**, 330–336.
Borst, P. (1986). *Annu. Rev. Biochem.* **55**, 701–732.
Bottomley, W., and Whitfeld, P. R. (1979). *Eur. J. Biochem.* **93**, 31–39.

Bottomley, W., Smith, H. J., and Bogorad, L. (1971). *Proc. Natl. Acad. Sci. U.S.A.* **68**, 2412–2416.
Bradley, D., and Gatenby, A. (1985). *EMBO J.* **4**, 3641–3648.
Bram, R. J., Young, A., and Steitz, J. A. (1980). *Cell* **19**, 393–401.
Briat, J. F., and Mache, R. (1980). *Eur. J. Biochem.* **111**, 503–509.
Briat, J. F., Laulhere, J. P., and Mache, R. (1979). *Eur. J. Biochem.* **98**, 285–292.
Briat, J. F., Droix, M., Loiseaux, S., and Mache, R. (1982). *Nucleic Acids Res.* **10**, 6865–6878.
Briat, J. F., Dron, M., and Mache, R. (1983). *FEBS Lett.* **163**, 1–5.
Briat, J. F., Lescure, A. M., and Mache, R. (1986). *Biochimie* **68**, 981–900.
Briat, J. F., Bisanz-Seyer, C., and Lescure, A. M. (1987). *Curr. Genet.* **11**, 259–263.
Bülow, S., Reiss, T., and Link, G. (1987). *Curr. Genet.* **12**, 157–159.
Burke, J. M., and RajBhandary, U. L. (1982). *Cell* **31**, 509–520.
Cannon, G., Heinhorst, S., and Weissbach, A. (1986). *Plant Physiol.* **80**, 601–603.
Cech, T. (1986). *Cell* **44**, 207–210.
Chang, S. H., Hecker, L. I., Brum, C. K., Schnabel, J. J., Heckman, J. E., Silberklang, M., RajBhandary, U. L., and Barnett, W. E. (1981). *Nucleic Acids Res.* **9**, 3199–3204.
Cozens, A. L., Walker, J. E., Phillips, A. L., Huttly, A. K., and Gray, J. C. (1986). *EMBO J.* **5**, 217–222.
Crossland, L. D., Rodermel, S. R., and Bogorad, L. (1984). *Proc. Natl. Acad. Sci. U.S.A.* **81**, 4060–4064.
Crouse, E. J., Bohnert, H. J., and Schmitt, J. M. (1984). *In* "Chloroplast Biogenesis" (R. J. Ellis, ed.), pp. 83–136. Cambridge Univ. Press, Cambridge, England.
Darnell, J. E., Jr. (1982). *Nature (London)* **297**, 365–371.
Deng, X. W., and Gruissem, W. (1987). *Cell* **49**, 379–387.
Deng, X. W., and Gruissem, W. (1988). *EMBO J.* **7**, 3301–3308.
Deng, X. W., Stern, D. B., Tonkyn, J. C., and Gruissem, W. (1987). *J. Biol. Chem.* **262**, 9641–9648.
Deno, H., Shinozaki, K., and Sugiura, M. (1984). *Gene* **32**, 195–201.
Deutscher, M. P. (1982). *Enzymes* **15**, 183–215.
Deutscher, M. P. (1984). *Cell* **40**, 731–732.
Dron, M., Robreau, S., and Legal, Y. (1979). *Exp. Cell Res.* **119**, 301–305.
Dyer, T. A., and Bedbrook, J. R. (1980). *In* "Genome Organization and Expression in Plants" (C. J. Leaver, ed.), pp. 305–311. Plenum, New York.
Ellis, R. J. (1981). *Annu. Rev. Plant Physiol.* **32**, 111–137.
Ellis, R. J. (1984). "Chloroplast Biogenesis." Cambridge Univ. Press, Cambridge, England.
Erickson, J. M., Rahire, M., and Rochaix, J. D. (1984). *EMBO J.* **3**, 2753–2762.
Erion, J. L., Tarnowski, J., Peacock, S., Caldwell, P., Redfield, B., Brot, N., and Weissbach, H. (1983). *Plant Mol. Biol.* **2**, 279–290.
Farnham, P. J., and Platt, PT. (1982). *Proc. Natl. Acad. Sci. U.S.A.* **79**, 998–1002.
Francis, M. A., and Dudock, B. S. (1982). *J. Biol. Chem.* **257**, 11195–11198.
Fukuzawa, H., Kohchi, T., Shirai, H., Ohyama, K., Umesono, K., Inokuchi, H., and Ozeki, H. (1986). *FEBS Lett.* **198**, 11–15.
Galli, G., Hofstetter, H., and Birnstiel, M. L. (1981). *Nature (London)* **294**, 626–631.
Gatenby, A. A., Castleton, J. A., and Saul, M. W. (1981). *Nature (London)* **291**, 117–122.
Ghirardi, M. L., and Melis, A. (1983). *Arch. Biochem. Biophys.* **224**, 19–28.
Glick, R. E., McCauley, S. W., Gruissem, W., and Melis, A. (1986). *Proc. Natl. Acad. Sci. U.S.A.* **83**, 4287–4291.
Graf, L., Kössel, H., and Stutz, E. (1980). *Nature (London)* **286**, 908–910.
Gray, P. W., and Hallick, R. B. (1978). *Biochemistry* **17**, 284–289.
Greenberg, B. M. (1985). Ph.D. thesis. University of Colorado, Boulder.
Greenberg, B. M., and Hallick, R. B. (1986). *Plant Mol. Biol.* **6**, 89–100.
Greenberg, B. M., Gruissem, W., and Hallick, R. B. (1984a). *Plant Mol. Biol.* **3**, 97–109.
Greenberg, B. M., Narita, J. O., DeLuca-Flaherty, C., Gruissem, W., Rushlow, K. A., and Hallick, R. B. (1984b). *J. Biol. Chem.* **259**, 14880–14887.

Gruissem, W. (1984). *Plant Mol Biol. Rep.* **2,** 15–23.
Gruissem, W., and Zurawski, G. (1985a). *EMBO J.* **4,** 1637–1644.
Gruissem, W., and Zurawski, G. (1985b). *EMBO J.* **4,** 3375–3383.
Gruissem, W., Prescott, D. M., Greenberg, B. M., and Hallick, R. B. (1982). *Cell* **30,** 81–92.
Gruissem, W., Greenberg, B. M., Zurawski, G., Prescott, D. M., and Hallick, R. B. (1983a). *Cell* **35,** 815–828.
Gruissem, W., Narita, J. O., Greenberg, B. M., Prescott, D. M., and Hallick, R. B. (1983b). *J. Cell. Biochem.* **22,** 31–46.
Gruissem, W., Greenberg, B. M., Hallick, R. B., and Zurawski, G. (1986a). *Methods Enzymol.* **118,** 253–270.
Gruissem, W., Elsner-Menzel, C., Latshaw, S., Narita, J. O., Schaffer, M. A., and Zurawski, G. (1986b). *Nucleic Acids Res.* **14,** 7541–7556.
Gruissem, W., Deng, X. W., Jones, H., Stern, D. B., and Tonkyn, J. (1987a). *In* "Plant Molecular Biology" (D. von Wettstein and N. H. Chua, eds.). Plenum, New York, pp. 135–148.
Gruissem, W., Callan, K., Lynch, J., Manzara, T., Meighan, M., Narita, J. O., Piechulla, B., Sugita, M., Thelander, M., and Wanner L. (1987b). *In* "Tomato Biotechnology" (D. J. Nevins and R. A. Jones, eds.), pp. 239–249. Alan R. Liss, Inc., New York.
Guerrier-Takada, C., Haydock, K., Allen, L., and Altman, S. (1983). *Cell* **35,** 849–857.
Hallick, R. B., Lipper, C., Richards, O. D., and Rutter, W. J. (1976). *Biochemistry* **15,** 3039–3045.
Hallick, R. B., Greenberg, B. M., Gruissem, W., Hollingsworth, M. J., Karabin, G. D., Narita, J. O., Nickoloff, J. A., Passavant, C. W., and Stiegler, G. L. (1983). *In* "Structure and Function of Plant Genomes" (O. Ciferri and L. Dure III, eds.), pp. 155–166. Plenum, New York.
Hallick, R. B., Hollingsworth, M. J., and Nickoloff, J. A. (1984). *Plant Mol. Biol.* **3,** 169–175.
Hallick, R. B., Gingrich, J. C., Johanningmeyer, U., and Passavant, C. W. (1985). *In* "Molecular Form and Function of the Plant Genome" (L. van Vloten-Doting, G. S. P. Groot, and T. C. Hall, eds.), pp. 211–220. Plenum, New York.
Hanley-Bowdoin, L. (1986). Ph.D. dissertation. Rockefeller University, New York.
Hanley-Bowdoin, L., and Chua, N. H. (1987). *Trends Biochem. Sci.* **12,** 67–70.
Hanley-Bowdoin, L., Orozco, E. M., Jr., and Chua, N. H. (1985). *Mol. Cell. Biol.* **5,** 2733–2745.
Harris, W. M., and Spurr, A. R. (1969a). *Am. J. Bot.* **56,** 369–379.
Harris, W. M., and Spurr, A. R. (1969b). *Am. J. Bot.* **56,** 380–389.
Harris, M. R., and Ellis, R. J. (1973). *Biochem. J.* **134,** 249–262.
Harris, M. R., and Ellis, R. J. (1979). *Eur. J. Biochem.* **96,** 301–309.
Hartley, M. R. (1979). *Eur. J. Biochem.* **96,** 311–320.
Hartley, M. R., and Head, C. (1979). *Eur. J. Biochem.* **96,** 301–309.
Hartley, M. R., Head, C. W., and Gardiner, J. (1977). *In* "Nucleic Acids and Protein Synthesis in Plants" (J. H. Weil and L. Bogorad, eds.), pp. 419–423. Centre National de la Recherche Scientifique, Paris.
Hawley, D. K., and McClure, D. W. (1983). *Nucleic Acids Res.* **11,** 2237–2255.
Heinemeyer, W., Alt, J., and Herrmann, R. G. (1984). *Curr. Genet.* **8,** 543–549.
Helling, R. B., El-Gewely, M. R., Lomax, M. I., Baumgartner, J. E., Schwartzbach, S. D., and Barnett, W. E. (1979). *Mol. Gen. Genet.* **174,** 1–10.
Herrmann, R. G., Westhoff, P., Alt, J., Tittgen, J., and Nelson, N. (1985). *In* "Molecular Form and Function of the Plant Genome" (L. van Vloten-Doting, G. S. P. Groot, and T. C. Hall, eds.), pp. 233–256. Plenum, New York.
Holben, W. E., and Morgan, E. A. (1984). *Proc. Natl. Acad. Sci. U.S.A.* **81,** 6789–6793.
Hollingsworth, M. J., and Martin, N. C. (1986). *Mol. Cell. Biol.* **6,** 193–199.
Holschuh, K., Bottomley, W., and Whitfeld, P. R. (1984). *Nucleic Acids Res.* **12,** 8819–8834.
Howe, C. J., Fearnley, I. M., Walker, J. E., Dyer, T. A., and Gray, J. C. (1985). *Plant Mol. Biol.* **4,** 333–345.

Jenni, B., and Stutz, E. (1978). *Eur. J. Biochem.* **88**, 127–134.
Job, C., Briat, J. F., Lescure, A. M., and Job, D. (1987). *Eur. J. Biochem.* **165**, 515–519.
Jolly, S. O., and Bogorad, L. (1980). *Proc. Natl. Acad. Sci. U.S.A.* **77**, 822–826.
Joussaume, M. (1973). *Physiol. Veg.* **11**, 69–82.
Keus, R. J. A., Dekker, A. F., van Roon, M. A., and Groot, G. S. P. (1983). *Nucleic Acids Res.* **11**, 6465–6474.
Kidd, G. H., and Bogorad, L. (1979). *Proc. Natl. Acad. Sci. U.S.A.* **76**, 4890–4892.
King, T. C., Sirdeskmukh, R., and Schlessinger, D. (1986). *Microbiol. Rev.* **50**, 428–451.
Kirsch, W., Seyer, P., and Herrmann, R. G. (1986). *Curr. Genet.* **10**, 843–855.
Klein, R. R., and Mullet, J. E. (1987). *J. Biol. Chem.* **262**, 4341–4348.
Koch, W., Edwards, K., and Kössel, H. (1981). *Cell* **25**, 203–213.
Koller, B., and Delius, H. (1982a). *Mol. Gen. Genet.* **188**, 305–308.
Koller, B., and Delius, H. (1984). *Cell* **36**, 613–622.
Koller, B., Delius, H., and Dyer, T. (1982). *Eur. J. Biochem.* **122**, 17–23.
Koller, B., Fromm, H., Galun, E., and Edelman, M. (1987). *Cell* **48**, 111–119.
Konarska, M. M., and Sharp, P. A. (1986). *Cell* **46**, 845–855.
Konarska, M. M., Padgett, R. A., and Sharp, P. A. (1985). *Cell* **42**, 165–171.
Koski, R. A., Bothwell, A. L. M., and Altman, S. (1976). *Cell* **9**, 101–116.
Kössel, H., Natt, E., Strittmatter, G., Fritzsche, E., Gozdzicka-Jozefiak, A., and Przybyl, D. (1985). *In* "Molecular Form and Function of the Plant Genome" (L. van Vloten-Doting, G. S. P. Groot, and T. C. Hall, eds.), pp. 183–198. Plenum, New York.
Kraus, M., and Hirsch, D. (1987). *Cell* **49**, 753–761.
Krebbers, E. T., Larrinua, I. M., McIntosh, L., and Bogorad, L. (1982). *Nucleic Acids Res.* **10**, 4985–5002.
Kreuz, K., Dehesh, K., and Apel, K. (1986). *Eur. J. Biochem.* **159**, 459–467.
Kruger, K., Grabowski, P. J., Zaug, A. J., Sands, J., Gottschling, D. E., and Cech, T. R. (1982). *Cell* **31**, 147–157.
Kück, U., Choquet, Y., Schneider, M., Dron, M., and Bennoun, P. (1987). *EMBO J.* **6**, 2185–2195.
Kung, S. D., and Lin, C. M. (1985). *Nucleic Acids Res.* **13**, 7543–7549.
Lam, E., and Chua, N. H. (1987). *Plant Mol. Biol.* **8**, 415–424.
Lamppa, G. K., and Bendich, A. J. (1979). *Plant Physiol.* **64**, 126–130.
Lang, B. F., Ahne, F., and Bonen, L. (1985). *J. Mol. Biol.* **184**, 353–366.
Lawrence, M. E., and Possingham, J. V. (1985). *Plant Physiol.* **81**, 708–710.
Lazowska, J., Jacq, C., and Slonimski, P. C. (1980). *Cell* **22**, 333–348.
Lerbs, S., Briat, J. F., and Mache, R. (1983). *Plant Mol. Biol.* **2**, 67–74.
Lerbs, S., Brautigam, E., and Parthier, B. (1985). *EMBO J.* **4**, 1661–1666.
Lerbs, S., Brautigam, E., and Mache, R. (1987). *Mol. Gen. Genet.* **211**, 459–464.
Lescure, A. M., Bisanz-Seyer, C., Pesey, H., and Mache, R. (1985). *Nucleic Acids Res.* **13**, 8787–8796.
Li, S. C., Squires, C. L., and Squires, C. (1984). *Cell* **38**, 851–860.
Link, G. (1981). *Planta* **152**, 379–380.
Link, G. (1982). *Planta* **154**, 81–86.
Link, G. (1984a). *EMBO J.* **3**, 1697–1704.
Link, G. (1984b). *Plant Mol. Biol.* **3**, 243–248.
Link, G., Coen, D. M., and Bogorad, L. (1978). *Cell* **15**, 725–731.
Macherel, D., Kobayashi, H., Valle, E., and Akazawa, T. (1986). *FEBS Lett.* **201**, 315–320.
MacIntosh, L., Poulsen, C., and Bogorad, L. (1980). *Nature* **288**, 556–560.
McCloskey, J. A., and Nishimura, S. (1977). *Acc. Chem. Res.* **10**, 403–410.
McClure, W. R. (1985). *Annu. Rev. Biochem.* **54**, 171–204.
Mazzara, G. P., Plunkett, G., III, and McClain, W. H. (1980). *In* "Cell Biology: A Comprehensive Treatise" (L. Goldstein and D. M. Prescott, eds.), pp. 439–545. Academic Press, New York.

Michel, F., and Dujon, B. (1983). *EMBO J.* **2**, 33–38.
Michel, F., and Lang, B. F. (1985). *Nature (London)* **316**, 641–643.
Miller, D. L., and Martin, N. (1983). *Cell* **34**, 911–917.
Montandon, P. E., Vasserot, A., and Stutz, E. (1986). *Curr. Genet.* **11**, 35–39.
Mott, J. E., Galloway, J. L., and Platt, T. (1985). *EMBO J.* **4**, 1887–1891.
Mullet, J. E., and Klein, R. R. (1987). *EMBO J.* **6**, 1571–1579.
Mullet, J. E., Orozco, E. M., Jr., and Chua, N. H. (1985). *Plant Mol. Biol.* **4**, 39–54.
Narita, J. O., Rushlow, K. E., and Hallick, R. B. (1985). *J. Biol. Chem.* **260**, 11194–11199.
Nelson, T., Harpster, M. H., Mayfield, S. P., and Taylor, W. C. (1984). *J. Cell Biol.* **98**, 558–564.
Neuhaus, H., and Link, G. (1987). *Curr. Genet.* **11**, 251–257.
Newbury, S. F., Smith, N. H., Robinson, E. C., Hiles, I. D., and Higgins, C. F. (1987). *Cell* **48**, 297–310.
Ohme, M., Kamogashira, T., Shinozaki, K., and Sugiura, M. (1985). *Nucleic Acids Res.* **13**, 1045–1056.
Ohme, M., Tanaka, M., Chunwongse, K., Shinozaki, K., and Sugiura, M. (1986). *FEBS Lett.* **200**, 87–90.
Ohyama, K., Fukuzawa, H., Kohchi, T., Shirai, H., Sano, T., Sano, S., Umesono, K., Shiki, Y., Takeuchi, M., Chang, Z., Aota, S., Inokuchi, H., and Ozeki, H. (1986). *Nature (London)* **322**, 572–574.
Ojala, D., Montoya, J., and Attardi, G. (1981). *Nature (London)* **290**, 470–474.
Orozco, E. M., Jr., Mullet, J. E., and Chua, N. H. (1985). *Nucleic Acids Res.* **13**, 1283–1302.
Orozco, E. M., Jr., Mullet, J. E., Hanley-Bowdoin, L., and Chua, N. H. (1986). *Methods Enzymol.* **118**, 232–253.
Osiewacz, H. D., and Esser, K. (1984). *Curr. Genet.* **8**, 299–305.
Palmer, J. D., Osorio, B., Watson, J. C., Edwards, H., Dodd, J., and Thompson, W. F. (1984). in "Biosynthesis of the Photosynthetic Apparatus: Molecular Biology, Development and Regulation" (J. P. Thornber, L. A. Staehelin, and R. B. Hallick, eds.), pp. 273–283. Liss, New York.
Peebles, C. L., Perlman, P. S., Mecklenburg, K. L., Petrillo, M. L., Tabor, J. H., Jarrell, K. A., and Cheng, H. L. (1986). *Cell* **44**, 213–223.
Piechulla, B., Chonoles-Imlay, K. R., and Gruissem, W. (1985). *Plant Mol. Biol.* **5**, 373–384.
Piechulla, B., Pichersky, E., Cashmore, A. R., and Gruissem, W. (1986). *Plant Mol. Biol.* **7**, 367–376.
Piechulla, B., Glick, R. E., Bahl, H., Melis, A., and Gruissem, W. (1987). *Plant Physiol.* **84**, 911–917.
Pirtle, R., Calagan, J., Pirtle, I., Kashdan, M. A., Vreman, H., and Dudock, B. S. (1981). *Nucleic Acids Res.* **9**, 183–188.
Platt, T. (1981). *Cell* **24**, 10–23.
Platt, T. (1986). *Annu. Rev. Biochem.* **55**, 339–372.
Polya, G. M., and Jagendorf, A. T. (1971). *Arch. Biochem. Biophys.* **146**, 635–648.
Poulsen, C. (1984). *Carlsberg Res. Commun.* **49**, 89–104.
Rawson, J. R. Y., Kushner, S. R., Vapnek, D., Kirby, N., Boerma, A., and Boerma, C. L. (1978). *Gene* **3**, 191–209.
Reed, R., and Maniatis, T. (1986). *Cell* **46**, 681–690.
Reisfeld, A., Gressel, J., Jakob, K. M., and Edelman, M. (1978). *Photochem. Photobiol.* **27**, 161–165.
Reiss, T., and Link, G. (1985). *Eur. J. Biochem.* **148**, 207–212.
Rochaix, J. D. (1981). *Experientia* **37**, 323–332.
Rochaix, J. D., Rahire, M., and Michel, M. (1985). *Nucleic Acids. Res.* **13**, 975–984.
Rock, C. D., Barkan, A., and Taylor, W. C. (1987). *Curr. Genet.* **12**, 69–77.
Rodermel, S. R., and Bogorad, L. (1985). *J. Cell Biol.* **100**, 463–476.
Rushlow, K. E., Orozco, E. M., Jr., Lipper, C., and Hallick, R. B. (1980). *J. Biol. Chem.* **255**, 3786–3792.

Ryan, T., and Chamberlin, M. J. (1983). *J. Biol. Chem.* **258,** 4690–4693.
Sasaki, Y., Sakihama, T., Kamikubo, T., and Shinozaki, K. (1983). *Eur. J. Biochem.* **133,** 617–620.
Schön, A., Krupp, G., Gough, S., Berry-Lowe, S., Kannangara, C. G., and Söll, D. (1986). *Nature (London)* **322,** 281–284.
Schuster, G., Ohad, I., Martineau, B., and Taylor, W. C. (1985). *J. Biol. Chem.* **260,** 11866–11873.
Schwarz, Z., Kössel, H., Schwarz, E., and Bogorad, L. (1981). *Proc. Natl. Acad. Sci. U.S.A.* **78,** 4748–4752.
Schwer, B., Visca, P., Vos, J. C., and Stunnenberg, H. G. (1987). *Cell* **50,** 163–169.
Scott, N. S., and Possingham, J. V. (1983). *J. Exp. Bot.* **34,** 1756–1767.
Scott, N. S., Tymms, M. J., and Possingham, J. V. (1984). *Planta* **161,** 12–19.
Sharp, P. A. (1985). *Cell* **42,** 397–400.
Sharp, P. A. (1987). *Cell* **50,** 147–148.
Sharp, J. A., and Platt, T. (1984). *J. Biol. Chem.* **259,** 2268–2273.
Sheen, J. Y., and Bogorad, L. (1985). *Plant Physiol.* **79,** 1072–1076.
Shinozaki, K., and Sugiura, M. (1982). *Nucleic Acids Res.* **10,** 4923–4934.
Shinozaki, K., Deno, H., Sugita, M., Kuramitsu, S., and Sugiura, M. (1986a). *Mol. Gen. Genet.* **202,** 1–5.
Shinozaki, K., Ohme, M., Tanaka, M., Wakasugi, T., Hayshida, N., Matsubayasha, T., Zaita, N., Chunwongse, J., Obokata, J., Yamaguchi-Shinozaki, K., Ohto, C., Torazawa, K., Meng, B. Y., Sugita, M., Deno, H., Kamogashira, T., Yamada, K., Kusuda, J., Takaiwa, F., Kata, A., Todoh, N., Shimada, H., and Sugiura, M. (1986b). *EMBO J.* **5,** 2043–2049.
Sijben-Müller, G., Hallick, B. B., Alt, J., Westhoff, P., and Herrmann, R. G. (1986). *Nucleic Acids Res.* **14,** 1029–1044.
Smith, S. M., and Ellis, R. J. (1981). *J. Mol. Appl. Genet.* **1,** 127–137.
Spencer, D., and Whitfeld, P. R. (1967). *Arch. Biochem. Biophys.* **121,** 336–345.
Steinmetz, A. A., Gubbins, E. J., and Bogorad, L. (1982). *Nucleic Acids. Res.* **10,** 3027–3037.
Steinmetz, A. A., Krebbers, E. T., Schwarz, Z., Gubbins, E. J., and Bogorad, L. (1983). *J. Biol. Chem.* **258,** 5503–5511.
Stern, D. B., and Gruissem, W. (1987). *Cell* **51,** 1145–1147.
Stirdivant, S. M., Crossland, L. D., and Bogorad, L. (1985). *Proc. Natl. Acad. Sci. U.S.A.* **82,** 4886–4890.
Strittmatter, G., and Kössel, H. (1984). *Nucleic Acids Res.* **12,** 7633–7647.
Strittmatter, G., Gozdzicka-Jozefiak, A., and Kössel, H. (1985). *EMBO J.* **4,** 599–604.
Sugita, M., and Sugiura, M. (1984). *Mol. Gen. Genet.* **195,** 308–313.
Sugita, M., Shinozaki, K., and Sugiura, M. (1985). *Proc. Natl. Acad. Sci. U.S.A.* **82,** 3557–3561.
Sugiura, M., Shinozaki, K., and Ohme, M. (1985). In "Molecular Form and Function of the Plant Genome" (L. van Vloten-Doting, G. S. P. Groot, and T. C. Hall, eds.), pp. 325–334. Plenum, New York.
Sun, E., Shapiro, D. R., and Tewari, K. K. (1986). *Plant Mol. Biol.* **6,** 201–210.
Takaiwa, F., and Sugiura, M. (1980). *Gene* **10,** 95–103.
Takaiwa, F., and Sugiura, M. (1982). *Nucleic Acids Res.* **10,** 2665–2676.
Tewari, K. K., and Goel, A. (1983). *Biochemistry* **22,** 2142–2148.
Thomas, F., Zeng, G. Q., Mache, R., and Briat, J. F. (1988). *Plant Mol. Biol.* **10,** 447–457.
Thompson, R. J., and Mosig, G. (1987). *Cell* **48,** 281–287.
Thompson, W. W., and Whatley, J. M. (1980). *Annu. Rev. Plant Physiol.* **31,** 375–394.
Tobin, E. M., and Silverthorne, J. (1985). *Annu. Rev. Plant Physiol.* **36,** 569–593.
Tohdoh, N., Shinozaki, K., and Sugiura, M. (1981). *Nucleic Acids Res.* **9,** 5399–5406.
Torazawa, K., Hayashida, N., Obokata, J., Shinozaki, K., and Sugiura, M. (1986). *Nucleic Acids Res.* **14,** 3143.
Tymms, M. J., Scott, N. S., and Possingham, J. V. (1983). *Plant Physiol.* **71,** 785–788.

Van Dyke, M. W., and Roeder, R. (1987). *Nucleic Acids Res.* **15,** 4365–4374.
Wakasugi, T., Ohme, M., Shinozaki, K., and Sugiura, M. (1986). *Plant Mol. Biol.* **7,** 385–392.
Wallace, D. (1982). *Microbiol. Rev.* **46,** 208–240.
Walter, G., Zillig, W., Palm, P., and Fuchs, E. (1967). *Eur. J. Biochem.* **3,** 194–201.
Wang, J. C. (1985). *Annu. Rev. Biochem.* **54,** 665–697.
Watson, J. C., and Surzycki, S. J. (1983). *Curr. Genet.* **7,** 201–210.
Westhoff, P. (1985). *Mol. Gen. Genet.* **201,** 115–123.
Westhoff, P., and Herrmann, R. G. (1988). *Eur. J. Biochem.* **171,** 551–564.
Westhoff, P., Alt, J., Nelson, N., Bottomley, W., Bünemann, H., and Herrmann, R. G. (1983). *Plant Mol. Biol.* **2,** 95–107.
Westhoff, P., Alt, J., Nelson, N., and Herrmann, R. G. (1985). *Mol. Gen. Genet.* **199,** 290–299.
Whitfeld, P. R., and Bottomley, W. (1983). *Annu. Rev. Plant Physiol.* **34,** 279–310.
Wong, H. C., and Chang, S. (1986). *Proc. Natl. Acad. Sci. U.S.A.* **83,** 3233–3237.
Yamaguchi-Shinozaki, K., Shinozaki, K., and Sugiura, M. (1987). *FEBS Lett.* **215,** 132–136.
Young, R. A., and Steitz, J. A. (1978). *Proc. Natl. Acad. Sci. U.S.A.* **75,** 3593–3597.
Zaita, N., Torazawa, K., Shinozaki, K., and Sugiura, M. (1987). *FEBS Lett.* **210,** 153–156.
Zech, M., Hartley, M. R., and Bohnert, H. J. (1981). *Curr. Genet.* **4,** 37–46.
Zenke, G., Edwards, K., Langridge, P., and Kössel, H. (1982). *In* "Cell Function and Differentiation" Part B (G. Akoyunoglou, A. E. Evangelopoulos, J. Georgalsos, G. Palaiologos, A. Trakatellis, and C. P. Tsiganos, eds.), pp. 309–319. Liss, New York.
Zhu, Y. S., Kung, S. D., and Bogorad, L. (1985). *Plant Physiol.* **79,** 371–376.
Zurawski, G., and Clegg, M. T. (1987). *Annu. Rev. Plant Physiol.* **38,** 391–418.
Zurawski, G., Perrot, B., Bottomley, W., and Whitfeld, P. R. (1981). *Nucleic Acids Res.* **9,** 3251–3270.
Zurawski, G., Bottomley, W., and Whitfeld, P. R. (1982). *Proc. Natl. Acad. Sci. U.S.A.* **79,** 6260–6264.
Zurawski, G., Clegg, M. T., and Brown, A. H. D. (1984). *Genetics* **106,** 735–749.
Zurawski, G., Whitfeld, P. R., and Bottomley, W. (1986). *Nucleic Acids Res.* **14,** 3975.

Protein Synthesis in Chloroplasts

ANDRÉ STEINMETZ
JACQUES-HENRY WEIL

I. Introduction
II. *In Organello* and *in Vitro* Synthesis of Chloroplast Proteins
III. Structure of Chloroplast Messenger RNAs
IV. Translation
    A. Aminoacylation of tRNAs
    B. The Ribosomal Components of the Translation Machinery
    C. The Ribosome Cycle in Protein Synthesis
    D. Genetic Code and Codon Usage in Chloroplasts
    E. Transfer RNAs and Codon Recognition in Chloroplasts
V. Maturation of Proteins
VI. Posttranscriptional Regulation of Chloroplast Gene Expression
VII. Concluding Remarks
    References

I. INTRODUCTION

In a plant cell, proteins are synthesized not only in the cytoplasm, but also in the organelles, namely the chloroplasts and the mitochondria. Chloroplasts of higher plants synthesize, in addition to the chloroplast rRNAs and tRNAs, approximately 100 different proteins (Ellis, 1981). The machinery involved in transcription and translation of chloroplast genes is chloroplast-specific and differs significantly from that used in the expression of nuclear genes. Several observations suggest that the chloroplast translation system is closely related to that of the bacteria: (1) the ribosomes are similar in size and in the number and structural features of their components (rRNAs and ribosomal proteins). (2) In both systems, initiation of protein synthesis involves N-formylmethionyl-tRNA^{Met}, whereas in eukaryotic cytoplasm the initiator methionyl-tRNA^{Met} is not formylated. (3) In many instances it has been shown that chloroplast tRNAs can be aminoacylated by bacterial aminoacyl-tRNA synthetases, and, conversely, that bacterial tRNAs can be charged by chloroplast enzymes. (4) Protein synthesis in chloroplasts and bacteria is inhibited by the same antibiotics (for instance chloramphenicol), but is not sensitive to antibiotics (for instance cycloheximide) which affect protein synthesis in eukaryotic cytoplasm.

The first step in gene expression, i.e., transcription of the chloroplast genes, is described by Gruissem in Chapter 4 of this volume. The present chapter will cover the posttranscriptional events in chloroplast protein synthesis and will deal more specifically with the structure of chloroplast mRNAs and the components of the translational machinery, the ribosome cycle in translation, and the maturation of chloroplast proteins. Similarities with *Escherichia coli* will be indicated. For additional information on prokaryotic and eukaryotic protein synthesis the reader should refer to reviews by Kozak (1983), Maitra *et al.* (1982), Marcus (1982), Moldave (1985). An excellent overview has also been given by Lewin (1985). Ribosome architecture has recently been reviewed by Wittmann (1983) and Lake (1985).

Translation of a mRNA into protein is the most critical step in protein synthesis and, despite large efforts during the past decade, it is still poorly understood. It is estimated that the number of protein and RNA species involved in the translation of a message is close to 150. This illustrates the extreme complexity of the system in which the various steps have to be carried out in a highly coordinated manner and with an absolute specificity, if the fidelity of translation is to be secured.

II. *IN ORGANELLO* AND *IN VITRO* SYNTHESIS OF CHLOROPLAST PROTEINS

Isolated, intact chloroplasts can synthesize proteins in a light-driven reaction. If chloroplasts are supplied with radioactively labeled amino acids, the newly synthesized proteins can be detected by autoradiography after electrophoretic fractionation on polyacrylamide gels (for a review, see Gray *et al.*, 1984; for methods, see also Nivison *et al.*, 1986). Approximately 100 radioactive spots could be resolved on two-dimensional gels.

A variety of prokaryotic and eukaryotic cell-free systems have been used to translate chloroplast mRNAs: the mRNA for the large subunit (LS) of ribulose bisphosphate carboxylase (RuBPCase) from spinach, *Euglena,* and maize chloroplasts has been translated in *E. coli* and wheat germ extracts, and rabbit reticulocyte lysates, respectively (Hartley *et al.*, 1975; Sagher *et al.*, 1976; Bedbrook *et al.*, 1978); rabbit reticulocyte lysates have also been used to translate mRNAs for thylakoid proteins, including the 32-kDa protein, from spinach chloroplasts (Driesel *et al.*, 1980; Westhoff *et al.*, 1981).

E. coli cells faithfully express the gene encoding the large subunit of RuBPCase if transformed with a plasmid carrying this gene (Gatenby et al., 1981). Chloroplast DNA and DNA fragments can also direct in vitro synthesis of chloroplast proteins in a cell-free transcription–translation system from E. coli (Bottomley and Whitfeld, 1978), in chloroplast extracts (Bard et al., 1985), and in mixed transcription (E. coli RNA polymerase)–translation (rabbit reticulocyte lysate) systems (Bogorad et al., 1978).

Many, but not all chloroplast protein genes can probably be expressed in bacteria and in bacterial extracts. Chloroplast genes that are likely to be not expressed or incorrectly expressed in these systems include the split genes for which a splicing apparatus is not available in eubacteria (see Section III).

III. STRUCTURE OF CHLOROPLAST MESSENGER RNAS

Approximately 50 protein genes have been identified to date on the chloroplast genome (Shinozaki et al., 1986b; Ohyama et al., 1986; see also Table I). Most of these genes are probably transcribed as polycistronic mRNAs. Large transcripts spanning several protein-coding regions have been identified, e.g., in the case of the two polypeptides of the photosystem I reaction center in maize (Rodermel and Bogorad, 1985) and spinach (Kirsch et al., 1986), the "51-kDa" protein of the photosystem II reaction center, apocytochrome b_6 and subunit 4 of the cytochrome b/f complex in spinach (Heine-

TABLE I

Protein Genes Identified on the Chloroplast Genome

Gene	Protein	References[a]
rpoA	RNA polymerase subunit α	1–3
rpoB	RNA polymerase subunit β	2, 3
rpoC	RNA polymerase subunit β'	2, 3
rps2	Ribosomal protein S2	2, 3
rps3	Ribosomal protein S3	2, 3
rps4	Ribosomal protein S4	2–4
rps7	Ribosomal protein S7	2, 3, 5
rps8	Ribosomal protein S8	2, 3
rps11	Ribosomal protein S11	1–3
rsp12	Ribosomal protein S12	2, 3, 5, 6
rps14	Ribosomal protein S14	2, 3
rps15	Ribosomal protein S15	2, 3
rps16	Ribosomal protein S16	2
rps18	Ribosomal protein S18	2, 3
rps19	Ribosomal protein S19	2, 3, 7

(continued)

TABLE I (*continued*)

Gene	Protein	References[a]
*rpl*2	Ribosomal protein L2	2, 3, 7
*rpl*14	Ribosomal protein L14	2, 3
*rpl*16	Ribosomal protein L16	2, 3
*rpl*20	Ribosomal protein L20	2, 3
*rpl*22	Ribosomal protein L22	2, 3
*rpl*23	Ribosomal protein L23	2, 3
*rpl*33	Ribosomal protein L33	2, 3
*inf*A	Initiation factor IF-1	1–3
*tuf*A	Elongation factor EF-Tu	8
*rbc*L	RuBPCase large subunit	2, 3, 9–12
*atp*A	ATPase subunit α	2, 3, 13
*atp*B	ATPase subunit β	2, 3, 13–16
*atp*E	ATPase subunit ε	2, 3, 13–16
*atp*F	ATPase subunit I	2, 3, 17
*atp*H	ATPase subunit III	2, 3, 18–20
*atp*I	ATPase subunit a	2, 3, 21
*psa*A	68-kDa polypeptide 1 "reaction center" PSI	2, 3, 22
*psa*B	68-kDa polypeptide 2 "reaction center" PSI	2, 3, 22
*psb*A	32-kDa (Qβ binding) protein PSII	2, 3, 23–27
*psb*B	51-kDa polypeptide "reaction center" PSII	2, 3, 28
*psb*C	44-kDa polypeptide "reaction center" PSII	2, 3, 29, 30
*psb*D	"D2-polypeptide" PSII	2, 3, 29, 30
*psb*F	10-kDa phosphoprotein PSII	2, 3, 31, 32
*psb*G	24-kDa protein PSII	2, 3, 33
*pet*A	Cytochrome f	2, 3, 34–37
*pet*B	Cytochrome b_6	2, 3, 38
*pet*C	Cytochrome b_{559}	2, 3, 39
*pet*D	Cytochrome b_6/f complex subunit 4	2, 3, 38
*ndh*A	NADH dehydrogenase ND1	2, 3
*ndh*B	NADH dehydrogenase ND2	2, 3
*ndh*C	NADH dehydrogenase ND3	2, 3
*ndh*D	NADH dehydrogenase ND4	2, 3
*ndh*E	NADH dehydrogenase ND4L	2, 3
*ndh*F	NADH dehydrogenase ND5	2, 3

[a] References: (1) Sijben-Müller *et al.*, 1986; (2) Shinozaki *et al.*, 1986b; (3) Ohyama *et al.*, 1986; (4) Subramanian *et al.*, 1983; (5) Montandon and Stutz, 1984; (6) Fromm *et al.*, 1986; (7) Zurawski *et al.*, 1984; (8) Montandon and Stutz, 1983; (9) McIntosh *et al.*, 1980; (10) Zurawski *et al.*, 1981; (11) Dron *et al.*, 1982; (12) Gingrich and Hallick, 1985; (13) Howe *et al.*, 1985; (14) Krebbers *et al.*, 1982; (15) Zurawski *et al.*, 1982b; (16) Zurawski and Clegg, 1984; (17) Bird *et al.*, 1985; (18) Howe *et al.*, 1982; (19) Alt *et al.*, 1983; (20) Passavant and Hallick, 1985; (21) Cozens *et al.*, 1986; (22) Fish *et al.*, 1985; (23) Zurawski *et al.*, 1982a; (24) Spielmann and Stutz, 1983; (25) Erickson *et al.*, 1984; (26) Karabin *et al.*, 1984; (27) Keller and Stutz, 1984; (28) Morris and Herrmann, 1984; (29) Alt *et al.*, 1984; (30) Holschuh *et al.*, 1984b; (31) Westhoff *et al.*, 1986; (32) Hird *et al.*, 1986; (33) Steinmetz *et al.*, 1986; (34) Alt and Herrmann, 1984; (35) Willey *et al.*, 1984a; (36) Willey *et al.*, 1984b; (37) Tyagi and Herrmann, 1986; (38) Heinemeyer *et al.*, 1984; (39) Herrmann *et al.*, 1984.

meyer *et al.*, 1984), subunits III, I, and α of wheat chloroplast ATPase (Bird *et al.*, 1985), subunits β and ε of maize chloroplast ATPase (Link and Bogorad, 1980), the 44-kDa chlorophyll *a* apoprotein and the "32-kDa-like" protein of photosystem II reaction center (Alt *et al.*, 1984). In most cases, more than one transcript could be detected on Northern blots: they may represent processing intermediates, or primary transcripts initiating or terminating at multiple sites. There is at least one example suggesting that oligocistronic mRNAs might be translated without being cleaved previously into monocistronic mRNAs. This is the case of the dicistronic mRNA coding for the β and ε subunits of ATPase in maize, spinach, tobacco, and wheat chloroplasts (Krebbers *et al.*, 1982; Zurawski *et al.*, 1982b; Shinozaki *et al.*, 1983; Howe *et al.*, 1985): these two genes have one nucleotide in common (the A in the initiation codon AUG of the second gene is also the last nucleotide of the last sense codon AAA of the first gene).

Some examples of monocistronically transcribed chloroplast mRNAs are also known: one of these is that of the large subunit of RuBPCase. In spinach, this gene is transcribed from a single promoter into a 1700-nucleotide-long mRNA which comprises the coding region and approximately 90 nucleotides downstream (Zurawski *et al.*, 1981). This mRNA is subsequently processed in its upstream region at several sites, thus generating a heterogeneous population of mRNAs (Mullet *et al.*, 1985b).

Chloroplasts, like bacteria, lack the enzyme guanylyl transferase that is known to add a cap structure to the unprocessed (therefore triphosphorylated) 5' ends of eukaryotic monocistronic mRNAs: chloroplast mRNAs are therefore uncapped (Mullet *et al.*, 1985b). As a result of 5' processing, most chloroplast mRNAs are monophosphorylated. Chloroplast mRNAs are not polyadenylated at their 3' end (Wheeler and Hartley, 1975). In this respect, these mRNAs resemble bacterial and yeast mitochondrial mRNAs, which are also uncapped and without poly(A) tails (Groot *et al.*, 1974). In mammalian mitochondria, mRNAs are uncapped, but they are polyadenylated (Clayton, 1984). In this case, since only a few mitochondrial mRNAs have a translational stop codon encoded in the mitochondrial DNA, the addition of As to mRNAs ending with U or UA produces a UAA stop codon.

Chloroplast mRNAs from higher plants also resemble eubacterial mRNAs in having a Shine–Dalgarno-like sequence upstream of the initiation codon (Table II). The most frequent chloroplast initiation codon is AUG; GUG appears to be used as initiation codon in the tobacco gene encoding ribosomal protein S19 (Sugita and Sugiura, 1983). Studies with eubacterial ribosomes have suggested that, in addition to the Shine–Dalgarno sequence and the initiation codon, the nucleotides immediately preceding and following the initiation codon contribute to recognition by prokaryotic ribosomes: a pyrimidine in position "-1" and a purine in position "+4" are most effective (see Kozak, 1983). This structure is also found in most chloroplast genes (Table II).

For some chloroplast protein genes, several potential translation start

TABLE II

Ribosome Binding Sites in Chloroplast mRNAs

Nucleotide sequence		Gene	References[a]
3'UUUCCUCCACUAGGUCGGCG ....5'		cp 16S rRNA	
A**AGAGGAGG**ACUUA	AUG AUU AUU	psaA maize	22
GCU**AGGAGG**AUUUGAAAGGCAUU	AUG GAA UUA	psaB maize	22
AAU**AGGAGG**AUCACU	AUG ACU AUA	psbD tobacco	2
AGU**AGGAGG**CAAACCUU	AUG CUA AAG	rpl22 tobacco	2
AAG**AGGAG**AGCAAAUUGAAGAA	AUG AAA UUA	atpE maize	14
GU**AGGGAGG**GACUUA	AUG UCA CCA	rbcL maize	9, 10
GACAC**GAGG**AACUACUCACC	AUG AAU CCA	atpH wheat	18
UUUCA**GAGG**GCAAGGCAAU	AUG AAU GUU	atpI tobacco	2
UAUA**AGAGG**AGAGCAU	AUG AAA AAU	atpF wheat	17
AAU**AGGAG**UAAGCUU	GUG ACA CGU	rps19 tobacco	2
AAA**AGGAG**UCUUC	AUG UCC GCU	rps4 maize	4
AAA**AGGAG**CCUUGGA	AUG GUC UUA	psbG maize	33
UAA**AGGGG**UAUUUCC	AUG GGU UUG	psbB spinach	28
UU**AAGGGGG**UUUUGUCUGGA	AUG AAA GAA	infA spinach	1
AAA**AGGGGG**GGUGUGGAGAGAA	AUG ACA AGA	rps2 tobacco	2
AAU**AGGA**AUUCAACCAUU	AUG GCA AGG	rps14 tobacco	2
mRNAs with poor ribosome binding site:			
UUUAU**GAGAUGA**AAAAAU	AUG GCA AAA	rps11 spinach	1
AAAAUUAU**GUGA**UAAUU	AUG AGA ACC	atpB maize	14
CAAU**GG**UUGGACU	AUG CAA ACU	petA spinach	34
U**AAAG**UCU	AUG AUU GGU	petB spinach	38
C**AAAG**UUUCAUUUAUUCA	AUG ACC AGA	rpl20 tobacco	2
C**AAAG**AACGGAUUAAAAAA	AUG AUU CAA	rpl14 tobacco	2
U**AAAG**AAAGAAUC	AUG GUA ACC	atpA tobacco	2
CC**AAG**AUUUUACC	AUG ACU GCA	psbA tobacco	2
AGGCCCCAAUAAUUUUAGUUCAUC	AUG GGU AGG	rps8 tobacco	2

[a] Numbers refer to references in Table I.

sites have been observed. In this case, the correct initiation site of translation can be determined using an *in vitro* dipeptide synthesis system (Bloom et al., 1986).

The presence of introns in some chloroplast genes of higher plants and green algae (*Euglena* and *Chlamydomonas*) is an extra feature not encountered in eubacteria. As a consequence, these pre-mRNAs, in addition to the processing at the 5' and 3' ends, have to undergo correct excision of the intervening sequences before they can carry out their structural (rRNAs), translational (tRNAs), or messenger (mRNAs) functions.

No information is presently available on the splicing machinery and on the splicing mechanism operating in chloroplasts. Initially, splicing was believed to be an exclusively intramolecular event, joining the exons from one precursor RNA molecule (*cis*-splicing). Recent reports indicate that intermolecular splicing (*trans*-splicing), involving two independent RNA transcripts, also exists in chloroplasts (Fromm *et al.*, 1986; Fukuzawa *et al.*, 1986; Torazawa *et al.*, 1986; Koller *et al.*, 1987).

In *Euglena*, introns make up to 20% of the chloroplast genome, and several genes with multiple introns (rbcL, psbA, psbC, tufA) have been identified and studied. In the case of the psbA gene, a heterogeneous population of pre-mRNAs was found (Hollingsworth *et al.*, 1984), apparently due to different splicing intermediates (Koller *et al.*, 1985).

The message carried by a split gene can be faithfully expressed only if the mRNA is correctly spliced before translation occurs. Since RNA splicing involves joining of primarily distant RNA regions, it is reasonable to assume that the splicing occurs only at the level of the fully transcribed RNA. Hence, in the case of the split genes, transcription and translation are probably separated in time and perhaps also in space. It is not clear whether this uncoupling of the two fundamental steps in chloroplast gene expression is restricted to the split mRNAs, or whether it can be considered a general feature of chloroplast mRNAs.

Uncoupling of transcription and translation rules out the possibility of regulation of gene expression by attenuation, an alternative mechanism regulating the expression of some eubacterial operons. It also raises the question as to how ribosomes are prevented from binding to or moving along an immature mRNA (see Section VI). In eukaryotes this situation is also found, but the problem is circumvented by the physical separation of transcription and translation by the nuclear membrane. An elegant way showing how T4 phage solves the problem in *E. coli*, in the case of its split thymidylate synthetase (td) gene, was recently described (Belfort *et al.*, 1985). The td gene (1875 bp long) is interrupted by a 1017-bp intron which starts with a UAA stop codon in phase with the sequence of the first exon. Translation of the first exon is coupled to transcription of the full-length td gene and produces a peptide that has dUMP and methylenetetrahydrofolate binding properties. After transcription of the gene is completed, self-splicing occurs on the full-length pre-mRNA (Chu *et al.*, 1986); the spliced mRNA is subsequently translated into the functional enzyme.

There is no evidence, at the moment, that in chloroplasts more than one polypeptide is synthesized from the same split gene. This suggests that translation occurs on a single message, the mature spliced mRNA. The finding that multiple introns are excised in a rather random order (Koller *et al.*, 1985) also supports this view.

IV. TRANSLATION

A. Aminoacylation of tRNAs

The formation of a peptide bond between two amino acids is thermodynamically unfavorable. This barrier is overcome by activating the carboxyl group of the precursor amino acid (amino acid ester). The activated intermediate serving in translation is the aminoacyl-tRNA.

The binding of a tRNA proceeds in two steps, both catalyzed by an aminoacyl-tRNA synthetase specific for a given amino acid and its corresponding isoacceptor tRNA(s). In the first step, the amino acid is activated by ATP to yield an aminoacyl-adenylate bound to the enzyme [Eq. (1)]. In the second step, the activated amino acid is transferred to its cognate tRNA in an esterification reaction involving a free hydroxyl group of the ribose moiety (2' or 3') at the 3' end of the tRNA (CCA-terminus) and the carboxyl group of the amino acid [Eq. (2)]:

$$\text{Amino acid}_x + \text{ATP} + \text{aminoacyl-tRNA synthetase}_x \longrightarrow \\ (\text{aminoacyl}_x\text{-adenylate}) - \text{aminoacyl-tRNA synthetase}_x + \text{PP}_i \tag{1}$$

$$(\text{Aminoacyl}_x\text{-adenylate}) - \text{aminoacyl-tRNA synthetase}_x + \text{tRNA}_x \longrightarrow \\ \text{aminoacyl}_x\text{-tRNA}_x + \text{AMP} + \text{aminoacyl-tRNA synthetase}_x \tag{2}$$

Discrimination of amino acids by aminoacyl-tRNA synthetases can occur at several stages: the correct amino acid may be bound in preference to most other amino acids; if an incorrect amino acid is bound, it may be proofread kinetically, by a conformational change induced by the tRNA, or chemically via an abortive transfer to tRNA.

The recognition, by an aminoacyl-tRNA synthetase, of its cognate tRNA(s) involves interactions between a few amino acids constituting the active site in the protein and the 3'-terminal region, the anticodon region, and the D stem of the tRNA. The recognition of the correct tRNA by the aminoacyl-tRNA synthetase is controlled at two levels: (1) Because the affinity of the aminoacyl-tRNA synthetase is greater for its cognate tRNA(s) than for an incorrect tRNA, binding of the correct tRNA occurs more readily; it can furthermore be stabilized by a conformational transition in the enzyme. (2) A slow aminoacylation rate of the incorrect tRNA increases the probability that synthetase and tRNA will dissociate.

The overall error rate in an aminoacylation reaction is lower than 1/10,000. *In vitro* and in heterologous systems, misactivations and/or misacylations have been observed with amino acid analogs (e.g., canavanine for arginine, azetidine 2-carboxylic acid for proline). In homologous systems where the amino acid analogs are naturally found, incorporation of the analogs into proteins does not occur, either because they are sequestered in vacuoles, or because the homologous aminoacyl-tRNA synthetase is able to discriminate between the correct amino acid and its analog.

1. Aminoacyl-tRNA Synthetases

Specific chloroplast aminoacyl-tRNA synthetases [L-amino acid:tRNA ligases (AMP forming), EC 6.1.1.x] that are different from their cytoplasmic counterparts have been characterized in *Euglena* (for tyrosine, valine, phenylalanine, leucine, lysine, serine, isoleucine, and histidine), in bean (for leucine, methionine, proline, lysine, and phenylalanine), in soybean (for tyrosine, phenylalanine, and tryptophan), in spinach (for leucine, isoleucine, proline, lysine, methionine, valine, and glycine), and in zucchini (for leucine, isoleucine, phenylalanine, and valine). Separation of chloroplast and cytoplasmic enzymes can usually be achieved by hydroxyapatite chromatography of total leaf protein extracts: in most cases the organellar enzyme is eluted before the cytoplasmic enzyme (for a review, see Weil and Parthier, 1982).

In some cases, chloroplast and cytoplasmic activities could not be separated. For instance, in protein extracts from zucchini leaves, only one peak of activity was found in the case of threonine, alanine, serine, and arginine. Since the charging specificities for most enzymes are different (see Steinmetz and Weil, 1986) one can differentiate between the chloroplast and the cytoplasmic enzyme. In the case of threonine, however, the two chloroplast tRNAs as well as the cytoplasmic threonine tRNA are charged by the enzyme isolated from purified chloroplasts (and *E. coli*), and by the cytoplasmic enzyme isolated from dark-grown hypocotyls. These observations suggest that the same threonyl-tRNA synthetase (Thr-RS) is present in the cytoplasm and in the chloroplasts, or that there are two enzymes that are very similar in their structure and their aminoacylation specificities. Another argument supporting this idea is the fact that borrelidin, and antibiotic that inhibits threonyl-tRNA formation in several microorganisms, inhibits the activity of the chloroplast and the cytoplasmic enzymes (Burkard *et al.,* 1970). That the same gene can code for a cytoplasmic and organellar aminoacyl-tRNA synthetase has recently been observed in the case of the yeast HTS1 gene, which codes for the cytoplasmic and mitochondrial histidyl-tRNA synthetase (His-RS) (Natsoulis *et al.,* 1986).

The physical properties and substrate specificities of some chloroplast and cytoplasmic aminoacyl-tRNA synthetases have been reviewed recently (Weil and Parthier, 1982; Steinmetz and Weil, 1986). Table III summarizes the molecular and structural features of the chloroplast and cytoplasmic enzymes studied so far. The most striking feature of this class of enzymes is the unusually high molecular weight (75 kDa at least). They are found as monomers (chloroplast and cytoplasmic Leu-RS and Val-RS, chloroplast Met-RS), as homodimers (chloroplast and cytoplasmic Tyr-RS), as heterodimers (chloroplast Phe-RS), or heterotetramers (cytoplasmic Phe-RS).

Immunochemical studies performed on chloroplast and cytoplasmic Leu-RS and Val-RS from *Euglena* (Colas *et al.,* 1982b), and on chloroplast and cytoplasmic Leu-RS from common bean (Dietrich *et al.,* 1983), have shown that the chloroplast and cytoplasmic enzymes have no common antigenic determinants, indicating that they are encoded by unrelated genes (which are located in the nucleus).

Aminoacyl-tRNA synthetases appear to have functional domains arranged linearly along the polypeptide chain(s). In the case of *E. coli* Ala-RS,

TABLE III

Chloroplast and Cytoplasmic Aminoacyl-tRNA Synthetases

Enzyme	Plant species	Cellular compartment	Molecular weight	Subunit structure	References
Glu-RS	Triticum aestivum	Chloroplast	110,000	α_2	Ratinaud et al. (1985)
		Cytoplasm	160,000	α_2	Ratinaud et al. (1985)
Leu-RS	Phaseolus vulgaris	Chloroplast	120,000	α	Souciet et al. (1982)
		Cytoplasm	130,000	α	Dietrich et al. (1983)
	Euglena gracilis	Chloroplast	100,000	α	Krauspe and Parthier (1974); Imbault et al. (1981); Colas et al. (1982a)
		Cytoplasm	116,000	α	Krauspe and Parthier (1974); Sarantoglou et al. (1981); Colas et al. (1982a)
Met-RS	Triticum aestivum	Chloroplast	75,000	α	Carias et al. (1981)
		Cytoplasm	165,000	α_2	Chazal et al. (1975)
Phe-RS	Phaseolus vulgaris	Chloroplast	78,000	$\alpha\beta$	Rauhut et al. (1986)
		Cytoplasm	260,000	$\alpha_2\beta_2$	Rauhut et al. (1984)
Tyr-RS	Glycine max	Chloroplast	98,000	α_2	Locy and Cherry (1978)
		Cytoplasm	126,000	α_2	Locy and Cherry (1978)
Val-RS	Euglena gracilis	Chloroplast	126,000	α	Imbault et al. (1979); Colas et al. (1982a)
		Cytoplasm	126,000	α	Sarantoglou et al. (1980); Colas et al. (1982a)

which is a homotetramer, three domains have been identified: the first domain (extending from amino acids 257 to 385) is responsible for the activation of the amino acid (aminoacyl-adenylate synthesis), the second domain (amino acids 404 to 461) for tRNA aminoacylation, and the third domain (amino acids 699 to 808) for oligomerization (Jasin et al., 1983).

2. Transfer RNAs

Since the discovery of a chloroplast-specific arginine tRNA in pea by Aliyev and Filippovitch in 1968, extensive studies have shown that chloroplasts contain specific tRNAs for each of the 20 amino acids. These tRNAs differ from their cytoplasmic and mitochondrial counterparts in their chromatographic and electrophoretic properties. They can also be charged selectively using either chloroplast or bacterial (*E. coli*) aminoacyl-tRNA synthetases, whereas cytoplasmic tRNAs are preferentially aminoacylated by cytoplasmic enzymes (Steinmetz and Weil, 1986). Exceptions to this general rule are chloroplast tRNA^{Ala}, one tRNA^{Val}, one tRNA^{Ile}, and both isoacceptors for threonine, which are charged by chloroplast and cytoplasmic enzymes.

The number of different chloroplast tRNAs has been estimated by two-dimensional polyacrylamide gel electrophoresis to be approximately 30 to 35 in spinach and the other higher plants studied (Steinmetz *et al.*, 1978; Mubumbila *et al.*, 1980). This is in agreement with the number of different tRNA genes identified on chloroplast genomes. This number, however, does not appear to be conserved in chloroplasts from different plant species. In tobacco, for instance, the analysis of the complete nucleotide sequence of the chloroplast genome has revealed 30 different tRNA genes (Shinozaki *et al.*, 1986b), whereas in liverwort 32 different tRNA genes were found (Ohyama *et al.*, 1986). This indicates that the complete set of chloroplast tRNAs is encoded by the chloroplast genome. It also indicates that several amino acids must have more than one tRNA. In spinach, multiple isoacceptors have been identified for leucine and serine (3 tRNAs for each), and arginine, glycine, isoleucine, methionine, threonine, and valine (2 tRNAs for each); each of the remaining amino acids has only one tRNA species. How the 61 sense codons may be recognized by the tRNAs available in chloroplasts will be discussed in Section IV,E of this chapter.

The sequences of 14 chloroplast tRNA species have been determined at the RNA level (see Table V). The analysis has shown relatively few modified nucleosides (5 to 10 per molecule) as compared to cytoplasmic tRNAs (about 15 per molecule). The positions at which the modified nucleosides are found are positions 12 (ac^4C), 18 (Gm), 19 (D), 20 (D), 25 (Ψ; $m^{2,2}G$), 26 (m^2G), 27 (Ψ), 28 (Ψ), 31 (Ψ), 34 (Cm; Um; mam^5s^2U), 35 (Ψ), 36 (m^7G), 37 (t^6A; mt^6A; ms^2i^6A; m^1G), 39 (Ψ), 46 (m^7G), 47 (acp^3U), 48 (m^5C), 54 (T), 55 (Ψ), 66 (Ψ). All of these modified nucleosides are also found in bacteria, except for $m^{2,2}G$, which is typically eukaryotic. The chemical nature and the properties of modified nucleosides have been reviewed by Dirheimer (1983).

Modified nucleosides can have an effect upon tRNA structure and codon–anticodon interaction (especially the modified nucleosides found in the anticodon region). In the case of chloroplast tRNA^{Ile}(*CAU), the modification of the C in the first position (wobble position) of the anticodon leads to the pairing with the isoleucine codon AUA (and not with codon AUG, which is the methionine codon). This situation has also been found in prokaryotes such as *E. coli* (Kuchino *et al.*, 1980), *Bacillus subtilis* (Green and Vold, 1983), *Mycoplasma* (Samuelsson *et al.*, 1985), *Spiroplasma* (Rogers *et al.*, 1984), and phage T4 (Fukada and Abelson, 1980).

Two different tRNA^{Met} species, with different functions, are found in the chloroplasts. One of them is used to initiate translation, the other one is used to read internal methionine codons. Both have the same unmodified anticodon (CAU). Both are charged by the same methionyl-tRNA synthetase. The chloroplast initiator tRNA^{Met}(f) has 80% homology with the *E. coli* tRNA^{Met}(f) and 70 to 76% homology with the higher plant mitochondrial tRNA^{Met}(f) (Maréchal et al., 1986). The bacterial, chloroplast, and mitochondrial initiator tRNAs have an unpaired nucleotide at the 5' end (contrary to all elongator tRNAs^{Met} in which this nucleotide is paired with a nucleotide from the 3' end), and an unmodified A residue adjacent to the last nucleotide of the anticodon. These features might be responsible for the interaction between the initiator tRNA and the initiation factor IF-2 during formation of the initiation complex.

In prokaryotes, initiation of translation involves a formylated methionyl-tRNA^{Met}(f). Participation of a formylated methionyl-tRNA in initiation of protein synthesis in chloroplasts was first demonstrated by Schwartz et al., (1967), who characterized N-formylmethionine in the proteins synthesized in a cell-free system prepared from *Euglena* chloroplasts and stimulated by f2 RNA. A formylatable methionyl-tRNA and a transformylase have subsequently been found in chloroplasts and mitochondria from higher plants (Burkard et al., 1969; Merrick and Dure, 1971; Guillemaut et al., 1972). In eukaryotes, the (cytoplasmic) initiator methionyl-tRNA is not formylated *in vivo*, but can be formylated *in vitro* by a heterologous (bacterial) transformylase. The cytoplasmic initiator tRNA from higher plants, however, is not formylated by the bacterial enzyme (Leis and Keller, 1970; Merrick and Dure, 1971; Guillemaut et al., 1972).

Three fractions of glutamate acceptor activity have been separated from barley chloroplast tRNA by high-performance liquid chromatographic (HPLC) fractionation (Schön et al., 1986). The first of these contained a tRNA^{Glu} species that is necessary for conversion of glutamate to δ-aminolevulinate, a precursor of chlorophyll. The nucleotide sequence of this tRNA^{Glu} shows high homology (94 to 100%) with the sequenced tRNA^{Glu} genes from chloroplasts of higher plants (see references in Table V). One distinctive feature of this tRNA is the presence of an A_{53}-U_{61} pair as the last base pair in the TΨC stem. It is not clear whether this tRNA^{Glu} can also function in protein synthesis. Since only one tRNA^{Glu} gene was found on chloroplast genomes, it is likely that the three isoacceptor tRNAs fractionated by HPLC differ in their posttranscriptional modifications.

B. The Ribosomal Components of the Translation Machinery

1. Chloroplast Ribosomes

Chloroplast ribosomes, discovered by Lyttleton in 1962, share a considerable number of features with prokaryotic ribosomes: their sedimentation coefficient is about 70S, and they consist of a small (30S) and a large (50S) subunit. The ribosomal subunits from chloroplasts and eubacteria are functionally interchangeable (Lee and Evans, 1971). Chloroplast protein synthesis is inhibited by antibiotics which also affect translation in prokaryotes and whose site of action has been identified as the ribosome. These antibiotics include chloramphenicol, erythromycin, kanamycin, streptomycin, lincomycin, and spectinomycin. Antibiotics such as cycloheximide, which inhibit translation on cytoplasmic ribosomes (which have a sedimentation coefficient of 80S and whose subunits are 40S and 60S), do not affect the function of chloroplast ribosomes. These observations indicate a close relationship between the structural components of the chloroplast and bacterial translation machineries, a relationship which will even become more evident below (see Table IV).

Chloroplast ribosomes are found free in the stroma, and associated with thylakoids. The presently available, limited data suggest that thylakoid-bound ribosomes synthesize intrinsic membrane polypeptides, while other polypeptides are made by both thylakoid and stroma ribosomes (Margulies, 1986).

2. Ribosomal RNAs

Chloroplast ribosomes from flowering plants contain four RNA species, three (23S, 5S, and 4.5S) in the large ribosomal subunit, and one (16S) in the small subunit. The sizes of the chloroplast 23S and 16S rRNAs have been estimated to be approximately 2850 and 1490 nucleotides, respectively. The 5S rRNA is about 120 nucleotides long. The 4.5S rRNA, which is missing in liverwort, moss, and fern (*Adianthum*) (Bowman and Dyer, 1979), and in algae such as *Euglena* and *Chlamydomonas,* is usually about 100 nucleotides long, except in broad bean and pea chloroplasts, where it is considerably shorter (only about 70 nucleotides). The 4.5S rRNA found in higher plant chloroplast ribosomes corresponds to the 3' end of the 23S rRNA in *E. coli* with which it shares 61% sequence homology. The large ribosomal subunit in *Chlamydomonas* chloroplasts has two extra rRNA species (7S and 3S) which appear to be structural equivalents of the 5' end of the 23S rRNA from higher plant chloroplasts and from bacteria. All rRNAs arise by specific endonucleolytic cleavages during the processing of a primary transcript which contains all rRNA species (Hartley, 1979). Ribosomal RNAs have a high degree of secondary and tertiary structure (Noller, 1984), and they provide the backbone to the complex ribosome architecture. Ribosome assembly is initiated by a coordinated binding of a number of ribosomal proteins to the ribosomal RNAs. Ribosomal RNAs are also involved directly in mRNA binding by a specific interaction of a sequence at the 3' end of the 16S rRNA with a complementary sequence on chloroplast mRNAs, located a few nucleotides upstream of the AUG initiation codon (see Table II). This sequence in the mRNA is called the Shine–Dalgarno sequence or ribosome binding site.

There is a high degree of structural homology between chloroplast and eubacterial ribosomal rRNAs (about 70% homology in the 23S rRNA and about 75% in the 16S rRNA). The large rRNAs from *E. coli* show a small number (less than 1%) of modified nucleosides; as the data available on the large chloroplast rRNAs are derived from DNA sequencing, the extent of base modification, as well as the nature and location of the modified nucleotides in chloroplast rRNAs are unknown.

Ribosomal RNAs can be involved in antibiotic resistance. In *Euglena* chloroplasts for instance, as well as in *E. coli*, the same point mutation in the 16S rRNA gene is linked to streptomycin resistance (Montandon *et al.*, 1985, 1986).

3. Ribosomal Proteins

Comparative studies of chloroplast and cytoplasmic ribosomal proteins by two-dimensional gel electrophoresis have shown, in the case of *Chlamydomonas*, that chloroplast proteins differ from their cytoplasmic counterparts in their number and in their electrophoretic mobilities (Hanson *et al.*, 1974). Similar experiments performed using tobacco led to the same observation (Capel and Bourque, 1982). These and other studies on chloroplast ribosomal proteins of pea (Eneas-Filho *et al.*, 1981) and spinach (Mache *et al.*, 1980; Posno *et al.*, 1984) have revealed a total number of chloroplast ribosomal proteins of approximately 60, with 22 to 25 proteins in the small subunit and 32 to 36 in the large subunit. This number is close to that found in the case of eubacterial ribosomes (55 spots could be resolved electrophoretically in the case of *E. coli*), but significantly lower than that found for cytoplasmic ribosomes from higher plants (73 to 80 proteins). When spinach chloroplast ribosomal proteins were first compared to those from *E. coli*, it could be shown that four chloroplast and bacterial proteins comigrated on a two-dimensional gel and that at least two bacterial proteins cross-reacted with antibodies directed against chloroplast ribosomal proteins (Dorne *et al.*, 1984).

E. coli ribosomes contain, as mentioned above, 55 proteins: 21 are present in the small ribosomal subunit (S1 to S21) and 34 in the large subunit (L1 to L34). The amino acid sequence of all these ribosomal proteins has been determined (Giri *et al.*, 1984). One protein of the small subunit (S20) and one from the large subunit (L26) are identical. This protein seems to lie at the interface of the two subunits. L7 and L12 differ only in their amino end, which is acylated in the case of L7. L7 and L12 are present in ribosomes as a tetramer of the complex L7/L12. L8 has been characterized and found to be a complex containing L7/L12 and L10. The number of different proteins of the *E. coli* ribosome is therefore 52.

The position of most of the individual proteins in the *E. coli* ribosome has been determined. One technique used to do this was deuterium labeling and neutron scattering (Ramakrishnan *et al.*, 1984). Immunochemical methods have shown that antigenic determinants of most ribosomal proteins are found on the surface of the ribosome.

The number of ribosomal proteins in chloroplasts appears to be slightly higher than in *E. coli*, but final conclusions await further analytical data on the individual proteins. Most information on their structure has been obtained by sequencing of chloroplast ribosomal protein genes. Approximately 20 to 25 ribosomal proteins appear to be encoded by the chloroplast genome (Eneas-Filho *et al.*, 1981; Posno *et al.*, 1984; Shinozaki *et al.*, 1986b) and the remainder are nuclear-encoded. The sequence homology at the amino acid level between the chloroplast and the corresponding *E. coli* proteins varies between 23 and 68%. Immunological cross-reactions with antibodies directed toward individual ribosomal proteins from *E. coli* have shown that a significant number (at least 13) of chloroplast ribosomal proteins share antigenic determinants with bacterial ribosomal proteins (Bartsch, 1985; see also Table IV). As a consequence of these structural similarities with *E. coli* ribosomal proteins, many chloroplast proteins have been named after their bacterial counterparts. Chloroplast ribosomal proteins have also been named on the basis of their electrophoretic mobilities (Mache *et al.*, 1980; Posno *et al.*, 1984). These two independent nomenclatures have introduced some confusion concerning the identity of a number of chloroplast ribosomal proteins.

E. coli ribosomal proteins contain some modified amino acids. Most of these modifications are methylations; some are acetylations. In chloroplasts, the extent and nature of amino acid modification in most of the ribosomal proteins is unknown. Whereas in *E. coli* no ribosomal protein seems to be phosphorylated (Gordon, 1971), at least two ribosomal proteins have been found to be phosphorylated *in vivo* in spinach chloroplasts (Posno *et al.*, 1984; Guitton *et al.*, 1984). These phosphorylations are light dependent. Phosphorylation of ribosomal proteins is more common in eukaryotic ribosomes (Wool and Stöffler, 1974); in soybean cotyledons, four polypeptides of the 40S ribosomal subunit could be phosphorylated by a cyclic AMP-independent protein kinase associated with ribosomes (Gowda and Pillay, 1980), and in the case of spinach 80S ribosomes four phosphorylated ribosomal proteins have been identified (Posno *et al.*, 1984).

Some chloroplast ribosomal proteins appear to be involved in antibiotic resistance. In *Chlamydomonas*, for instance, erythromycin-resistant mutants were shown to contain altered chloroplast ribosomal proteins (Mets and Bogorad, 1972; Davidson *et al.*, 1974). In *E. coli*, erythromycin resistance is related to alterations in ribosomal proteins L4 and L22 (Wittmann *et al.*, 1973). Streptomycin resistance and dependence were found to be associated in bacteria with a mutation in ribosomal protein S12. Alterations at the C-terminus of ribosomal protein S4 (deletion and frame shift) lead to the suppression of the streptomycin dependence which is caused by the mutation in protein S12 (Wittmann and Wittmann-Liebold, 1974).

4. Translation Factors

Specific protein factors, which bind transiently to ribosomes or RNAs, are required in the various steps of protein synthesis (Weissbach, 1980). They include initiation factors, elongation factors, and termination factors.

Eubacteria have three protein factors that promote initiation: IF-1, IF-2, and IF-3. Three factors are involved in the elongation step, two of which (EF-Tu and EF-Ts) function in aminoacyl-tRNA binding, and the third factor, EF-G, in translocation. Three factors, RF-1, RF-2, and RF-3, are re-

TABLE IV

Ribosomal Proteins and Translation Factors

Protein	Molecular weight		Homology with *E. coli* (%)	Immunological cross-reaction with *E. coli*[c]
	Chloroplast[a]	*E. coli*[b]		
Ribosomal proteins				
S1		61,159		+[d]
S2	26,943	26,613	33	
S3	25,085	25,852	32	−
S4	23,420	23,137	39	−
S5		17,515		−
S6		15,704		−
S7	17,386	19,732	38	+
S8	15,790	13,996	30	−
S9		14,569		+
S10		11,736		−
S11	14,883	13,728	52	+
S12	13,764	13,606	68[e]	+
S13		12,968		
S14	11,744	11,063	45	−
S15	10,445	10,001	30	
S16	9,921	9,191	34	−
S17		9,573		−
S18	12,052	8,896	30	
S19	10,411	10,299	56	+
S20		9,553		−
S21		8,369		
L1		24,599		+
L2	30,010	29,416	47	+
L3		22,258		+
L4		22,087		−
L5		20,171		−
L6		18,832		+
L7		12,220		
L8		=L7/L12 + L10		
L9		15,531		−
L10		17,737		−
L11		14,874		−
L12	13,576[f]	12,176	49	+
L13		16,019		+
L14	13,738	13,541	55	
L15		14,981		
L16	15,214	15,296	56	
L17		14,364		+
L18		12,770		−
L19		13,002		−
L20	15,541	13,366	45	

(*continued*)

TABLE IV (continued)

Protein	Molecular weight		Homology with E. coli (%)	Immunological cross-reaction with E. coli[c]
	Chloroplast[a]	E. coli[b]		
L21		11,565		
L22	17,769	12,227	25	–
L23	10,763	11,013	23	
L24		11,185		–
L25		10,694		
L26		9,553		
L27		8,993		
L28		8,875		
L29		7,274		
L30		6,411		
L31		6,971		
L32		6,315		
L33	7,693	6,255	26	
L34		5,183		
Initiation factors				
IF-1	9,107[g]	8,119	43	
IF-2		118,000[h]		
IF-3	50,000[i]	20,695		
Elongation factors				
EF-Tu	45,011[j]	43,225	70	
EF-Ts		30,000[h]		
EF-G	77,000[k]	80,000[h]		
Termination factors				
RF-1		36,000[l]		
RF-2		38,000[l]		
RF-3		46,000[l]		
RRF		23,500[m]		

[a] The molecular weights of all chloroplast ribosomal proteins, except for L12, are derived from sequencing data of the corresponding genes from tobacco.

[b] The molecular weight values of all *E. coli* ribosomal proteins, initiation factors IF-1 and IF-3, and elongation factor EF-Tu are taken from Wittmann (1982).

[c] All data, except for S1, are taken from Bartsch (1985).

[d] V. Hahn, R. Mache, and P. Stiegler, personal communication.

[e] *Euglena* chloroplast ribosomal protein S12 (Montandon and Stutz, 1984).

[f] Ribosomal protein L12 from spinach chloroplasts: molecular weight derived from protein sequencing data (Bartsch *et al.*, 1982).

[g] Initiation factor IF-1 from spinach chloroplasts derived from gene sequence (Sijben-Müller *et al.*, 1986).

[h] From Weissbach (1980).

[i] Initiation factor IF-3 from *Euglena* chloroplasts (Kraus and Spremulli, 1986).

[j] Elongation factor EF-Tu from *Euglena* chloroplasts (Montandon and Stutz, 1983).

[k] From Tiboni and Ciferri (1986).

[l] From Watson *et al.* (1987).

[m] From Ryoji *et al.* (1981).

sponsible for chain termination and are called release factors. RF-1 and RF-2 have been shown to share common antigenic determinants (Ratcliffe and Caskey, 1977). A ribosome release factor, RRF, releases the ribosome from the mRNA and tRNAs.

In eukaryotes, translation factors are complex in number and in structure. There are at least eight initiation factors (eiF-1, eiF-2, eiF-3, eiF-4A, eiF-4B, eiF-4C, eiF-4D, and eiF-5), four elongation factors (eEF-Tu, eEF-Ts, eEF-2, and eEF-3), but only one termination factor (RF). For more details see Moldave (1985).

The chloroplast translation factors thus far characterized appear to share significant homology with the factors from *E. coli*. *Euglena* chloroplast elongation factor EF-Tu shares 70% homology in amino acid sequence (Montandon and Stutz, 1983) and spinach chloroplast initiation factor IF-1 has 43% homology (Sijben-Müller *et al.*, 1986) with its bacterial counterpart. Spinach chloroplast EF-Tu activity is affected by kirromycin, and antibiotic that specifically interacts with EF-Tu from many eubacteria (Tiboni and Ciferri, 1986). Spinach chloroplast initiation factors IF-1 and IF-2 and elongation factors EF-T and EF-G are active with *E. coli* ribosomes and vice versa (Tiboni *et al.*, 1976). *E. coli* initiation factors IF-2 and IF-3 stimulate the binding of formylmethionyl-tRNA to *Euglena* ribosomes (Graves and Spremulli, 1983). IF-3 from *Euglena* chloroplasts facilitates dissociation of *E. coli* ribosomes and stimulates the formation of the initiation complex on *E. coli* ribosomes (Kraus and Spremulli, 1986); IF-2 from *Euglena* chloroplasts cannot, however, replace the bacterial factor during initiation of protein synthesis on bacterial ribosomes (Gold and Spremulli, 1985). These observations suggest that, in higher plant chloroplasts, and maybe to a lesser extent in *Euglena* chloroplasts, the mechanisms of initiation, elongation, and termination are similar to those known for *E. coli*.

C. The Ribosome Cycle in Protein Synthesis

1. Initiation

A mRNA cannot bind to an intact (70S) ribosome. The first step in initiation is therefore the dissociation of the ribosomes into ribosomal subunits, which appears to be promoted by the binding to the 70S ribosome of initiation factor IF-3 and, to a minor extent, of IF-1. In a subsequent step, IF-3 is translocated onto the inner surface of the 30S subunit and acts as an antiassociation factor. Two proteins are involved in the binding of the mRNA to the 30S subunit of the ribosome: IF-3 and ribosomal protein S1. The latter protein is also responsible for in-phase positioning of the mRNA, i.e., the precise fitting of the AUG codon in the P site of the ribosome (Subramanian, 1985). The binding of the initiator, formylmethionyl-tRNA, to this complex is mediated by IF-2, which forms a binary complex with this tRNA. The last step in initiation is the binding of the 50S ribosomal subunit to the initiation complex with a concomitant hydrolysis of GTP (by IF-2), and subsequent release of all initiation factors. After this step, the formylmethionyl-tRNA finds itself placed in the P site of the 70S ribosome, while the A site is available for the attachment of the second aminoacyl-tRNA, which will lead to the formation of the first peptide bond between the methionine and the second amino acid.

2. *Elongation*

Elongation proceeds in three steps: (1) binding of the aminoacyl-tRNA to the A site, (2) peptide bond formation, and (3) translocation. Binding of the next aminoacyl-tRNA (selected according to its anticodon, which must be complementary to the next codon) to the A site is promoted by its association with elongation factor EF-Tu, the active form of the latter carrying GTP. After the aminoacyl-tRNA is positioned on the ribosome, which involves hydrolysis of GTP, EF-Tu.GDP is released. This complex is inactive and is dissociated by the binding of factor EF-Ts to EF-Tu. EF-Ts is then displaced by GTP to restore the active complex EF-Tu.GTP.

Peptide bond formation is catalyzed by a peptidyltransferase, with ribosomal proteins L2, L3, L4, L15, and L16 involved in the reaction. In bacteria, chloroplasts, and mitochondria, peptidyltransferase activity is inhibited by chloramphenicol. During peptide bond formation, the growing polypeptide chain, initially attached to the tRNA in the P site, is transferred to the aminoacyl-tRNA present in the A site, leaving the P site with an uncharged tRNA. The uncharged tRNA is expelled in a translocation step, which moves the peptidyl-tRNA from the A site to the P site, leaving the ribosome with an empty A site accessible to the next aminoacyl-tRNA. The translocation step is triggered by the association of another elongation factor, EF-G, with the ribosome. This step also requires hydrolysis of GTP.

3. *Termination*

Termination of protein synthesis occurs when ribosomes are stalled at codons for which there is no tRNA with a complementary anticodon. These stop codons (UAA, UAG, and UGA) are recognized by release factors RF-1 (for codons UAA and UAG) and RF-2 (for codons UAA and UGA). The presence of a release factor on a nucleotide triplet leads to the cleavage of the ester bond between the polypeptide and the tRNA in the P site. A third release factor, RF-3, appears to stimulate the action of the other two. Following the release of the polypeptide, the ribosome release factor (RRF) interacts with the ribosome and removes the last tRNAs and the mRNA (Ryoji *et al.*, 1981).

D. Genetic Code and Codon Usage in Chloroplasts

The correspondence between codons and amino acids is determined by comparing the DNA-derived amino acid sequence of a protein with data obtained by direct sequencing of the protein. In the case of higher plant chloroplasts, when the proteins and the available corresponding genes are analyzed (large subunit of RuBPCase from tobacco, subunit III of spinach ATPase, and some partially sequenced chloroplast proteins), only 58 sense codons are represented, all in agreement with the universal code. Three codons, namely AGG, UCG, and CUC, could not be assigned an amino acid on the basis of the above-described comparisons. These codons are found in chloroplast genes, but at a low frequency (see Table V). Thus is is likely that the chloroplast genetic code does not deviate from the universal code; in mitochondria, several deviations have been reported (Barrell *et al.*, 1980; Heckman *et al.*, 1980; Bonitz *et al.*, 1980).

To illustrate codon usage in chloroplast protein genes, the frequency of the individual codons in a total of 19 chloroplast protein genes from three different plants (maize, tobacco, and spinach) has been studied and the results are summarized in Table V. In a total of 7283 codons considered there is a strong preference for U or A in the wobble position of each codon family; the only frequently used codons terminating with C are serine codon UCC, threonine codon ACC, isoleucine codon AUC, and the only frequently used codon terminating with G is leucine codon UUG. In fact, 39.6% of the codons terminate with U, 29.4% with A, 15.6% with C, and 15.4% with G. A similar codon usage pattern was found in yeast mitochondria (Bonitz *et al.*, 1980; Sibler *et al.*, 1986). In human mitochondria, codons ending in A or C predominate (36.4 and 41%, respectively), while those ending in U or G are used less frequently (16.3 and 6.4%, respectively) (Anderson *et al.*, 1981).

A nonrandom usage of synonymous codons is also found in genes from other organisms, prokaryotes as well as eukaryotes. In *E. coli*, the pattern of nonrandom codon usage varies with levels of gene expression. A strong positive correlation between codon usage and tRNA content was observed and the extent of this correlation was found to relate to the level of production of the proteins encoded by the individual genes studied (Ikemura, 1985). In chloroplasts, a strong correlation exists between the amounts of certain tRNAs and the frequency of the corresponding codons (Pfitzinger *et al.*, 1987).

E. Transfer RNAs and Codon Recognition in Chloroplasts

Higher plant chloroplasts contain 30 to 35 tRNA species. This number is lower than that in bacteria (39 species), but significantly higher than in mammalian (22 species) or yeast (24 species) mitochondria. Since at least 32 species are required, according to the wobble hypothesis, to read the complete set, chloroplast tRNAs would be expected to show no deviation from the classical wobble theory. Recent studies on the structure of chloroplast tRNAs and their genes have, however, revealed that codon recognition in chloroplasts is more complex than expected.

For the present discussion, the case of tobacco is considered. The tobacco chloroplast genome carries genes for 30 different tRNAs (Shinozaki et al., 1986b). If nucleotide modification in the anticodon is not considered (in spite of the fact that it has been shown—at least in one case—to occur and to cause a change in codon recognition, as discussed above), the 30 different genes which have been sequenced and which account for the complete set of chloroplast tRNAs, encompass only 28 anticodons, a figure clearly too low to read all the codons. With a non-Watson–Crick base pair (U-G) allowed in the wobble position, the minimum number of different anticodons required would be 31 (the minimum number of different tRNAs would be 32 because of the initiator tRNA^{Met}). Additional new base pairings must therefore be considered if the complete set of 61 sense codons is to be read by the known anticodons.

A possible candidate for multiple interactions in the wobble position is U (in the tRNA). In chloroplasts a single tRNA^{Ala} is able to read all four codons for alanine. The anticodon of this tRNA, as derived from its gene structure, is UGC (U in the wobble position). There is also a single tRNA^{Pro}, reading all four codons for proline. In this case, the nucleotide in the wobble position in the tRNA is also a U (modified). If this situation were common in chloroplasts, only one tRNA^{Thr} (that with the anticodon UGU) would read the four threonine codons, one tRNA^{Val} (that with anticodon UAC) the four valine codons, one tRNA^{Gly} (anticodon UCC) the four glycine codons, two tRNA^{Ser} the six serine codons, etc. The presence of two tRNAs for threonine, valine, and glycine and three tRNAs for serine and leucine suggests that additional factors are involved in codon–anticodon recognition, a point known for a long time. In fact, in some cases a U in the wobble position of a tRNA can only read A and G in the wobble position of the codon if correct translation is to be secured. An example is tRNA^{Leu}(UAA), which can read the leucine codons UUA and UUG, but not the codons UUU and UUC, which are phenylalanine codons. Similarly, tRNA^{Lys}(UUU) reads only the lysine codons AAA and AAG, but not the codons AAU and AAC, which are asparagine codons.

It has been shown that some modified nucleosides at position 34 in the tRNA (wobble position) increase the wobble capacity to four nucleotides (reading a family of four codons for the same amino acid); other modifications restrict the wobble interaction to a single nucleotide (Björk, 1984; Yokoyama et al., 1985). In chloroplasts, 6 of the 14 tRNAs sequenced have modified nucleotides (U or C) in the wobble position. Their effect on codon recognition, however, is not understood.

TABLE V
Codon Recognition

Amino acid	Anticodon[a]	References[b] tRNA	References[b] tDNA	Codons recognized[c]
Ala	UGC		1–4	GCU (4.24), GCA (2.31), GCC (1.34), GCG (0.75)
Arg	UCU		3, 5, 6	AGA (1.17), AGG (0.36)
	ACG		3, 5, 6, 7	CGU (1.80), CGA (0.96), CGC (0.63), CGG (0.20)
	CCG		8	CGG[d]
Asn	GUU		3, 5, 6, 9	AAU (2.42), AAC (1.17)
Asp	GUC		3, 9–11	GAU (3.13), GAC (1.00)
Cys	GCA		3, 7, 9, 11	UGU (0.51), UGC (0.27)
Gln	UUG		3, 7, 12	CAA (2.60), CAG (0.82)
Glu	*UUC	13	3, 6, 7, 10, 14–16	GAA (4.02), GAG (1.21)
Gly	GCC		3, 7, 11, 17	GGU (3.65), GGC (1.29)
	UCC		3, 7, 16, 17	GGA (3.06), GGG (1.47)
	GUG		3, 6, 7, 18–20	CAU (1.96), CAC (0.81)
His	GAU	21	1–4	AUU (3.56), AUC (1.55)
Ile	*CAU	22	3, 23	AUA (1.46)
Leu	*CAA	24, 25	3, 6, 26, 27	UUG (2.14)
	*UAA	24, 25	3, 7, 28–30	UUA (3.49), UUG (2.14)
	UAG*	24, 25, 31	3, 7	CUU (2.11), CUA (1.37), CUG (0.60), CUC (0.36)
Lys	UUU		3, 32	AAA (2.75), AAG (0.89)
Met	CAU (i)	33–35	3, 7, 16, 17	AUG (0.25), GUG (0.01)
	CAU	36	3, 7, 26, 37	AUG (2.14)
Phe	GAA	30, 37, 38	3, 7, 26, 27	UUU (3.49), UUC (1.99)

Pro	*UGG	40	3	CCU (2.24), CCA (1.06), CCC (0.88), CCG (0.52)
Ser	GCU		3, 7	AGU (1.23), AGC (0.37)
	GGA		3, 26	UCU (1.72), UCC (1.11)
	UGA		3, 17, 41, 42	UCA (0.73), UGC (0.62)
Thr	GGU	43	3, 14–16, 44	ACU (2.58), ACC (1.45)
	UGU		3, 7, 25	ACA (1.43), ACG (0.48)
Trp	CCA	45	3, 7	UGG (2.09)
Tyr	GUA	46	3, 6, 7, 10, 14–16	UAU (2.26), UAC (0.85)
Val	GAC		3, 6, 47, 48	GUU (2.59), GUC (0.52)
	*UAC	49	3, 7, 37, 41	GUA (2.87), GUG (0.78)

[a] Asterisk indicates a modified nucleotide.

[b] References: (1) Graf et al., 1980; (2) Koch et al., 1981; (3) Wakasugi et al., 1986; (4) Schneider and Rochaix, 1986; (5) Keus et al., 1984; (6) Shapiro and Tewari, 1986; (7) Hallick et al., 1984; (8) Ohyama et al., 1986; (9) Holschuh et al., 1983; (10) Rasmussen et al., 1984; (11) Quigley et al., 1985; (12) Steinmetz et al., 1985; (13) Schön et al., 1986; (14) Kuntz et al., 1984; (15) Holschuh et al., 1984a; (16) Quigley and Weil, 1985; (17) Oliver and Poulsen, 1984; (18) Schwarz et al., 1981a; (19) Spielmann and Stutz, 1983; (20) Zurawski et al., 1984; (21) Guillemaut and Weil, 1982; (22) Francis and Dudock, 1982; (23) Kashdan and Dudock, 1982b; (24) Osorio-Almeida et al., 1980; (25) Pillay et al., 1984; (26) Steinmetz et al., 1983; (27) Bonnard et al., 1985; (28) Steinmetz et al., 1982; (29) Bonnard et al., 1984; (30) Ma and Doebley, 1986; (31) Canaday et al., 1980a; (32) Sugita et al., 1985; (33) Calagan et al., 1980; (34) Canaday et al., 1980b; (35) McCoy and Jones, 1980; (36) Pirtle et al., 1981; (37) Zurawski and Clegg, 1984; (38) Chang et al., 1976; (39) Guillemaut and Keith, 1977; (40) Francis et al., 1982; (41) Krebbers et al., 1984; (42) Holschuh et al., 1984b; (43) Kashdan et al., 1980; (44) Kashdan and Dudock, 1982a; (45) Canaday et al., 1981; (46) Green and Jones, 1985; (47) Schwarz et al., 1981b; (48) Briat et al., 1982; (49) Sprouse et al., 1981.

[c] The figures in parentheses indicate the relative frequencies (in %) of the various codons in chloroplast protein genes. The genes considered include the tobacco genes $rbcL$, $psbA$, $rps19$, $atpA$, $atpB$, $atpE$; the spinach genes $rbcL$, $petA$, $petB$, $petD$, $atpH$, $psbB$, $psbC$, $psbD$; and the maize genes $rbcL$, $atpB$, $rps4$, $psaA$, $psaB$ (a total of 7283 codons).

[d] Gene only found in liverwort, not in tobacco chloroplasts.

Whereas individual tRNAs are able to recognize several codons, it is not clear whether the same codon can be read by different isoacceptor tRNAs. This question can be asked in the case of the leucine codon UUG. This codon is likely to be read by the tRNA^{Leu}(UAA) and tRNA^{Leu}(CAA). Since tRNA^{Leu}(UAA) is required to read the codon UUA, and since it can also read the codon UUG, the tRNA^{Leu}(CAA) would be dispensable. However, tRNA^{Leu}(CAA) is present in all the plant species studied, and has recently been found on polyribosomes isolated from purified chloroplasts (P. Guillemaut, personal communication). It appears therefore that tRNA^{Leu}(CAA) is also required to read the UUG codons in chloroplast mRNAs.

U and G are the most common bases in the wobble position of tRNAs (12 tRNAs in each case). C is found in this position in five chloroplast tRNAs. Only one chloroplast tRNA has an A in the wobble position: tRNA^{Arg}(ACG). If A is unmodified, it should read only the arginine codon CGU. As the second arginine isoacceptor, tRNA^{Arg}(UCU), can only read the two arginine codons AGA and AGG (AGC and AGU are serine codons), half of the six arginine codons would not be read. This reveals another unusual situation, where an extension of the wobble theory must be considered. In eukaryotes, an A in the wobble position is usually modified into inosine (I) which can pair with U, C, and A, but not with G. If we assume therefore that in chloroplasts A is modified into I, the arginine codon CGG would still be left unread. How this codon is read remains unclear, not only in tobacco (and other) chloroplasts, but also in yeast mitochondria, where A was shown to be unmodified (Sibler *et al.*, 1986). The only plausible explanation would be in this case a "two out of three" mechanism, strengthened by the fact that two G-C base pairs could confer enough stability to the codon–anticodon interaction (Lagerqvist, 1978).

In liverwort (*Marchantia polymorpha*) chloroplast DNA, two other tRNA genes have been identified in addition to the 30 tRNA genes also found on the tobacco chloroplast genome (Ohyama *et al.*, 1986). One of these two genes corresponds to that of a tRNA^{Arg}(CCG) which would be required to read the arginine codon CGG. The second gene corresponds to a tRNA^{Pro}(GGG) gene, with, however, an incomplete acceptor stem structure and therefore may be a pseudogene.

The codon recognition pattern shown in Table V has been constructed by considering optimal base pairings between universal codons and the anticodons available in chloroplasts from three higher plants (maize, spinach, and tobacco). It can be observed that

1. When U (modified or not) is the wobble base in the anticodon, the preferential order for the codon wobble bases is, for a four codon family, U > A > C > G; for a two codon family, A > G.

2. In the case of the only tRNA having A in the wobble position (arginine anticodon ACG), the same preferential order for the codon wobble bases is observed as in 1: U > A > C > G.

3. For G in the anticodon wobble position the preferential order is U > C.
4. C in the anticodon wobble position recognizes G, except in the case of tRNA^{Ile}(*CAU), where the C with an undetermined modification pairs with A.

These observations show that the most stable codon–anticodon interaction is not always the most frequently used. A too strong or a too weak interaction between codon and anticodon could in fact interfere with a fast and accurate translation.

V. MATURATION OF PROTEINS

Primary translation products generally undergo a number of modifications before reaching their final form. These modifications include removal of the formyl group of the N-terminal methionine, or of the N-terminal methionine (these generally occur before the polypeptide chain is completed), proteolytic digestion involving removal of short oligopeptides at the amino- or the carboxy-terminus, and amino acid modifications.

Nuclear-encoded chloroplast proteins, which are synthesized on cytoplasmic ribosomes and are subsequently taken up by the chloroplast, contain transit peptides which allow passage through the chloroplast membrane (Chua and Schmidt, 1979; Schmidt et al., 1979) and also determine the subsequent localization of the proteins within the organelle (thylakoid lumen or stroma) (Smeekens et al., 1986). The transit peptide is removed, either during or after the transport, by a specific peptidase. Most chloroplast-encoded proteins stay in the chloroplast where they can be either free (in the stroma), engaged in organellar structures (such as the ribosomes), or membrane-bound (to the thylakoids).

Only a few chloroplast proteins have been purified to date and studied at the level of their amino acid sequence. Removal of the N-terminal methionine has been observed for spinach cytochrome b_{559} (Widger et al., 1984), and for the ε subunit of wheat chloroplast ATPase (Howe et al., 1985), but not for subunit III of spinach chloroplast ATPase, which even remains formylated (Sebald and Wächter, 1980). Proteolytic digestion removes 12 to 16 amino acid residues at the C-terminus of the 32-kDa protein (Marder et al., 1982). In the large subunit of tobacco RuBPCase the N-terminus is heterogeneous as a result of two different proteolytic cleavages, one removing a tetrapeptide and the second removing a 14-residue-long oligopeptide (Amiri et al., 1984). In the case of spinach chloroplast cytochrome f, a 35-amino acid-long polypeptide is removed from the N-terminus (Alt and Herrmann, 1984).

The only type of amino acid modification characterized so far in chloroplast proteins is phosphorylation. At least two ribosomal proteins are phosphorylated on serine residues (Guitton et al., 1984; Posno et al., 1984) and the nuclear-encoded light-harvesting chlorophyll a/b proteins are modified on threonine residues (Clark et al., 1985); the amino acids phosphorylated in the nuclear-encoded pyruvate,orthophosphate dikinase [located in the stromal fraction of maize mesophyll chloroplasts (Foyer, 1984)] and in the 10-kDa protein of photosystem II from spinach chloroplasts (Farchaus and Dilley, 1986) are unidentified. Protein kinase activities have been found associated with the outer membrane (Soll and Buchanan, 1983), the soluble stromal fraction (Bennett, 1984), and the thylakoid membrane (Bennett, 1979). A protein kinase is also present on chloroplast ribosomes (Guitton et al., 1984).

Glycosylation of a chloroplast-encoded protein has not been observed. It is believed that this modification occurs in the cytoplasm and facilitates transport of nuclear-encoded proteins into the chloroplasts.

Very few polypeptides synthesized in chloroplasts carry out their function as monomeric independent units. Most become part of complex structures such as the ribosomes, photosystem I complex, photosystem II complex, ATPase complex (nine subunits), cytochrome b_6/f complex, and RuBPCase (which in higher plants exists as a hexadecamer of eight large subunits and eight small subunits). In many cases, the incorporation into a complex structure is associated with a processional step.

VI. POSTTRANSCRIPTIONAL REGULATION OF CHLOROPLAST GENE EXPRESSION

Gene expression can be controlled at various levels in the series of reactions leading from transcription of a gene to the final functional protein. The best known case of regulation of a chloroplast protein at the level of transcription is that of the large subunit (LS) of RuBPCase in maize (a C_4 plant): this gene is transcribed in bundle sheath but not in mesophyll cells (Link et al., 1978). In pea, Shinozaki et al. (1982) reported a 20- to 30-fold increase in LS mRNA after illumination of dark-grown plants for 48 hr. These data are not consistent with those published by Thompson et al. (1983) and Inamine et al. (1985), who showed that the hybridizable amount of LS mRNA, already significant in the plastids of dark-grown plants, increases only about threefold, relative to total plant RNA, when dark-grown plants are illuminated. This increase in LS mRNA could be accounted for by a similar increase in genome copy number. The increase in specific activity of RuBPCase upon exposure to light, however, was shown to be due to increased synthesis of both subunits of the enzyme and not to light activation of preexisting enzyme present in the dark-grown cells. Similar observations were made by Cushman et al. (1985) in proplastids and chloroplasts from *Euglena* and by Van Grinsven et al. (1985) in green suspension cultures of *Petunia hybrida*.

Light activation (or dark inactivation) of a preexisting enzyme was shown to exist for RuBPCase from chloroplasts of bean and other species (Seemann *et al.*, 1985). In fact, the amount of RuBPCase (determined by immunoprecipitation of the enzyme radioactively labeled with carboxyarabinitol-1,5-bisphosphate) was shown by these authors not to differ significantly in light-grown plants and plants kept in the dark overnight, but they noticed a considerable loss of activity in the dark-grown plants, due to the presence of an inhibitor. This nocturnal inhibitor has recently been identified in potato (Gutteridge *et al.*, 1986) and in bean (Berry *et al.*, 1987): its structure is that of 2-carboxy-D-arabinitol-1-phosphate. A loss of RuBPCase activity in the dark was not observed in a number of plant species including pea.

Posttranscriptional control of gene expression has also been described for the 32-kDa protein of *Spirodela* chloroplasts (Fromm *et al.*, 1985). In tissue containing mature chloroplasts, light–dark regimes have no substantial effect on the levels of the psbA transcript, but the synthesis of the protein decreases considerably in the dark. Rapid turnover of the 32-kDa membrane protein under high light intensities has been noted, and is considered to be the cause of photoinhibition of photosynthesis in higher plants (Kyle, 1985).

An important factor in posttranscriptional regulation of gene expression is the stability of mRNAs, and differential stability of mRNAs has been shown to occur in the photosynthesis operon in *Rhodopseudomonas capsulata* (Belasco *et al.*, 1985).

Translation of mRNAs can be affected by the secondary structure of the mRNA, masking the ribosome binding site(s) which is (are) therefore unavailable for the interaction with the 16S rRNA. A specific endonucleolytic cleavage upstream of this site can uncover a ribosome binding site by destroying the secondary structure and subsequently promote initiation of translation. Alternatively, long upstream regions may be required for efficient translation of the mRNA: Mullet *et al.* (1985a) observed that in pea and spinach chloroplasts the highest rate of large subunit synthesis is found in plastid populations enriched in the largest rbcL transcript.

In prokaryotes, some proteins have been identified that bind in a highly specific manner to one or a few mRNAs, thus preventing translation (e.g., ribosomal proteins as translational repressors in *E. coli;* see Nomura *et al.*, 1984). On the other hand, RNAs complementary to the mRNA (antisense RNAs, transcribed from the opposite DNA strand) could impede translation by base-pairing with the upstream or coding region of the mRNA. This mechanism has been proposed to explain translational regulation in the case of the transposase of IS10 (Simons and Kleckner, 1983) and the OmpF protein (Mizuno *et al.*, 1984). Overlapping, divergently transcribed genes are candidates for this type of control.

In the case of the ATPase operon of *E. coli*, the synthesis of the three subunits encoded by the *unc* operon in stoichiometric amounts is determined by the efficiency of translation initiation (McCarthy *et al.*, 1985). Translation can also be regulated at the level of polypeptide elongation: tRNA content, unusual codon usage pattern, and codon context are believed to modulate translation in bacteria (Ikemura, 1985; Yarus and Folley, 1985).

VII. CONCLUDING REMARKS

The aim of the present chapter was to review basic features of protein synthesis and the components of the translation apparatus in chloroplasts. In retrospect, it is clear that, although much information has been obtained, our knowledge of the subject is still incomplete. Most of the components involved in protein synthesis in chloroplasts have been identified, but very few have been purified in amounts allowing further characterization. Molecular cloning and gene sequencing have allowed the determination of primary structures of RNAs (rRNAs and tRNAs) and some (chloroplast-encoded) proteins, but this approach does not substitute for a detailed analysis of the final components themselves.

Whereas data are available on the chloroplast-encoded components of the translation machinery, this is not the case for nuclear-encoded proteins, including the entire set of chloroplast aminoacyl-tRNA synthetases, most translation factors, and about two-thirds of the ribosomal proteins.

Understanding the interactions between the various components during protein synthesis also requires knowledge of their three-dimensional structure. Such studies require purification and crystallization of the individual components, and are presently under way in organisms such as yeast and *E. coli*. The rules governing RNA and protein interactions derived from these studies will no doubt be useful to understand many features of chloroplast protein synthesis.

Chloroplast mRNAs are fairly well characterized. Most, but not all, have Shine–Dalgarno sequences which are presumed to interact with initiation factor IF-3 and ribosomal protein S1 in the binding of the mRNA to the ribosome during the initiation step. How mRNAs lacking these sequences interact with ribosomes is not known.

RNA processing, which includes specific cleavages in precursor RNAs, as well as nucleotide modifications (especially in tRNAs), has not extensively been studied in chloroplasts. The most fascinating aspect of RNA processing is the splicing of RNAs transcribed from split genes. Chloroplast introns have been classified in three groups (Shinozaki *et al.*, 1986a) and it has been proposed that different mechanisms may be used for splicing these introns, but so far the mechanisms have not been studied in detail. Do chloroplasts contain structures similar to the "spliceosomes" described in the case of yeast and other pre-mRNAs (Brody and Abelson, 1985)?

That the amount and activity of proteins in a cell or organelle can be regulated at levels other than transcription is well documented. The extent to which mRNA structure and stability, codon usage of individual genes, maturation and turnover of proteins are involved in regulating protein synthesis in chloroplasts, remains to be established. It is reasonable to believe that many of these questions will be answered during the next few years.

ACKNOWLEDGMENTS

We are grateful to P. Guillemaut, P. Stiegler, and V. Hahn for communicating results prior to publication; to A. Subramanian for providing data on bacterial and chloroplast ribosomal proteins. We thank P. Guillemaut, M. Keller, M. Kuntz, R. Martin, and P. Stiegler for critical reading of the manuscript.

REFERENCES

Aliyev, K. A., and Filippovitch, I. I. (1968). *Mol. Biol.* **2,** 364.
Alt, J., and Herrmann, R. G. (1984). *Curr. Genet.* **8,** 551.
Alt, J., Winter, P., Sebald, W., Moser, J. G., Schedel, R., Westhoff, P., and Herrmann, R. G. (1983). *Curr. Genet.* **7,** 129.
Alt, J., Morris, J., Westhoff, P., and Herrmann, R. G. (1984). *Curr. Genet.* **8,** 597.
Amiri, I., Salnikow, J., and Vater, J. (1984). *Biochim. Biophys. Acta* **784,** 116.
Anderson, S., Bankier, A. T., Barrell, B. G., de Bruijn, M. H. L., Coulson, A. R., Drouin, J., Eperon, I. C., Nierlich, D. P., Roe, B. A., Sanger, F., Schreier, P. H., Smith, A. J. H., Staden, R., and Young, I. G. (1981). *Nature (London)* **290,** 457.
Bard, J., Bourque, D. P., Hildebrandt, M., and Zaitlin, D. (1985). *Proc. Natl. Acad. Sci. U.S.A.* **82,** 3983.
Barrell, B. G., Anderson, S., Bankier, A. T., de Bruijn, M. H. L., Chen, E., Coulson, A. R., Drouin, J., Eperon, I. C., Nierlich, D. P., Roe, B. A., Sanger, F., Schreier, P. H., Smith, A. J. H., Staden, R., and Young, I. G. (1980). *Proc. Natl. Acad. Sci. U.S.A.* **77,** 3164.
Bartsch, M. (1985). *J. Biol. Chem.* **260,** 237.
Bartsch, M., Kimura, M., and Subramanian, A. R. (1982). *Proc. Natl. Acad. Sci. U.S.A.* **79,** 6871.
Bedbrook, J. R., Link, G., Coen, D. M., Bogorad, L., and Rich, A. (1978). *Proc. Natl. Acad. Sci. U.S.A.* **75,** 3060.
Belasco, J. G., Beatty, J. T., Adams, C. W., von Gabain, A., and Cohen, S. N. (1985). *Cell* **40,** 171.
Belfort, M., Pedersen-Lane, J., West, D., Ehrenmann, K., Maley, G., Chu, F., and Maley, F. (1985). *Cell* **41,** 375.
Bennett, J. (1979). *FEBS Lett.* **103,** 342.
Bennett, J. (1984). *Physiol. Plant.* **60,** 583.
Berry, J. A., Lorimer, G. H., Pierce, J., Seemann, J. R., Meek, J., and Freas, S. (1987). *Proc. Natl. Acad. Sci. U.S.A.* **84,** 734.
Bird, C. R., Koller, B., Auffret, A. D., Huttly, A. K., Howe, C. J., Dyer, T. A., and Gray, J. C. (1985). *EMBO J.* **4,** 1381.
Björk, G. R. (1984). *In* "Processing of RNA" (D. Apirion, ed.), pp. 291–330. CRC Press, Boca Raton, Florida.
Bloom, M., Brot, N., Cohen, B. N., and Weissbach, H. (1986). *Methods Enzymol.* **118,** 309.

Bogorad, L., Bedbrook, J. R., Coen, D. M., Kolodner, R., and Link, G. (1978). *In* "Chloroplast Development" (G. Akoyunoglou and J. Argyroudi-Akoyunoglou, eds.), pp. 541–551. Elsevier/North-Holland, Amsterdam.
Bonitz, S. G., Berlani, R., Coruzzi, G., Li, M., Macino, G., Nobrega, F. G., Thalenfeld, B. E., and Tzagoloff, A. (1980). *Proc. Natl. Acad. Sci. U.S.A.* **77**, 3167.
Bonnard, G., Michel, F., Weil, J. H., and Steinmetz, A. (1984). *Mol. Gen. Genet.* **194**, 330.
Bonnard, G., Weil, J. H., and Steinmetz, A. (1985). *Curr. Genet.* **9**, 417.
Bottomley, W., and Whitfeld, P. (1978). *In* "Chloroplast Development" (G. Akoyunoglou and J. Argyroudi-Akoyunoglou, eds.), pp. 657–662. Elsevier/North-Holland, Amsterdam.
Bowman, C. M., and Dyer, T. A. (1979). *Biochem. J.* **183**, 605.
Briat, J. F., Dron, M., Loiseaux, S., and Mache, R. (1982). *Nucleic Acids Res.* **10**, 6865.
Brody, E., and Abelson, J. (1985). *Science* **228**, 963.
Burkard, G., Eclancher, B., and Weil, J. H. (1969). *FEBS Lett.* **4**, 485.
Burkard, G., Guillemaut, P., and Weil, J. H. (1970). *Biochim. Biophys. Acta* **224**, 184.
Calagan, J. L., Pirtle, R. M., Pirtle, I. L., Kashdan, M. A., Vreman, H. J., and Dudock, B. S. (1980). *J. Biol. Chem.* **255**, 9981.
Canaday, J., Guillemaut, P., Gloeckler, R., and Weil, J. H. (1980a). *Plant Sci. Lett.* **20**, 57.
Canaday, J., Guillemaut, P., and Weil, J. H. (1980b). *Nucleic Acids Res.* **8**, 999.
Canaday, J., Guillemaut, P., Gloeckler, R., and Weil, J. H. (1981). *Nucleic Acids Res.* **9**, 47.
Capel, M. S., and Bourque, D. P. (1982). *J. Biol. Chem.* **257**, 7746.
Carias, J. R., Mouricout, M., and Julien, R. (1981). *Biochim. Biophys. Res. Commun.* **98**, 735.
Chang, S. H., Brum, C. K., Silberklang, M., RajBhandary, U. L., Hecker, L. I., and Barnett, W. E. (1976). *Cell* **9**, 717.
Chazal, P., Thomes, J. C., and Julien, R. (1975). *FEBS Lett.* **56**, 268.
Chu, F. K., Maley, G. F., West, D. K., Belfort, M., and Maley, F. (1986). *Cell* **45**, 157.
Chua, N. H., and Schmidt, G. W. (1979). *J. Cell Biol.* **81**, 461.
Clark, R. D., Hind, G., and Bennett, J. (1985). *In* "Molecular Biology of the Photosynthetic Apparatus" (K. E. Steinback, S. Bonitz, C. J. Arntzen, and L. Bogorad, eds.), pp. 259–267. Cold Spring Harbor Lab., Cold Spring Harbor, New York.
Clayton, D. A. (1984). *Annu. Rev. Biochem.* **53**, 573.
Colas, B., Imbault, P., Sarantoglou, V., Boulanger, Y., and Weil, J. H. (1982a). *Biochim. Biophys. Acta* **697**, 71.
Colas, B., Imbault, P., Sarantoglou, V., and Weil, J. H. (1982b). *FEBS Lett.* **141**, 213.
Cozens, A. L., Walker, J. E., Phillips, A. L., Huttly, A. K., and Gray, J. C. (1986). *EMBO J.* **5**, 217.
Cushman, J. C., Barrasso, D. S., and Price, C. A. (1985). *Proc. Int. Congr. Plant Mol. Biol., 1st* p. 54 (Abstr.).
Davidson, J. N., Hanson, M. R., and Bogorad, L. (1974). *Mol. Gen. Genet.* **132**, 119.
Dietrich, A., Souciet, G., Colas, B., and Weil, J. H. (1983). *J. Biol. Chem.* **258**, 12386.
Dirheimer, G. (1983). *Recent Results Cancer. Res.* **84**, 15.
Dorne, A. M., Eneas-Filho, J., Heizmann, P., and Mache, R. (1984). *Mol. Gen. Genet.* **193**, 129.
Driesel, A., Speirs, J., and Bohnert, H. J. (1980). *Biochim. Biophys. Acta* **610**, 297.
Dron, M., Rahire, M., and Rochaix, J. D. (1982). *J. Mol. Biol.* **162**, 775.
Ellis, R. J. (1981). *Annu. Rev. Plant Physiol.* **32**, 111.
Eneas-Filho, J., Hartley, M. R., and Mache, R. (1981). *Mol. Gen. Genet.* **184**, 484.
Erickson, J. M., Rahire, M., and Rochaix, J. D. (1984). *EMBO J.* **3**, 2753.
Farchaus, J., and Dilley, R. A. (1986). *Arch. Biochem. Biophys.* **244**, 94.
Fish, L. E., Kück, U., and Bogorad, L. (1985). *J. Biol. Chem.* **260**, 1413.
Foyer, C. (1984). *Biochem. J.* **222**, 247.
Francis, M. A., and Dudock, B. S. (1982). *J. Biol. Chem.* **257**, 11195.
Francis, M. A., Kashdan, M., Sprouse, H., Otis, L., and Dudock, B. (1982). *Nucleic Acids Res.* **10**, 2755.

Fromm, H., Devic, M., Fluhr, R., and Edelman, M. (1985). *EMBO J.* **4,** 291.
Fromm, H., Edelman, M., Koller, B., Goloubinoff, P., and Galun, E. (1986). *Nucleic Acids Res.* **14,** 883.
Fukada, K., and Abelson, J. (1980). *J. Mol. Biol.* **139,** 377.
Fukuzawa, H., Kohchi, T., Shirai, H., Ohyama, K., Umesono, K., Inokushi, H., and Ozeki, H. (1986). *FEBS Lett.* **198,** 11.
Gatenby, A. A., Castleton, J. A., and Saul, M. W. (1981). *Nature (London)* **291,** 117.
Gingrich, J. C., and Hallick, R. B. (1985). *J. Biol. Chem.* **260,** 16156.
Giri, L., Hill, W. E., and Wittmann, H. G. (1984). *Adv. Protein Chem.* **36,** 1.
Gold, J. C., and Spremulli, L. (1985). *J. Biol. Chem.* **260,** 14897.
Gordon, J. (1971). *Biochem. Biophys. Res. Commun.* **44,** 579.
Gowda, S., and Pillay, D. T. N. (1980). *Plant Cell Physiol.* **21,** 1357.
Graf, L., Kössel, H., and Stutz, E. (1980). *Nature (London)* **286,** 908.
Graves, M. C., and Spremulli, L. L. (1983). *Arch. Biochem. Biophys.* **222,** 192.
Gray, J. C., Phillips, A. L., and Smith, A. G. (1984). In "Chloroplast Biogenesis" (J. Ellis, ed.), pp. 137–163. Cambridge Univ. Press, Cambridge, England.
Green, G. A., and Jones, D. S. (1985). *Nucleic Acids Res.* **13,** 1659.
Green, C. J., and Vold, B. S. (1983). *Nucleic Acids Res.* **11,** 5763.
Groot, G. S. P., Flavell, R. A., van Ommen, G. J. B., and Grivell, L. A. (1974). *Nature (London)* **252,** 167.
Guillemaut, P., and Keith, G. (1977). *FEBS Lett.* **84,** 351.
Guillemaut, P., and Weil, J. H. (1982). *Nucleic Acids Res.* **10,** 1653.
Guillemaut, P., Burkard, G., and Weil, J. H. (1972). *Phytochemistry* **11,** 2217.
Guitton, C., Dorne, A. M., and Mache, R. (1984). *Biochem. Biophys. Res. Commun.* **121,** 297.
Gutteridge, S. M., Parry, M. A. J., Burton, S., Keys, A. J., Mudd, A., Feeney, J., Servaites, J. C., and Pierce, J. (1986). *Nature (London)* **324,** 274.
Hallick, R. B., Hollingsworth, M. J., and Nickoloff, J. (1984). *Plant Mol. Biol.* **3,** 169.
Hanson, M. R., Davidson, J. N., Mets, L. J., and Bogorad, L. (1974). *Mol. Gen. Genet.* **132,** 105.
Hartley, M. R. (1979). *Eur. J. Biochem.* **96,** 311.
Hartley, M. R., Wheeler, A. M., and Ellis, R. J. (1975). *J. Mol. Biol.* **91,** 67.
Heckman, J. E., Sarnoff, J., Alzner-DeWeerd, B., Yin, S., and RajBhandary, U. L. (1980). *Proc. Natl. Acad. Sci. U.S.A.* **77,** 3159.
Heinemeyer, W., Alt, J., and Herrmann, R. G. (1984). *Curr. Genet.* **8,** 543.
Herrmann, R. G., Alt, J., Schiller, B., Widger, W. R., and Cramer, W. A. (1984). *FEBS Lett.* **176,** 239.
Hird, S. M., Dyer, T. A., and Gray, J. C. (1986). *FEBS Lett.* **209,** 181.
Ho, K. K., and Krogmann, D. W. (1980). *J. Biol. Chem.* **255,** 3855.
Hollingsworth, M. J., Johanningsmeyer, U., Karabin, G. D., Stiegler, G. L., and Hallick, R. B. (1984). *Nucleic Acids Res.* **12,** 2001.
Holschuh, K., Bottomley, W., and Whitfeld, P. R. (1983). *Nucleic Acids Res.* **11,** 8547.
Holschuh, K., Bottomley, W., and Whitfeld, P. R. (1984a). *Plant Mol. Biol.* **3,** 313.
Holschuh, K., Bottomley, W., and Whitfeld, P. R. (1984b). *Nucleic Acids Res.* **12,** 8819.
Howe, C. J., Auffret, A. D., Doherty, A., Bowman, C. M., Dyer, T. A., and Gray, J. C. (1982). *Proc. Natl. Acad. Sci. U.S.A.* **79,** 6903.
Howe, C. J., Fearnley, I. M., Walker, J. E., Dyer, T. A., and Gray, J. C. (1985). *Plant Mol. Biol.* **4,** 333.
Ikemura, T. (1985). *Mol. Biol. Evol.* **2,** 13.
Imbault, P., Sarantoglou, V., and Weil, J. H. (1979). *Biochem. Biophys. Res. Commun.* **88,** 75.
Imbault, P., Colas, B., Sarantoglou, V., Boulanger, Y., and Weil, J. H. (1981). *Biochemistry* **20,** 5855.
Inamine, G., Nash, B., Weissbach, H., and Brot, N. (1985). *Proc. Natl. Acad. Sci. U.S.A.* **82,** 5690.

Jasin, M., Regan, L., and Schimmel, P. (1983). *Nature (London)* **306**, 441.
Karabin, G. D., Farley, M., and Hallick, R. B. (1984). *Nucleic Acids Res.* **12**, 5801.
Kashdan, M., and Dudock, B. (1982a). *J. Biol. Chem.* **257**, 1114.
Kashdan, M., and Dudock, B. (1982b). *J. Biol. Chem.* **257**, 11191.
Kashdan, M., Pirtle, R. M., Pirtle, I. L., Calagan, J., Vreman, H. J., and Dudock, B. (1980). *J. Biol. Chem.* **255**, 8831.
Keller, M., and Stutz, E. (1984). *FEBS Lett.* **175**, 173.
Keus, R. J. A., Stam, N. J., Zwiers, T., de Heij, H. T., and Groot, G. S. P. (1984). *Nucleic Acids Res.* **12**, 5639.
Kirsch, W., Seyer, P., and Herrmann, R. G. (1986). *Curr. Genet.* **10**, 843.
Koch, W., Edwards, K., and Kössel, H. (1981). *Cell* **25**, 203.
Koller, B., and Delius, H. (1984). *Cell* **36**, 613.
Koller, B., Clarke, J., and Delius, H. (1985). *EMBO J.* **4**, 2445.
Koller, B., Fromm, H., Galun, E., and Edelman, M. (1987). *Cell* **48**, 111.
Kozak, K. (1983). *Microbiol. Rev.* **47**, 1.
Kraus, B. L., and Spremulli, L. L. (1986). *J. Biol. Chem.* **261**, 4781.
Krauspe, R., and Parthier, B. (1974). *Biochem. Physiol. Pflanz.* **165**, 18.
Krebbers, E. T., Larrinua, I. M., McIntosh, L., and Bogorad, L. (1982). *Nucleic Acids Res.* **10**, 4985.
Krebbers, E. T., Steinmetz, A., and Bogorad, L. (1984). *Plant Mol. Biol.* **3**, 13.
Kuchino, Y., Watanabe, S., Harada, F., and Nishimura, S. (1980). *Biochemistry* **19**, 2085.
Kuntz, M., Weil, J. H., and Steinmetz, A. (1984). *Nucleic Acids Res.* **12**, 5037.
Kyle, D. J. (1985). In "Molecular Biology of the Photosynthetic Apparatus" (K. E. Steinback, S. Bonitz, C. J. Arntzen, and L. Bogorad, eds.), pp. 33–38. Cold Spring Harbor Lab., Cold Spring Harbor, New York.
Lagerqvist, U. (1978). *Proc. Natl. Acad. Sci. U.S.A.* **75**, 1759.
Lake, J. A. (1985). *Annu. Rev. Biochem.* **54**, 507.
Lee, S. G., and Evans, W. R. (1971). *Science* **173**, 241.
Leis, J. P., and Keller, E. B. (1970). *Proc. Natl. Acad. Sci. U.S.A.* **67**, 1593.
Lewin, B. (1985). "Genes II." Wiley, New York.
Link, G., Coen, D. M., and Bogorad, L. (1978). *Cell* **15**, 725.
Locy, R. O., and Cherry, J. H. (1978). *Phytochemistry* **17**, 19.
Lyttleton, J. W. (1962). *Exp. Cell Res.* **26**, 312.
McCarthy, J. E. G., Schairer, H. U., and Sebald, W. (1985). *EMBO J.* **4**, 519.
McCoy, J. M., and Jones, D. S. (1980). *Nucleic Acids Res.* **8**, 5089.
McIntosh, L., Poulsen, C., and Bogorad, L. (1980). *Nature (London)* **288**, 556.
Mache, R., Dorne, A. M., and Batlle, R. M. (1980). *Mol. Gen. Genet.* **177**, 333.
Maitra, U., Stringer, E. A., and Chaudhury, A. (1982). *Annu. Rev. Biochem.* **51**, 869.
Marcus, A. (1982). *Encycl. Plant Physiol., New Ser.* **14A**, 113.
Marder, J. B., Goloubinoff, P., and Edelman, M. (1982). *J. Biol. Chem.* **259**, 3900.
Maréchal, L., Guillemaut, P., Grienenberger, J. M., Jeannin, G., and Weil, J. H. (1986). *Plant Mol. Biol.* **7**, 245.
Margulies, M. M. (1986). In "Regulation of Chloroplast Differentiation" (G. Akoyunoglou and H. Senger, eds.), pp. 171–180. Liss, New York.
Merrick, W. C., and Dure, L. (1971). *Proc. Natl. Acad. Sci. U.S.A.* **68**, 641.
Mets, L. J., and Bogorad, L. (1972). *Proc. Natl. Acad. Sci. U.S.A.* **69**, 3779.
Mizuno, T., Chou, M. Y., and Inouye, M. (1984). *Proc. Natl. Acad. Sci. U.S.A.* **81**, 1966.
Moldave, K. (1985). *Annu. Rev. Biochem.* **54**, 1109.
Montandon, P. E., and Stutz, E. (1983). *Nucleic Acids Res.* **11**, 5877.
Montandon, P. E., and Stutz, E. (1984). *Nucleic Acids Res.* **12**, 2851.
Montandon, P. E., Nicolas, P., Schurmann, P., and Stutz, E. (1985). *Nucleic Acids Res.* **13**, 4299.

Montandon, P. E., Wagner, R., Nicolas, P., and Stutz, E. (1986). *Proc. Int. Symp. Plant Mol. Biol.* Abstr. C26.
Morris, J., and Herrmann, R. G. (1984). *Nucleic Acids Res.* **12,** 2837.
Mubumbila, M., Burkard, G., Keller, M., Steinmetz, A., Crouse, E. J., and Weil, J. H. (1980). *Biochim. Biophys. Acta* **609,** 31.
Mullet, J. E., Boyer, S., and Klein, R. R. (1985a). *Proc. Int. Cong. Plant Mol. Biol., 1st* p. 54 (Abstr.).
Mullet, J. E., Orozco, E. M., Jr., and Chua, N. H. (1985b). *Plant Mol. Biol.* **4,** 39.
Natsoulis, G., Hilger, F., and Fink, G. R. (1986). *Cell* **46,** 235.
Nivison, H. T., Fish, L. E., and Jagendorf, A. T. (1986). *Methods Enzymol.* **118,** 282.
Noller, H. F. (1984). *Annu. Rev. Biochem.* **53,** 119.
Nomura, M., Gourse, R., and Baughman, G. (1984). *Annu. Rev. Biochem.* **53,** 75.
Ohyama, K., Fukuzawa, H., Kohchi, T., Shirai, H., Sano, T., Sano, S., Umesono, K., Shiki, Y., Takeushi, M., Chang, Z., Aota, S., Inokushi, H., and Ozeki, H. (1986). *Plant Mol. Biol. Rep.* **4,** 149.
Oliver, R. P., and Poulsen, C. (1984). *Carlsberg Res. Commun.* **49,** 647.
Osorio-Almeida, M. L., Guillemaut, P., Keith, G., Canaday, J., and Weil, J. H. (1980). *Biochem. Biophys. Res. Commun.* **92,** 102.
Passavant, C. W., and Hallick, R. B. (1985). *Plant Mol. Biol.* **4,** 347.
Pfitzinger, H., Guillemaut, P., Weil, J. H., and Pillay, D. T. N. (1987). *Nucleic Acids Res.* **15,** 1377.
Pillay, D. T. N., Guillemaut, P., and Weil, J. H. (1984). *Nucleic Acids Res.* **12,** 2997.
Pirtle, R., Calagan, J., Pirtle, I., Kashdan, M., Vreman, H., and Dudock, B. (1981). *Nucleic Acids Res.* **9,** 183.
Posno, M., van Noort, M., Débise, R., and Groot, G. S. P. (1984). *Curr. Genet.* **8,** 147.
Quigley, F., and Weil, J. H. (1985). *Curr. Genet.* **9,** 495.
Quigley, F., Grienenberger, J. M., and Weil, J. H. (1985). *Plant Mol. Biol.* **4,** 305.
Ramakrishnan, V., Capel, M., Kjeldgaard, M., Engelman, D. M., and Moore, P. B. (1984). *J. Mol. Biol.* **174,** 265.
Rasmussen, O. F., Stummann, B. M., and Henningsen, K. W. (1984). *Nucleic Acids Res.* **12,** 9143.
Ratcliffe, J. C., and Caskey, C. T. (1977). *Arch. Biochem. Biophys.* **181,** 671.
Ratinaud, M. R., Thomes, J. C., and Julien, R. (1985). *Eur. J. Biochem.* **135,** 471.
Rauhut, R., Gabius, H. J., and Cramer, F. (1984). *Hoppe-Seyler's Z. Physiol. Chem.* **365,** 289.
Rauhut, R., Gabius, H. J., and Cramer, F. (1986). *J. Biol. Chem.* **261,** 2799.
Rodermel, S. R., and Bogorad, L. (1985). *J. Cell Biol.* **100,** 463.
Rogers, M. J., Steinmetz, A., and Walker, R. T. (1984). *Isr. J. Med. Sci.* **20,** 768.
Ryoji, M., Karpen, J. W., and Kaji, A. (1981). *J. Biol. Chem.* **256,** 5798.
Sagher, D., Grosfeld, H., and Edelman, M. (1976). *Proc. Natl. Acad. Sci. U.S.A* **73,** 722.
Samuelsson, T., Elias, P., Lustig, F., and Guindy, Y. S. (1985). *Biochem. J.* **232,** 223.
Sarantoglou, V., Imbault, P., and Weil, J. H. (1980). *Biochem. Biophys. Res. Commun.* **93,** 134.
Sarantoglou, V., Imbault, P., and Weil, J. H. (1981). *Plant Sci. Lett.* **22,** 291.
Schmidt, G. W., Devillliers-Thiery, A., Desruisseaux, H., Blobel, G., and Chua, N. H. (1979). *J. Cell Biol.* **86,** 615.
Schneider, M., and Rochaix, J. D. (1986). *Plant Mol. Biol.* **6,** 265.
Schön, A., Krupp, G., Gough, S., Berry-Lowe, S., Kannangara, C. G., and Söll, D. (1986). *Nature (London)* **322,** 281.
Schwartz, J., Meyer, R., Eisenstadt, J., and Brawerman, G. (1967). *J. Mol. Biol.* **25,** 571.
Schwarz, Z., Jolly, S. O., Steinmetz, A., and Bogorad, L. (1981a). *Proc. Natl. Acad. Sci. U.S.A.* **78,** 3423.
Schwarz, Z., Kössel, H., Schwarz, E., and Bogorad, L. (1981b). *Proc. Natl. Acad. Sci. U.S.A* **78,** 4748.

Sebald, W., and Wächter, E. (1980). *FEBS Lett.* **122**, 307.
Seemann, J. R., Berry, J. A., Freas, S. M., and Krump, M. A. (1985). *Proc. Natl. Acad. Sci. U.S.A.* **82**, 8024.
Shapiro, D. R., and Tewari, K. K. (1986). *Plant Mol. Biol.* **6**, 1.
Shinozaki, K., Sasaki, Y., Sakihama, T., and Kamikubo, T. (1982). *FEBS Lett.* **144**, 73.
Shinozaki, K., Deno, H., Kato, A., and Sugiura, M. (1983). *Gene* **24**, 147.
Shinozaki, K., Deno, H., Sugita, M., Kuramitsu, S., and Sugiura, M. (1986a). *Mol. Gen. Genet.* **202**, 1.
Shinozaki, K., Ohme, M., Tanaka, M. Wakasugi, T., Hayashida, N., Matsubayashi, T., Zaita, N., Chunwongse, J., Obokata, J., Yamagushi-Shinozaki, K., Ohto, C., Torazawa, K., Meng, B. Y., Sugita, M., Deno, H., Kamogashira, T., Yamada, K., Kusuda, J., Takaiwa, F., Kato, A., Tohdoh, N., Shimada, H., and Sugiura, M. (1986b). *Plant Mol. Biol. Rep.* **4**, 111.
Sibler, A. P., Dirheimer, G., and Martin, R. P. (1986). *FEBS Lett.* **194**, 131.
Sijben-Müller, G., Hallick, R. B., Alt, J., Westhoff, P., and Herrmann, R. H. (1986). *Nucleic Acids Res.* **14**, 1029.
Simons, R. W., and Kleckner, N. (1983). *Cell* **34**, 683.
Sissakian, N., Filippovitch, I., Svetailo, E., and Aliyev, A. (1965). *Biochim. Biophys. Acta* **95**, 474.
Smeekens, S., Bauerle, C., Hageman, J., Keegstra, K., and Weisbeek, P. (1986). *Cell* **46**, 365.
Soll, J., and Buchanan, B. B. (1983). *J. Biol. Chem.* **258**, 6668.
Souciet, G., Dietrich, A., Colas, B., Razafimahatratra, P., and Weil, J. H. (1982). *J. Biol. Chem.* **257**, 9598.
Spielmann, A., and Stutz, E. (1983). *Nucleic Acids Res.* **11**, 7157.
Sprouse, H. M., Kashdan, M., Otis, L., and Dudock, B. (1981). *Nucleic Acids Res.* **9**, 2543.
Steinmetz, A., and Weil, J. H. (1986). *Methods Enzymol.* **118**, 212.
Steinmetz, A., Mubumbila, M., Keller, M., Burkard, G., Weil, J. H., Driesel, A. J., Crouse, E. J., Gordon, K., Bohnert, H. J., and Herrmann, R. G. (1978). *In* "Chloroplast Development" (G. Akoyunoglou and J. Argyroudi-Akoyunoglou, eds.), pp. 573–580. Elsevier/North-Holland, Amsterdam.
Steinmetz, A., Gubbins, E. J., and Bogorad, L. (1982). *Nucleic Acids Res.* **10**, 3027.
Steinmetz, A. A., Krebbers, E. T., Schwarz, Z., Gubbins, E. J., and Bogorad, L. (1983). *J. Biol. Chem.* **258**, 5503.
Steinmetz, A., Bonnard, G., Kuntz, M., Green, G. A., Mubumbila, M., Crouse, E. J., and Weil, J. H. (1985). *In* "Molecular Biology of the Photosynthetic Apparatus" (K. E. Steinback, S. Bonitz, C. J. Arntzen, and L. Bogorad, eds.), pp. 279–284. Cold Spring Harbor Lab., Cold Spring Harbor, New York.
Steinmetz, A. A., Castroviejo, M., Sayre, R. T., and Bogorad, L. (1986). *J. Biol. Chem.* **261**, 2485.
Subramanian, A. R. (1985). *In* "Essays in Biochemistry," pp. 45–85. The Biochem. Soc., London.
Subramanian, A. R., Steinmetz, A., and Bogorad, L. (1983). *Nucleic Acids Res.* **11**, 5277.
Sugita, M., and Sugiura, M. (1983). *Nucleic Acids Res.* **11**, 1913.
Sugita, M., Shinozaki, K., and Sugiura, M. (1985). *Proc. Natl. Acad. Sci. U.S.A.* **82**, 3557.
Thompson, W. F., Everett, M., Polans, N. O., Jörgensen, R. A., and Palmer, J. D. (1983). *Planta* **158**, 487.
Tiboni, O., and Ciferri, O. (1986). *Methods Enzymol.* **118**, 296.
Tiboni, O., Di Pasquale, G., and Ciferri, O. (1976). *Plant Sci. Lett.* **6**, 419.
Torazawa, K., Hayashida, N., Obokata, J., Shinozaki, K., and Sugiura, M. (1986). *Nucleic Acids Res.* **14**, 3143.
Tyagi, A. K., and Herrmann, R. G. (1986). *Curr. Genet.* **10**, 481.
Van Grinsven, M., Zethof, J., Gielen, J., and Kool, A. J. (1985). *Proc. Int. Congr. Plant Mol. Biol., 1st* p. 126 (Abstr.).

Wakasugi, T., Ohme, M., Shinozaki, K., and Sugiura, M. (1986). *Plant Mol. Biol.* **7,** 385.
Watson, J. D., Hopkins, N. H., Roberts, J. W., Steitz, J. A., and Weiner, A. M. (1987). "Molecular Biology of the Gene." Benjamin/Cummings, Menlo Park, California.
Weil, J. H., and Parthier, B. (1982). *Encycl. Plant Physiol., New Ser.* **14A,** 65.
Weissbach, H. (1980). *In* "Ribosomes: Structure, Function and Genetics" (G. Chambliss, G. R. Craven, J. Davies, K. Davis, L. Kahan, and M. Nomura, eds.), pp. 377–411. Univ. Park Press, Baltimore, Maryland.
Westhoff, P., Nelson, N., Bühnemann, H., and Herrmann, R. G. (1981). *Curr. Genet.* **4,** 109.
Westhoff, P., Farchaus, J. W., and Herrmann, R. G. (1986). *Curr. Genet.* **11,** 165.
Wheeler, A. M., and Hartley, M. R. (1975). *Nature (London)* **257,** 66.
Widger, W. R., Cramer, W. R., Hermodson, M., Meyer, D., and Gullifor, M. (1984). *J. Biol. Chem.* **259,** 3870.
Willey, D. L., Auffret, A. D., and Gray, J. C. (1984a). *Cell* **36,** 555.
Willey, D. L., Howe, C. J., Auffret, A. D., Bowman, C. M., Dyer, T. A., and Gray, J. C. (1984b). *Mol. Gen. Genet.* **194,** 416.
Wittmann, H. G. (1982). *Annu. Rev. Biochem.* **51,** 155.
Wittmann, H. G. (1983). *Annu. Rev. Biochem.* **52,** 35.
Wittmann, H. G., and Wittmann-Liebold, B. (1974). *In* "Ribosomes" (M. Nomura, A. Tissières, and P. Lengyel, eds.), pp. 115–140. Cold Spring Harbor Lab. Cold Spring Harbor, New York.
Wittmann, H. G., Stöffler, G., Apirion, D., Rosen, L., Tanaka, K., Tamaki, M., Takata, R., Dekio, S., Otaka, E., and Osawa, S. (1973). *Mol. Gen. Genet.* **127,** 175.
Wool, I. G., and Stöffler, G. (1974). *In* "Ribosomes" (M. Nomura, A. Tissières, and P. Lengyel, eds.), pp. 417–460. Cold Spring Harbor Lab., Cold Spring Harbor, New York.
Yarus, M., and Folley, L. S. (1985). *J. Mol. Biol.* **182,** 529.
Yokoyama, S., Watanabe, T., Murao, K., Ishikura, H., Yamaizumi, Z., Nishimura, S., and Miyazawa, T. (1985). *Proc. Natl. Acad. Sci. U.S.A.* **82,** 4905.
Zurawski, G., and Clegg, M. T. (1984). *Nucleic Acids Res.* **12,** 2549.
Zurawski, G., Perrot, B., Bottomley, W., and Whitfeld, P. R. (1981). *Nucleic Acids Res.* **9,** 3251.
Zurawski, G., Bohnert, H. J., Whitfeld, P. R., and Bottomley, W. (1982a). *Proc. Natl. Acad. Sci. U.S.A.* **79,** 7699.
Zurawski, G., Bottomley, W., and Whitfeld, P. R. (1982b). *Proc. Natl. Acad. Sci. U.S.A.* **79,** 6260.
Zurawski, G., Bottomley, W., and Whitfeld, P. R. (1984). *Nucleic Acids Res.* **12,** 6547.

The Plant Mitochondrial Genome

6

DAVID M. LONSDALE

I. Introduction
II. Physical Parameters of Mitochondrial DNA
    A. Base Composition
    B. Mitochondrial Genome Size
III. Mitochondrial Genome: Composition and Organization
    A. Large Circular DNA Species
    B. Small Circular DNA Plasmids
    C. Small Linear DNA Replicons
    D. The S, R, and D Episomes of Maize
    E. R1- and R2-Related Sequences of the Maize N-Cytoplasms
    F. R1, S1, and S2 Sequences of the Maize S-Cytoplasms
    G. The 2.3-kb Linear Plasmid of Maize
    H. Double and Single-Stranded RNAs
IV. Cells–Mitochondria–Mitochondrial DNA
V. Promiscuous DNA
    A. Chloroplast Sequences in Mitochondrial DNA of Maize
    B. Chloroplast Sequences in the Mitochondrial Genomes of Species Other Than Maize
VI. Genetic Complexity of Plant Mitochondrial Genomes
    A. Conserved Sequences
    B. Protein Synthesis by Isolated Mitochondria
VII. Mitochondrial Coding Sequences
    A. Mitochondrial Ribosomal RNA Genes
    B. Mitochondrial Transfer RNA Genes
    C. Polypeptide Genes
    D. Unusual Transcribed Sequences
    E. Introns
VIII. Gene Copy Number and Mapping
IX. Transcription
X. Ribosome Binding and Translation Initiation
XI. Genomic Reorganizations
XII. Conclusions
    References

I. INTRODUCTION

The cytoplasm of higher plants contains two extranuclear genetic systems, the mitochondrion and chloroplast, both of which are unique and distinct from the nuclear–cytosol system. Efficient interaction between these three genetic systems within the plant cell is essential for normal plant development. Alterations in the way the three organelles interact or changes in their genetic information may have profound effects on plant phenotype (Lonsdale, 1987a,b).

The chloroplast genome has been studied extensively and in two species, tobacco (*Nicotiana tabacum*) and the liverwort (*Marchantia polymorpha*), the entire genome has been sequenced (Ohyama *et al.*, 1986; Shinozaki *et al.*, 1986). Its small size and its similarity with prokaryotic systems (for example, *Escherichia coli*) has made it an easy and amenable system (Palmer, 1985; also Steinmetz and Weil, Chapter 5, this volume). In contrast, the plant mitochondrial genome is exceptionally diverse in size and in some species has in addition low-molecular-weight, circular and linear DNAs and single and double-stranded RNAs. There appears to be no relationship between mitochondrial genome size and genetic complexity. In addition, the mitochondrial genetic code may exhibit at least one difference from the universal one. These factors complicate studies on plant mitochondrial genome organization and expression. This review concentrates primarily on the molecular biology of the mitochondrial genome. Cytoplasmic male sterility (Hanson and Conde, 1985; Lonsdale, 1987b) and the genetic engineering of higher plant cytoplasms (Lonsdale, 1987a) are dealt with elsewhere.

II. PHYSICAL PARAMETERS OF MITOCHONDRIAL DNA

A. Base Composition

The base composition of mitochondrial DNA (mtDNA) based on buoyant density and melting temperature determinations is remarkably constant across a wide range of higher plants; values of 46 to 48% G + C have been obtained (see Table I; Pring and Lonsdale, 1985), the only reported exception is *Oenothera berteriana* (evening primrose) for which a G + C value of 51% has been calculated (Brennicke, 1980). There appears to be little or no heterogeneity in base sequence composition as evidenced by several criteria: absence of heavy and light satellite fractions in neutral CsCl, little or no band dispersion in alkaline CsCl, and single sharp transitions of thermal melting profiles (Suyama and Bonner, 1966; Kolodner and Tewari, 1972; Wong and Wildman, 1972; Vedel and Quetier, 1974). Unusual bases have not been detected in mtDNAs, nor is there any evidence for a significant level of

TABLE I

Physical Parameters of Plant Mitochondrial DNA

Species		Density[a] (g/cm³)	Kinetic complexity[b] (Da × 10⁶)	References[c]
Common name	Latin name			
Broad bean	Vicia faba	1.705		7
Broad bean	Vicia faba		190	14
Broad bean	Vicia faba	1.706		17
Cucumber	Cucumis sativus	1.705	1000	11[d], 13
Evening primrose	Oenothera berteriana	1.710		3
Flax	Linum usitatissum	1.706		17
Hairy vetch	Vicia villosa		250	14
Jimson weed	Datura innoxia	1.706		17
Lettuce	Lactuca sativa	1.706	104	7
Maize	Zea mays	1.706		9, 16
Maize	Zea mays	1.705	320	13
Mung bean	Phaseolus aureus	1.706		1, 17
Muskmelon	Cucumis melo	1.705	1600	13
Onion	Allium cepa	1.706		1, 12
Pea	Pisum sativum	1.706	73.8	2
Pea	Pisum sativum	1.706	240	13
Petunia	Petunia hybrida	1.706		17, 18
Potato	Solanum tuberosum	1.706	99	6, 11[d]
Red bean	Phaseolus vulgaris	1.707		8
Sorghum	Sorghum bicolor	1.705		10
Soybean	Glycine max	1.706		5, 17
Spinach	Spinacea oleracea	1.706		7
Sweet potato	Ipomoea batatas	1.706		1
Swiss chard	Beta vulgaris	1.705		12
Turnip	Brassica rapa	1.706		1
Tobacco	Nicotiana tabacum	1.705	60	4
Tobacco	Nicotiana tabacum	1.706		17
Tobacco	Nicotiana tabacum	1.707		15
Virginia creeper	Parthenocissus tricuspidata	1.706		11[d]
Watermelon	Citrullus vulgaris	1.706	220	13
Wheat	Triticum aestivum	1.706		11[d]
Zucchini squash	Cucurbita pepo	1.705	560	13

[a] The calculated densities are based on the following standards, *Bacillus subtilis* (N¹⁵), 1.740 (1); *Pseudomonas aeruginosa*, 1.737 (3); *Micrococcus luteus*, 1.731 (5); φE, 1.742 (6); SPOI, 1.742 (8); *Micrococcus lysodeikticus*, 1.731 (2, 4, 7, 12, 13). With the exception of Ward *et al.* (1981), it can only be assumed that the densities of the standards are relative to *E. coli* 1.710.

[b] The kinetic complexities are based on T4, 106×10^6 Da, (2, 6); T4, 108×10^6 Da (4); T4, 130×10^6 Da (2, 7); *E. coli* (7); T7 and *Bacillus subtilis* (13).

[c] References (1) Suyama and Bonner (1966); (2) Kolodner and Tewari (1972); (3) Brennicke (1980); (4) Wong and Wildman (1972); (5) Synenki *et al.* (1978); (6) Vedel and Quetier (1974); (7) Wells and Birnstiel (1969); (8) Wolstenholme and Gross (1968); (9) Pring and Levings (1978); (10) Pring *et al.* (1982); (11) Quetier and Vedel (1977); (12) Wells and Ingle (1970); (13) Ward *et al.* (1981); (14) Bendich (1982); (15) Sparks and Dale, 1980; (16) Shah and Levings, 1974; (17) Bailey-Serres *et al.* (1987); (18) Kool *et al.* (1985).

[d] Data not shown.

methylation based on the use of restriction endonuclease isoschizomers that differentiate between methylated and unmethylated restriction sites (Bonen and Gray, 1980; Ward *et al.*, 1981; Borck and Walbot, 1982; Bailey-Serres *et al.*, 1987).

B. Mitochondrial Genome Size

Estimations of mitochondrial genome sizes have been attempted by three separate methods: electron microscopy, renaturation kinetics, and summations of the molecular weights of restriction fragments. None of these methods is entirely satisfactory, owing to the large size and multipartite organization of mitochondrial genomes.

1. Electron Microscopy Studies

These have mostly revealed a heterogeneous population of linear molecules (Suyama and Miura, 1968; Wolstenholme and Gross, 1968; Mikulska *et al.*, 1970; Vedel and Quetier, 1974; Kim *et al.*, 1982; Kool *et al.*, 1985; Manna *et al.*, 1985; Bailey-Serres *et al.*, 1987). In a few studies a low proportion of circular DNA molecules has been observed (see Section III). None of the electron microscopy studies have provided evidence that plant mitochondria contain a uniform population of molecules as is the case for mammalian mitochondria (see review by Borst, 1972). The one exception is the report of Kolodner and Tewari (1972) who found that 55% of the DNA released from osmotically shocked mitochondria of pea had a circular configuration of monomer length 30.3 μm. This circular species also constituted 25% of the DNA in deproteinized DNA preparations. The circular contour length was in apparent agreement with their kinetic complexity measurements. However, these experiments have not been independently confirmed and they contrast with an earlier study which demonstrated only linear molecules with an average size of 10.1 μm (Mikulska *et al.*, 1970).

Rarely, exceptionally large molecules have been observed (Suyama and Miura, 1968; Fontarnau and Hernandez-Yago, 1982; Negruk *et al.*, 1986); these may represent an intact genome though circularity was not proven and accurate size measurements could not be made due to their highly tangled nature.

2. DNA Renaturation Studies

These provide a direct estimate of the size of the unique sequence component of the genome, relative to known standards such as *E. coli*. Early studies relied on T4 coliphage as a standard, an inappropriate choice as its molecular weight was not accurately known and the variable extent of glucosylation and 5-hydroxymethylcytosine affected its rate of renaturation (Mandel and Marmur, 1968; Christiansen *et al.*, 1973). Despite this, molecular weight estimations based on renaturation kinetics indicated a large genome size (Table I). More recent studies (Ward *et al.,* 1981) have demonstrated that mitochondrial genome sizes can be exceptionally large, up to 2400 kb in the case of muskmelon and that a small reiterative component can be detected in some species (Ward *et al.,* 1981; De Heij *et al.,* 1985).

3. Restriction Fragment Analysis

Restriction endonuclease profiles of plant mtDNA preparations are more complex than those of chloroplast DNAs. Restriction bands can be observed in both submolar and supramolar amounts (Quetier and Vedel, 1977; Pring and Levings, 1978; Ward *et al.,* 1981). In estimating the molecular weight of mitochondrial genomes from restriction profile data, the stoichiometry of individual restriction bands has to be taken into account. Widely differing values for genome size can be obtained, depending on which restriction band is taken to represent a single copy sequence (Borck and Walbot, 1982). For this reason, molecular weight estimates of plant mitochondrial genome size based on restriction fragment molecular weight summations are not included in this review. Only molecular weights based on restriction mapping data are given (see Table II).

III. MITOCHONDRIAL GENOME: COMPOSITION AND ORGANIZATION

In many plant species, the mitochondria contain, besides a high-molecular-weight DNA component which represents the main genomic DNA, smaller circular and linear molecules of either DNA or RNA.

In order to discuss these various components of the mitochondrial genome in their correct context, I have distinguished between large circular species and small plasmids. The larger circular species have rarely been observed by electron microscopy and have been mainly predicted from physical mapping studies. Such molecules arise from the high-molecular-weight mitochondrial chromosome. The small plasmids apparently have the capacity to self-replicate and show no homology to the mitochondrial chromosome. The division between these two groups is unclear, particularly for the circles of intermediate size which are less well characterized.

A. Large Circular DNA Species

In addition to the small circular DNAs that characterize some mtDNA preparations (Section III,B), some mtDNA preparations when viewed in the electron microscope have additional larger circular species which have not been detected by other methods of analysis. The existence of such circles as components of the mitochondrial genome, present at low stoichiometric values, was one explanation for the observed restriction profile heterogene-

TABLE II
Plants for Which Restriction Maps of the Mitochondrial Genomes Are Complete

Species		Genome size (kb)	Number of repeats	Size[a] (kb)	SGC[b]	References[c]
Common name	Latin name					
White mustard	*Brassica hirta*	208	None	—	—	1
Turnip	*Brassica campestris*	218	1	2	135, 83	2
Cabbage	*Brassica oleracea*	219	1	2	170, 49	1, 3[d]
Oilseed rape	*Brassica napus*	221	1	2	123, 98	1
Black mustard	*Brassica nigra*	231	1	7	135, 96	8
Radish	*Raphanus sativa*	242	1	10	139, 103	8
Spinach	*Spinacea oleracea*	327	1	6	234, 93	4
Sugar beet	*Beta vulgaris*[e]	386	5	ND[b]	Many	7
Wheat	*Triticum aestivum*	440	10	ND	Many	5[c]
Maize	*Zea mays* (Wf9-N)	570	6	14, 12, 10, 5, 0.6, 0.5	Many	6

[a] Repeat sizes are approximate and are based mainly on restriction mapping data.
[b] SGC, subgenomic circles; ND, not determined.
[c] References: (1) Palmer and Herbon (1987); (2) Palmer and Shields (1984); (3) Chetrit *et al.* (1984); (4) Stern and Palmer (1986); (5) Quetier *et al.* (1985); (6) Lonsdale *et al.* (1984); (7) T. Brears and D. M. Lonsdale (unpublished results); (8) Palmer and Herbon (1986).
[d] The reported restriction maps of the master circle are not colinear.
[e] The "Owen" sterile cytoplasm of 82 RW 181 was used for these studies.
[f] Reported map is based on the analysis of recombinant clones with a single restriction endonuclease. Three of the four 18S rRNA repeats have been located on the master circle, suggesting an unlikely genome arrangement.

ity (Spruill *et al.*, 1980; Sederoff *et al.*, 1981; Bendich, 1982; Borck and Walbot, 1982; Dale *et al.*, 1983). Hybridization and restriction mapping studies provided direct evidence for sequence duplications and another explanation for restriction profile heterogeneity (Spruill *et al.*, 1980; Bonen and Gray, 1980; Lonsdale *et al.*, 1981; Borck and Walbot, 1982). In addition, many of these repeated DNA sequences appear to be involved in homologous recombination events. The four unique sequences flanking the repeat DNA element could be found in four pair-wise combinations (Fig. 1; Lonsdale *et al.*, 1983b, 1984; Palmer and Shields, 1984; Stern and Palmer, 1984a, 1986). The relative orientation of the repeated DNA sequences therefore leads to two possible genome resolutions. Recombination between repeats which have an opposite or inverted orientation leads to sequence inversion (flip-flop; Fig. 1-I). This process is a common feature of the chloroplast genomes of algae and higher plants (Bohnert and Loffelhardt, 1982; Palmer, 1983, 1985; Aldrich *et al.*, 1985; Brears *et al.*, 1986). Where the repeats have the same relative orientation, recombination leads to the formation of two circular species (Fig. 1-II). It is this process of recombination between repeated DNA elements, leading to either sequence inversions or the resolution of circular species into smaller subgenomic circles which accounts for

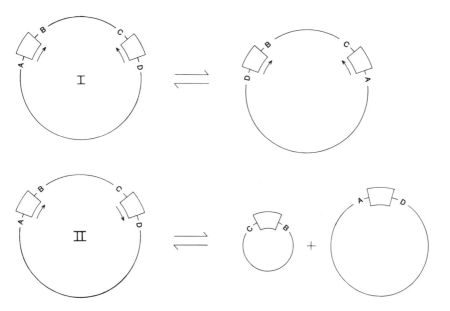

Fig. 1. Recombination between inverted and directly repeated sequences. Four unique sequences represented by A, B, C, and D flank a repetitive DNA sequence element (open box). The relative orientations of the repetitive elements are indicated (arrows). I, recombination between inverted repeats leads to sequence inversion ("flip-flop"). II, recombination between direct repeats leads to the formation of subgenomic circles ("loop-out").

the mixed stoichiometries of fragments observed in restriction endonuclease profiles of higher plant mtDNA.

The structure and organization of few plant mitochondrial genomes have been elucidated and published owing to problems associated with restriction mapping such large complex molecules. The plant mitochondrial genomes that have been mapped are given in Table II.

The simplest mitochondrial genome described is that of *Brassica hirta* (Palmer and Herbon, 1987). It is a simple circular molecule of 208 kb with no repeated DNA elements and therefore is analogous to the chloroplast genomes of the *Leguminosae* (Palmer, 1985). Other members of the *Cruciferae* including *Raphanus sativa* (radish) and *Spinacea oleracea* (spinach: *Chenopodiaceae*) have small mitochondrial genomes (200–300 kb) with a relatively simple genomic organization. These mitochondrial genomes are characterized by having a single repeated DNA element, both copies having the same relative orientation. Physical mapping has identified the four paired combinations of unique sequences which flank the repeats. The mitochondrial genomes can therefore be predicted to exist as single circular molecules, "master circles," or as two subgenomic circles. If homologous recombination is continually occurring between the copies of the repeated DNA elements, then the master circle and the two subgenomic circles would exist in a dynamic equilibrium.

The mitochondrial genomes of maize (Lonsdale *et al.*, 1984) and sugar beet (T. Brears and D. M. Lonsdale, unpublished results) can also be organized as master circles. The master circles of the sugar beet and maize mitochondrial genomes have five and six repeated DNA elements, respectively. Many of these repeated DNA elements, including those of other species such as wheat, where genome organization is less certain (Quetier *et al.*, 1985), appear to be involved in homologous recombination events (Lonsdale *et al.*, 1983b, 1984; Stern and Palmer, 1984a; Falconet *et al.*, 1984, 1985) and therefore the genomes can be visualized as complex multicircular populations (Fig. 2).

If recombination is truly homologous, then intermolecular recombination will occur between circles having the same sequence or sharing sequence elements. This is the probable origin of the multimeric forms of mitochondrial plasmids and for the multimeric forms of the subgenomic circles of tobacco (Dale *et al.*, 1983). Homologous recombination will therefore lead to a very large number of circular species, but whether these molecules exist *in vivo* remains enigmatic. In the absence of an active recombination system, all the circles must be independently maintained in the same relative stoichiometry. This is a necessary prediction because of the stability of mtDNA restriction profiles both in tissue culture material and in plants (Brennicke and Schwemmle, 1984; Oro *et al.*, 1985). Considering the large number of discrete molecular species which are known to exist in some cytoplasms, their independent maintenance would seem extremely improbable.

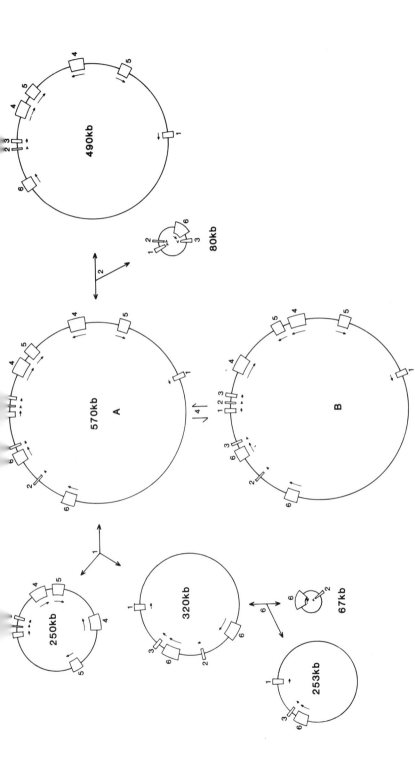

Fig. 2. Multicircular organization of the maize mitochondrial genome. The master circle of 570 kb has six pairs of repeats (open boxes), numbered 1 to 6. The relative orientation of the repeats is indicated (arrows). The master circle of 570 kb can exist in two isomeric forms: A and B; these result from recombination between the two copies of repeat 4, which have an inverted orientation. The resulting sequence inversion between the two copies of repeat 4 alters the relative orientation of the two copies of repeat 5. Recombination between the copies of repeats 1, 2, and 6 illustrates the formation of subgenomic circles.

The models of mitochondrial genome organization mentioned above: (1) a dynamic recombinational equilibrium between a master circle and a population of subgenomic circles and multimers thereof, and (2) a population of self-replicating molecules maintained in a strict stoichiometric relationship which recombine infrequently but nevertheless sufficiently often to maintain the lack of variation seen at the base sequence level between copies of the repeated sequences (Hiesel et al., 1987; Houchins et al., 1986) are not the only possibilities. A third type of genome organization is possible. This combines the features of (1) and (2) above and is illustrated by *Oenothera berteriana*. The mitochondrial genome is multicircular; some of the smaller circles can be visualized on agarose gels (Brennicke and Blanz, 1982) and have been arbitrarily designated nos. 1 to 8 in ascending size. The high-molecular-weight DNA has been designated as band 9. Circle 3, 7.5 kb, contains the 3' portion of the 26S rRNA gene. This circle arose following a recombination between two identical decanucleotide DNA sequences (5'-GGAAGCAGCC), one of which is within 26S rRNA coding sequence, the other being approximately 7 kb 3' to 26S rRNA. The other predicted product of this recombination is a subgenomic circle containing the 5'-portion of the 26S rRNA gene; it has not been detected. Interestingly, the copy of the decanucleotide repeat on circle 3, 5'-GGTTGCAGCC, is no longer identical to the two copies on the master circle. In all probability, the recombination event leading to the formation of circle 3 occurred once in evolution and its ability to self-replicate has ensured its survival (Manna and Brennicke, 1986). The sequence divergence of the decanucleotide repeat can be seen as sequence drift or, as the sequence was determined from a single clone, it may not be representative of the circle 3 population. Homologous recombination between circle 3 and the 26S rDNA region of the main genome probably occurs. This would lead to tandem reiterations of this 7.5-kb sequence. However, this would not allow copy correction of the decanucleotide repeat to occur.

A lack of recombination between the two copies of the *Petunia hybrida* ATP9 gene has been reported (Rothenberg and Hanson, 1987). If recombination were to occur, the ATP9 gene would be flanked by four unique sequence combinations. Only the arrangements A-*atp*9-B and C-*atp*9-D have been detected; A-*atp*9-D and C-*atp*9-B were not detected. Surprisingly, however, recombination can occur between the two copies of the *atp*9 gene during somatic hybridization. The lack of observable recombination between the two *atp*9 genes of *P. hybrida* is therefore unexpected. It is possible that recombination between the copies of this gene may lead to the formation of subgenomic circles which are selected against.

It is apparent from the studies on *Petunia* and *Oenothera* that mitochondrial genome organization may differ from the models developed for *Brassica* (Fig. 1, II), maize (Fig. 2) and sugar beet.

B. Small Circular DNA Plasmids

Small circular DNA species, sometimes referred to as minicircular DNAs (Sederoff, 1984; Pring and Lonsdale, 1985), are associated with the mitochondria of some fungal (Nargang, 1985) and plant species (Table III). In plants they are by far the most abundant extrachromosomal elements associated with plant mitochondria. They are best visualized by running unrestricted DNA preparations on agarose gels, where they exhibit the properties of extrachromosomal plasmids, having three topological states: supercoiled (form I), relaxed circular (form II), and linear (form III). In

TABLE III

Circular Plasmid DNAs in Higher Plant Mitochondria

Species	Size (bp)[a] kb	References
Zea mays (maize)		
N, cms-C,T,S	1.9 (1913)[b]	Kemble and Bedbrook (1980)
	1.4	Dale *et al.* (1981)
cms-C	1.5	
cms-C	1.4	Ludwig *et al.* (1985)[b]
Beta vulgaris (sugar beet)	1.3 (1308)[c]	Powling (1981)
	1.4 (1440)[d]	Hansen and Marcker (1984)[d]
	1.45	
	1.5 (1620)[c]	Thomas (1986)[c]
CMS 01I13M4	7.3	
Phaseolus vulgaris (red bean)	1.9	Dale *et al.* (1981)
Vicia faba (broad bean)	1.42 (1476)[e]	Goblet *et al.* (1983)
	1.7 (1704)[e]	Goblet *et al.* (1985)
	1.7 (1695)[e]	Nikiforova and Negruk (1983)
(cms 350)	1.54	Negruk *et al.* (1985)
		Wahleithner and Wolstenholme (1987)[e]
Helianthus annus (sunflower)		
HA89 cms	1.45	Leroy *et al.* (1985)
Sorghum bicolor (sorghum)	2.3	Chase and Pring (1985)
	1.7	
	1.36	
Oryza sativa (rice)		
BT-cytoplasm	1.5	Yamaguchi and Kakiuchi (1983)
	1.2	
Zhen Shan 97B	1.39	Mignouna *et al.* (1987)
	1.38	
	0.9	
Zhen Shan 97A (WA-cytoplasm)	2.1	

[a] Size in base pairs given where plasmid sequence has been reported.
[b-e] References which report plasmid sequence.

addition, oligomeric forms of the plasmids have been detected in several plant species (Dale, 1981; Dale *et al.,* 1981; Powling, 1981; Yamaguchi and Kakiuchi, 1983; Abbott *et al.,* 1985; Bendich, 1985; Thomas, 1986; Mignouna *et al.,* 1987). The occurrence of these plasmids is species-specific and in some species appears to be cytoplasm-specific, for example in sunflower the 1.45-kb plasmid has only been reported in cytoplasms which are associated with male sterility (Leroy *et al.,* 1985). Similarly, the 1.4- and 1.5-kb plasmids of maize have only been reported for the cms-C cytoplasm (Kemble and Bedbrook, 1980). The 1.7-kb plasmid of broad bean was thought to be unique to cytoplasms associated with male sterility (Goblet *et al.,* 1983, 1985); however, it was isolated and sequenced from a fertile accession (Wahleithner and Wolstenholme, 1987). Similarly, the distribution of the 1.5-kb sugar beet plasmid is not restricted to male sterile accessions as once though (C. M. Thomas, personal communication). The distribution of these plasmids can best be explained by stochastic segregation and maternal evolution (Birky, 1983).

Several plasmids have been sequenced (Table III) and sequence comparisons demonstrate a lack of extensive homology between the broad bean, sugar beet, and maize plasmids. Homology does exist, however, within the sugar beet plasmids and between plasmid 1 and plasmid 2 of broad bean. Where extensive homology exists, the plasmids have perhaps arisen from a common ancestor or from an intermolecular recombination between two unrelated plasmids (Thomas, 1986).

Within the common region of broad bean plasmid 1 and 2 are six sets of inverted repeats. Plasmid 3 also contains four sets of inverted repeats, though they are unrelated to those of plasmid 1 and 2. Inverted repeats capable of forming stable hairpin structures have also been identified in the sugar beet plasmids (Thomas, 1986). These hairpin structures are similar to the inverted repeat sequences of the yeast and HeLa replication origins (Bernardi *et al.,* 1980; Attardi *et al.,* 1978); however, their function as far as the mitochondrial plasmids are concerned is not known as no replication studies have yet been carried out. In maize no such hairpin structures have been identified, though three 11-bp sequences that have homology to the yeast ARS element consensus sequence, 5'-A/TTTTATPuTTTA/T-3', have been identified, though their ability to function as ARS elements in yeast has not been tested (Stinchcomb *et al.,* 1980).

The 1913-bp plasmid of maize, the sugar beet plasmids, and plasmid 2 of broad bean hybridize to specific RNA species. RNA transcripts homologous to plasmids 1 and 3 of broad bean have not been detected. In sugar beet the RNA transcripts homologous to the 1308- and 1440-bp plasmids have only been detected in mitochondria from male-fertile cytoplasms; no transcripts have been detected in mitochondria from cytoplasms associated with male sterility (Thomas, 1986). The RNA transcripts generally do not contain large open reading frames which could possibly be translated. The open reading frame sufficient to code for a protein of 111 amino acids with homology to NAD4 on the 1620-bp plasmid of sugar beet is not transcribed (C. M. Thomas, personal communication). Therefore, in the absence of any identified function for the transcripts of plasmids, they must be considered as arising fortuitously as a result of sequence motifs which are recognized by the mitochondrial RNA polymerase.

Without exception, the mitochondrial plasmids which characterize individual cytoplasms are unrelated to the main mitochondrial chromosome, though sequences homologous to the 1913-bp plasmid of maize have been detected in the nuclear genome (Abbott *et al.*, 1985). The lack of homology of these plasmids to the mitochondrial chromosomal DNA and to other plasmids makes their origins and genetic function uncertain, supporting the idea that these types of plasmids make no genetic contribution either to the mitochondrion or to the plant (Pring and Lonsdale, 1985). Similar observations and conclusions have been made for the linear plasmids of *Sorghum bicolor* and *Brassica* spp. (Chase and Pring, 1986; Palmer *et al.* 1983; Kemble *et al.* 1986b).

In the light of current studies, it would seem that plant mitochondrial plasmids may represent selfish replicons. Whether or not their maintenance is preserved on a copy number basis (see Birky, 1983, for discussion on stochastic segregation and maintenance of small replicons) or whether it is controlled by the nuclear genotype has not yet been ascertained. These considerations are important should such plasmids be considered for use as vectors for introducing foreign genetic material into mitochondria (see Lonsdale, 1987a).

C. Small Linear DNA Replicons

Small linear replicons, apparently with no virus etiology or transposon characteristics, have been identified in the filamentous Gram-positive bacterium, *Streptomyces rochei;* in the ascomycetes, *Kluyveromyces lactis* and *Gaeumannomyces graminis;* and in a limited number of plant species, *Sorghum, Zea,* and *Brassica* (see Table IV).

The main characteristic feature of these linear replicons is a terminal inverted repeated sequence with a polypeptide covalently linked to the terminal 5'-phosphates. In the adenoviruses and *Bacillus* phages the terminal protein is part of the replication complex, DNA replication occurring bidirectionally by strand displacement (Tamanoi and Stillman, 1983). A similar function can be assumed for the terminal polypeptides of the plant linear replicons. Electron microscopy studies have shown that in the absence of proteinase treatment the S and R episomes of maize have a circular configuration, replication proceeding bidirectionally from the terminal protein complex (Sederoff and Levings, 1985). It remains unclear from these studies whether there is one or two terminal polypeptides per molecule (Sederoff

TABLE IV
Linear DNA Replicons[a]

Species	Replicon	kbp	TIR[b] (bp)	5'-Polypeptide	References
Fungi					
Gaeumannomyces graminis[c] (Ha-01)	E1	8.4	ND	Yes	Honeyman and Currier (1986)
	E2	7.2	ND	Yes	
Streptomyces rochei	pSLA1	17	614	Yes	Hirochika et al. (1984)
	pSLA2[d]	17	614	Yes	
Kluyveromyces lactis	pGKL1	8.9	202	Yes	Gunge et al. (1981)
	pGKL2	13.4	184	Yes	Hishinuma et al. (1984)
Saccharomyces kluyveri	pSKL	14.2	483	Yes	Kitada and Hishinuma (1987)
Plants					
Brassica spp.		11.3	325	Yes	Palmer et al. (1983)
					Erickson et al. (1985)
					Turpen et al. (1987)
Sorghum bicolor					
male-sterile accession	N1	5.8	ND	Yes	Chase and Pring (1986)
IS1112C	N2	5.4	ND	Yes	
Zea mays:	2.3L	2.3	170	Yes	Bedinger et al. (1986)
	2.1L[d]	2.15	170	Yes	
S-cytoplasms	S1	6.4	208	Yes	Levings and Sederoff (1983)
	S2	5.4	208	Yes	Paillard et al. (1985)
RU-cytoplasms	R1	7.4	ND[e]	Yes	Weissinger et al. (1982)
	R2	5.4	ND[e]	Yes	
Zea diploperennis	D1	7.4	ND	Yes	Timothy et al. (1983)
	D2	5.4	ND	Yes	

[a] This table does not include the giant linear plasmids of *Streptomyces* (Kinashi et al., 1987) or the large genomic linear replicons of the mitochondrial genomes of *Zea* species (Lonsdale et al., 1988).
[b] TIR, terminal inverted repeat; ND, not determined.
[c] Strains contain one or two linear molecules. Up to four different sizes, including E1 and E2, have been observed in different strains.
[d] Have arisen following a deletion.
[e] The TIR size has not been determined directly by sequencing, but sequencing of the integrated versions would

and Levings, 1985). There is some evidence that the terminal polypeptide attached to the killer plasmids KL1 and KL2 of the yeast *Kluyveromyces lactis* may differ. This is based on the different mobilities of the two identically sized terminal restriction fragments after proteinase K treatment (Kikuchi *et al.* 1984), the mobility differences being attributed to differences between the terminal amino acids remaining after proteinase K treatment. This is unusual and, more importantly, it may suggest an ability to control the replication of the plasmids independently. In maize, the relative proportions of the S1 and S2 plasmids of the S-cytoplasm can vary with the nuclear background (Laughnan *et al.*, 1981). Differential DNA replication can be easily achieved if the terminal polypeptides are the products of different nuclear genes. In *Brassica* varieties which contain the 11.3-kb linear plasmid, the relative amount of this linear molecule varies over 100-fold (Palmer *et al.*, 1983; Kemble *et al.*, 1986b). It can only be concluded at present that the nuclear genome must control the relative levels of these linear replicons by an as yet unidentified mechanism.

D. The S, R, and D Episomes of Maize

S1 (6397 bp; Paillard *et al.*, 1985) and S2 (5453 bp; Levings and Sederoff, 1983) are characteristic components of mtDNA from the S-cytoplasm of maize (Pring and Levings, 1978). In addition, mitochondria from the Ecuadorian race of maize, Racimo de Uva (RU), and from *Zea diploperennis* (ZD) contain two linear molecules, R1/R2 and D1/D2, respectively (Weissinger *et al.*, 1982; Timothy *et al.*, 1983). Because these linear molecules have related sequences in the main mtDNA and have been shown to integrate, these replicons have been termed episomes (Schàrdl *et al.*, 1984). S2, R2, and D2 appear to be identical, though R2 contains a *Bgl*I restriction site not found in S2 (Sederoff and Levings, 1985). R1 and D1, which are approximately 7.5 kb, also appear to be identical. Restriction mapping, and heteroduplex and sequencing studies (Timothy *et al.*, 1983; Levings *et al.*, 1983; Elmore-Stamper and Levings, 1986) strongly suggest that S1 arose from an intermolecular recombination event between an R1-type molecule and an R2-type molecule (Fig. 3). Such a recombination leading to the formation of S1 would result in the deletion of approximately 2.6 kb of the R1 sequence. The terminal part of this sequence, designated R1*, has been identified in Wf9-S and Vg-S cytoplasms as a component of the high-molecular-weight genome (Schardl *et al.*, 1985b; Houchins *et al.*, 1986). This provides evidence that the maize S-cytoplasms originally contained R1- and R2-type molecules. The R and D episome sequences are therefore ancient in relation to the evolution of the maize cytoplasm (Fig. 4). The origins of the R/D episomes are obscure: apart from the integrated sequences, there is no sequence homology to other mitochondrial sequences. This suggests an exogenous origin, perhaps fungal or viral (see Table IV; Lonsdale, 1986). This

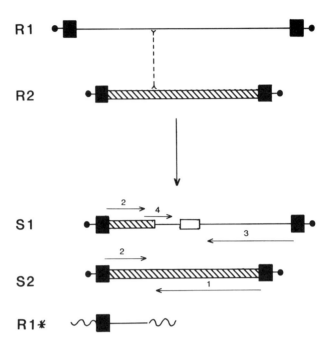

Fig. 3. Relationships between the R and S episomes of maize. The structures of the linear R and S episomes are illustrated. They are characterized by having a polypeptide covalently linked to the 5′-phosphate (●) and a short terminal inverted repeat (■). The unique sequences of R1 (unbroken line) and R2 (hatched box) are shown. The S1 and S2 episomes have been sequenced and their open reading frames (numbered arrows) and the sequence homologous to chloroplast *psbA* (open box) are shown. All of these features, apart from open reading frame 4, exist in R1 and R2 but are not shown. S1 and S2 have arisen from a recombination between R1 and R2 as shown. The terminal part of the R1 sequence lost during the creation of S1 is found as an integrated sequence (R1*) in the S-mitochondrial genome; see text and Fig. 5.

hypothesis is supported by their G + C content which, at 37–39%, is significantly less than the 47% G + C content of the mitochondrial genome and its associated plasmids.

The Latin American RU cytoplasms, in addition to containing the R1 and R2 mitochondrial episomes, have mtDNA restriction profiles almost indistinguishable from those of the current North American fertile lines (Timothy *et al.*, 1983; Weissinger *et al.*, 1982, 1983). It is probable that the North American fertile lines have arisen from these Latin American RU cytoplasms; the integrated R1 and R2 episomes became fixed by deletion of one of their two terminal inverted repeats with the concomitant loss of the free replicating episomes. Interestingly, the positions at which R1 and R2 integrate and partially linearize the RU mitochondrial genome are close to the positions of the fixed homologous sequences in the current North American lines (Lonsdale *et al.*, 1988).

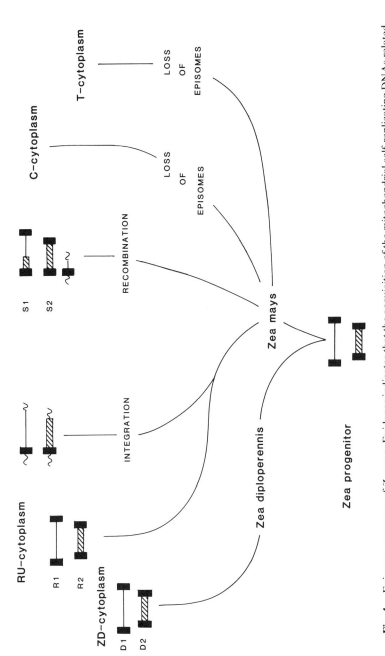

Fig. 4. Episome sequences of *Zea* spp. Evidence indicates that the acquisition of the mitochondrial self-replicating DNAs related to the R and D episomes preceded the evolutionary divergence of *Zea diploperennis* and *Zea mays*. Cytoplasmic evolution led to five recognized cytoplasmic types of *Zea mays*. Only in the RU and S-cytoplasms are free-replicating episomes maintained. In the N, C, and T cytoplasms free-replicating episomes have been lost (Thompson *et al*., 1980). Only in the N-cytoplasm are sequences related to R1 and R2 present, though the C- and T-cytoplasm may contain short sequences related to the terminal inverted repeats.

The linear replicons, N1 and N2, of sorghum exhibit no sequence homology to the high-molecular-weight mitochondrial genome (Chase and Pring, 1986). Similarly, homology between the 11.3-kb linear replicon of *Brassica* and the mitochondrial genomes of several *Brassica* species, including some which did not contain the plasmid, was not observed by Palmer *et al.* (1983) or Erickson *et al.* (1986). However, a more recent publication reported a weak homology between the 11.3-kb plasmid and the mitochondrial and chloroplast genomes (Turpen *et al.*, 1987). The origins of this homology were not investigated and therefore the D, R, and S episomes of *Zea spp.* remain unusual in having considerable sequence homology to the high-molecular-weight mitochondrial genome. These related sequences will be described, as they have been studied in detail in both the S-cytoplasm and in the normal (N) fertile cytoplasm.

E. R1- and R2-Related Sequences of the Maize N-Cytoplasms

Sequences related to the S1 and S2 episomes of maize reside adjacent to a 5270-bp repeated DNA sequence (Repeat 1; Fig. 2). Restriction analysis, heteroduplex studies, and sequencing all strongly suggest that these integrated sequences derive from the R1 and R2 replicons which characterize the related RU cytoplasms of maize (Fig. 4; Levings *et al.*, 1983; Houchins *et al.*, 1986) rather than from the S1 and S2 replicons of the S-cytoplasm. Unlike S1 and S2, R1 and R2 replicons share no sequence homology apart from their terminal inverted repeats. In the integrated sequences of the N-cytoplasm the terminal inverted repeat is 187 bp in length and forms the junction of the 5-kb repeat (Repeat 1, Fig. 2) with the unique sequences homologous to the R1 and R2 episomes. The terminal inverted repeats, including a small segment of the unique sequence of both R1 and R2, distal to the 5-kb repeat have been deleted (Levings *et al.*, 1983; McNay *et al.*, 1983; Houchins *et al.*, 1986). The R1 and R2 replicons presumably integrated into the mitochondrial master chromosome and became fixed following deletion of one of their two terminal inverted repeats, concomitant with the loss of the free replicons. Such a process has been observed in the male sterile M825 accessions of the S-cytoplasm, where reversion to fertility is associated with the loss of the S1 and S2 replicons and deletion of one of the two terminal inverted repeats of the integrated copy of S2 (Schardl *et al.*, 1985b).

F. R1, S1, and S2 Sequences of the Maize S-Cytoplasms

S1 and S2 are characteristic mitochondrial components of the S-cytoplasms of maize. In the main mitochondrial genome, restriction mapping and sequencing studies have identified a 187-bp sequence which, apart from the terminal A, is identical to the first 187 bp of the 208-bp terminal inverted repeats of S1 and S2 (Isaac *et al.*, 1985b; Schardl *et al.*, 1985b; Braun *et al.*,

1986). Sequence comparisons reveal that the adjacent 533 bp were virtually identical to the sequence adjacent to the 5-kb repeat (Repeat 1, Fig. 2) in the N-cytoplasm (Houchins *et al.*, 1986) which has been identified as the unique sequence of the R1 replicon (Levings *et al.*, 1983; Houchins *et al.*, 1986).

The homology to the R1 replicon in the S-mitochondrial genome, designated R1* (see Fig. 3), occurs at two loci in the main genome. This repeat of approximately 1 kb is flanked by four unique sequences which have been designated σ (sigma), σ' (sigma-prime), ψ (psi), and ψ' (psi-prime) (Fig. 5). Recombination between the two copies of R1* generates the four pairwise combination of the flanking sequences, σ-σ', σ-ψ', ψ-ψ', and ψ-σ'. Further recombinations involving the episomes occur as a result of the 187 bp of homology to the terminal inverted repeats of the S1 and S2 episomes. These latter recombinations, which can involve either of the two terminal inverted repeats of S1 and S2, cause a presumably circular genome to linearize (Fig. 5; Schardl *et al.*, 1984) and create numerous linear sequence configurations (Schardl *et al.*, 1985b). In the absence of S1 and S2, the organization of the S-mitochondrial genome would presumably be multicircular with the homology to R1* occurring on some but not all subgenomic circles, analogous to the N-genome organization (Fig. 2). The linear molecules generated as a

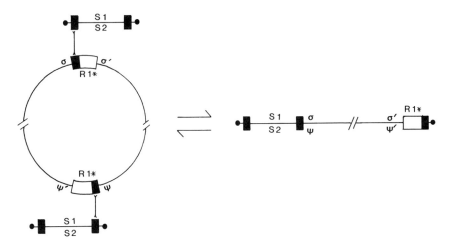

Fig. 5. Linearization of the S-mitochondrial genome by S1 and S2. A fragment of the R1 episome (R1*, see Fig. 3), including a sequence homologous to the first 187 bp of S1 and S2, is found at two locations in the S-mitochondrial genome. The two R1* sequences are flanked by four unique sequences which are labelled σ, σ' ψ, and ψ' in order to differentiate between them. Recombination occurs between the terminal 187 bp of the S1/S2 episomes and the R1* homologous sequences. This leads directly to the formation of large linear structures, having S1 or S2 attached to one terminus and the R1* sequence at the other. All the linear molecules have 5'-blocked termini (-●) and can be considered to be large linear replicons. A full description of these is given in Schardl *et al.* (1985b).

result of recombination between the terminal inverted repeats of the S1 and S2 episomes and the homologous 187 bp of the integrated partial copy of R1 have S1 or S2 attached to one end and R1* at the other (Fig. 5). Like the free episomes, these large linear molecules have a terminal inverted repeat with their 5'-phosphates covalently linked to a polypeptide.

The ability of the nuclear genotype to control the relative stoichiometry of S1 and S2 (Laughnan et al., 1981) must, and does, affect the number of linear chromosomes having S1 or S2 at one of their termini (Escote et al., 1986). This partially linear genome organization is, however, unaffected by nuclear fertility restorer genes (Schardl et al., 1985a). In cytoplasmic fertile revertants which retain the S episomes, genome organization has not yet been investigated (Escote et al., 1985) but it seems improbable that the genome organization will be anything other than partially linear and that genome organization is unrelated to the fertility status of the plant.

G. The 2.3-kb Linear Plasmid of Maize

This exists in two forms, one of 2.3 kb and one of 2.15 kb. The smaller form results from a deletion which maps to one end of the plasmid (Bedinger et al., 1986). This plasmid is found in all races of maize with the exception of the teosintes (for example, *Zea diploperennis* and *Zea luxurians*). Though these species lack the 2.3-kb species, they have homology to the plasmid in their main mitochondrial genomic DNA. This homology, however, appears to be restricted to a homology to the chloroplast *Bam*HI fragment, 15', which contains the genes for tRNA^{Trp} and tRNA^{Pro} (see Section V,A,5; Larrinua et al., 1983; Bedinger et al., 1986). The plasmid also shows homology to the internal coding sequence of the URF1 gene of the S2 episome from which it may be derived. Like the R/D and S episomes, it has a terminal inverted repeat of 170 bp, the terminal 16/17 bp of which reveals a striking homology to the terminal sequences of the S episome (Fig. 6). This homology may be sufficient to allow the 2.1/2.3-kb linear plasmid to integrate at low frequency into the main mitochondrial genome in the N- and S-cytoplasms via homologous recombination. If this occurred in the N-cytoplasm, it would result in a partial linear genomic organization similar to that resulting with S1 and S2 in the S-cytoplasm; however, there is no evidence to suggest that this occurs (Bedinger et al., 1986).

Fig. 6. DNA sequence homology between the termini of S1 and S2 and the 2.1/2.3-kb linear plasmid.

H. Double- and Single-Stranded RNAs

Both double- and single-stranded RNAs have been found associated with the mitochondrial fraction of some plant species (Schuster et al., 1983; Sisco et al., 1984; Powling, 1981; Pring and Lonsdale, 1985). The best described are those which are associated with the mitochondria of the S- and RU-cytoplasms of maize. Two double-stranded RNA species with sizes of approximately 2900 and 900 bp have been identified in the LBN-cytoplasm (an S-type) and designated LBN1 and LBN2, respectively (Sisco et al., 1982, 1984; Schuster et al., 1983). Two single-stranded RNAs have also been identified in other cytoplasms of the S group and in RU, which by restriction profile appears to be related to the N-cytoplasm types. These single-stranded RNAs have been designated S/RU-RNA-a and S/RU-RNA-b.

By isolating mitochondria from the S(B73)-cytoplasm of maize, and incubating them in medium which supports RNA synthesis, in the presence of actinomycin D, two double-stranded RNA species are predominantly labeled. These are equivalent in size to the LBN1 and LBN2 double-stranded RNAs described for the LBN-cytoplasm. A low level of incorporation into the more abundant single-stranded RNAs was also observed, suggesting that this labeling arose from replication or transcription of the double-stranded RNAs (Finnegan and Brown, 1986).

Hybridization and heat denaturation experiments demonstrate quite clearly that S/RU-RNA-b hybridizes to LBN2 but not to itself. Following 60°C formamide denaturation, the majority of LBN2 comigrates with and hybridizes to S/RU-RNA-b (Schuster et al., 1983). The implication of these results is that LBN2 is the replicative double-stranded form of S/RU-RNA-b. It is probable that a similar relationship exists between LBN1 and S/RU-RNA-a. The structure of these RNAs is not known.

These RNAs, LBN1 and LBN2, are not related to other mitochondrial plasmid or genomic sequences (Schuster et al., 1983). Neither are they associated only with the male-sterile phenotype of the S-cytoplasm, as they have been identified in R369, a fertile revertant derived from the inbred Vg-S (Singh and Laughnan, 1972), and in the RU-cytoplasm, a fertile Latin American cytoplasm. There is a correlation, however, with the presence of episome-related sequences in that these RNAs have only been identified in cytoplasms which contain either the free replicating S1/R1, S2/R2 episomes, or in the case of R369-cytoplasm, integrated sequences of the S/R episomes (Schardl et al., 1985b). However, this association is not absolute as they have not been detected in N-cytoplasms which also contain integrated sequences related to the R episomes (Houchins et al., 1986).

In sugar beet, RNA species were detected in some but not all mitochondrial preparations from plants of the same breeding line; this is suggestive of virus contamination (Powling, 1981). Virus or viruslike particles tend to associate with the organelles in infected tissue (Reisner and Gross, 1985) and would probably copurify with them. Perhaps one conclusion that can be drawn from such a close association is that the mitochondrial RNA replicons are relics of abortive viral or other plant pathogen (fungal) infections (Lonsdale, 1986).

A mechanism for importing RNA into mitochondria exists in mammalian cells; the RNA component of the mitochondrial endonuclease which cuts the mitochondrial replication primer of the point of transition to DNA synthesis is nuclear encoded (Chang and Clayton, 1987). In *Tetrahymena pyriformis* (Suyama, 1986) and *Chlamydomonas reinhardtii* (Gray and Boer, 1988), the majority of the tRNAs required to support mitochondrial protein synthesis are nuclear encoded. These examples of RNA import into mitochondria may not be isolated and a ubiquitous mechanism may exist in which case the utilization of this "RNA-import" pathway by RNA species of invading organisms is therefore a possibility. A similar mechanism for the acquisition of linear DNA replicons may be operative. For example, in certain *Neurospora* strains the Kalilo and Maranhar plasmids (Bertrand *et al.*, 1985; reviewed in Jacobs and Lonsdale, 1987) replicate in the nucleus; in certain strains these linear plasmids transfer into the mitochondrion, where they insert into the mitochondrial genome causing senescence.

Alternatively, the DNA/RNA replicons of plant mitochondria may have been acquired following mitochondrial fusion. It is interesting, in this respect, to note that some plant pathogens, for example *Ophiostroma ulmi*, the fungus associated with Dutch elm disease (Rogers *et al.*, 1987) and *Gaeumannomyces graminis,* the fungal agent of the disease, Take-All, in wheat (Honeyman and Currier, 1986), have mitochondrial RNA and DNA linear replicons.

IV. CELLS–MITOCHONDRIA–MITOCHONDRIAL DNA

There are very few published studies on relationships between mtDNA content of cells, the number of mitochondria per cell, and the cell type. Differences in mitochondria, particularly in the density of their cristae, can be observed in different cell types (Malone *et al.*, 1974). Studies on the organelle population in cells of the apical meristematic region of *Zea mays* roots showed dramatic differences in the number of mitochondria between different cell types, ranging from less than 50 mitochondria per cell in root cap cells to over 10^3 mitochondria per cell some 400 μm behind the cap (Clowes and Juniper, 1964; Juniper and Clowes, 1965). These figures are in the same range as those determined for the number of mitochondria in Timothy grass (*Phleum pratense*), watermelon, and muskmelon cells (Avers, 1962; Bendich and Gauriloff, 1984). In differentiating or differentiated cells the number of mitochondria tends to increase. A 20- to 40-fold increase in the number of mitochondria per cell has been recorded in developing anther tissue of maize (Warmke and Lee, 1978). Similarly, during oogenesis in the liverwort (*Sphaerocarpus donnellii*), there is an increase in the number of mitochondria in the oocyte. However, this relates directly to the increase in volume of the developing oocyte rather than to an absolute increase in the number of mitochondria per unit volume (Diers, 1966). If mtDNA content of individual mitochondria is constant, the amount of mtDNA per cell should vary. In pea, mtDNA content per cell varied from 0.3% to over 1.5%, depending on the tissue from which the mtDNA has been isolated (Lamppa and Bendich, 1984). In the cucurbits (watermelon, zucchini squash, cucumber, and muskmelon) electron micrograph studies suggest that in differential cells the number of mitochondria may exceed the number of mitochondrial genomes (Bendich and Gauriloff, 1984). In mung bean (*Phaseolus aureus*) the amount of mtDNA per mitochondrion has been estimated at 5×10^{-10} μg. Similar values were obtained for turnip (*Brassica rapa*), sweet potato (*Ipomoea batatas*), and onion (*Allium cepa*) (Suyama and Bonner, 1966). Based on this value, the turnip, whose genome can be expected to be about 218 kb, has two genome copies per mitochondrion. Interestingly and perhaps fortuitously, this is equivalent to one master circle and the two subgenomic circles per mitochondrion (Palmer and Shields, 1984).

In considering the relationship between the number of mitochondria per cell and the number of mitochondrial genomes, the complexity and stability of mitochondrial DNA restriction profiles need to be taken into account. If one accepts that recombination is homologous and that mitochondrial restriction profiles represent the steady-state equilibrium generated by random homologous recombination, then the inevitable conclusion is that every mitochondrion contains sufficient mtDNA to generate a state of recombinational equilibrium, or that every cell contains a single mitochondrion, or that the mitochondrial population is in a continuous state of fusion and division, a dynamic syncitium, and the mitochondrial genome population is panmictic (Lonsdale *et al.*, 1988). It is this latter possibility which appears to be the most attractive alternative and it is supported by the observation that, in fused cells, mitochondrial fusion and mtDNA recombination readily occur (see Lonsdale, 1987a, and Section XI).

V. PROMISCUOUS DNA

The organelles within the eukaryotic cell have originated in one of two ways: either the mitochondrion and chloroplast descended from free living prokaryotes which entered into an endosymbiotic relationship within a host cell having a nuclear genome, or the progenitor nuclear genome segregated, becoming physically compartmentalized and functionally specialized within a single cell. Arguments relating to these two evolutionary hypotheses are discussed elsewhere (Gray and Doolittle, 1982) and will not be presented here, though it does appear that the endosymbiont hypothesis is the more attractive and more plausible.

The close proximity of these three genetic compartments within a single cell raises the possibility of genetic transfer. Such a process is a necessary prediction of the endosymbiont hypothesis, in that much of the original genetic material of the chloroplast and mitochondrion must have been transferred to the nucleus in order to account for the current size and coding capacity of the mitochondrial and chloroplast genomes. This concept of genetic transfer is supported by the fact that for both organelles the majority of the structural polypeptides are nuclear encoded.

This genetic transfer appears to be an on-going evolutionary process as exemplified by the ATPase 9 gene (DCCD-binding protein) of fungi. In yeast (*Saccharomyces cerevisiae*), it is a mitochondrial gene (Orian et al., 1981) whereas in *Neurospora crassa*, also an ascomycete, the active gene for the DCCD-binding protein is in the nucleus (Jackl and Sebald, 1975); the protein is synthesized on cytoplasmic ribosomes and transported into the mitochondrion. Surprisingly, the *N. crassa* mitochondrion also contains a gene for the DCCD-binding protein, though it appears to be inactive (van den Boogaart et al., 1982). In photosynthetic eukaryotic cells, genetic transfer from the chloroplast to the nucleus has also occurred. One possible example involves one of the two subunits of ribulose bisphosphate carboxylase (RuBPCase). In the single-celled alga, *Cyanophora paradoxa,* the gene for the small subunit of RuBPCase is found in the cyanelle (chloroplast) genome immediately 3' to the gene encoding the large subunit of RuBPCase (Heinhorst and Shively, 1983), whereas in higher plants, the small subunit polypeptide is now coded for by a mutigene family in the nuclear genome (Cashmore, 1983). Similarly, the gene for the elongation factor Tu, which is located in the *Euglena* chloroplast DNA, is not found in either the liverwort (*Marchantia polymorpha*) or tobacco chloroplast genomes. In addition, a comparison of the liverwort and tobacco chloroplast genome sequences (Ohyama et al., 1986; Shinozaki et al., 1986) demonstrates minor variation in gene content. It can only be assumed that the "missing genes" are located in the nucleus and that their products are imported into chloroplasts. Proteins that are imported into chloroplasts are synthesized as precursors on cytoplasmic ribosomes. The precursor has an amino terminal extension (or target peptide) which ensures that the polypeptide is imported into the chloroplast with the concomitant removal of the target peptide (Ellis, 1983). It is perhaps rare that relocated genes acquire the necessary additional genetic information to ensure not only their correct transcription but also that the gene product is targeted back to the organelle from which the gene presumably originated. Mistakes have probably been made by nature in this respect, and it is not unreasonable to assume that a "chloroplast gene" could acquire a mitochondrial targeting sequence. A possible example of such a mistake is the detection of the small subunit polypeptide of RuBPCase within the mitochondria of the photosynthetic alga *Ochromonas danica* (Lacoste-Royal and Gibbs, 1985), though it is also possible that the mitochondria contain an active copy or partial copy of the small subunit gene.

Examination of nuclear DNA sequences of many eukaryotic species using mitochondrial DNA probes has provided evidence that transfer of sequences has occurred in recent evolutionary time (see Lonsdale, 1985). These apparently random transfers of DNA may still be effecting the transfer of functional genes, as is perhaps the case for the ATPase 9 gene of *N. crassa*. In plants, the majority of the mitochondrial sequences with homology in the nuclear genome appear to be exclusively from the mitochondrial plasmids. So far such transfers, based on hybridization results, have been described for the S1 episome (Kemble *et al.*, 1983), the 2.3-kb linear plasmid (Bedinger *et al.*, 1986), and the 1.9-kb circular plasmid (Abbott *et al.*, 1985) of maize. In these instances the direction of transfer cannot be established. In *Oenothera berteriana*, a clear example of sequence transfer from the nucleus to the mitochondrion has been described. Part of the nuclear gene coding for the cytoplasmic 18S rRNA has been identified in the mitochondrial genome (Schuster and Brennicke, 1987a, 1988a).

The transfer of genetic information has also been demonstrated to occur from the chloroplast to the nucleus. This genetic transfer appears to be truly promiscuous as it has occurred on an apparently enormous scale (Timmis and Scott, 1985). In spinach, up to five chloroplast genome equivalents per haploid nuclear genome have been reported (Scott and Timmis, 1984), and therefore it is not surprising that similar chloroplast sequences have been detected in both the nuclear and mitochondrial genomes in this species (Timmis and Scott, 1983).

The chloroplast genomes of higher plants apparently do not contain DNA sequences related to the mitochondrion or nucleus, apart from those sequences which have sequence homology due to related function, for example the ribosomal RNAs and tRNAs (see Table VI). This lack of promiscuous DNA may relate to the density of functional genes within the chloroplast genome (see Ohyama *et al.*, 1986; Shinozaki *et al.*, 1986). The insertion of foreign sequences would in all probability cause lethal mutations; these would be directly selected against. This density of genetic information is not observed in the mitochondrial and nuclear DNAs, where functional gases are widely dispersed, and therefore these organelles are more capable of retaining inserted sequences.

A. Chloroplast Sequences in Mitochondrial DNA of Maize

Mitochondrial restriction fragments with homology to chloroplast DNA sequences have been studied in detail from a number of plant species but the best characterized examples are from the mitochondrial genome of maize:

1. 16S Ribosomal RNA Gene and Its Flanking Sequences

An uninterrupted sequence of 12 kb from the inverted repeat of the chloroplast genome was identified as part of the maize mitochondrial genome (Stern and Lonsdale, 1982). This sequence starts within the intron of *trn*A (tRNA^{Ala}), and contains all the coding sequences 3' to this junction. These include tRNA^{Ile}, 16S rRNA, tRNA^{Val}, and presumably the coding sequences of ORF115 and 131, RPS7, 3'exon of RPS12, NDHB, and tRNA^{Leu}, based on this region's colinearity to the chloroplast genomes of tobacco and liverwort (Ohyama *et al.*, 1986; Shinozaki *et al.*, 1986).

2. Large Subunit Gene of Ribulose-Bisphosphate Carboxylase (rbcL)

The gene (*rbc*L) for the large subunit of RuBPCase including both 5' and 3' flanking sequences is found within a 3.35-kb mitochondrial *Hin*dIII restriction fragment. The authentic chloroplast *rbc*L directs the synthesis of a 54,000-Da polypeptide in *E. coli* either *in vivo* or in an *in vitro* transcription–translation system (Gatenby *et al.*, 1981). The mitochondrial *rbc*L, when added to an *E. coli in vitro* coupled transcription–translation system, directs the synthesis of a 21,000-Da polypeptide which can be immunoprecipitated using wheat RuBPCase antibodies (Lonsdale *et al.*, 1983c). This clearly demonstrates that the mitochondrial sequence is no longer recognized as an authentic chloroplast sequence and is incorrectly processed by the *E. coli* transcription–translational system.

3. Chloroplast 5S Ribosomal RNA

This sequence is contained in a chloroplast DNA segment of not more than 2 kb located on a 5-kb *Bam*HI mitochondrial fragment. When the mitochondrial sequence is used as a probe, it hybridizes to the chloroplast *Hin*dIII fragments of 1269 and 740 bp. This region, besides coding for the ribosomal 5S RNA, codes for the 3' end of the ribosomal 23S RNA, the 4.5S RNA, tRNA^{Arg}, and tRNA^{Asp} (Larrinua *et al.*, 1983; Selden *et al.*, 1983; Ohyama *et al.*, 1986; Shinozaki *et al.*, 1986). Sequencing of the mitochondrial DNA identified a 1270-bp sequence with 98% homology to the chloroplast sequence containing the 3' end of the 23S rRNA, the 4.5S and 5S RNAs, and a nearly complete copy of tRNA^{Arg} (W. W. Hauswirth, personal communication; Braun and Levings, 1986).

4. Homology to the Chloroplast psbA Gene

The central region of the S1 episome contains a 420-bp segment of the chloroplast *psb*A (Paillard *et al.*, 1985; Sederoff and Levings, 1985; Sederoff *et al.*, 1986). The DNA sequence of maize *psb*A has not been published but the comparison to other *psb*A sequences demonstrates a high degree of homology (>90%). The *psb*A sequence homology lies within the noncoding region between S1-*orf3* and S1-*orf4* (Fig. 3). In all probability, it is this chloroplast sequence which results in the apparent homology of S1 to nuclear DNA sequences (Kemble et al., 1983).

The *psb*A homology is also found in the homologous D1 and R1 episomes and in the integrated R1 sequence in the N-cytoplasm. This suggests that the original integration events of the *psb*A sequence preceded the evolutionary divergence of the current day *Zea* cytoplasms (Fig. 4).

5. Homology to Chloroplast tRNA^{Trp} and tRNA^{Pro}

The 2.3 kb/2.1-kb linear plasmid contains a sequence homologous to the chloroplast *Bam*HI fragment 15' (Larrinua et al., 1983; Bedinger et al., 1986). This region of the chloroplast genome of maize, tobacco, and liverwort contains the coding sequences for tRNA^{Pro}, tRNA^{Trp}, and *psb*E in addition to several ORFs. Probing mtDNA with an oligonucleotide derived from bean mitochondrial tRNA^{Trp} confirmed the presence of this homology on the 2.3/2.1-kb linear plasmid (Maréchal et al., 1987). No homology to other tRNA^{Trp}-like sequences was detected on the main chromosomal DNA, except in the teosintes, for example, *Zea diploperennis*, which lack the free replicating form of this plasmid (Bedinger et al., 1986). Sequencing established that both tRNA^{Trp} and tRNA^{Pro} sequences were present, separated by 140 bp of spacer sequence. This is virtually identical to organization of these genes in the chloroplast genome from where this fragment presumably originated.

6. Homology to Chloroplast tRNA^{His}

The entire coding sequence of chloroplast tRNA^{His} has been found in the mitochondrial genome, though its location is not known. In the chloroplast genome, tRNA^{His} is located on *Bam*HI fragments 6 and 8 within the inverted repeat close to the junction of the inverted repeat with the long single copy sequence (Larrinua et al., 1983; Selden et al., 1983). The full extent of this chloroplast homology has not been determined as only one of the two sequence junctions with mtDNA has been identified (Iams et al., 1985).

7. Homology to Chloroplast tRNA^{Cys} and tRNA^{Arg}

A tRNA gene identified as tRNA^{Cys} from its sequence (Wintz et al., 1988) shows nearly complete homology with a wheat chloroplast tRNA^{Cys} gene (Quigley et al., 1985). There is only one nucleotide difference in the amino acid stem between the two sequences, suggesting that it may be part of a chloroplast DNA insertion into the mitochondrial genome. A sequence related to the chloroplast tRNA^{Arg} forms the carboxy-terminus of *orf25* (Dewey et al., 1986). This chloroplast-related sequence is not found at the carboxy-terminus of the *orf25* sequence of tobacco (Stamper et al., 1987).

Many other smaller homologies have been identified in maize, the majority of which derive from the chloroplast inverted repeat but these have not been studied in any detail (Stern and Palmer, 1984b).

B. Chloroplast Sequences in the Mitochondrial Genomes of Species Other Than Maize

In other higher plant species, chloroplast homologous sequences located in the mitochondrion have been studied in detail and a few of these are worth mentioning.

1. *Cauliflower* (Brassica oleracea)

A 550-bp sequence virtually 100% identical to a region of the chloroplast genome and containing a tRNA^{Leu} gene has been identified (Dron *et al.*, 1985). The border sequences between the mitochondrial DNA and the chloroplast insert were not identified.

2. *Mung Bean* (Phaseolus aureus)

Sequence homology to the chloroplast *atp*B and *atp*E genes accounts for the major cross-hybridizations (Stern and Palmer, 1984b).

3. *Spinach* (Spinacea oleracea)

The physical maps of both the chloroplast and mitochondrial genomes have been determined (Stern and Palmer, 1986). Cross-hybridization studies demonstrated that 11 of the 12 chloroplast DNA clones hybridized with mtDNA. The homologies to the chloroplast DNA clones were dispersed throughout the mitochondrial genome, similar to the chloroplast homologies on the maize mitochondrial genome. Details such as function, length, and degree of homology were not determined, though it was evident that both multiple transfer events and possible rearrangements of the mitochondrial genome after genetic transfer may have contributed to the present-day chloroplast DNA sequence distribution.

4. *Evening Primrose* (Oenothera berteriana)

Part of the chloroplast rRNA operon has been identified in the mitochondrial genome. This sequence of 2081 bp contains the 3' end of the 23S rRNA and the 4.5S rRNA separated by the intergenic spacer region. The first 298 bp of the homology to the 23S rRNA gene are also located immediately adjacent to part of an open reading frame with homology to the ribosomal protein S13 (Schuster and Brennicke, 1987b). In maize, the homology to the 23S rRNA is shorter, but extends further 3' to genes encompassing not only the 4.5S rRNA but the 5S rRNA and much of tRNA^{Arg}. Comparison of the chloroplast genome sequences in maize, tobacco, and *Oenothera* and the mitochondrial sequences in maize and *Oenothera* suggests that these two sequences were transferred independently.

An homology to part of chloroplast gene encoding ribosomal protein S4, the transcribed spacer and tRNA^{Ser} has been identified (Schuster and Brennicke, 1987a). This homology has not been directly compared to the

Oenothera chloroplast sequence, but a comparison to the tobacco chloroplast sequence reveals that only the tRNA sequence is highly conserved, showing only one nucleotide change.

Interestingly, as in maize and wheat, a fragment containing the sequences of tRNA^{Pro} and tRNA^{Trp} is present in the *Oenothera* mitochondrial genome. Comparison of the *Oenothera* chloroplast and mitochondrial sequences reveals that only the tRNA^{Trp} sequence is conserved; the intergenic region and the tRNA^{Pro} sequences have diverged. In *Oenothera,* maize, and wheat, comparison of the mitochondrial sequences shows that the intergenic region, though divergent, is less divergent than an equivalent comparison of this region in chloroplast genomes. This strongly suggests that this particular chloroplast fragment was transferred to the mitochondrial genome prior to the separation of higher plants into monocotyledonous and dicotyledonous species (Schuster and Brennicke, 1988b).

The identification of an open reading frame with homology to reverse transcriptase in *Oenothera* (Schuster and Brennicke, 1987a) raises the possibility that the chloroplast sequences found in higher plant mitochondrial genomes are the result of RNA import and reverse transcription. However, in considering all the fragments of chloroplast DNA in mitochondrial genomes it appears improbable that reverse transcription has played or plays any significant role in interorganelle sequence transposition.

The widespread, perhaps ubiquitous, nature of these chloroplast homologous sequences in higher plant mitochondrial genomes and the apparent randomness of the sequences transferred in individual species make it unlikely that the majority of these sequences play a biological role within the mitochondrion. This is particularly evident in cases where the chloroplast sequence contains a gene whose product is normally part of a multisubunit enzyme, for example, the 16S rRNA gene. However, chloroplast tRNA genes may provide functional transcripts. The mitochondrial tRNA fraction from maize, radioactively labeled using terminal nucleotidyl transferase, does not hybridize to the mitochondrial DNA region which contains the chloroplast 12-kb inverted repeat homology and the tRNA genes for tRNA^{Leu}, tRNA^{Val}, tRNA^{Ala}, and tRNA^{Ile} (Wintz *et al.,* 1988). In contrast, the mitochondrial fragments containing the chloroplast-related tRNA^{Cys} and tRNA^{Trp} genes hybridize, indicating that they are transcribed or that the mitochondrial tRNA preparations are significantly contaminated with specific chloroplast tRNAs, though this latter possibility seems unlikely. In the case of tRNA^{Trp}, studies on the mitochondrial tRNA species in mung bean (Maréchal *et al.,* 1985b) and soybean (Swamy and Pillay, 1982) have identified only one tRNA^{Trp} species. The sequence of the mung bean tRNA^{Trp} species was found to be virtually identical to its chloroplast homology and also to the wheat and maize tRNA^{Trp} gene sequences. Therefore, it would appear that intracellular promiscuity has contributed to mitochondrial function in this instance.

The transcription of chloroplast tRNA genes in the mitochondrion may be serendipitous, but the highly conserved secondary and tertiary structure of the nascent tRNA transcript may allow correct processing to produce mature functional tRNA species (Peebles et al., 1983). The contribution of chloroplast sequences, particularly tRNAs, to mitochondrial biogenesis remains to be determined.

VI. GENETIC COMPLEXITY OF PLANT MITOCHONDRIAL GENOMES

The only logical way to determine the complete genetic complexity of a plant mitochondrial genome is to sequence it. However, given the enormity of sequencing an entire mitochondrial genome, several other approaches have been used to assess the genetic complexity or the number of genes in plant mitochondrial DNA. These include DNA–DNA and RNA–DNA hybridization studies and *in vitro* protein synthesis in isolated, intact mitochondria.

A. Conserved Sequences

Experiments using purified and radioactively labeled mitochondrial DNA to probe restriction endonuclease digests of other mitochondrial genomes showed that a set of conserved sequences existed that appeared to correlate with the transcribed fraction of the mitochondrial genome (Stern et al., 1983; Stern and Newton, 1985a). In maize, heterologous hybridizations with radioactively labeled *Brassica* and sugar beet mtDNAs identify a relatively low number of restriction fragments, many of the genomic locations of which correlated with the positions of known mitochondrial genes (D. M. Lonsdale, unpublished data). Similar experiments using cloned fragments of the *Brassica campestris* (218 kb) mitochondrial genome to probe RNA detected 24 abundant nonoverlapping transcripts totalling some 60 kb. This number of transcripts is in reasonable agreement with the observation that approximately 20 polypeptides are synthesized in isolated mitochondria (Makaroff and Palmer, 1987).

RNA excess hybridization studies suggest that a significant proportion of the mitochondrial genome is transcribed, for example up to 70% (225 kb) of the 330-kb watermelon genome and 30% (480 kb) of the 2400-kb muskmelon genome appear to be transcribed (Bendich, 1985). Using radioactively labeled RNA synthesized by isolated mitochondria, hybridization to separated restriction fragments of *Brassica* and maize mitochondrial DNA provided evidence that the entire genomes were transcribed (Carlson et al., 1968a,b). Such results suggest that, if the majority of these transcripts are from unique

DNA sequences, a significant proportion of the transcribed RNA must derive from noncoding regions of the genome, given the apparently low coding potential of higher plant mitochondrial genomes (see Section IX; Bendich, 1985; Stern and Newton, 1985a). The transcription of chloroplast-related sequences and nonfunctional sequences in *Brassica* and maize mitochondrial genomes (Carlson *et al.*, 1986a,b) contradicts more critical experiments which demonstrate that, with the exception of specific tRNAs, such sequences are not transcribed at easily detectable levels (Houchins *et al.*, 1986; Makaroff and Palmer, 1987; Wintz *et al.*, 1988). It would appear that the use of RNA synthesized in isolated mitochondria leads to artifactually high hybridization results, possibly due to the loss of specificity by the RNA polymerase.

B. Protein Synthesis by Isolated Mitochondria

When provided with an osmoticum, oxidizable substrate, and a radioactive amino acid (for example, [^{35}S]methionine), isolated mitochondria synthesize ^{35}S-labeled polypeptides (Forde and Leaver, 1980; see Fig. 7). The number of polypeptide bands which can be visualized on a one-dimensional sodium dodecyl sulfate (SDS)-polyacrylamide gel is approximately 15–20 for all plant species investigated and seems to be unrelated to mitochondrial genome size. *In vivo* protein synthesis in the presence of antibiotics which inhibit cytosolic and chloroplast protein synthesis have demonstrated that the *in vivo* and *in vitro* products of mitochondrial protein synthesis are very similar (Forde *et al.*, 1978; Newton and Walbot, 1985), sustaining the conclusion that plant mitochondrial genomes do indeed synthesize only a small number of polypeptides. Two-dimensional gel electrophoresis reveals between 30 and 50 discrete products, indicative of perhaps a larger number of genes (Hack and Leaver, 1983). However, many of these additional products may arise from proteolytic cleavage or secondary modification of the polypeptides. This is certainly the case in mammalian and fungal systems where, for example, *in vitro* protein synthesis using isolated intact mammalian mitochondria results in the synthesis of approximately 24 polypeptides though the genome only contains 13 known reading frames (Anderson *et al.*, 1981).

In plants, mitochondrial polypeptide profiles appear to be remarkably similar across a wide range of species, suggesting an evolutionary conservation of genes. However, the polypeptide profiles do differ enough to enable closely related species and even different cytoplasms of the same species to be distinguished. Excellent examples of such differences are those between fertile and male-sterile cytoplasms of maize, sorghum, and sugar beet (Forde and Leaver, 1980; Boutry *et al.*, 1984; Bailey-Serres *et al.*, 1986) which reveal additional "variant" polypeptides (Fig. 7). The role of these extra

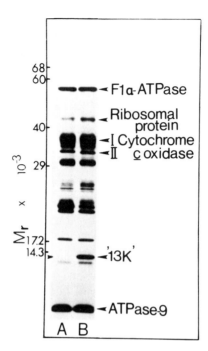

Fig. 7. Polypeptide synthesis by isolated mitochondria. Mitochondria isolated from the normal fertile (A) and male-sterile T-cytoplasm (B) were incubated in a medium containing [^{35}S]methionine and an energy-generating system. Mitochondrial polypeptides were fractionated by SDS-polyacrylamide gel electrophoresis and the radioactively labeled polypeptides detected by autoradiography (Forde and Leaver, 1980). Several of the radioactively labeled polypeptides have been identified, including the "13K" polypeptide which is coded for by T-*urf13* (see text and Fig. 8A). Reproduced with the permission of C. J. Leaver.

products as determinants of the male-sterile phenotype is uncertain, but evidence is accumulating that the 13,000-Da polypeptide of the T-cytoplasm of maize may have such a role (Rottmann *et al.*, 1987; Lonsdale, 1987b).

Mitochondria isolated from different tissues and at different times of development exhibit both quantitative and qualitative differences (Boutry *et al.*, 1984; Newton and Walbot, 1985). Such differences, particularly where apparently novel translation products have been detected, may reflect alterations in polypeptide processing or modification in addition to developmentally regulated or tissue-specific expression of additional genes.

VII. MITOCHONDRIAL CODING SEQUENCES

The plant mitochondrial genome, like all mitochondrial genomes, codes for the large (26S) and small (18S) ribosomal RNAs (Pring and Thornbury,

1975; Chao et al., 1983). In addition, and only in plant mitochondria, has a small (5S) rRNA component been detected (Leaver and Harmey, 1976). The mitochondrial genome presumably codes for a complete set of tRNAs though this is by no means certain. Recent sequencing of the tobacco chloroplast genome demonstrates that the genes coding for the tRNAs having the anticodons G/AAG (leucine), G/AGG (proline), G/AGC (alanine), and CGC/C/U/GCGA/G (arginine) are absent (Wakasugi et al., 1986), though their existence is predicted from codon usage. Although the lack of specific tRNAs can be compensated for by increased codon recognition of other tRNA species, in at least two protists, *Tetrahymena pyriformis* (Suyama, 1986) and *Chlamydomonas reinhardtii* (Gray and Boer, 1988), the mitochondrial genomes do not code for sufficient tRNAs to support protein synthesis and therefore it can only be assumed that tRNA import occurs. This is not an unreasonable assumption based on the recent demonstration that the RNA moiety of the mammalian mitochondrial RNase MRP is imported (Chang and Clayton, 1987).

Last, and in addition to the expected complement of genes, including polypeptide genes whose products are involved in electron transport and ATP synthesis, mitochondria contain a collection of unidentified open reading frames, gene chimeras, nonfunctional genes, and nonfunctional transcribed sequences.

A. Mitochondrial Ribosomal RNA Genes

The size and organization of mitochondrial rRNA genes vary widely among eukaryotes. In mammalian mitochondria, the RNAs from the large and small subunits of the ribosome have sedimentation coefficients of 16S and 12S, respectively. In yeast, the rRNAs are 21S and 15S, while in higher plants they are 26S and 18S, representing the largest mitochondrial rRNAs known.

In mammals the mitochondrial rRNAs are separated by a single tRNA gene (Anderson et al., 1981); in yeast they are some 40 kb apart (Borst and Grivell, 1978). In higher plants, for example maize, they are 16 kb apart and transcribed in opposite directions (Iams and Sinclair, 1982; Stern et al., 1982; Dawson et al., 1986a). However, in other species the spatial relationship and the relative orientation of the two genes depend on the genome organization.

Only in plant mitochondria has a small (5S) rRNA component been detected (Leaver and Harmey, 1976). The gene encoding the 5S rRNA is closely linked to the 18S rRNA gene in all higher plants studied to date (Bonen and Gray, 1980; Huh and Gray, 1982; Stern et al., 1982), and is located 3' to the 18S gene. The intergenic spacer region between the 3' end of the 18S rRNA and the 5' end of the 5S rRNA varies between species; in maize it is 108 bp, in soybean 150 bp, and in *Oenothera berteriana* it is 582 bp, containing a pseudo-tRNA gene (Brennicke *et al.,* 1985). This organization of the 18S–5S rRNA genes appears to be peculiar to higher plants.

Ribosomal RNA genes have been sequenced in wheat and maize, which are representative of the monocotyledons, and in *Oenothera* and soybean, which are dicotyledonous species (Table V). There is a remarkable degree of sequence conservation between the ribosomal rRNA genes: approximately 85% between two groups and 95% within the groups. It is evident that, compared to mitochondrial genes of other organisms, the rate of mutational change in plants is much slower (Grabau, 1985). Terminal heterogeneity at both the 5' and 3' ends of the mature RNAs is evident (Spencer *et al.,* 1981, 1984; Schnare and Gray, 1982). In addition, the length variations between the different rRNAs (Table V) are the result of small additions and deletions located in the variable loop regions of the derived secondary structures. Such length variations have been noted previously between different plant species (Leaver and Harmey, 1973), and, more recently, evidence for two different size classes of the 26S rRNA transcripts in watermelon, muskmelon (Stern and Newton, 1985a), and mung bean has been obtained (Stern and Newton, 1985b). Filter hybridization data suggest that in watermelon only a single 26S rRNA gene exists. Therefore it is probable that the larger of the two transcripts is a precursor of the mature 26S rRNA. In muskmelon, where multiple genes for the 26S rRNA exist (Stern and Newton, 1985a), the two transcripts may be from different genes, which are either different sizes or have primary transcripts that are processed differently.

The 3' ends of sequenced mitochondrial 18S genes do not have a recognizable Shine–Dalgarno sequence (CCUCC) which has been identified at the 3'

TABLE V

Sequenced Ribosomal RNA Genes of Higher Plants

Species	Length (bp)	Reference
26S rRNA		
Maize	3546	Dale *et al.* (1984)
Oenothera berteriana	3268	Manna and Brennicke (1985)
18S rRNA		
Soybean	1990	Grabau (1985)
Wheat	1995	Spencer *et al.* (1984)
Maize	1968	Chao *et al.* (1984)
Zea diploperennis	1966	Gwynn *et al.* (1987)
Oenothera berteriana	1901	Brennicke *et al.* (1985)
5S rRNA		
Soybean	118	Morgens *et al.* (1984)
Wheat	122	Spencer *et al.* (1981)
Maize	126	Chao *et al.* (1983)
Zea diploperennis	126	Gwynn *et al.* (1987)
Oenothera berteriana	118	Brennicke *et al.* (1985)

ends of the *E. coli* and chloroplast 16S rRNAs. However, another sequence (UGAAU) may have an equivalent role in plant mitochondria (see Section X and Table IX).

B. Mitochondrial Transfer RNA Genes

Sequencing studies on isolated tRNAs and on tRNA genes have revealed certain interesting features. First, sequenced genes do not encode the terminal CCA, and second, they can (with one exception) be classified into two groups: *chloroplast-homologous genes,* those exhibiting 90-100% homology to their equivalent chloroplast genes, and *mitochondrial genes.* The latter group exhibits approximately 70-80% homology to the equivalent chloroplast genes. The exception is one of the methionine tRNAs of maize. This gene, which is duplicated in the maize mitochondrial genome, exhibits only 41% homology to its equivalent maize chloroplast sequence and to the mitochondrial tRNA^{Met} of *Phaseolus vulgaris* (Table VI). This gene as well as the sequences homologous to the formyl methionine initiator tRNA and elongator tRNA species of *Phaseolus vulgaris,* have been localized at different positions on the maize mitochondrial map (see Fig. 10).

The chloroplast-homologous tRNA genes fall into two classes: transcribed and nontranscribed. The latter group is illustrated by those sequences homologous to tRNA^{Val}, tRNA^{Leu}, and tRNA^{Ile}. In the maize mitochondrial genome these tRNA genes are located on the 12-kb chloroplast-homologous fragment (see Section V, A, 1; Stern and Lonsdale, 1982). Hybridization studies demonstrate that transcripts of these genes are present in the tRNA fraction of chloroplasts but not mitochondria. In contrast, transcripts of tRNA^{Trp} and tRNA^{Cys} are present in both the chloroplast and mitochondrial tRNA fractions (Maréchal *et al.,* 1987; Wintz *et al.,* 1988). This suggests that chloroplast sequences can provide functional transcripts in the mitochondrion. Supporting this is the fact that, in *Phaseolus vulgaris,* both the mitochondrial and chloroplast tRNA^{Trp}-CCA species have been purified and sequenced. They show a high degree of sequence homology only differing at five positions (Maréchal *et al.,* 1985b). Interestingly, the *P. vulgaris* mitochondrial tRNA^{Trp}-CCA species does not promote readthrough of the UGA universal stop codon as do the *E. coli* and yeast tRNA^{Trp}-CCA species (Guillemaut *et al.,* 1986). An alteration in the genetic code, CGG = tryptophan instead of arginine, is predicted from the sequence of several mitochondrial polypeptide genes (Section VII,C,1), though as yet no gene or tRNA species has been identified which will translate the arginine codon CGG as tryptophan.

The plant mitochondrial tRNA genes therefore constitute an unusual and unique system, where chloroplast sequences containing tRNA genes have apparently integrated into the mitochondrial genome. Some of these chloro-

TABLE VI
Sequenced Mitochondrial tRNAs and tRNA Genes

Sequenced tRNAs

Species	tRNA[a]	Anticodon	Length in nucleotides	% Homology to chloroplast[b]	Reference
Phaseolus vulgaris (French bean)	Trp	CCA	76	97 (p)	Marechal et al. (1985b)
	Met	CAU	76	93.3 (t)	Marechal et al. (1986)
	f-Met	CAU	76	76.6 (s)	Marechal et al. (1986)
	Phe	GAA	76	75 (p)	Marechal et al. (1985a)
	Pro	UGG	78	72.0 (+)	Runeberg-Roos et al. (1987)
	Tyr-1	NUA	76	70.9 (s)	Marechal et al. (1985c)
	Tyr-2	NUA	76	70.9 (s)	Marechal et al. (1985c)
	Leu-1	NAG	76	57 (p)	Green et al. (1987)
	Leu-2	NAG	76	57 (p)	Marechal-Drouard et al. (1988)

Sequenced tRNA genes

Species	Gene	Anticodon	Length in nucleotides	% Homology to chloroplast	Reference
Triticum aestivum (wheat)	φtrnP	—	23	100 (w)	Marechal et al. (1987)
	trnW	CCA	73	97 (w)	Marechal et al. (1987)
	trnP	UGG	75	75 (t)	Joyce et al. (1988)
	trnfM	CAU	74	73 (s)	Gray & Spencer et al. (1983)

	trnD	GUC	74	70.3 (t)	Joyce et al. (1988)
	trnY	GUA	83	69.1 (t)	Joyce et al. (1988)
	trnS	GCU	88	68.1 (t)	Joyce et al. (1988)
	trnS	UGA	87	73.7 (t)	Joyce et al. (1988)
Zea mays (maize)	trnH	GUG	76	100 (m)	Iams et al. (1985)
	trnC	GCA	71	99 (w)	Wintz et al. (1988b)
	trnW	CCA	73	96 (w)	Marechal et al. (1987)
	φtrnP	UGG	74	96 (w)	Marechal et al. (1987)
	φtrnF	UAA	119	93 (p)	Wintz et al. (1988b)
	φtrnR	ACG	68	93 (t)	Dewey et al. (1986)
	trnfM	CAU	74	72 (s)	Parks et al. (1984)
	trnD	GUC	74	70 (+)	Parks et al. (1985)
	trnS	GCU	88	68 (+)	Wintz et al. (1988b)
	trnM	CAU	74	43 (m)	Parks et al. (1984)
Oenothera berteriana (evening primrose)	trnW	CCA	74	100 (o)	Schuster et al. (1988b)
	trnS	GGA	87	99 (t)	Schuster & Brennicke (1987a)
	φtrnP	UGG	70	89 (o)	Schuster et al. (1988b)
	trnfM	CAU	74	73 (s)	Gottschalk & Brennicke (1985)
Glycine max (soybean)	trnM	CAU	69	95 (g)	Wintz et al. (1988a)
	trnfM	CAU	74	73 (+)	Grabau (1987)
	trnE	UUC	72	65 (+)	Wintz et al. (1987)
Brassica	trnL	CAA	85	100 (b)	Dron et al. (1985)
Lupinus luteus (lupin)	trnfM	CAU	73	nd	Borsuk et al. (1986)
	trnG	GCC	73	nd	Bartnick & Borsuk (1986)
Arabidopsis thaliana	trnM	CAU	69	95 (g)	Wintz et al. (1988a)

[a] Includes terminal CCA
[b] s, spinach; t, tobacco; w, wheat; m, maize; p, Phaseolus; b, Brassica; g, soybean; o, Oenothera; +, unspecified; nd, not determined.

plast tRNA genes are transcribed and probably provide functional tRNA species which may supplement or replace mitochondrial tRNAs.

C. Polypeptide Genes

1. Identification

Genes encoding polypeptides in higher plant mtDNA, listed in Table VII, have been identified by two procedures:

1. Clone banks of plant mitochondrial DNA restriction fragments or cDNAs (Hiesel and Brennicke, 1987; Hiesel et al., 1987) were hybridized with total radioactively labeled mitochondrial RNA. The clones which hybridized were then screened to eliminate those containing homology to mitochondrial rRNA and tRNAs and previously sequenced genes. The selected clones were then sequenced. The sequences were compared to known eukaryotic mitochondrial polypeptide sequences held in DNA sequence data banks such as "Genebank." This procedure identified, for example, the coding sequences for α-subunit of the F_1 portion of ATP synthase (Braun and Levings, 1985) and subunits 6 and 9 of the F_o portion of ATP synthase (Dewey et al., 1985a,b).

2. The second and more direct way of identifying particular polypeptide genes is to use heterologous probes (Dawson et al., 1986b). For example, the yeast *oxi-1* gene (Fox, 1979) was used to isolate the equivalent maize gene, *coxII* (Fox and Leaver, 1981). This more directed procedure has identified the genes for cytochrome oxidase subunits I, II, and III (Fox and Leaver, 1981; Isaac et al., 1985b; Schuster and Brennicke, 1986; McCarty and Hauswirth, 1988), apocytochrome *b* of the *bc'* complex (Dawson et al., 1984), and the α-subunit of the F_1 portion of F_o–F_1 ATP synthase (Isaac et al., 1985a).

2. Genetic Code and Codon Usage

Comparison of the codon and amino acid sequence of plant mitochondrial genes to other equivalent eukaryotic sequences predicts that CGG codes for tryptophan instead of arginine (Fox and Leaver, 1981). This is supported by comparisons of plant mitochondrial genes in different species where TGG and CGG are optional in highly conserved domains of the proteins (Hiesel and Brennicke, 1983).

Analysis of the protein coding genes in plant mtDNAs shows a consistent bias in favor of codons ending in T and A residues ($T > A > G > C$), similar to the bias in fungal mitochondrial genes (Waring et al., 1981), though it contrasts with mammalian mitochondrial genes where A and C predominate in the third position (Anderson et al., 1981). There are no obvious omissions in the codons used, suggesting that mitochondria require a full complement of tRNAs (Schuster and Brennicke, 1985). The translation of plant mito-

TABLE VII
Mitochondrial Sequences with Homology to Known Proteins

Gene	Species[a]	Predicted length of coding sequence (bp)	Terminal codon	Intron No.	Intron length (bp)	Type	Reference
Cytochrome oxidase complex							
Subunit I	*coxI*						
	Sorghum : 9E	1893	TAG	—	—	—	Bailey-Serres *et al.* (1986)
	Sorghum : Milo	1590	TAA	—	—	—	Bailey-Serres *et al.* (1986)
	Maize	1584	TAG	—	—	—	Isaac *et al.* (1985b)
	Obe	1584	TAA	—	—	—	Hiesel *et al.* (1987)
	Soybean	1581	TAA	—	—	—	Grabau (1986)
	Wheat	1572	TAA	—	—	—	Bonen *et al.* (1987)
Subunit II	*coxII*						
	Rice	783	TAA	1	1265	II	Kao *et al.* (1984)
	Soybean	783	TAA	—	—	—	Grabau (1987)
	Maize	780	TAA	1	794	II	Fox and Leaver (1981)
	Wheat	780	TAA	1	1217	II	Bonen *et al.* (1984)
	Zdi	780	TAA	1	794	II	Gwynn *et al.* (1987)
	Pea	777	TAA	—	—	—	Moon *et al.* (1985)
	Obe	777	TAA	—	—	—	Hiesel and Brennicke (1983)
Subunit III	*coxIII*						
	Obe	795	TGA	—	—	—	Hiesel *et al.* (1987)
Cytochrome bc_1 complex							
Apocytochrome *b*	*cob*						
	Wheat	1194	TAG	—	—	—	Boer *et al.* (1985)
	Obe	1182	TGA	—	—	—	Schuster and Brennicke (1985)
	Broad bean	1176	TGA	—	—	—	Wahleithner & Wolstenholme (1988)
	Maize	1164	TAG	—	—	—	Dawson *et al.* (1984)
F_o–F_1 ATPase complex							
Subunit (F_1)	*atpA*						
	Obe	1533	TAG	—	—	—	Schuster and Brennicke (1986)
	Maize	1524	TGA	—	—	—	Isaac *et al.* (1985a)
	Maize : cms-T	1524	TGA	—	—	—	Braun and Levings (1985)
	Pea	1521	TAA	—	—	—	Morikami & Nakamura (1987)

(*continued*)

TABLE VII (continued)

	Gene	Species[a]	Predicted length of coding sequence (bp)	Terminal codon	Intron No.	Intron length (bp)	Type	Reference
Subunit 6 (F$_o$)	atp6	Tobacco	1185	TAG	—	—	—	Bland et al. (1987)
		Obe	948	TAG	—	—	—	Schuster & Brennicke (1987c)
		Maize : cms-T	873	TAG	—	—	—	Dewey et al. (1985a)
		Soybean	873	TAG	—	—	—	Grabau et al. (1988)
Subunit 9 (F$_o$)	atp9	Tobacco	231	TAG	—	—	—	Bland et al. (1986)
		Petunia	231	TAG	—	—	—	Young et al. (1986)
		Maize	222	TAA	—	—	—	Dewey et al. (1985b)
		Broad bean	222	TAA	—	—	—	Wahleithner & Wolstenholme (1988)
NAD : Q1 complex								
Subunit 1	nad1	Watermelon	?	?	1	52	?	Stern et al. (1986)
					2	1430	II	
					3	177	?	
					4	1022	?	
Subunit 3	nad3	Maize	354	TAA	—	—	—	Gualberto et al. (1988)
		Wheat	354	TAA	—	—	—	Gualberto et al. (1988)
Subunit 5	nad5	Obe	1446	TAG	1	850	II	Wissinger et al. (1988)
					2	357	?	
Ribosomal proteins								
Small subunit	rps12	Maize	375	TGA	—	—	—	Gualberto et al. (1988)
		Wheat	375	TGA	—	—	—	Gualberto et al. (1988)
	rps13	Maize : cms-C	387	TGA	—	—	—	Bland et al. (1986)
		Tobacco	348	TGA	—	—	—	Bland et al. (1986)
		Wheat	348	TGA	—	—	—	Bonen (1987)
		Obe	342	TGA	—	—	—	Schuster & Brennicke (1987b)
	rps14	Broad bean	300	TAG	—	—	—	Wahleithner & Wolstenholme (1988)

[a] Obe, Oenothera berteriana, the evening primrose; Zdi, Zea diploperennis.

chondrial mRNAs is presumably initiated at AUG using tRNA^{fMet}. Termination of translation would appear to occur at the universal code triplets TAA, TGA, and TAG. The use of TGA as a stop codon is unusual in that in all other mitochondria TGA codes for tryptophan. No amino acid sequence of a plant mitochondrial gene product is yet available and therefore a comparison between predicted and actual amino acid sequences cannot be made.

D. Unusual Transcribed Sequences

In addition to those genes for which a function has been assigned, several other classes or groups of genes can be identified. These are as follows.

1. Unassigned and Open Reading Frames (URFs and ORFs)

Open reading frames (ORFs) may be transcribed, but this cannot be assumed to imply that such transcripts are pre-mRNAs or are functional mRNAs. Unassigned reading frames (URFs) are open reading frames that are transcribed and translated and for which the function of the polypeptide which the gene encodes is not known, as no equivalent or related sequence exists in yeast or mammalian mitochondrial genomes. Such genes have been identified both in the main mitochondrial genome and in plasmids, for example, the four ORFs of the S1 and S2 episomes of the S-cytoplasm of maize. Of these S1-*urf3* and S2-*urf1* have been shown to have protein products (Manson *et al.*, 1986; Zabala and Walbot, 1988). S1-*urf3*, has limited homology to a viral-type DNA polymerase (Kuzmin and Levchenko, 1987) while the S2-*urf1* has homology to the RNA polymerase of yeast mitochondria (Kuzmin *et al.*, 1988). In *Oenothera* an ORF with homology to reverse transcriptase has been identified (Schuster and Brennicke, 1987a). These three ORFs represent the only sequences so far identified with homology to known DNA/RNA polymerases in plant mitochondria. Other transcribed ORFs with no known or obvious functional homologies have been identified, for example, the ORFs in the intron of the COXII gene of wheat (Bonen *et al.*, 1984) and in the 5' regions of the COXI, COXII, and COXIII genes of *Oenothera* (Hiesel and Brennicke, 1985; Hiesel *et al.*, 1987). In maize, an ORF sufficient to encode a polypeptide of approximately 25,000 Da, *orf25*, is found as a single-copy transcribed sequence of the N, C, T, and S mitochondrial genomes (Stamper *et al.*, 1987). It is also a transcribed sequence of the tobacco mitochondrial genome, and homologies to this sequence can be detected in other species including bean, wheat, pea, rice (Stamper *et al.*, 1987). As *orf25* occurs in a number of diverse plant species, it seems probable that there may be polypeptide product, though it has yet to be identified.

Transcription of DNA containing ORFs may be fortuitous based on an advantageously placed RNA transcription promoter motif (see Section IX). Such transcripts may be untranslatable, for example, due to lack of ribosome binding sites or other as yet unidentified translational signals, and such

ORFs are unlikely to be conserved between species. Therefore, the conservation of transcribed ORFs between diverse species is probably an indicator of evolutionary conservation and therefore function.

2. Chimeric Genes

These are transcribed ORFs which contain a sequence or sequences derived from other genes. The *orf25* sequence of maize has a homology to chloroplast tRNA^{Arg}, located at its 3' terminus (Dewey *et al.*, 1986). This homology is found in the *orf25* genes of the N-, C-, S-, and T-cytoplasms of maize but is absent from the tobacco sequence (Stamper *et al.*, 1987). A more extreme example of a chimeric gene is T-*urf13*. This gene is located immediately 5' to the *orf25* sequence in the mitochondrial genome of the T-cytoplasm and it encodes the 13-kDa polypeptide (Forde and Leaver, 1980; Wise *et al.*, 1987). T-*urf13* is composed of a mosaic of sequences derived from the flanking and coding regions of the 26S rRNA gene (Fig. 8A; Dewey *et al.*, 1986; Rottmann *et al.*, 1987). A similar complex gene, S-*pcf*, has been identified in the mitochondrial genome of male-sterile petunia (Young and Hanson, 1987; Hanson *et al.*, 1988). The 5' flanking sequences are identical to the 5' flanking sequence of *atp9*. The coding region has sequence homology to *atp9*, *coxII*, and *urfS* (Fig. 8B); the latter is an URF found in the mitochondrial genomes of both fertile and male-sterile cytoplasms.

Both T-*urf13* and S-*pcf* appear to be causally linked to the male-sterile phenotype in maize and petunia, respectively. In addition, deletion of the T-*urf13* gene in the fertile revertants of the male-sterile T-cytoplasm is associated with the loss of susceptibility to the T-toxin of *Helminosporium maydis* race T (Rottmann *et al.*, 1987). The expression of T-*urf13* in *E. coli* confers susceptibility to the *H. maydis* toxin, strongly implicating the T-*urf13* gene product as the receptor protein for the pathotoxins of *H. maydis* race T (Levings and Dewey, 1988; D. M. Lonsdale and W. H. Rottmann, unpublished data). These chimeric genes, T-*urf13* and S-*pcf*, are the only mitochondrial genes which are known to have a direct effect on the plant phenotype.

In the C-cytoplasm of maize, the mitochondrial genes for *atp6* and *coxII* are chimeric (Fig. 8C). For example, the first 24 codons of *atp6* have been replaced with the amino terminal 13 codons of *atp9* followed by 147 codons from an ORF of 441 bp which has significant homology to the chloroplast genome (Fig. 8C). The *coxII* sequence has immediately 5' to its ATG codon the 5' flanking sequence and the first 65 amino acids of the predicted ATP6 polypeptide. However, the reading frame of *atp6* extends 5' for a further 408 bp, with alternative methionine codons occurring at -294, -324, -348, -351, -354, and -357 bp upstream of the predicted initiation codon of the *atp6* (Dewey *et al.*, 1985a; Levings and Dewey, 1988). Amino-terminal extensions and carboxy-terminal extensions are not unusual features of plant mitochondrial genes, though at present there is no evidence to suggest that

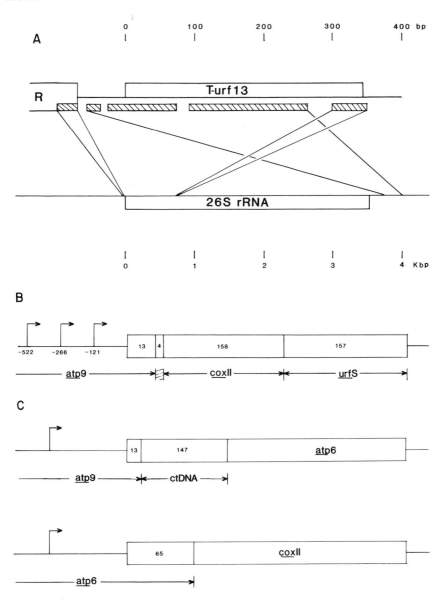

Fig. 8. Chimeric genes. (A) The T-URF13 gene from the T-cytoplasm of maize. The sequences of T-*urf13* derived from the mitochondrial 26S rRNA coding region, those exhibiting >90% sequence homology are shown (hatched boxes). A repeated sequence (R) contains the transcriptional promoter sequence; this sequence is also found 5′ to the ATP6 gene. (B) The S-PCF gene from the male-sterile cytoplasm of petunia. The sequences of S-*pcf* origin are from *atp9*, *coxII*, and *urfS*. A spacer consisting of four codons between the *atp9* and *coxII* sequences (hatched) has no obvious origins. The gene is transcribed from the three *atp9* transcription promoters (arrows). (C) The ATP6 and COXII genes from the C-cytoplasm of maize. The numbers within the boxed coding regions indicate number of codons.

amino-terminal extension sequences are translated to form chimeric proteins. However, the carboxy-terminal extension of 101 amino acids associated with the COXI gene of the sorghum 9E cytoplasm is translated, giving a chimeric protein (Bailey-Serres *et al.*, 1986).

It cannot be ascertained whether these types of chimeric genes having amino-terminal extensions or carboxy-terminal extensions, as in the case for the sorghum COXI gene, are responsible for aberrant plant phenotypes such as male-sterility.

3. Nonfunctional Genes (Pseudogenes)

a. Inactive Genes. Gene inactivation can occur by a variety of processes, two of which have been identified in plant mitochondria and which affect the same gene: the large ORF of the S2 episome (S2-*urf1*) of the S-cytoplasm of maize (Levings and Sederoff, 1983). This gene transcribed in the male-sterile S-cytoplasm, the RU-cytoplasm, and the normal fertile (N) cytoplasm of maize (Houchins *et al.*, 1986; Traynor and Leavings, 1986). In male-sterile maize plants having the S-cytoplasm and the M825 nuclear genotype, spontaneous reversion to fertility has been observed (Laughnan and Gabay-Laughnan, 1983). This fertile phenotype is maternally inherited and is associated with alterations in the mtDNA (Levings *et al.*, 1980), specifically the loss of the free S1 and S2 episomes and the deletion of part of the terminal inverted repeat of the integrated S2 episome. This deletion includes the region involved in transcription initiation which effectively abolishes transcription of S2-*urf1* (Schardl *et al.*, 1985b). In the normal fertile cytoplasm the R2/S2-URF1 gene is transcribed in an identical manner to S2-*urf1* of the male-sterile S-cytoplasm. However, sequencing studies on the 5' coding region of R2-*urf1* have indicated that frame shift mutations have occurred, resulting in the introduction of multiple termination codons into the amino-terminal region of the reading frame (Houchins *et al.*, 1986). Western blotting of mitochondrial polypeptides using antibodies against a *lac*Z/S2-*urf1* fusion polypeptide has detected a protein of approximately 125,000 kDa in mitochondria from the male-sterile S-cytoplasm but confirmed the absence of cross-reacting polypeptides in mitochondria from normal fertile cytoplasms (Manson *et al.*, 1986).

b. Gene Fragments. It is not uncommon in probing restriction endonuclease digests of mtDNA with specific probes to identify not only the restriction fragments carrying the major homology to the probe but also other restriction fragments which hybridize less intensely. It is evident from the preceding sections that chimeric genes such as T-*urf13* and S-*pcf* or partial gene/sequence duplications can account for these minor hybridizations. Other instances of sequence duplications have been described, for example in the wheat mitochondrial genome, 193 nucleotides of the *coxII* 5' exon have been duplicated (Bonen *et al.,* 1984). These are nucleotides 118–310, which correspond to the transmembrane domain of the protein. In maize, 122 bp of the coding domain of ATP6 (amino acids 25 to 65) are duplicated and immediately precede the COXII gene in the N- and T-cytoplasms (Dewey *et al.,* 1985a). In *Oenothera berteriana* the 3' region of the 26S rRNA is found on circle 3 (Manna and Brennicke, 1986). In addition, the ATPA gene exists in several nontranscribed forms at different genomic locations and on different molecules (Schuster and Brennicke, 1986). In the SB-1 tissue culture line of soybean, part of a transcribed ORF has been fused to the only copy of the 5S rRNA gene; the 5S rRNA is subsequently processed from a larger transcript (Morgens *et al.,* 1984).

It is probable that these small sequence duplications are the direct result of nonreciprocal recombination events. Continuing and active recombination between such small sequence duplications has not been demonstrated, though it seems probable that such recombinations do occur, albeit at subliminal levels based on our current understanding of mitochondrial genome organization. Recombination between sequence duplications of this type and subsequent amplification of one of the recombination products would lead to a rapid rearrangement in genome organization. Possible examples of such events have been identified in the analysis of the *atp*A sequences in different maternal lineages of maize (Small *et al.,* 1987; Leaver *et al.,* 1988), and in the T-revertants (Rottmann *et al.,* 1987) and S-revertants (Schardl *et al.,* 1985b; Lonsdale *et al.,* 1988) of maize.

E. Introns

Introns have been identified in only three mitochondrial genes, *coxII*, *nad1*, and *nad5* (Table VII). The intron of *coxII* is present in species such as wheat, rice, maize, rye (Bonen *et al.,* 1984), sunflower, sugar beet (M. Lund and K. Marcker, personal communication), and carrot (Turano *et al.,* 1987) and absent from *Oenothera berteriana,* soybean, pea, cucumber, and broad bean (Bonen *et al.,* 1984). The sequence of the *coxII* intron has only been determined for three monocotyledonous species, namely maize, rice, and wheat, and in these three species it is highly conserved. In these species the intron begins 391 bp after the start of the gene. In maize the intron is 794 bp long. In wheat this is expanded to 1216 by way of a 422-bp insertion, and in rice a further 52-bp insertion occurs within the wheat insertion sequence (Fig. 9). Both insertion sequences are flanked by almost perfect short direct repeats. The insertions occur within a nonconserved region of the intron, based on the secondary structure model for type II introns proposed by Michel and Dujon (1983).

ORFs exist within the introns of the sequenced COXII genes (Fox and Leaver, 1981; Bonen *et al.,* 1984; Kao *et al.,* 1984). These ORFs have no homology to yeast group I intron ORFs that may be associated with

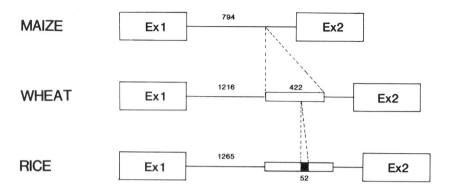

Fig. 9. Introns of cytochrome oxidase subunit II genes. The COXII genes of maize, wheat, and rice are illustrated (see text for details). The total length of the intron is given in base pairs and is inclusive of the insertion element sizes which are also shown.

maturase functions (Bonitz *et al.*, 1980; Lazowska *et al.*, 1980) and there is no evidence at present to suggest that these ORFs are expressed in mitochondria.

The *nad1* gene of watermelon has been predicted to have five exons and four introns (Stern *et al.*, 1986) of 52, 1430, 177, and 1022 bp. Intron 2 (1430 bp), which separates the two most highly conserved exons of this gene, exhibits homology with other group II introns. However, intron 1 (52 bp) and intron 3 (177 bp) have no homology to either the group II or group I introns (Davies *et al.*, 1982), though intron 4 could be classified as a group I-type intron (Davies *et al.*, 1982). Interestingly, exon I and intron 1 have not been identified in the *nad1* gene of maize and tobacco (Bland *et al.*, 1986). Though *trans*-splicing cannot be excluded, the small size and the absence of direct evidence to show that this small intron is excised raise doubts as to the proposed gene structure and to the functionality of this gene.

VIII. GENE COPY NUMBER AND MAPPING

The multipartite structure of higher plant mitochondrial genomes results from recombinations across duplicated sequences. In some instances, these repeats contain genes which are indispensable for mitochondrial function. In such cases the gene copy number is raised to two, for example, the 18S/5S and 26S rRNAs of wheat reside within different repeated sequences (Falconet *et al.*, 1984, 1985). In spinach, sugar beet, and two species of pokeweed (*Phytolacca americana* and *P. heterotepala*) only the 26S rRNA gene is duplicated (Stern and Palmer, 1984a, 1986; T. Brears and D. M. Lonsdale, unpublished data). In *Brassica campestris, B. oleracea*, and *B.*

The Plant Mitochondrial Genome 275

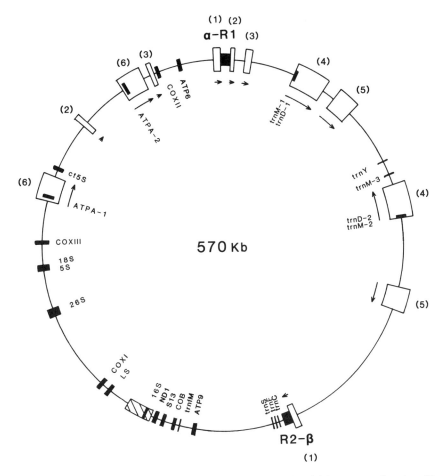

Fig. 10. Physical and genetic map of the Wf9-N mitochondrial genome of maize. The approximate sizes and orientations (inner arrows) of repeated sequences (open boxes), numbered 1 to 6, are shown. The black bars represent sequences whose origins and functions are known. Gene designations are given in Tables VI and VII. These sequences include the chloroplast sequences, the 23S–5S rRNA homology (ct5S), large subunit gene of RuBisCo (LS), and the 12-kb sequence homologous to part of the inverted repeat (hatched box; see Section V). The sequences flanking repeat 1 have been labeled α, β, R1, and R2, according to notation used previously (Lonsdale *et al.*, 1983b; Houchins *et al.*, 1986). R1 and R2 are integrated sequences related to the R1 and R2 episomes (see text, Section III,E).

napus, the 2-kb repeated sequence contains *coxII*, while in maize, *atpA* is duplicated as part of the 12-kb repeated sequence (repeat 6) in some of the fertile cytoplasms (Fig. 10; Isaac *et al.*, 1985a; Dawson *et al.*, 1986a; Small *et al.*, 1987). Also in maize, elongators tRNA^{Met} and tRNA^{Asp} (Parks *et al.*,

1984) are duplicated in maize, since they lie within the 14-kb repeat (repeat 4, Fig. 10). However, not all repeated sequences contain functional genes: examples are the "5" kb, "1" kb, and "2" kb repeats of maize, repeats 1, 2, and 3, respectively (Houchins *et al.*, 1986; D. M. Lonsdale, unpublished data) and a 657-bp repeated sequence in *Oenothera* (Hiesel *et al.*, 1987). The relative positions of genes with respect to repeated sequences that are recombinationally active can create problems in assessing the gene copy number from Southern blot analysis. For example, the proximity of *coxI* in the mitochondrial genome of the S-cytoplasm of maize to a small repeated sequence, RI*, that is related to the terminal inverted repeats of the S1 and S2 episomes (see Section III,F; Schardl *et al.*, 1984, 1985b; Isaac *et al.*, 1985b) can create numerous restriction fragments containing *coxI* homology, even though *coxI* is a single-copy gene. The origin and function of repeated sequences, particularly where gene duplications are involved, need to be resolved. In particular, where gene duplication has occurred, questions arise as to whether both copies of the gene are active and whether both are required for normal mitochondrial biogenesis. For example, *atpA*, the gene coding for the α-subunit of the F_1 ATPase is present in two copies in some fertile and male-sterile cytoplasms and in only one copy in others (Braun and Levings, 1985; Small *et al.*, 1987). This perhaps suggests that duplication is fortuitous and is not used as a mechanism for controlling the level of gene expression.

Comparisons of the relative positions of genes between maize (Dawson *et al.*, 1986a), spinach (Stern and Palmer, 1986), turnip (Makaroff and Palmer, 1987), and unpublished studies on the male-sterile T-cytoplasm of maize (C. M.-R. Fauron and D. M. Lonsdale, unpublished data) and the Owen cytoplasm of sugar beet (T. Brears and D. M. Lonsdale, unpublished data) have shown that the order and relative orientation of individual genes are not conserved. This is similar to fungal mitochondrial DNA organization (see Sederoff, 1984), but contrasts with the fixed organization of animal mitochondrial genomes (Anderson *et al.*, 1981, 1982; Bibb *et al.*, 1981; Roe *et al.*, 1985).

Apart from the 18S and 5S rRNA genes and, in wheat, the tRNA^{fMet}–18S–5S rRNA genes, most genes appear to be well dispersed (Fig. 10). In maize, for example, *cob* and *coxII* are separated by 226 kb on the 570-kb master circle; whereas on the 503-kb circle, which derives from the 570-kb circle by excision of the 67-kb circle, these two genes are 160 kb apart. The ATPA genes can be mapped to the 67- and 80-kb circles, while the COXII and the ATP6 genes can reside both on the 47-kb circle as well as on the 80-kb circle along with the gene for ATPA. The subgenomic circles, without exception, do not carry the full mitochondrial genetic complement; in this respect, they can be considered equivalent to the ρ mitochondrial genomes of yeast.

IX. TRANSCRIPTION

Transcription initiation is dependent on sequences which have the ability to interact with RNA polymerase(s). Because of the exceptionally large size of plant mitochondrial genomes, sequences related to true promoters may be randomly scattered throughout the genome not necessarily located immediately 5' to genes. Transcription would be initiated at these sequences, terminating at any fortuitously located termination signal if such signals exist. Such transcription initiation sequences are absent from the broad bean plasmids 1 and 3, as these plasmids do not have detectable transcripts (Wahleithner and Wolstenholme, 1987). Transcription promotion and termination sequences have, however, been mapped on broad bean plasmid 2 (Wahleithner and Wolstenholme, 1987) and on the p0 plasmid of sugar beet (Hansen and Marcker, 1984). It seems highly improbable, based on analysis of the DNA sequence, that such transcripts constitute functional mRNAs. This supports the concept of fortuitous transcription, though it does not rule out the possibility that such transcripts may have an as-yet-unidentified role in mitochondrial biogenesis. A similar case exists for the 2720- and 2520-nucleotide transcripts of the 5-kb repeat of maize, where no long ORFs can be identified (Houchins *et al.*, 1986). However, the ability of specific DNA sequences to hybridize to specific mitochondrial transcripts does not necessarily imply that that particular DNA sequence is itself transcribed. Small sequence duplications may well reflect the transcriptional profile of the gene from which they were derived (see Section VII,D,3).

Where numerous transcripts can be observed for single genes, not all the observed transcripts need necessarily arise from RNA processing events. As discussed above, some may be from unrelated genes and may even represent chloroplast transcripts. For example, the immediate 5' region of the maize *coxII* gene is found to comprise 122 bp of nucleotide sequence which is derived from the coding region of the ATP6 gene. Probes used to identify *coxII* transcripts which include this ATP6 homology will also identify transcripts of the ATP6 gene and vice versa. In *Oenothera berteriana*, sequences upstream of the *coxI* and *coxIII* form a short repeat sequence (Hiesel *et al.*, 1987); hybridization probes containing sequences from this repeat will identify both the *coxI* and *coxIII* transcripts. These considerations should always be borne in mind when analyzing plant mitochondrial RNA transcripts (Fig. 11).

The lack of close clustering of genes would suggest that most genes are within their own transcriptional domain. However, several examples of multigene transcripts have been described. In wheat, the gene for tRNA^{fMet} is separated by only 1 bp from the 5' end of the 18S rRNA gene, suggesting that they are cotranscribed (Gray and Spencer, 1983). The highly ordered structure of the tRNA may provide the RNA-processing signal, similar to

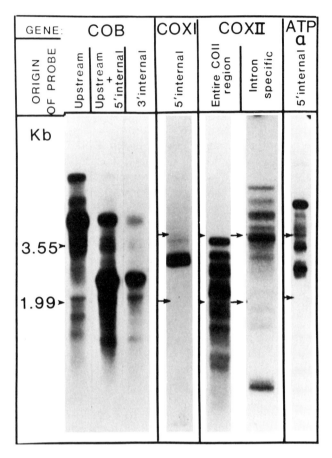

Fig. 11. Transcripts of maize mitochondrial genes. The transcripts detected by Northern hybridization using various regions of genes illustrate the complexity of transcription. The distances migrated by the 26S and 18S rRNA are indicated (arrows). From Dawson et al. (1986b), published with permission.

the processing of the mammalian mitochondrial mRNAs (Ojala et al., 1981). The 5S rRNA may also be part of the same primary transcript, as it is located close to the 3' end of the 18S rRNA gene and, in Oenothera berteriana, this transcript may also include nad5 (Wissinger et al., 1987). In the soybean tissue culture cell line SB-1, the 5S rRNA appears to be part of a larger 800-nucleotide transcript which initiates close to or at the mature 5' end of the 5S rRNA (Morgens et al., 1984). In tobacco, atp9 and rps13 are separated by 478 bp and appear to be cotranscribed, sharing sequence homology to 8 RNA species (Bland et al., 1986). In the T-cytoplasm of maize two ORFs, T-urf13 and orf25, are cotranscribed initially as a 3900-nucleotide transcript which is subsequently processed (Dewey et al., 1986; Rottmann et al., 1987).

The 5' termini of several gene transcripts have been identified either by S1 mapping or by primer extension (Table VIII); guanyltransferase capping experiments have not yet been reported. With a few exceptions, transcripts appear to initiate on either of the first two A residues within the consensus TAAG$\overset{\text{A}}{\text{T}}$GA. This consensus is similar to yeast promoter consensus ATATAAGTA (Osinga and Tabak, 1982; Osinga *et al.*, 1984; Biswas *et al.*, 1985; Schinkel *et al.*, 1987), particularly in that they both contain a highly conserved tetranucleotide sequence: TAAG. However, there are several notable exceptions to the plant mitochondrial consensus sequence: the mapped 5' ends of the broad bean plasmid 2 transcripts, and the maize and sorghum *coxI* site 1 transcripts (Table VIII). Whether these sites are RNA polymerase initiation sites or possible RNA-processing sites is not at all clear in the absence of guanyltransferase capping data. There is therefore evidence to suggest that the mitochondrial RNA polymerase can initiate at several sites, and this provides an explanation for the many low-abundance large transcripts that are observed using 5' gene probes (see Fig. 11).

The 3' nontranslated sequences of genes show little sequence conservation between species and no conserved motifs have yet been reported. Mapping of the 3' termini of transcripts of genes which have more than one transcript demonstrates that the multiple transcripts may all terminate at the same position, as is the case for the transcripts of the *coxII* gene of pea (Moon *et al.*, 1985), or that the multiple transcripts may have several 3' termini. This is illustrated by the transcripts of sorghum (Milo) *coxI*, where transcriptional termination occurs 0.2, 0.8, 1.0, and 1.4 kb 3' to the translation stop codon (Bailey-Serres *et al.*, 1986). Sequence analysis of cDNAs (Wissinger *et al.*, 1987) and the failure to detect poly(A) in RNA preparations suggest that the 3' termini of mitochondrial mRNAs are not polyadenylated, despite a recent report to the contrary (Carlson *et al.*, 1986a) whose results using oligo(dT) cellulose chromatography highlight the problems of nonspecific binding.

Stem-loop structures can invariably be identified in the 3' noncoding region of genes. For example, in *Oenothera berteriana* complex stem-loop structures have been found immediately preceding the transcript 3' termini of the *coxII* and *atpA* genes (Fig. 12; Schuster *et al.*, 1986) and immediately 3' to the end tRNA^{fMet} gene (Gottschalk and Brennicke, 1985); in addition, short stem-loop structures have been identified immediately preceding the postulated mature 3' ends of the 5S, 18S, and 26S rRNAs (Manna and Brennicke, 1985). Such stem-loop structures, whatever their size and stability, may act as RNA-processing signals rather than being involved in transcription termination. Whether or not transcription termination occurs as a specific process in plant mitochondria has not been established. The functional relationship between the DNA sequence and putative transcription initiation and transcription termination signals needs to be further defined by

TABLE VIII
DNA Sequences Surrounding the Known 5' Termini of Plant Mitochondrial Transcripts

Gene[a]		5' Sequence[b]	Length[c]	Reference
		RNA →		
Obe coxII	−239	TGGGCTCTTTTACCTCTAACTAAAAATCTCGTATGAGAATCAAAAGAATCTG	207	Hiesel and Brennicke (1985)
Maize coxIII site 1	−349	TCAAGTAACGATGAGAATTGAGCTATTCAAAGTGAA	324	W. W. Hauswirth (personal communication)
Petunia O4 atp9 9-1/2 site 1	−153	GGGTGTGGTTCAGTGTACCGCTTGTCTAGCCTATGCTTTGCATGAAC	121	Young et al. (1986)
Petunia O4 atp9 9-1 site 3	−560	TGTACCATGCTCTCTATTTGATGTAATATAGTATAGAGGCTG	528	Young et al. (1986)
Petunia O4 atp9 9-2 site 2	−272	TTCGCTTTATAAGAAGAAAGCTTTTT	252	Rothenberg and Hanson (1987)
Soybean 5S rRNA	865	TTTCCTTCTGTTCATCAATCGAAATCAAGCAAACCGGCACTACG	0	Morgens et al. (1984)
Obe 5S rRNA	2449	TTTCCTTCTGTTCATCAATCGGAAATCAAGACAAACCGGCACTACG	5	Brennicke et al. (1985)
Sorghum Milo coxI site 2	−437	TTTCTTATAAAGATGAAAGTGGGCTGCGCTCAAGAACTAGTGAAGTAAATC	405	Bailey-Serres et al. (1986)
Maize S-plasmids		5'-pAAAAGTATACAAGCACATGTCCAATCTACATAAAGATACCAACCAGGTATCT		
Maize 18S rRNA	−265	TCCGTTTGTTTGTTTTGAATTGACATAGATAAATTCTATCGCGTTCCCTT	234	Mulligan et al. (1988)
Maize 26S rRNA	−217	TGTTTTAGTTGGACGAAAAGAAAATCGTATAAAAATCAAGCAAGAGGATGC	175	Mulligan et al. (1988)
Maize coxI site 2	−180	ACTTTTGCACCGAAGAAACTCATAAGTAATCCAATTTCCGAGG	152	Isaac et al. (1985b)
Obe atp6	−240	GATCGGATATATTATCATAAGTGAGGAG	222	Schuster and Brennicke (1987c)
Obe 18S rRNA	−155	AACAAAGACAAAAAACAGATTGAAATGTCATAAGTGATGTTCGAAATCGCT	118	A. Brennicke (personal communication)
Petunia O4 atp9 9-1 site 2	−298	CATCCTGCTTCTCTTCTACAAAAGAAATTTCATAAGATAAGAGAGATGAGG	266	Young et al. (1986)
Obe atpA	−245	AGTTCCTTTTCTAAAGAAAAGTTGATAAATCATAAGAAGCAAAGTCCCTAG	213	A. Brennicke (personal communication)
Sugarbeet 1.3 kb		973 AAATAACCATAAGTGAC		C. M. Thomas (personal communication)
Sugarbeet 1.44 kb plasmid	944	CAAGAGAATAGCAGCTAATTAGCTAAAAATCATAAGTGATATCCGAAATCACT		Hansen and Marcker (1984)
Sugarbeet 1.62 kb plasmid		373 AAATATCGTAAGTGAC		C. M. Thomas (personal communication)
Obe coxI[d]	−469	CATTGCGTAGTCTTCCTGTTCAACAATTGCGTAAGTGAGGGTGGAGGACTCG	337	Hiesel et al. (1987)
Obe coxIII[d]	−1505	CATTGCGTAGTCTTCCTGTTCAACAATTGCGTAAGTGAGGGTGGAGGACTCG	1473	Hiesel et al. (1987)
Petunia O4 atp9 9-2 site 3	−324	ATCCAACTATCAATCTCGTAAGAGAATAAA	301	Rothenberg and Hanson (1987)
Pea coxII site 1	−295	GAAATCACGTAAGTGATAGAAAGAATCGTT	285	Moon et al. (1985)
Pea coxII site 2	−334	GAATCCTGAATCCCTATTCTATAAATTTACTAAGAAG	302	Moon et al. (1985)
Obe 26S rRNA	−34	TATGCGAAGCAAGCCTTATTTATAAGGTTAGTAAGGTTAGGGGGGTACAAGA	25	Manna and Brennicke (1985)

CONSENSUS   AAAT (X$^{1-6}$) TAAG$_A^T$GA

Broad bean plasmid 2 minor		1176 TGGTTTCGGATACTTTGAGACCGATGTC 1149	51	Wahleithner and Wolstenholme (1986)
Sorghum Milo/9E coxI site 1	−72	TCTAACTCCTCTCTTTCCATCCAGCCTTAGTT	365	Bailey-Serres et al. (1986)
Petunia O4 atp9 9-2 site 4	−390	AATAAGAGCCTTCTTAACTTCACTCTTTAATT	57	Rothenberg and Hanson (1987)
Maize coxI site 1	−76	TTTTCACTGCCCTTCCATTCTTTGGCCTGC		Isaac et al. (1985b)
Broad bean plasmid 2 major		1193 TGGTTGTTTATCGTAGGTGGTTTCGAT 1166		Wahleithner and Wolstenholme (1986)
Maize coxIII site 2	−392	TCGAAGTTACAGATGAGAAATGACGTATCTTACGTATCGAATC	361	W. W. Hauswirth (personal communication)
Maize coxIII site 3	−490	AAGTAGGCTTTTTTCATCTCCGAAACCTCCCAAAACTCCAC	459	W. W. Hauswirth (personal communication)

[a] Obe, Oenothera berteriana
[b] Sequence around mapped 5' termini (•). Transcription starts given below the consensus appear to be unusual. Those above consensus have been grouped based on sequence relatedness around the transcription start.
[c] Length in base pairs of untranslated 5' leader sequence.
[d] Genes share same promoter.

The Plant Mitochondrial Genome

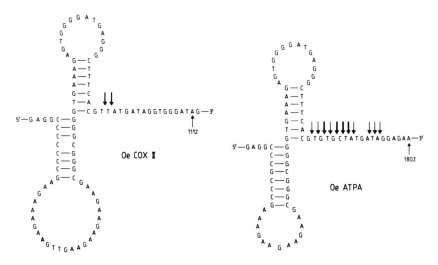

Fig. 12. Putative transcription termination signals. Stem-loop structures located 3' to the *Oenothera berteriana* (Oe) cytochrome oxidase subunit II gene and the α-subunit of the ATPase gene have been suggested to function as transcription terminators. The downward arrows show the locations and relative abundance of the termini. From Schuster *et al.* (1986), published with permission.

in vitro site-directed mutagenesis. Until these kinds of experiments are performed, transcription initiation and termination sequence motifs, such as those illustrated in Table VIII and Fig. 12, must be treated cautiously.

X. RIBOSOME BINDING AND TRANSLATION INITIATION

Many, but not all prokaryotic, eukaryotic, and chloroplast mRNAs contain identifiable Shine–Dalgarno sequences at their 5' end. This sequence is complementary to a sequence at the 3' end of the small subunit rRNA and is considered to be important for ribosome binding to the mRNA prior to the initiation of translation (Shine and Dalgarno, 1974; Kozak, 1983). The *E. coli* Shine–Dalgarno sequence, 5'-CCUCCU is unrelated to the mature end of the mitochondrial 18S rRNA 5'-UGAAUCC in wheat (Schnare and Gray, 1982), and the terminal redundancy of mitochondrial 18S rRNA, involving both C residues, makes it unlikely that these bases are involved in ribosome binding.

Analysis of sequences immediately preceding the proposed AUG methionine initiation codon for the maize COB gene revealed a limited homology to the 3' end of the ribosomal 18S RNA (Table IX). Dawson *et al.*, (1984) proposed that this homology represented a Shine–Dalgarno-like se-

TABLE IX

Possible Ribosome Binding Sites Associated with Higher Plant Mitochondrial Transcripts

Plant mitochondrial 18S rRNA		3'				C[a]	C	U	A	A	G	U		U	A				
Maize	cob	5'	g	g	a	G	u	u	g	U	C	A	(15 nucleotides)	A	U	G	a		
Obe[b]	cob	5'	a	g	a	G	u	u	g	U	C	A	(15 nucleotides)	A	U	G	g		
Maize	coxI	5'	a	a	g	G	u	u	U	U	C	A	(13 nucleotides)	A	U	G	a		
Sorghum	coxI	5'	u	u	u	G	a	A	a	U	C	A	(13 nucleotides)	A	U	G	a		
Maize	coxII	5'	u	c	c	u	a	c	U	U	C	u	(10 nucleotides)	A	U	G	a		
Obe[b]	coxII	5'	c	g	g	a	G	A	g	U	C	A	(17 nucleotides)	A	U	G	a		
Pea	coxII	5'	g	a	g	c	a	g	U	U	a	A	(17 nucleotides)	A	U	G	a		
Rice	coxII	5'	g	g	a	G	c	A	g	a	C	g	(21 nucleotides)	A	U	G	a		
Obe[b]	atpA	5'	a	u	c	G	a	A	U	U	g	A	(3 nucleotides)	A	U	G	g		
Tobacco	atp6	5'	g	a	g	G	G	A	a	U	C	u	(16 nucleotides)	A	U	G	u		
Tobacco	rps13	5'	g	a	g	G	G	A	a	U	C	g	(21 nucleotides)	A	U	G	u		

[a] Terminal C-OH residue present in 80% of wheat mitochondrial 18S rRNA species (Schnare and Gray, 1982).

[b] Obe, *Oenothera berteriana*

quence. Similar sequences could be found in some other mitochondrial genes but not all. For example, the COB gene of wheat (Boer et al., 1985) and the ORFs of the S1 and S2 episomes from the male-sterile S-cytoplasm of maize (Levings and Sederoff, 1983; Paillard et al., 1985) have no sequences with homology to the 3' end of the 18S rRNA immediately preceding the proposed methionine initiation codon.

Ribosome binding may therefore occur by some other as-yet-unrecognized mechanism. In mammalian mitochondrial mRNAs, 5' untranslated leader sequences are absent, ribosome recognition must therefore rely on other factors. In yeast, nuclear-encoded gene products which interact with the 5' untranslated leader sequence are required for translation to occur (Costanzo et al., 1986; Poutre and Fox, 1987; Fox et al., 1988). Though such requirements for translation of plant mitochondrial genes have not been demonstrated, it seems likely that specific proteins in addition to the ribosomes will be involved in mRNA recognition and translation initiation. Transcripts without the correct 5' signals would not compete for ribosomes and would not form part of the stable mRNA population. It is interesting in this respect that some plant mitochondrial genes share the same 5' untranslated leader sequence, for example T-*urf13* and *atp6*, S-*pcf* and *atp9*, supporting the idea that these sequences have a role in translation initiation.

Many mitochondrial genes have 5' extensions to the ORF of the gene. These in-frame ORFs in many instances are transcribed and contain AUG codons, providing the opportunity for translation initiation and the formation of a chimeric gene product. The *coxII* gene of maize, for example, has two possible initiation codons (Fox and Leaver, 1981). Comparison to other sequenced *coxII* genes, particularly that of wheat (Bonen *et al.*, 1984), indicates that the second AUG is the probable initiator codon, based on the start of sequence homology between two genes. The first AUG lies within the sequence that is derived from the internal coding sequence of *atp6*, amino acid 54 (Dewey *et al.*, 1985a). Ribosome binding to sequences immediately preceding either of the AUG codons must therefore involve the internal sequences of the ATP6 gene. Is there a possibility that ribosomes locate and initiate translation from any suitable AUG?

In the absence of protein sequencing, the amino-terminal methionine codon cannot be positioned. It remains a possibility that any in-frame AUG codon could be used to initiate translation. In other organisms AUG is not an obligatory initiation codon; in mammalian mitochondria, AUA and AUG are used (Anderson *et al.*, 1981; Bibb *et al.*, 1981); in *Aspergillus ridulans*, UUA or AUU; in *Neurospora crassa* UCA or AUA have been suggested as the authentic initiation codons (Ziaie and Suyama, 1987). Therefore, the use of unusual codons to initiate translation cannot be excluded in plants at the present time.

The amino-terminal reading frame extensions observed in some plant mitochondrial genes may be required to ensure the correct targeting of the protein to the intermembrane space or into the inner membrane, as has been suggested for *coxI* of *Aspergillus, Neurospora, Paramecium,* and *Tetrahymena* (Ziaie and Suyama, 1987) and *coxII* of yeast (Mannhaupt *et al.*, 1985; Pratje and Guiard, 1986). Amino-terminal extensions, if not involved in targeting, may be tolerated or removed by the matrix protease that is involved in the removal of the amino-terminal targeting sequences of nuclear-encoded genes.

XI. GENOMIC REORGANIZATIONS

In an evolutionary time scale, large genomic reorganizations appear to constitute the major pathway of both mitochondrial and chloroplast genome evolution. In chloroplast genomes, sequence inversions of up to 50 kb, particularly in the long unique region of the genomes, characterize many groups of plants (Palmer, 1985). In mitochondria, significantly fewer studies have been done because of the complexity of the genome organization. In *Brassica*, the evolution of the entire genus can be described based on the organization of the mitochondrial genomes (J. D. Palmer, unpublished data). For example, three major sequence inversions can be predicted to have occurred in the evolution of the *B. napus* and *B. campestris* mitochondrial genomes from their common progenitor (Fig. 13; J. D. Palmer and L. A. Herbon, unpublished data). In maize, a similar but more complex set of rearrangements can be predicted to have occurred during the evolution of N and T mitochondrial genomes from their common progenitor (Lonsdale *et al.*,

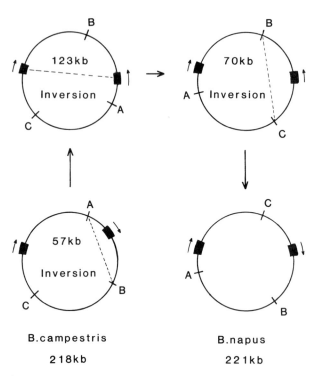

Fig. 13. Mitochondrial genomic relationship between *Brassica campestris* and *Brassica napus*. In evolutionary terms the *B. campestris* mitochondrial genome requires three major sequence inversions in order that its sequence organization reflects that found in *B. napus*. The first involves a 57-kb inversion between A and B. This inversion alters the orientation (arrows) of the repeated sequence (black box). Isomerization between the repeated sequence can occur (123-kb inversion). Last, a further sequence inversion of 70 kb between B and C restores the direct orientation of the repeats. From J. Palmer and L. Hebron, published with permission.

1983a, and unpublished data). Analysis of mitochondrial genome organization in different maternal lineages of one cytoplasmic type also illustrates the capacity of genomes to rearrange by this mechanism (Small et al., 1987). Such alterations in genome organization probably involve homologous recombination between small repeated sequences with the subsequent selection and amplification of novel forms of the genome.

Regeneration of plants from callus appears to promote genetic variability. This variability, termed somaclonal variation (Karp and Bright, 1985), in some instances includes the selection and amplification of variant forms of the mitochondrial genome (Kemble and Shepard, 1984). Such forms may preexist in the plant mitochondrial population at a low level and may be the consequence of restriction site polymorphism or hypervariability at restricted loci.

Genomic rearrangements have not been observed in tobacco plants regenerated from protoplasts (Nagy *et al.*, 1983); however, in long-term cultures of tobacco cells, heterogeneity in the mtDNA population is observed (Dale *et al.*, 1981; Grayburn and Bendich, 1987). Similarly, in nonregenerable callus cultures of wheat, extensive heterogeneity has been reported and in one culture line a complete loss of one of the two copies of the 26S rRNA gene has occurred (Rode *et al.*, 1987).

Plants regenerated from callus of the male-sterile T-cytoplasm of maize in some instances exhibited reversion to fertility and resistance to the T-toxin of *Helminthosporium maydis*, a fungal pathogen (Umbeck and Gengenbach, 1983). These phenotypic changes correlated with a discrete alteration in the mitochondrial DNA (Umbeck and Gengenbach, 1983). Detailed analysis has shown that a homologous recombination event involving a small sequence reiteration has led to a genomic reorganization and loss of the T-URF13 gene (Rottmann *et al.*, 1987).

Spontaneous mitochondrial genome reorganization is also associated with the phenomenon of fertility reversion in the S-cytoplasm of maize. Plants having the S-cytoplasm are male-sterile in nuclear backgrounds which are nonrestoring. In a few cytoplasm–nucleus combinations, particularly Vg-S cytoplasm + M825 nucleus, reversion to fertility can occur; this phenotypic alteration is stable and maternally inherited (Laughnan and Gabay, 1973, 1975, 1978; Laughnan and Gabay-Laughnan, 1983). Analysis of the mtDNA from male-sterile parents and male-fertile revertants demonstrates that alterations in the structure of the mitochondrial genome have occurred in the fertile revertants (Levings *et al.*, 1980; Schardl *et al.*, 1985b). These changes are consistent with a recircularization of the partially linear genome, promoted by the loss of the S1 and S2 episomes and by deletion of part of the integrated S2 sequence (Schardl *et al.*, 1985b).

The full nature and consequences of these deletions on mitochondrial genome structure have not been investigated in detail, though it is evident that where a loss of sequence occurs it must be from the master circle and involve the whole mitochondrial population; it cannot arise by the selective loss of one of the subgenomic circles, as such a loss does not delete sequence information; the lost subgenomic circle would be reformed by recombination between the repeated sequence elements of the master circle.

Cytoplasmic hybridization through protoplast fusion, either interspecific or intergeneric, can lead to the formation of novel mitochondrial genomes (Belliard *et al.*, 1978, 1979; Galun *et al.*, 1982; Izhar *et al.*, 1983; Pelletier *et al.*, 1983; Fluhr *et al.*, 1983; Kemble *et al.*, 1986a; Vedel *et al.*, 1986). This process occurs as a result of the ability of the mitochondria to fuse and the ability of mitochondrial genomes to recombine. Such promiscuous recombination and the formation of novel genotypes is not, however, a feature of chloroplast genomes in cell fusion hybrids, where only one recombinant-type molecule has ever been described (Medgyesy *et al.*, 1985).

It is apparent that mitochondrial genomes have the capacity to reorganize. Reorganization may be spontaneous, may be the result of stress, or may be promoted by the nuclear genotype. Whatever the outcome, the final organization must contain a complete genetic complement and must be stable.

XII. CONCLUSIONS

The current literature on plant mitochondria covers many facets of the biology and the molecular biology of this organelle. Studies on mitochondrial genome recombination in cybrids and somatic hybrids, mitochondrial genome size and organization, gene content, gene structure, transcription, and translation suggest that the biology of plant mitochondria and the molecular biology of their genomes are mirrored in fungal systems.

In fungi, as in plants, the mixing of cytoplasms promotes mitochondrial fusion and recombination between mitochondrial genomes. Recombination between repeated sequences, creating subgenomic circles is also a feature of both groups of organisms, though only in the fungi can amplification of the subgenomes occur, giving phenotypic mutants. Introns are found in fewer genes in plants than in fungi, and as in fungi these introns are "optional." Transcription, in plant mitochondria, appears to be promoted by a sequence which is very similar to the yeast transcription promoter consensus; it would not be at all surprising if the mitochondrial RNA polymerases of plants and fungi were also highly related.

The lack of mitochondrial mutants in higher plants precludes genetic analysis. However, novel nuclear–cytoplasmic combination in many plant species can be obtained either from sexual crosses or by regenerating plants from protoplast fusions. In some species such novel nuclear–cytoplasmic combinations are reflected by an inability to produce functional pollen (cytoplasmic male sterility). Similarly, crosses between two sexual compatible strains of yeast, *Saccharomyces cerevisiae* and *Saccharomyces douglasii*, give rise to hybrids which are infertile and, in addition, respiratory deficient (Kotylak *et al.*, 1985).

There is little doubt that an inability to develop an *in vitro* protein synthesis system and the inability to transform mitochondria either *in vitro* or *in vivo* are the Achilles' heels of mitochondrial studies not only in plants, but in all eukaroytes. Until these problems are solved and with the lack of mitochondrial genetics, plant mitochondrial molecular biology will remain descriptive. The last major remaining objective in this field must be the sequencing of a mitochondrial genome.

ACKNOWLEDGMENTS

I would like to thank all my colleagues for their help in the compilation of this review, particularly Dr. William Rottmann and Sara Melville. In addition, I would like to thank Dr. Axel Brennicke (Tübingen), Dr. Jean Michel Grienenberger (Strasbourg), Professor Chris Leaver (Edinburgh), and Dr. Jeff Palmer (Ann Arbor) for their helpful comments and criticisms of this review. Last, I would like to thank Karen Parr for her patience and help in typing and correcting this review. This review was completed in January 1988; Tables VI, VII, and VIII were updated in August 1988.

REFERENCES

Abbott, A. G., O'Dell, M., and Flavell, R. B. (1985). *Plant Mol. Biol.* **4**, 233–240.
Aldrich, J., Cherney, B., Merlin, E., Williams, C., and Mets, L. (1985). *Curr. Genet.* **9**, 233–238.
Anderson, S., Bankier, A. T., Barrell, B. G., de Bruijn, M. H. L., Coulson, A. R., Drouin, J., Eperon, I. C., Nierlich, D. P., Roe, B. A., Sanger, F., Schreier, P. H., Smith, A. J. H., Staden, R., and Young, I. G. (1981). *Nature (London)* **290**, 457–465.
Anderson, S., de Bruijn, M. H. L., Coulson, A. R., Eperon, I. C., Sanger, F., and Young, I. G. (1982). *J. Mol. Biol.* **156**, 683–717.
Attardi, G., Crews, S. T., Nishiguchi, J., Ojala, D. K., and Posakony, J. W. (1978). *Cold Spring Harbor Symp. Quant. Biol.* **43**, 179–192.
Avers, C. J. (1962). *Am. J. Bot.* **49**, 996–1003.
Bailey-Serres, J., Hanson, D. K., Fox, T. D., and Leaver, C. J. (1986). *Cell* **47**, 567–576.
Bailey-Serres, J., Leroy, P., Jones, S. S. Wahleithner, J. A., and Wolstenholme, D. R. (1987). *Curr. Genet.* **12**, 49–53.
Bartnik, E., and Borsuk, P. (1986). *Nucleic Acids Res.* **14**, 2407.
Bedinger, P., de Hostos, E. L., Leon, P., and Walbot, V. (1986). *Mol. Gen. Genet.* **205**, 206–212.
Belliard, G., Pelletier, G., Vedel, F., and Quetier, F. (1978). *Mol. Gen. Genet.* **165**, 231–237.
Belliard, G., Vedel, F., and Pelletier, G. (1979). *Nature (London)* **281**, 401–403.
Bendich, A. J. (1982). In "Mitochondrial Genes" (P. Slonimski, P. Borst, and G. Attardi, eds.), pp. 477–481. Cold Spring Harbor Lab., Cold Spring Harbor, New York.
Bendich, A. J. (1985). In "Plant Gene Research" (B. Hohn and E. S. Dennis, eds.), Vol. 2, pp. 111–138. Springer-Verlag, New York.
Bendich, A. J., and Gauriloff, L. P. (1984). *Protoplasma* **119**, 1–7.
Bernardi, G., *et al.* (1980). *Dev. Genet.* **2**, 21–31. Amsterdam.
Bertrand, H., Chan, B. S.-S., and Griffiths, A. J. F. (1985). *Cell* **41**, 877–884.
Bibb, M. J., Van Etten, R. A., Wright, C. T., Walberg, M. W., and Clayton, D. A. (1981). *Cell* **26**, 167–180.
Birky, C. W., Jr. (1983). *Science* **222**, 468–475.
Biswas, T. K., Edwards, J. C., Rabinowitz, M., and Getz, G. S. (1985). *Proc. Natl. Acad. Sci. U.S.A.* **82**, 1954–1958.
Bland, M. M., Levings, C. S., III, and Matzinger, D. F. (1987). *Curr. Genet.* **12**, 475–481.
Bland, M. M., Levings, C. S., III, and Matzinger, D. F. (1986). *Mol. Gen. Genet.* **204**, 8–16.
Boer, D. H., McIntosh, J. E., Gray, M. W., and Bonen, L. (1985). *Nucleic Acids Res.* **13**, 2281–2292.
Bohnert, H. J., and Loffelhardt, W. (1982). *FEBS Lett.* **150**, 403–406.
Bonen, L. (1987). *Nucleic Acids Res.* **15**, 10393–10404.

Bonen, L., and Gray, M. W. (1980). *Nucleic Acids Res.* **8**, 319–335.
Bonen, L., Boer, P. H., and Gray, M. W. (1984). *EMBO J.* **3**, 2531–2536.
Bonen, L., Boer, P. H., McIntosh, J. E., and Gray, M. W. (1987). *Nucleic Acids Res.* **15**, 6734.
Bonitz, S. G., Coruzzi, G., Thalenfeld, B. E., Tzagoloff, A., and Macino, G. (1980). *J. Biol. Chem.* **255**, 11927–11941.
Borck, K. S., and Walbot, V. (1982). *Genetics* **102**, 109–128.
Borst, P. (1972). *Annu. Rev. Biochem.* **41**, 333–376.
Borst, P., and Grivell, L. A. (1978). *Cell* **15**, 705–723.
Borsuk, P., Sirko, A., and Bartnik, E. (1986). *Nucleic Acids Res.* **14**, 7508.
Boutry, M., Faber, A.-M., Charbonnier, M., and Briquet, M. (1984). *Plant Mol. Biol.* **3**, 445–452.
Braun, C. J., and Levings, C. S., III (1985). *Plant Physiol.* **79**, 571–577.
Braun, C. J., and Levings, C. S., III (1986). *Maize Genet. Coop. News Lett.* **60**, 111.
Braun, C. J., Sisco, P. H., Sederoff, R. R., and Levings, C. S., III (1986). *Curr. Genet.* **10**, 625–630.
Brears, T., Schardl, C. L., and Lonsdale, D. M. (1986). *Plant Mol. Biol.* **6**, 171–177.
Brennicke, A. (1980). *Plant Physiol.* **65**, 1207–1210.
Brennicke, A., and Blanz, P. (1982). *Mol. Gen. Genet.* **187**, 461–466.
Brennicke, A., and Schwemmle, B. (1984). *Z. Naturforsch. C* **39**, 191–192.
Brennicke, A., Moller, S., and Blanz, P. A. (1985). *Mol. Gen. Genet.* **198**, 404–410.
Carlson, J. E., and Kemble, R. J. (1985). *Plant Mol. Biol.* **4**, 117–123.
Carlson, J. E., Brown, G. L., and Kemble, R. J. (1986a). *Curr. Genet.* **11**, 151–160.
Carlson, J. E., Erickson, L. R., and Kemble, R. J. (1986b). *Curr. Genet.* **11**, 161–163.
Cashmore, A. R. (1983). *In* "Genetic Engineering of Plants: An Agricultural Perspective" (T. Kosuge, C. P. Meredith, and A. Hollaender, eds.), p. 29. Plenum, New York.
Chang, D. D., and Clayton, D. A. (1987). *Science* **235**, 1178–1184.
Chao, S., Sederoff, R. R., and Levings, C. S., III (1983). *Plant Physiol.* **71**, 190–193.
Chao, S., Sederoff, R. R., and Levings, C. S., III (1984). *Nucleic Acids Res.* **12**, 6629–6644.
Chase, C. D., and Pring, D. R. (1985). *Plant Mol. Biol.* **5**, 303–312.
Chase, C. D., and Pring, D. R. (1986). *Plant Mol. Biol.* **6**, 53–64.
Chetrit, P., Mathieu, C., Muller, J. P., and Vedel, F. (1984). *Curr. Genet.* **8**, 413–421.
Christiansen, C., Christiansen, G., and Bak, A. L. (1973). *Biochem. Biophys. Res. Commun.* **52**, 1426–1433.
Clowes, F. A. L., and Juniper, B. E. (1964). *J. Exp. Bot.* **15**, 622–630.
Costanzo, M. C., Seaver, E. C., and Fox, T. D. (1986). *EMBO J.* **5**, 3637–3641.
Dale, R. M. K. (1981). *Proc. Natl. Acad. Sci. U.S.A.* **78**, 4453–4457.
Dale, R. M. K., Duesing, J. H., and Keene, D. (1981). *Nucleic Acids Res.* **9**, 4583–4593.
Dale, R. M. K., Wu, M., and Kiernan, M. C. C. (1983). *Nucleic Acids Res.* **11**, 1673–1685.
Dale, R. M. K., Mendu, N., Ginsburg, H., and Kridl, J. C. (1984). *Plasmid* **11**, 141–150.
Davies, R. W., Waring, R. B., Ray, J. A., Brown, T. A., and Scazzocchio, C. (1982). *Nature (London)* **300**, 719–724.
Dawson, A. J., Jones, V. P., and Leaver, C. J. (1984). *EMBO J.* **3**, 2107–2113.
Dawson, A. J., Hodge, T. P., Isaac, P. G., Leaver, C. J., and Lonsdale, D. M. (1986a). *Curr. Genet.* **10**, 561–564.
Dawson, A. J., Jones, V. P., and Leaver, C. J. (1986b). *Methods Enzymol.* **118**, 470–485.
De Heij, H. T., Lustig, H., Van Ee, J. H., Vos, Y. J., and Groot, G. S. P. (1985). *Plant Mol. Biol.* **4**, 219–224.
Dewey, R. E., Levings, C. S., III, and Timothy, D. H. (1985a). *Plant Physiol.* **79**, 914–919.
Dewey, R. E., Schuster, A. M., Levings, C. S., III, and Timothy, D. H. (1985b). *Proc. Natl. Acad. Sci. U.S.A.* **82**, 1015–1019.
Dewey, R. E., Levings, C. S., III, and Timothy, D. H. (1986). *Cell* **44**, 439–449.
de Zamaroczy, M., Faugeron-Fonty, G., and Bernardi, G. (1983). *Gene* **21**, 193–202.
Diers, L. (1966). *J. Cell Biol.* **28**, 527–543.

Dron, M., Hartmann, C., Rode, A., and Sevignac, M. (1985). *Nucleic Acids Res.* **13,** 8603–8609.
Ellis, J. (1983). *Nature (London)* **304,** 308–309.
Elmore-Stamper, S., and Levings, C. S., III (1986). *Maize Genet. Coop. News Lett.* **60,** 110–111.
Erickson, L., Beversdorf, W. D., and Pauls, K. P. (1985). *Curr. Genet.* **9,** 679–682.
Erickson, L., Grant, I., and Beversdorf, W. (1986). *Theor. Appl. Genet.* **72,** 151–157.
Escote, L. J., Gabay-Laughnan, S. J., and Laughnan, J. R. (1985). *Plasmid* **14,** 264–267.
Escote, L. J., Gabay-Laughnan, S., and Laughnan, J. R. (1986). *Maize Genet. Coop. News Lett.* **60,** 127–128.
Falconet, D., Lejeune, B., Quetier, F., and Gray, M. W. (1984). *EMBO J.* **3,** 297–302.
Falconet, D., Delorme, S., Lejeune, B., Sevignac, M., Delcher, E., Bazetouz, S., and Quetier, F. (1985). *Curr. Genet.* **9,** 169–174.
Finnegan, P. M., and Brown, G. G. (1986). *Proc. Natl. Acad. Sci. U.S.A.* **83,** 5175–5179.
Fluhr, R., Aviv, D., Edelman, M., and Galen, E. (1983). *Theor. Appl. Genet.* **65,** 289–294.
Fontarnau, A., and Hernandez-Yago, J. (1982). *Plant Physiol.* **70,** 1678–1682.
Forde, B. G., and Leaver, C. J. (1980). *Proc. Natl. Acad. Sci. U.S.A.* **77,** 418–422.
Forde, B. G., Oliver, R. J. C., and Leaver, C. J. (1978). *Proc. Natl. Acad. Sci. U.S.A.* **75,** 3841–3845.
Fox, T. D. (1979). *J. Mol. Biol.* **130,** 63–82.
Fox, T. D., and Leaver, C. J. (1981). *Cell* **26,** 315–323.
Fox, T. D., Costanzo, M. C., Strick, C. A., Marykwas, D. L., Seaver, E. C., and Rosenthal, J. K. (1988). *Philos. Trans. R. Soc. London, Ser. B,* **319,** 97–105.
Galun, E., Arzee-Gonen, P., Fluhr, R., Edelman, M., and Aviv, D. (1982). *Mol. Gen. Genet.* **186,** 50–56.
Gatenby, A. A., Castleton, J. A., and Saul, M. W. (1981). *Nature (London)* **291,** 117–121.
Goblet, J.-P., Boutry, M., Duc, G., and Briquet, M. (1983). *Plant Mol. Biol.* **2,** 305–309.
Goblet, J.-P., Flamand, M.-C., and Briquet, M. (1985). *Curr. Genet.* **9,** 423–426.
Gottschalk, M., and Brennicke, A. (1985). *Curr. Genet.* **9,** 165–168.
Grabau, E. A. (1985). *Plant Mol. Biol.* **5,** 119–124.
Grabau, E. A. (1986). *Plant Mol. Biol.* **7,** 377–384.
Grabau, E. A. (1987). *Curr. Genet.* **11,** 287–293.
Grabau, E., Havlik, M., and Gesteland, R. (1988). *Curr. Genet.* **13,** 83–89.
Gray, M. W., and Boer, P. H. (1988). *Philos. Trans. R. Soc. London, Ser. B,* **319,** 135–147.
Gray, M. W., and Doolittle, W. F. (1982). *Microbiol. Rev.* **46,** 1–42.
Gray, M. W., and Spencer, D. F. (1983). *FEBS Lett.* **161,** 323–327.
Grayburn, W. S., and Bendich, A. J. (1987). *Curr. Genet.* **12,** 257–261.
Gualberto, J. M., Wintz, H., Weil, H.-H. and Grienenberger, J. M. (1988). *Mol. Gen. Genet.,* in press.
Gunge, N., Tamaru, A., Ozawa, F., and Sakaguchi, K. (1981). *J. Bacteriol.* **145,** 382–390.
Guillemaut, P., Dietrich, A., Marechal, L., and Weil, J.-H. (1986). *Nucleic Acids Res.* **14,** 6775.
Gwynn, B., Dewey, R. E., Sederoff, R. R., Timothy, D. H. and Levings, C. S. III (1987). *Theor. Appl. Genet.* **74,** 781–788.
Hack, E., and Leaver, C. J. (1983). *EMBO J* **2,** 1783–1789.
Hansen, B. M., and Marcker, K. A. (1984). *Nucleic Acids Res.* **12,** 4747–4756.
Hanson, M. R., and Conde, M. F. (1985). *Int. Rev. Cytol.* **94,** 213–267.
Hanson, M. R., Young, E. G., and Rothenberg, M. (1988). *Philos. Trans. R. Soc. London, Ser. B,* **319,** 199–208.
Heinhorst, S., and Shively, J. M. (1983). *Nature (London)* **304,** 373–374.
Hiesel, R., and Brennicke, A. (1983). *EMBO J.* **2,** 2173–2178.
Hiesel, R., and Brennicke, A. (1985). *FEBS Lett.* **193,** 164–168.
Hiesel, R., and Brennicke, A. (1987). *Plant Sci.* **51,** 225–230.
Hiesel, R., Schobel, W., Schuster, W., and Brennicke, A. (1987). *EMBO J.* **6,** 29–34.

Hirochika, H., Nakamura, K., and Sakaguchi, K. (1984). *EMBO J.* **3,** 761–766.
Hishinuma, F., Nakamura, K., Hirai, K., Nishizawa, R., Gunge, N., and Maeda, T. (1984). *Nucleic Acids Res.* **12,** 7581–7597.
Honeyman, A. L., and Currier, T. C. (1986). *Appl. Environ. Microbiol.* **52,** 924–929.
Houchins, J. P., Ginsburg, H., Rohrbaugh, M., Dale, R. M. K., Lonsdale, D. M., Hodge, T. P., and Schardl, C. L. (1986). *EMBO J.* **5,** 2781–2788.
Huh, T. Y., and Gray, M. W. (1982). *Plant Mol. Biol.* **1,** 245–249.
Iams, K. P., and Sinclair, J. H. (1982). *Proc. Natl. Acad. Sci. U.S.A.* **79,** 5926–5929.
Iams, K. P., Heckman, J. E., and Sinclair, J. H. (1985). *Plant Mol. Biol.* **4,** 225–233.
Isaac, P. G., Brennicke, A., Dunbar, S. M., and Leaver, C. J. (1985a). *Curr. Genet.* **10,** 321–328.
Isaac, P. G., Jones, V. P., and Leaver, C. J. (1985b). *EMBO J.* **4,** 1617–1623.
Izhar, S., Schlichter, M., and Swartzberg, D. (1983). *Mol. Gen. Genet.* **190,** 468–474.
Jackl, G., and Sebald, W. (1975). *Eur. J. Biochem.* **54,** 97–106.
Jacobs, H. T., and Lonsdale, D. M. (1987). *Trends Genet.* **3,** 337–341.
Joyce, P. B. M., Spencer, D. F., Bonen, L. and Gray, M. W. (1988). *Plant Mol. Biol.* **10,** 251–262.
Juniper, B. E., and Clowes, F. A. L. (1965). *Nature (London)* **208,** 864–865.
Kao, T.-H., Moon, E., and Wu, R. (1984). *Nucleic Acids Res.* **12,** 7305–7315.
Karp, A., and Bright, S. W. J. (1985). *Oxford Surv. Plant Mol. Cell Biol.* **2,** 199–234.
Kemble, R. J., and Bedbrook, J. R. (1980). *Nature (London)* **284,** 565–566.
Kemble, R. J., and Shepard, J. F. (1984). *Theor. Appl. Genet.* **69,** 211–216.
Kemble, R. J., Mans, R. J., Gabay-Laughnan, S., and Laughnan, J. R. (1983). *Nature (London)* **304,** 744–747.
Kemble, R. J., Barsby, T. L., Wong, R. S. C., and Shepard, J. F. (1986a). *Theor. Appl. Genet.* **72,** 787–793.
Kemble, R. J., Carlson, J. E., Erickson, L. R., Sernyk, L., and Thompson, D. J. (1986b). *Mol. Gen. Genet.* **205,** 183–185.
Kikuchi, Y., Hirai, K., and Hishinuma, F. (1984). *Nucleic Acids Res.* **12,** 5685–5692.
Kim, B. D., Lee, L. J., and DeBusk, A. G. (1982). *FEBS Lett.* **147,** 231–234.
Kinashi, H., Shimaji, M., and Sakai, A. (1987). *Nature (London)* **328,** 454–456.
Kitada, K., and Hishinuma, F. (1987). *Mol. Gen. Genet.* **206,** 377–381.
Kolodner, R., and Tewari, K. K. (1972). *Proc. Natl. Acad. Sci. U.S.A.* **69,** 1830–1834.
Kool, A. J., de Haas, J. M., Mol, J. N. M., and van Marrewijk, G. A. M. (1985). *Theor. Appl. Genet.* **69,** 223–233.
Kotylak, Z., Lazowski, J., Hawthorne, D. C., and Slonimski, P. P. (1985). *In* "Achievements and Perspectives of Mitochondrial Research: II. Biogenesis" (E. Quagliariello *et al.*, eds.), pp. 1–19. Elsevier, Amsterdam.
Kozak, M. (1983). *Microbiol. Rev.* **47,** 1–45.
Kuzmin, E. V., and Levchenko, I. V. (1987). *Nucleic Acids Res.* **15,** 6758.
Lacoste-Royal, G., and Gibbs, S. P. (1985). *Proc. Natl. Acad. Sci. U.S.A.* **82,** 1456–1459.
Lamppa, G. K., and Bendich, A. J. (1984). *Planta* **162,** 463–468.
Larrinua, I. M., Muskavitch, K. M. T., Gubbins, E. J., and Bogorad, L. (1983). *Plant Mol. Biol.* **2,** 129–140.
Laughnan, J. R., and Gabay, S. J. (1973). *Theor. Appl. Genet.* **43,** 109–116.
Laughnan, J. R., and Gabay, S. J. (1975). *In* "Genetics and the Biogenesis of Cell Organelles" (C. W. Birky, Jr., P. S. Perlman, and T. J. Byers, eds.), pp. 330–349. Ohio State Univ. Press, Columbus.
Laughnan, J. R., and Gabay, S. J. (1978). *In* "Maize Breeding and Genetics" (D. B. Walden, ed.), pp. 427–446. Wiley, New York.
Laughnan, J. R., and Gabay-Laughnan, S. (1983). *Annu. Rev. Genet.* **17,** 27–48.
Laughnan, J. R., Gabay-Laughnan, S., and Carlson, J. E. (1981). *Stadler Genet. Symp.* **13,** 93–114.

Lazowska, J., Jacq, C., and Slonimski, P. P. (1980). *Cell* **22**, 333–348.
Leaver, C. J., and Harmey, M. A. (1973). *Biochem. Soc. Symp.* **38**, 175–193.
Leaver, C. J., and Harmey, M. A. (1976). *Biochem. J.* **157**, 175–177.
Leaver, C. J., Isaac, P. G., Small, I. D., Bailey-Serres, J., Liddell, A. D., and Hawkesford, M. J. (1988). *Philos. Trans. R. Soc. London, Ser. B,* **319**, 165–176.
Leroy, P., Bazetouz, S., Quetier, F., Delbut, J., and Berville, A. (1985). *Curr. Genet.* **9**, 242–251.
Levings, C. S., III, and Dewey, R. E. (1988). *Philos. Trans. R. Soc. London, Ser. B,* **319**, 177–185.
Levings, C. S., III, and Sederoff, R. R. (1983). *Proc. Natl. Acad. Sci. U.S.A.* **80**, 4055–4059.
Levings, C. S., III, Shah, D. M., Hu, W. W. L., Pring, D. R., and Timothy, D. H. (1979). In "Extrachromosomal DNA" (D. J. Cummings, P. Borst, I. B. Dawid, S. M. Weissmann, and C. F. Fox, eds.), p. 63–73. Academic Press, New York.
Levings, C. S., III, Kim, B. D., Pring, D. R., Conde, M. F., Mans, R. J., Laughnan, J. R., and Gabay-Laughnan, S. J. (1980). *Science* **209**, 1021–1023.
Levings, C. S., III, Sederoff, R. R., Hu, W. W. L., and Timothy, D. H. (1983). In "Structure and Function of Plant Genomes" (O. Ciferri and L. Dure, eds.), pp. 363–371. Plenum, New York.
Lonsdale, D. M. (1984). *Plant Mol. Biol.* **3**, 201–206.
Lonsdale, D. M. (1985). In "Genetic Flux in Plants" (B. Hohn and E. S. Dennis, eds.), pp. 51–60. Springer-Verlag, New York.
Lonsdale, D. M. (1986). *Nature (London)* **323**, 299.
Lonsdale, D. M. (1987a). *Genet. Eng.* **6**, 47–102.
Lonsdale, D. M. (1987b). *Plant Physiol. Biochem.* **25**, 265–271.
Lonsdale, D. M., Thompson, R. D., and Hodge, T. P. (1981). *Nucleic Acids Res.* **9**, 3657–3669.
Lonsdale, D. M., Fauron, C. M.-R., Hodge, T. P., Pring, D. R., and Stern, D. B. (1983a). In "Genetic Rearrangement" (K. F. Chater, A. Cullis, D. A. Hopwood, A. W. B. Johnson, and H. W. Woolhouse, eds.), pp. 183–206. Croon-Helm, London.
Lonsdale, D. M., Hodge, T. P., Fauron, C. M.-R., and Flavell, R. B. (1983b). *UCLA Symp. Mol. Cell. Biol., New Ser.* **12**, 445–456.
Lonsdale, D. M., Hodge, T. P., Howe, C. J., and Stern, D. B. (1983c). *Cell* **34**, 1007–1014.
Lonsdale, D. M., Hodge, T. P., and Fauron, C. M.-R. (1984). *Nucleic Acids Res.* **12**, 9249–9261.
Lonsdale, D. M., Brears, T., Hodge, T. P., Melville, S. E., and Rottmann, W. H. (1988). *Philos. Trans. R. Soc. London, Ser. B,* **319**, 149–163.
Ludwig, S. R., Pohlman, R. F., Vieira, J., Smith, A. G., and Messing, J. (1985). *Gene* **38**, 131–138.
McCarty, J., and Hauswirth, W. W. (1988). Personal communication.
McNay, J. W., Pring, D. R., and Lonsdale, D. M. (1983). *Plant Mol. Biol.* **2**, 177–187.
Makaroff, C. A., and Palmer, J. D. (1987). *Nucl. Acids Res.* **15**, 5141–5146.
Malone, C., Koeppe, D. E., and Miller, R. J. (1974). *Plant Physiol.* **53**, 918–927.
Mandel, M., and Marmur, J. (1968). *Methods Enzymol.* **12**, 195–206.
Manna, E., and Brennicke, A. (1985). *Curr. Genet.* **9**, 505–515.
Manna, E., and Brennicke, A. (1986). *Mol. Gen. Genet.* **203**, 377–381.
Manna, F., Del Giudice, L., Massaro, D. R., Schreil, W. H., Cermola, M., Devreux, M., and Wolf, K. (1985). *Curr. Genet.* **9**, 411–415.
Mannhaupt, G., Beyreuther, K., and Michaelis, G. (1985). *Eur. J. Biochem.* **150**, 435–439.
Manson, J. C., Liddell, A. D., Leaver, C. J., and Murray, K. (1986). *EMBO J.* **5**, 2775–2780.
Maréchal, L., Guillemaut, P., Grienenberger, J.-M., Jeannin, G., and Weil, J.-H. (1985a). *FEBS Lett.* **184**, 289–293.
Maréchal, L., Guillemaut, P., Grienenberger, J.-M., Jeannin, G., and Weil, J.-H. (1985b). *Nucleic Acids Res.* **13**, 4411–4416.
Maréchal, L., Guillemaut, P., and Weil, J. H. (1985c). *Plant Mol. Biol.* **5**, 347–351.

Maréchal, L., Guillemaut, P., Grienenberger, J.-M., Jeannin, G., and Weil, J.-H. (1986). *Plant Mol. Biol.* **7,** 245–253.
Maréchal, L., Runeberg-Roos, P., Grienenberger, J. M., Colin, J., Weil, J.-H., Lejeune, B., Quetier, F., and Lonsdale, D. M. (1987). *Curr. Genet.* **12,** 91–98.
Marechal-Drouard, L., Weil, J.-H. and Guillemaut, P. (1988). *Nucleic Acids Res.* **16,** 4777–4788.
Medgyesy, P., Fejes, E.,and Maliga, P. (1985). *Proc. Natl. Acad. Sci. U.S.A.* **82,** 6960–6964.
Michel, F., and Dujon, B. (1983). *EMBO J.* **2,** 33–38.
Mignouna, H., Virmani, S. S., and Briquet, M. (1987). *Theor. Appl. Genet.* **74,** 666–669.
Mikulska, E., Odintsova, M. S., and Turischeva, M. S. (1970). *J. Ultrastruct. Res.* **32,** 258–267.
Moon, E., Kao, T.-H. and Wu, R. (1985). *Nucleic Acids Res.* **13,** 3195–3212.
Morgens, P. H., Grabau, E. A., and Gesteland, R. F. (1984). *Nucleic Acids Res.* **12,** 5665–5684.
Morikami, A., and Nakamura, K. (1987). *J. Biochem.* **101,** 967–976.
Mulligan, M. R., Maloney, A. P., and Walbut, V. (1988). *Mol. Gen. Genet.* **211,** 373–380.
Nagy, F., Lazar, G., Menczel, L., and Maliga, P. (1983). *Theor. Appl. Genet.* **66,** 203–207.
Nargang, F. E. (1985). *Exp. Mycol.* **9,** 285–293.
Negruk, V. I., Goncharova, N. P., Gelnin, L. G., and Mardamshin, A. G. (1985). *Mol. Gen. Genet.* **198,** 486–490.
Negruk, V. I., Eisner, G. I., Redichkina, T. D., Dumanskaya, N. N., Cherny, D. I., Alexandrov, A. A., Shemyakin, M. F., and Butenko, R. G. (1986). *Theor. Appl. Genet.* **72,** 541–547.
Newton, K. J., and Walbot, V. (1985). *Proc. Natl. Acad. Sci. U.S.A.* **82,** 6879–6883.
Nikiforova, I. D., and Negruk, V. I. (1983). *Planta* **157,** 81–84.
Ohyama, K., Fukuzawa, H., Kohchi, T., Shirai, H., Sano, T., Sano, S., Umesono, K., Shiki, Y., Takeuchi, M., Chang, Z., Aota, S., Inokuchi, H., and Ozeki, H. (1986). *Nature (London)* **322,** 572–574.
Ojala, D., Montoya, J., and Attardi, G. (1981). *Nature (London)* **290,** 470–474.
Orian, J. M., Murphy, M., and Marzuki, S. (1981). *Biochim. Biophys. Acta* **652,** 234–239.
Oro, A. E., Newton, K. J., and Walbot, V. (1985). *Theor. Appl. Genet.* **70,** 287–293.
Osinga, K. A., and Tabak, H. F. (1982). *Nucleic Acids Res.* **10,** 3617–3626.
Osinga, K. A., De Vries, E., Van der Horst, G. T. J., and Tabak, H. F. (1984). *Nucleic Acids Res.* **12,** 1889–1900.
Paillard, M., Sederoff, R. R., and Levings, C. S., III (1985). *EMBO J.* **4,** 1125–1128.
Palmer, J. D. (1983). *Nature (London)* **301,** 92–93.
Palmer, J. D. (1985). *In* "Monographs in Evolutionary Biology: Molecular Evolutionary Genetics" (R. J. MacIntyre, ed.), pp. 131–240. Plenum, New York.
Palmer, J. D., and Herbon, L. A. (1986). *Nucleic Acids Res.* **14,** 9755–9764.
Palmer, J. D., and Herbon, L. A. (1987). *Curr. Genet.* **11,** 565–570.
Palmer, J. D., and Shields, C. R. (1984). *Nature (London)* **307,** 437–440.
Palmer, J. D., Shields, C. R., Cohen, D. B., and Orton, T. J. (1983). *Nature (London)* **301,** 725–728.
Parks, T. D., Dougherty, G., Levings, C. S., III, and Timothy, D. H. (1984). *Plant Physiol.* **76,** 1079–1082.
Parks, T. D., Dougherty, G., Levings, C. S., III, and Timothy, D. H. (1985). *Curr. Genet.* **9,** 517–519.
Peebles, C. L., Gegenheimer, P., and Abelson, J. (1983). *Cell* **32,** 525–536.
Pelletier, G., Primard, C., Vedel, F., Chetrit, P., Remy, R., Rousselle, and Renard, M. (1983). *Mol. Gen. Genet.* **191,** 244–250.
Poutre, C. G., and Fox, T. D. (1987). *Genetics* **115,** 637–647.
Powling, A. (1981). *Mol. Gen. Genet.* **183,** 82–84.
Pratje, E., and Guiard, B. (1986). *EMBO J.* **5,** 1313–1317.
Pring, D. R., and Levings, C. S., III (1978). *Genetics* **89,** 121–136.
Pring, D. R., and Lonsdale, D. M. (1985). *Int. Rev. Cytol.* **97,** 1–46.

Pring, D. R., and Thornbury, D. W. (1975). *Biochim. Biophys. Acta* **383,** 140–146.
Pring, D. R., Conde, M. F., Schertz, K. F., and Levings, C. S., III (1982). *Mol. Gen. Genet.* **186,** 180–184.
Quetier, F., and Vedel, F. (1977). *Nature (London)* **268,** 365–368.
Quetier, F., Lejeune, B., Delorme, S., Falconet, D., and Jubier, M. F. (1985). *In* "Molecular Form and Function of the Plant Genome" (L. van Vloten-Doting, G. S. P. Groot, and T. C. Hall, eds.), pp. 413–420. Plenum, New York.
Quigley, F., Grienenberger, J.-M., and Weil, J.-H. (1985). *Plant Mol. Biol.* **4,** 305–310.
Reisner, D., and Gross, H. J. (1985). *Annu. Rev. Biochem.* **54,** 531–564.
Rode, A., Hartmann, C., Falconet, D., Lejeune, B., Quetier, F., Benslimane, A., Henry, Y., and de Buyser, J. (1987). *Curr. Genet.* **12,** 369–376.
Roe, B. A., Din-Pow, M., Wilson, R. K., and Wong, J. F.-H. (1985). *J. Biol. Chem.* **260,** 9759–9774.
Rogers, H. J., Buck, K. W., and Brasier, C. M. (1987). *Nature (London)* **329,** 558–560.
Rothenberg, M., and Hanson, M. R. (1987). *Curr. Genet.* **12,** 235–240.
Rottmann, W. H., Brears, T., Hodge, T. P., and Lonsdale, D. M. (1987). *EMBO J.* **6,** 1541–1546.
Runeberg-Roos, P., Grienenberger, J. M., Guillemaut, P., Marechal, L., Gruber, V., and Weil, J. H. (1987). *Plant Mol. Biol.* **9,** 237–246.
Schardl, C. L., Lonsdale, D. M., Pring, D. R., and Rose, K. R. (1984). *Nature (London)* **301,** 292–296.
Schardl, C. L., Lonsdale, D. M., and Pring, D. R. (1985a). *UCLA Symp. Mol. Cell. Biol.* **35,** 575–583.
Schardl, C. L., Pring, D. R., and Lonsdale, D. M. (1985b). *Cell* **43,** 361–368.
Schinkel, A. H., Groot Koerkamp, M. J. A., Stuiver, M. H., Van der Horst, G. T. J., and Tabak, H. F. (1987). *Nucleic Acids Res.* **15,** 5597–5612.
Schnare, N. M., and Gray, M. W. (1982). *Nucleic Acids Res.* **10,** 3921–3932.
Schuster, A. M., Sisco, P. H., and Levings, C. S., III (1983). *UCLA Symp. Mol. Cell. Biol. New Ser.* **12,** 437–444.
Schuster, W., and Brennicke, A. (1985). *Curr. Genet.* **9,** 157–163.
Schuster, W., and Brennicke, A. (1986). *Mol. Gen. Genet.* **204,** 29–35.
Schuster, W., and Brennicke, A. (1987a). *EMBO J.* **6,** 2857–2863.
Schuster, W., and Brennicke, A. (1987b). *Mol. Gen. Genet.* **210,** 44–51.
Schuster, W., and Brennicke, A. (1987c). *Nucleic Acids Res.* **15,** 9092.
Schuster, W., and Brennicke, A. (1988a). *Plant Sci.* **54,** 1–10.
Schuster, W., and Brennicke, A. (1988b). *Nucleic Acids Res.* (in press).
Schuster, W., Hiesel, R., Isaac, P. G., Leaver, C. J., and Brennicke, A. (1986). *Nucleic Acids Res.* **14,** 5943–5954.
Scott, N. S., and Timmis, J. N. (1984). *Theor. Appl. Genet.* **67,** 279–288.
Sederoff, R. R. (1984). *Adv. Genet.* **22,** 1–108.
Sederoff, R. R., and Levings, C. S., III (1985). *Plant Gene Res.* **2,** 91–109.
Sederoff, R. R., Levings, C. S., III, Timothy, D. H., and Hu, W. W. L. (1981). *Proc. Natl. Acad. Sci. U.S.A* **78,** 5953–5957.
Sederoff, R. R., Ronald, P. Bedlinger, P., Rivin, C., Walbot, V., Bland, M., and Levings, C. S., III (1986). *Genetics* **113,** 469–482.
Selden, R. F., Steinmetz, A. McIntosh, L., Bogorad, L., Burkard, G., Mubumbila, M., Kuntz, M., Crouse, E. J., and Weil, J. (1983). *Plant Mol. Biol.* **2,** 141–153.
Shah, D. M., and Levings, C. S., III (1974). *Crop Sci.* **14,** 852–853.
Shine, J., and Dalgarno, L. (1974). *Proc. Natl. Acad. Sci. U.S.A.* **71,** 1342–1346.
Shinozaki, K., Ohme, M., Tanaka, M., Wakasugi, T., Hayshida, N., Matsubayasha, T., Zaita, N., Chunwongse, J., Obokata, J., Yamaguchi-Shinozaki, K., Ohto, C., Torazawa, K., Meng, B. Y., Sugita, M., Deno, H., Kamogashira, T., Yamada, K., Kusuda, J., Takaiwa, F., Kata, A., Todoh, N., Shimada, H., and Sugiura, M. (1986). *EMBO J.* **5,** 2043–2049.

Singh, A., and Laughnan, J. R. (1972). *Genetics* **71**, 607–620.
Sisco, P. H., Gracen, V. E., Manchester, C. E., and Everett, H. L. (1982). *Maize Genet. Coop. News Lett.* **56**, 80.
Sisco, P. H., Garcia-Arenal, F., Zaitlin, M., Earle, E., and Gracen, V. (1984). *Plant Sci. Lett.* **34**, 127–134.
Small, I. D., Isaac, P. G., and Leaver, C. J. (1987). *EMBO J.* **6**, 865–869.
Sparks, R. B., Jr., and Dale, R. M. K. (1980). *Mol. Gen. Genet.* **180**, 351–355.
Spencer, D. F., Bonen, L., and Gray, M. W. (1981). *Biochemistry* **20**, 4022–4029.
Spencer, D. F., Schnare, M. N., and Gray, M. W. (1984). *Proc. Natl. Acad. Sci. U.S.A.* **81**, 493–497.
Spruill, W. M., Jr., Levings, C. S., III, and Sederoff, R. R. (1980). *Dev. Genet.* **1**, 363–378.
Stamper, S. E., Dewey, R. E., Bland, M. M., and Levings, C. S., III (1987). *Curr. Genet.* **12**, 457–463.
Stern, D. B., and Lonsdale, D. M. (1982). *Nature (London)* **299**, 698–702.
Stern, D. B., and Newton, K. J. (1985a). *Curr. Genet.* **9**, 395–405.
Stern, D. B., and Newton, K. J. (1985b). *Methods Enzymol.* **118**, 488–496.
Stern, D. B., and Palmer, J. D. (1984a). *Nucleic Acids Res.* **12**, 6141–6157.
Stern, D. B., and Palmer, J. D. (1984b). *Proc. Natl. Acad. Sci. U.S.A.* **81**, 1946–1950.
Stern, D. B., and Palmer, J. D. (1986). *Nucleic Acids Res.* **14**, 5651–5666.
Stern, D. B., Dyer, T. A., and Lonsdale, D. M. (1982). *Nucleic Acids Res.* **10**, 3333–3340.
Stern, D. B., Palmer, J. D., Thompson, W. F., and Lonsdale, D. M. (1983). *UCLA Symp. Mol. Cell. Biol., New Ser.* **12**, 467–477.
Stern, D. B., Bang, A. G., and Thompson, W. F. (1986). *Curr. Genet.* **10**, 857–869.
Stinchcomb, D. T., Thomas, M., Kelly, J., Selker, E., and Davis, R. M. (1980). *Proc. Natl. Acad. Sci. U.S.A.* **77**, 4559–4563.
Suyama, Y. (1986). *Curr. Genet.* **10**, 411–420.
Suyama, Y., and Bonner, W. D., Jr. (1966). *Plant Physiol.* **41**, 383–388.
Suyama, Y., and Miura, K. (1968). *Proc. Natl. Acad. Sci. U.S.A.* **60**, 235–242.
Swamy, G. S., and Pillay, D. T. N. (1982). *Plant Sci. Lett.* **25**, 73–84.
Synenki, R. M., Levings, C. S., III, and Shah, D. M. (1978). *Plant Physiol.* **61**, 460–464.
Tamanoi, F., and Stillman, B. W. (1983). *Proc. Natl. Acad. Sci. U.S.A.* **80**, 6446–6450.
Thomas, C. M. (1986). *Nucleic Acids Res.* **14**, 9353–9370.
Thompson, R. D., Kemble, R. J., and Flavell, R. (1980). *Nucleic Acids Res.* **8**, 1999–2008.
Timmis, J. N., and Scott, N. S. (1983). *Nature (London)* **305**, 65–67.
Timmis, J. N., and Scott, N. S. (1985). *Plant Gene Res.* **2**, 61–78.
Timothy, D. H., Levings, C. S., III, Hu, W. W. L., and Goodman, H. H. (1983). *Maydica* **28**, 139–149.
Traynor, P. L., and Levings, C. S., III (1986). *Plant Mol. Biol.* **7**, 255–263.
Turano, F. J., Debonte, L. R., Wilson, K. G., and Matthews, B. F. (1987). *Plant Physiol.* **84**, 1074–1079.
Turpen, T., Garger, S. J., Marks, M. D., and Grill, L. K. (1987). *Mol. Gen. Genet.* **209**, 227–233.
Umbeck, P. F., and Gengenbach, B. G. (1983). *Crop Sci.* **23**, 584–588.
van den Boogaart, D., Samallo, J., and Asgsteribb, E. (1982). *Nature (London)* **298**, 187–189.
Vedel, F., and Quetier, F. (1974). *Biochim. Biophys. Acta* **340**, 374–387.
Vedel, F., Chetrit, P., Mathieu, C., Pelletier, G., and Primard, C. (1986). *Curr. Genet.* **11**, 17–24.
Wahleithner, J. A., and Wolstenholme, D. R. (1987). *Curr. Genet.* **12**, 55–67.
Wahleithner, J. A., and Wolstenholme, D. R. (1988). *Nucleic Acids Res.* **16**, 6897–6913.
Wakasugi, T., Ohme, M., Shinozaki, K., and Sugiura, M. (1986). *Plant Mol. Biol.* **7**, 385–392.
Ward, B. L., Anderson, R. S., and Bendich, A. J. (1981). *Cell* **25**, 793–803.
Waring, R. B., Davies, R. W., Lee, S., Grisi, E., McPhail-Berks, M., and Scazzocchio, C. (1981). *Cell* **27**, 4–11.

Warmke, H. E., and Lee, S.-L. J. (1978). *Science* **200,** 561–563.
Weissinger, A. K., Timothy, D. H., Levings, C. S., III, Hu, W. W. L., and Goodman, M. M. (1982). *Proc. Natl. Acad. Sci. U.S.A.* **79,** 1–5.
Weissinger, A. K., Timothy, D. H., Levings, C. S., III, and Goodman, M. M. (1983). *Genetics* **104,** 365–379.
Wells, R., and Birnstiel, M. (1969). *Biochem. J.* **112,** 777–786.
Wells, R., and Ingle, J. (1970). *Plant Physiol.* **46,** 178–179.
Wintz, H., Chen, H.-C. and Pillay, D. T. N. (1987). *Nucleic Acids Res.* **15,** 10588.
Wintz, H., Chen, H.-C. and Pillay, D. T. N. (1988a). *Curr. Genet.* **13,** 255–260.
Wintz, H., Grienenberger, J.-M., Weil, J.-H., and Lonsdale, D. M. (1988b). *Curr. Genet.* **13,** 247–254.
Wise, R. P., Fliss, A. E., Pring, D. R., and Gengenbach, B. G. (1987). *Plant Mol. Biol.* **9,** 121–126.
Wissinger, B., Hiesel, R., Schuster, W., and Brennicke, A. (1988). *Mol. Gen. Genet.* **212,** 56–65.
Wolstenholme, D. R., and Gross, N. J. (1968). *Proc. Natl. Acad. Sci. U.S.A.* **61,** 245–252.
Wong, F. Y., and Wildman, S. G. (1972). *Proc. Natl. Acad. Sci. U.S.A.* **69,** 1830–1834.
Yamaguchi, H., and Kakiuchi, H. (1983). *Jpn. J. Genet.* **58,** 607–611.
Young, E. G., and Hanson, M. R. (1987). *Cell* **50,** 41–49.
Young, E. G., Hanson, M. R., and Dierks, P. M. (1986). *Nucleic Acids Res.* **14,** 7995–8006.
Zabala, G., and Walbot, V. (1988). *Mol. Gen. Genet.* **211,** 386–392.
Ziale, Z., and Suyama, Y. (1987). *Curr. Genet.* **12,** 357–368.

The Biochemistry and Molecular Biology of Seed Storage Proteins

7

MARK A. SHOTWELL
BRIAN A. LARKINS

I. Globulin Storage Proteins
 A. Soybean Storage Globulins
 B. Globulin Storage Proteins of Other Species
 C. Structure of Globulin Storage Proteins
II. Synthesis and Deposition of Storage Globulins
III. Organization and Structure of Storage Globulin Genes
IV. Regulation of Globulin Gene Expression
V. Prolamine Storage Proteins
 A. Wheat and Barley Prolamines
 B. Maize Prolamines
VI. Synthesis and Deposition of Cereal Prolamines
VII. Organization and Structure of Prolamine Genes
 A. Wheat and Barley Prolamines
 B. Maize Zein Prolamines
VIII. Regulation of Prolamine Gene Expression
IX. Summary
 References

Seeds are one of the richest sources of plant proteins. Among major crop plants, seeds contain 10 to 50% protein, and most of this is storage protein. The storage proteins have no enzymatic activity, and simply provide a source of amino acids, nitrogen, and carbon skeletons for the developing seedling. Storage proteins are deposited in the seed in an insoluble form in protein bodies and survive desiccation for long periods of time. They characteristically contain high levels of amide amino acids (glutamine and asparagine) and are deficient in others. Cereal storage proteins are limiting in lysine, while legume storage proteins are limiting in methionine and cysteine. Because seeds provide an important source of protein for human and livestock nutrition, much research has been devoted to increasing the content of the essential amino acids and improving the nutritional quality of seed proteins (Nelson, 1979).

Research by Osborne (1908) and Daniellson (1949) provided a fundamental understanding of the chemistry of storage proteins, and a review of the general properties of storage proteins is presented by Larkins in Chapter 11, Volume 6 of this series. Over the past 5 years these proteins have been characterized at the molecular level, and we now know the structures of a number of storage proteins as well as the mechanisms by which they are synthesized and deposited in the seed. We also know details of the organization and structure of the genes encoding storage proteins and the nature of the DNA sequences regulating their expression. This chapter presents an overview of recent findings pertaining to the structure, biosynthesis, and genetic regulation of the major seed proteins, the storage globulins and prolamines. Information pertaining to phytohemagglutinins and protease inhibitors is not included, although these may also account for a significant portion of the seed protein. Details of recent studies on these proteins are available in Higgins (1985) and Shannon and Chrispeels (1985). Other recent reviews on storage proteins are those by Altschul and Wilcke (1985), Gottschalk and Muller (1983), and Pernollet (1985).

I. GLOBULIN STORAGE PROTEINS

Globulin storage proteins are separated into classes by cryoprecipitation, differential salt solubility, or ultracentrifugation (see Chapter 11, Volume 6 of this series). Most storage globulins fall into two major groups with sedimentation coefficients of 11–12S and 7–8S. The seeds of many plant species contain both 11S and 7S storage proteins, although in most cases one or the other predominates (Derbyshire *et al.*, 1976). Storage globulins have best been characterized in the legumes, and in these genera the 11S proteins are sometimes collectively referred to as legumins and the 7S proteins as vicilins. In many plants, the two classes are given trivial names derived from the genus of the plant. For example, the 11S and 7S proteins of soybean (*Glycine max*) are referred to as glycinin and conglycinin. To a large extent, the molecular structures of the 11S and 7S globulins are conserved among species. For illustrative purposes we therefore consider in detail storage proteins of only one species, soybean. For additional information of storage globulins, see Gatehouse *et al.*, (1984), Nielsen (1985b), and Casey *et al.*, (1986).

A. Soybean Storage Globulins

Soybean seeds contain about 40% protein by weight. Initial studies showed four major proteins: glycinin (11S) and α, β, and γ-conglycinin (7S). Glycinin and β-conglycinin are the most abundant and account for about 70% of the total seed protein. The proportions of glycinin and β-conglycinin vary, with ratios of 1:1 to 3:1 in different soybean cultivars. On the average, the ratio of glycinin to conglycinin is 1.6:1 (Nielsen, 1984). Both 11S and 7S proteins are low in sulfur amino acids (1.8% for glycinin and 0.6% for β-conglycinin; Nielsen, 1985a).

Glycinin is isolated from seeds as a hexameric complex of M_r 360,000, composed of six nonidentical subunits with M_r averaging 60,000 (Fig. 1, lane 2). The subunits, which are not glycosylated, are made up of a polypeptide of approximately M_r 40,000 with an acidic pI and an M_r 20,000 polypeptide with a basic pI, covalently linked by a single disulfide bond (Fig. 1, lane 6). The glycinin hexamer is assembled from five different subunit molecules that fall into two classes: (1) the group I glycinins, M_r 58,000, which have five to eight methionine residues and consist of two members; and (2) the group II glycinins, M_r 62,000 to 69,000, which have three methionine residues and consist of two members (Nielsen, 1985a). There is about 80% identity in amino

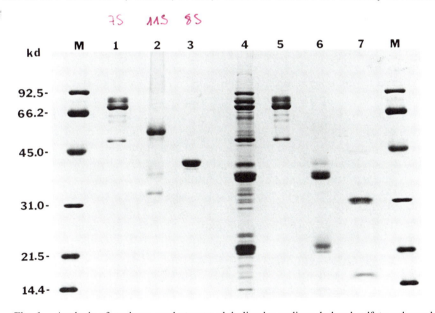

Fig. 1. Analysis of soybean seed storage globulins by sodium dodecyl sulfate-polyacrylamide gel electrophoresis (SDS-PAGE). Purified proteins (lanes 1–3 and 5–7) and total soybean protein (lane 4) from the variety Century were dissolved in SDS in the absence (lanes 1–3) and presence (lanes 4–7) of 2-mercaptoethanol. Samples are as follows: lanes 1 and 5, 7S (conglycinin); lanes 2 and 6, 11S (glycinin); lanes 3 and 7, 8S basic globulin.

acid sequence between members of the same group, but only 40 to 50% sequence identity between members of different groups (Moreira et al., 1981).

Glycinin is isolated from extracts of total soybean protein by isoelectric precipitation at pH 6.6 at 4°C (Moreira et al., 1979). After the protein is denatured in 6 M urea in the presence of 2-mercaptoethanol, the acidic and basic polypeptides are fractionated by anion-exchange chromatography. The initial protein fraction contains a mixture of basic polypeptides, while subsequent fractions contain acidic polypeptides. The mixture of basic polypeptides is fractionated by anion-exchange chromatography to obtain five different polypeptides.

β-Conglycinin is generally isolated as a trimer of M_r 180,000. At low ionic strength, however, it dimerizes to form hexameric complexes. β-Conglycinin exists in at least seven isomeric forms. It is made up of three major subunits, α, α', and β (Thanh and Shibasaki, 1976), of M_r 76,000, 72,000, and 53,000 (Meinke et al., 1981), as well as several minor components (Coates et al., 1985) (Fig. 1, lane 1). In contrast to glycinin, these subunits contain a single polypeptide (Fig. 1, cf. lanes 1, 5). The subunit molecules are structurally equivalent, and β-conglycinin complexes arise by random assembly (Nielsen, 1984). Unlike glycinin, β-conglycinin is glycosylated with about 5% sugar by weight. The sugar component is covalently linked to one or more asparagine residues and is made up of 2 central N-acetylglucosamine residues to which are attached 7 to 10 mannose residues (Yamauchi and Yamagishi, 1979).

β-Conglycinin is precipitated from total seed globulin after removal of glycinin by lowering the pH to 4.8, with subsequent removal of lower-molecular-weight proteins by gel filtration (Coates et al., 1985). Further purification is obtained on a concanavalin A-Sepharose affinity column to separate β-conglycinin from nonglycosylated protein contaminants. Finally, the three major subunit polypeptides, α, α', and β, are separated by a combination of anion- and cation-exchange chromatography.

A third class of globulin, sedimenting at 8.2S, has been characterized in soybean (Hu and Esen, 1982; Yamauchi et al., 1984; Lilley and Nielsen, 1987). This protein is a relatively minor component and accounts for 5 to 10% of the seed protein, depending on the cultivar. This globulin is a tetramer of M_r 168,000 composed of four identical M_r 42,000 subunits (Fig. 1, lane 3), and, in contrast to the 7S and 11S globulins, is basic (pI 9.5). Like the 11S globulins, each subunit is composed of a large polypeptide (M_r 30,000) disulfide-bonded to a small polypeptide (M_r 16,000) (Fig. 1, lane 7). Unlike the 11S globulin, the large polypeptide is basic and the small polypeptide is acidic. The 8S globulin has higher levels of methionine and cysteine than the 11S and 7S proteins of soybean, but is lower in acidic amino acids, accounting for its higher isoelectric point (Lilley and Nielsen, 1987).

B. Globulin Storage Proteins of Other Species

Storage proteins with structures similar to the soybean 7S and 11S globulins have been characterized in many dicots and monocots and have been most intensively studied in the legume family. Besides soybean, these proteins have been examined in pea (*Pisum sativum*) (Derbyshire *et al.*, 1976), French bean (*Phaseolus vulgaris*) (Sun and Hall, 1975; Pusztai and Watt, 1974), broad bean (*Vicia faba*) (Bailey and Boulter, 1970, 1974), peanut (*Arachis hypogea*) (Neucere and Ory, 1970), castor bean (*Ricinus communis*) (Sharief and Li, 1982), jack bean (*Canavalia ensiformis*) (Smith *et al.*, 1982), cowpea (*Vigna unguiculata*) (Carasco *et al.*, 1978; Khan *et al.*, 1980), lupin (*Lupinus agustifolius*) (Blagrove and Gillespie, 1975), and broom (*Cytisus scoparius*) (Citharel and Citharel, 1985). Whereas the 11S globulins thus far characterized are similar to one another, the 7S globulins have more varied structures. In soybean (Meinke *et al.*, 1981) and French bean (Brown *et al.*, 1981), the 7S trimers are composed of three major polypeptides. Pea vicilin contains more than 10 polypeptides from M_r 12,000 to 75,000, which arise from extensive proteolytic processing (Spencer *et al.*, 1983). The relative amounts of the 7S and 11S proteins differ among species. For example, the ratio of 11S to 7S in broad bean is 4:1 (Wright and Boulter, 1972), while in French bean the ratio is 1:9 (Derbyshire and Boulter, 1976).

Less information is available for nonleguminous dicot species, but most of those examined contain 11S-type storage proteins. In rape (*Brassica napus*) seed, 60% of the protein in the mature seed is a 12S globulin of M_r 300,000 named cruciferin (Crouch and Sussex, 1981). A similar protein has been found in other members of the mustard family, including white mustard (*Sinapus alba*) (Kirk and Pyliotis, 1976), radish (*Raphanus sativus*) (Laroche *et al.*, 1984), and *Arabidopsis thaliana* (Heath *et al.*, 1986). The seeds of pumpkin (*Cucurbita pepo*) and other members of the cucurbit family contain an 11S globulin, cucurbitin, as their only storage protein (Hara *et al.*, 1976).

In cotton (*Gossypium hirsutum*), a member of the Malvaceae, the storage globulins fall into two classes designated α and β (Dure and Galau, 1981). The α-globulin family comprises M_r 46,500 nonglycosylated proteins and M_r 52,000 glycosylated proteins, and the β-globulin family is made up of M_r 58,000 proteins that are cleaved twice and linked by disulfide bonds (Chlan *et al.*, 1986). Amino acid sequence analysis has shown that the cottonseed α-globulins are related to the 7S globulins of legumes and the β-globulins are related to the 11S globulins (Borroto and Dure, 1987).

The major storage proteins in oat (*Avena sativa*) seeds are also globulins. This sets oats apart from most cereals, which store predominantly prolamines. Oat seed globulins fall into three classes with sedimentation coefficients of 12S, 7S, and 3S (Burgess *et al.*, 1983), with the 12S component by far the most abundant. Rice (*Oriza sativa*) is also unusual among cereals because its seeds contain very low levels of prolamines (Juliano, 1972). About 80% of the rice seed protein is designated glutelin because of the requirement of NaOH, SDS, or urea for solubility (Juliano and Boulter, 1976). The rice glutelin is structurally similar to the 11S-type globulins in legumes (Zhao *et al.*, 1983) and is antigenically related to them (Roberts *et al.*, 1985).

In addition to 11S and 7S proteins, a third storage globulin has been characterized in lupin. This minor component, conglutin γ, is isolated from lupin seeds as an M_r 280,000 hexamer that sediments at about 10S (Blagrove *et al.*, 1980). The hexamer is composed of six M_r 47,000 subunits, each consisting of an M_r 30,000 polypeptide, p*I* 8.0, disulfide bonded to an M_r 17,000 polypeptide, p*I* 6.9. Although no enzymatic activity has been demonstrated for this protein, Blagrove and Gillespie (1975) questioned whether conglutin γ might have a metabolic rather than a storage function.

C. Structure of Globulin Storage Proteins

The physical structures of the 11S and 7S storage globulins have been investigated using a number of techniques. Based on electron microscopy (EM) of sunflower (*Helianthus annuus*) 11S protein, helianthin, Reichelt and co-workers (1980) concluded that each 11S complex is composed of six spherical subunits arranged in two trimers stacked one atop the other but offset by 60°. The quaternary structure thus takes the form of a trigonal antiprism (see Fig. 4). The pattern of penetration of a hydrophilic staining agent indicated that the three subunits making up each trimer may be held together primarily by hydrophobic forces, whereas the two trimers making up each hexamer may be joined by hydrogen bonds or electrostatic interactions.

Extending the EM results, Plietz and co-workers (1983a) analyzed helianthin by small-angle X-ray scattering and found the complex to be almost spherical with a radius of gyration of 3.95 nm. As in the EM study, the six structurally identical subunits were arranged as a trigonal antiprism. By this method, rape seed 11S globulin was found to have the same structure but with a radius of 4.1 nm.

Phaseolin, the 7S globulin from French bean, has also been studied using these physical techniques. The phaseolin molecule was found to be composed of three Y-shaped subunits arranged around a threefold symmetry axis and separated by deep solvent clefts (Pleitz *et al.*, 1983b) (see Fig. 3). Based on small-angle X-ray scattering and quasi-elastic light scattering measurements, the trimeric complex was found to approximate an oblate ellipsoid 12.5 × 12.5 × 3.75 nm (Plietz *et al.*, 1983c).

Comparisons have also been made of amino acid sequences between different members of the 11S or 7S group in a given species, as well as between 11S and 7S proteins from different species. Using a computer program to compare sequences on the basis of amino acid physical characteristics, Argos *et al.* (1985) found structural similarities between 11S and 7S storage proteins of soybean, pea, and French bean. Based on alignment of regions of similar predicted structure, they proposed a model in which the polypeptide is divided into three structural domains (Fig. 2). Domain I is the NH_2-terminal quarter of the subunit molecule and differs significantly between 11S and 7S proteins. It is predicted to consist mostly of helical and turn conformations whose distribution is not conserved between the two proteins. Domain II is the central portion and contains regions of common predicted secondary structure. It is proposed to consist of a mixture of helical, beta sheet, and turn conformations. Domain III is the COOH-terminal half of the molecule and is the most highly conserved of the domains. It is predicted to consist largely of beta sheet and turn conformations and to be quite hydrophobic.

Based on this model, a major difference between the 11S and 7S subunits is the presence between domains II and III in the 11S subunits of a hypervariable region containing inserts of different length. Variation in this region accounts for the size differences between different members of the group I glycinins and the group II glycinins. The insertions, consisting largely of repeated aspartate and glutamate residues, are very acidic and are predicted to exist in a helical conformation.

Note in Fig. 2 that domains I and II together correspond to the acidic polypeptide of 11S subunits and domain III to the basic polypeptide. Figure 2 also shows that the site of proteolytic cleavage is at the carboxyl side of the hypervariable region. This highly charged region would be expected to be exposed to the solute at the surface of the molecule. Accordingly, Argos *et al.* (1985) suggested these repeat regions might act to make this site accessi-

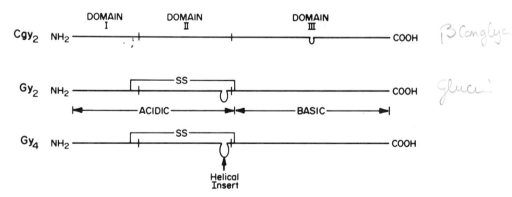

Fig. 2. Relationship of predicted domain organization between 7S and 11S globulins. Cgy_2 corresponds to the β-conglycinin subunit; Gy_2 and Gy_4 correspond to soybean 11S subunits. Sites of peptide inserts are shown as loops. From Argos *et al.* (1985) with permission of the author and publisher.

ble to the enzyme that cleaves it into acidic and basic polypeptides. They further suggested the hydrophobic nature of domain III may result in its being buried within the molecule and that the single disulfide bond linking domain I with domain III contributes to the maintenance of the conformation of the proteolytically processed subunit.

In comparing predicted amino acid sequences from different pea legumin cDNAs, Lycett *et al.* (1984b) found three tandem repeats of 18 amino acid residues in a part of the sequence corresponding to the hypervariable region at the end of domain II. Like those in soybean glycinin, these repeats contain a high proportion of polar, mainly acidic, residues. In cruciferin, no insert exists at the end of domain II, and instead an insert of 38 amino acids was found between domain I and domain II. This insert has a very different composition from the highly acidic repeats in legume 11S proteins and is composed almost entirely of glutamine and glycine (Simon *et al.*, 1985). Sequence analysis of several cDNA clones corresponding to the oat 12S globulin revealed the presence of an 8-amino acid block that was repeated either four or five times (Walburg and Larkins, 1986; Shotwell *et al.*, 1988). These repeats are in the same position as the glycinin inserts, but have the consensus sequence "Gln-Tyr-Gln-Val/Glu-Gly-Gln-Ser-Thr."

Subunits of the 7S proteins also contain peptide insertions. A large insert in the first domain accounts for the size difference between the α- and α'-subunits in soybean (Doyle *et al.*, 1986). Somewhat smaller inserts occur in domain III, the most highly conserved region of the molecule. These inserts, which are considerably shorter than those in 11S subunits, are also rich in glutamate and aspartate and are assumed to reside at the surface of the molecule.

It is thus apparent that inserts of polar amino acids are a common structural feature of both 11S and 7S globulin storage proteins. These inserts can vary considerably in length, amino acid composition, and location in the subunit molecule, both within species and between species. This variability has attracted attention to the insertion regions as possible targets for the genetic engineering of altered amino acid sequences. The introduction of sulfur-containing amino acids into these regions by the techniques of site-directed mutagenesis is a likely future approach in the improvement of storage globulins as protein sources.

II. SYNTHESIS AND DEPOSITION OF STORAGE GLOBULINS

A general model has emerged for the mechanism by which storage globulins are transported from their site of synthesis, the rough endoplasmic reticulum (RER), to their site of deposition, the vacuole (Figs. 3 and 4). The proteins are synthesized by membrane-bound polysomes as precursor poly-

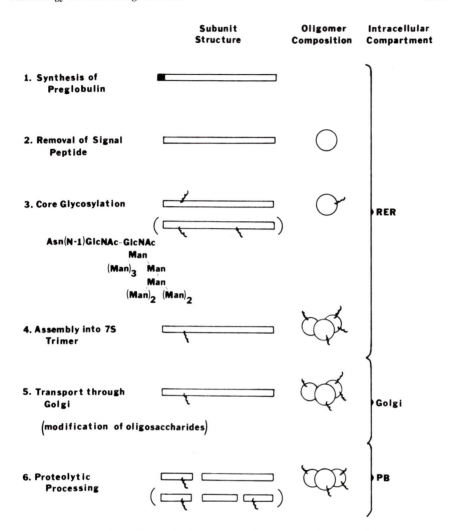

Fig. 3. Pathway for synthesis and processing of 7S seed storage globulins.

peptides with NH_2-terminal signal sequences. The signal peptide directs the translocation of the nascent polypeptide into the lumen of the RER and is cotranslationally removed. Some precursor polypeptides are cotranslationally glycosylated by addition of a mannose-containing core oligosaccharide. Soon after translation is complete, the 11S and 7S precursors appear to assemble into trimers within the ER and are then transported to the central vacuole via the Golgi apparatus, where oligosaccharide chains of the glycoproteins may be modified by addition or removal of sugar residues. Transport from the Golgi to the central vacuole is thought to involve small

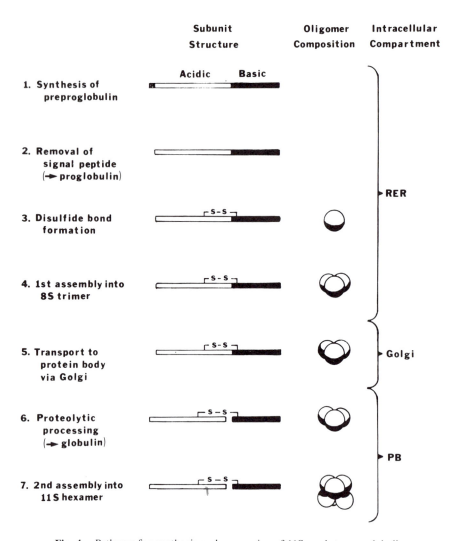

Fig. 4. Pathway for synthesis and processing of 11S seed storage globulins.

membrane vesicles. After deposition in the vacuole, the 11S precursors are cleaved into acidic and basic polypeptides, which remain linked by a disulfide bond. The 7S precursors are also subject to proteolytic processing, but the sites of cleavage are species-specific. After deposition in the vacuole and proteolytic processing, the 11S-type trimers assemble into hexamers. With further accumulation of storage protein the vacuole subdivides to form protein bodies, small spherical structures composed of a protein matrix surrounded by a single membrane.

Electron microscopy studies clearly implicated the RER in the synthesis of globulin storage protein (Harris, 1979), but evidence that transport to the RER involved signal peptides came from *in vitro* translations of mRNAs in cell-free systems, where the products obtained were M_r 1000 to 3000 larger than the mature storage proteins (Barton *et al.*, 1982; Tumer *et al.*, 1982; Brinegar and Peterson, 1982; Walburg and Larkins, 1983; Adeli and Altosaar, 1983; Yamagata *et al.*, 1982). Subsequent cDNA and gene sequencing revealed the coding sequence for a signal peptide preceding the codon for the first amino acid of the mature protein (Lycett *et al.*, 1983, 1984a; Bassüner *et al.*, 1984). In most instances, the predicted signal sequence is 20 to 30 amino acids long and consists of a conserved core of hydrophobic amino acid residues followed by a region of variable composition where cleavage occurs (Nielsen, 1985a; Doyle *et al.*, 1986).

In addition to the removal of the signal peptide, some globulin subunits undergo cotranslational glycosylation. Studies of proteins from a number of species have shown that glycosylation is primarily restricted to subunits of the 7S globulins (Hurkman and Beevers, 1982) and that it involves the transfer of an N-acetylglucosamine-mannose core oligosaccharide (GlnNAcGln-NAc(Man)$_9$GlcGlc) to an asparagine residue (Davies and Delmer, 1981). Glycosylation of phaseolin was shown to be a cotranslational event by Chrispeels and co-workers, who isolated glycosylated polypeptides after run-off synthesis with polysomes isolated from RER (Bollini *et al.*, 1983). These investigators also found that some phaseolin polypeptides are glycosylated at a second site. The amino acid sequences deduced from cDNA clones of 7S globulins generally contain the N-glycosylation signal "Asn-X-Ser/Thr" one to three times in different regions of the polypeptide (Doyle *et al.*, 1986). The transfer of the core oligosaccharide to nascent 7S globulins is prevented by tunicamycin, but, interestingly, the inhibition of glycosylation does not affect the synthesis, assembly, or transport of the protein (Badenoch-Jones *et al.*, 1981). For this reason, and because 11S globulins and canavalin, the 7S globulin of jack bean (Smith *et al.*, 1982), are not glycosylated, the physiological significance of the glycosylation of globulin storage proteins is unclear. Doyle *et al.* (1986) suggested that glycosylation near the COOH-terminus of 7S globulins may be involved in proteolytic processing or storage of these proteins.

The core oligosaccharide of 7S globulins may be further modified posttranslationally during the intracellular transport of the protein either by removal or addition of sugar residues. In a study of the high-mannose core oligosaccharide of phaseolin, Herman *et al.* (1986), found that a minority of the glycosylated molecules underwent subsequent removal of four mannose residues while the protein was still in the RER. In the Golgi vesicles two additional mannose residues were removed and two terminal GlcNAcs added. Finally, in the protein bodies these GlcNAc residues were cleaved and a single terminal xylose was attached. For unknown reasons, in the majority of phaseolin molecules little processing occurred, and the molecules were deposited in protein bodies with oligosaccharide side chains containing seven or nine mannose residues. These differences in the glycosylation pattern of phaseolin subunits partially account for the variation in molecular weights estimated by polyacrylamide gel electrophoresis (PAGE).

The assembly of subunit polypeptides into oligomers has been studied in a number of plants. In pea it was shown that the formation of 7S trimers of vicilin occurs in the RER (Chrispeels et al., 1982). The assembly of pea legumin, on the other hand, was found to be a two-step process, with some 8S trimers first formed in the RER in an assembly analogous to that of vicilin. Only after transport to the protein bodies do the 11S hexamers completely assemble. A similar two-step assembly mechanism has been proposed for soybean glycinin (Barton et al., 1982) and pumpkin cucurbitin (Hara-Nishimura et al., 1985). This pattern of assembly correlates with the sequence of dissociation of 11S globulins ($11S \rightarrow 2 \times 7S \rightarrow 6 \times 3S$) reported by Derbyshire et al. (1976) and also agrees with the EM studies summarized earlier showing the 11S hexamer to be composed of two stacked subunit trimers (Reichelt et al., 1980).

Proteolytic processing of globulin precursors also occurs in the protein body (Chrispeels et al., 1982). Pulse-chase labeling experiments showed that cleavage of legumin precursors to generate disulfide-bonded acidic and basic polypeptides occurs within 2 hr after synthesis (Spencer and Higgins, 1980). In some 11S globulins, including pea legumin (Lycett et al., 1984b), there is a single proteolytic cleavage giving rise to the acidic and basic polypeptide. In others, however, there appear to be two cleavage points so that processing yields an acidic polypeptide, a short linker peptide, and a basic polypeptide. For example, two of the five types of soybean glycinin are thought not to have linkers, and three are thought to have linker peptides of 3, 4, and 17 amino acid residues (Nielsen, 1985a). The sites of proteolytic processing of glycinin precursors are believed to be specified by paired basic residues in the amino acid sequence. In one glycinin precursor there is an additional processing site in the acidic polypeptide. Cleavage at this site gives rise to acidic polypeptides A_5 and A_4 and gives a final subunit composition of $A_5A_4B_3$ (Nielsen, 1984). A short peptide was also found to be removed from the COOH-terminus of pea legumin (Lycett et al., 1983).

The 7S globulin polypeptides of some legumes are not proteolytically processed. This has been shown for β-conglycinin in soybean (Meinke et al., 1981) and phaseolin in French bean (Brown et al., 1981). In other species, the 7S globulins comprise a complex mixture of processed and unprocessed subunits. This has been found for jack bean (Smith et al., 1982) and lupin (Blagrove and Gillespie, 1975) and has best been characterized for pea vicilin (Spencer et al., 1983). Fractionation of pea vicilin by PAGE yields components of M_r 75,000, 70,000, 50,000, 34,000, 30,000, 25,000, 18,000, 14,000, 13,000, and 12,000 (Spencer et al., 1983). In addition, the M_r 50,000 band contains several polypeptides that give a complex pattern on isoelectric focusing (IEF) gels. On the basis of antigenic relationships and amino acid sequence comparisons, Spencer and co-workers deduced that vicilin subunit polypeptides smaller than M_r 50,000 result from proteolytic cleavage of the M_r 50,000 molecules. Cleavage was found to occur at either or both of two processing sites (Spencer et al., 1983).

The role of the Golgi apparatus in globulin transport was established by two experimental approaches (Chrispeels, 1985). In the first, the storage proteins were radioactively labeled and their appearance monitored in different subcellular fractions over a period of time. In this way, newly synthesized phytohemagglutinin labeled with [^{3}H]fucose was followed from Golgi cisternal stacks to Golgi vesicles and ultimately to the protein bodies (Chrispeels, 1983). Similarly, pumpkin cucurbitin labeled with [^{35}S]-labeled methionine was found unprocessed in the Golgi fraction after 30 min and in a mature form in protein bodies after 1 to 2 hr (Hara-Nishimura et al., 1985). The second experimental approach involved the immunologic tagging of the storage proteins and localization by EM in thin sections of developing cotyledons. By this technique Craig and colleagues observed that immunogold-labeled vicilin was specifically associated with RER, Golgi vesicles, and protein bodies in pea cotyledons (Craig and Goodchild, 1984; Craig and Miller, 1984) (Figs. 5 and 6). Using similar methods, others observed the same distribution of vicilin and legumin in broad bean (Zur Nieden et al., 1984) and phaseolin and phytohemagglutinin in French bean (Chrispeels and Greenwood, 1985). Additional evidence that the intracellular transport of seed storage protein is mediated by the Golgi apparatus comes from studies on the effects of ionophores on protein movements in pea cotyledon cells. Treatment of developing cotyledons with monensin and nigericin, which both disrupt movement of Golgi vesicles, redirected transport of newly synthesized vicilin from the central vacuole to the plasmalemma, where the protein was released from the cells (Craig and Goodchild, 1984).

Craig and colleagues studied the deposition of storage globulins of pea using light and electron microscopy and showed that legumin and vicilin first appear as small clumped deposits at the periphery of the large central vacuoles about 8 days after flowering (DAF) (Craig et al., 1980). Serial sections of storage parenchyma cells at later times in development revealed that the central vacuole, which initially has numerous complex protrusions, becomes further convoluted, and by 20 DAF is fragmented into many small, discrete, spherical vacuoles, now called protein bodies (Craig et al., 1980). During the period 8 to 20 DAF, the total vacuole surface area per cell increases by a factor of 100, and the amount of protein in the vacuole increases to the point that the vacuoles stain uniformly. Double-labeling of storage proteins showed that the protein bodies of pea contain both legumin and vicilin in the same protein deposits (Craig and Millerd, 1981).

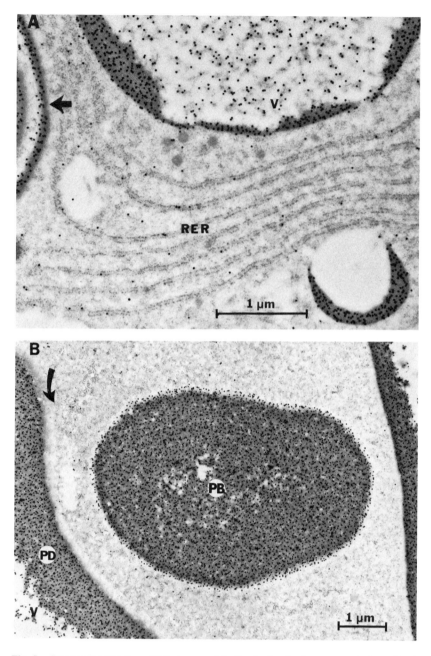

Fig. 5. Immunolocalization of 7S storage globulins in developing pea cotyledons. Sections of tissue were first treated with anti-rabbit IgG directed against the M_r 50,000 subunit complex of vicilin, followed by treatment with goat anti-rabbit IgG adsorbed to colloidal gold. Subcellular structures include rough endoplasmic reticulum (RER), vacuole (V), peripheral deposit (PD), and protein body (PB). (A) Vicilin is labeled in RER, vacuoles, and electron-dense cisterna (arrow); (B) few gold particles are found over cytoplasm (arrow). From Craig and Miller (1984) with permission of the author and publisher.

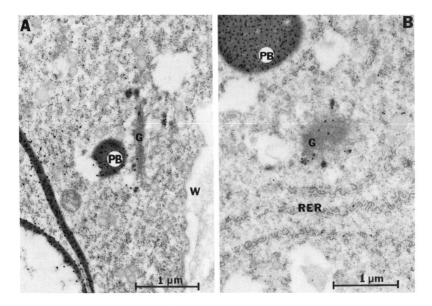

Fig. 6. Immunolocalization of 7S storage globulin in developing pea cotyledons. Tissue sections prepared and treated as described in Fig. 5. In sections normal to the Golgi (G) cisternal stack, antibody appears to bind preferentially to one face, probably the *trans*, and to peripheral vesicles (A). When the organelles are sectioned parallel to their membrane stack (B), gold is bound over the cisternal profile. Protein body (PB); cell wall (W). From Craig and Miller (1984) with permission of the author and publisher.

III. ORGANIZATION AND STRUCTURE OF STORAGE GLOBULIN GENES

Cloned cDNAs corresponding to mRNAs encoding 7S and 11S storage globulins have allowed the analysis of the primary amino acid sequence of the storage proteins, the number of genes encoding the proteins, the expression of these genes during seed development, and the isolation and determination of the structure of the genes themselves. Globulin seed storage proteins are encoded by gene families varying from a few to as many as 20 or more members. For example, group I glycinins of soybean are encoded by 3 or 4 genes (Fischer and Goldberg, 1982), and group II glycinins are encoded by 2 genes (Scallon, 1986). The conglycinin gene family of soybean consists of 15 to 20 members (Ladin *et al.*, 1984). In pea it was estimated that there are 8 genes for legumin, 11 for vicilin, and 1 for convicilin (Domoney and Casey, 1985). In French bean, 7 genes appear to encode phaseolin (Talbot *et al.*, 1984). The oat 12S globulin gene family appears to contain 6 to 8 members (Shotwell *et al.*, 1988).

The organization of storage globulin genes varies among species. By restriction fragment length polymorphisms, Domoney *et al.* (1986) demonstrated that there are three subfamilies of legumin genes in pea. In two, on chromosome 7 near the *r* locus and chromosome 1 near the *a* locus, the genes are tightly clustered, whereas in the third, also on chromosome 1, they are less closely spaced. Pea 7S globulin genes map to six different loci. The structural genes (or gene) for convicilin are restricted to a single locus on chromosome 2 (near the *k* locus) (Matta and Gatehouse, 1982), whereas the genes for vicilin are at five loci with each containing several copies (Ellis *et al.*, 1986). Three of the vicilin loci map to chromosome 7, and the legumin locus appears to map between two of these.

Genes encoding soybean globulins are more scattered throughout the genome. Of the five glycinin genes, two, Gy_1 and Gy_2, are tightly linked (Fischer *et al.*, 1988), whereas Gy_3 and Gy_5 segregate independently (N. C. Nielsen, personal communication). The regions around Gy_1–Gy_2 and Gy_3 are similar and contain genes expressed in the leaf in analogous positions (Fischer *et al.*, 1988). By heteroduplex formation it was found that the Gy_1–Gy_2 region differs from Gy_3 by a simple deletion or duplication of 4.3 kb of DNA. It appears that group I glycinins arose from an ancestral gene that gave rise to the Gy_1–Gy_2 cluster by duplication. Duplication of the genome (tetraploidization) and deletion of DNA could account for the two glycinin domains, one containing two genes (Gy_1–Gy_2) and the other one gene (Gy_3). The 15 or more genes encoding β-conglycinin are in three linkage groups in different regions of the soybean genome (Harada *et al.*, 1988). Some genes contain a 556-bp insertion that accounts for the two size classes of β-conglycinin mRNA and protein.

In some cases the number of genes estimated to encode a particular class of storage globulins may seem too small to account for the large number of polypeptides resolved by SDS-PAGE. This discrepancy may be due in part to the proteolytic processing of subunits that gives rise to several polypeptides, as described earlier for pea vicilin. Moreover, charge variants of a single-sized polypeptide are often observed after IEF analysis, as seen with the β-conglycinin subunit polypeptides in soybean (Ladin *et al.*, 1984). It is often unclear whether the charge variants of these proteins are encoded by different genes or whether they are the result of posttranslational events. After introducing a gene encoding a β-conglycinin subunit into petunia, Beachy and collaborators found a number of charged isomers of the soybean protein in seeds of regenerated plants (Beachy *et al.*, 1985). This indicates that a protein encoded by a single storage protein gene can give rise to multiple charged isomers of the protein.

Genes for 11S globulins of several species have been found to have similar sequence organization. The genes in pea and soybean have four exons separated by three introns (Fig. 7). The three introns are in similar positions and interrupt the coding sequence of the acidic polypeptide twice and the basic polypeptide once (Nielsen, 1985a; Lycett *et al.*, 1984a). A similar arrange-

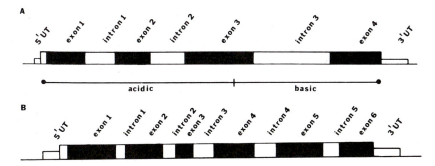

Fig. 7. Generalized structure of genes encoding 11S (A) and 7S (B) seed storage globulins. The positions of protein-coding regions (exons) are dark, and noncoding sequences (introns) are white. Regions corresponding to the acidic and basic polypeptides of the 11S subunit are indicated (A).

ment has been found in the 12S globulin genes of oats (M. A. Shotwell and B. A. Larkins, unpublished results). Variants of this structure have been found in pea (R. Casey, personal communication), broad bean (Bäumlein *et al.*, 1986), and sunflower (T. Thomas, personal communication). One of the genes in both pea and broad bean lacks the first intron, and the third intron is absent in a gene from sunflower. The introns in these genes vary in length and range up to 600 bp in soybean to less than 100 bp in pea.

The two genes for 7S globulin storage proteins characterized thus far have similar structures. Genes for a β-subunit of phaseolin and an α'-subunit of β-conglycinin consist of six exons separated by five small (100 to 200 bp) introns (Doyle *et al.*, 1986) (Fig. 7). Doyle and co-workers compared the coding sequences of these two genes with cDNA sequences corresponding to a number of other 7S storage proteins and observed that the degree of sequence similarity varies considerably in different regions. Within the exons, regions of high homology are separated by completely diverged regions. On the basis of these patterns, they suggested that evolutionary conservation operates on units smaller than exons. Likewise, the computer-generated pattern of predicted hydropathy transcended the exon boundaries, adding to evidence that exons do not coincide with structural or evolutionary domains. Doyle *et al.* (1986) concluded that legume storage proteins have not arisen by splicing together of exons coding for structural domains, as has been proposed for other proteins.

IV. REGULATION OF GLOBULIN GENE EXPRESSION

In attempts to delimit nucleotide sequences responsible for regulating the expression of storage globulin genes, Lycett and co-workers (1985) analyzed the 5' flanking regions of three pea legumin genes. In one of these, an 82-bp sequence was repeated twice in the 5' upstream region, and each repeat contained a pair of 21-bp inverted repeats. These authors proposed that this region of the gene could form either of two mutually exclusive stem-loop structures, one of which with a potentially stable 42-bp stem. They further suggested that switching between these alternative structures may play a role in the control of the expression of this legumin gene. In a slightly different approach, Bäumlein and colleagues (1986) compared the sequences of 11S genes of broad bean, pea, and soybean. They found a region in the 5' flanking DNA of these genes that was more highly conserved than most of the coding region. The region of highest conservation was in a 28-bp sequence about 80 bp upstream of the mRNA cap site in which 25 of the 28 nucleotides were invariant. This consensus sequence, tentatively termed the "legumin box," could not be located in any genes encoding 7S proteins, leading these authors to suggest that it serves a specific regulatory function.

Transformation with the Ti plasmid of *Agrobacterium tumefaciens* has also been used to study the expression of cloned seed storage protein genes. Because efficient transformation and regeneration have mainly been achieved with members of the Solanaceae (Horsch *et al.*, 1985), these analyses have primarily involved tobacco (*Nicotiana tabacum*) and petunia (*Petunia hybrida*). Sengupta-Gopalan *et al.* (1985) introduced a gene encoding the β-subunit of the 7S protein of French bean into tobacco and immunologically detected phaseolin in seeds of regenerated plants, where it accounted for about 1% of the total seed protein. By immunocytochemical techniques, phaseolin was found in protein bodies in six of seven tobacco embryos but in only one of seven endosperms (Greenwood and Chrispeels, 1985). The phaseolin was glycosylated, but it was proteolytically cleaved differently than in bean seeds. Very little of the protein was found in transformed calli or seedlings, indicating that expression of the gene was both tissue-specific and developmentally regulated. A single copy of the phaseolin gene was present in the transformed plants, and it was inherited as a simple Mendelian dominant trait.

In similar studies, Beachy *et al.* (1985) introduced a soybean gene for the α'-subunit of β-conglycinin into petunia. Transcripts of the soybean gene were detected in immature embryos but not in leaves of transgenic plants. The soybean α'-subunit accumulated in parallel with the petunia seed storage protein and was estimated to be between 0.1 and 1.0% of the total seed protein. The majority of the α'-subunit polypeptides assembled into multimeric proteins with sedimentation coefficients of 7–8S, with smaller amounts existing in 9–11S complexes.

Extending these studies, Chen *et al.* (1986) constructed a series of deletion mutants containing the soybean α'-subunit coding region with different lengths of 5' noncoding DNA. When flanked by 159 nucleotides of DNA upstream from the mRNA cap site, the gene was expressed only at low levels in seeds of transgenic petunias. With 208 nucleotides of flanking DNA, gene expression was at least 16 times as high as with 159 nucleotides. The 49-nucleotide difference between these two constructs contains three repeats of the sequence $A_A^G CCCA$. In addition, the gene with 159 bp of 5' flanking sequence contains a single repeat (AACCCA). Interestingly, the level of expression of the α'-subunit gene was not significantly increased by an additional 8 kb of 5' flanking sequence, which includes a sequence identical to the core enhancer of SV40 560 bp upstream of the transcribed region.

In addition to the *cis*-acting factors that regulate the developmental expression of storage protein genes, environmental conditions, such as mineral nutrition, have been shown to affect storage protein gene expression. In peas grown with a suboptimal sulfate supply, the accumulation of legumin is greatly reduced (Randall et al., 1979). Likewise, sulfur deficiency suppresses the synthesis of conglutin α, the major 11S globulin of lupin (Gillespie et al., 1978). In pea the reduced protein accumulation results from a 90% reduction in legumin mRNA rather than from a decrease in legumin mRNA translatability or the degradation of legumin polypeptides. In addition, sulfur deficiency leads to an increase in both the relative level and the duration of synthesis of M_r 50,000 vicilin polypeptides. The increased accumulation of vicilin correlates with elevated vicilin mRNA in these seeds (Chandler et al., 1984), leading to the conclusion that control of storage protein synthesis by sulfur availability results from modulation of mRNA levels by transcription, posttranscriptional processing, or turnover.

Legumin polypeptides have higher amounts of sulfur-containing amino acids (1.0% cysteine and 0.7% methionine; Casey and Short, 1981) than vicilin polypeptides, which have very few or no cysteine and methionine residues. Thus, a lack of sulfur shifts storage protein synthesis away from relatively sulfur-rich proteins to sulfur-poor proteins. The effect is reversible, as restoring optimal sulfur levels causes a shift back to a normal protein profile (Chandler et al., 1984). In addition to sulfate, compounds such as methionine, cysteine, glutathione, and mercaptoethanol also cause increased legumin synthesis in sulfur-deficient seeds, leading to suggestions that a reduced sulfur compound mediates this response. The mechanism by which sulfur availability modules mRNA levels is unknown, but apparently the effect is not regulated by cysteinyl- and methionyl-tRNA levels, which are unchanged under sulfur-deficient conditions (Macnicol, 1983).

In soybeans the accumulation of the three major 7S subunit polypeptides, α, α', and β, has also been found to be influenced by sulfur metabolism. In cotyledon cultures supplemented with methionine, the β-subunit is not detected (Holowach et al., 1984). The β-subunit contains no methionine (Thanh and Shibasaki, 1977) and is the only polypeptide whose accumulation is suppressed by exogenous methionine. Suppression of protein accumulation is completely reversible, so the effect of methionine cannot be to accelerate the degradation of the β-subunit. By *in vitro* translation assays, functional β-subunit mRNA was shown to be absent in tissues that do not accumulate the protein (Holowach *et al.*, 1986), and nontranslatable mRNA could not be detected by Northern blot hybridization. The inhibition of β-subunit accumulation by methionine must therefore act transcriptionally, during mRNA synthesis, processing, or turnover.

V. PROLAMINE STORAGE PROTEINS

The major storage proteins in most cereals are alcohol-soluble prolamines (Osborne, 1908; see Chapter 11, Volume 6 of this series). In genera such as *Triticum* (wheat), *Hordeum* (barley), *Secale* (rye), *Zea* (maize), *Sorghum* (sorghum or milo), and *Pennisetum* (millet), the prolamines account for 50 to 60% of the total endosperm protein. Rice and oats are unusual because their prolamine fractions account for only 5 to 10% of the endosperm protein, and most of the storage protein is similar to the 11S globulins previously described.

Cereal prolamines are characterized by a high content of proline and glutamine (hence the name prolamine) and a low level of charged amino acids, especially lysine. The low lysine content is primarily responsible for the poor nutritional quality of high prolamine-containing cereals. The prolamines of most cereals are complex mixtures of polypeptides that occur in protein bodies. Some of these proteins can be cross-linked by disulfide bonds, thus forming high-molecular-weight complexes. Differences in the solubility and the potential of some to cross-link and associate with other proteins have made it difficult to isolate and characterize prolamine proteins by traditional protein chemistry. The analysis of recombinant DNA clones corresponding to prolamine mRNAs and genes, however, has allowed the deduction of the primary amino acid sequences of many prolamines.

Cereals are members of the family Gramineae and are divided into two subfamilies: the Festucoideae, which contains the genera *Triticum, Secale, Hordeum, Avena,* and *Oryza,* and the Panicoideae, which contains *Zea* and the related genus *Tripsacum,* as well as *Sorghum* and *Pennisetum.* From NH_2-terminal sequence analysis of mixtures of prolamines, Bietz (1982) showed that while prolamines from a given species are complex, there is homology among them. Furthermore, there is conservation of prolamine NH_2-terminal protein sequences among members of the various cereal tribes. For example, prolamine sequences of the Triticeae (wheat, rye, and barley) are more closely related to one another than they are to the Oryzeae (rice) and the Aveneae (oats). The prolamines of the Andropogoneae (maize, *Tripsacum,* and sorghum) are more closely related to one another than to those of the Paniceae (millet). The sequence relationships among this heterogeneous group of proteins have prompted several reviewers to suggest that prolamines originated through duplication and subsequent mutation of an ancestral prolamine gene (Kasarda *et al.*, 1976; Bietz *et al.*, 1977), and this hypothesis has been largely supported by recent molecular analyses of storage protein genes.

In this review we consider the prolamine proteins in wheat, barley, and maize. The proteins in these species have been characterized to the greatest extent in recent years and are illustrative of the prolamines of the Festucoideae and Panicoideae. Additional information on the biochemistry and molecular biology of cereal prolamines may be obtained from several recent reviews (Bright and Shewry, 1983; Wilson, 1983; Porceddu *et al.*, 1983; Kreis *et al.*, 1985b).

A. Wheat and Barley Prolamines

Wheat storage proteins are traditionally divided into gliadins and glutenins, which together form the gluten complex responsible for the breadmaking properties of wheat. Glutenin is believed to impart elasticity to dough, while gliadin makes it viscous and provides extensibility. Typically, wheat storage proteins consist of about 50% gliadin, 10% high-molecular-weight (HMW) glutenin, and 40% low-molecular-weight (LMW) glutenin subunits (Payne *et al.*, 1984).

Traditionally, gliadins are distinguished as being soluble in 70% ethanol at room temperature, while glutenins are insoluble. Glutenins will dissolve in dissociating media such as dilute acid, urea, or ionic detergents (Bietz and Wall, 1975; Huebner, 1977). Payne *et al.* (1984) classified the two prolamine groups according to their state of aggregation in dissociating media rather than solely on their solubility: the gliadins are a mixture of single polypeptides, while the glutenins are aggregated, primarily through disulfide bonds. Treatment of glutenin with a reducing agent, such as 2-mercaptoethanol, causes dissociation into two groups of polypeptides classified as low-molecular-mass (LMM) and high-molecular-mass (HMM) subunits. With this system of separation about 10% of the gliadin, variously referred to as HMM gliadin (Beckwith *et al.*, 1966), aggregated gliadin (Shewry *et al.*, 1984), LMM glutenin (Nielson *et al.*, 1968), and gluteinin III (Graveland *et al.*, 1982), is transferred to the glutenin fraction.

Two-dimensional PAGE techniques separate gliadins and glutenins into subunits (Wrigley and Shepherd, 1973; Jackson *et al.*, 1983; Payne *et al.*, 1985). Most of the gliadin fraction consists of polypeptides of M_r 32,000 to 44,000 that are distinguished as α-, β-, and γ-gliadins, based on decreasing mobility at acidic pI (Fig. 8). The ω-gliadins have the lowest mobility at acidic pI and exhibit M_r of 50,000 to 70,000. The size and charge heterogeneity of gliadin proteins is variable among different wheat varieties, and this provides one method to differentiate between them (Payne *et al.*, 1982). Following reduction of disulfide bonds, the HMM glutenins can be resolved

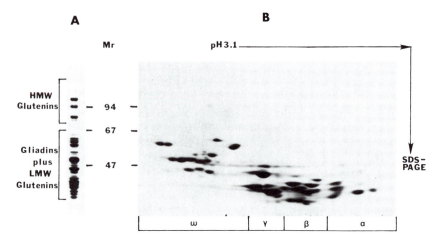

Fig. 8. Analysis of wheat storage proteins by SDS-PAGE. (A) Separation of high-molecular-weight (HMW) glutenins, low-molecular-weight (LMW) glutenins, and gliadins by molecular weight. (B) Two-dimensional separation of gliadin proteins by charge in aluminum lactate buffer (pH 3.1) (left to right) and by molecular weight (top to bottom). The positions of α-, β-, γ-, and ω-gliadins are indicated.

into polypeptides with M_r ranging from 80,000 to 150,000. Under similar conditions, the LMM glutenins migrate with M_r similar to the γ-gliadins. These proteins are optimally separated from the γ-gliadins by a combination of equilibrium and nonequilibrium IEF (Jackson et al., 1983; Payne et al., 1984). The LMM glutenins can then be resolved into two sets of polypeptides with either acidic or basic pIs.

Miflin and co-workers (Shewry et al., 1984; Kreis et al., 1985b) proposed an alternative method for the classification of wheat prolamines based not on aggregation, but rather on M_r and sulfur amino acid content. These investigators distinguished three groups of proteins called sulfur-rich (S-rich), sulfur-poor (S-poor), and high-molecular-weight (HMW). The S-rich wheat prolamines, which contain 2.5 to 3.5% cysteine plus methionine, consist of the α-, β-, and γ-gliadins as well as the "aggregated gliadins" (LMM glutenins of Payne et al., 1984). The S-poor wheat prolamines, containing 0 to 0.2% cysteine plus methionine, are composed of the ω-gliadins. In addition to having higher M_r than the S-rich prolmaines, the S-poor proteins contain higher percentages of glutamine, proline, and phenylalanine. These three amino acids are present in a ratio of 4:3:1 and account for about 80% of the total (Kreis et al., 1985a). The HMW prolamines (HMM glutenins of Payne et al., 1984) have a sulfur amino acid content that is intermediate between the S-rich and S-poor fractions (0.5 to 1.9%) and are cross-linked by disulfide bonds.

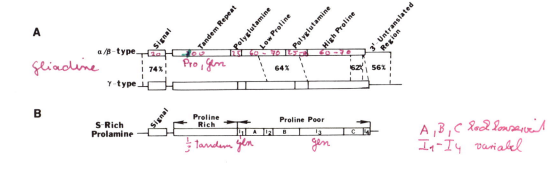

Fig. 9. Structural organization of wheat storage proteins. (A) General structure of wheat gliadin proteins. From Okita *et al.* (1985) with permission of the author and publisher. (B) General structure of wheat and barley sulfur S-rich prolamines. From Kreis *et al.* (1985b) with permission of the author and publisher. (C) Generalized structure of high-molecular-weight (HMW) gluteinin proteins.

At least partial amino acid sequence information is available for many of the gliadin and glutenin proteins (Kreis *et al.*, 1985a), and the complete amino acid sequence is known for examples of the α-, β-, and γ-gliadins (Kasarda *et al.*, 1984; Rafalski *et al.*, 1984; Okita *et al.*, 1985) and several HMW glutenins (Forde *et al.*, 1985a; Thompson *et al.*, 1985; Sugiyama *et al.*, 1985). The α-, β-, and γ-gliadins are structurally related (Fig. 9A) and can be divided into several distinct domains (Kasarda *et al.*, 1984) or modules (Reeck and Hedgcoth, 1985). The gliadins that have been characterized to date are around 300 amino acids long. The mature protein is preceded by a signal peptide of approximately 20 amino acids that directs the synthesis of the protein into the lumen of the RER (Blobel, 1977). Immediately following the signal sequence is a region of 90 to 100 amino acids that contains tandemly repeated peptides composed mainly of proline and glutamine. Because these repeats are tandem and slightly degenerate, different investigators have deduced somewhat different consensus repeats. Kasarda *et al.* (1984) and Sumner-Smith *et al.* (1985) recognized nine copies of a 12-amino acid repeat (Pro-Gln-Pro-Gln-Pro-Phe-Pro-Pro-Gln-Gln-Pro-Tyr) in an α/β-type gliadin, although these authors differed in the assignment of the first amino acid in the repeat. A shorter 7- or 8-amino acid repeat (Pro-Gln-Gln-Pro-Phe-Pro-Leu/Gln-Gln) was determined in this region of the γ-gliadin (Okita *et al.*, 1985; Sheets *et al.*, 1985). The remainder of the protein is divided into four distinct regions. The first and third are segments of 25 to 30 residues that are mostly polyglutamine. The second and fourth are regions of 60 to 70 amino acids that are designated "unique" (Kasarda et al., 1984). These regions are deficient in three of the four amino acids that constitute the repetitive peptide region (Phe, Gln, and Pro) and contain the majority of the Thr, Ala, Ser, Cys, Val, Arg, and Ile (Okita, 1984).

Kreis et al. (1985b) distinguished a somewhat different organization of the primary amino acid sequence of the S-rich wheat prolamines (Fig. 9B). Following the signal peptide the protein was divided into a proline-rich and a proline-poor domain. The proline-rich region, which constitutes the first one-third of the molecule, corresponds to the region of tandemly repeated peptides. The proline-poor domain consists of seven regions that are designated I_1, A, I_2, B, I_3, C, and I_4. Regions A, B, and C are strongly conserved among all the S-rich wheat prolamines, while regions I_1 to I_4 are more variable (Kreis et al., 1985b). The I_1 region corresponds to the first polyglutamine-rich tract (Fig. 9A), and I_3 corresponds to most of the second polyglutamine tract.

The complete primary amino acid sequences of several HMW glutenin subunits have also been determined (Forde et al., 1985a; Thompson et al., 1985; Sugiyama et al., 1985). The lengths of the mature proteins vary from 504 to 817 amino acids, but each appears to be organized into three domains: an NH_2-terminal region of around 100 residues, a middle repetitive region of 400 to 670 residues, and a COOH-terminal region of around 40 residues (Fig. 9C). Two consensus peptides appear in the repetitive domain of these molecules. One is a hexapeptide (Pro-Gly-Gln-Gly-Gln-Gln) and one is a nonapeptide (Gly-Tyr-Tyr-Pro-Tyr-Ser-Leu-Gln-Gln). The repetitive peptides are interspersed such that copies of the nonapeptide are located between regions containing several tandemly arranged copies of the hexapeptide repeat. The nonrepetitive regions have some homology to the nonrepetitive regions of the S-rich gliadin genes (Kreis et al., 1985b). These proteins contain relatively low amounts of cysteine, four to six residues, which are clustered in the NH_2-terminal and COOH-terminal regions of the protein. Interestingly, several lysine residues are present in the NH_2-terminal and COOH-terminal regions.

Most of the information pertaining to the secondary and tertiary structures of wheat prolamines comes from circular dichroism (CD) measurements or from computer-generated models based on the conformation of model proteins whose structures are known from crystallographic analyses. However, there are limitations to interpretations of structural predictions with either approach. To analyze the CD spectrum the proteins must be dissolved in alcohol solutions or dissociating media, and whether or not the polypeptides assume the same conformation in these solutions as they do in the cell is unknown. Computer-generated models are inexact because most of the model proteins upon which the structural predictions are based are hydrophilic rather than hydrophobic. Presumably, predictions based on a combination of these procedures have a measure of validity.

Based on CD analysis and Chou and Fasman (1978) predictions, the α-gliadins contain predominantly alpha helix (35%) and beta turn (35%) with the remainder being beta sheet and random coil (Tatham et al., 1985). These data are consistent with the CD analysis of Kasarda et al. (1968) and the optical rotary dispersion analysis of Cluskey and Wu (1971), although there is poor agreement with the estimates of α-helix and β-sheet determined by the predictive method of Garnier et al. (1978). Most of the α-helical regions are presumed to be associated with polyglutamine stretches, while the β turns are in the proline-rich regions: the repetitive domain (see Fig. 9B) and the COOH-terminal domain. The CD spectra of the β- and γ-gliadins are similar to the α-gliadins, suggesting that these proteins have similar conformations.

The CD analysis of ω-gliadins reveals little α-helix or β-sheet structure, with most of the molecule composed of β turn structure (Tatham et al., 1985). A preponderance of β-turn structure is also deduced from CD analysis of the HMW subunits of glutenin, and it is thought to result from the many repeated peptides in the central region of the protein (Tatham et al., 1985). The NH_2-terminal and COOH-terminal regions, which contain most of the cysteine residues, are thought to be α-helical. It is proposed that cross-linking of cysteine residues in these regions causes the HMW glutenins to assemble into linear polymers (Tatham et al., 1984; 1985), which are responsible for the elastic nature of bread dough (Ewart, 1977). The presence of multiple cysteines in the NH_2-terminal region may allow branching and cross-linking with the LMW subunits of glutenin.

Although complete amino acid sequences of the ω-gliadins are unknown, partial amino acid sequencing and comparison with similar proteins in barley suggest that these proteins contain a significant proportion of repeated peptides similar to the HMW glutenins (Kreis et al., 1985b).

The prolamines of barley, hordeins, are structurally related to the wheat prolamines. Hordeins are classified into three main groups of polypeptides designated B, C, and D, based on separation of SDS-PAGE (Fig. 10). The A group is heterogeneous and does not show characteristics of prolamine proteins (Salcedo et al., 1980). The group B hordeins, which have M_r of 30,000 to 50,000 and account for 80 to 90% of the prolamine fraction, are structurally similar to the S-rich wheat prolamines. Like the gliadins, the region of the protein following the signal peptide is proline-rich and contains a series of tandemly repeated peptides with the sequence Pro-Gln-Gln-Pro-Pro making up the core of the repeat (Forde et al., 1985b,c). Several variations of this repeat can be recognized, all consisting of the tetrapeptide Pro-Gln-Gln-Pro with the addition of Phe, Ile, Val, Pro, Phe-Pro, or Gln-Pro-Tyr. This region encompasses approximately the first one-fourth of the polypeptide. The remainder of the protein is relatively deficient in proline, has no repeated peptides, and contains most of the cysteine residues. Within the COOH-

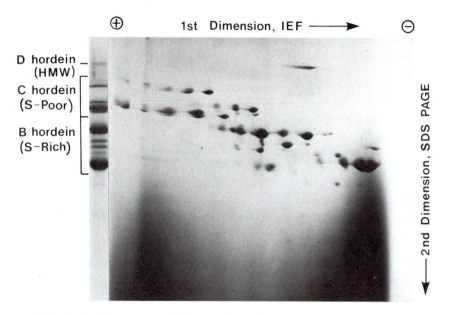

Fig. 10. Analysis of barley prolamine proteins. Reduced and pyridylethylated total hordein from the variety Sundance was separated by SDS-PAGE (left) or by two-dimensional separation with IEF between pH 5 and 9 and SDS-PAGE (right). The positions of the S-rich B, S-poor C, and high-molecular-weight (HMW) D hordeins are indicated. From Shewry *et al.* (1984) with permission of the author and publisher.

terminal portion of the protein are three regions, designated A, B, and C, that are highly conserved between the B hordeins and the α/β- and γ-gliadins (see Fig. 9). The extent of similarity between these regions clearly indicates a common origin for the proteins (Fig. 11).

Only partial amino acid sequences are available for the C hordeins (Forde *et al.*, 1985b), but these proteins appear to be structurally related to the S-poor wheat prolamines, the ω-gliadins. The C hordeins, which have M_r of 60,000 to 80,000 and compose 10 to 20% of the prolamine, contain a repeated peptide (Pro-Gln-Gln-Pro-Phe-Pro-Gln-Gln) that is thought to account for most of the protein sequence. The first four amino acids of this repeat are identical with the repeated peptides in the NH_2-terminal portion of the B hordein, and it may be that they have a common origin (Kreis *et al.*, 1985b).

The D hordein accounts for about 5% of the barley prolamine fraction and consists of a few polypeptides with apparent M_r of 105,000. The amino acid sequences of the D hordeins have not been determined, but they are thought to be related to the HMW subunits of wheat glutenin (Kreis *et al.*, 1985b). Barley flour does not have the same rheological qualities for breadmaking as wheat flour, so there may be some structural differences between these proteins.

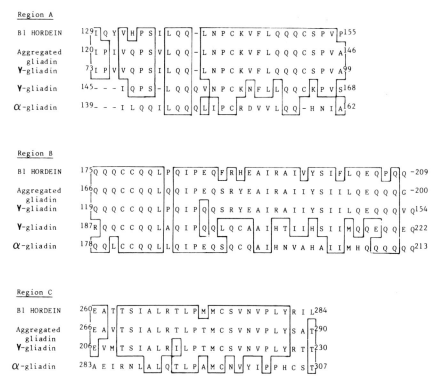

Fig. 11. Comparison of amino acid sequence homologies between the A, B, and C regions (see Fig. 9B) of the S-rich prolamines of barley and wheat. Amino acids: cysteine (C), histidine (H), methionine (M), serine (S), valine (V), alanine (A), glycine (G), proline (P), threonine (T), phenylalanine (F), arginine (R), tyrosine (Y), tryptophan (W), leucine (L), isoleucine (I), aspartic acid (D), asparagine (N), glutamic acid (E), glutamine (Q), lysine (K). From Kreis *et al.* (1985b) with permission of the author and publisher.

B. Maize Prolamines

Maize prolamines, zeins, are composed of several proteins that differ in alcohol solubility, depending on the presence of reducing agents. Various investigators have used slightly different procedures for isolating zeins, leading to a somewhat complicated nomenclature (Wilson, 1983; Esen, 1986). The total zein fraction can be extracted from maize endosperm in 70% ethanol or 55% 2-propanol containing 1% 2-mercaptoethanol. Separation by SDS-PAGE resolves components with apparent M_r of 27,000, 22,000, 19,000, 15,000, 14,000, and 10,000 (Fig. 12). Estimates of molecular weight for these groups of polypeptides vary, but the patterns of protein separation are very similar (Bunce *et al.*, 1985). Esen (1986) fractionated zeins into groups designated α, β, and γ on the basis of their solubility in 2-propanol in the presence or absence of reducing agent. The proteins within these groups

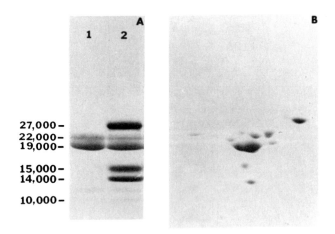

Fig. 12. Analysis of maize zein proteins by SDS-PAGE. (A) A one-dimensional separation of proteins extracted from mature endosperm with 70% ethanol (lane 1) or 70% ethanol plus 2-mercaptoethanol (lane 2). Apparent M_r are indicated on the left. (B) A two-dimensional separation of zein proteins from protein bodies of developing endosperm; proteins extracted with 70% ethanol containing 1% 2-mercaptoethanol. The nonequilibrium IEF gradient between pH 9 and 4 is from left to right; apparent M_r correspond to values left of A. From Larkins et al. (1984) with permission of the author and publisher.

are structurally distinct, so this appears to be a valid basis on which to distinguish among them.

The α-zeins, which typically account for about 70% of the total zein fraction, are composed of four or five polypeptides with apparent M_r ranging from 19,000 to 24,000. Each of these can be further separated into differently charged species by IEF (Hastings et al., 1984; Wilson, 1985). The IEF heterogeneity is genotype-specific and is inherited codominantly in a simple Mendelian fashion (Righetti et al., 1977). The heterogeneity has been useful in distinguishing among various maize genotypes (Hastings et al., 1984; Wilson, 1985).

Zeins are made as preproteins with an NH_2-terminal signal peptide of 20 or 21 amino acids (Geraghty et al., 1981; Marks and Larkins, 1982). For the α-zeins, the length of the mature protein is variable and ranges from 210 to 245 amino acids. All of the α-zeins have high contents of glutamine (25%), leucine (20%), alanine (15%), and proline (11%), and none has been identified that contains lysine. A distinguishing feature of the α-zeins is the presence of tandemly repeated peptides of approximately 20 amino acids in the central region of the protein (Geraghty et al., 1981; Pedersen et al., 1982; Spena et al., 1982). These repeats vary slightly in length (Fig. 13A), but can be averaged to obtain a consensus repeated peptide sequence (Pedersen et al., 1982). Based on CD analysis and computer-generated structural models (Argos et al., 1982), it was proposed that the repeated peptides are α-helices that interact through hydrogen bonding to fold the proteins into rod-shaped molecules (Fig. 13B). Hydrogen bonding of polar amino acids on the surface of the repeats, as well as between glutamines at the ends of the repeats, may contribute to aggregation of α-zeins within protein bodies (Argos et al., 1982).

The β-zeins are proteins of M_r 14,000 to 16,000 that account for around 15% of the zein fraction. These polypeptides show less charge heterogeneity than the α-zeins on IEF analysis (Hurkman et al., 1981; Marks et al., 1985a), with each component composed of only one or two polypeptides. Amino acid sequence analysis of an M_r 15,000 zein (Pedersen et al., 1986) shows that the mature protein following the signal peptide contains 160 amino acids. This protein has less glutamine (16%), leucine (10%), and proline (9%) than the α-zeins but contains significantly more of the sulfur amino acids, methionine (11%) and cysteine (4%) (Esen et al., 1985; Pedersen et al., 1986). Unlike the α-zeins, the β-zein contains no repetitive peptides. Based on CD and computer-generated structural analysis, the proteins have very little α-helical structure and are mostly composed of β strand and turn structure (Pedersen et al., 1986).

The γ-zein, also referred to as the "reduced-soluble protein" (Wilson et al., 1981), alcohol-soluble reduced glutelin (Paulis and Wall, 1971), or glutelin-2 (Prat et al., 1985), generally accounts for 20% of the total zein (Esen, 1986), although it can make up as much as 50% of this fraction (Ortega and Bates, 1983). In addition to being soluble in alcohol solutions, the γ-zein is soluble in saline solution, and this has led to its extraction in several different protein fractions. Although charge variants of the γ-zein have been isolated (Esen et al., 1982), the protein appears to be primarily a single species (Prat et al., 1985; Wang and Esen, 1986). Prat et al. (1985) recognized five distinct, consecutive regions in the 180-amino acid protein sequence: an NH$_2$-terminal segment of 11 amino acids, a repetitive peptide region composed of 8 tandem copies of the hexapeptide Pro-Pro-Pro-Val-His-Leu, an alternating Pro-X sequence between residues 70 and 91, and a COOH-terminal part rich in Gln that can be divided into two parts. The fourth region, comprising residues 92–148, is cysteine-rich and has some internal homology: the heptapeptide Gln-Cys-X-Glx-X-Leu-Arg is repeated twice and the related sequence Gln-Cys-Gln-Ser-Leu-Arg also appears once in this region. The fifth region, beginning with residue 149, has no apparent internal homology. While the γ-zein is structurally distinct from both the α- and β-zeins, it does have two regions of internal homology to the M_r 15,000 β-zein: there are matches of 10 residues each between amino acids 34 to 49 of the β-zein and

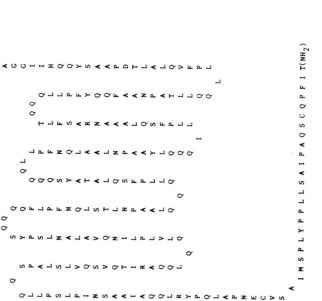
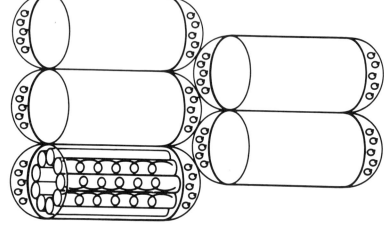

Fig. 13. Proposed structural organization of α-zein polypeptides. (A) Primary amino acid sequence of M_r 19,000 α-zein; following the NH$_2$-terminus the repeated peptides are illustrated folding back and forth on one another. (B) Structural model for α-zein. Repeated peptides form α-helices that wind up and down and fold the protein into a rod-shaped molecule. (See legend to Fig. 11 for single-letter amino acid designations.) From Argos et al. (1982) with permission of the author and publisher.

112 to 125 of the γ-zein, and between amino acids 61 to 73 of the β-zein and 130 to 141 of the γ-zein. These regions of homology probably account for the small degree of cross-reactivity between the M_r 15,000 β-zein and antisera directed against the γ-zein. Little is known about the structure of the γ-zein. The fact that it migrates more slowly on SDS-PAGE than the larger α-zeins may be a reflection of an unusual secondary structure associated with its high proline content (25%).

The M_r 10,000 zein protein contains a high proportion of sulfur amino acids and has an amino acid composition resembling that of the M_r 15,000 β-zein (Gianazza et al., 1977). However, it differs somewhat in alcohol solubility from the β-zeins (Esen, 1986) and may represent a structurally distinct fourth type of maize storage protein.

VI. SYNTHESIS AND DEPOSITION OF CEREAL PROLAMINES

Prolamines are synthesized on RER and subsequently deposited into protein bodies. In electron micrographs of maize and sorghum endosperm, protein bodies are observed at the middle and ends of RER cisternae and as distinct membrane-enclosed bodies (Khoo and Wolf, 1970; Larkins et al., 1979a; Taylor et al., 1985). In maize the mRNAs associated with the RER are identical to those on membranes surrounding protein bodies (Larkins and Hurkman, 1978); thus the two membrane systems appear to be equivalent and continuous. Injection of zein mRNAs into *Xenopus laevis* oocytes resulted in the synthesis and transport of proteins into membrane vesicles that sediment with densities similar to protein bodies from maize endosperm (Larkins et al., 1979b; Hurkman et al., 1981). This suggests that assembly of zeins into protein bodies is mediated primarily by the structure of the proteins rather than the transport system.

Studies of protein body formation in other cereals have shown that some prolamines aggregate into protein bodies within the RER, while others are transported into vacuoles. In rice, protein bodies containing prolamines form within the RER and are easily distinguished from the larger and more abundant protein bodies in vacuoles that contain the globulin-type storage proteins (Krishnan et al., 1986). However, in oats both prolamine and globulin proteins are transported into the vacuole where aggregates of avenin, the prolamine, are surrounded by the 11S globulin (C. R. Lending and B. A. Larkins, unpublished observations). The mechanism by which wheat and barley prolamines are deposited in protein bodies appears more complex. Investigation of developing wheat endosperm by electron microscopy (e.g., Campbell et al., 1981; Parker, 1982; Bechtel et al., 1982) has implicated the RER in the synthesis and transport of wheat prolamines, but the extent to which these proteins undergo further transport is not clear. Several studies

(Parker, 1982; Bechtel et al., 1982; Bechtel and Barnett, 1986) noted Golgi bodies adjacent to protein deposits within the RER and enlargements of the Golgi vesicles containing material resembling storage protein. Stero-pair images of thick endosperm sections showed interconnections between cisternal ER and tubular ER, with elements of the tubular ER associated with Golgi (Parker and Hawes, 1982); however, interconnections between the tubular ER and Golgi were not evident. These studies suggest that the wheat storage proteins are transported to protein bodies by a mechanism similar to that in legumes (Parker and Hawes, 1982). More recent analysis of wheat endosperm using freeze-etch electron microscopy indicated connections between Golgi and large sheets of ER, and it was concluded that protein body formation may occur by fusion of ER and Golgi-derived vesicles (Bechtel and Barnett, 1986). The role of Golgi in the transport of wheat gliadins was also recently demonstrated by Kim et al. (1988), who detected gliadins in Golgi vesicles by immunolocalization. It is sometimes difficult to see continuous membranes surrounding protein bodies in developing wheat and barley endosperm (Cameron-Mills and von Wettstein, 1980; Parker, 1982; Bechtel et al., 1982). Because these protein bodies are sensitive to protease treatment, it was suggested that the storage protein aggregates may disrupt protein body membranes (Miflin and Burgess, 1982).

Protein transport through the Golgi apparatus is often associated with glycosylation, but there is only indirect evidence that cereal prolamines are glycosylated. Most prolamines react negatively in glycoprotein tests, although Burr and Burr (1979) and Smith et al. (1985) suggested zein polypeptides contain a single glucose residue. Since 11S storage globulins are not glycosylated, this is apparently not a prerequisite for transport through Golgi.

There is evidence that prolamines do not associate randomly into protein bodies. Electron micrographs of developing barley endosperm show differential staining of proteins within protein bodies (Munck and von Wettstein, 1976; Cameron-Mills and von Wettstein, 1980). Barley protein bodies have a homogeneous matrix surrounded by irregular deposits of electron-dense material (Fig. 14A). Within the vacuoles, spherical bodies of the homogeneous matrix as well as electron-dense spheres are clustered, and the spaces between them are filled with a fibrillar material. Protein bodies from the barley mutant Risφ 56 have considerably more fibrillar material, and the homogeneous spherical deposits are less than half the diameter of the normal type (Fig. 14B). These morphological differences correlate with a significant reduction in the B hordeins and a near doubling of the C hordeins (Kreis et al., 1983b). Protein bodies in the Risφ 1508 mutant, which contains less B and C hordein but more D hordein (Hopp et al., 1983; Kreis et al., 1983b), are more fibrillar with larger and more common electron-dense spheres than in the normal genotype.

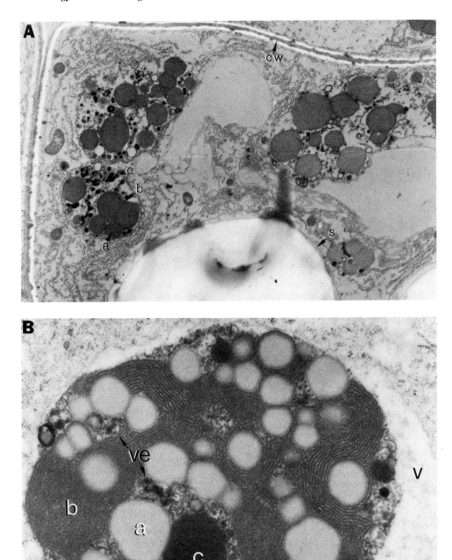

Fig. 14. (A) Protein bodies in developing endosperm of Carlsberg II barley 20 days after pollination. Subcellular structures include starch (s) and cell walls (cw). Three storage protein components (a, b, and c) are found in small vacuoles. Magnification: ×6000. (B) Protein body in 19-day endosperm of barley mutant Risø 56, derived from Carlsberg II. The three components (a, b, and c) of the wild-type protein bodies can be observed, but the proportion of fibrillar matrix to homogeneous component is considerably increased. v, Vacuole; ve, vesicle. Magnification: ×30,000. From Cameron-Mills and von Wettstein (1980) with permission of the author and publisher.

Protein bodies in maize appear more homogeneous than those in barley (Khoo and Wolf, 1970; Larkins and Hurkman, 1978), but zeins are also asymmetrically distributed. Studies by Ludevid *et al.* (1984) showed the γ-zein was primarily on the periphery of the protein body. More recently, it was shown that the α-zeins make up the core of the protein body (Lending *et al.*, 1988), with both the γ- and β-zeins making up the outer region of the protein body. In the *floury*-2 mutant of maize the α-, β-, and γ-zeins are not symmetrically organized, and the protein bodies appear to coalesce within the RER (C. R. Lending and B. A. Larkins, unpublished observations). The factors responsible for the assembly of the α-, β-, and γ-zeins into protein bodies and the extent to which it affects the endosperm texture (floury versus vitreous) are unknown.

VII. ORGANIZATION AND STRUCTURE OF PROLAMINE GENES

A. Wheat and Barley Prolamines

The size and charge heterogeneity among wheat and barley prolamines, as well as the diversity of genetic stocks of these species, has made it possible to map the loci controlling their synthesis (Wrigley and Shepherd, 1973; Payne *et al.*, 1984; Shewry *et al.*, 1984). Barley and wheat differ with regard to ploidy: barley is diploid with 14 pairs of chromosomes, while common wheat is allohexaploid with 42 chromosomes. The three genomes of common wheat are designated A, B, and D, with each contributing seven pairs of chromosomes. Chromosomes from the A, B, and D genomes are related but nonpairing (homoeologous) and are given similar numerical designations, e.g., 1A, 1B, and 1D. The seven chromosomes of barley are related to those in wheat: chromosome 1 is homoeologous to chromosome 7 of wheat, 2 to 2, 3 to 3, 4 to 4, 5 to 1, 6 to 6, and 7 to 5.

In wheat, genes controlling prolamine synthesis occur at complex loci on two chromosomes. The *Glu-1* triplicated locus on the long arm of chromosomes 1A, 1B, and 1D (Fig. 15) controls the synthesis of the HMW glutenins. The *Gli-1* locus on the short arm of homoeologous chromosomes 1A, 1B, and 1D controls the synthesis of the ω-gliadins, most of the γ-gliadins, the LMW glutenins (aggregated gliadins), and some of the β-gliadins. Other S-rich prolamines, the α/β-gliadins and some of the γ-gliadins are controlled by the *Gli-2* locus, which is on the short arm of chromosomes 6A, 6B, and 6D. Recently, the D subunits of the LMWglutenins have been mapped on the short arm of chromosome 1B proximal to the *Gli-1* locus, and this locus has been termed *Glu-B2* (Jackson *et al.*, 1985).

All three loci controlling synthesis of the barley hordeins map to chromosome 5. The D hordein, which is thought to be related to the HMW glutenins (Kreis *et al.*, 1985a), is controlled by the *Hor-3* locus on the long arm of

WHEAT — Chromosome 1A, 1B, 1D

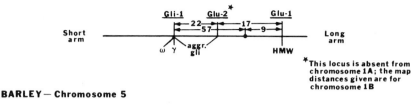

BARLEY — Chromosome 5

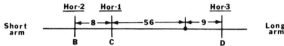

WHEAT — Chromosome 6A, 6B, 6D

Fig. 15. Chromosomal locations of structural genes for prolamines in wheat and barley. From Kreis *et al.* (1985b) with permission of the author and publisher.

chromosome 5 (Shewry *et al.*, 1983). The loci controlling the S-rich B hordeins (*Hor-1*) and S-poor C hordeins (*Hor-2*) are located on the short arm of chromosome 5. Like the *Gli-1* and *Gli-2* loci in wheat, the *Hor-1* and *Hor-2* loci of barley are complex and control the synthesis of multiple polypeptides (Jensen *et al.*, 1980; Shewry and Miflin, 1982). From an examination of eight cultivars, Faulks *et al.* (1981) identified 47 major B hordein polypeptides with 8 to 16 proteins present in each variety. Between 4 and 18 C hordein polypeptides were mapped at the *Hor-1* locus.

The similarity in positions of loci controlling prolamines on barley chromosome 5 and the homoeologous chromosomes 1 of wheat suggests an evolutionary relationship (Shewry *et al.*, 1985). The *Hor-1* and *Hor-3* loci on barley chromosome 5 are similarly spaced with respect to the centromere as the *Gli-1* and *Glu-1* loci on the homoeologous chromosomes 1 of wheat (Fig. 15). Kreis *et al.* (1985b) noted, however, that a comparison of the two major loci for S-rich (*Hor-1*) and S-poor (*Hor-2*) barley prolamines with the complex *Gli-1* locus may be an oversimplification.

Little is currently known regarding the molecular organization of loci encoding wheat and barley prolamines. Analyses of cDNA clones corresponding to prolamine mRNAs (Okita *et al.*, 1985; Forde *et al.*, 1983; Thompson *et al.*, 1983) confirm that numerous genes encode these proteins. The cDNA clones corresponding to different prolamine types, such as the α/β- and γ-gliadins, have varying degrees of sequence similarity as determined by cross-hybridization. The extent to which one type of sequence can be distinguished from another depends on the criterion (T_m) of the hybridization reaction. At a high stringency ($T_m - 8°C$), Okita et al. (1985) were able to distinguish five classes of α/β-gliadin mRNAs and three classes of γ-gliadin mRNAs. Similar heterogeneity has been detected among hordein mRNAs (Kreis et al., 1983a; Rasmussen et al., 1983; Forde et al., 1985b). Two major types of B hordein mRNAs were distinguished that differed in nucleotide sequence by 16%. The variation in nucleotide sequence among the mRNAs is generally in the third or "wobble" nucleotide of a codon, so the encoded amino acid is conserved. A detailed analysis of the structural relationships among these mRNAs has been presented (Kreis et al., 1985b).

Hordein and gliadin cDNA clones hybridize to multiple DNA fragments in restriction enzyme digests of genomic DNA (Anderson et al., 1984; Kreis et al., 1983a; Shewry et al., 1985), demonstrating that there are multiple genes homologous to the mRNA sequences. Some DNA fragments have hybridization intensities greater than single genes, but it is not apparent whether they contain more than one gene or correspond to multiple copies of a similarly sized DNA fragment.

All of the genomic clones examined to date contain only a single coding sequence, so it is unclear whether the genes are physically tightly clustered. The glutenin (Forde et al., 1985a; Thompson et al., 1985; Sugiyama et al., 1985), gliadin (Rafalski et al., 1984; Sumner-Smith et al., 1985; Anderson et al., 1984), and hordein clones (Forde et al., 1985c) that have been characterized contain no introns in their coding sequences. DNA sequences responsible for the transcriptional and developmental regulation have not been determined, but their 5' flanking regions contain canonical eukaryotic promoter sequences such as TATA and CAAT boxes (Benoist et al., 1980), and their 3' ends contain sequences directing polyadenylation (AATAAA).

B. Maize Zein Prolamines

Salamini and his colleagues (Soave and Salamini, 1984; Hastings et al., 1984) mapped loci controlling zein synthesis to several maize chromosomes. The genes controlling a number of the α-zeins are on chromosomes 4, 7, and 10 (Fig. 16). Seven genes corresponding to zeins of M_r 19,000 and 20,000 have been mapped to the short arm of chromosome 7, and linked with these are two regulatory loci, opaque-2 and De B-30, that exert qualitative and quantitative effects on zein synthesis (Soave and Salamini, 1984). A set of nine genes corresponding to M_r 20,000 and 22,000 zeins was mapped to the short and long arms of chromosome 4, and associated with these is the regulatory locus floury-2. A third locus corresponding to an M_r 22,000 zein is on the long arm of chromosome 10, and associated with it is the regulatory locus opaque-7. Although the gene or genes encoding the γ-zein have not been mapped, a gene encoding an M_r 15,000 β-zein has recently been

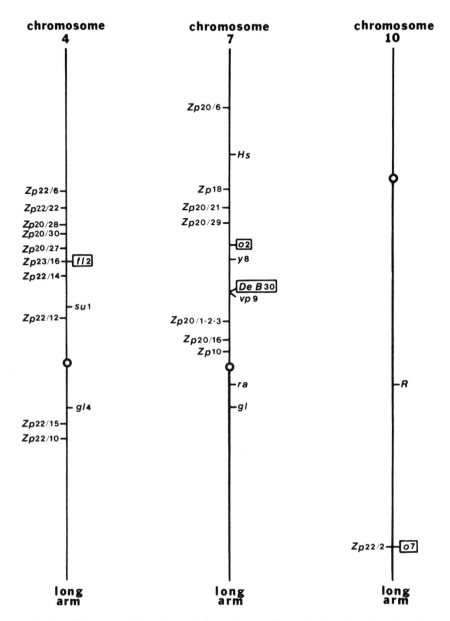

Fig. 16. Chromosomal locations of structural genes for prolamines in maize. From Soave and Salamini (1984) with permission of the author and publisher.

mapped to the long arm of chromosome 6 (J. Cramer, Agrigenetics Corporation, personal communication). The gene corresponding to the M_r 10,000 zein protein has been mapped to the short arm of chromosome 7 (Hastings *et al.*, 1984).

The complexity among α-zein proteins detected by IEF analysis has also been demonstrated among the genes that encode these proteins. There are multiple families of cDNA clones corresponding to α-zein mRNAs that can be distinguished from one another at different hybridization criteria (Burr *et al.*, 1982; Marks and Larkins, 1982; Viotti *et al.*, 1985). The heterogeneity among these sequences results from variation in the number of elements encoding repetitive peptides and from high frequencies of base substitutions, insertions, and deletions. The cDNA clones hybridize to multiple coding sequences in the genome, leading to estimates of between 75 and 150 genes (Hagen and Rubenstein, 1981; Wilson and Larkins, 1984); however, there appears to be only one or two copies of genes encoding the β- and γ-zeins (Wilson and Larkins, 1984; Boronat *et al.*, 1986).

Although little is known about the molecular organization of the zein genes, there is evidence of close linkage among some of the α-zein genes. Several genomic clones have been isolated that contain multiple genes spaced within a few kilobase pairs (Spena *et al.*, 1983; Marks *et al.*, 1984; Kriz *et al.*, 1987). The genes are tandemly arranged 5' to 3', and the DNA sequences flanking the coding regions are conserved. Analysis of one gene family of α-zeins suggested the region of homology extends from about 1 kb on the 5' end to 0.3 kb on the 3' end (Kriz *et al.*, 1987). Outside these conserved regions the DNA sequences are moderately to highly repetitive. The conserved regions flanking the genes contain canonical eukaryotic promoters and polyadenylation sequences. The coding sequences of zein genes contain no introns. Several genomic clones have been isolated in which the coding sequences have altered initiation codons or premature stop codons (Spena *et al.*, 1983; Kridl *et al.*, 1984; Marks *et al.*, 1984). These are thought to be pseudogenes, but it has been suggested that transcripts from genes with premature stop codons may be translated (Viotti *et al.*, 1985).

VIII. REGULATION OF PROLAMINE GENE EXPRESSION

A number of genetic and environmental factors influence the expression of cereal prolamine genes. Thus, the application of nitrogen fertilizer greatly enhances prolamine synthesis. This has not been investigated at the molecular genetic level, but it is likely that mineral nutrition affects prolamine gene expression as it does for storage globulin genes (see Section IV).

Mutations that reduce prolamine synthesis have been identified in the diploid cereals maize, barley, and sorghum (Nelson, 1979). These mutations are generally recessive, so it is not surprising that they have not been found in genera that are allopolyploid, e.g., *Triticum*. The nutritional quality of the seed protein in the mutants is improved as a consequence of the reduction in prolamine content. In particular, the relative content of the essential amino acid lysine is increased, so these are sometimes referred to as "high lysine" mutants.

One of the first of these mutants to be characterized was *opaque*-2 in maize (Mertz *et al.*, 1964). The *opaque*-2 mutation is recessive and causes about a 50% reduction in zein synthesis. Seeds of this genotype have a floury endosperm that does not transmit light, giving rise to the name "opaque." Most zein groups are affected by *opaque*-2 mutations, and many of its alleles almost completely eliminate the M_r 22,000 α-zeins (Lee *et al.*, 1976). Other nonallelic opaque mutants, notably *opaque*-6 and *opaque*-7, also cause a significant reduction in zein synthesis. In *opaque*-6 all zein groups are equally affected, while in *opaque*-7 the M_r 19,000 α-zeins are disproportionately reduced (DiFonzo *et al.*, 1980). Another mutation, *floury*-2, affects each zein group equally and causes a stepwise reduction in zein content depending on dosage (Jones, 1978).

The mechanisms by which these mutations cause a reduction in zein synthesis are unknown. When zein cDNA clones were used to analyze genomic DNA from the mutants, no major changes in gene number or organization were detected (Burr and Burr, 1982), so it appears the mutations affect regulatory functions rather than coding sequences. Several studies have shown that zein mRNA levels are reduced in the mutants relative to the normal genotypes (Pedersen *et al.*, 1980; Burr and Burr, 1982), but it is unknown whether this results from altered gene transcription or mRNA stability.

Feix and co-workers showed that some zein transcripts may be two to four times the length of the mature mRNA and that the level of these large RNAs is reduced in *opaque*-2 (Langridge *et al.*, 1982; Langridge and Feix, 1983). The large RNAs appear to result from transcription that begins several thousand nucleotides before the coding sequence, but it is not apparent whether these large RNAs are translated, are processed to the mature mRNA, or are products arising from transcription at minor start sites (Kriz *et al.*, 1987).

Products of zein regulatory loci have not been identified. A soluble protein of M_r 32,000 is absent in *opaque*-2 and *opaque*-6 mutants (Soave *et al.*, 1981). This protein is synthesized in developing endosperm coincident with zein accumulation, and although it was initially thought to be the product of the *opaque*-6 locus, more recent experiments indicate this is not the case (N. DiFonzo, personal communication). A protein of M_r 70,000 associated with protein bodies increases in *floury*-2 and several other mutants (Soave and Salamini, 1984), but its function is also unknown (Galante *et al.*, 1983).

Several mutations in barley have been identified that cause a reduction in hordein synthesis (Miflin and Shewry, 1979). One was found to occur naturally (Munck *et al.*, 1970), and the others were identified following mutagenesis. Mutant Risϕ 56, which was generated by gamma irradiation of the variety Carlsberg II, maps near the *Hor*-2 locus (Doll, 1980); others map to different sites in the genome and are thought to be regulatory. In Risϕ 56 there is about a 30% reduction in hordein accumulation, mostly corresponding to the major B hordeins. In this mutant the C hordeins are nearly doubled, but the D hordein is unaffected. The reduction of hordein synthesis appears to result from a major deletion in the *Hor*-2 locus (Kreis et al., 1983b). Hybridization of a B hordein clone to genomic DNA of Risø 56 and Carlsberg II revealed that around 13 copies of B hordein genes and their accompanying DNA sequences are missing in the mutant. It is estimated that this may correspond to a deletion at the *Hor*-2 locus of 80 to 90 kb of DNA sequence. Analysis of mRNA from the developing seeds of the mutant shows a corresponding absence of B hordein mRNA sequences and an increase in C hordein mRNA sequences (Hopp et al., 1983; Kreis et al., 1983b).

The mechanisms by which other "high lysine" mutations affect hordein synthesis are unclear. In the mutant Risø 1508, the B and C hordeins are reduced by 20 and 7%, respectively, and the D hordein is increased fourfold. This mutation, which maps to chromosome 7, reveals no apparent alteration in the structure and organization of hordein genes (Kreis et al., 1984). In this mutant the levels of hordein mRNAs are reduced commensurate with the reduction in storage proteins (Hopp et al., 1983; Kreis et al., 1983b). There are only trace amounts of C hordein mRNA, the two families of B hordein mRNA sequences are reduced to 40 and 5% of those in wild type, and the D hordein mRNA is increased twofold. Cameron-Mills and Ingversen (1978) reported that functional microsomes from the mutant could not be reconstituted *in vitro* and suggested the reduction of mRNA in this mutant may be due to instability as a consequence of altered protein synthesis. However, the possibility that this mutation alters gene transcription cannot be ruled out (Hopp et al., 1983; Kreis et al., 1984).

The DNA sequences responsible for developmental regulation of prolamine gene expression are less well characterized than those for storage globulins. This is a reflection of the slower development of methods to transform and regenerate monocots. The Ti plasmid, an effective vector for dicots, transforms monocots inefficiently (Hooykas-Van Slogteren et al., 1984). Maize zein genes have been transferred to dicots via the Ti plasmid (Matzke et al., 1984; Goldsbrough et al., 1986), and while accurate gene transcription was detected, developmental regulation could not be studied.

Several investigators noted a conserved region in the 5' flanking region of prolamine genes from different cereal species with sequence similarity to the SV40 transcriptional enhancer region (Freeling and Bennet, 1985; Kreis et al., 1985b). This region appears to be important for zein gene transcription (Roussell et al., 1988), but whether it plays a role in developmental regulation is unknown. Delimiting the regions that regulate the expression of prolamine genes will be an active area of future research.

IX. SUMMARY

Two basic types of storage proteins can be distinguished in seeds: the globulins, which are found in the embryos of most seeds, and the prolamines, which occur in the endosperms of many cereals. It would not be surprising if they evolved from common ancestral genes, but as they now exist there are significant structural differences between these proteins and between the genes that encode them. Although both globulins and prolamines have a high content of amide amino acids, their primary and secondary structures are quite different. Subunits of storage globulins are spherical and composed of three domains. Insertions and deletions of short peptides account for most of the structural variation among globulin storage proteins of different species. Although less is known about the tertiary and quaternary structure of prolamine proteins, the best studied examples have more extended and rodlike conformations. Duplication of short peptides appears to have occurred in the evolution of prolamine proteins. The segmented nature of some prolamines, e.g., those of wheat and barley, suggests they may have evolved from parts of other proteins. This idea is supported by the analysis of Kreis *et al.* (1985b), showing homology of regions in nonrepetitive domains of wheat and barley prolamines with seed globulins and protease inhibitors. There is insufficient evidence to evaluate the role that may have been played in the evolution of these proteins by intron/exon structures. All of the globulin genes examined thus far contain introns, while the prolamine genes completely lack introns. The size of introns in the 7S and 11S globulin genes in legumes is variable, but their number and organization are generally conserved. The recent observation that certain introns are absent in some of these genes may indicate that the genes evolved by addition of introns to an ancestral gene.

The mechanisms by which storage proteins are synthesized and deposited in cells of embryo or endosperm tissue are conserved among seed plants. In all cases they are made as preproteins that are transported through the endomembrane system; they become concentrated and associate into aggregates called protein bodies. In maize, sorghum, and rice, aggregation occurs directly within the lumen of the RER, but for the prolamines of other cereals and the globulins of most dicots, transport continues through the ER and Golgi apparatus to the vacuole. During their transport, globulin storage proteins are subject to proteolytic processing. Cleavage of 11S globulins occurs between the acidic and basic parts of subunit polypeptides and may play a role in the association of 7–8S complexes into 11S complexes. It is less clear what function proteolytic digestion plays in the processing and assembly of 7S globulins. Cleavage of these proteins is more variable among species. The fact that the same protein can undergo digestion at different sites in different plants suggests that the cleavage may not be important in processing and assembly. Since the 7S complexes form within the RER immediately following synthesis, the peptide cleavages may not affect the structure of the molecule.

The role of glycosylation in the transport and assembly of storage proteins is also unclear. Glycosylation is apparently not involved in transport or assembly of 11S globulins or cereal prolamines. However, many, though not all, 7S storage globulins undergo a series of sugar addition and subtraction reactions. Present evidence suggests that these carbohydrate moieties are not necessary for protein transport. It is possible that they are involved in assembly within the protein body or in proteolytic digestion during seed germination.

The ability to transform plants with storage protein genes provides a tool to address important questions pertaining to gene regulation as well as the synthesis and processing of these proteins. The DNA sequences responsible for developmental regulation of storage globulins appear to be relatively short *cis*-acting elements that precede the genes by several hundred base pairs. Although equivalent sequences remain to be identified in cereals, the conservation of DNA sequences flanking prolamine genes suggests that they are similarly situated. There must nevertheless be some difference in the regulation of prolamine and globulin genes, since one group is expressed in the embryo and the other in the endosperm. Future experiments will address these questions as well as the transcriptional, posttranscriptional, and translational mechanisms responsible for storage protein synthesis in developing seeds. Modifying the coding sequences of storage protein genes will provide an approach to studying the proteolytic processing and glycosylation of these proteins and to determining the potential of creating storage proteins with higher levels of essential amino acids.

ACKNOWLEDGMENTS

We thank J. S. Lindell for help in preparation of the manuscript and figures, and N. C. Nielsen, M. A. Hermodson, and C. E. Bracker for careful reading of the manuscript. Supported in part by a grant from NIH (GM36970-02) to B.A.L. M.A.S. was the recipient of a postdoctoral fellowship from NSF. This is a journal paper No. 11,874 from the Purdue Agricultural Experiment Station.

REFERENCES

Adeli, K., and Altosaar, I. (1983). *Plant Physiol.* **73,** 949–955.
Altschul, A. M., and Wilcke, H. L., eds. (1985). "New Protein Foods, Vol. 5." Academic Press, New York.
Anderson, O. D., Litts, J. C., Gautier, M.-F., and Green, F. C. (1984). *Nucleic Acids Res.* **12,** 8129–8144.

Argos, P., Pedersen, K., Marks, M. D., and Larkins, B. A. (1982). *J. Biol. Chem.* **257,** 9984–9990.
Argos, P., Narayana, S. V. L., and Nielsen, N. C. (1985). *EMBO J.* **4,** 1111–1117.
Badenoch-Jones, J., Spencer, D., Higgins, T. J. V., and Millerd, A. (1981). *Planta* **153,** 201–209.
Bailey, C. J., and Boulter, D. (1970). *Eur. J. Biochem.* **17,** 460–466.
Bailey, C. J., and Boulter, D. (1974). *Phytochemistry* **11,** 59–64.
Barton, K. A., Thompson, J. F., Madison, J. T., Rosenthal, R., Jarvis, N. P., and Beachy, R. N. (1982). *J. Biol. Chem.* **257,** 6089–6095.
Bassüner, R., Wobus, U., and Rapoport, T. (1984). *FEBS Lett.* **166,** 314–320.
Bäumlein, H., Wobus, U., Pustell, J., and Kafatos, F. C. (1986). *Nucl. Acids Res.* **14,** 2707–2720.
Beachy, R. N., Chen, Z.-L., Horsch, R. B., Rogers, S. B., Hoffmann, N. J., and Fraley, R. T. (1985). *EMBO J.* **4,** 3047–3053.
Bechtel, D. B., and Barnett, B. D. (1986). *Cereal Chem.* **63,** 232–240.
Bechtel, D. B., Gaines, R. L., and Pomeranz, Y. (1982). *Ann. Bot.* **50,** 507–518.
Beckwith, A. C., Nielson, H. C., Wall, J. S., and Huebner, F. R. (1966). *Cereal Chem.* **43,** 14–27.
Benoist, C., O'Hare, K., Breathnach, R., and Chambon, P. (1980). *Nucleic Acids Res.* **8,** 127–142.
Bietz, J. A. (1982). *Biochem. Genet.* **20,** 1039–1053.
Bietz, J. A., and Wall, J. S. (1975). *Cereal Chem.* **52,** 145–155.
Bietz, J. A., Huebner, F. R., Sandersen, J. E., and Wall, J. S. (1977). *Cereal Chem.* **54,** 1070–1083.
Blagrove, R. J., and Gillespie, J. M. (1975). *Aust. J. Plant Physiol.* **2,** 13–27.
Blagrove, R. J., Gillespie, J. M., Lilley, G. G., and Woods, E. F. (1980). *Aust. J. Plant Physiol.* **7,** 1–13.
Blobel, G. (1977). *In* "International Cell Biology 1976–1977" (B. R. Brinkley and K. R. Porter, eds.), pp. 318–325. Rockefeller Univ. Press, New York.
Bollini, R., Vitale, A., and Chrispeels, M. J. (1983). *J. Cell Biol.* **96,** 999–1007.
Boronat, A., Martinez, M. C., Reina, M., Puigdomenech, and Palau, J. (1986). *Plant Sci.* **47,** 95–102.
Borroto, K., and Dure, L., III (1987). *Plant Mol. Biol.* **8,** 113–131.
Bright, S. W. J., and Shewry, P. R. (1983). *Crit. Rev. Plant Sci.* **1,** 49–92.
Brinegar, A. C., and Peterson, D. M. (1982). *Plant Physiol.* **70,** 1767–1769.
Brown, J. W. S., Bliss, F. A., and Hall, T. C. (1981). *Theor. Appl. Genet.* **60,** 251–258.
Bunce, N. A. C., White, R. P., and Shewry, P. R. (1985). *J. Cereal Sci.* **3,** 131–142.
Burgess, S. R., Shewry, P. R., Matlashewski, G. J., Altosaar, I., and Miflin, B. J. (1983). *J. Exp. Bot.* **147,** 1320–1332.
Burr, F. A., and Burr, B. (1979). *In* "The Plant Seed: Development, Preservation, and Germination" (I. Rubenstein, R. L. Phillips, C. E. Green, and B. G. Gengenbach, eds.), pp. 27–48. Academic Press, New York.
Burr, F. A., and Burr, B. (1982). *J. Cell Biol.* **94,** 201–206.
Burr, B., Burr, F. A., St. John, T. P., Thomas, M., and Davis, R. W. (1982). *J. Mol. Biol.* **154,** 33–49.
Cameron-Mills, V., and Ingversen, J. (1978). *Carlsberg Res. Commun.* **43,** 471–489.
Cameron-Mills, V., and von Wettstein, D. (1980). *Carlsberg Res. Commun.* **45,** 577–594.
Campbell, W. P., Lee, J. W., O'Brien, T. P., and Smart, M. G. (1981). *Aust. J. Plant Physiol.* **8,** 5–19.
Carasco, J. F., Croy, R., Derbyshire, E., and Boulter, D. (1978). *J. Exp. Bot.* **29,** 309–323.
Casey, R., Domoney, C., and Ellis, N. (1986). *Oxford Surv. Plant Mol. Cell Biol.* **3,** 1–95.
Casey, R., and Short, M. N. (1981). *Phytochemistry* **20,** 21–23.
Chandler, P. M., Spencer, D., Randall, P. J., and Higgins, T. J. V. (1984). *Plant Physiol.* **75,** 651–657.

Chen, Z.-L., Schuler, M. A., and Beachy, R. N. (1986). *Proc. Natl. Acad. Sci. U.S.A.* **83**, 8560–8564.
Chlan, C. A., Pyle, J. B., Legocki, A. B., and Dure, L., III (1986). *Plant Mol. Biol.* **7**, 475–489.
Chou, P. Y., and Fasman, G. D. (1978). *Annu. Rev. Biochem.* **47**, 251–256.
Chrispeels, M. J. (1983). *Planta* **158**, 140–151.
Chrispeels, M. J. (1985). *Oxford Surv. Plant Mol. Cell Biol.* **2**, 43–68.
Chrispeels, M. J., and Greenwood, J. S. (1985). *Planta* **164**, 295–302.
Chrispeels, M. J., Higgins, T. J. V., and Spencer, D. (1982). *J. Cell Biol.* **93**, 306–313.
Citharel, L., and Citharel, J. (1985). *Planta* **166**, 39–45.
Cluskey, J. E., and Wu, Y. V. (1971). *Cereal Chem.* **48**, 203–211.
Coates, J. B., Medeiros, J. S., Thanh, V. H., and Nielsen, N. C. (1985). *Arch. Biochem. Biophys.* **243**, 184–194.
Craig, S., and Goodchild, D. J. (1984). *Protoplasma* **122**, 35–44.
Craig, S., and Miller, C. (1984). *Cell Biol. Int. Rep.* **8**, 879–886.
Craig, S., and Millerd, A. (1981). *Protoplasma* **105**, 333–339.
Craig, S., Millerd, A., and Goodchild, D. J. (1980). *Aust. J. Plant Physiol.* **7**, 339–351.
Crouch, M. L., and Sussex, I. M. (1981). *Planta* **153**, 64–74.
Danielsson, C. E. (1949). *Biochem. J.* **44**, 387–400.
Davies, H. M., and Delmer, D. P. (1981). *Plant Physiol.* **68**, 284–291.
Davies, C. S., Coates, J. B., and Nielsen, N. C. (1985). *Theor. Appl. Genet.* **71**, 351–358.
Derbyshire, E., and Boulter, D. (1976). *Phytochemistry* **15**, 411–414.
Derbyshire, E., Wright, D. J., and Boulter, D. (1976). *Phytochemistry* **15**, 3–24.
DiFonzo, N., Fornasari, E., Salamini, F., Reggiani, R., and Soave, C. (1980). *J. Hered.* **71**, 397–402.
Doll, H. (1980). *Heridtas* **93**, 217–222.
Domoney, C., and Casey, R. (1985). *Nucleic Acids Res.* **13**, 687–699.
Domoney, C., Ellis, T. H. N., and Davies, D. R. (1986). *Mol. Gen. Genet.* **202**, 280–285.
Doyle, J. J., Schuler, M. A., Godette, W. D., Zenger, V., Beachy, R. N., and Slightom, J. L. (1986). *J. Biol. Chem.* **261**, 9228–9238.
Dure, L., III, and Galau, G. A. (1981). *Plant Physiol.* **68**, 187–194.
Eaton-Mordas, C. A., and Moore, K. G. (1978). *Phytochemistry* **17**, 619–621.
Ellis, T. H. N., Domoney, C., Castleton, J., Cleary, W., and Davies, D. R. (1986). *Mol. Gen. Genet.* **205**, 164–169.
Esen, A. (1986). *Plant Physiol.* **80**, 623–627.
Esen, A., Bietz, J. A., Paulis, J. W., and Wall, J. S. (1982). *Nature (London)* **296**, 678–679.
Esen, A., Beitz, J. A., Paulis, J. W., and Wall, J. S. (1985). *J. Cereal Sci.* **3**, 143–152.
Ewart, J. A. D. (1977). *J. Sci. Food Agric.* **28**, 191–199.
Faulks, A. J., Shewry, P. R., and Miflin, B. J. (1981). *Biochem. Genet.* **19**, 841–858.
Fischer, R. L., and Goldberg, R. B. (1982). *Cell* **29**, 651–660.
Fischer, R. L., Sims, T. L., Drews, G. N., and Goldberg, R. B. (1988). Submitted for publication.
Forde, J., Forde, B. G., Fry, R. P., Kreis, M., Shewry, P. R., and Miflin, B. J. (1983). *FEBS Lett.* **162**, 360–366.
Forde, J., Malpica, J.-M., Halford, N. G., Shewry, P. R., Anderson, O. D., Greene, F. C., and Miflin, B. J. (1985a). *Nucleic Acids. Res.* **13**, 6817–6832.
Forde, B. G., Kreis, M., Williamson, M. S., Fry, R. P., Pywell, J., Shewry, P. R., Bunce, N., and Miflin, B. J. (1985b). *EMBO J.* **4**, 9–15.
Forde, B. G., Heyworth, A., Pywell, J., and Kreis, M. (1985c). *Nucleic Acids Res.* **13**, 7327–7339.
Freeling, M., and Bennet, D. C. (1985). *Annu. Rev. Genet.* **19**, 297–323.
Galante, E., Vitale, A., Manzocchi, L., Soave, C., and Salamini, F. (1983). *Mol. Gen. Genet.* **192**, 316–321.
Garnier, J., Osguthorpe, D. J., and Robson, B. (1978). *J. Mol. Biol.* **120**, 97–120.

Gatehouse, J. A., Croy, R. R. D., and Boulter, D. (1984). *Crit. Rev. Plant Sci.* **1,** 287–314.
Geraghty, D., Peifer, M. A., Rubenstein, I., and Messing, J. (1981). *Nucleic Acids Res.* **9,** 5163–5174.
Gianazza, E., Viglienghi, V., Righetti, P. G., Salamini, F., and Soave, C. (1977). *Phytochemistry* **16,** 315–317.
Gillespie, J. M., Blagrove, R. J., and Randall, P. J. (1978). *Aust. J. Plant Physiol.* **5,** 641–650.
Goldsbrough, P. B., Gelvin, S. B., and Larkins, B. A. (1986). *Mol. Gen. Genet.* **202,** 374–381.
Gottschalk, W., and Muller, H. P. (1983). "Seed Proteins, Biochemistry, Genetics, and Nutritive Value." Nijhoff/Junk, The Hague, The Netherlands.
Graveland, A., Bosveld, P., Lichtendonk, W. J., Moonen, H. H. E., and Scheepstra, A. (1982). *J. Sci. Food Agric.* **33,** 1117–1128.
Greenwood, J. S., and Chrispeels, M. J. (1985). *Plant Physiol.* **79,** 65–71.
Hagen, G., and Rubenstein, I. (1981). *Gene* **13,** 239–249.
Hara, I., Wada, K., Wakabayashi, S., and Matsubara, H. (1976). *Plant Cell Physiol.* **17,** 799–814.
Harada, J. J., Barker, S., and Goldberg, R. B. (1988). Submitted for publication.
Hara-Nishimura, I., Nishimura, M., and Akazawa, T. (1985). *Plant Physiol.* **77,** 747–752.
Harris, N. (1979). *Planta* **146,** 63–69.
Hastings, H., Bonanomi, S., Soave, C., DiFonzo, N., and Salamini, F. (1984). *Genet. Agrar.* **38,** 447–464.
Heath, J. D., Weldon, R., Monnot, C., and Meinke, D. W. (1986). *Planta* **169,** 304–312.
Herman, E. M., Leland, M. S., and Chrispeels, M. J. (1986). *In* "Molecular Biology of Seed Storage Proteins and Lectins" (L. M. Shannon and M. J. Chrispeels, eds.). Am. Soc. Plant Physiol., Rockville, Maryland.
Higgins, T. V. J. (1985). *Annu. Rev. Plant Physiol.* **35,** 191–221.
Holowach, L. P., Thompson, J. F., and Madison, J. T. (1984). *Plant Physiol.* **74,** 576–583.
Holowach, L. P., Madison, J. T., and Thompson, J. F. (1986). *Plant Physiol.* **80,** 561–567.
Hooykaas-Van Slogteren, G. M. S., Hooykaas, P. J. J., and Schilperoort, R. A. (1984). *Nature (London)* **311,** 763–764.
Hopp, H. E., Rasmussen, S. K., and Brandt, A. (1983). *Carlsberg Res. Commun.* **48,** 201–216.
Horsch, R. B., Fry, J. E., Hoffman, N. L., Eichholtz, D., Rogers, S. G., and Fraley, R. T. (1985). *Science* **227,** 1229–1231.
Hu, B., and Esen, A. (1982). *J. Agric. Food Chem.* **30,** 21–25.
Huebner, F. R. (1977). *Baker's Dig.* **51,** 25–31.
Hurkman, W. J., and Beevers, L. (1982). *Plant Physiol.* **69,** 1414–1417.
Hurkman, W. J., Smith, L. D., Richter, J., and Larkins, B. A. (1981). *J. Cell Biol,* **89,** 292–299.
Jackson, E. A., Holt, L. M., and Payne, P. I. (1983). *Theor. Appl. Genet.* **66,** 29–37.
Jackson, E. A., Holt, L. M., and Payne, P. I. (1985). *Genet. Res.* **46,** 11–17.
Jensen, J., Jorgensen, J. H., Jensen, H. P., Giese, H., and Doll, H. (1980). *Theor. Appl. Genet.* **58,** 27–31.
Jones, R. A. (1978). *Biochem. Genet.* **16,** 27–38.
Juliano, B. O. (1972). *In* "Rice Chemistry and Technology" (D. F. Houston, ed.), pp. 16–74. Am. Assoc. Cereal Chem., St. Paul, Minnesota.
Juliano, B. O., and Boulter, D. (1976). *Phytochemistry* **15,** 1601–1606.
Kasarda, D. D., Bernardin, J. E., and Gaffield, W. (1968). *Biochemistry* **7,** 3950–3957.
Kasarda, D. D., Bernardin, J. E., and Nimmo, C. C. (1976). *Adv. Cereal Sci. Technol.* **1,** 158–236.
Kasarda, D. D., Okita, T. W., Bernardin, J. E., Baecker, P. A., Nimmo, C. C., Lew, E. J.-L., Dietler, M. D., and Green, F. C. (1984). *Proc. Natl. Acad. Sci. U.S.A.* **81,** 4712–4716.
Khan, M. R. I., Gatehouse, J. A., and Boulter, D. (1980). *J. Exp. Bot.* **31,** 1599–1611.
Khoo, U., and Wolf, M. J. (1970). *Am. J. Bot.* **57,** 1042–1050.
Kim, W. T., Franceschi, V. R., Krishnan, H. B., and Okita, T. W. (1988). *Planta* (in press).

Kirk, J. T. O., and Pyliotis, N. A. (1976). *Aust. J. Plant Physiol.* **3**, 731–746.
Kreis, M., Rahman, S., Forde, B. G., Pywell, J., Shewry, P. R., and Miflin, B. J. (1983a). *Mol. Gen. Genet.* **191**, 194–200.
Kreis, M., Shewry, P. R., Forde, B. G., Rahman, S., and Miflin, B. J. (1983b). *Cell* **34**, 161–167.
Kreis, M., Shewry, P. R., Rahman, S., Bahramian, M. B., and Miflin, B. J. (1984). *Biochem. Genet.* **22**, 231–255.
Kreis, M., Forde, B. G., Rahman, S., Miflin, B. J., and Shewry, P. R. (1985a). *J. Mol. Biol.* **183**, 499–502.
Kreis, M., Shewry, P. R., Forde, B. G., Forde, J., and Miflin, B. J. (1985b). *Oxford Surv. Plant Mol. Cell Biol.* **2**, 253–317.
Kridl, J. C., Vieira, J., Rubenstein, I., and Messing, J. (1984). *Gene* **28**, 112–118.
Krishnan, H. B., Franceschi, V. R., and Okita, T. W. (1986). *Planta* **169**, 471–480.
Kriz, A., Boston, R. S., Slightom, J. L., and Larkins, B. A. (1987). *Mol. Gen. Genet.* **207**, 90–98.
Ladin, B. F., Doyle, J. J., and Beachy, R. N. (1984). *J. Mol. Appl. Genet.* **2**, 372–380.
Langridge, P., and Feix, G. (1983). *Cell* **34**, 1015–1022.
Langridge, P., Pintor-Toro, J. A., and Feix, G. (1982). *Planta* **156**, 166–170.
Larkins, B. A., and Hurkman, W. J. (1978). *Plant Physiol.* **62**, 256–263.
Larkins, B. A., Pearlmutter, N. L., and Hurkman, W. J. (1979a). *In* "The Plant Seed: Development, Preservation and Germination" (I. Rubenstein, R. L. Phillips, C. E. Green, and B. G. Gengenbach, eds.), pp. 27–48. Academic Press, New York.
Larkins, B. A., Pedersen, K., Handa, A. K., Hurkman, W. J., and Smith, L. D. (1979b). *Proc. Natl. Acad. Sci. U.S.A.* **76**, 6448–6452.
Larkins, B. A., Pedersen, K., Marks, M. D., and Wilson, D. R. (1984). *Trends Biochem. Sci.* **9**, 306–308.
Laroche, M., Aspart, L., Delseny, M., and Penon, P. (1984). *Plant Physiol.* **74**, 487–493.
Lee, K. H., Jones, R. A., Dalby, A., and Tsai, C. Y. (1976). *Biochem. Genet.* **14**, 641–650.
Lending, C. R., Kriz, A. L., Larkins, B. A., and Bracker, C. E. (1988). *Protoplasma* **143**, 51–62.
Lilley, G. G., and Nielsen, N. C. (1987). Submitted for publication.
Ludevid, M. C., Torrent, M., Martinez-Izquierdo, J. A., Puigdomenech, P., and Palau, J. (1984). *Plant Mol. Biol.* **3**, 227–234.
Lycett, G. W., Delauney, A. J., Gatehouse, J. A., Gilroy, J., Croy, R. R. D., and Boulter, D. (1983). *Nucleic Acids Res.* **11**, 2367–2380.
Lycett, G. W., Croy, R. R. D., Shirsat, A. H., and Boulter, D. (1984a). *Nucleic Acids Res.* **12**, 4493–4506.
Lycett, G. W., Delauney, A. J., Zhao, W., Gatehouse, J. A., Croy, R. R. D., and Boulter, D. (1984b). *Plant Mol. Biol.* **3**, 91–96.
Lycett, G. W., Croy, R. R. D., Shirsat, A. H., Richards, D. M., and Boulter, D. (1985). *Nucleic Acids Res.* **13**, 6733–6743.
Macnicol, P. K. (1983). *FEBS Lett.* **156**, 55–57.
Marks, M. D., and Larkins, B. A. (1982). *J. Biol. Chem.* **257**, 9976–9983.
Marks, M. D., Pedersen, K., Wilson, D. R., DiFonzo, N., and Larkins, B. A. (1984). *Curr. Top. Plant Biochem. Physiol.* **3**, 9–18.
Marks, M. D., Lindell, J. S., and Larkins, B. A. (1985a). *J. Biol. Chem.* **260**, 16445–16450.
Matta, N. K., and Gatehouse, J. A. (1982). *Heredity* **48**, 383–392.
Matzke, M. A., Susani, M., Binns, A. N., Lewis, E. D., Rubenstein, I., and Matzke, A. J. M. (1984). *EMBO J.* **3**, 1525–1531.
Meinke, D. W., Chen, J., and Beachy, R. N. (1981). *Planta* **153**, 130–139.
Mertz, E. T., Bates, L. S., and Nelson, O. E. (1964). *Science* **145**, 279–280.
Miflin, B. J., and Burgess, S. R. (1982). *J. Exp. Bot.* **33**, 251–260.
Miflin, B. J., and Shewry, P. R. (1979). *In* "Recent Advances in the Biochemistry of Cereals" (D. Lardman and R. G. Wyn Jones, eds.), pp. 239–273. Academic Press, London.

Moreira, M. A., Hermodson, M. A., Larkins, B. A., and Nielsen, N. C. (1979). *J. Biol. Chem.* **254,** 9921–9926.
Moreira, M. A., Hermodson, M. A., Larkins, B. A., and Nielsen, N. C. (1981). *Arch. Biochem. Biophys.* **210,** 633–642.
Munck, L., and von Wettstein, D. V. (1976). *In* "Genetic Improvement of Seed Proteins," pp. 71–79. Natl. Acad. Sci., Washington, D.C.
Munck, L., Karlsson, K. E., Hagberg, A., and Eggum, B. O. (1970). *Science* **168,** 985–987.
Nelson, O. E. (1979). *Adv. Cereal Sci. Technol.* **3,** 41–71.
Neucere, N. J., and Ory, L. (1970). *Plant Physiol.* **45,** 616–619.
Nielsen, N. C. (1984). *Philos. Trans. R. Soc. London Ser. B* **204,** 287–296.
Nielsen, N. C. (1985a). *J. Am. Oil Chem. Soc.* **62,** 1680–1686.
Nielsen, N. C. (1985b). *New Protein Foods* **5,** 27–64.
Nielson, H. C., Beckwith, A. C., and Wall, J. S. (1968). *Cereal Chem.* **45,** 37–47.
Okita, T.W. (1984). *Plant Mol. Biol.* **3,** 325–332.
Okita, T. W., Cheesbrough, V., and Reeves, C. D. (1985). *J. Biol. Chem.* **260,** 8203–8213.
Ortega, E. I., and Bates, L. S. (1983). *Cereal Chem.* **60,** 107–111.
Osborne, T. B. (1908). *Science* **28,** 417–427.
Osborne, T. B., and Campbell, G. F. (1898). *J. Am. Chem. Soc.* **20,** 348–362.
Parker, M. L. (1982). *Plant Cell Environ.* **5,** 37–43.
Parker, M. L., and Hawes, C. R. (1982). *Planta* **154,** 277–283.
Paulis, J. W., and Wall, J. S. (1971). *Biochim. Biophys. Acta* **251,** 57–69.
Payne, P. I., Holt, L. M., Lawrence, G. J., and Law, C. N. (1982). *Qual. Plant–Plant Foods Hum. Nutr.* **31,** 229–241.
Payne, P. I., Holt, L. M., Jackson, E. A., and Law, C. N. (1984). *Philos. Trans. R. Soc. London, Ser. B* **304,** 359–371.
Payne, P. I., Holt, L. M., Jarvis, M. G., and Jackson, E. A. (1985). *Cereal Chem.* **62,** 319–326.
Pedersen, K., Bloom, K. S., Anderson, J. N., Glover, D. V., and Larkins, B. A. (1980). *Biochemistry* **19,** 1644–1650.
Pedersen, K., Devereux, J., Wilson, D. R., Sheldon, E., and Larkins, B. A. (1982). *Cell* **29,** 1015–1026.
Pedersen, K., Argos, P., Narayana, S. V. L., and Larkins, B. A. (1986). *J. Biol. Chem.* **261,** 6279–6284.
Pernollet, J. C. (1985). *Physiol. Veg.* **23,** 45–59.
Plietz, P., Damaschun, G., Müller, J. J., and Schwenke, K.-D. (1983a). *Eur. J. Biochem.* **130,** 315–320.
Plietz, P., Damaschun, G., Müller, J. J., and Schlesier, B. (1983b). *FEBS Lett.* **162,** 43–46.
Plietz, P., Damaschun, G., Zirwer, D., Gast, K., and Schlesier, B. (1983c). *Int. J. Biol. Macromol.* **5,** 356–360.
Porceddu, E., Lafiandra, D., and Scarascia-Mugnozza, G. T. (1983). *In* "Seed Proteins" (W. Gottschalk and H. P. Muller, eds.), pp. 77–141. Nijhoff/Junk, The Hague, The Netherlands.
Prat, S., Cortadas, J., Puigdomenech, P., and Palau, J. (1985). *Nucleic Acids Res.* **13,** 1493–1504.
Pusztai, A., and Watt, W. B. (1974). *Biochim. Biophys. Acta* **365,** 57–71.
Rafalski, J. A., Scheets, K., Metzler, M., Peterson, D. M., Hedgcoth, C., and Söll, D. G. (1984). *EMBO J.* **3,** 1409–1415.
Randall, P. J., Thomson, J. A., and Schroeder, H. E. (1979). *Aust. J. Plant Physiol.* **6,** 11–24.
Rasmussen, S. K., Hopp, H. E., and Brandt, A. (1983). *Carlsberg Res. Commun.* **48,** 187–199.
Reeck, G. R., and Hedgcoth, C. (1985). *FEBS Lett.* **180,** 291–294.
Reichelt, R., Schwenke, K.-D., König, T., Pähtz, W., and Wangermann, G. (1980). *Biochem. Physiol. Pflanz.* **175,** 653–663.
Righetti, P. G., Gianazza, E., Viotti, A., and Soave, C. (1977). *Planta* **136,** 115–123.
Roberts, L. S., Nozzolillo, C., and Altosaar, I. (1985). *Biochim. Biophys. Acta* **329,** 19–26.

Roussell, D., Boston, R. S., Goldsbrough, P. B., and Larkins, B. A. (1988). *Molec. Gen. Genet.* **211**, 202-209.
Salcedo, G., Sanchez-Monge, R., Argamentaria, A., and Aragoncillo, C. (1980). *Plant Sci. Lett.* **19**, 109-119.
Scallon, B. J. (1986). Ph.D. thesis. Purdue University, West Lafayette, Indiana.
Scheets, K., Rafalski, J. A., Hedgcoth, C., and Söll, D. (1985). *Plant Sci. Lett.* **37**, 221-225.
Sengupta-Gopalan, C., Reichert, N. A., Barker, R. F., Hall, T. C., and Kemp, J. D. (1985). *Proc. Natl. Acad. Sci. U.S.A.* **82**, 3320-3324.
Shannon, L. M., and Chrispeels, M. J., eds. (1986). "Molecular Biology of Seed Storage Proteins and Lectins." Am. Soc. Plant Physiol., Rockville, Maryland.
Sharief, F. S., and Li, S. S.-L. (1982). *J. Biol. Chem.* **257**, 14753-14759.
Shewry, P. R., and Miflin, B. J. (1982). *Qual. Plant.-Plant Foods Hum. Nutr.* **31**, 251-267.
Shewry, P. R., Finch, R., Parmer, S., Franklin, J., and Miflin, B. J. (1983). *Heredity* **50**, 179-189.
Shewry, P. R., Miflin, B. J., and Kasarda, D. D. (1984). *Philos. Trans. R. Soc. London, Ser. B* **304**, 297-308.
Shewry, P. R., Bunce, N. A. C., Kreis, M., and Forde, B. G. (1985). *Biochem. Genet.* **23**, 391-404.
Shotwell, M. A., Afonso, C., Davies, E., Chesnut, R. L. S., and Larkins, B. A. (1988). *Plant Physiol.* **87**, 698-704.
Simon, A. E., Tenbarge, K. M., Scofield, S. R., Finkelstein, R. R., and Crouch, M. L. (1985). *Plant Mol. Biol.* **5**, 191-201.
Smith, J. A., Rottmann, W. L., and Rubenstein, I. (1985). *Plant Sci.* **38**, 93-98.
Smith, S. C., Johnson, S., Andrews, J., and McPherson, A. (1982). *Plant Physiol.* **70**, 1199-1209.
Soave, C., and Salamini, F. (1984). *Philos. Trans. R. Soc. London Ser. B* **304**, 341-347.
Soave, C., Tardani, R., DiFonzo, N., and Salamini, F. (1981). *Cell* **27**, 403-410.
Spena, A., Viotti, A., and Pirrotta, V. (1982). *EMBO J.* **1**, 1589-1594.
Spena, A., Viotti, A., and Pirrotta, V. (1983). *J. Mol. Biol.* **169**, 779-811.
Spencer, D., and Higgins, T. J. V. (1980). *Biochem. Int.* **1**, 502-509.
Spencer, D., Chandler, P. M., Higgins, T. J. V., Inglis, A. S., and Rubira, M. (1983). *Plant Mol. Biol.* **2**, 259-267.
Staswick, P. E., Hermodson, M. A., and Nielsen, N. C. (1981). *J. Biol. Chem.* **256**, 8752-8755.
Sugiyama, T., Rafalski, A., Peterson, D., and Söll, D. (1985). *Nucleic Acids Res.* **13**, 8729-8737.
Sumner-Smith, M., Rafalski, A., Sugiyama, T., Stoll, M., and Söll, D. (1985). *Nucleic Acids Res.* **13**, 3905-3916.
Sun, S. M., and Hall, T. C. (1975). *J. Agric. Food Chem.* **23**, 184-189.
Talbot, D. R., Adang, M. J., Slightom, J. L., and Hall, T. C. (1984). *Mol. Gen. Genet.* **198**, 42-49.
Tatham, A. S., Shewry, P. R., and Miflin, B. J. (1984). *FEBS Lett.* **177**, 205-208.
Tatham, A. S., Miflin, B. J., and Shewry, P. R. (1985). *Cereal Chem.* **62**, 405-412.
Taylor, J. R. N., Schüssler, L., and Liebenberg, N. v. d. W. (1985). *S. Afr. J. Bot.* **51**, 35-40.
Thanh, V. H., and Shibasaki, K. (1976). *Biochim. Biophys. Acta* **439**, 326-338.
Thanh, V. H., and Shibasaki, K. (1977). *Biochim. Biophys. Acta* **490**, 370-384.
Thompson, R. D., Bartels, D., Harberd, N. P., and Flavell, R. B. (1983). *Theor. Appl. Genet.* **67**, 87-96.
Thompson, R. D., Bartels, D., and Harberd, N. P. (1985). *Nucleic Acids Res.* **13**, 6833-6846.
Tumer, N. E., Richter, J. D., and Nielsen, N. C. (1982). *J. Biol. Chem.* **257**, 4016-4018.
Viotti, A., Cairo, G., Vitale, A., and Sala, E. (1985). *EMBO J.* **4**, 1103-1110.
Walburg, G., and Larkins, B. A. (1983). *Plant Physiol.* **72**, 161-165.
Walburg, G., and Larkins, B. A. (1986). *Plant Mol. Biol.* **6**, 161-169.
Wang, S.-Z., and Esen, A. (1986). *Plant Physiol.* **81**, 70-74.

Wilson, C. M. (1983). *In* "Seed Proteins, Biochemistry, Genetics, Nutritive Value" (W. Gottschalk and H. P. Muller, eds.), pp. 271–307. Nijhoff/Junk, The Hague, The Netherlands.
Wilson, C. M. (1985). *Biochem. Genet.* **23,** 115–124.
Wilson, D. R., and Larkins. B. A. (1984). *J. Mol. Evol.* **29,** 330–340.
Wilson, C. M., Shewry, P. R., and Miflin, B. J. (1981). *Cereal Chem.* **58,** 275–281.
Wright, D. J., and Boulter, D. (1972). *Planta* **105,** 60–65.
Wrigley, C. W., and Shepherd, K. W. (1973). *Ann. N.Y. Acad. Sci.* **209,** 154–162.
Yamagata, H., Sugimoto, T., Tanaka, K., and Kasai, Z. (1982). *Plant Physiol.* **70,** 1094–1100.
Yamauchi, F., and Yamagishi, T. (1979). *Agric. Biol. Chem.* **43,** 505–510.
Yamauchi, F., Saio, K., and Yamagishi, T. (1984). *Agric. Biol. Chem.* **48,** 645–650.
Zhao, W.-M., Gatehouse, J. A., and Boulter, D. (1983). *FEBS Lett.* **162,** 96–102.
Zur Nieden, J., Manteuffel, R., Weber, E., and Neumann, D. (1984). *Eur. J. Cell Biol.* **34,** 9–17.

Stress-Induced Proteins: Characterization and the Regulation of Their Synthesis

8

TUAN-HUA DAVID HO
MARTIN M. SACHS

I. Introduction
II. Temperature Stress
   A. Alteration of Gene Expression during Heat Stress
   B. Potential Physiological Roles of HSPs
   C. Cold Temperature-Induced Gene Expression
III. Drought and Salt Stress-Induced Proteins
IV. Anaerobic Stress
V. Response to Ultraviolet Light Exposure
VI. Heavy Metal-Induced Proteins and Peptides
VII. Biological Stress
VIII. Summary and Perspective
   References

I. INTRODUCTION

Field-grown plants are constantly subject to adverse environmental conditions, such as drought, flooding, extreme temperatures, excessive salts, heavy metals, high-intensity irradiation, and infection by pathogenic agents. Because of their immobility, plants have to make necessary metabolic and structural adjustments to cope with the stress conditions. To this end, the genetic program in normal plants is altered by the stress stimuli to produce specific protein and activate biochemical pathways that ensure survival. For example, flooded plant tissues synthesize alcohol dehydrogenase (ADH) to catalyze ethanol formation coupled to the oxidation of NADH, thereby maintaining glycolysis in the anaerobically stressed cells (Sachs and Ho, 1986). Some stress-induced proteins may play a role in structural alterations that protect cells from being damaged by the stress conditions. The cell wall hydroxyproline-rich glycoproteins (HRGP) are produced in mechanically wounded tissues (Showalter and Varner, Chapter 12, this book). Because of their rigid structure, it is conceivable that the elevated levels of these proteins in the cell walls may help to seal off the tissues from further injury.

Several questions are addressed in the study of stress-induced proteins: (1) How is the stress signal perceived by the cells? (2) How does the perceived stress signal alter the expression of genes? (3) What are the physiological roles of the stress-induced proteins? Studies designed to answer these questions usually begin with the finding of new proteins in stressed tissues, usually by gel electrophoretic techniques. This initial observation is followed by purification of the stress proteins, and the cloning and characterization of their genes. In several cases, putative promoter sequences regulating the expression of these genes have been identified. Research on the physiological roles of stress proteins has been progressing, although many stress proteins remain unidentified. Very little is presently known about the molecular mechanism underlying the perception of stress signal.

II. TEMPERATURE STRESS

Because of seasonal changes, almost all plants are affected by temperature fluctuations during their life cycles. Very high ambient temperatures have been reported in many arid zones around the world, and the lack of effective transpiration in plants located in these areas causes the temperatures inside these plants to be significantly higher than ambient (Levitt, 1980a). Chilling or subfreezing temperatures are even more common in many areas of the world. In temperate zones, most plants can encounter a wide temperature range, from higher than 40°C in summer to subfreezing temperatures in winter. A rapid temperature upshift on sunny winter mornings, or a downshift in temperature as would occur after sunset in a desert, make the condition even more stressful. Three types of adjustments are expected in temperature-related stress conditions. First, macromolecules such as proteins are denatured at high temperatures; thus, denatured proteins have to be removed and their replacements synthesized. Second, metabolic pathways are often affected by temperature perturbations, causing the accumulation or depletion of certain metabolites. Third, some physical properties of membranes such as lipid fluidity are influenced by temperature shifts.

A. Alteration of Gene Expression during Heat Stress

The most readily observable response to heat stress in many organisms is the induction of heat-shock proteins (HSPs) (for reviews, see Schlesinger *et al.*, 1982; Craig, 1986; Burdon, 1986). This phenomenon was first investigated in the fruit fly, *Drosophila melanogaster*. When *Drosophila* cells are rapidly shifted from their normal growth temperature (25°C) to an increased temperature (37°C), there is a cessation of normal protein synthesis with the concomitant synthesis of a novel set of HSPs. This alteration of gene expression is accompanied by the regression of old polytene chromosome puffs and the generation of new ones (for review, see Ashburner and Bonner, 1979). The HSPs are not detectable at 25°C except in embryonic tissues, and their induction at 37°C is rapid; they appear within a few minutes after heat shock begins (Ashburner and Bonner, 1979). In *Drosophila*, the mRNAs encoding the HSPs result from *de novo* transcription and are selectively translated during heat shock.

The induction of HSPs has been studied in several higher plants, including soybean (Key *et al.*, 1981), pea (Barnett *et al.*, 1980), tobacco (Altschuler and Mascarenhas, 1982), tomato (Scharf and Nover, 1982), and maize (Baszczynski *et al.*, 1982; Cooper and Ho, 1983). Three size groups of HSPs have been observed: (1) the large HSP group ranging in size from 68 to 104 kDa (this group is ubiquitous among all organisms studied, including bacteria, animals and plants), (2) the intermediate-size group between 20 and 33 kDa, and (3) the small HSP group, about 15–18 kDa in size (this group is unique to higher plants). All of these proteins appear to be coordinately expressed when the tissue is under heat stress. The optimal condition for HSP induction in higher plants is a drastic temperature upshift to 39–42°C. However, a drastic temperature surge (shock) is not absolutely required for HSP induction; these proteins are also induced if there is a gradual temperature rise such as a 2.5°C per hour increase, a condition closer to what occurs in the field (Altschuler and Mascarenhas, 1982). Furthermore, HSPs and their mRNAs have been detected in field-grown plants under heat stress (Burke *et al.*, 1985; Kimpel and Key, 1985). Thus it is more appropriate to call these proteins "heat-stress proteins" rather than the commonly used "heat-shock proteins," although the abbreviation "HSP" remains the same. HSP synthesis can be detected within 20 min of heat stress, and the increase in transcript levels of some *Hsp* genes is noted within 3–5 min (Schöffl and Key, 1982; Figs. 1 and 2). However, the induction of HSP appears to be transient, lasting for only a few hours despite the continuous presence of heat-stress temperatures (Schöffl and Key, 1982). In maize roots, the HSPs are synthesized during the first 4 hr followed by a quick decline upon prolonged heat stress (Cooper and Ho, 1983; Fig. 1).

The induction of HSPs correlates with the increase of the levels of their transcripts. Liquid hybridization studies conducted by Schöffl and Key (1982) using cloned cDNA probes have revealed that about 20 different species of HSP18 mRNA accumulate to 19,000 copies per cell within 2 hr of heat stress in soybean cells (Fig. 2). When a heat-stressed tissue is returned to normal temperatures (e.g., 28°C) the synthesis of HSPs decreases over the next few hours with a concomitant decline in the levels of gene transcripts. Several plant *Hsp* genes have been cloned and sequenced, including

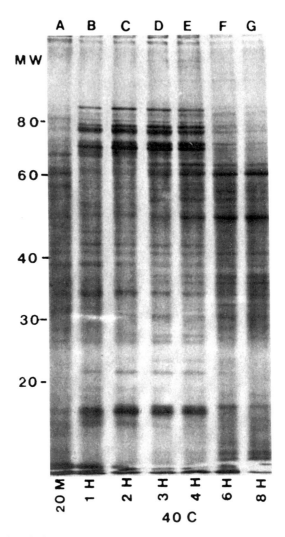

Fig. 1. SDS gel analysis of proteins synthesized by excised maize roots incubated at continuous 40°C. Roots of 3-day-old maize seedlings were excised and incubated at 40°C for increasing times [20 min (20M) to 8 hr (8H)] as indicated. Labeling with [^{35}S]methionine was carried out in the final 20 min of the incubation. Proteins were visualized by fluorography. The size distribution of kilodaltons is indicated at left. From Cooper and Ho (1983).

the genes for maize HSP70 (Rochester et al., 1986) and several genes encoding soybean HSP18 (Schöffl and Key, 1983; Czarnecka et al., 1985). The maize *Hsp70* gene appears to be very similar to the *Drosophila* counterpart, with 75% sequence homology in the coding region (Rochester et al., 1986). The three soybean *Hsp18* genes share greater than 90% homology in their

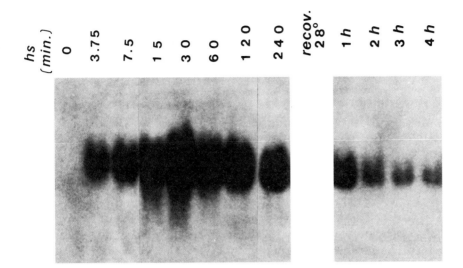

Fig. 2. Time course of accumulation of HSP mRNA. One microgram of poly(A)$^+$ RNA isolated from soybean hypocotyls after different times of incubation at 42.5°C (hs) or at additional times after transfer back to 28°C after 4 hr at the elevated temperature (recov), were electrophoresed in formaldehyde agarose gels. Blots of these gels were hybridized with a mixture of four cDNAs encoding small soybean HSPs ranging from 15 to 23 kDa. From Schöffl and Key (1982).

deduced amino acid sequences. They also share similarities with the *Drosophila* small HSPs (22–27 kDa) in hydropathy profiles (Schöffl *et al.*, 1984). Besides the TATA box (Goldberg, 1979), sequences related to the *Drosophila* heat-shock consensus regulatory element (CTnGAAnnTTCnAG; Pelham and Bienz, 1982) are found −48 to −62 bp 5' to the start of transcription (Schöffl *et al.*, 1984). In addition, there are secondary heat-shock consensus elements (decameric palindromic sequences: AGAAATTTTCT; CTnGAAnnTTCnAG) located further upstream (Nagao *et al.*, 1985). The DNA sequence analysis of these plant *Hsp* genes supports the view that the molecular mechanism involved in the induction of *Hsp* genes is highly conserved among eukaryotes. To further analyze the promoter regions, Gurley *et al.* (1986) have introduced a soybean small *Hsp* gene into primary sunflower tumors via Ti plasmid-mediated transformation. The soybean *Hsp* gene containing 3.25 kb of 5' flanking sequences is strongly transcribed in a thermoinducible manner in sunflower tumors. The 5' flanking region can be deleted to −192 base with no effect on the levels of thermoinduction, but the deletion results in a large increase in basal transcription. Additional deletion to position −95 reduces both the thermoinducible and basal transcription. Further analysis could pinpoint more precisely the sequences that are important for the induction of *Hsp* genes.

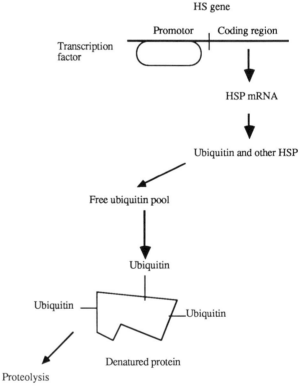

Fig. 3. Postulated mechanism for HSP induction as suggested by Munro and Pelham (1985). See text for details.

Although it is not clear how the heat-stress condition causes the induction of HSPs, Munro and Pelham (1985) have proposed a model to explain the molecular mechanism of HSP induction (Fig. 3). The key elements of the model are the induction of ubiquitin by heat stress (Bond and Schlesinger, 1985) and a transcription factor which binds to the promotor of *Hsp* genes (Wu *et al.*, 1987; Parker and Topol, 1984; Weiderrecht *et al.*, 1987). Ubiquitin is a 76-amino acid protein that forms conjugates with aberrant proteins through the ε-amino group of a lysine residue, and tags these proteins for degradation (for review, see Rechsteiner, 1985). It has also been shown that the function of histones can be altered by ubiquitination. Munro and Pelham (1985) speculate that, by analogy with histones, the *Hsp* gene transcription cofactor is normally in an inactive form conjugated with ubiquitin (Fig. 3A). When cells are under heat stress, many cellular proteins are thermally denatured and become ubiquitinated so that they could be removed by proteolysis. The depletion of the free ubiquitin pool causes the dissociation of ubiquitin from the transcription factor, allowing it to bind to the *Hsp* genes (Fig. 3B). Since ubiquitin has been shown to be one of the HSPs, the replenishment of the free ubiquitin pool will again lead to the ubiquitination of the HSP transcription factor, rendering it inactive and thus preventing further expression of *Hsp* genes. Although ubiquitinated HSP transcription factor has not been found in plants, the heat-stress-induced ubiquitin mRNA accumulation has been demonstrated in carrot root disks (Fig. 4). An alternative model of HSP induction involves the accumulation of special metabolites during heat stress. Lee *et al.* (1983) have found that heat stress in *Escherichia coli* causes the accumulation of a rare compound, P^1,P^4-diadenosine tetraphosphate (AppppA). They suggest that this type of compound is an "alarmone" responsible for the induction of HSPs and other heat-stress-induced processes. Guedon *et al.* (1985) microinjected *Xenopus laevii* oocytes with AppppA and found no induction of HSPs in the absence of heat stress. When this compound was microinjected after a mild heat stress, HSP70 synthesis was enhanced whereas the synthesis of other HSPs was inhibited. These authors suggest that AppppA may be involved in the eventual termination of the heat-stress response.

Besides heat stress, many other factors, including osmotic stress, high salt, 2,4-dinitrophenol, arsenite, anaerobiosis, and high concentrations of abscisic acid (ABA), ethylene, or auxins, also induce the synthesis of certain HSPs (Czarnecka *et al.*, 1984; Key *et al.*, 1985). In soybean, arsenite and cadmium induce a normal spectrum of HSPs, yet some of the other factors induce HSP27 specifically. In maize, ABA induces HSP70 (Heikkila *et al.*, 1984). Although the physiological significance of these inductions is not known, these factors can certainly be used as additional tools to investigate the molecular mechanisms underlying *Hsp* gene expression.

The induction of HSPs is accompanied by the alteration in the expression of other genes. In *Drosophila,* HSPs are essentially the only proteins that are

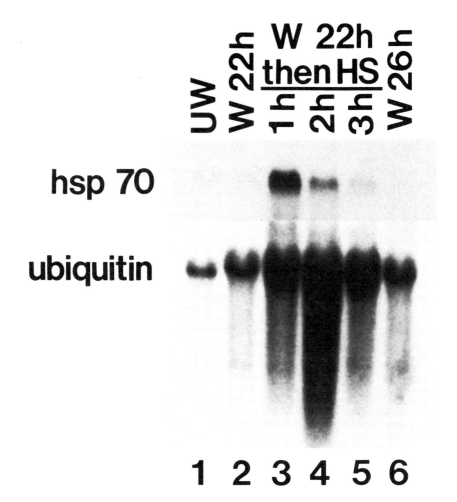

Fig. 4. Heat-stress induction of ubiquitin gene in carrot. Carrot root disks were mechanically wounded for 22 hr before heat stress (40°C) was imposed for various lengths of time. Total RNA was extracted and analyzed by Northern blot techniques. The probes used were maize HSP70 genomic clone and chicken fibroblast ubiquitin genomic clone. UW, Unwounded; W, wounded. Note that ubiquitin is expressed at low level before and after wounding, yet heat stress drastically enhances the level of its expression. From M. R. Brodl, M. Tierney, and T.-H. D. Ho (unpublished data).

synthesized during heat stress. The expression of normal proteins is suppressed and many preexisting mRNAs are sequestered rather than being degraded (Storti, 1980). When returned to a normal temperature the synthesis of normal proteins resumes even in the absence of new transcription (Lindquist, 1981). Thus, while the induction of HSPs in *Drosophila* is at the transcriptional level, the repression of the synthesis of normal proteins is at the translational level. In yeast, the preexisting mRNAs are not sequestered but undergo normal turnover (Lindquist, 1981). Hence, in the absence of continuous synthesis of these preexisting mRNAs their levels decrease gradually during heat stress. In higher plants, the effect of heat stress on the synthesis of normal proteins appears to be diverse. Vierling and Key (1985) have examined the effect of heat stress on the synthesis of ribulose-1,5-bisphosphate carboxylase in a soybean cell culture line. The synthesis of both the large and small subunits of this enzyme is decreased by 80% during heat stress. The levels of mRNA for the small subunit (encoded in the nuclear genome) also decrease during heat stress. In contrast, changes in the synthesis of the large subunit (encoded in the chloroplast genome) show little relationship to the corresponding mRNA levels; large subunit mRNA levels remain relatively unchanged by heat stress. Since the expression of the small subunit genes is most likely suppressed during heat stress, one cannot ascertain if the decrease in the levels of this mRNA is the consequence of normal turnover or of heat-stress-enhanced degradation. In cereal plants, unlike the case in soybean, the synthesis of many normal proteins continues during heat stress (Cooper and Ho, 1983). However, because the rate of total protein synthesis is unchanged, there appears to be suppression of the synthesis of some of the normal proteins to compensate for the increase in HSP synthesis. Belanger et al. (1986) used the barley aleurone layer to study the effect of heat stress on the synthesis of normal proteins. This tissue synthesizes α-amylase and protease in response to gibberellic acid (GA_3) treatment at normal temperatures (e.g., 25°C; Ho et al., 1987), and it was observed that heat stress quickly suppresses the synthesis of these two enzymes as well as a number of other secretory proteins. The suppression appears to occur not at the translational level, but at the level of mRNA stability, with the normally stable α-amylase mRNA being degraded during the heat stress. Concomitantly, there is a substantial destruction of the endoplasmic reticulum (ER). Upon recovery from heat stress, the synthesis of α-amylase resumes but only in the presence of new transcription. Since all the secretory proteins are synthesized on the ER, the disruption of this structure during heat stress could be related to the decrease in the synthesis of these proteins and the degradation of their mRNAs. Different from both soybean leaves and barley aleurone layers, the developing legume seeds synthesize some storage proteins even more rapidly during heat stress (Mascarenhas and Altschuler, 1985). In developing *Phaseolus* seeds, Chrispeels and Greenwood (1987) have shown that although the synthesis of phytohemagglutinin is enhanced, the posttranslational processing of this protein is retarded during heat stress. In contrast to *Drosophila* and soybean, both maize and barley appear to have proteins whose synthesis is neither suppressed nor enhanced by heat stress. It is not yet known why the synthesis of certain proteins is preserved while the synthesis of other proteins is quickly suppressed by heat stress.

B. Potential Physiological Roles of HSPs

Although the function of most of the HSPs in plants has not been elucidated, some of these proteins have been identified in other organisms: ubiquitin in chicken embryo fibroblasts (Bond and Schlesinger, 1985), lysyl-tRNA synthetase (Hirshfield *et al.*, 1981) and ATP-dependent protease in *E. coli* (Philips *et al.*, 1984), an isozyme of enolase in yeast (Iida and Yahara, 1985), and the uncoating ATPase that releases triskelia from coated vesicles in mammals (Ungewickell, 1985). Other properties of HSPs include the binding of these proteins to poly(A)$^+$ RNA in Hela cells (Schönfelder *et al.*, 1985), to fatty acids in rat (Guidon and Hightower, 1986), and to collagen in chicken embryo fibroblasts (Nagata *et al.*, 1986). The association of HSPs with cellular components could be beneficial to the heat-stressed cells. Minton *et al.* (1982) have shown that mixing HSPs with several enzymes *in vitro* actually protects the enzymes from being thermally denatured. To date, only ubiquitin has been identified as a HSP in plants (Fig. 4). Although the role of ubiquitin in plants is still under investigation (Vierstra, 1985), it is conceivable that ubiquitin plays a role in tagging proteins for degradation as it does in animal cells. It has been shown that the induction of HSPs allows the cells to establish thermotolerance in many organisms (Schlesinger *et al.*, 1982), i.e., to survive at a temperature that is normally lethal. Lin *et al.* (1984) have shown that briefly subjecting soybean seedlings to 40°C followed by incubation at 28°C results in the induction of HSPs and a concomitant establishment of thermotolerance (Fig. 5). In maize roots the temperature regime and time course of HSP induction also correlate closely with those for the establishment of thermotolerance (Cooper, 1985). The thermotolerant maize plants not only survive at the lethal temperature but slightly outgrow the control plants when they are returned to normal temperatures.

All the tissues in maize plants, with the exception of germinating pollen, synthesize HSPs during heat stress (Cooper *et al.*, 1984; Xiao and Mascarenhas, 1985). It is known that germinating pollen is more sensitive to high temperature than other tissues (Herrero and Johnson, 1980), a phenomenon that also suggests that the ability to synthesize HSPs is related to thermotolerance. However, this view is questioned by the observation that it is possible to establish thermotolerance in germinating *Tradescantia* pollen without the induction of HSPs (Xiao and Mascarenhas, 1985).

Since most of the chemical inducers of HSPs are either sulfhydryl reagents or compounds affecting oxygen metabolism, Nieto-Sotelo and Ho (1986) reasoned that HSPs may be involved in the metabolism of thiols in heat-stressed cells. They found that heat stress in maize root causes an enhancement in the biosynthesis of glutathione, an important thiol reductant in cells. It is conceivable that glutathione is used in heat-stressed cells to reduce "over-oxidized" cellular components to restore their function. It is not yet known which of the HSPs is involved in glutathione biosynthesis.

Stress-Induced Proteins

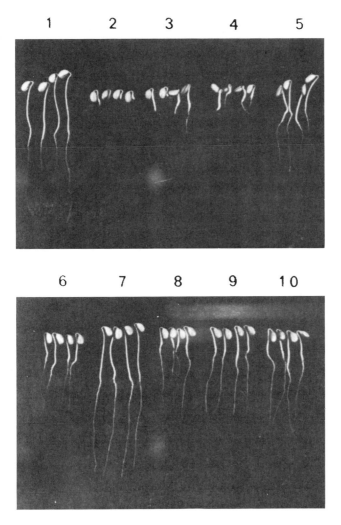

Fig. 5. Development of thermotolerance to 45°C in soybean seedlings by a 40°C pretreatment. Seedlings shown were incubated for 72 hr at 28°C after returning from the following treatments: (1) 28°C, 2 hr; (2) 45°C, 2 hr; (3) 40°C, 15 min → 45°C, 2 hr; (4) 40°C, 30 min → 45°C, 2 hr; (5) 40°C, 1 hr → 45°C, 2 hr; (6) 40°C, 2 hr → 45°C, 2 hr; (7) 40°C, 2 hr; (8) 40°C, 10 min → 28°C, 4 hr → 45°C, 2 hr; (9) 40°C, 30 min → 28°C, 2 hr → 45°C, 2 hr; and (10) 40°C, 2 hr → 28°C, 4 hr → 45°C, 2 hr. From Lin *et al.* (1984).

One approach to investigating the potential role of HSPs is to specify the cellular location of these proteins, which could be related to the function of specific organelles. Lin *et al.* (1984) have found in soybean that HSP15-18 and HSP69-70 are associated with both nuclei and ribosomes, HSP21 and 27 with mitochondrial fractions which may also contain plastids, and HSP84 and 92 with the postribosomal supernatant. They have also observed that the HSPs induced by other factors, such as arsenite, are not associated with these organelles, indicating that the association of HSPs with specific organelles is heat-mediated. Cooper and Ho (1987) determined the cellular location of maize HSPs using sucrose gradient centrifugation to fractionate a whole-cell extract. HSP29 (possibly equivalent to the soybean HSP27) is found to be associated with mitochondria, HSP18 and 70 with the plasma membranes, HSP25 and 72 with the ER, and the higher-molecular-weight HSPs such as HSP79–83 in the soluble fractions. The association of HSPs with membrane fractions suggests a role of HSPs in the altered membrane functions which likely occur at elevated temperatures. Nover *et al.* (1983) have presented evidence to suggest that a considerable amount of small HSPs in tomato and maize forms granular aggregates in the cytoplasm. However, the function of these heat-shock granules is unknown. Since some HSPs are highly conserved among many organisms ranging from bacteria to higher eukaryotes, it would be intriguing to investigate whether organelles such as mitochondria and chloroplasts still retain the ability to synthesize HSPs. Sinibaldi and Turpen (1985) and Nebiolo and White (1985) have reported that isolated mitochondria of maize synthesize a 62-kDa HSP when incubated at heat-stress temperature. However, Nieto-Sotelo and Ho (1987) have presented evidence to suggest that the 62-kDa HSP is likely produced by bacteria contaminating plant materials not prepared under strict aseptic conditions. These authors have shown that neither mitochondria nor plastids (i.e., chloroplasts and proplastids) can synthesize any HSP. However, possibilities exist that these organelles can take up HSPs encoded by the nuclear genome. Kloppstech *et al.* (1985) have shown that HSP22 can be transported *in vitro* to the membrane fraction of chloroplasts isolated from heat-shocked pea and *Clamydomonas*. Vierling *et al.* (1986) have also demonstrated that in soybean, pea, and maize, HSP21 and 27 can be taken up by isolated chloroplasts *in vitro*. It is apparent that a well-defined role for any of the plant HSPs has not yet been elucidated. To this end a clear-cut investigation on the cellular location of HSPs using immunohistochemical methods with monospecific antibodies would reveal useful information.

C. Cold Temperature-Induced Gene Expression

Low, nonfreezing, temperatures are essential for important physiological processes such as vernalization, stratification, and cold acclimation for freezing tolerance. Low temperatures have been shown to cause changes in protein content, enzyme activities, and membrane structures (Levitt, 1980a). The increase of freezing tolerance of plants during cold acclimation has been suggested to be the consequence of physiological and metabolic changes dependent on gene expression (Weiser, 1970). Exposure of spinach plants to 5°C induced a greater tolerance to freezing. Guy *et al.* (1985) have analyzed the changes in the mRNA population during cold acclimation of spinach by *in vitro* translation followed by SDS-gel electrophoresis. The levels of mRNAs encoding 82- and 180-kDa proteins are enhanced within 2 days of cold treatment, with the timing of this change correlating with the increase in freezing tolerance. More extensive changes in the levels of several other mRNAs are also observed in longer cold treatment. With the exception of two proteins, with M_r of 31 and 19 kDa, the cold-induced proteins appear to be different from HSPs. Thus, the change in mRNA levels induced by cold treatment is a separate response from that induced by heat stress. More recently, Guy and Haskell (1987) have shown that the induction of freezing tolerance in spinach is associated with the synthesis of cold acclimation-induced proteins. Analyzing the newly synthesized proteins in 0°C-treated rapeseed seedlings, Meza-Basso *et al.* (1986) found that protein synthesis continues at 0°C, and some proteins preferentially accumulate at this temperature. On the other hand, the synthesis of several other proteins is repressed while many are insensitive to cold treatment. These changes are probably regulated at the level of mRNA because *in vitro* translation primed with mRNA revealed the same pattern of changes. Among the proteins whose synthesis is suppressed by low temperature treatment is the small subunit of ribulose-1,5-bisphosphate carboxylase.

III. DROUGHT AND SALT STRESS-INDUCED PROTEINS

Several physiological changes induced by drought have been documented, including an increase in ABA levels, the closure of stomata, and the increase in cellular osmolarity. The increase in ABA levels is probably due to the *de novo* synthesis of this hormone (Zeevaart, 1980), and the process requires transcription (Guerrero and Mullet, 1986). The increase in ABA appears to be related to the closure of stomata to avoid further water loss. To date, the enzyme system involved in ABA biosynthesis has not been well investigated. The most commonly observed cellular components enhanced by drought are proline and betaine. A 10- to 100-fold increase of free proline content occurs in leaf tissues of many plants during moderate water deficit (Hanson and Hitz, 1982). It is not clear, however, whether changes in enzyme activity or gene expression are involved in the increase in proline. The increase in betaine is probably more related to salinity stress. One of the enzymes involved in the biosynthesis of betaine has been purified (A. D. Hanson, personal communication), which should facilitate the investigation into the molecular mechanisms underlying the regulation of betaine synthesis. The elevated levels of proline and betaine during drought appear to serve as osmotica to prevent further water loss from the stressed cells. It has also been reported that osmotic shock enhances the biosynthesis of polyamines (Flores and Galston, 1982). However, the concentrations of polyamines seem to be too low to account for any significant increase in osmolarity.

Drought also has a profound effect on protein synthesis. In many plant tissues, a reduced water potential causes a reduction of total protein synthesis and a rapid dissociation of polyribosomes. The latter has been shown not to be the consequence of increase in ribonuclease activity (Hsiao, 1973; Dhindsa and Bewley, 1976). The synthesis of a few proteins, however, is enhanced by low water potential. Jacobsen et al. (1986) have shown in barley leaves that water stress enhances the synthesis of one of the α-amylase isozymes. Using cDNA probe, they found that water-stressed leaves contained much more α-amylase mRNA than unstressed plants. Bozarth and Boyer (1987) have reported the accumulation of a 28-kDa cell wall glycoprotein in soybean upon prolonged exposure to low water potential. Lin and Ho (1987) have found that sorbitol or salt treatment results in the synthesis of a 29-kDa protein. It is not known whether these two proteins share common features.

Low water potential has a profound effect on chloroplast activities, including the decreases in electron transport and photophosphorylation (Mayoral et al., 1981) and the changes in conformation of the thylakoid and of the coupling factor (Younis et al., 1979). Some of these effects are similar to exposing thylakoid or coupling factor to Mg^{2+} above 5 mM (Younis et al., 1983), thus leading to the suggestion that Mg^{2+} concentration increase in chloroplasts due to water loss could cause these changes (Kaiser et al., 1986).

Salinity is another major limiting factor in agriculture, affecting a large area of cultivated land and, with increasing irrigation, salinity stress has become more widespread. In recent years there has been increased interest in the introduction of halophytes and in the breeding of salt-resistant crop plants to be used in coastal areas and lands that require heavy irrigation. Many culture cell lines adapted to high salt conditions have been isolated. As an initial attempt to study the biochemical mechanism of salt adaptation, Singh et al. (1985) have examined the protein profiles in a NaCl-adapted (salt-adapted) line of Nicotiana tabacum and a non-NaCl-adapted line. They found several new or enhanced protein bands (M_r of 58, 37, 35.5, 34, 26, 21, 19.5 and 18 kDa) with increasing levels of NaCl adaptation, while the intensities of other protein bands (54, 52, 17.5, and 16.5 kDa) were reduced. The 21-kDa protein has 4% hydroxyproline and 11.3% proline, unusually high levels for these amino acids, yet less than those in the cell wall HRGP (Ericson and Alfinito, 1984). The 26-kDa protein is the best characterized salt-stress-induced protein. It has a pI greater than 8.2 and it is also induced by ABA, whose level is enhanced in salt-adapted cells (Fig. 6). However, the synthesis of ABA-induced 26-kDa protein is transient unless the cultured cells are simultaneously exposed to NaCl stress. Singh et al. (1987) suggest that ABA

Stress-Induced Proteins

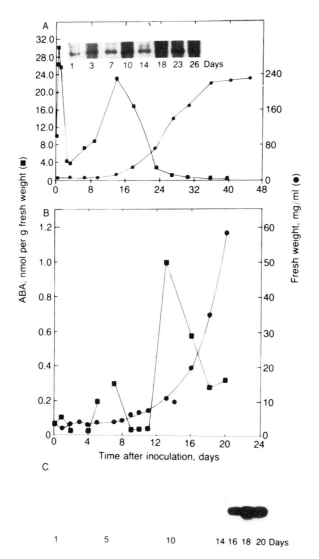

Fig. 6. Endogenous levels of ABA during adaptation of cultured tobacco cells to medium containing NaCl (10 g/liter) in the presence (A) and absence (B) of exogenously applied ABA. The synthesis of immunoprecipitable 26-kDa protein in cells treated with ABA is shown in A (inset) and in cells not treated with ABA in C. From Singh *et al.* (1987).

is involved in the normal induction of the synthesis of this 26-kDa protein and that the presence of NaCl is necessary for the protein to accumulate. In tobacco plants many tissues synthesize the 26-kDa protein in response to ABA treatment, but the highest level of expression of this protein was observed in the outer stem tissue. Although exogenously applied ABA induces the synthesis of an immunologically cross-reactive 26-kDa protein in cultured cells of several plant species, the physiological role of this 26-kDa protein remains unknown. King et al. (1986) found that tobacco suspension culture accumulates this protein even in the absence of NaCl as the cells approach stationary phase, but the accumulation never reaches the level seen in the salt-adapted cells. This protein can also be induced by other agents that lower the water potential, such as PEG and KCl, but no increase in the levels of this protein is seen after heat stress or heavy metal ($CdCl_2$) treatment. Although this protein accumulates with salt stress in hydroponically grown tomato plants, the levels of this protein do not seem to correlate with natural salt tolerance in wild tomato species. Therefore, it is speculated that the 26-kDa protein plays a role in responding to lowered water potential in plants instead of being related to salt tolerance. The induction of new mRNAs has also been reported in salt-stressed monocots. In wheat, salt stress induces 10 new mRNAs and suppresses 8 mRNAs in root tissue, but not in shoot tissues (Gulick and Dvorak, 1987).

An intriguing stress-induced alteration in gene expression occurs in a succulent plant, *Mesembryanthemum crystallinum,* which switches its primary photosynthetic CO_2 fixation pathway from C_3 to CAM upon salt or drought stress (Winter, 1974). Ostrem et al. (1987) have shown that the pathway switching involves the increase in the level of mRNA encoding phosphoenolpyruvate carboxylase, a key enzyme in CAM photosynthesis.

IV. ANAEROBIC STRESS

Anaerobic treatment results in a drastic alteration in the pattern of protein synthesis in maize seedlings (Sachs et al., 1980). Preexisting (aerobic) protein synthesis is repressed while selective synthesis of new polypeptides is initiated (Sachs et al., 1980). This is most likely a plant's natural response to flooding.

Studies on the maize anaerobic response stemmed from the extensive analysis of the maize ADH system by Drew Schwartz and co-workers (reviewed in Freeling and Bennett, 1985; Freeling and Birchler, 1981; Gerlach et al., 1986). Initially it was shown that ADH activity in maize seedlings increases as a result of flooding (Hageman and Flesher, 1960). Freeling (1973) later reported that ADH activity increased at a zero order rate between 5 and 72 hr of anaerobic treatment, reflecting a simultaneous expression of two unlinked genes, *Adh1* and *Adh2*. Schwartz (1969) showed that ADH activity is required to allow the survival of maize seeds and seedlings during flooding. ADH is the major terminal enzyme of fermentation in plants and is responsible for recycling NAD^+ during anoxia. It has been suggested that ethanolic fermentation permits tight cytoplasmic pH regulation, thus preventing acidosis from competing with lactic fermentation (Roberts et al., 1984a,b, 1985).

Except for one possible overlap, anaerobiosis induces a different set of proteins in maize than is induced by heat stress (Cooper and Ho, 1983; Kelley and Freeling, 1982; Sachs et al., 1980). In soybeans there appears to be a ~27- to 28-kDa HSP that is induced by a number of different environmental insults, including both heat stress and anaerobiosis (Czarnecka et al., 1984). In maize seedlings, as is the case for many heat-stress systems (cf. Schlesinger et al., 1982), repression of preexisting (aerobic) protein synthesis and the induction of a new set of proteins occurs very shortly after being subjected to anoxia (e.g., an argon atmosphere: Sachs et al., 1980).

The induction of anaerobic polypeptide synthesis occurs in two phases. During the first few hours of anaerobic treatment there is a transition period during which there is a rapid increase in the synthesis of a class of polypeptides of approximately 33 kDa (Fig. 7). These have been referred to as the transition polypeptides (TPs) as they represent most of the protein synthesis occurring in early anaerobiosis.

After approximately 90 min of anoxia, the induced synthesis of an additional group of ~20 polypeptides is first detected. This group of 20 anaerobic polypeptides (ANPs) represents greater than 70% of the total labeled amino acid incorporation after 5 hr of anaerobiosis (Fig. 7). By this time the synthesis of the TPs is at a minimal level; however, these polypeptides accumulate to a high level during early anaerobiosis and have been shown by pulse–chase experiments to be very stable. The synthesis of the ANPs continues at a constant rate for up to ~72 hr of anaerobic treatment (depending on which maize line is being examined), at which time protein synthesis decreases concurrently with the onset of cell death (Sachs et al., 1980). It has been shown in maize primary roots that hypoxia causes the induction of the ANPs, but does not cause the complete repression of preexisting protein synthesis (Kelley and Freeling, 1982). In addition, a novel set of polypeptides, not normally observed under aerobic or anaerobic conditions, is synthesized under hypoxic conditions (Kelley and Freeling, 1982).

The identities of some of the ANPs are now known. The isozymes of alcohol dehydrogenase, encoded by the *Adh1* and *Adh2* genes, have been identified as ANPs through the use of genetic variants (Ferl et al., 1979; Sachs and Freeling, 1978). More recently, glucose-6-phosphate isomerase (Kelley and Freeling, 1984a), fructose-1,6-diphosphate aldolase (Kelley and Freeling, 1984b; Kelley and Tolan, 1986), and sucrose synthase (Springer et al., 1986) have been identified as ANPs. Pyruvate decarboxylase activity has also been shown to be induced by anaerobiosis (Laszlo and St. Lawrence, 1983; Wignarajah and Greenway, 1976) and therefore is probably one of the ANPs. The identities and functions of the remaining ANPs or of the TPs are unknown. It may be noted, however, that five of the ANPs so far identified

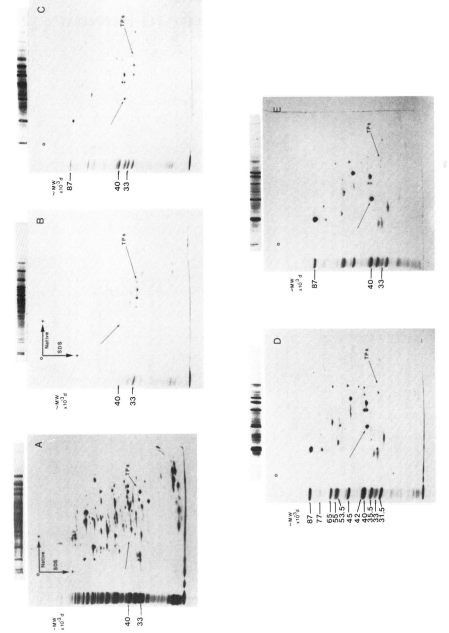

Fig. 7. Protein synthesis in a maize primary root during (A) 1-hr pulse labeling with [³H]leucine under aerobic conditions; (B–D) pulse labeling with [³H]leucine under anaerobic conditions; (B) 0–1 hr; (C) 4–5 hr; (D) 12–17 hr; (E) 50–55 hr. The arrow labeled "TPs" indicates the position of the transition polypeptides. The unlabeled arrow indicates the position of alcohol dehydrogenase 1 (ADH1).

are glycolytic enzymes, and sucrose synthase is also involved in glucose metabolism. In light of the inability of maize seedlings to survive 5 hr of flooding in the absence of an active *Adh1* gene (Schwartz, 1969), it appears that at least one function of the anaerobic response is to enable the plant to produce as much ATP as possible when there is an oxygen deficit, as would occur in water-logged soils.

In the presence of air, the roots, coleoptile, mesocotyl, endosperm, scutellum, and anther wall of maize synthesize a tissue-specific spectrum of polypeptides. The scutellum and endosperm of the immature kernel synthesize many or all of the ANPs constitutively, along with many other proteins under aerobic conditions. Under anaerobic conditions, all of the above organs selectively synthesize only the ANPs. Moreover, except for a few characteristic qualitative and quantitative differences, the patterns of anaerobic protein synthesis in these diverse organs are remarkably similar (Okimoto *et al.*, 1980). On the other hand, maize leaves, which have emerged from the coleoptile, do not incorporate labeled amino acids under anaerobic conditions, and do not survive even a brief exposure to anaerobiosis (Okimoto *et al.*, 1980).

The shift in pattern of protein synthesis during anaerobiosis has been observed in root tissue of many other plant species including rice (Bertani *et al.*, 1981; Pradet *et al.*, 1985), sorghum, barley, pea, and carrot (M. Freeling, unpublished data). In anaerobically treated barley aleurone cells, lactate dehydrogenase (LDH) activity increases (Hanson and Jacobsen, 1984) as do enzyme activity and mRNA levels for ADH (Hanson *et al.*, 1984). In the barley aleurone system, gibberellic acid (GA_3) induction of α-amylase synthesis and the accumulation of its mRNA are suppressed by anaerobiosis. The "anaerobic" pattern of protein synthesis is perturbed only slightly by GA_3 when aleurone cells incorporate labeled amino acids in a nitrogen atmosphere (Hanson and Jacobsen, 1984).

The rapid repression of preexisting protein synthesis caused by anaerobic treatment is correlated with a near complete dissociation of polysomes in primary roots of soybeans (Lin and Key, 1967) and maize (E. S. Dennis and A. J. Pryor, personal communication). This does not result from degradation of "aerobic" mRNAs, because the mRNAs encoding the preexisting proteins remain translatable in an *in vitro* system at least 5 hr after anaerobic treatment is initiated (Sachs *et al.*, 1980). This is in agreement with the observation that the polysomes, dissociated by anaerobiosis, rapidly reform up to 80–90% of their pretreatment levels, even in the absence of new RNA synthesis, when soybean seedlings are returned to air (Lin and Key, 1967).

The molecular basis of the maize anaerobic response has been analyzed with cDNA clones from high-molecular-weight poly(A)$^+$ RNA of anaerobically treated maize seedling roots (Gerlach *et al.*, 1982). Cloned anaerobic-specific cDNAs were identified by colony hybridization analysis, using differential hybridization to labeled cDNA of mRNA from anaerobic and aerobic roots. The anaerobic-specific cDNA clones were grouped into families on the basis of cross-hybridization, and several of the families were analyzed by hybrid-selected translation and by RNA gel blot (Northern) hybridization. The *Adh1* and *Adh2* cloned cDNA families were subsequently identified from this anaerobic-specific cDNA clone library, and the cDNA clone families and the genes encoding them were analyzed extensively (Dennis *et al.*, 1984, 1985; Sachs *et al.*, 1986).

The anaerobic-specific cDNA clones were used as probes to measure gene expression in maize seedling roots and shoots. In both tissues, the levels of mRNA hybridizable to the cDNAs increase during anaerobic treatment. This has been quantified in the case of *Adh1* and *Adh2;* with the kinetics of mRNA increase being the same for both mRNAs. The mRNA level first begins to increase after 90 min of anaerobic treatment, and continues to increase until it plateaus at 50-fold above the aerobic level at 5 hr of anaerobiosis. The level is maintained until after 48 hr in the case of *Adh1* but starts declining after 10 hr in the case of *Adh2* (Dennis *et al.*, 1985). This pattern of mRNA level increase and decrease is reflected in the previously described rates of *in vivo* anaerobic protein synthesis for ADH1 and ADH2 (Sachs *et al.*, 1980). Sucrose synthase (ANP87) also appears to have the same kinetics of mRNA increase (Hake *et al.*, 1985). *In vitro* run-off transcription experiments (Rowland and Strommer, 1985, 1986; L. Beach, personal communication) show that, in root cells, there is an increase in the transcription rate of the *Adh1* gene during anoxia, indicating that the increase in the levels of anaerobic-specific mRNAs is due to induced transcription of the *Anp* genes.

A comparison of the regions of the *Adh1* and *Adh2* genes upstream from the site of transcription initiation reveals only a few islands of homology. These include a 11-bp region that includes the "TATA box" (Goldberg, 1979) and three additional 8-bp regions of homology. One or more of these sequences might account for the anaerobic control of these genes (Dennis *et al.*, 1985), much as the 10-bp consensus sequence determined in the heat-stress system appears to be responsible for the high-temperature induction of *Hsp* gene transcription (Pelham and Bienz, 1982). *In vitro* mutagenesis coupled with transformation and gene expression studies as well as the analysis of the 5' regions of the other anaerobic genes will be necessary to determine which, if any, of these homologous regions might be important in regulating the induced expression of the anaerobic response genes.

Up to now, there has not been any reliable method of DNA transformation in maize (Cocking and Davey, 1987). Recently, electroporation has been used to introduce DNA into maize protoplasts (Fromm *et al.*, 1985, 1986; Howard *et al.*, 1987). Using this method together with *in vitro* mutagenesis of the *Adh1* promoter, Walker *et al.* (1987) found two regions upstream from the site of transcription initiation which appear to control the anaerobic induction of mRNA accumulation. Sequences homologous to those detected by Walker *et al.* (1987) are found in the *Adh1* and *Adh2* genes of maize and an *Adh1*-like gene from pea (Llewellyn *et al.*, 1987). However, there are no obviously comparable regions upstream of the *Adh* gene of *Arabidopsis thaliana* (Chang and Meyerowitz, 1986). Transformation by electroporation of DNA into maize protoplasts is a promising general system for *in vitro* analysis of gene expression, but this system has drawbacks. First, maize protoplasts cannot yet be regenerated into intact plants. Second, the cell culture systems used to make protoplasts for these studies are derived from maize embryo (scutellum), a tissue where ADH1 and other ANPs are synthesized at a high constitutive level under aerobic conditions.

Ellis *et al.* (1987), reported a similar study using the *Agrobacterium* system to introduce the maize *Adh1* promoter, associated with a marker gene, into tobacco. In this case, some incompatibilities were found in expression of the maize (a monocot) promoter in tobacco (a dicot), as the maize promoter was functional in transgenic tobacco plants only when it was augmented with an enhancer-like sequence from a gene constitutively expressed in tobacco. It would be preferable to use an homologous maize system where intact transgenic plants could be obtained. Also, it is necessary to test the expression of anaerobically stimulated promoters in tissues where they are normally inducible by anoxic treatment (e.g., the seedling root and shoot). Recent reports indicate that several methods of transformation are possible in maize, including the use of *Agrobacterium* infection (Graves and Goldman, 1986; Grimsley *et al.*, 1987), direct DNA transfer into pollen (De Wet *et al.*, 1986; Ohta, 1986), high-velocity microprojectiles for DNA transfer into intact cells which can be regenerated into whole plants (Klein *et al.*, 1987), and direct injection of DNA into premeiotic plants (de la Peña *et al.*, 1987).

V. RESPONSE TO ULTRAVIOLET LIGHT EXPOSURE

Plants can be damaged by the ultraviolet rays of intense sunlight. As a defense, plants produce flavonoid pigments that absorb the UV irradiation, minimizing damage to them. Cell suspension cultures of parsley produce and accumulate flavonoids when irradiated with ultraviolet light (Hahlbrock, 1981). It was found that UV treatment causes the coordinate induction of phenylalanine ammonia-lyase, 4-coumarate–CoA ligase, chalcone synthase, and UDPapiose synthase, all enzymes required for flavonoid biosynthesis. It was shown, using cDNA probes, that the levels of mRNAs encoding these enzymes increase during UV treatment (Kuhn *et al.*, 1984) and that this is the result of increased transcription rates of the genes (Chappell and Hahlbrock, 1984).

In contrast to the induction of the formation of flavonoid glycosides by UV irradiation, a cell wall fraction ("elicitor") from a fungal pathogen induces the formation of furanocoumarins (antimicrobial substances or "phytoalexins," see details below; Fig. 8) in parsley cell suspension cultures. Both responses induce the synthesis of three enzymes (phenylalanine ammonia-lyase, cinnamate 4-hydroxylase, and 4-coumarate–CoA ligase) to convert phenylalanine to 4-coumaroyl-CoA, a key intermediate of the phenylpropanoid metabolism in higher plants. Thereafter, the induction patterns diverge in accordance with the metabolic end product. UV induces the synthesis of a second group of at least 13 enzymes (including the aforementioned chalcone synthase and UDPapiose synthase) that allow the synthesis of flavonoids from 4-coumaroyl-CoA (Hahlbrock, 1981). Elicitor induces a series of enzymes that produce phytoalexins from the same intermediate (Hahlbrock *et al.*, 1985). Induction of phytoalexin synthesis by elicitor is correlated with a rapid and dramatic increase in the amounts of mRNAs encoding a number of small proteins having low isoelectric points (Hahlbrock *et al.*, 1985).

VI. HEAVY METAL-INDUCED PROTEINS AND PEPTIDES

Heavy metals, such as cadmium and copper, contaminate the environment because of increasing activities in mining and industrial waste disposal. Most of these heavy metals are toxic to plant metabolism, and plants have

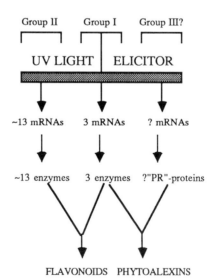

Fig. 8. Comparison of the two signal–reaction chains leading either to the UV light-induced formation of flavonoids or to the elicitor-induced formation of furanocoumarins and related compounds with antimicrobial activity. From Hahlbrock *et al.* (1985). "PR" proteins are pathogenesis-related proteins.

developed both a strategy of avoiding uptake of these toxic metal ions and an ability to synthesize proteins and peptides that can tightly bind and sequester these metals. A few cell lines have been selected for ability to grow in a normally toxic concentration of cadmium (Cd). In *Datura innoxia,* resistance to Cd is correlated with the ability to synthesize one or more cysteine-rich, metal-binding peptides (Jackson *et al.*, 1985). Similar Cd-binding peptides have been isolated from several other plant species (Rauser, 1984; Wagner, 1984; Reese and Wagner, 1987). The size of these peptides ranges from 2 to 10 kDa, and they are highly acidic with an apparent p*I* lower than 4.0. Amino acid analyses indicate that they are composed of predominantly Glu/Gln, Cys, and Gly. Although some similarities exist, these plant Cd-binding peptides appear to be different from metallothioneins, the mammalian heavy metal-binding proteins (Webb, 1979). Grill *et al.* (1985) have elucidated the structure of a group of cysteine-rich peptides capable of binding heavy metal ions via thiolate coordination. These peptides, named phytochelatins, have a general structure of $[\gamma\text{Glu-Cys}]_n$-Gly ($n = 2$–8), and the primary form has a molecular weight of 3600. Phytochelatins are induced by a wide range of metal ions including Cd^{2+}, Zn^{2+}, Pb^{2+}, Ag^+, AsO_4^{3-}, and SeO_3^{2-}. In cultured cells of several plant species, the synthesis of phytochelatins is induced within an hour after the addition of 200 μM $Cd(NO_3)_2$ (Fig. 9). The synthesis of phytochelatins can be inhibited by buthionine sulfoximine, a specific inhibitor of γ-glutamylcysteine synthetase. Concomitant with the induction of phytochelatins, a decrease in the cellular pool of glutathione takes place (Grill *et al.*, 1987; Fig. 9). This observation and other kinetic studies indicate that phytochelatins are probably synthesized from glutathione or its precursor γ-glutamylcysteine in a sequential manner, generating the set of homologous peptides. It is apparent that small peptides, such as phytochelatins, are more efficient than metallothioneins in sequestering heavy metal ions.

VII. BIOLOGICAL STRESS

Besides the physical stress conditions discussed above, plants are constantly subjected to invasion by insects and animals and infection by microorganisms. To cope with these biological stresses, plants produce specific proteins or metabolites that discourage the invading agents from further damaging the rest of the plants. One of the best examples of this type of response is the accumulation of serine proteinase inhibitor proteins in the leaves of plants from the Solanaceae and Leguminosae families when severely damaged by attacking insects or other mechanical agents (Nelson *et al.*, 1983). Cell wall fragments released from the damaged tissue can trigger a systemic response in other parts of plants (Ryan *et al.*, 1985). In response to this "wound hormone," the plant expresses two small gene families of

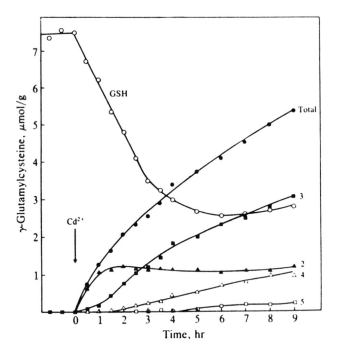

Fig. 9. Time course of phytochelatin induction and glutathione consumption after administration of 200 μM Cd(NO$_3$)$_2$ to *Rauwolfa serpentina* cell suspension culture. Quantities of glutathione (GSH, ○), total phytochelatin (●), and individual phytochelatins with n (number of γ-glutamylcysteine units per molecule) = 2 (▲), 3 (■), 4 (△), or 5 (□) are expressed as micromoles of γ-glutamylcysteine per gram (dry weight) of cells. From Grill *et al.* (1987).

serine proteinase inhibitors (Fig. 10). The accumulation of these inhibitors constitutes a defensive response that interferes with the digestive process of attacking agents.

Bishop *et al.* (1984) have isolated a pectic polysaccharide (M_r 6,000) from tomato leaves that can elicit the induction of proteinase inhibitors. Upon further hydrolysis, this pectic polysaccharide yields oligogalacturanan (degree of polymerization from 2 to 6) that still possesses the proteinase inhibitor inducing activities. Cell wall fragments from cultured sycamore cells can also elicit the induction of proteinase inhibitors in tomato plants, indicating the wide distribution of similar oligosaccharides among different species. Both inhibitors I and II from tomato plants have been characterized. Inhibitor I is initially synthesized as a 111-amino acid polypeptide (Graham *et al.*, 1985a), with the first 42 amino acids near the N-terminus apparently being removed during maturation. Inhibitor II has 148 amino acids, including a 25-amino acid signal sequence. It exhibits two domains, one a trypsin inhibitory site and the other a chymotrypsin inhibitory site, apparently evolved from a smaller gene by a process of gene duplication and elongation (Graham *et al.*,

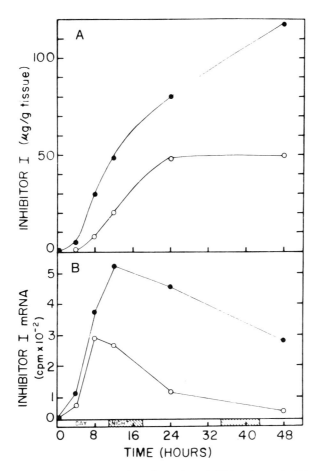

Fig. 10. Wound-induced accumulation of proteinase inhibitor I protein (A) and mRNA (B) in tomato leaves following a single wound (○) or four wounds (●) inflicted at 1-hr intervals. From Ryan *et al.* (1985).

1985b). Both inhibitors exhibit significant sequence homology to potato proteinase inhibitors.

Mechanical wounding also induces the synthesis of HRGP which are deposited in cell walls (Showalter and Varner, Chapter 12 this volume), such proteins apparently making the walls more rigid. The recent discovery that the seed coat is a rich source of HRGP suggests a protective role for these proteins against mechanical damages to the seeds (Cassab *et al.*, 1985).

In addition to forming passive physical barriers such as cuticles and rigid cell walls, there are at least two other means by which plants can defend themselves against fungal infection: the infected plants produce specific secondary metabolites (phytoalexins) which are toxic to the invading fungus (for review, see Bailey and Mansfield, 1982), or the plants synthesize enzymes which can hydrolyze fungal cell walls to stop the further migration of fungal hyphae. Most of the phytoalexins are derivatives of phenylalanine (Hahlbrock *et al.*, 1985; see Fig. 8), and phenylalanine ammonia-lyase and two other enzymes in the same pathway, chalcone synthase and chalcone isomerase, can be induced not only by infection but also by the fungal cell wall elicitors (Lawton *et al.*, 1983). Using thiouridine incorporation to isolate newly synthesized RNA from elicitor-treated cells, Cramer *et al.* (1985) have demonstrated that the induction of these enzymes is part of a rapid and extensive change in the pattern of mRNA synthesis directing production of a set of proteins associated with expression of disease resistance.

Fungal elicitors can also induce cell wall degrading enzymes such as chitinase and β-1,3-glucanase, which damages the fungal cells. Both of these enzymes have been purified and characterized in several plant species. Chitinase is around 30 kDa and β-1,3-glucanase is 33 kDa in size (Boller, 1985). Chappell *et al.* (1984) have shown that fungal elicitor can induce ethylene and the ethylene-forming enzyme, 1-aminocyclopropane-1-carboxylic acid (ACC) synthase, several hours before the induction of chitinase (Fig. 11). He also demonstrated that exogenously applied ethylene can induce both chitinase and β-1,3-glucanase. However, the action of elicitors in the induction of these enzymes is unlikely to be mediated by ethylene because the potent ethylene biosynthesis inhibitor, aminoethoxyvinylglycine (AVG), has no effect on the elicitor-induced chitinase (Boller, 1985). Indeed, ethylene and fungal infection appear to be separate, independent, stimuli for the induction of these enzymes.

In addition to the proteins mentioned above, there have been many reports of the induction of "pathogenesis-related" proteins (PRPs) in a number of plant species infected by viruses, viroids, bacteria, or fungi (Van Loon, 1985). It has been shown that, in cultured parsley cells, increased transcription of two PRP genes occurs within a few minutes of exposure to fungal elicitors (Somssich *et al.*, 1986). Some of the PRPs can be also induced by chemicals. For example, spraying tobacco with salicylic acid induces a number of PRPs, as well as resistance to tobacco mosaic virus infection (White, 1979). Although several PRPs from tobacco have been purified and their cDNAs isolated (Hooft van Huijsduijnen *et al.*, 1986), the exact action of these proteins remains unclear.

VIII. SUMMARY AND PERSPECTIVE

Because of a constantly changing environment, field-grown plants are always under the influence of stressful conditions. In order to fulfill the genetic potential, plants have adopted many strategies to cope with the stress. Most of the stress proteins discussed in this chapter have a role in

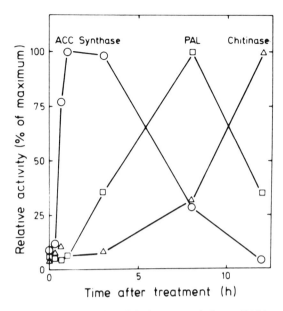

Fig. 11. Induction of chitinase, phenylalanine ammonia-lyase (PAL), and ACC synthase in parsley cell cultures by an elicitor from a fungal pathogen, *Phytophthora megasperma* (40 μg/ml). From Chappell *et al.* (1984).

either helping the plants survive, or in minimizing the effectiveness of the stress agent. In helping plants survive under stress conditions, the stress proteins perform the following functions: (1) maintenance of the basic metabolism in the stressed cell, e.g., the induction of ADH and some glycolysis enzymes in anaerobic stressed cells; (2) protection of cellular components from being damaged by the stressful condition, e.g., the association of HSPs with the enzymes or organelles, resulting in tolerance to thermodenaturation; and (3) removal of damaged cellular components, e.g., the tagging of denatured proteins by ubiquitin for proteolysis in heat-stressed cells. Although the stressed cells are usually not metabolically active, the induction of stress proteins can keep the cells from being killed, and the cells can recover once the stress condition is relieved. In minimizing the effectiveness of the stress agents, the stress proteins take up a more active set of functions: (1) the physical blockage of entry of stress agents such as the induction of cell wall HRGP and lignin synthesis by fungal elicitors; (2) the sequestration of stressful agents, as in the induction of phytochelatin to chelate heavy metal ions; and (3) the impairment of the biological stress agents, such as the induction of proteinase inhibitors and enzymes capable of hydrolyzing fungal cell walls, and the synthesis of phytoalexins.

The functions of many of the stress proteins, including many HSPs, TPs induced in anaerobic stress, P26 in salt-adapted cells, and many PRPs induced by pathogens, remain unclear at this point. On the other hand, proteins (enzymes) responsible for several physiological responses to stress conditions have not been isolated and studied. For example, ABA is well known for its role in stomatal closure during water stress, yet the enzyme catalyzing the rate-limiting step of ABA biosynthesis is yet to be identified.

An interesting and notable feature in plant stress response is the similarity among proteins induced by different stress conditions. In soybean, HSP27 can be induced by various stress conditions in addition to heat stress. In maize roots, the TPs induced by anaerobic stress are very similar in size and in charge to HSP33 in the same tissue. Phenylalanine ammonia-lyase is induced by both UV and fungal elicitors. Both heavy metal toxicity and heat stress affect the metabolism of glutathione. Therefore, it is not inconceivable that the diverse stress responses may share some common regulatory step(s).

Besides physiological and biochemical approaches, genetic analysis will be essential in the establishment of causal relationships between the induction of a stress protein and the establishment of tolerance to the stress condition. In most cases, it is not difficult to detect the induction of new proteins during stress. However, the induction of new proteins does not necessarily establish stress resistance; it may well be the consequence of damages caused by stress conditions. Thus, genetic mutants will be necessary to test the physiological role of a stress protein. Perhaps one of the best examples is the demonstration that an ADH null mutant in maize fails to survive during anaerobic stress (Schwartz, 1969). With the advances in genetic transformation techniques, it is likely that genes encoding stress proteins could be introduced into plants which are normally sensitive to the stress conditions. This approach would not only reveal DNA sequences which are responsive to stress stimuli, but also pave the way for constructing "tailor-made" stress-resistant plants.

ACKNOWLEDGMENTS

We thank Mark Brodl, David Hirsch, Jorge Nieto-Sotelo, and Douglas Russell for their critical reading of this chapter. The authors' research is supported by National Science Foundation: DCB 8702299 to T-H.D.H. and National Institutes of Health: 5 R01 GM34740 to M.M.S.

REFERENCES

Altschuler, M., and Mascarenhas, J. P. (1982). *Plant Mol. Biol.* **1**, 103–115.
Ashburner, M., and Bonner, J. J. (1979). *Cell* **17**, 241–254.
Bailey, J. A., and Mansfield, J. W. (eds.) (1982). "Phytoalexins." Blackie, Glasgow, Scotland.
Barnett, T., Altschuler, M., McDaniel, C. N., and Mascarenhas, J. P. (1980). *Dev. Genet.* **1**, 331–340.
Baszczynski, C. L., Walden, D. B., and Atkinson, B. G. (1982). *Can. J. Biochem.* **60**, 569–579.
Belanger, F. C., Brodl, M. R., and Ho, T.-H. D. (1986). *Proc. Natl. Acad. Sci. U.S.A.* **83**, 1354–1358.
Bertani, A., Menegus, F., and Bollini, R. (1981). *Z. Pflanzenphysiol.* **103**, 37–43.

Bishop, P. D., Pearce, G., Bryant, J. E., and Ryan, C. A. (1984). *J. Biol. Chem.* **259,** 13172–13177.
Boller, T. (1985). *In* "Cellular and Molecular Biology of Plant Stress" (J. L. Key and T. Kosuge, eds.), pp. 247–262. Academic Press, New York.
Bond, U., and Schlesinger, M. J. (1985). *Mol. Cell. Biol.* **5,** 949–956.
Bozarth, C. S., and Boyer, J. S. (1987). *Plant Physiol.* **83,** Abstr. 282.
Burdon, R. H. (1986). *Biochem. J.* **240,** 313–324.
Burke, J. J., Hatfield, J. L., Klein, R. R., and Mullet, J. E. (1985). *Plant Physiol.* **78,** 394–398.
Cassab, G. I., Nieto-Sotelo, J., Cooper, J. B., Van Holst, G.-J., and Varner, J. E. (1985). *Plant Physiol.* **77,** 532–535.
Chang, C., and Meyerowitz, E. M. (1986). *Proc. Natl. Acad. Sci. U.S.A.* **83,** 1408–1412.
Chappell, J., and Hahlbrock, K. (1984). *Nature (London)* **311,** 76–78.
Chappell, J., Hahlbrock, K., and Boller, T. (1984). *Planta* **161,** 475–480.
Chrispeels, M. J., and Greenwood, J. S. (1987). *Plant Physiol.* **83,** 778–784.
Cocking, E. C., and Davey, M. R. (1987). *Nature (London)* **236,** 1259–1262.
Cooper, P. (1985). Ph.D. dissertation. University of Illinois, Urbana.
Cooper, P., and Ho, T.-H. D. (1983). *Plant Physiol.* **71,** 215–222.
Cooper, P., and Ho, T.-H. D. (1987). *Plant Physiol.* **84,** 1197–1203.
Cooper, P., Ho, T.-H. D., and Hauptmann, R. M. (1984). *Plant Physiol.* **75,** 431–441.
Craig, E. A. (1986). *CRC Rev. Biochem.* **18,** 239–280.
Cramer, C. L., Ryder, T. B., Bell, J. N., and Lamb, C. J. (1985). *Science* **227,** 1240–1242.
Czarnecka, E., Edelman, L., Schöffl, F., and Key, J. L. (1984). *Plant Mol. Biol.* **3,** 45–58.
Czarnecka, E., Gurley, W. B., Nagao, R. T., Mosquera, L., and Key, J. L. (1985). *Proc. Natl. Acad. Sci. U.S.A.* **82,** 3726–3730.
de la Peña, A., Lörz, H., and Schell, J. (1987). *Nature (London)* **235,** 274–276.
Dennis, E. S., Gerlach, W. L., Pryor, A. J., Bennetzen, J. L., Inglis, A., Llewellyn, D., Sachs, M. M., Ferl, R. J., and Peacock, W. J. (1984). *Nucleic Acids Res.* **12,** 3983–4000.
Dennis, E. S., Sachs, M. M., Gerlach, W. L., Finnegan, E. J., and Peacock, W. J. (1985). *Nucleic Acids Res.* **13,** 727–743.
De Wet, J. M. J., De Wet, A. E., Brink, D. E., Hepburn, A. G., and Woods, J. A. (1986). *In* "Biotechnology and Ecology of Pollen" (D. L. Mulcahy, G. B. Mulcahy, and E. Ottaviano, eds.), pp. 59–64. Springer-Verlag, New York.
Dhindsa, R. S., and Bewley, J. D. (1976). *Science* **191,** 181–182.
Ellis, J. G., Llewellyn, D. J., Dennis, E. S., and Peacock, W. J. (1987). *EMBO J.* **6,** 11–16.
Ericson, M. C., and Alfinito, S. H. (1984). *Plant Physiol.* **74,** 506–509.
Ferl, R. J., Dlouhy, S. R., and Schwartz, D. (1979). *Mol. Gen. Genet.* **169,** 7–12.
Flores, H. E., and Galston, A. W. (1982). *Science* **217,** 1259–1260.
Freeling, M. (1973). *Molec. Gen. Genet.* **127,** 215–227.
Freeling, M., and Bennett, D. C. (1985). *Annu. Rev. Genet.* **19,** 297–323.
Freeling, M., and Birchler, J. A. (1981). *In* "Genetic Engineering Principles and Methods" (J. K. Setlow and A. Hollaender, eds.), Vol. 3, pp. 223–264. Plenum, New York.
Fromm, M., Taylor, L. P., and Walbot, V. (1985). *Proc. Natl. Acad. Sci. U.S.A.* **82,** 5824–5828.
Fromm, M., Taylor, L. P., and Walbot, V. (1986). *Nature (London)* **319,** 791–793.
Gerlach, W. L., Pryor, A. J., Dennis, E. S., Ferl, R. J., Sachs, M. M., and Peacock, W. J. (1982). *Proc. Natl. Acad. Sci. U.S.A.* **79,** 2981–2985.
Gerlach, W. L., Sachs, M. M., Llewellyn, D., Finnegan, E. J., and Dennis, E. S. (1986). *In* "Plant Gene Research: A Genetic Approach to Plant Biochemistry" (A. D. Blonstein and P. J. King, eds.), Vol. 3, pp. 73–100. Springer-Verlag, New York.
Goldberg, M. L. (1979). Ph.D. dissertation. Stanford University, Stanford, California.
Graham, J. E., Pearce, G., Merryweather, J., Titani, K., Ericsson, L., and Ryan, C. A. (1985a). *J. Biol. Chem.* **260,** 6555–6560.
Graham, J. E., Pearce, G., Merryweather, J., Titani, K., Ericsson, L., and Ryan, C. A. (1985b). *J. Biol. Chem.* **260,** 6561–6564.

Graves, A. C. F., and Goldman, S. L. (1986). *Plant Mol. Biol.* **7**, 43–50.
Grill, E., Winnacker, E.-L., and Zenk, M. H. (1985). *Science* **230**, 674–676.
Grill, E., Winnacker, E.-L., and Zenk, M. H. (1987). *Proc. Nat. Acad. Sci. U.S.A.* **83**, 439–443.
Grimsley, N., Hohn, T., Davies, J. W., and Hohn, B. (1987). *Nature (London)* **325**, 177–179.
Guedon, G., Sovia, D., Ebel, J. P., Befort, N., and Remy, P. (1985). *EMBO J.* **4**, 3743–3749.
Guerrero, F., and Mullet, J. E. (1986). *Plant Physiol.* **80**, 588–591.
Guidon, P. T., and Hightower, L. E. (1986). *Biochemistry* **25**, 3231–3239.
Gulick, P., and Dvorak, J. (1987). *Proc. Natl. Acad. Sci. U.S.A.* **84**, 99–103.
Gurley, W. B., Czarnecka, E., Nagao, R. T., and Key, J. L. (1986). *Mol. Cell. Biol.* **6**, 559–565.
Guy, C. L., and Haskell, D. (1987). *Plant Physiol.* **84**, 872–878.
Guy, C. L., Niemi, K. J., and Brambl, R. (1985). *Proc. Natl. Acad. Sci. U.S.A.* **82**, 3673–3677.
Hageman, R. H., and Flesher, D. (1960). *Arch. Biochem. Biophys.* **87**, 203–209.
Hahlbrock, K. (1981). *In* "The Biochemistry of Plants" (E. E. Conn, ed.), Vol. 7, pp. 425–429. Academic Press, New York.
Hahlbrock, K., Chappell, J., Jahnen, W., and Walter, M. (1985). *In* "Molecular Form and Function of the Plant Genome" (L. van Vloten-Doting, G. S. P. Groot, and T. C. Hall, eds.), pp. 129–140. Plenum, New York.
Hake, S., Kelley, P. M., Taylor, W. C., and Freeling, M. (1985). *J. Biol. Chem.* **260**, 5050–5054.
Hanson, A. D., and Hitz, W. D. (1982). *Annu. Rev. Plant Physiol.* **33**, 163–203.
Hanson, A. D., and Jacobsen, J. V. (1984). *Plant Physiol.* **75**, 566–572.
Hanson, A. D., Jacobsen, J. V., and Zwar, J. A. (1984). *Plant Physiol.* **75**, 573–581.
Heikkila, J. J., Papp, J. E. T., Schultz, G. A., and Bewley, J. D. (1984). *Plant Physiol.* **76**, 270–274.
Herrero, M. P., and Johnson, R. R. (1980). *Crop Sci.* **20**, 796–800.
Hirshfield, I. N., Bloch, P. L., Van Bogelen, R. A., and Neidhardt, F. C. (1981). *J. Bacteriol.* **146**, 345–351.
Ho, T.-H. D., Nolan, R. C., Lin, L.-S., Brodl, M. R., and Brown, P. H. (1987). *In* "Molecular Biology of Plant Growth Control" (M. Jacobs and J. E. Fox, eds.), pp. 35–49. Liss, New York.
Hooft van Huijsduijnen, R. A. M., Van Loon, L. C., and Bol, J. F. (1986). *EMBO J.* **5**, 2057–2061.
Howard, E. A., Walker, J. C., Dennis, E. S., and Peacock, W. J. (1987). *Planta* **170**, 535–540.
Hsiao, T. C. (1973). *Annu. Rev. Plant Physiol.* **24**, 519–570.
Iida, H. I., and Yahara, I. (1985). *Nature (London)* **315**, 688–690.
Jackson, P. J., Naranjo, C. M., McClure, P. R., and Roth, E. J. (1985). *In* "Cellular and Molecular Biology of Plant Stress" (J. L. Key and T. Kosuge, eds.), pp. 145–160. Liss, New York.
Jacobsen, J. V., Hanson, A. D., and Chandler, P. C. (1986). *Plant Physiol.* **80**, 350–359.
Kaiser, W. M., Schröppel-Meier, G., and Wirth, E. (1986). *Planta* **167**, 292–299.
Kelley, P. M., and Freeling, M. (1982). *In* "Heat Shock from Bacteria to Man" (M. J. Schlesinger, M. Ashburner, and A. Tissières, eds.), pp. 315–319. Cold Spring Harbor Lab., Cold Spring Harbor, New York.
Kelley, P. M., and Freeling, M. (1984a). *J. Biol. Chem.* **259**, 673–677.
Kelley, P. M., and Freeling, M. (1984b). *J. Biol. Chem.* **259**, 14180–14183.
Kelley, P. M., and Tolan, D. R. (1986). *Plant Physiol.* **82**, 1076–1080.
Key, J. L., Lin, C. Y., and Chen, Y. M. (1981). *Proc. Natl. Acad. Sci. U.S.A.* **78**, 3526–3530.
Key, J. L., Kimpel, J. A., Lin, C. Y., Nagao, R. T., Vierling, E., Czarnecka, E., Gurley, W. B., Roberts, J. K., Mansfield, M. A., and Edelman, L. (1985). *In* "Cellular and Molecular Biology of Plant Stress" (J. L. Key and T. Kosuge, eds.), pp. 161–179. Liss, New York.
Kimpel, J. A., and Key, J. L. (1985). *Plant Physiol.* **79**, 672–678.
King, G. J., Hussey, C. E., Jr., and Turner, V. A. (1986). *Plant Mol. Biol.* **7**, 441–449.
Klein, T. M., Wolf, E. D., Wu, R., and Sanford, J. C. (1987). *Nature (London)* **327**, 70–73.
Kloppstech, K., Meyer, G., Schuster, G., and Ohad, I. (1985). *EMBO J.* **4**, 1901–1909.

Kuhn, D. N., Chappell, J., Boudet, A., and Hahlbrock, K. (1984). *Proc. Natl. Acad. Sci. U.S.A.* **81**, 1102–1106.
Laszlo, A., and St. Lawrence, P. (1983). *Mol. Gen. Genet.* **192**, 110–117.
Lawton, M. A., Dixon, R. A., Rowell, P. M., Bailey, J. A., and Lamb, C. J. (1983). *Eur. J. Biochem.* **129**, 593–601.
Lee, P. C., Bochner, B. R., and Ames, B. N. (1983). *Proc. Natl. Acad. Sci. U.S.A.* **80**, 7496–7500.
Levitt, J. (1980a). "Response of Plants to Environmental Stress," Vol. 1, pp. 166–222. Academic Press, New York.
Levitt, J. (1980b). "Response of Plants to Environmental Stress," Vol. 1, pp. 347–470. Academic Press, New York.
Lin, C. Y., and Key, J. L. (1967). *J. Mol. Biol.* **26**, 237–247.
Lin, C. Y., Roberts, J. K., and Key, J. L. (1984). *Plant Physiol.* **74**, 152–160.
Lin, L.-S., and Ho, T.-H. D. (1987). *Plant Physiol.* **83**, Abstr. 288.
Lindquist, S. (1981). *Nature (London)* **294**, 311–314.
Llewellyn, D. J., Finnegan, E. J., Ellis, J. G., Dennis, E. S., and Peacock, W. J. (1987). *J. Mol. Biol.* **195**, 115–123.
Mascarenhas, J. P., and Altschuler, M. (1985). In "Changes in Eukaryotic Gene Expression in Response to Environmental Stress" (B. G. Atkinson and D. B. Walden, eds.), pp. 315–326. Academic Press, Orlando, Florida.
Mayoral, M. L., Atsmon, D., Gromet-Elhanan, Z., and Shimshi, D. (1981). *Aust. J. Plant Physiol.* **8**, 385–394.
Meza-Basso, L., Alberdi, M., Raynal, M., Ferrero-Cadinanos, M.-L., and Delseny, M. (1986). *Plant Physiol.* **82**, 733–738.
Minton, K. W., Karmin, P., Hahn, G. M., and Minton, A. P. (1982). *Proc. Natl. Acad. Sci. U.S.A.* **79**, 7107–7111.
Munro, S., and Pelham, H. (1985). *Nature (London)* **317**, 477–478.
Nagao, R. T., Czarnecka, E., Gurley, W. B., Schöffl, F., and Key, J. L. (1985). *Mol. Cell. Biol.* **5**, 3417–3428.
Nagata, K., Saga, S., and Yamada, K. M. (1986). *J. Cell Biol.* **103**, 223–229.
Nebiolo, C. M., and White, E. M. (1985). *Plant Physiol.* **79**, 1129–1132.
Nelson, C., Walker-Simmons, M., Makus, D., Zuroske, G., Graham, J., and Ryan, C. A. (1983). In "Plant Resistance to Insects," (P. K. Hedin, ed.), *ACS Symposium Series* **208**, 103–122.
Nieto-Sotelo, J., and Ho, T.-H. D. (1986). *Plant Physiol.* **82**, 1031–1039.
Nieto-Sotelo, J., and Ho, T.-H. D. (1987). *J. Biol. Chem.* **262**, 12288–12292.
Nover, L., Scharf, K. D., and Neumann, D. (1983). *Mol. Cell. Biol.* **3**, 1648–1655.
Ohta, Y. (1986). *Proc. Natl. Acad. Sci. U.S.A.* **83**, 715–719.
Okimoto, R., Sachs, M. M., Porter, E. K., and Freeling, M. (1980). *Planta* **150**, 89–94.
Ostrem, J. A., Olson, S. W., Schmitt, J. M., and Bohnert, H. (1987). *Plant Physiol.* **84**, 1270–1275.
Parker, C. S., and Topol, J. (1984). *Cell* **36**, 357–369.
Pelham, H. R. B., and Bienz, M. (1982). *EMBO J.* **1**, 1473–1477.
Philips, T. A., Van Bogelen, R. A., and Neidhardt, F. C. (1984). *J. Bacteriol.* **159**, 283–287.
Pradet, A., Mocquot, B., Raymond, P., Morisset, C., Aspart, L., and Delseny, M. (1985). In "Cellular and Molecular Biology of Plant Stress" (J. L. Key and T. Kosuge, eds.), pp. 227–245. Liss, New York.
Rauser, W. E. (1984). *Plant Physiol.* **74**, 1025–1029.
Rechsteiner, M. (1985). *Curr. Top. Plant Biochem. Physiol.* **4**, 15–24.
Reese, R. N., and Wagner, G. J. (1987). *Biochem. J.* **241**, 641–647.
Roberts, J. K. M., Callis, J., Jardetzky, O., Walbot, V., and Freeling, M. (1984a). *Proc. Natl. Acad. Sci. U.S.A.* **81**, 6029–6033.
Roberts, J. K. M., Callis, J., Wemmer, D., Walbot, V., and Jardetzky, O. (1984b). *Proc. Natl. Acad. Sci. U.S.A.* **81**, 3379–3383.

Roberts, J. K. M., Andrade, F. H., and Anderson, I. C. (1985). *Plant Physiol.* **77,** 492–494.
Rochester, D. E., Winter, J. A., and Shah, D. M. (1986). *EMBO J.* **5,** 451–458.
Rowland, L. J., and Strommer, J. N. (1985). *Proc. Natl. Acad. Sci. U.S.A.* **82,** 2875–2879.
Rowland, L. J., and Strommer, J. N. (1986). *Mol. Cell. Biol.* **6,** 3368–3372.
Ryan, C. A., Bishop, P. D., Walker-Simmons, M., Brown, W., and Graham, J. S. (1985). *In* "Cellular and Molecular Biology of Plant Stress" (J. L. Key and T. Kosuge, eds.), pp. 319–334. Liss, New York.
Sachs, M. M., and Freeling, M. (1978). *Mol. Gen. Genet.* **161,** 111–115.
Sachs, M. M., and Ho, T.-H. D. (1986). *Annu. Rev. Plant Physiol.* **37,** 363–376.
Sachs, M. M., Freeling, M., and Okimoto, R. (1980). *Cell* **20,** 761–767.
Sachs, M. M., Dennis, E. S., Gerlach, W. L., and Peacock, W. J. (1986). *Genetics* **113,** 449–467.
Scharf, K. D., and Nover, L. *Cell* **30,** 427–437.
Schlesinger, M. J., Ashburner, M., and Tissières, A. (1982). "Heat Shock from Bacteria to Man." Cold Spring Harbor Lab., Cold Spring Harbor, New York.
Schöffl, F., and Key, J. L. (1982). *J. Mol. Appl. Genet.* **1,** 301–314.
Schöffl, F., and Key, J. L. (1983). *Plant Mol. Biol.* **2,** 269–278.
Schöffl, F., Raschke, E., and Nagao, R. T. (1984). *EMBO J.* **3,** 2491–2497.
Schönfelder, M., Horsch, A., and Schmid, H.-P. (1985). *Proc. Natl. Acad. Sci. U.S.A.* **82,** 6884–6888.
Schwartz, D. (1969). *Am. Nat.* **103,** 479–481.
Singh, N. K., Handa, A. K., Hasegawa, P. M., and Bressan, R. A. (1985). *Plant Physiol.* **79,** 126–137.
Singh, N. K., LaRosa, C., Handa, A. K., Hasegawa, P. M., and Bressan, R. A. (1987). *Proc. Natl. Acad. Sci. U.S.A.* **84,** 739–743.
Sinibaldi, R. M., and Turpen, T. (1985). *J. Biol. Chem.* **260,** 15382–15385.
Somssich, I. E., Schmeizer, E., Bollmann, J., and Hahlbrook, K. (1986). *Proc. Natl. Acad. Sci. U.S.A.* **82,** 6551–6555.
Springer, B., Werr, W., Starlinger, P., Bennett, D. C., Zokolica, M., and Freeling, M. (1986). *Mol. Gen. Genet.* **205,** 461–468.
Storti, R. V., Scott, M. P., Rich, A., and Pardue, M. L. (1980). *Cell* **22,** 825–834.
Ungewickell, E. (1985). *EMBO J.* **4,** 3385–3391.
Van Loon, L. C. (1985). *Plant Mol. Biol.* **4,** 111–116.
Vierling, E., and Key, J. L. (1985). *Plant Physiol.* **78,** 155–162.
Vierling, E., Mishkind, M. L., Schmidt, G. W., and Key, J. L. (1986). *Proc. Natl. Acad. Sci. U.S.A.* **83,** 361–365.
Vierstra, R. D. (1985). *J. Biol. Chem.* **260,** 12015–12021.
Wagner, G. J. (1984). *Plant Physiol.* **76,** 797–805.
Walker, J. C., Howard, E. A., Dennis, E. S., and Peacock, W. J. (1987). *Proc. Natl. Acad. Sci. U.S.A.,* **84,** 6624–6628.
Webb, M. (1979). *In* "The Chemistry, Biochemistry, and Biology of Cadmium" (M. Webb, ed.), pp. 195–266. Elsevier/North-Holland, Amsterdam.
Weiderrecht, G., Shuey, J. D., Kibbe, W. A., and Parker, C. S. (1987). *Cell* **48,** 507–515.
Weiser, C. J. (1970). *Science* **169,** 1269–1278.
White, R. F. (1979). *Virology* **99,** 410–412.
Wignarajah, K., and Greenway, H. (1976). *New Phytol.* **77,** 575–584.
Winter, K. (1974). *Planta* **121,** 147–153.
Wu, C., Wilson, S., Walker, B., Dawid, I., Paisley, T., Zimarino, V., Veda, H. (1987). *Science* **238,** 1247–1253.
Xiao, C.-M., and Mascarenhas, J. P. (1985). *Plant Physiol.* **78,** 887–889.
Younis, H. M., Boyer, J. S., and Govindjee. (1979). *Biochim. Biophys. Acta* **548,** 328–340.
Younis, H. M., Weber, G., and Boyer, J. S. (1983). *Biochemistry* **22,** 2505–2512.
Zeevaart, J. A. D. (1980). *Plant Physiol.* **66,** 672–678.

The Thaumatins 9

H. VAN DER WEL
A. M. LEDEBOER

I. Introduction
II. Isolation and Characterization of Thaumatins
III. Biochemistry and Physiology
IV. Molecular Genetics of the Thaumatins
V. Study of the Natural Genes Encoding Thaumatin
VI. Production of Thaumatin by Microorganisms
    A. Expression of the Thaumatin Gene in *Escherichia coli*
    B. Expression of the Thaumatin Gene in Yeasts
    C. Expression of the Thaumatin Gene in *Kluyveromyces lactis*
    D. Prospects for the Production of Thaumatin by Microorganisms
VII. Expression of Thaumatin in Plants Other Than *Thaumatococcus daniellii*
VIII. Conclusions
    References

I. INTRODUCTION

Thaumatins are sweet proteins originating from the fruits of *Thaumatococcus daniellii* Benth (fam. Marantaceae) and were isolated for the first time by van der Wel and Loeve (1972).

The plant has been used extensively in tropical Africa for decades by inhabitants, who used the sweet material to make overfermented palm wine palatable, and the fruits were eaten by children (Daniell, 1855). There is a jelly around the seeds that may swell to 10 times its own weight and which can be used as a substitute for agar (Onwueme *et al.*, 1979). This tasteless transparent gel resembles gelatin (Adesina and Higginbotham, 1977). In West Africa, the leaves of the plant are used for roofing, its lamina for wrapping, and its petioles for weaving mats and baskets. The sweet ingredients of the fruit, the thaumatins, are intensively sweet, so they might be used as natural low-calorie sweeteners. Replacing sucrose by low-calorie sweeteners may reduce cariogenicity and diabetogenicity and lower caloric intake.

The *Thaumatococcus* plant grows in West African rainforest zones from Sierra Leone to Zaire but also in Angola. The plants mostly grow in areas of good, well-drained soil and where the annual rainfall is 1500–2000 mm. They prefer shade from tall trees in secondary forests. The papery leaves on petioles, which may be as high as 3 m, are large and oval-shaped. Its purple-pink flowers form pairs of small spikes near the ground. Usually, 20 to 40% bear clusters of mature fruits, which are mostly covered by leaves and soil debris. The fruit is trigonal in shape, fleshy, and the color changes from dark-green through brown to a bright red during maturation.

The fruit contains one to three hard black seeds; on the bottom there is a soft aril containing the sweet proteins. Normally, flowering occurs from March to June and again from August to October. Ripening takes about 3–4 months. The weight of a ripe fruit is between 9 and 40 g, with 16 g typical.

Small-scale surveys made in Ghana (Higginbotham, 1979) to determine fruit yield showed 2–8 tonnes per hectare. According to Higginbotham (1979), production under controlled conditions yields over 30 tonnes per hectare per annum. The recovery of thaumatin per kilogram of fruit could be greatly improved from 0.9 g/kg (van der Wel and Loeve, 1972) to 6 g/kg (Higginbotham, 1979) by applying a more efficient extraction process. It is remarkable that the fruits only grow on plants in their natural environment and not in other climates, nor in greenhouses despite vigorous flowering (Higginbotham, 1979). One approach to circumvent these agriculturally related problems is the biotechnological production of thaumatin in genetically engineered microorganisms, and Edens *et al.* (1982, 1984) have begun investigations into thaumatin production by microbial fermentation (Unilever European patent applications 54330 and 54331).

Higginbotham *et al.* (1981) and Higginbotham (1983) showed that thaumatins may also act as flavor enhancers/potentiators. To improve the taste quality (more sucrose-like), mixtures of thaumatin, amino acids, citric acid, and succinic acid have been formulated (Ochi, 1980; Higginbotham, 1983). Before a new sweetener can be marketed, extensive toxicological research is necessary. In an overview of such data, Higginbotham *et al.* (1983) indicate that there are no toxicological indications. In Japan, thaumatin is already being used in several food products.

II. ISOLATION AND CHARACTERIZATION OF THAUMATINS

The sweet-tasting thaumatins are extracted from the arils by homogenization with water, centrifugation, and freeze-drying of the aqueous solution after ultrafiltration. The proteins are further purified by ion-exchange chromatography (van der Wel and Loeve, 1972). Besides the two major proteins, thaumatins I and II, thaumatins III, b, and c were also identified (Higginbotham, 1979).

The number of thaumatins present in the fruit strongly depends on the origin of the fruits. The fruits from Nigeria contained about 70% thaumatin I, 20% thaumatin II, and 10% of the others. Some fruits from Ghana contain as much as 100% thaumatin I.

All thaumatins have a molecular mass of around 22,000 Da, an isoelectric point between pH 11.5 and 12.5, and a very similar amino acid composition with alanine as N-terminus.

Thaumatins I and II consist of 207 amino acid residues and have identical amino acid sequences, except for five residues (Iyengar *et al.*, 1979; Edens *et al.*, 1982). The thaumatins have 16 cysteine residues forming 8 disulfide bonds, only 1 methionine residue, and no histidine residue.

Specific chemical modification of lysine residues by acetylation and methylation and of arginine residues by treatment with cyclohexanedione shows that the isoelectric point of the molecule as such is not essential for sweetness sensation but that the lysine residues are important (van der Wel and Bel, 1976; van der Wel, 1983).

The three-dimensional structure of thaumatin I was elucidated by X-ray analysis at 3.1 Å resolution (De Vos *et al.*, 1985). The structure of thaumatin comprises three domains (Fig. 1). Domain I is formed by a long β-sheet folded in such a way that the β-strands of one flat part of the sheet are almost parallel to and on top of those of the other part of the sheet. The 11 strands are antiparallel to each other, except the amino- and carboxyl-terminal strands, which are parallel to each other. This domain is a so-called flattened "β-barrel." Attached to this domain are two small domains, rich in disulfide bonds. Domain II, from amino acids 128 to 177, contains a small α-helix and the disulfide bonds between Cys 126–177, Cys 134–145, and Cys 159–164. Domain III, from amino acids 54 to 80, contains the disulfide bonds Cys 56–66 and those between Cys 71 and 77. This backbone structure enables a precise assignment of the eight disulfide bridges, which agrees with the bonds determined by enzymatic hydrolysis of the intact molecule with trypsin and pepsin (van der Wel *et al.*, 1984). There are only two exceptions, namely the bonds between Cys 134–145 and Cys 149–158, instead of Cys 134–149 and Cys 145–158.

Small loops in domains II and III are stabilized by the disulfide bridges. This structure is analogous to that found in a class of small disulfide-rich proteins with a four-disulfide core (Drenth *et al.*, 1980). These proteins all bind to membrane-bound receptors.

De Vos *et al.* (1985) suggest that domain II and/or III may play a role in the binding of thaumatin to the taste-receptor membrane. Results of selective chemical modifications (van der Wel, 1980, 1983) suggest that one of the lysine residues in these regions does indeed play a role in the binding of the protein.

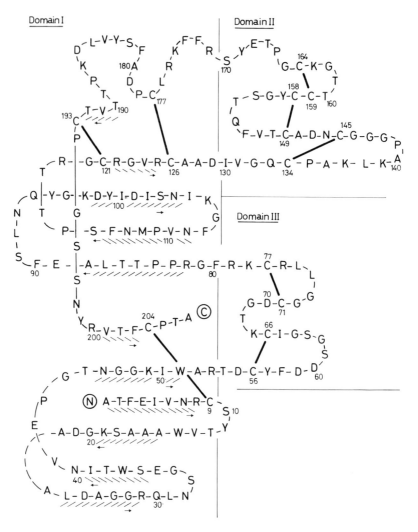

Fig. 1. Schematic diagram of thaumatin I. The vertical bars are the disulphide bonds. The β-strands of the top β-sheet are marked ////→ and the β-strands of the bottom β-sheet are marked ←\\\\. Ⓝ and Ⓒ are the N- and C-termini.

III. BIOCHEMISTRY AND PHYSIOLOGY

The thaumatin molecule is attractive for studying structural aspects connected with sweetness perception (van der Wel, 1980, 1983). Aqueous solutions of thaumatin have a sweetness 5500 times that of sucrose when compared with a 0.6% sucrose solution. On a molar basis, this is around 100,000 times sweeter than sucrose. In salt solutions and especially in phosphate, the sweetness is greatly reduced (van der Wel, 1981; Higginbotham and Hough, 1977). Considering the intensely sweet taste (threshold value of 10^{-8} mol/liter), it is a useful tool of chemoreceptor studies. The sweet sensation, elicited by the binding of the molecule to the receptor located on the taste cells is comparable to hormone receptor binding. Thus for a hormone, 50% binding to the receptor usually takes place at concentrations of around 10^{-8} to 10^{-11} mol/liter. For sucrose, a concentration of 10^{-1} to 10^{-3} mol/liter is required to elicit a sweet taste sensation. At this high concentration, a high degree of nonspecific binding occurs. With the thaumatins, studies similar to those with hormones can be carried out.

Another important question is the identification of the active center of the thaumatins responsible for the sweetness of the molecule. Helpful here is our knowledge of monellin, a sweet-tasting protein from another source (van der Wel, 1980). The tertiary structure of both groups of proteins is essential for sweetness and there are a number of similarities between thaumatins and monellins, such as equal sweetness intensity on a molar basis, immunological cross-reactivity, and local structural similarities in amino acid sequence.

Another interesting aspect is the phylogenetic implication of the taste response to thaumatin and monellin. Electrophysiological and behavioral experiments (Brouwer *et al.*, 1973; Glaser *et al.*, 1978; Hellekant *et al.*, 1981) revealed that besides man, only old-world monkeys, the catarrhine, can taste thaumatin. Only the Cercopithecidae, Hylobatidae, and Pongidae respond to this protein as man does and prefer this substance to water. These data may indicate there are different sweet receptor sites for different sweeteners, and this aspect is important for a further study of the transduction mechanism in taste.

IV. MOLECULAR GENETICS OF THE THAUMATINS

Molecular biological techniques enabled a detailed study of the genetic background of the thaumatins. Intriguing questions in this respect are

Are the thaumatins encoded by one gene and subsequently modified, or do they form part of a multigene family?

How is the expression of the genes regulated? They are strongly developmentally expressed in the arils, where they form 50% of the aril dry weight. Moreover, thaumatin I and II predominate and often constitute more than 20% of the aril dry weight each.

To elucidate these phenomena and to consider the feasibility of thaumatin production by microbial fermentation, isolation and analysis of the thaumatin-encoding genes were undertaken by Edens *et al.* (1982).

Total RNA was isolated from deep-frozen, homogenized arils and enriched for mRNA by passages on oligo (dT)-cellulose followed by a denaturing polyacrylamide gel. This mRNA preparation, which was shown to synthesize mainly thaumatin protein in a cell-free translation system, was used for the construction of a cDNA clone bank in *Escherichia coli*.

From 106 Tc^r Amp^s transformants obtained, 21 were found using a hybrid release screening procedure, to contain at least part of the thaumatin-encoding sequence. The DNA sequence of two clones was determined completely, one containing part of the structural gene up to the poly(A) tail and the other comprising the complete structural gene sequence (but missing parts of the 3' and 5' nontranslated sequences). The sequence of the 5' nontranslated region that was not cloned was determined directly on the mRNA. Comparison of the deduced amino acid sequence with the known amino acid sequence of thaumatin I (Iyengar *et al.*, 1979) shows a number of differences. At the amino-terminal end of the protein, there is a 22-amino acid-long hydrophobic presequence that closely resembles a signal sequence involved in the process of cotranslational secretion (Blobel and Dobberstein, 1975; Von Heijne, 1983). The carboxy-terminus has a six-amino acid-long, very acidic extension. Since the carboxy-terminal amino acid analysis of each of the thaumatins reveals no acidic extension, this prosequence must also be cleaved off during maturation of the protein in the plant. It is suggested that this sequence serves either as a recognition signal for compartmentalization or that it compensates for the extremely basic character of thaumatin, thereby facilitating membrane passage (Edens *et al.*, 1982; Edens and van der Wel, 1985). Apart from these preprosequences, the DNA-derived amino acid sequence differs in five positions from the thaumatin I sequence. Considering the effect that these differences would have on a tryptic map, it was concluded that the gene encoding thaumatin II had been cloned.

Of interest in the structure of the RNA molecule is the fairly short 5' nontranslated sequence which is only 31 nucleotides long. The mRNA also contains one of the few 5' nontranslated sequences that does not have a purine at the −3 position (Kozak, 1984). The 3' nontranslated sequence is 190 nucleotides long, contains three translational stop codons in phase and three (5') AAU AAA (3') sequences generally considered to function as a polyadenylation signal (Proudfoot and Brownlee, 1976).

V. STUDY OF THE NATURAL GENES ENCODING THAUMATIN

The cDNA clone was used to study the genes encoding thaumatin (Ledeboer *et al.*, 1984). Southern blots of *Hin*dIII-cleaved DNA from arils and leaves revealed at least seven different bands. From a *Hin*dIII clone bank in phage λ, five different clones were isolated. By hybridization with different ends of the cDNA clone, three were proved to contain virtually complete genes. Restriction enzyme analysis showed that all five clones were strongly related, correlating with the presence of several different thaumatins and suggesting that the different forms of thaumatin are the result of the presence of different genes, i.e., a gene family. Nuclease S1 mapping showed that all clones contained two small introns, which could not, however, be revealed by heteroduplex mapping. These very small introns resemble the situation in seed storage protein gene families, such as the genes for soybean glycinin (Fischer and Goldberg, 1982) or French bean phaseolin (Slightom *et al.*, 1983). Sometimes, introns may even be absent, e.g., in the case of hordein (Forde *et al.*, 1985).

VI. PRODUCTION OF THAUMATIN BY MICROORGANISMS

Since the production of thaumatin by industrial fermentation of genetically engineered microorganisms could be an attractive alternative to agricultural production and provide a constant supply essential for use as a food additive, a study of the expression of the gene in several microbial hosts was carried out.

A. Expression of the Thaumatin Gene in *Escherichia coli*

To study the expression of the gene in *E. coli*, the genes for the different maturation forms of thaumatin were brought under the transcriptional control of the *lac UV-5* or *trp* regulon. Expression was monitored by an enzyme-linked immunosorbent assay and by electrophoresis of [^{35}S]cysteine-labeled immunoprecipitated proteins on a sodium dodecyl sulfate-polyacrylamide gel. All maturation forms were expressed in *E. coli*, but yields were always low, up to about 10^3 molecules per cell for preprothaumatin. This calculation is based on the immunological detection of the proteins assuming that the epitopes of the natural protein and the *E. coli*-produced protein are alike. The expected nonoptimal folding will influence these calculations greatly. The different maturation forms were not processed to the mature form (Edens *et al.*, 1982).

B. Expression of the Thaumatin Gene in Yeasts

The yeasts *Saccharomyces cerevisiae* and *Kluyveromyces lactis* are GRAS (generally recognized as safe) organisms and are, therefore, more attractive for the production of food-grade proteins than *E. coli*. Moreover, *S. cerevisiae* can produce and excrete various heterologous proteins in an active configuration (Smith *et al.*, 1985; Kingsman *et al.*, 1985). The redox potential of the cells will lead to the correct folding in *S. cerevisiae*, which makes an additional renaturation step unnecessary.

To effect the transcription of the thaumatin genes, they were brought under control of the regulon of the gene encoding glyceraldehyde-3-phosphate dehydrogenase (GAPDH) (Edens *et al.*, 1984). This enzyme accounts for up to 5% of the dry weight of commercial bakers' yeast (Krebs, 1953) and the mRNA represents 2–5% of the total mRNA (Holland and Holland, 1978). There are three genes encoding GAPDH and one of them, recently named *tdh2* (McAlister and Holland, 1985), was isolated from a *S. cerevisiae* clone bank in *E. coli*. Either a fragment reaching up to position −843 or a fragment reaching up to −279 was fused to the different maturation forms of the thaumatin gene in *E. coli–S. cerevisiae* shuttle plasmids. These plasmids consist of the origin of replication and the Amp^r gene of pBR322 and of the 2μ origin of replication and the defective *leu2* gene from pMP81 (Hollenberg, 1982). A maximum of $3–5 \times 10^4$ molecules per cell was obtained when the preprothaumatin gene was fused to the longest form of the GAPDH regulon (Fig. 2A, pURY 528-03).

In the construction of pURY 528-03, the amino-terminal signal sequence was correctly cleaved off, as shown by the amino-terminal sequence analysis of the yeast-produced protein (Edens *et al.*, 1984). This proves that *S. cerevisiae* is able to recognize a plant signal sequence and that this processing step proceeds according to a universal mechanism. No data were obtained on a possible processing at the carboxy-terminal end of the protein. Other forms of either the regulon or the thaumatin gene lead to a substantial reduction in expression level.

Plasmids in which the defective *leu2* gene is used as a selection marker must be present in high copy numbers to be able to produce enough enzyme. This leads to a genetically unstable situation which can be shown by the loss of thaumatin production upon growth in a chemostat, even under selective conditions. Therefore, the defective *leu2* gene was replaced by the complete *leu2* gene from pYeleu10 (Ratzkin and Carbon, 1977). When *S. cerevisiae* containing this plasmid was grown in a chemostat, the amount of thaumatin molecules produced per cell dropped to 10^4, but this production remained stable under selective conditions. Under nonselective conditions, however, the plasmids are still unstable. No thaumatin was detected after 20 generations of growth in a chemostat on nonselective medium.

Another option for the stabilization of heterologous genetic information may be the transfer to the chromosome. An elegant procedure, based on site-specific chromosomal integration by a one-step gene disruption, has been worked out for *S. cerevisiae* by Rothstein (1983). Following this procedure, a plasmid has been constructed in which the preprothaumatin gene, fused to the GADPH regulon from plasmid pURY 528-03, is integrated into

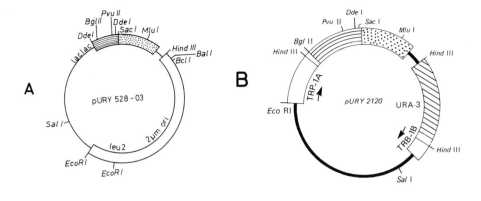

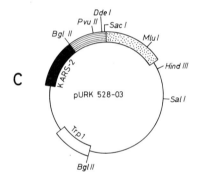

Fig. 2. Schematic representation of thaumatin-encoding (A) *E. coli–S. cerevisiae* (pURY 528-03), (B) *S. cerevisae* integrating plasmid pURY 2120, and (C) *E. coli–K. lactis* (pURK 528-03) shuttle plasmids. Dotted areas represent thaumatin-encoding sequences, hatched areas GAPDH promoter sequences. (A) pURY 528-03 was constructed by ligating first the 804-bp long *Dde*I fragment of the GAPDH promoter (positions −843 to −39), provided with *Eco*RI linkers, into *Eco*RI-cleaved pUR 528 (Edens *et al.*, 1982, 1984). The original GAPDH promoter was reconstructed by insertion of a synthetic DNA fragment containing the nucleotides −38 to −1. For further constructions, a *Sac*I site was designed into this fragment. The plasmid obtained was called pUR 528-03. From pMP81 (Hollenberg, 1982), a 5.4-kb *Hin*dIII–*Sal*I fragment was isolated, containing the 2μ ori and the *S. cerevisiae* leu2 gene. This fragment was inserted into *Hin*dIII–*Sal*I-cleaved pUR 528-03 to give pURY 528-03. (B) pURY 2120 was constructed by inserting the *Bgl*II–*Bam*HI fragment of pUR 528-03, containing the GAPDH promoter fused to the thaumatin gene, into the unique *Bgl*II site of pURY 2114. In this plasmid, the *ura3* gene is inserted into the *trp1* gene of *S. cerevisiae* on a 2μ-based plasmid (Overbeeke, 1989). (C) pEK2-7 is the *E. coli–S. cerevisiae* shuttle plasmid YRP7 (Stinchomb *et al.* 1979), containing the 1.2-kb KARS2 fragment, which allows autonomous replication in *K. lactis*. pEK2-7 is described in the European patent application 96430. A *Bgl*II fragment, containing KARS2 and the *trp1* gene, was isolated from pEK2-7 and inserted into the unique *Bgl*II site of pUR 528-03 to give pURK 528-03.

the *trpl* gene of *S. cerevisiae*, using the *ura3* gene as a selectable marker (Fig. 2B). trp^-, ura^+ Transformants of a trp^+, ura^- strain were shown to contain the thaumatin gene situated in the trp gene on chromosome four. Thaumatin production proved to be stable at a level of about 10^3 molecules per cell (without further selection) for at least 100 generations (Overbeeke, 1989).

C. Expression of the Thaumatin Gene in *Kluyveromyces lactis*

The yeast *K. lactis* is used in the industrial production of β-galactosidase, which is used to split milk sugar, and is, therefore, highly suited as a host for the production of a food-grade protein. Plasmids based on the origin of replication of the 2μ plasmid from *S. cerevisiae* do not replicate in *K. lactis*. To transfer the preprothaumatin gene to this yeast, use was made of an autonomously replicating sequence, KARS2, isolated from *K. lactis* by Das and Hollenberg (1982). Behind the GAPDH regulon and using the *trpl* gene as a selectable marker (Fig. 2C), the production of thaumatin-like protein was comparable to that in *S. cerevisiae*.

D. Prospects for the Production of Thaumatin by Microorganisms

Production of thaumatin by microbial fermentation is only attractive if it can compete with agricultural production. Because cultivation on plantations is not always possible, production by microbial fermentation could be a good alternative. Up to now, production yields in microorganisms have been low, which may be attributed to ineffective transcription termination, the need for integration in proper chromosomal surroundings, or an unstable or wrongly structured mRNA or protein. When translation or protein properties are considered, a clear deviation is found between the preferred codon usage of the organism from which the gene was isolated and the microbial hosts used. Moreover, the protein itself might interfere with the cellular metabolism of the hosts as a result of its high isoelectric point. Excretion as performed for chymosin (Smith *et al.*, 1985) could be a solution. The protective role of a pre- or prosequence might be deduced from the fact that the nonprocessed forms are expressed more effectively than the processed forms. The importance of an effective transcription termination or codon usage is indicated in a recent patent application, in which up to 10% of the SDS-soluble protein was claimed to be thaumatin, when a synthetic thaumatin gene in yeast-preferred codons was equipped with the PGK promoter and terminater of *S. cerevisiae* (European patent application 139501).

VII. EXPRESSION OF THAUMATIN IN PLANTS OTHER THAN *Thaumatococcus daniellii*

In recent years, the transfer of genes into plant cells and their expression in these cells has become feasible using the *Agrobacterium* Ti-vector system (Fraley *et al.*, 1983; Caplan *et al.*, 1984; Hoekema *et al.*, 1983). Although there are some indications of developmentally regulated organelle-specific expression, such as the expression of the French bean β-phaseolin gene in tobacco seeds (Sengupta-Gopalan *et al.*, 1985), we are still far from the developmentally controlled, high-level expression and organelle routing required for the commercial agricultural production of genetically engineered crops. Gene transfer for most commercial crops is not yet well established either, although much progress, including gene transfer into the commercially important monocots (Hooykaas-Van Slogteren *et al.*, 1984), has been made.

In this light, the transfer of thaumatin genes into plant cells and the study of their expression is an interesting system.

The localized high expression of thaumatin in the arils of the fruits of *T. daniellii* may provide further knowledge of the regulation that accomplishes these properties. This knowledge may enable the accomplishment of a high expression and the targeting of other genes of interest to seeds in other plants.

The thaumatin gene could be introduced into other plant species that are more suitable for the agricultural production of thaumatin under nontropical conditions.

In addition to sweetness, thaumatin has flavor-enhancing properties. This might contribute to the flavor improvement of existing crops and fruits, even if expression of the gene is low.

VIII. CONCLUSIONS

The plant protein thaumatin is of particular interest because of its extremely sweet properties. Although these properties are attractive for industrial exploitation, further development is still hampered for the following reasons: (1) *T. daniellii* is a shrub that only bears fruit under highly specific tropical conditions; (2) the thaumatins have some negative properties, such as a rather low thermostability and a too long-lasting sweet taste.

Production via microbial fermentation of genetically engineered microorganisms will circumvent the specific tropical conditions. Although reported production levels are still low, examples in the field of genetic engineering suggest that scale-up is only a matter of time. The highly basic properties of the protein could be a problem, which may be circumvented by excretion. The expression in other crops, more suitable for agricultural production, is still in an exploratory stage.

Moreover, the study of the highly localized expression may help to gain a certain insight into the complex process of developmental expression. The cloning of the thaumatin gene has made the protein accessible to modern protein engineering techniques, by which negative properties might be changed. The three-dimensional model of the protein will also facilitate the further identification of the essential regions.

ACKNOWLEDGMENTS

The authors thank Dr. N. Overbeeke for making his results available before publication. They are also grateful to Drs. J. Maat and N. Overbeeke for critically reading the manuscript.

REFERENCES

Adesina, S. K., and Higginbotham, J. D. (1977). *Carbohydr. Res.* **59,** 517–524.
Blobel, G., and Dobberstein, D. (1975). *J. Cell Biol.* **67,** 835–851.
Brouwer, J. N., Hellekant, G., Kasahara, Y., Van der Wel, H., and Zotterman, Y. (1973). *Acta Physiol. Scand.* **89,** 550–557.
Caplan, A., Herrera-Estrella, L., Inzé, D., Van Haute, E., Van Montagu, M., Schell, J., and Zambryski, P. (1984). *Science* **222,** 815–821.
Daniell, W. F. (1855). *Pharm. J.* **14,** 158.
Das, S., and Hollenberg, C. P. (1982). *Curr. Genet.* **6,** 123–128.
De Vos, A. M., Hatada, M., Van der Wel, H., Krabbendam, H., Peerdeman, A. F., and Sung-Han, K. (1985). *Proc. Natl. Acad. Sci. U.S.A.* **82,** 1406–1409.
Drenth, J., Low, B., Richardson, J., and Wright, C. (1980). *J. Biol. Chem.* **255,** 2652–2655.
Edens, L., and Van der Wel, H. (1985). *Trends Biotechnol.* **3,** 1–4.
Edens, L., Heslinga, L., Klok, R., Ledeboer, A. M., Maat, J., Toonen, M. Y., Visser, C., and Verrips, C. T. (1982). *Gene* **18,** 1–12.
Edens, L., Bom, I., Ledeboer, A. M., Maat, J., Toonen, M. Y., Visser, C., and Verrips, C. T. (1984). *Cell* **37,** 629–633.
Fischer, R. L., and Goldberg, R. B. (1982). *Cell* **29,** 651–660.
Forde, B. G., Heyworth, A., Pywell, J., and Kreis, M. (1985). *Nucleic Acids Res.* **13,** 7327–7338.
Fraley, R. T., Rogers, S. G., Horsch, R. B., Sanders, P. R., Flick, J. F., Adams, S. P., Bittner, M. L., Brand, L. A., Fink, C. L., Fry, J. S., Galluppi, G. R., Goldberg, S. B., Hoffmann, N. L., and Woo, S. C. (1983). *Proc. Natl. Acad. Sci. U.S.A.* **80,** 4803–4807.
Glaser, D., Hellekant, G., Brouwer, J. N., and Van der Wel, H. (1978). *Folia Primatol.* **29,** 56–63.
Hellekant, G., Glaser, D., Brouwer, J. N., and Van der Wel, H. (1981). *Chem. Senses* **6,** 165–173.
Higginbotham, J. D. (1979). *In* "Developments in Sweeteners" (C. A. M. Hough, K. J. Parker, and A. J. Vlitos, eds.), Vol. 1, pp. 86–123. Appl. Sci. Publ., London.

Higginbotham, J. D. (1983). *In* "Developments in Sweeteners" (T. H. Grenby, K. J. Parker, and M. G. Lindley, eds.), Vol. 2, pp. 119-130. Appl. Sci. Publ., London.
Higginbotham, J. D., and Hough, C. A. M. (1977). *In* "Sensory Properties of Foods" (G. C. Birch, J. G. Brennam, and K. J. Parker, eds.), pp. 129-149. Appl. Sci. Publ., London.
Higginbotham, J. D., Lindley, M. G., and Stephens, J. P. (1981). *In* "The Quality of Foods and Beverages" (G. Charalambous and G. E. Inglett, eds.), pp. 91-111. Academic Press, New York.
Higginbotham, J. D., Snodin, D. J., Eaton, K. K., and Daniel, J. W. (1983). *Food Chem. Toxicol.* **21,** 815-822.
Hoekema, A., Hirsch, P., Hooykaas, P., and Schilperoort, R. (1983). *Nature (London)* **303,** 179-180.
Holland, M. J., and Holland, J. P. (1978). *Biochemistry* **17,** 4900-4907.
Hollenberg, C. P. (1982). *Curr. Top. Microbiol. Immunol.* **96,** 119-144.
Hooykaas-Van Slogteren, G. M. S., Hooykaas, P. J. J., and Schilperoort, R. A. (1984). *Nature (London)* **311,** 763-764.
Iyengar, R. B., Smits, P., Van der Wel, H., Van der Ouderaa, F. J. G., Van Brouwershaven, J. H., Ravenstein, P., Richters, G., and Van Wassenaar, P. D. (1979). *Eur. J. Biochem.* **96,** 193-204.
Kingsman, S. M., Kingsman, A. J., Dobson, M. J., Mellor, J., and Roberts, N. A. (1985). *Biotechnol. Genet. Eng. Rev.* **3,** 377-416.
Kozak, M. (1984). *Nucleic Acids Res.* **12,** 857-872.
Krebs, E. G. (1953). *J. Biol. Chem.* **200,** 471-478.
Ledeboer, A. M., Verrips, C. T., and Dekker, B. M. M. (1984). *Gene* **30,** 23-32.
McAlister, L., and Holland, M. J. (1985). *J. Biol. Chem.* **260,** 15019-15027.
Ochi, T. (1980). *New Food Ind.* **23,** 13-28.
Onwueme, I. C., Onochine, B. E., and Sofowora, E. A. (1979). *World Crops* **May/June,** 106-111.
Overbeeke, N. (1989). *In* "Yeast Biotechnology" (P. J. Barr, A. J. Brake and P. Valenzuela, eds.). Butterworths, Stoneham (USA). In press.
Proudfoot, N. J., and Brownlee, G. G. (1976). *Nature (London)* **263,** 211-214.
Ratzkin, B., and Carbon, J. (1977). *Proc. Natl. Acad. Sci. U.S.A.* **74,** 487-491.
Rothstein, R. J. (1983). *Methods Enzymol.* **101,** 202-211.
Sengupta-Gopalan, C., Reichert, N. A., Barker, R. F., Hall, T. C., and Kemp, J. D. (1985). *Proc. Natl. Acad. Sci. U.S.A.* **82,** 3320-3324.
Slightom, J. L., Sun, S. M., and Hall, T. C. (1983). *Proc. Natl. Acad. Sci. U.S.A.* **80,** 1897-1901.
Stinchomb, D. T., Struhl, K., and Davis, R. W. (1979). *Nature (London)* **282,** 39-43.
Smith, R. A., Duncan, M. J., and Moir, D. T. (1985). *Science* **229,** 1219-1223.
van der Wel, H. (1980). *Trends Biochem. Sci.* **5,** 122-123.
van der Wel, H. (1981). *In* "Criteria of Food Acceptance" (J. Solms and R. L. Hall, eds.), pp. 292-295. Foster Verlag, Zurich, Switzerland.
van der Wel, H. (1983). *Chem. Ind.* **1,** 19-22.
van der Wel, H., and Bel, W. J. (1976). *Chem. Senses* **2,** 211-218.
van der Wel, H., and Loeve, K. (1972). *Eur. J. Biochem.* **31,** 221-225.
van der Wel, H., Iyengar, R. B., Van Brouwershaven, J., Van Wassenaar, P. D., Bel, W. J., and Van der Ouderaa, F. J. G. (1984). *Eur. J. Biochem.* **144,** 41-45.
Von Heijne, G. (1983). *Eur. J. Biochem.* **133,** 17-21.

Cytoskeletal Proteins and Their Genes in Higher Plants

10

DONALD E. FOSKET

I. The Cytoskeleton—A Definition
II. The Structure of Cytoskeletal Elements
    A. Microtubules
    B. Microfilaments
    C. Intermediate Filaments
III. Dynamics of the Cytoskeleton in Plant Cells
    A. The Cortical Cytoskeleton
    B. The Preprophase Band
    C. The Mitotic Spindle
    D. The Phragmoplast
    E. Cytoskeletal Derivatives
IV. Microtubule Proteins
    A. Isolation of Higher Plant Tubulin
    B. Microtubule Protein Isolation by Cyclic Rounds of Assembly
    C. Electrophoretic Mobility of Plant Tubulin Monomers
    D. Tubulins Are Encoded by Multigene Families
    E. Amino Acid Sequence of Tubulin Monomers
    F. Tubulin Domain Organization
    G. Tubulin Isotypes and Isoforms
    H. Colchicine Binding to Tubulins
    I. Herbicides as Plant Antimicrotubule Agents
    J. Tubulin Mutants
    K. Microtubule-Associated Proteins
    L. Tubulin Self-Assembly
    M. Cold Stability of Plant Microtubules
    N. Tubulin Synthesis in Relation to Microtubule Formation
V. Actin and Other Microfilament Proteins
    A. Types of Actins and Their Relationships
    B. Higher Plant Actins and Actin Genes
    C. Some Biochemical Properties of Actin
    D. Polymerization of G-Actin
    E. Actin-Binding Proteins

VI. Intermediate Filament Proteins
   A. Types of Intermediate Filament Proteins
   B. Structural Characteristics Common to Intermediate Filament Proteins
   C. A Common Structure from Different Proteins
   D. Higher Plant Intermediate Filaments
      References

The literature on cytoskeletal proteins has grown dramatically in the past few years. Several recent books deal exclusively with the cytoskeleton (Borisy *et al.,* 1984; Dustin, 1984; Schliwa, 1986; Bershadsky and Vasiliev, 1987), including one devoted to the cytoskeleton of higher plants (Lloyd, 1982). In addition, numerous contemporary reviews on specific aspects of the cytoskeleton have appeared, many of which are cited below. The following represents an up-to-date summary of the more important characteristics of cytoskeletal proteins, in to which the unique features and the challenges presented by the higher plant cytoskeleton have been integrated.

I. THE CYTOSKELETON—A DEFINITION

Interphase cells of nearly all eukaryotes exhibit an interconnected network of filamentous proteins, collectively known as the cytoskeleton. In addition, the mitotic and meiotic spindles of all eukaryotic cells are constructed of cytoskeletal proteins. These structures, therefore, can be considered to be mitotic or meiotic cytoskeletons. The principal functions of the cytoskeleton are (1) to determine cell shape, (2) to organize the cytoplasm, (3) to transport cellular structures, such as vesicles and chromosomes, and (4) to bring about cell motility. The cytoskeleton of vertebrate cells is composed of three filamentous polymers; microtubules, microfilaments, and intermediate filaments. Microtubules (MTs) are assembled chiefly from subunits of the heterodimeric protein, tubulin, and microfilaments (MFs) are linear polymers composed principally of actin subunits. Intermediate filaments (IFs) also are polymers assembled from specific structural protein monomers, but the nature of these proteins is different in different cell types. All three cytoskeletal components may contain additional specific proteins which contribute to their function and connect them with each other or with other cellular structures.

The cytoskeleton of higher plant cells is less well characterized than that of vertebrates. There is a large body of literature on the ultrastructural characteristics of plant MTs, and both MF and MT arrays have been visualized by specific immunochemical and/or fluorescent staining. Furthermore, plant tubulin and actin genes have been cloned and sequenced. However, the biochemical characterization of higher plant MT and MF proteins is just beginning. Although there is at least one report of an immunochemical demonstration of IFs in plant cells utilizing an antibody raised against a vertebrate IF protein (Dawson *et al.*, 1985), little is known about the distribution of these cytoskeletal structures in higher plants, and nothing is known about the biochemical characteristics of their component proteins. The functions of the plant cytoskeleton include those listed above for vertebrate cells, except that the cytoskeleton is only indirectly responsible for determining cell shape, and of course, higher plant cells do not move about, although cytoskeletal elements are responsible for moving organelles and chromosomes within the cell.

II. THE STRUCTURE OF CYTOSKELETAL ELEMENTS

Cytoskeletal components were first identified by electron microscopy and the characteristics of individual cytoskeletal elements, and their distribution within cells, continue to be investigated with this technique (see Gunning and Hardham, 1982, for a review of EM studies dealing with plant MTs, and Parthasarathy and Pesacreta, 1980, for a review of ultrastructural studies on plant MF). More recently, light microscopic procedures have been developed for visualizing cytoskeletal proteins using fluorescence-labeled probes, and/or various immunochemical procedures. This approach gives a more complete picture of the orientation and extent of a particular cytoskeletal array. Furthermore, it provides us with the means of identifying particular immunological determinants within the cytoskeleton. Investigations of the higher plant cytoskeleton using these procedures have been reviewed by Lloyd (1987).

A. Microtubules

The structure of the MT is highly conserved in evolution. In the electron microscope, MTs are tubular structures, 24 nm in diameter, which can be seen to be composed of subunits of the dumbbell-shaped heterodimeric protein, tubulin. The tubulin dimers within the MT are aligned end-to-end to form protofilaments. Microtubules from a diverse array of organisms contain 13 protofilaments, which associate laterally to form the wall of the tubule. The subunits in one protofilament are out of register with those in the adjacent protofilaments, creating a shallow helix (Amos, 1979). *In vitro*-assembled brain tubulin forms MTs with either 13 or 14 protofilaments. In either case, the subunits are arranged in a 3-start helix (Mandelkow *et al.*, 1986).

In thin sections, MTs frequently are seen to possess lateral appendages by which they seem to be connected to other MTs or to other cellular organelles. These appear to represent a specific class of microtubule-associated proteins (MAPs). The MAPs do not form an integral part of the MT. Typically, MAPs are linear rather than globular proteins, and they associate with the surface of the MT to form the lateral projections. The nature and position of the MAPs possibly determines which structures the MT may interact with and how this interaction will occur. They may be essential for some types of MT function, though they are not necessary for the formation of the basic MT structure.

B. Microfilaments

These are 5–7 nm diameter fibers composed of the globular protein actin. Like tubulin, actin also is capable of self-assembly to form the actin filament. Each actin MF is a two-stranded helix. Microfilaments usually are associated with other filamentous proteins which participate in forming the MF cytoskeleton or in its function. Some of these, such as fimbrin and tropomycin, are closely associated with the actin filaments and can be said to be part of the MF. Others, such as myosin, are integral parts of the functioning MF network, but only associate briefly with the actin filaments.

C. Intermediate Filaments

As the name implies, IFs are smaller in diameter than MTs but larger than MFs. Their structure in plant cells has not been described, if indeed they occur in these organisms. In vertebrates, they are 7–11 nm in diameter and show a lateral periodicity of 21 nm after they have been exposed to glycerol and heavy metal shadowing (Aebi *et al.*, 1983; Henderson *et al.*, 1982). While the morphology of these structures is similar in most vertebrate cells, their protein composition tends to vary with the cell type in which they are found. Some IF proteins are self-assembly competent while others can assemble *in vitro* only through an interaction with other IF proteins.

III. DYNAMICS OF THE CYTOSKELETON IN PLANT CELLS

The cytoskeleton is a highly dynamic entity, particularly in dividing cells where individual arrays appear, conduct their function, and are succeeded by other arrays as the cells progress through the division cycle. Higher plant cells exhibit four different cytoskeletal arrays: (1) the cortical array, (2) the preprophase band, (3) the mitotic spindle, and (4) the phragmoplast. Each of these consists of MTs, which most likely play the key role in the dynamics of the array. However, there is increasing evidence that all four also contain MFs. The contribution of the MFs to the organization and function of the plant cytoskeletal arrays, however, is not known.

A. The Cortical Cytoskeleton

Walled cells exhibit a peripheral cytoskeletal array during interphase in which a network of MTs run through the cortical cytoplasm (Wick et al., 1981). This cortical array may be required for the progression of cells through the mitotic cycle (Hahne and Hoffmann, 1984). In elongating cells, cortical MTs tend to be aligned at right angles to the direction of elongation, either as hoops or in helical arrays (Wick et al., 1981; Lloyd, 1982, 1984). Numerous studies have shown that disruption of the cortical MTs results in disordered deposition of cellulose microfibrils and a loss of the polarity of cell elongation (summarized by Robinson and Quader, 1982, and Lloyd, 1984). These studies emphasize the importance of MTs in determining the orientation of the deposition of cellulose microfibrils. In turn, the orientation of the cellulose microfibrils appears to be a major factor in plant cell and tissue morphogenesis (Green, 1985). Cortical MTs also appear to be involved in determining the pattern of wall deposition in differentiating tracheary elements (Pickett-Heaps, 1967; Hepler and Fosket, 1971; Falconer and Seagull, 1985, 1986). Since cellulose deposition occurs outside the cell from a plasmalemma-bound cellulose synthetase complex (Delmer, 1987), the mechanism by which MTs control cellulose microfibril orientation is not immediately obvious, although a number of theories have been proposed to account for this (summarized by Heath and Seagull, 1982).

In recent years, MF arrays have been visualized in interphase plant cells by means of immunochemical methods (Metcalf et al., 1980), or more often, specific fluorescent-labeled chemical probes (Parthasarathy, 1985; Seagull et al., 1987). Some of these MFs appear to be localized in the cortical cytoplasm, along with the MTs, while other MFs form a diffuse network throughout the cytoplasm. Seagull et al. (1987) reported that the MFs in elongating cells initially were transverse to the axis of elongation and parallel to the cortical MTs. MFs also may be associated with the cortical MTs which precede and accompany the deposition of the secondary wall (Quader et al., 1986; Kobayashi et al., 1987). The relationship between the MFs and the cortical MTs, as well as the possible contribution of the MFs to the orientation of the deposition of cellulose microfibrils, is not known.

B. The Preprophase Band

A dense array of MTs appears in the cortical cytoplasm as a ring around the nucleus when many plant cells prepare to enter mitosis. This is known as the preprophase band (Pickett-Heaps and Northcote, 1966; Gunning and Wick, 1985). Recently, Palevitz (1987) demonstrated that microfilaments also occur in the preprophase band. The formation of the preprophase band appears to be coincident with the loss of the more generalized cortical MT array (Wick and Duniec, 1984). It is thought to determine the position in which the cell plate forms, although the mechanism for this is not evident since the preprophase band disappears before or during prophase, while the phragmoplast does not form until telophase (Gunning and Wick, 1985; Simmonds, 1986).

C. The Mitotic Spindle

The mitotic spindle forms in late prophase, after the preprophase band has disappeared. Microtubules are the most extensively studied, and perhaps the most significant component of the mitotic spindle (Bajer and Molé-Bajer, 1986). However, there is evidence that actin filaments also occur in the spindle (Forer and Jackson, 1979; Seagull et al., 1987). The function of the mitotic spindle is, of course, the separation of the chromatids and their transport to opposite poles. Although subjected to intense investigation, the mechanism for the movement of the chromosomes in mitosis and meiosis is not known (but see Pickett-Heaps et al., 1982, for a discussion of some of the possibilities).

D. The Phragmoplast

The phragmoplast, the fourth cytoskeletal array in higher plant cells, forms in late telophase, as the spindle disappears. It transports vesicles containing cell wall materials to the developing cell plate. The phragmoplast consists of two sets of parallel MTs, both oriented at right angles to the division plane and overlapping at their tips within it (Hepler and Jackson, 1968). Euteneuer and McIntosh (1980) demonstrated that the growing (+) ends of the two sets of phragmoplast MTs were both within the division plane where they overlap. In addition to MTs, the phragmoplast also contains MFs (Clayton and Lloyd, 1984; Gunning and Wick, 1985) and components that react with an antibody to IF proteins (Dawson et al., 1985).

E. Cytoskeletal Derivatives

The cytoskeletal elements of the interphase cell or the dividing cell are highly dynamic and contain some degree of disorder. In contrast, cilia, flagella, centrioles, basal bodies, and muscle sarcomeres all contain comparatively stable, and much more highly ordered cytoskeletal derivatives. It seems reasonable to suppose that these structures arose in evolution after the cytoskeleton itself through the acquisition of additional MT- and/or MF-associated proteins. Flagella, centrioles, and basal bodies are found throughout the algae and are present in some of the reproductive cells of lower tracheophytes. None of these structures are present in angiosperm or most conifer cells.

IV. MICROTUBULE PROTEINS

The biochemistry and molecular biology of plant tubulins have been reviewed recently (Morejohn and Fosket, 1986; Dawson and Lloyd, 1987). The following section presents some results or perspectives not covered in these reviews.

A. Isolation of Higher Plant Tubulin

Tubulin was first isolated from cultured vertebrate cells by ion-exchange chromatography on DEAE-cellulose (Weisenberg et al., 1968). The acidic tubulin protein binds more tightly than most cellular proteins, but it can be eluted with high salt. The highly purified tubulin isolated from vertebrate brain tissue by this procedure is capable of self-assembly to form microtubules in vitro (Lee and Timasheff, 1977). The first isolations of assembly-competent tubulin from higher plant cells utilized modifications of this procedure (Morejohn and Fosket, 1982; Mizuno, 1985). It is important to protect against proteolysis of the tubulins by endogenous proteases during their isolation from some species of higher plants. α-Tubulin is especially susceptible to degradation (Morejohn et al., 1985).

Tubulin isolated by the procedure of Morejohn and Fosket (1982) is 85% pure, as judged by Coomassie blue-stained polyacrylamide gel electrophoretic (PAGE) separations. Other chromatographic procedures have been used to further purify plant tubulin initially isolated by ion-exchange chromatography. Some of these, particularly Sephadex G-200 chromatography, substantially improve tubulin purity, but they markedly decrease the tubulin yield (Kato et al., 1985; Mizuno, 1985; Mizuno et al., 1985). Alternatively, taxol-induced assembly of the tubulin, followed by collection of the assembled MTs by centrifugation, is a fast and highly effective method for the further purification of plant tubulin initially isolated by ion-exchange chromatography (Morejohn and Fosket, 1982; Cyr et al., 1987a).

B. Microtubule Protein Isolation by Cyclic Rounds of Assembly

Weisenberg (1972) first demonstrated that tubulin in brain homogenates could self-assemble in vitro. Subsequently, this was developed into a method for the isolation of MT proteins (for example, see Shelanski et al., 1973). Tissue homogenates, in a pH 6.8 buffer containing GTP and ethylene glycol bis(β-aminoethyl ether-N,N'-tetracetic acid (EGTA), are warmed (24–37°C) to induce MT assembly. The in vitro-assembled MTs are collected by centrifugation, resuspended in ice-cold buffer to disassemble the MTs, and the homogenate is centrifuged to remove particulate matter and cold-stable MTs. The homogenate is warmed again to induce another round of MT assembly, and the assembled MTs are collected by centrifugation. These cycles of warm assembly followed by cold-induced disassembly are repeated several times to give a preparation that is almost pure MT protein. However, it is only 70–90% tubulin. The major additional proteins maintain a constant stoichiometric relationship with tubulin through the isolation procedure, so they are unlikely to be contaminants. In some cases, the proteins coisolated with tubulin by this procedure have been shown to be MAPs (Dentler et al., 1975; Sloboda and Rosenbaum, 1979a).

Although this represents the most commonly used method for isolating MT proteins from vertebrate tissues, it has not been successful for higher plant tubulin isolation. There are several reasons for this, and the reader is referred to Morejohn and Fosket (1986) for a discussion of these reasons. One of these reasons, however, deserves some additional comments. The tubulin concentration in most plant cell homogenates is too low to permit tubulin self-assembly. In order to form MTs, tubulin must exceed the critical concentration for assembly, which can be as high as 4 mg/ml (Sloboda et al., 1975). Because plant cells frequently have a comparatively small cytoplasmic volume and a large central vacuole, the total protein concentration in many plant cell homogenates is low. Furthermore, tubulin usually represents less than 1% of the total protein in plant cells (R. J. Cyr, unpublished observations). Thus, the tubulin concentration in plant homogenates is likely to be less than the critical concentration for self-assembly.

C. Electrophoretic Mobility of Plant Tubulin Monomers

Electrophoresis of tubulin from many sources under denaturing conditions separates the tubulin monomers and indicates apparent molecular weights of 56,000 (α) and 53,000 (β) (Olmsted et al., 1971). Higher plant tubulins exhibit electrophoretic mobilities similar to, but not identical with, those of vertebrate tubulin monomers. The magnitude of this difference, however, is not clear. Some studies have found that plant β-tubulin comigrates with vertebrate brain tubulin, while plant α-tubulin migrates faster than vertebrate α subunit, but slower than plant β-tubulin (Morejohn and Fosket, 1982; Yadav and Filner, 1983; Morejohn et al., 1984). Other studies have reported that plant α-tubulins migrate more rapidly than plant β-tubulin under conditions where the relative mobilities of vertebrate brain α- and β-tubulins are not affected (Mizuno et al., 1985; Hussey and Gull, 1985; Fukuda, 1987). Cyr et al. (1987a) compared the relative electrophoretic mobilities of reduced and akylated rat brain and carrot tubulin subunits in two different sodium dodecyl sulfate (SDS)-acrylamide gel electrophoretic systems, one a modification of the Laemmli (1970) procedure, and the second a modification of the Studier (1973) procedure. In the Studier-type system, both the rat and carrot subunits were well separated. The carrot β-tubulin comigrated with the rat brain β-tubulin, while the carrot α-tubulin migrated more rapidly than the brain α-, but more slowly than the carrot β-tubulin. In the Laemmli-type gel system, the plant tubulin subunits were poorly separated, and the α-tubulin migrated more rapidly than the β-tubulin subunit, while the relative electrophoretic mobilities of the rat brain tubulin subunits were not changed. When the carrot α-tubulin subunit initially separated by the Laemmli procedure was eluted from the gel and reelectrophoresed in the Studier system, it migrated more slowly than the β-tubulin, demonstrating that the subunit was not permanently altered. Since these two gel systems differ principally in the pH at which the separation occurs, this result indicates that pH may markedly influence the interaction of SDS with plant, but not vertebrate α-tubulin.

The apparent molecular weights of tubulins determined by SDS-gel electrophoresis are 10–12% greater than their true molecular weights as determined from their amino acid sequence (Pontingl *et al.*, 1981; Krauhs *et al.*, 1981; Silflow *et al.*, 1987). Since vertebrate α- and β-tubulins are both about 50,000, it should not be possible to separate them on the basis of molecular weight differences under the conditions usually used for SDS-PAGE. In fact, Bryan and Wilson (1971) demonstrated that animal tubulin monomers are separated on the basis of charge and not molecular weight differences. This indicates that some factor, possibly the highly charged C-terminal domain, interferes with the binding of SDS to both α- and β-tubulin, but that this interference is greater with α- than with β-tubulin. Many microorganisms also have been shown to exhibit altered relative mobilities of their tubulin subunits in SDS gels (Maekawa and Sakai, 1978; Clayton and Gull, 1982; White *et al.*, 1983).

D. Tubulins Are Encoded by Multigene Families

Tubulins are encoded by a multigene family in most eukaryotes, with the number of α- and β-tubulin genes varying from 2 to over 20 in some vertebrates (Cleveland and Sullivan, 1985). The human genome contains more tubulin pseudogenes than functional genes; however, humans, along with most mammals, appear to have six functional β- and six functional α-tubulin genes (Villasante *et al.*, 1986; Wang *et al.*, 1986). Chickens have seven expressed β-tubulins (Sullivan *et al.*, 1986). The amino acid sequences of five of the six mammalian β-tubulins are over 95% homologous, while the sixth mammalian β-tubulin is 78% homologous to the other five. The alga *Chlamydomonas* has two α- and two β-tubulin genes (Brunke *et al.*, 1984). The two α-tubulin genes encode slightly different proteins, while the two β-tubulin genes encode identical proteins (Silflow *et al.*, 1985; Youngblom *et al.*, 1984). Both the α- and β-tubulins of *Chlamydomonas* are nearly 90% homologous at the amino acid level with the most prevalent, comparable vertebrate tubulins.

Only in the case of *Arabidopsis thaliana* do we have anything approaching a complete picture of the size of tubulin multigene families in higher plants, although β-tubulin sequences have been published for both soybean and *Zea mays*. Southern blots of *Arabidopsis* genomic DNA, hybridized with 5′ and 3′ tubulin coding sequence probes, demonstrated that this Angiosperm with an unusually small genome contains at least seven β-tubulin genes (Marks et al., 1987; Oppenheimer et al., 1988) and 4 α-tubulin genes (Ludwig et al., 1987a). The sequence of two of the *Arabidopsis* α-tubulin genes has been published (Ludwig et al., 1987a,b), as has the sequence of four of the *Arabidopsis* β-tubulin genes (Marks et al., 1987; Silflow et al., 1987; Oppenheimer et al., 1988). Furthermore, these six tubulin genes have been shown to be transcribed and to encode different, although highly homologous, proteins. The nucleotide sequence of the coding region of the *Arabidopsis* α-tubulin gene designated *tua3* has 72% homology with the *Chlamydomonas* α-tubulin genes. This *Arabidopsis* α-tubulin gene would encode a 450-amino acid protein whose sequence is 90% homologous to the *Chlamydomonas* α-tubulin protein, and 83% homologous to a human α-tubulin (Ludwig et al., 1987a). The *tua3* gene is abundantly expressed in all tissues of the *Arabidopsis* plant. In contrast, the more divergent *Arabidopsis* α-tubulin gene, *tua1*, is expressed only in flowers, and possibly only in pollen (Ludwig et al., 1987b). The *Arabidopsis* β-tubulin gene *tub1* also exhibits a restricted pattern of expression. Its transcript was found primarily in roots (Oppenheimer et al., 1988).

The β-tubulin genes have been isolated from a soybean genomic library and sequenced (Guiltinan et al., 1987a,b). They encode β-tubulin proteins that are approximately 80–85% homologous to vertebrate tubulins. Both genes contain two introns in exactly the same positions and the first intron is in the same position as the third intron of the *Chlamydomonas* β-tubulins. The sequenced *Zea mays* β-tubulin has a single intron at exactly the same position as the first intron of the soybean and the third intron of the *Chlamydomonas* β-tubulin genes. The *Arabidopsis tub1* and *tub5* genes both encode an additional internal amino acid (Ala), inserted after residue 40, that is not present in the β-tubulins of other organisms. Allowing for this additional codon, all four sequenced *Arabidopsis* β-tubulin genes also have two introns inserted into their coding sequences in exactly the same position as the two soybean β-tubulins.

Although the two soybean β-tubulins exhibit very similar codon usage, they are as divergent from each other in amino acid sequence as they are from *Chlamydomonas* or vertebrate tubulins. Although both genes exhibit a similar degree of divergence with regard to the number of amino acid substitutions, as compared to the most prevalent vertebrate β-tubulin, in one case the substitutions are nearly all conservative, while nearly 30% of the substitutions are nonconservative in the other β-tubulin. These appear to be the only two classes of β-tubulin genes in the soybean genome, although it is likely that each is a member of a family of related β-tubulin genes.

E. Amino Acid Sequence of Tubulin Monomers

The amino acid sequence of many tubulins has been determined, either by direct protein sequencing in the case of porcine α- and β-tubulins (Krauhs *et al.*, 1981; Ponstingl *et al.*, 1981), or more often, these sequences were deduced from genomic or cDNA nucleotide sequences. The derived amino acid sequences of six higher plant β-tubulins is shown in Fig. 1, along with the plant β-tubulin consensus sequence. These sequences are presented along with β-tubulins from representative animals, fungi and protists. Analysis of these sequences demonstrates a high degree of tubulin amino acid sequence conservation during evolution. There are 222 positions for which there is absolute conservation of a particular amino acid in all the tubulins compared. Some regions of the protein (e.g., residues 400 to 430) contain long sequences that are nearly 100% homologous. This striking degree of homology is even more apparent when fungi are omitted from the comparison. Much of the divergence is clustered in specific regions. The carboxy-terminal 14–19 residues of both α- and β-tubulin, for example, are highly divergent, although there is a conservation of charged residues in this region. Positions 431–441 are all occupied by acidic residues (either aspartate or glutamate) in the β-tubulin consensus sequence, although any particular tubulin may have from one to three uncharged residues in this region. Between residue 442 and the carboxy terminus there is no apparent conservation in the β-tubulin sequence, and even the length varies from 442 amino acids (*Trypanosoma brucei*) to 458 amino acids (*S. cerevisae*). The carboxy terminus of α-tubulins also exhibits numerous charged residues, but they are more uniform in size (448–450 amino acids) while most, including the two *Arabidopsis* α-tubulins, have an encoded terminal tyrosine residue.

In comparing mouse, chicken, human, rat and pig β-tubulins, Cleveland (1987) noted four different patterns of evolutionarily conserved isotypes, each of which had a characteristic sequence in the variable C-terminus. Some of these isotypes were ubiquitous, while others were expressed in specific tissues. Individual isotypes were highly conserved among the vertebrates, approaching 100% (Sullivan and Cleveland, 1986). At this point too few higher plant tubulins have been sequenced to know if these organisms also exhibit interspecies conservation of isotypes.

One surprising difference in the β-tubulins of plants and animals is the strong conservation of a particular residue in some positions in the plant sequence, where there is an equally strong conservation of a different residue in the same position in the β-tubulins of other organisms. For example, all six known plant β-tubulin sequences have leucine in position 5, whereas all the animal tubulins examined have valine occupying this same position. There are at least 37 internal positions in β-tubulin (not taking into consideration the highly variable C-terminal 15–20 residues) in which there are differences in the consensus amino acids between plants and animals (Table I).

```
PLANT         1                                                              50
Consensus     MREILHiQgG QCGNQIGaKF WEViCdEHGi D*TGrY*Gds dLQLERinVY
  Soybeta1    ------V-A- -------G-- ---M------ -A---N-V-NF H-----V---
  Soybeta2    ---------- ---------- ---V-A---- -P----G--- E---------
  A.t.TUB1    ------V--- -------S-- ---I-----V -P----N--- ----------
  A.t.TUB4    ---------- ---------- ---------- -H---Q-V--- P------D--
  A.t.TUB5    ---------- -------S-- ---------- -S----S--T ----------
  Z.mays10    ---------- ---------- ---V-A---- -A----G--- ------V---

ANIMAL
  Humbeta2    ----V---A- ---------- ----S----- -P--T-H--- ----------
  Mousetu5    ----V---A- ---------- ----S----- -P--T-H--- -------S--
  Chktub4b    ----V---A- ---------- ----S----- -PS-N-V--- -------S--
  Chktub5b    ----V---A- -------T-- ----S----- -PA-G-V--- A---------
  Drostub2    ----V---A- -------G-- ----S---C- -A--T-Y--- ----------

PROTIST
  Chlamytb    ----V---G- ---------- ---VS----- -P--T-H--- ----------
  Trypanos    ----VCV-A- -------S-- ----S----- -P--T-Q--- ----------

FUNGI
  Yeastsp     ----Vl--A- -----V--A- -ST-A----L -SA-I-H-TS EA-H--L---
  Yeastsc     ----I--SA- -Y------A- -ET-CG---L -FN-T-H-HD DI-K--L---
  Neurosp     ----V-L-T- --------A- -QT-SG---L -AS-V-N-T- E-----M---


PLANT         51                                                             100
Consensus     yNEASgGryV PRAVLMDLEP GTMDS*RSGP fGqIFRPDNF VFGQSGAGNN
  Soybeta1    ---------- ---------- -----L---- --K------- ----N-----
  Soybeta2    -----C--F- ---------- -----V---- Y--------- ----------
  A.t.TUB1    ---------- ---------- -----I---- Y--------- ----------
  A.t.TUB4    F------K-- ---------- -----L---- ---------- ----------
  A.t.TUB5    ---------- ---------- -----I---- ---------- ----------
  Z.mays10    -----C---- ---------- -----V---- Y-H------- ----------

ANIMAL
  Humbeta2    ----T--K-- -----V---- -----V---- ---------- ----------
  Mousetu5    ----T--K-- ---I-V---- -----V---- ---------- ----------
  Chktub4b    -----SHK-- ---I-V---- -----V---A --HL------ I---------
  Chktub5b    ---S-SQK-- -----V---- -----V---- ---L------ I---T-----
  Drostub2    ----T-AK-- ---I-V---- --------R- ---------- ----------

PROTIST
  Chlamytb    F---T----- ---I------ -----V---- Y--------- ----T-----
  Trypanos    FD--T----- --S--I---- -----V-A-- Y--------- I---------

FUNGI
  Yeastsp     F---A--K-- -----V---- ----AVK--K --NL-----I IY-------I
  Yeastsc     F----S-KW- --SINV---- W-I-AV-NSA I-NL-----Y I----S---V
  Neurosp     F-----NK-- -----V---- ----AV-A-- ---L------ ----------
```

Fig. 1. The derived amino acid sequences of six higher plant β-tubulins were compared and a consensus sequence was generated. The consensus sequence is given, using the standard single-letter amino acid code. When all six β-tubulins had the same amino acid at a given position, a capital letter was used to designate it in the consensus sequence. Where there was disagreement, the amino acid residue found in the majority of the polypeptides at that position was given as a lower case letter. Where there was no agreement, that position in the consensus sequence was marked with an asterisk. When a given position in the sequence of a particular β-tubulin was in agreement with the consensus sequence, that position was marked with a dash. When the residue at a given position in a particular β-tubulin disagreed with the consensus, that position was marked with a capital letter representing the amino acid found at that position. The *Arabidopsis* β-tubulins encoded by *tub1* and *tub5* contain an additional amino acid after consensus position 40. The additional amino acid was deleted from these sequences to obtain maximum agreement among the six plant β-tubulins. The plant sequences compared include two from soybean, designated *Soybeta1* and *Soybeta2* (Guiltinan et al., 1987), three from *Arabi-*

```
          PLANT      101                                                          150
          Consensus  wAKGHYTEGA ELIDsVLDVV RKEAENcDCL QGFQvCHSLG GGTGSgMGTL
          Soybeta1   ---------- ---------- ---------- ---------- ----------
          Soybeta2   ---------- ---------- ---------- ----I----- ----------
          A.t.TUB1   ---------- ----A----- ---------- ---------- ----------
          A.t.TUB4   ---------- ---------- ------S--- ---------- ----------
          A.t.TUB5   ---------- ----A----- ---------- ---------- ----------
          Z.mays10   S--------- ---------- ---------- ---------- -----A----

          ANIMAL
          Humbeta2   ---------- --V------- -----S---- ----LT---- ----------
          Mousetu5   ---------- --V------- -----S---- ----LT---- ----------
          Chktub4b   ---------- --V------- ---C------ ----LT---- ----------
          Chktub5b   ---------- --V------- ---C-H---- ----LT---- ----------
          Drostub2   ---------- --V------- ---S-G---- ----LT---- ----------

          PROTIST
          Chlamytb   ---------- ---------- -----S---- ----V----- ----------
          Trypanos   ---------- --------C  C----s---- ---------- ----------

          FUNGI
          Yeastsp    ---------- --AVA----- -R---a--A- ----LT---- ----------
          Yeastsc    ---------- --V---M--I -R---g--S- -----T---- ----------
          Neurosp    ---------- --V-Q----- -R---g---- -----T---- ----A-----


          PLANT      151                                                          200
          Consensus  LISKIREEYP DRMM1TFSVF psPkvSDTvV EPYNATLSVH qLVENADECm
          Soybeta1   ---------- ---------- AV-EG----- ---------- ----------
          Soybeta2   ---------- ---------- -------E-- ------S--- D--------S
          A.t.TUB1   ---------- ---------- ---------- ---------- ----------
          A.t.TUB4   ---------- ----M----- ---------- ---------- ----------
          A.t.TUB5   ---------- ---------- ---------- ---------- ----------
          Z.mays10   ---------- ---------- ---------- ---------- ----------

          ANIMAL
          Humbeta2   ---------- --I-N----V ---------- ---------- -----T--TY
          Mousetu5   ---------- --I-N----V ---------- ---------- -----T--TY
          Chktub4b   ----V----- --I-N----V ---------- --------I- -----T--TY
          Chktub5b   ---------- --I-N----M ---------- ---------- -----T--TY
          Drostub2   ---------- --I-N----V ---------- ---------- -----T--TY

          PROTIST
          Chlamytb   ---------- ---------V ---------- ---------- ---------M
          Trypanos   ----L--Q-- --I-M----II ---------- ----T----- -----S--SM

          FUNGI
          Yeastsp    -L-------- ----A----A -A-KS----- --------M- -----S--TF
          Yeastsc    -F---K--L- ----A----L ---KT----- ---------- ----HS--TF
          Neurosp    --------F- ----A-Y--V ---K------ ---------- -----S--TF
```

Fig. 1. (*Continued*)

dopsis thaliana, designated *A.t.TUB1* (Oppenheimer *et al.*, 1988), *A.t.TUB4* (Marks *et al.*, 1987), and *A.t.TUB5* (Silflow *et al.*, 1987), and one from maize designated *Z.mays10* (Silflow *et al.*, 1987). These higher plant β-tubulin sequences were compared to five animal, two protist, and three fungal β-tubulins. *Humbeta2*, a human β-tubulin, designated 5B, that is expressed in neural tissues (Gwo-Shu Lee, *et al.*, 1984). *Mousetu5*, the ubiquitously expressed β-5-tubulin from Swiss mice (Wang *et al.*, 1986). *Chktub4b*, the neuronal tissue specific β-4-tubulin from chicken (Sullivan and Cleveland, 1984), while *Chktub5b* is also from chicken, but is the β-5-tubulin expressed only in non-neuronal tissues (Sullivan *et al.*, 1986b). *Drostub2*, the testis-specific β-2-tubulin of *Drosophila melanogaster* (Rudolph *et al.*, 1987). *Chlamytb*, the β-tubulin of *Chlamydomonas* (Youngblom *et al.*, 1984). *Trypanos*, the β-tubulin from *Trypanosoma brucei* (Kimmel *et al.*, 1985). *Yeastsp*, the *Schizosaccharomyces pombe* β-tubulin (Hiraoka *et al.*, 1984). *Yeastsc*, the *Saccharomyces cerevisiae* (Neff *et al.*, 1983). *Neurosp*, the β-tublin of *Neurospora crassa* (Orbach *et al.*, 1986).

```
PLANT        201                                                        250
Consensus    VLDNEALYDI CFRTLKL*tP sfGDLNHLIS aTMSGVTCcL RfPgQLNSDL
Soybeta1     ---------- -------TN- ---------- T--------- ----------
Soybeta2     ---------- -------T-- -C-------- ---------- -N-S------
A.t.TUB1     ---------- -------S-- ---------- -------S-- ----------
A.t.TUB4     ---------- -------AN- T--------- ---------- ----------
A.t.TUB5     ---------- -------S-- ---------- -------S-- ----------
Z.mays10     ---------- -------T-- ---------- ---------- ----------

ANIMAL
Humbeta2     CI-------- -------T-- TY------V- -------T-- -------A--
Mousetu5     CI-------- -------T-- TY------V- -------T-- -------A--
Chktub4b     CI-------- -------A-- TY------V- -------TS- -------A--
Chktub5b     CI-------- -------T-- TY------V- -------TS- -------A--
Drostub2     CI-------- -------T-- TY------V- -------T-- -------A--

PROTIST
Chlamytb     ---------- -------T-- T--------- -V---I---- -------A--
Trypanos     CI-------- -------T-- T-------V- -VV------- ----------

FUNGI
Yeastsp      CI-----SS- IAN---IKS- -YD-----V- -V-A---TSF ----E-----
Yeastsc      CI-------- -Q-----NQ- -Y----N-V- SV-----TS- -Y--------
Neurosp      CI-------- -M-----SN- -Y------V- -V-----VS- ----------


PLANT        251                                                        300
Consensus    RKLAVNLIPF PRLHFFMVGF APLtSRGSQQ Y*aLtvPELT QQMWDakNMM
Soybeta1     ---------- ---------- ---------- -RS--I---- ------R---
Soybeta2     ---------- ---------- ---A------ -R--S----- -----S----
A.t.TUB1     ---------- ---------- ---------- -IS------- ----------
A.t.TUB4     ---------- ---------- ---------- -S--S----- ----------
A.t.TUB5     ---------- ---------- ---------- -IS------- ----------
Z.mays10     ---------- ---------- ---------- -R-------Y ----------

ANIMAL
Humbeta2     ------MV-- -------P-- ---------- -R-------- ---F------
Mousetu5     ------MV-- -------P-- ---------- -R-------- --VF------
Chktub4b     ------MV-- -------P-- ----R----- -R-------- ---F------
Chktub5b     ------MV-- -------P-- ----A----- -R-------- ---F------
Drostub2     ------MV-- -------P-- ---------- -R-------- ---F------

PROTIST
Chlamytb     ---------- ---------- T--------- -R-------- ----------
Trypanos     -------V-- ------*-M- ---------- -RG-S----- ---F------

FUNGI
Yeastsp      ------MV-- ---------- ---AAI--SS FQ-VS----- ---F--N---
Yeastsc      -------V-- ---------Y ----AI---S FRS------- ---F------
Neurosp      ------MV-- ---------- ------AHH FR-VS----- ---F-P----
```

Fig. 1. (*Continued*)

Twenty-four of these are conservative differences. That is, the amino acids are different in the plant and animal tubulins, but they are of the same functional type. However, thirteen of these differences are nonconservative. Twelve of these are polar/nonpolar differences. In six cases the residue occupying the position in the plant β-tubulin is polar, whereas the animal β-tubulin has a nonpolar residue in this same position. While overall the number of polar and nonpolar residues is the same in the plant and animal sequences, four of these nonpolar substitutions occur near position 200 in the plant β-tubulin. This creates a hydrophobic pocket in this position that is

```
PLANT        301                                                         350
Consensus    CAADPRHGRY LTaSAmFRGk mSTKEVDeQm *NvQNKNSSY FVEWIPNNVK
  Soybeta1   ---------- ---------- -------Q-- I--------- ----------
  Soybeta2   ---------- ---------- ---------- I--------- ------H---
  A.t.TUB1   ---------- ---------- ---------I L--------- ----------
  A.t.TUB5   ---------- -----I---Q ---------I L-I------- ----------
  A.t.TUB4   ---------- --R--V---- L--------- M-I------- ----------
  Z.mays10   ---------- ---------- ---------- L--------- ----------

ANIMAL
  Humbeta2   A-C------- --VA-V---R --M------- L--------- ----------
  Mousetu5   A-C------- --VA-V---R --M------- L--------- ----------
  Chktub4b   A-C------- --VATV---R --M------- LAI-S----- ----------
  Chktub5b   A-C------- --VATV---P --M------- LAI------- ----------
  Drostub2   A-C------- --VA-I---R --M------- L-I------F --------C-

PROTIST
  Chlamytb   ---------- -----L---R --T------- L--------- ----------
  Trypanos   Q--------- -----L---R --T------- L--------- -I------I-

FUNGI
  Yeastsp    V--------- --VA-L---- V-M------I RS--T---A- ------D--L
  Yeastsc    A-----N--- --VA-F---- V-V---EDE- HK--S---D- ---------Q
  Neurosp    A-S-F-N--- --C--I---- V-M---ED-- R--------- ---------Q


PLANT        351                                                         400
Consensus    SsVCDIpP*G lkMaSTFiGN STSIQEMFRR VSEQFTAMFR RKAFLHWYTG
  Soybeta1   --------T- -S-S---M-- ---------- ------V--- ----------
  Soybeta2   -T------T- -R-------- ---------- ---------- ----------
  A.t.TUB1   --------T- I------V-- ---------- ---------- ----------
  A.t.TUB4   ------A-K- ---------- ---------- ---------- ----------
  A.t.TUB5   --------K- ----A--V-- ---------- ---------- ----------
  Z.mays10   -T------H- ---------- ---------- ---------- ----------

ANIMAL
  Humbeta2   TA------R- ----A----- --A---L-K- I--------- ----------
  Mousetu5   TA------R- ----V----- --A---L-K- I--------- ----------
  Chktub4b   VA------R- ---------- --A---L-K- I--------- ----------
  Chktub5b   VA------R- ---------- --A---L-K- I----S---- -------F--
  Drostub2   TA------R- ----A----- --A---L-K- ---------- -------F--

PROTIST
  Chlamytb   --------K- ----A----- --A----K-- ---------- ----------
  Trypanos   --------K- ----V----- N-C------- -G----L--- ----------

FUNGI
  Yeastsp    KA--SV--KD ----A----- ------I--- LGD--S---- -N--------
  Yeastsc    TA--SVA-Q- -D--A---A- ------L-K- -GD--S---k ---------S
  Neurosp    TAL-S---R- -------V-- --A---L-K- IG-------k ----------
```

Fig. 1. (*Continued*)

not present in animal β-tubulins, which could account for some of the unique pharmacological properties of the plant tubulins.

Another unusual feature of the plant β-tubulins is that there are seven internal positions for which there is no consensus. Four different amino acids were found in positions 32 and 36, while three different residues occurred at positions 76, 218, 282, 331, and 359 among the six plant β-tubulin sequences compared. Residues 32 and 36 are in a hypervariable region of the β-tubulin molecule in which many divergent residues are clustered. Two of the *Arabidopsis* β-tubulins have an additional amino acid (alanine) inserted

```
PLANT         401                                              450
Consensus     EGMDEMEFTE aesNMNDLVa EYQQYQDATA dee*e*eeEe e********-
Soybeta1      ---------- VRA------- ---------- VDDH-D-D-D -AMAA.....
Soybeta2      ---------- --------S- ---------- ---EYE---- -EMFAQHDM.
A.t.TUB1      ---------- --------S- ---------- ---D-YD--- -QVYES....
A.t.TUB4      ---------- ---------- ---------- G--EYE---- -YET......
A.t.TUB5      ---------- ---------- ---------- ---G-YDV-- -EEGDYET..
Z.mays10      ---------- --------S- ---------- ---G-Y---- GDLQD.....

ANIMAL
Humbeta2      ---------- ---------- ---------- E--G-F---A -EEVA.....
Mousetu5      ---------- ---------- ---------- E--EDFG--A -EEA......
Chktub4b      ---------- ---------- ---------- E--G-MY-D- -EESEQGAK.
Chktub5b      ---------- ---------- ------E--- NDGE-AF-D- -EEINE....
Drostub2      ---------- ---------- -A----E--- AD-EGEFD-- -EGGGDE...

PROTIST
Chlamytb      ---------- ---------- --------S- E--G-F-G-- -EA.......
Trypanos      ---------- ---------- ---------I E--G-FD--- QY........

FUNGI
Yeastsp       ---------- ---------- ------E-GI --GD-DY-I- -EKEPLDY..
Yeastsc       -----L--S- ---------- ------E--V E--E-VDENG DFGAPQNQDE PITENFEB
Neurosp       ---------- ---------- --------GV ---E-EY--- APLEGEE...
```

Fig. 1. (*Continued*)

in this hypervariable region. However, the remaining plant variable positions occur in regions of the molecule that exhibit very high degrees of sequence conservation in other groups, particularly in the animal sequences. For example, there is a clear consensus of proline at residue 32 in animals and protists, and valine occurs at residue 76 in all other organisms compared.

The secondary structure of the tubulin monomers is not known for any organism. However, some of the amino acid differences that are noted between the β-tubulins of different organisms could be expected to have important consequences for secondary structure. This assumption is supported by a comparison of the secondary structures predicted by the algorithm of Chou and Fasman (1978) for the different β-tubulins. Not unexpectedly, tubulins with a high degree of homology tend to give similar predicted secondary structures. Many of the highly homologous mammalian tubulins exhibited a predicted secondary structure such as that illustrated in Fig. 2. These structures exhibit relatively short regions of alpha helical and beta sheet configurations, separated by longer regions of random coil. All of the plant β-tubulins examined exhibited a somewhat similar pattern of beta sheet, alpha helical and random coil configuration, but differed from the mammalian tubulins primarily in the number and positioning of beta turns. The plant β-tubulins exhibited an array of predicted secondary structures, some of which were not very different from that of the mammalian β-tubulins (Fig. 3). However, others, such as the soybean SB1, exhibited rather different predicted secondary structures (Fig. 4).

TABLE I
Unique Residues in Plant and Animal β-Tubulin Sequences[a]

Residue number	Higher plant	Animal	Chlamydomonas	Type of substitution
5	LEU (6/6)	VAL (5/5)	VAL	C
9	GLY (5/6)	ALA (5/5)	GLY	C
35	CYS (6/6)	SER (5/5)	SER	C
58	ARG (5/6)	LYS (5/5)	ARG	C
66	MET (6/6)	VAL (5/5)	MET	C
113	ILE (6/6)	VAL (5/5)	ILE	C
135	VAL (5/6)	LYS (5/5)	VAL	N/B
136	CYS (6/6)	THR (5/5)	CYS	C
163	MET (6/6)	ILE (5/5)	MET	C
165	LEU (5/6)	ASN (5/5)	LEU	N/P
170	PHE (6/6)	VAL (4/5)	VAL	C
196	ALA (6/6)	THR (5/5)	ALA	N/P
199	CYS (6/6)	TYR (5/5)	CYS	C
200	MET (5/6)	TYR (5/5)	MET	N/P
201	VAL (6/6)	CYS (5/5)	VAL	N/P
202	LEU (6/6)	ILE (5/5)	ILE	C
221	SER (5/6)	THR (5/5)	THR	C
222	PHE (5/6)	TYR (5/5)	PHE	N/P
229	ILE (6/6)	VAL (5/5)	ILE	C
238	CYS (6/6)	THR (5/5)	CYS	C
248	SER (6/6)	ALA (5/5)	ALA	P/N
257	LEU (6/6)	MET (5/5)	LEU	C
258	ILE (6/6)	VAL (5/5)	ILE	C
268	VAL (6/6)	PRO (5/5)	VAL	C
294	TRP (6/6)	PHE (5/5)	TRP	C
301	CYS (6/6)	ALA (5/5)	CYS	P/N
303	ALA (6/6)	CYS (5/5)	ALA	N/P
313	ALA (5/6)	VAL (5/5)	ALA	C
314	SER (6/6)	ALA (5/5)	SER	P/N
316	MET (4/6)	VAL (4/5)	LEU	C
320	LYS (5/6)	ARG (4/5)	ARG	C
323	THR (6/6)	MET (5/5)	THR	P/N
352	SER (4/6)	ALA (5/5)	SER	P/N
373	SER (6/6)	ALA (5/5)	ALA	P/N
377	MET (6/6)	LEU (5/5)	MET	C
379	ARG (6/6)	LYS (5/5)	LYS	C
381	VAL (6/6)	ILE (4/5)	VAL	C

[a] The 37 internal positions in which there is a difference in the consensus amino acid between higher plant and animal β-tubulin sequences. The numbers in parentheses are the number of β-tubulins with the indicated amino acid over the total number of β-tubulin sequences examined. The sequences used in this comparison are listed in Fig. 1.

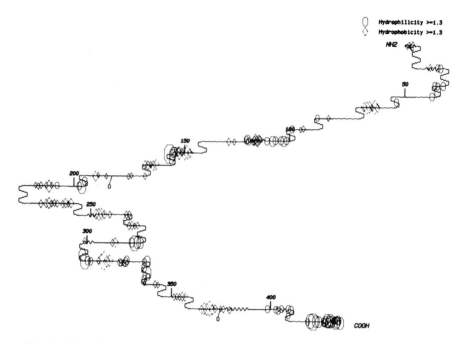

Fig. 2. Secondary structure of the β-5-tubulin of mouse (Wang *et al.*, 1986) predicted according to the algorithm of Chou and Fasman (1978), with the aid of a computer program known as "Plotstructure" developed by the University of Wisconsin Genetics Computer Group. The continuous line represents the sequence of amino acids, with numbered flags giving the positions of landmark residues. A sharp saw tooth pattern represents regions of predicted β-sheet structure, while the large amplitude sine wave represents regions of predicted alpha helix. The low amplitude wave represents regions of predicted random coil. Beta turns are indicated by a change in the direction of the line. Superimposed on the line are diamonds in regions of hydrophobicity, while octagons are placed over hydrophylic regions. The size of the diamond or octagon indicate the relative strength of the hydrophobic or hydrophylic character.

F. Tubulin Domain Organization

Tubulin exhibits a number of distinct biochemical properties. In addition to the tubulin dimer interactions which are necessary for self-assembly, tubulin binds other molecules, such as MAPs, GTP, and Ca^{2+}, and after the subunits have assembled into MTs, tubulin hydrolyzes bound GTP. Recent evidence indicates that many of these activities are segregated into different domains of the tubulin monomers. Mild proteolysis of tubulin by subtilisin cleaves a C-terminal peptide from β-tubulin to give a dimer containing an unmodified α- and a truncated β-tubulin (called α-βS). Further proteolysis cleaves a C-terminal peptide from α-tubulin as well to give a tubulin-S dimer in which both subunits are truncated. Both the tubulin-S and the α-βS dimers assemble into MTs more readily than intact tubulin dimers (Serrano *et*

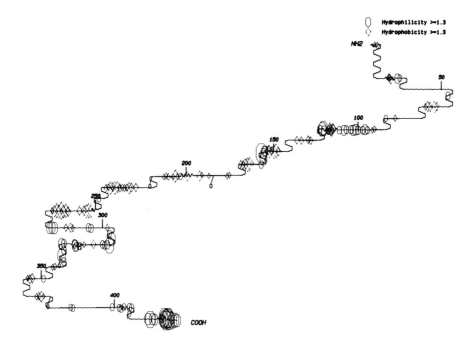

Fig. 3. Secondary structure of the *Arabidopsis* β-tubulin protein encoded by the gene designated *tub4* (Marks *et al.*, 1987) determined by the algorithm of Chou and Fasman (1978). The significance of the characters is described in the legend for Fig. 2.

al., 1984a). Microtubules composed of tubulin-S are more resistant to depolymerization by Ca^{2+}, while MTs formed by α-βS are more resistant to depolymerization by cold, GDP, and drugs as well (Bhattacharyya *et al.*, 1985). Serrano *et al.* (1986) localized the high affinity Ca^{2+} binding sites of both α- and β-tubulins to the C-terminal region that is removed by subtilisin. MTs assembled from tubulin-S do not bind MAPs (Serrano *et al.*, 1984b, 1985), while the C-terminal peptide released by subtilisin itself binds MAPs (Serrano *et al.*, 1984a). Littauer *et al.* (1986) examined the binding of MAP2 and *tau* factor to peptides derived from porcine brain α- and β-tubulins. In addition to a *tau*-binding site near the N-terminus of α-tubulin, both *tau* and MAP2 showed high affinity for the C-terminal peptides. These findings were confirmed by Maccioni *et al.* (1988) using somewhat different methods. In the latter study, the region of β-tubulin containing residues 422–434 and of α-tubulin containing residues 430–441 were identified as having a high affinity for MAP2 and *tau* factor. Brietling and Little (1986) demonstrated that the C-terminus is located on the surface of the MT, based on the interactions of MTs with different monoclonal antibodies.

α and β-tubulins each also exhibit a single internal highly protease-sensitive cleavage site. Trypsin cleaves proteins after arginine or lysine residues.

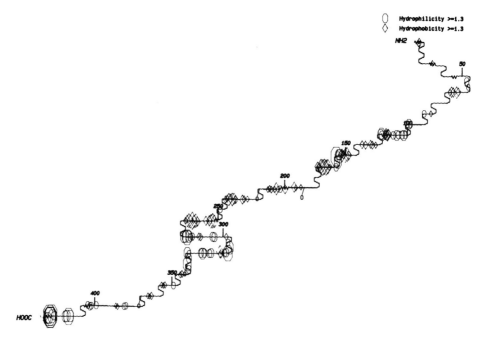

Fig. 4. Secondary structure of the soybean β-tubulin encoded by the gene *soybeta1* (Guiltinan *et al.*, 1987) determined by the algorithm of Chou and Fasman (1978). The significance of the characters is described in the legend for Fig. 2.

Although there are many potential trypsin cleavage sites in denatured tubulin molecules, trypsin readily cleaves the native tubulin dimer only after Arg-339. Similarly, chymotrypsin cleaves the native tubulin dimer after the Try-281 of β-tubulin (Mandelkow *et al.*, 1985; Sackett and Wolff, 1986). In either case, the monomer is cleaved into a large fragment containing the N-terminus and approximately two-thirds of the mass of the polypeptide and a smaller fragment composed of the C-terminal one-third of the mass, but not the C-terminus itself. There is another chymotryptic cleavage site within 1 kDa of the C-terminus in β-tubulin. Most likely these sites are readily available to the proteases because they are on the surface of the molecule. Frequently it has been shown that such highly protease-sensitive sites represent bridges between compact structural domains.

Together, these data indicate that both α- and β-tubulins are organized into two compact structural domains which are closely associated to form a globular structure, from which a short linear tail projects (Fig. 5). The N-terminus is found in the large globular domain, while the carboxy-terminal approximately 2 kDa of each subunit is the linear projection domain (Sackett and Wolff, 1986). Kirchner and Mandelkow (1985) used chemical cross-linking and limited proteolysis to show that tubulin dimer formation oc-

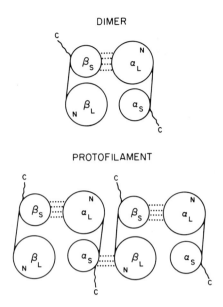

Fig. 5. Model for the domain structure of tubulin monomers and for their interaction to form tubulin dimers and protofilaments. The model of Sackett and Wolff (1986) was modified to incorporate the data of Kirchner and Mandelkow (1985) and Serrano et al. (1986). The circles represent the large and small compact structural domains of the α and β subunits, respectively, while C represents the linear carboxy terminal domain. N, the location of the amino terminus. Dotted lines illustrate the binding sites that hold the monomers together to form the dimer and the dimers together to form the protofilament.

curred as a result of a bond between the large domain of α-tubulin and the small domain of β-tubulin, while the bond responsible for protofilament formation occurred between the large domain of β-tubulin and the small domain of α-tubulin. Szasz et al. (1986) demonstrated that three basic residues in α-tubulin, Lys-394, His-393 and Arg-390 played an important role in microtubule assembly. Together, the large and small domains are capable of self-assembly. The exchangeable GTP binding site has been localized in the N-terminal domain of β-tubulin (Nath and Himes, 1986). The carboxy terminal domain represents the regulatory component, to which other molecules bind to regulate subunit assembly, or to determine the interaction of MTs with other cellular components.

While plant tubulins have not yet been subjected to a similar analysis, the amino acid sequences available show that both Arg-339 in α-tubulin and Tyr-281 of β-tubulin are conserved in higher plant tubulins. Furthermore, Chou–Fasman secondary structure predictions suggest that these residues would reside in a hydrophylic region of the polypeptides and thus they would be expected to be found at the surface of the molecules where they would be accessible for proteolytic cleavage. Similarly, the C-terminus of all β-tubulins examined is highly hydrophylic and, although the last 10 to 15 residues of the C-terminus are highly diverged, the sequence in the regions identified by Maccioni et al. (1988) as important for the interaction of microtubule-associated proteins is largely conserved between plants and animals. In addition, the basic residues of α-tubulin identified by Szasz et al. (1986) as important in microtubule assembly are conserved in all the higher plant α-sequences currently known (Silflow et al., 1987). These considerations suggest that plant tubulins may have a domain organization similar to that of vertebrate tubulins. It also suggests that plant tubulin assembly may be regulated as it is invertebrates.

G. Tubulin Isotypes and Isoforms

Multiple α- and β-tubulins can be detected in most organisms by two-dimensional or isoelectric focusing gel electrophoretic separations of isolated tubulins, or upon probing Western blots of electrophoretically separated whole-cell proteins with antibodies to tubulins. For example, Field et al. (1984) detected 6 α- and 12 β-tubulin isoforms after separating brain tubulin by isoelectric focusing gel electrophoresis. Multiple tubulin isoforms also have been reported in higher plants (Dawson and Lloyd, 1985; Hussey and Gull, 1985; Mizuno et al., 1985; Cyr et al., 1987b; Hussey et al., 1988). Heterogeneity can come about either as a result of genetically encoded differences in amino acid sequence or as a result of posttranslational modifications. Tubulins are encoded by small gene families in most organisms, including the higher plants that have been examined (see Section IV,D above). The two sequenced β-tubulin genes of soybean encode proteins that differ sufficiently in net charge to be separated by isoelectric focusing gel electrophoresis (Guiltinan et al., 1987b). However, the relationship between the number of tubulin genes and the number of detectable isoforms has not yet been examined rigorously in any higher plant. In *Physarum*, it has been shown that each isoform is encoded by a different gene and the number of genes can account for all of the isoforms observed (Burland et al., 1983; Schedl et al., 1984). Although this is a simple relationship in *Physarum*, in many organisms the number of genes cannot be inferred from the number of isoforms because they also arise as a result of the posttranslational modification of tubulin precursors. A number of different mechanisms have been described for the posttranslational modification of tubulins, including detyrosination of α-tubulin (Kumar and Flavin, 1981), phosphorylation of β-tubulin (Goldenring et al., 1983), and the ε-amino acetylation of a lysine residue in α-tubulin (L'Hernault and Rosenbaum, 1985).

The significance of multiple tubulins is obscure. Fulton and Simpson (1976) proposed that the different MT arrays in eukaryotes were composed of different tubulin isotypes. A rigorous interpretation of this multitubulin hypothesis proposes that a particular tubulin isotype performs some essential function in a particular MT array for which other tubulin isotypes cannot substitute. There are many examples of the specific developmental expression of particular tubulin isotypes (see Cleveland and Sullivan, 1985, for a review). However, in very few cases has it been demonstrated that a specific tubulin isotype is utilized preferentially to form a given MT array. For example, *Drosophila* has four α- and four β-tubulin genes, some of which are expressed constituitively, while others exhibit tissue-specific expression (Kemphues *et al.*, 1980; Theurkauf *et al.*, 1986). One of the β-tubulin genes (*β-2*) is expressed exclusively in the testis where its product is the dominate form of β-tubulin. However, even though this gene is expressed in a highly regulated, tissue-specific manner, the product of that gene is utilized indiscriminately for all the MT arrays found in this tissue. These include the mitotic and meiotic spindles, a MT array known as the manchette that is involved in nuclear shaping during spermatogenesis, and the MTs of the sperm flagellar axoneme (Kemphues *et al.*, 1982). In some microorganisms it is clear that individual tubulin isotypes could not be MT array-specific since it has been demonstrated that the products of one gene can substitute for another without affecting viability. *Saccharomyces cerevisiae* has two α-tubulin genes which encode proteins that are about 10% divergent from each other, yet either α-tubulin gene can substitute for the other when the expression of one is eliminated by nul mutations (Shatz *et al.*, 1986). Also, in *Aspergillus nidulans*, one β-tubulin gene normally is expressed only during conidiation, but it can be replaced by the products of β-tubulin genes usually expressed in vegetative cells (Weatherbee *et al.*, 1985).

There are a few examples of MT array-specific tubulin isotypes. In chickens, a specific, divergent β-tubulin isotype is expressed only in erythrocytes and thrombocytes (Murphy and Wallis, 1983). An immunofluorescent study has shown that the erythrocyte-specific β-tubulin first appears at the onset of hemoglobin synthesis in developing erythroblasts, and that it is the only β-tubulin utilized in the formation of the marginal band of MTs that characterizes the mature erythrocytes (Murphy *et al.*, 1986). The assembly characteristics of the erythrocyte tubulin (in which 100% of the β subunits are the divergent erythrocyte-specific β-tubulin) are quite different from those of chick brain tubulin (Murphy and Wallis, 1985). Thus, this may represent an example of a MT array-specific tubulin isotype that performs an essential functional role in the marginal band. Mammals also have been shown to have a highly divergent β-tubulin expressed only in erythroblasts, with the unique β-tubulin utilized to form a marginal band of MTs (Wang *et al.*, 1986). In an effort to evaluate the possibility that this β-tubulin is function-specific, the gene encoding it has been transfected into cultured mammalian cells, where it was expressed and the protein was incorporated into all MT arrays which continued to function normally (Lewis *et al.*, 1987). This result argues against the exclusive functioning of the erythrocyte-specific β-tubulin in the marginal band, but it still may have an essential function in that MT array. At the same time it is clear from this and other studies that the mitotic spindle and interphase MT arrays of cultured mammalian cells can incorporate divergent tubulins into their MTs and continue to function (Bond et al., 1986). It is still possible that individual tubulin isotypes perform specific and unique functional roles in MT arrays of specialized cells. This is suggested by the observation that there is strong selection pressure for the conservation of several specific tubulin sequences in vertebrate evolution (Cleveland, 1987).

A unique α-tubulin isoform (α-3) is found in the flagellar axoneme in *Chlamydomonas*. α-3 arises posttranslationally by acylation of a lysine ε-amino group in one of the two *Chlamydomonas* α-tubulin isotypes (L'Hernault and Rosenbaum, 1985). Although the assembly–disassembly kinetics of MTs constructed with tubulin dimers having acetylated and unacetylated α subunits do not differ (Maruta et al., 1986), one of the major differences between flagellar axoneme and cytoplasmic microtubules is their stability. The enhanced stability of axoneme microtubules may result from their interaction with other flagellar proteins and it is possible that acetylation represents a signal for the binding of these accessory flagellar proteins with the axoneme microtubules (J. L. Rosenbaum, personal communication). The acetylated residue is Lys-40 of the flagellar α-tubulin (LeDizet and Piperno, 1987). The sequence of α-tubulin is highly conserved in this region, and Lys-40 is found in many vertebrate α-tubulins. Acetylated α-tubulin occurs in a subset of the cytoplasmic MTs that is more stable in some cultured mammalian cells (Bulinski et al., 1988). Thus, acetylation may be a general mechanism for marking sets of microtubules for interaction with stabilizing MAPs.

Tyrosination/detyrosination of α-tubulin may be another mechanism to create a subset of MTs with different functional properties. Tyrosine is the genetically encoded carboxy-terminal residue in most (but not all) α-tubulins, including the two *Arabidopsis* α-tubulins. In vertebrates, the terminal tyrosine may be removed by a unique carboxypeptidase, exposing a glutamate residue (Kumar and Flavin, 1981). Vertebrates also have a ligase which adds tyrosine to the detyrosinated tubulin (Barra et al., 1974; Schroder et al., 1985). Using highly specific antibodies for the tyrosinated and detyrosinated α-tubulins, Gundersen et al. (1984) demonstrated that the detyrosinated tubulin was confined to a subset of interphase MT array, and that it was not present in the MTs of the mitotic spindle or the astral fibers. The MTs containing the detyrosinated tubulin have been shown to turn over much more slowly in cultured mammalian cells. Whereas the MTs containing predominately tyrosinated tubulin turn over with a half-life measured in minutes, MTs enriched in detyrosinated tubulin persist for hours (Kreis, 1987; Webster et al., 1987). In some cell lines, the MTs enriched in detyrosinated tubulin also are enriched in acetylated tubulin (Bulinski et al., 1988). Wehland et al. (1984) observed that all MT arrays in onion root cells reacted with an antibody specific for the tyrosinated α-tubulin. However, it is not known if a mechanism exists for the reversible removal of the genetically encoded tyrosine from the C-terminus of higher plant α-tubulins.

H. Colchicine Binding to Tubulins

One of the chief differences between mammalian brain tubulin and the higher plant tubulins is their relative affinities for certain alkaloids, chiefly colchicine. Colchicine is a secondary metabolite of *Colchicium autumnale*, and is a drug with a potent ability to inhibit mitosis. While it has a relatively minor effect on the duration of the various phases of the cell cycle, it will block mitosis at metaphase due to the destruction of the spindle (see Dustin, 1984, for a review). This phenomenon occurs in both plant and animal cells, although plant cells may require concentrations of colchicine three orders of magnitude higher than those that are effective in mammalian cells. Many animal cells are blocked in metaphase by colchicine concentrations in the nanomolar to low micromolar range, whereas the same effect in higher plant cells may require millimolar levels of colchicine. Microtubules rapidly disassemble in the presence of substoichiometric levels of colchicine *in vivo* in mammalian systems. Tubulin binds colchicine to form a tubulin–colchicine complex and the tubulin–colchicine complex is incorporated into MTs at their ends, poisoning further assembly (Margolis and Wilson, 1977; Bergen and Borisy, 1983). Plant tubulin assembly *in vitro* is inhibited by colchicine only at millimolar concentrations, while brain tubulin assembly is completely blocked by micromolar colchicine levels (Morejohn *et al.*, 1984). Furthermore, although tubulin was first isolated as the colchicine-binding protein of mammalian cells, and was shown to be the only cellular component that colchicine bound to with high affinity (Weisenberg *et al.*, 1968), attempts to isolate the colchicine-binding protein from plants have not been successful in most cases (see Hart and Sabnis, 1976, for a review of early work on this problem). However, Okamura *et al.* (1984) and Kato *et al.* (1985) have succeeded in isolating the colchicine-binding activity of carrot and demonstrating that it was tubulin.

The difference in the response of plant and animal cells to colchicine is likely due to the differing affinities of vertebrate and higher plant tubulins for colchicine. Each mammalian tubulin dimer has a single high-affinity colchicine-binding site with a binding constant of approximately $2 \times 10^6 \, M^{-1}$ (Margolis and Wilson, 1977). The binding of colchicine to mammalian tubulin is a slow, temperature-dependent reaction that may involve a change in the conformation of the protein (Andreu and Timasheff, 1982b). The colchicine-binding site decays rapidly in purified tubulin, but it is stabilized by a number of agents, including sucrose, vinblastine sulfate, and organic acids (Wilson *et al.*, 1976). Colchicine binding to plant tubulin is similar in some respects. Rose tubulin also has a single class of colchicine-binding sites; the binding is slow and the binding site appears to be stabilized by sucrose or organic acids. However, colchicine binding to rose tubulin is not strongly temperature dependent, binding is not enhanced by vinblastine sulfate, and the affinity constant of rose tubulin for colchicine is low, $K = 9.7 \times 10^2 \, M^{-1}$ (Morejohn et al., 1987a). Morejohn et al. (1984) noted that plant tubulins differ in their affinity for colchicine. Although neither rose nor hibiscus tubulins exhibited high-affinity colchicine binding, carrot tubulin does, but colchicine binding to carrot tubulin is still less than that of brain tubulin. Okamura and co-workers measured an affinity constant of $5.7 \times 10^5 \, M^{-1}$ for carrot tubulin in the presence of 1 M tartrate, which prevents the rapid decay of colchicine binding in the isolated protein. Under the same conditions, brain tubulin exhibited a 10-fold higher colchicine affinity constant (Okamura et al., 1984; Kato et al., 1985).

The differences in the colchicine-binding site of higher plant and vertebrate tubulins are further illustrated by the effects of competitive inhibitors of colchicine binding in the two systems. Colchicine consists of three hydrocarbon rings, designated A, B, and C. The A ring is trimethoxybenzene. Other compounds with a trimethoxybenzene ring, such as *N*-acetylmescaline, bind to brain tubulin. Binding is rapid, although the affinity of brain tubulin for these compounds is low. *N*-acetylmescaline also is a competitive inhibitor of colchicine binding, suggesting that colchicine binds, in part, through an interaction of its A ring with tubulin (Cortese et al., 1977; Andreu and Timasheff, 1982a). The C ring of colchicine has a tropolone structure and tropolone itself has been shown to bind to tubulin and to be a competitive inhibitor of colchicine binding to brain tubulin (Andreu and Timasheff, 1982a). Since colchicine analogs lacking the B ring, such as colcemid, also bind to animal tubulin and inhibit its colchicine binding, it is assumed that colchicine is a bifunctional ligand, binding to tubulin through its A and C rings (Andreu and Timasheff, 1982b; Andreu et al., 1984). However, podophyllotoxin, which has an A and a B ring and may bind to tubulin through both its A and B rings, also is an effective inhibitor of colchicine binding to mammalian tubulin (Cortese et al., 1977).

The data on the effects of these inhibitors on colchicine binding to plant tubulins do not present a complete picture, but they establish that the colchicine-binding site of plant tubulins differs significantly from that of animal tubulin. Ludueña et al. (1980) reported that podophyllotoxin inhibited colchicine binding to fern (*Marsilea*) sperm axoneme outer doublet tubulin, although they present data showing that this tubulin exhibited a low-affinity interaction with colchicine, as has been found to be the case with rose and hibiscus tubulins. However, podophyllotoxin has no effect on colchicine binding to carrot or rose tubulin (Okamura, 1983; Okamura et al., 1984; Kato et al., 1985; Morejohn et al., 1987a). This would indicate that higher plant tubulin lacks the binding site for the A ring and that colchicine interacts with plant tubulin primarily through its C ring. However, Morejohn *et al.* (1987a) also found that tropolone did not inhibit colchicine binding to the rose tubulin. Thus, it is possible that colchicine interacts with plant tubulin primarily through its B ring.

I. Herbicides as Plant Antimicrotubule Agents

Two classes of herbicides have colchicine-like antimitotic effects on seedling roots. These are the phosphoric amide and the dinitroaniline herbicides. In either case, the cytological effects are very similar to those induced by colchicine, except that these herbicides are active at concentrations in the micromolar range, whereas millimolar levels of colchicine are required to produce the same effects. The effects include the swelling of the root tip, blockage of mitosis in metaphase, destruction of cytoplasmic and spindle MTs, and a loss of the polarity of growth (see Ashton and Crafts, 1981, and Fedtke, 1982, for reviews of this literature).

1. Phosphoric Amide Herbicides

The phosphoric amide herbicides, chiefly amiprophos methyl (APM), bring about the destruction of cytoplasmic microtubules and produce mitotic arrest in metaphase in a variety of plant tissues. Micromolar concentrations of APM depolymerize spindle MTs and stop chromosome movement in *Hemanthus* endosperm. These effects are fully reversed when the herbicide is washed out of the cells (Bajer and Molé-Bajer, 1986). Being an herbicide, APM is selective in its action, and the MTs of some plant species are almost completely resistant to depolymerization by APM (Schroeder *et al.*, 1985; Falconer and Seagull, 1987). Nevertheless, for those plants that are susceptible, APM is a fast, effective, reversible antimitotic agent. Furthermore, it appears to have no effect on MTs of vertebrate cells (Bajer and Molé-Bajer, 1986).

There is now good evidence that these anti-MT effects are brought about by the binding of the herbicide to plant tubulin, which blocks its assembly into MTs. Morejohn and Fosket (1984a) demonstrated that micromolar levels of APM block taxol-driven rose tubulin assembly *in vitro*. The action of APM appeared to be preferentially on the nucleation step since, as the concentration of APM in the assembly mixture was increased, the number of MTs formed decreased. APM had no effect on the taxol-driven assembly of brain tubulin. The affinity of APM for tubulins was not determined because radiolabeled APM is not available and alternative methods for determining APM binding have not been worked out.

2. Dinitroaniline Herbicides

This group includes a number of related chemicals, of which oryzalin and trifluralin are the most thoroughly studied. Morejohn *et al.* (1987b) showed that 10 nM oryzalin slowed anaphase chromosome movement in *Hemanthus* endosperm cells, while 0.1 μM stopped it completely within minutes. Treatment of *Hemanthus* endosperm cells for 2 min or less with 0.1 μM oryzalin caused virtually all of the MTs to depolymerize in both the mitotic spindle and in interphase cells. This effect was not readily reversible. In contrast, exposure of primary cultures of *Xenopus leavis* heart cells to oryzalin at concentrations up to 10 μM for 10 hr had no effect on chromosome movement, on the mitotic spindle, or on interphase MTs.

Scatchard analysis of the binding of dinitroaniline herbicides is difficult for technical reasons. These herbicides also bind to glass surfaces, their solubility in aqueous solution is low, and, although radiolabeled herbicides are available, their specific activity is low. Thus, the binding constants obtained thus far are only rough estimates of the affinity for tubulins for the dinitroaniline herbicides. *Chlamydomonas* flagellar axoneme tubulin has been shown to bind both trifluralin (Hess and Bayer, 1977) and oryzalin (Strachen and Hess, 1983). The latter study obtained an apparent binding constant of $2.08 \times 10^5 \ M^{-1}$ for oryzalin, while Morejohn *et al.* (1987b) obtained a similar apparent constant, $k = 1.19 \times 10^5 \ M^{-1}$, for oryzalin binding to rose tubulin.

There are conflicting reports about the effect of oryzalin and trifluralin on the assembly of mammalian tubulin. Bartels and Hilton (1973) report no effect on assembly, while Robinson and Herzog (1977) report that the herbicides depolymerize assembled MTs and completely block brain tubulin assembly. Hertel and Marmé (1983) observed that the herbicides partially inhibit the assembly of brain tubulin. Morejohn *et al.* (1987b) found that oryzalin has no effect on brain tubulin assembly *in vitro* at any herbicide concentration tested, but it is highly effective in blocking taxol-driven rose tubulin assembly under the same conditions. Oryzalin also rapidly depolymerizes rose MTs when added to the assembly reaction at steady state, but it does not depolymerize assembled brain MTs. The reasons for the discrepancies are not immediately obvious. However, the study of Morejohn *et al.* (1987b) clearly shows a marked difference in response of brain and rose tubulin assemblies to these herbicides. Morejohn *et al.* (1987b) also noted that there are some nonspecific oryzalin binding to brain tubulin at a level nearly 20-fold lower than that giving the specific binding observed with rose tubulin. The nonspecific binding of the herbicide to brain tubulin may affect its ability to assemble under some conditions, possibly by increasing the critical concentration for assembly. The reported data on the effects of the dinitroaniline herbicides on brain tubulin assembly are not in sufficient detail to evaluate these possibilities.

J. Tubulin Mutants

Bolduc *et al.* (1988) isolated two independent *Chlamydomonas* mutants which are more resistant to colchicine. Both mutations were shown to be in one of the β-tubulin genes, resulting in the appearance of an acidic β-tubulin isoform. In addition to increased colchicine resistance, these mutants show cross-resistance to a number of additional antimicrotubule agents, including oryzalin, trifluralin and APM. Recently, Mudge *et al.* (1984) reported the identification of a trifluralin-resistant biotype of a weed known as goosegrass (*Eleusine indica*) which is cross-resistant to other dinitroaniline herbicides and to APM, but not to colchicine (Vaughn, 1986a; Vaughn *et al.*, 1987). Vaughn (1986b) also reported that the electrophoretic mobility of the β-tubulin subunit of the resistant biotype exhibited an altered electrophoretic mobility, suggesting that the resistance may result from a mutation in a β-tubulin gene. This is an important finding, since molecular genetics could contribute substantially to the understanding of the plant cytoskeleton, and this is the first indication of a mutant plant tubulin.

The power of this approach is illustrated by recent advances in understanding benomyl resistance in fungi. Benomyl inhibits the growth of many fungi by binding to tubulin and destabilizing their MTs (Davidse and Flach, 1977; Quinlan *et al.*, 1980). Benomyl-resistant mutants have been isolated from *Aspergillus nidulans* (Sheir-Neiss *et al.*, 1978), *Schizosaccharomyces pombe* (Hiraoka *et al.*, 1984), *Saccharomyces cerevisiae* (Thomas *et al.*, 1985), and *Neurospora crassa* (Orbach *et al.*, 1986). In each of these cases, the resistant mutations were demonstrated to result from changes in the genes encoding β-tubulin. Both *N. crassa* and *S. cerevisiae* appear to have only one β-tubulin gene (Neff *et al.*, 1983; Orbach *et al.*, 1986), but *A. nidulans* has two functional β-tubulin genes (Weatherbee and Morris, 1984). The *N. crassa* benomyl-resistant β-tubulin gene has been used as a dominant selectable marker in transformation studies (Orbach *et al.*, 1986). As a result, it is likely that a mutation in one member of a β-tubulin multigene family would confer resistance. Orbach *et al.* (1986) also showed that a single amino acid change, the substitution of a tyrosine for phenylalanine at position 167 in the β-tubulin of *N. crassa* was responsible for benomyl resistance. Thomas *et al.* (1985) also found that benomyl resistance resulted from a single amino acid change in the β-tubulin of *S. cerevisiae*, but in this case a histidine was substituted for arginine at position 241. While it is possible that these analyses of the benomyl-resistant mutants have identified the benomyl-binding site, it is also possible that the amino acid substitutions result in a conformational change in the β-tubulin, altering a benomyl-binding site remote from the mutations. Nevertheless, these studies have begun the molecular analysis of the benomyl-binding site and they are beginning to determine the changes necessary to obtain benomyl resistance.

K. Microtubule-Associated Proteins

There is no compelling biochemical evidence that higher plant MTs have microtubule-associated proteins (MAPs). However, EM micrographs are highly suggestive of the presence of MAPs on plant MTs *in vivo*. Furthermore, the data on animal MAPs show that these proteins play vital roles in the formation, organization, and function of MT arrays. The various functions that have been demonstrated or indicated for animal MAPs include (1) enhancement of MT stability, (2) cross-bridging MTs to other cellular structures, such as the plasma membrane, MFs, and IFs, (3) transport of particles along MTs, and (4) enhancing MT formation by acting as MT nucleation sites. Even if plant MAPs are not homologous with animal MAPs, it is difficult to envision plant MT arrays forming and functioning without analogous proteins. It is likely that higher plant MAPs have not been isolated for technical reasons. Presumably these difficulties will be overcome in the near future.

1. *MAP1, 2, 3, and Tau Factor*

There are several recent reviews of MAPs (Sloboda and Rosenbaum, 1979a; Vallee *et al.*, 1984; Wiche, 1985; Olmsted, 1986; Matus, 1988). The most thoroughly characterized MAPs are those that associate with flagellar axoneme MTs. These include the ATPase, dynein, which plays an important role in flagellar movement. Several classes of MAPs have been isolated from animal cells and the nature of the MAP or putative MAP depends not only on the tissue from which the proteins are isolated, but also upon the method used for their isolation. Two types of proteins copurify with tubulin when it is isolated by cyclic rounds of assembly from vertebrate brain homogenates. The first, with masses near or over 200 kDa, include the MAP1 group, consisting of at least three different proteins (300–350 kDa), the MAP2 group, consisting of at least two proteins (270–300 kDa), (Sloboda *et al.*, 1975, Binder *et al.*, 1984), and MAP3 (180 kDa) (Huber *et al.*, 1985). The second type of copurifying protein, designated *tau* factor, consists of at least four phosphoproteins of between 55 and 68 kDa (Cleveland *et al.*, 1977). The amino acid sequence of two *tau* proteins from mouse brain was determined from cDNA clones (Lee *et al.*, 1988). Both proteins contain an 18-residue sequence that is repeated three times which was postulated to represent the *tau* MT binding site. *Tau* has been shown to be a linear, elastic protein (Lichtenberg *et al.*, 1988). After binding to specific sites on the tubulin dimer, MAPs project from the MT surface with regular spacings of from 21 to over 100 nm, depending upon the study (Voter and Erikson, 1982; de la Torre *et al.*, 1986). Limited proteolysis of MAP2 cleaves the protein into two fragments, a 240-kDa arm, to which a cAMP-dependent protein kinase may associate, and a 40-kDa MT-binding peptide (Vallee, 1980). The MT-binding fragment also binds calmodulin (Lee and Wolff, 1984).

2. *Other MAPs*

A fraction of the MTs assembled *in vitro* from crude extracts are insensitive to cold-induced disassembly. These cold-stable MTs, which disassemble in the presence of calmodulin and low levels of calcium ions, have a special class of MAPs associated with them called STOPs (for stable tubule only peptides) (Margolis and Rauch, 1981; Job *et al.*, 1982). STOPs are proteins of 56, 70 to 82, and 145 kDa. The association of these proteins with MTs has been correlated with their enhanced stability *in vivo*, as well as their resistance to cold-induced disassembly *in vitro* (Job *et al.*, 1985; Margolis *et al.*, 1986).

Although MAPs have been isolated from tissues other than brain by cyclic rounds of assembly (Bulinski and Borisy, 1979), in most cases this approach has not been successful. Instead, other procedures have been devised. For example, MAPs have been isolated from a number of different animal cells by first isolating taxol-stabilized MTs, and then removing the proteins bound to the MTs by increasing the ionic strength of the solution (Vallee, 1982; Vallee and Bloom, 1983; Goldstein *et al.*, 1986). Many, but not all, of the proteins isolated by this procedure have been shown to be MAPs, and some of the MAPs thus isolated have been shown to be identical with, or related to MAP1 or 2 or *tau* factor (Wiche *et al.*, 1984; Bloom *et al.*, 1985).

3. Chartins

Solomon and his co-workers devised yet another approach for the isolation of MAPs in which the cytoskeleton or mitotic spindle is first isolated and then proteins associated with it are released with drugs that depolymerize MTs (Solomon *et al.*, 1979). They have isolated a number of different proteins (which they have termed chartins) by this procedure and demonstrated that they were microtubule-associated. These have included a 150-kDa protein, which appears to be unique to the mitotic spindle (Zieve and Solomon, 1982), and a pair of proteins found only in neural tissues (Duerr *et al.*, 1981; Zieve and Solomon, 1984).

4. Kinesin

Perhaps the most exciting discovery made in this field in recent years grew out of the demonstration that particles are transported along MTs in the axioplasm of squid giant axons (Allen *et al.*, 1982, 1985). Vale (1985a) showed that isolated axoplasmic organelles moved along *in vitro* assembled, MAP-free microtubules at a rate similar to that observed in dissected axioplasm. This led to the isolation of the conditional MAP, kinesin, from the squid axioplasm, which associates with cellular particles and transports them along MTs. Experimentally, kinesin-coated latex beads are transported from the (−) ends of MTs toward the (+) ends. Kinesin has also been isolated from vertebrate brain tissues and shown to have similar properties (Kuznetsov and Gelfand, 1986). It appears to be a pentameric protein consisting of one peptide of approximately 135 kDa, plus four peptides ranging from 45 to 70 kDa. Kinesin is a MT-activated ATPase. It associates with the MT only when it has bound ATP, and is released from the MT after ATP hydrolysis. It can be coisolated with taxol-stabilized brain MTs, prepared in the presence of a nonhydrolyzable ATP analog, which prevents the release of kinesin from the MTs (Vale *et al.*, 1985b). Kinesin is a motor which may attach to particles and transport them along MTs. It only transports particles in one direction along MTs, although it is known that transport is bidirectional on a single MT, suggesting that another MAP must be responsible for the retrograde movement (Schnapp *et al.*, 1985; Vale *et al.*, 1985a). Leslie *et al.* (1987) showed that kinesin associates with the mitotic spindles of sea urchin embryos. Furthermore, using a highly specific antibody for kinesin, they observed that kinesin also associates with remnants of the spindle that persist after the MTs are disrupted. Other MT-based motors have been isolated from various neuronal tissues. These include a variant of MAP1 known as MAP1c (Paschal *et al.*, 1987) and a 292-kDa protein designated vesikin that appears to be involved in vesicle transport (Do *et al.*, 1988). The literature in kinesin and other MT-based motors has been reviewed (Vale, 1987).

L. Tubulin Self-Assembly

Tubulin self-assembly has been reviewed recently by Hill and Kirschner (1982) and Purich and Kristofferson (1984). Kirschner and Mitchison (1986) also reviewed much of the pertinent literature on tubulin assembly *in vitro* and *in vivo,* and in addition discuss the ways in which their "dynamic instability" hypothesis (discussed below) can account for the formation of different MT arrays and for morphogenesis of animal cells.

1. Assembly Kinetics

Microtubule assembly is similar to other self-assembly processes, such as formation of viruses from components, or more appropriately, formation of actin microfilaments from G-actin monomers. Typically, MT formation *in vitro* is monitored by following changes in turbidity as the tubulin dimers polymerize. A plot of the time course of brain tubulin assembly is sigmoidal. A brief lag phase is followed by a phase of exponential increase in turbidity, and then a stable plateau. The total amount of polymer formed is a function of the initial tubulin concentration. However, there is a critical tubulin concentration, (Cc), below which no MTs are formed. Above Cc, the amount of polymer formed is proportional to total tubulin concentration, but all the tubulin dimers do not assemble. The MTs behave as though they are in equilibrium with a pool of unassembled free tubulin dimers, the concentration of which equals Cc.

Tubulin self-assembly is a two-step process in which a comparatively slow nucleation step is followed by a rapid elongation reaction. Nucleation is the process in which aggregates are formed that have a greater probability of persisting than disaggregating. In living cells, MTs usually are nucleated by microtubule organizing centers (MTOCs). The most important MTOC in animal cells is the centrosome (reviewed by McIntosh, 1983; Brinkley, 1985). The centrosome contains the centriole or centriole pair, plus some amorphous electron-dense material known as pericentriolar granules. Although both the centriole and the pericentriolar granules can nucleate MT formation *in vitro,* it is the pericentriolar material that nucleates the interphase cytoskeletal MTs. Mitchison and Kirschner (1984a) showed that isolated centrosomes nucleate a fixed number of MTs, and that they permit MT formation at tubulin concentrations below the Cc. Because of this, most MTs originate from the centrosome within a living animal cell.

Most analyses of the elongation reaction are based on the assumption that subunits add only at the ends of the growing MT. Recently, this assumption was confirmed experimentally (Soltys and Borisy, 1985; Schulze and Kirschner, 1986). The rate of subunit addition to the MT ends, k_1, is proportional to the free subunit concentration (c). Subunits also are lost from the ends of the MTs at a rate k_2 that is independent of the free subunit concentration. As a result, the net assembly rate at one end of the MT is

$$dn/dt = ck_1 - k_2$$

This equation also defines the Cc, where $dn/dt = 0$, as

$$Cc = k_2/k_1$$

Assembly results in the progressive reduction of the free tubulin dimer concentration so that the reaction should come to equilibrium and the net assembly rate should reach zero when the free tubulin dimer concentration is reduced to Cc.

2. GTP Hydrolysis and Tubulin Assembly

MTs are not true equilibrium polymers because tubulin hydrolyzes GTP. Tubulin binds 2 moles of GTP/mole of tubulin dimer. One of the GTPs is exchangeable with GTP in solution, while the other is not exchangeable. After assembly, GTP bound at the exchangeable site is hydrolyzed to GDP by a GTPase activity of the tubulin itself. While GTP must be bound to tubulin if it is to assemble, GTP hydrolysis is not necessary for assembly. There is evidence that the dissociation of GTP–tubulin occurs at a slower rate than that of GDP–tubulin, and it has been proposed that GTP–tubulin provides a kinetic cap on the MTs (Carlier *et al.,* 1984; Chen and Hill, 1985; Kirschner and Mitchison, 1986). According to this model, GTP hydrolysis lags behind assembly, particularly at high assembly rates. However, GTP hydrolysis is a stocastic process and there is a probability that GTP hydrolysis will occur in all or most of the subunits at the MT end so that the cap is lost. Once the MTs have lost their GTP–tubulin caps, the exposed GDP–tubulin would rapidly disassemble. Once the GDP–tubulin is released from the MT, the bound GDP is free to exchange with GTP in solution, thus regenerating GTP–tubulin which can again participate in the assembly reaction. It is assumed that there is a much greater probability for the GTP–tubulin to be added to capped than uncapped MTs. As a result, the capped MTs continue to grow, while the uncapped MTs completely depolymerize.

3. Treadmilling

Margolis and Wilson (1978, 1981) noted that, if the disassembly rate were greater at the (−) end than at the (+) end of MTs, and if the assembly rate were greater at the (+) end, the net result could be (+) end MT assembly with simultaneous (−) end disassembly. Since the assembly rates are a function of the tubulin dimer concentration, at high relative tubulin concentrations there will be a net addition of subunits to both ends of the MT. At low free tubulin concentrations there will be a net loss of subunits from both ends, and the MT will shrink. At steady state, however, the net assembly rate at the (+) end would equal net disassembly rate at the (−) end, and subunits would move through the MTs, or treadmill, while MT length remained constant. Margolis and Wilson (1981) proposed this mechanism as a way to explain the substoichiometric poisoning of MT assembly by colchicine. Colchicine only binds to tubulin dimers, but the amount of colchicine required to dissociate MTs is far less than the tubulin concentration in animal cells. If the colchicine–tubulin complex binds to the (+) end of the MT and prevents further assembly, according to the treadmilling hypothesis, the MT would rapidly disassemble from the (−) end. While some important parts of this hypothesis appear to be correct, Bergen and Borisy (1983) demonstrated that the tubulin–colchicine complex binds to both the (+) and the (−) ends of the MT. While substoichiometric levels of tubulin–colchicine complex reduced the association rate constants at both the (+) and (−) ends, there was no effect of the tubulin–colchicine complex on the dissociation rate constants at any concentration they tested. In other words, the tubulin–colchicine complex blocked assembly at both ends, while permitting subunit disassociation to continue.

4. Dynamic Instability

Treadmilling is an intriguing hypothesis which has important implications for MT function in living cells. While there is evidence that it occurs with *in vitro*-assembled MTs (Wilson *et al.*, 1985; Farrell *et al.*, 1986), it has not been demonstrated unambiguously *in vivo*. Also, some *in vitro* studies that support the hypothesis suggest that the rate of flux of tubulin dimers through the MTs is low (Bergen and Borisy, 1980). Mitchison and Kirschner (1984a,b) suggest that the apparent steady state of *in vitro*-assembled MTs does not represent a true steady state. They observed that, although the turbidity of the solution remained constant at "steady state," in reality some MTs were growing slowly, while others were rapidly shrinking. As a result, the numbers of MTs decreased while the total mass of assembled MTs remained constant. They termed this behavior "dynamic instability," and suggested that it was brought about by the lag between the assembly of GTP–tubulin and the subsequent hydrolysis of its bound GTP. As in the model of Carlier *et al.* (1984), the growing MTs are postulated to have a kinetic cap of GTP–tubulin, and the loss of this cap due to GTP hydrolysis results in the rapid disassembly of the exposed GDP–tubulin subunits.

Recently, Horio and Hotani (1986), using dark-field microscopy, observed directly the instability of MTs assembled to "steady state" *in vitro*. As predicted by Mitchison and Kirschner (1984a,b), most MTs grew slowly, while a minority rapidly shortened and some even disassembled completely. The observed shortening rate, which was greater at the (−) end of the MT, was 10- to 15-fold greater than the growth rate. Although these results tend to confirm the dynamic instability hypothesis, Horio and Hotani (1986) also noted that a given MT end alternated between the growing and a shrinking phase, something that is predicted to be highly improbable in the Mitchison–Kirschner model.

Farrell *et al.* (1986) confirmed the observations of Mitchison and Kirschner (1984a,b), i.e., that there is an increase in the average MT length as MT number decrease at "steady state." However, this occurred only in the assembly of MAP-depleted tubulin, and only after the MTs were sheared mechanically. If the MTs were undisturbed, there was little change in average MT length when the MTs were assembled from either MAP-rich or MAP-depleted tubulin. Shearing the MTs brought about a redistribution of MT length in either case, but a new steady state was reached quickly and a substantial change in MT length occurred only when the MTs were assembled from MAP-depleted tubulin. Horio and Hotani (1986) also noted that MAPs stabilized MTs, reducing both the growth and shrinking rates at steady state.

Since nucleated MT assembly in living cells is unlikely to occur without MAPs participation, it is not certain to what degree the dynamic stability hypothesis is relevant to MT formation *in vivo*. In any case, it is clear that there is a pool of unassembled tubulin dimers in cells and that the MTs of any given array will be in pseudoequilibrium with that pool. In the mitotic spindle, individual MTs have half-lives measured in seconds, while interphase MTs have half-lives measured in minutes to at most a few hours (Kirschner and Mitchison, 1986).

5. *Higher Plant Tubulin Assembly* in Vitro

Very little is known about the kinetics of plant MT assembly either *in vitro* or *in vivo*. Morejohn and Fosket (1982) showed that DEAE-purified rose tubulin would assemble *in vitro* either in the presence of the taxane alkaloid taxol, or in the presence of glycerol. Taxol promotes tubulin assembly by binding to assembled tubulin dimers, thereby decreasing the dissociation constants of the dimer from either end of the MT (Kumar, 1981). With saturating amounts of taxol, rose tubulin assembles with sigmoidal kinetics, reaching an apparent steady state in about 1 hr. Rose tubulin assembly in the presence of glycerol or dimethyl sulfoxide (DMSO), without taxol, is not efficient and only a few long MTs are formed. Taxol-induced brain tubulin assembly exhibits hyperbolic kinetics, reaching steady state in a few minutes. The apparent Cc for rose tubulin was 2.1 μM, while the Cc for DEAE-purified brain tubulin, assembled in the presence of saturating levels of taxol, was 3.3 μM (Morejohn and Fosket, 1984b). Together, these data indicate that plant tubulin nucleates much more slowly than brain tubulin, but that other aspects of the plant tubulin assembly are similar to those of brain tubulin. Mizuno (1985) demonstrated that tubulins isolated from seedling tissues of several different higher plant species can be assembled *in vitro* in the presence of DMSO, but he has not subjected this assembly to kinetic analysis.

6. MT Formation in Plant Cells

Given the ability of plant tubulin to assemble in a manner analogous to brain tubulin, we may assume that MTs in plant cells are in equilibrium with a cytoplasmic free tubulin dimer pool, as is the case in animal cells. The formation of MTs at any particular site within the cell would be determined by the appearance of factors that nucleate MT assembly and reduce the Cc locally, much as the centrosome nucleates MT assembly in animal cells. The progressive appearance and disappearance of specific MT arrays thus could be explained by the appearance of different nucleating sites, or by their relocation or modification.

A human autoimmune disease has been identified in which individuals produce antibodies to the pericentriolar material of the centrosome. Antiserum from such an individual has been used to follow the behavior of the centrosome during the cell cycle of mammalian cells (Brenner and Brinkley, 1982). Pickett-Heaps (1969) proposed that osmiophilic amorphous material observed at the poles of cells in anastral division was analogous to the pericentriolar material of the centrosome of cells exhibiting astral division, and he suggested the term microtubule organizing centers (MTOCs) for both. The validity of this proposal was demonstrated by the findings of Clayton *et al.* (1985) that the sclerodoma antiserum containing antibodies to human pericentriolar material also stained the amorphous polar granules in mitotic onion cells. The pericentriolar-like material was only detected in cells that were dividing, or had divided recently. There was no evidence that these same MTOCs nucleated the cortical MT array. Immediately after division, pericentriolar-like material was detected around the cytoplasmic face of the nuclear envelope of the daughter nuclei, where it nucleated MTs that radiated out toward the cell surface. Later this array was replaced with the cortical array, which appeared without any detectable relationship to the pericentriolar-like material. Higher plant cells without walls, such as endosperm cells of many monocots and male meiocytes, exhibit a persistent radial array of MTs which projects from electron-dense material surrounding the nuclear envelope (Franke et al., 1977; De Mey et al., 1982; Bajer and Molé-Bajer, 1986; Sheldon and Dickinson, 1986). Perhaps this radial MT array is a fifth type of plant cytoskeletal array, since it is apt to be functionally as well as spatially and temporally distinct from the cortical array.

It is possible that the cellulose synthetic machinery, which is localized in the plasma membrane, contains transmembrane elements which act as MTOCs for the cortical MT array in walled cells. In the absence of cellulose synthesis, the radial array would persist, with nucleation by the pericentriolar-like material at the surface of the nuclear envelope. However, the onset of cellulose synthesis would present competing MT nucleating centers on the inner surface of the plasma membrane that nucleate the MTs of the cortical array.

M. Cold Stability of Plant Microtubules

Many higher plant MTs are comparatively resistant to cold depolymerization. Temperate zone plants must be able to withstand a wide range of temperatures, and some plants are capable of growing at temperatures that would depolymerize the MTs of homothermic animals. Cold stable MTs may arise by two mechanisms in vertebrates. In mammalian brains, a group of microtubule associated proteins called STOPs has been shown to bind to MTs at substoichiometric levels to increase their resistance to cold-induced depolymerization, at least in MTs assembled *in vitro* from crude homogenates (Job et al., 1982). In addition, Detrich and Overton (1986) have shown that tubulin isolated from fishes that have evolved to live in polar waters is able to form cold-stable MTs, without the aid of MAPs. The MTs of Arctic and Antarctic fishes must be able to assemble and function at temperatures that would depolymerize mammalian MTs. A major change in the structure of the tubulin dimer may not be necessary to induce cold stability or lability. Mutations in the gene encoding β-tubulin have been shown to result in a cold-sensitive growth phenotype in both *Saccharomyces* and *Schizosaccharomyces,* and in the case of *Saccharomyces,* the mutation was shown to be a single amino acid substitution (Hiraoka et al., 1984; Thomas et al., 1985).

Some plants clearly exhibit cold-sensitive MTs *in vivo* (Lambert and Bajer, 1977; Juniper and Lawton, 1979; Ilker et al., 1979). The sensitivity of plant MTs to cold depolymerization may be inversely related to their cold-hardiness (Rikin et al., 1980, 1983). Interestingly, plant hormone treatments have been shown to increase both the resistance of sensitive plants to chilling injury and the resistance of their MTs to cold-induced depolymerization (Rikin et al., 1983; Mita and Shibaoka, 1984).

N. Tubulin Synthesis in Relation to Microtubule Formation

Many developing systems exhibit sudden and dramatic changes in MT number. The tubulin dimer pool has been estimated to be approximately 30% of the total tubulin in cultured mouse fibroblasts. Since MTs are in equilibrium with the tubulin dimer pool, changes in the total amount of assembled MTs would be expected to increase or decrease the tubulin dimer pool by a corresponding amount. How does the tubulin synthesis rate respond to changes in the size of the tubulin dimer pool? Conversely, could an increase in the tubulin dimer pool lead to an increase in the number of MTs? The data available at this point indicate that in some systems the rate of tubulin synthesis is regulated by a mechanism that is sensitive to the size of the tubulin dimer pool, whereas in other systems additional factors or regulatory mechanisms make this relationship difficult to observe, if it exists. Two examples in which tubulin synthesis has been shown to be coordinated with the size of the tubulin dimer pools are flagellar development in microorganisms such as *Chlamydomonas,* and in cultured, undifferentiated animal cells. The two systems behave differently in important respects. In many types of cultured animal cells, tubulin synthesis is regulated by a posttranscriptional mechanism, while in microorganisms exhibiting flagellar outgrowth or regrowth, both transcriptional and posttranscriptional mechanisms control tubulin synthesis rates. In other cases, principally differentiating neuroblastoma cells, MT assembly, with the concomitant depletion of the tubulin dimer pool, appears to have no significant effect on the rate of tubulin synthesis.

1. Autoregulation of Tubulin Synthesis in Cultured Animal Cells

This topic has been reviewed by Cleveland and Sullivan (1985) and Cleveland (1986). Essentially, the synthesis of tubulin dimers is regulated by a mechanism that is sensitive to the size of the tubulin dimer pool, a phenomenon that has been termed autoregulation. Ben Ze'ev *et al.* (1979) noted that treatment of cultured mammalian cells with colchicine at a level sufficient to depolymerize the MTs, thereby increasing the pool of free tubulin dimers, specifically depressed the rate of tubulin synthesis. Other drugs that affect the tubulin dimer–MT equilibrium also affect tubulin synthesis. These include nocodazole, an antitumor drug with a mode of action similar to that of colchicine (De Brabander *et al.,* 1976), vinblastine, which decreases both the assembled MTs and the tubulin dimer pool by binding to tubulin, following which the tubulin–vinblastine complex crystallizes (Bryan, 1971), and taxol, which decreases the tubulin dimer pool by increasing MT stability (Schiff and Horowitz, 1980). All of these drugs were shown to affect tubulin synthesis in a manner consistent with the autoregulatory hypothesis (Cleveland *et al.,* 1981; Cleveland and Havercroft, 1983; Lau *et al.*, 1985). Direct injection of tubulin dimers into cultured cells also has been shown to depress tubulin synthesis (Cleveland *et al.*, 1983). The reduction in tubulin synthesis is paralleled by a concomitant reduction in tubulin mRNA, as determined by probing Northern blots of cellular RNA with cloned tubulin sequences (Cleveland and Havercroft, 1983). However, *in vitro* run-off transcription assays with nuclei isolated from control and colchicine-treated cells revealed no difference in the rate of tubulin gene transcription (Cleveland and Havercroft, 1983), indicating that autoregulation is not mediated at the transcriptional level. Pittenger and Cleveland (1985) and Caron *et al.* (1985) found that colchicine treatment also depresses tubulin synthesis in cytoplasts, cells lacking nuclei. Recently, Gay *et al.* (1987) demonstrated that a 49-bp sequence encoding the amino-terminus of a chicken β-tubulin was sufficient for autoregulation to occur when this sequence was inserted in a thymidine kinase gene and transfected into mouse cells. Their results showed that autoregulation of tubulin synthesis most likely involves a mechanism affecting tubulin mRNA stability. Since untranslated cytoplasmic mRNAs decay more rapidly than translated mRNAs, it is possible that the mechanism operates at the level of tubulin mRNA translation.

2. Tubulin Synthesis during Flagellar Outgrowth in Chlamydomonas

The biflagellated, unicellular alga *Chlamydomonas* can be deflagellated by mechanical or pH shock. The deflagellated cells respond by regenerating their flagella, partially from an endogenous pool of precursors, but also from components that are synthesized in response to deflagellation. The response has been the subject of numerous, elegant investigations, summarized in a recent review (Lefebvre and Rosenbaum, 1986). Tubulin synthesis is induced within a few minutes of deflagellation, reaches a peak about 1 hr later, and then declines to the basal level over the course of the next 2 hr (Weeks and Collis, 1976). The transient increase in the rate of tubulin synthesis is accompanied by a concomitant increase in tubulin mRNA (Silflow and Rosenbaum, 1981; Schloss *et al.*, 1984) that results from the coordinate expression of all four tubulin genes (Brunke *et al.*, 1982). Using the technique of *in vitro* run-off transcription from isolated nuclei, Keller *et al.* (1984) demonstrated that deflagellation stimulated the rate of tubulin gene transcription from 4- to 10-fold. Since the induced level of cytoplasmic tubulin mRNA is 10- to 14-fold greater than the basal level, it is likely that there is an enhancement of tubulin mRNA stability as well as the increased rate of tubulin gene transcription (Baker *et al.*, 1984). The enhancement of tubulin synthesis, however, is not initiated in response to the depletion of the endogenous tubulin dimer pool, since tubulin synthesis is stimulated to the same degree by deflagellation in the presence of sufficient colchicine to prevent axoneme MT formation (Lefebvre *et al.*, 1980).

3. Microtubule Formation in Neuroblastoma Cells

Neuroblastoma cells are derived from tumors of neural tissue and they can be grown indefinitely in culture, where they have a somewhat rounded morphology. The neuroblastoma cells known as rat PC12 (pheochromocytoma) cells average 9 μm in diameter. However, if they are cultured in a medium that contains nerve growth factor (NGF), they will extend neurites from the cell body which may reach over 500 μm in length. The neurites are packed with closely bundled MTs. Other neuroblastoma cell lines can be induced to differentiate with a variety of factors, including cAMP. This phenomenon raises a number of important questions about the factors controlling MT formation. Since MT assembly occurs from a soluble tubulin dimer pool, how does the formation of the extensive MT array in the neurites affect the synthesis of the tubulin monomers? Also, what factors are responsible for the initiation of MT assembly during neurite outgrowth, and how do these initiation sites come to be localized in the regions of neurite outgrowth? It is possible that the answers to these questions are not relevant to MT formation in plants, but they are at least suggestive of the type of phenomena that could be responsible for the organization and orientation of the different plant MT arrays.

Olmsted (1981) used a radioimmunoassay (RIA) to determine the tubulin levels in mouse neuroblastoma cells both before and after cAMP-induced differentiation. She found no change in the total amount of tubulin per cell over the week during which neurite outgrowth occurred. In either case, the cells have approximately 4 pg of tubulin/cell. In the undifferentiated cells, only 11–16% of the tubulin was in the assembled fraction with 84–89% of the tubulin in the free tubulin dimer pool. In the differentiated cells, 48–63% of the tubulin was in the assembled MT fraction. Thus, there was a 4- to 5-fold increase in the assembled MTs, with no increase in the total tubulin. Olmsted *et al.* (1984) found enhanced MT assembly during neurite outgrowth to be correlated with the synthesis and accumulation of MAP4. Kirschner's group has studied nerve growth factor-induced PC12 neuroblastoma cell differentiation, which is slightly different in that there is some increase in tubulin during neurite outgrowth (Drubin *et al.*, 1984, 1985). However, they found that, after a lag of approximately 3 days, both *tau* factor and MAP1 increased much more dramatically and this increase paralleled the increase in neurite length and number of assembled MTs. Thus, in this system, MT formation appears to be determined by the appearance in the cell of proteins which act both as sites of initiation for MT assembly and factors which enhance the stability of the MTs.

4. Tubulin Synthesis during Carrot Somatic Embryogenesis

Somatic embryogenesis is induced in cultured carrot cells upon transfer to fresh medium lacking auxin (Steward *et al.*, 1964; Halperin and Wetherell,

1965; Sung and Okimoto, 1981). The relatively disordered pattern of growth of the cultured cells is replaced by the comparatively precise growth patterns of somatic embryogenesis, a process similar in developmental sequence to zygotic embryogenesis. As the cultured cells initiate embryogenesis, the average cell size is reduced, largely as a result of the elimination of the central vacuole, and the cells initiate a series of regular divisions to form the globular stage embryo. The transition from an irregular growth pattern in the cultured somatic cells to the ordered growth of the globular stage embryo has been shown by electron microscopy to be accompanied by a large increase in the number of MTs in the cortical array (Halperin and Jensen, 1967; Wochok, 1973). During later stages of somatic embryogenesis, a marked polarization of growth occurs as the cells become elongate and some vascularization is initiated (torpedo and plantlet stages).

Recently, Cyr et al. (1987b) examined changes in tubulin protein and tubulin mRNAs during carrot somatic embryogenesis. A RIA was developed using an antiserum to soybean α-tubulin and a monoclonal antibody to vertebrate β-tubulin, respectively. Tubulin mRNA levels were estimated by densitometric scans of Northern blots of carrot poly(A)+ RNA, probed with soybean α- or β-tubulin sequences. The data indicated no increase in the amount of tubulin/cell as the somatic carrot cells initiated embryogenesis. During late stages of embryogenesis, however, where cell elongation and differentiation are occurring (torpedo/plantlet stages), there was approximately a 5-fold increase in tubulin protein/cell. Northern analysis demonstrated that the cellular levels of tubulin mRNA varied concomitantly with the tubulin protein level. These results indicate that MT formation early in embryogenesis is not limited by the size of the pool of free tubulin dimers. Although the actual tubulin synthesis rates were not examined, it is clear that the large increase in cortical MTs that has been observed as the cultured cells initiate embryogenesis takes place with no change in the total amount of tubulin/cell. At this stage the regulation of MT formation in carrot somatic embryogenesis appears to be analogous to that observed in cultured neuroblastoma cells. However, during the later stages of embryogenesis, the parallel increase in tubulin protein and tubulin mRNA indicates that MT formation is coordinated with tubulin synthesis. The data also indicate that there is a coordination of tubulin synthesis and MT formation with cell elongation and/or vascular differentiation.

5. *Tubulin Synthesis and MT Formation during Tracheary Element Differentiation*

Cortical MTs may be involved in orienting the deposition of cellulose microfibrils in the formation of the secondary wall during tracheary element differentiation (Hepler and Fosket, 1971). Recent work has shown that, in addition to the orientation of the cortical MTs, there is a substantial increase in the number of MTs in the cortical array as cells initiate tracheary element differentiation. Fukuda (1987) has quantified this increase by electron microscopy and has shown a nearly 6-fold increase in the density of cortical MTs (number of MTs/micrometer of cell surface in cross-sections) in the differentiating xylem elements as compared to cells before the initiation of differentiation. His investigations were conducted in cultured *Zinnia* leaf mesophyll cells, where 40% of the cells differentiated more or less synchronously, many without first dividing, within the first 72 hr after the start of the culture (see Fukuda and Komamine, 1985, for a review of this system). Fukuda (1987) also measured changes in the amount of tubulin and in its synthesis as tracheary elements formed from the cultured mesophyll cells. Total tubulin levels were measured with an immunoblot assay (Fukuda and Iwata, 1986), using monoclonal antibodies to chick brain α- and β-tubulins, while tubulin synthesis was estimated from *in vivo* incorporation of [^{35}S]methionine into whole-cell proteins, separated by two-dimensional gel electrophoresis. The mobilities of the tubulins were determined by comparison with authentic plant tubulins and by Western blotting with immunodetection. There was a 6-fold increase in both α- and β-tubulins, relative to other cellular proteins, which reached a peak at 48 hr after the start of the culture period. This increase was paralleled by an increase in the observed rate of tubulin synthesis, and blocking protein synthesis with inhibitors of RNA or protein synthesis also prevented the increase in tubulin content. Since the increase in tubulin protein synthesis paralleled the increase in the density of MTs in the differentiating tracheary elements, this result suggests a possible relationship between the regulation of tubulin synthesis and the size of the tubulin dimer pool in this system.

V. ACTIN AND OTHER MICROFILAMENT PROTEINS

Actin and actin-binding proteins have been the subject of several recent reviews (Korn, 1982; Stossel *et al.*, 1985; Pollard and Cooper, 1986; Staiger and Schliwa, 1987). Also, Jackson (1982) has reviewed this field from the perspective of plant biology. Actin polymers or MFs play many different roles within cells and exhibit strikingly different morphologies, depending upon the nature of the cells in which they are found. The actin MFs in cytoskeletal arrays have two functions: First, they play an important role in structuring the cytoplasm. Second, they are involved in movement. It is difficult to say that any particular MF array has a purely structural or a purely motor role in nonmuscle cells.

A. Types of Actins and Their Relationships

There are six different kinds of actin in vertebrate cells. Five of these actins are found in specific kinds of muscle: cardiac, skeletal, smooth, enteric, and vascular. In addition, there is a cytoplasmic actin that is present in nonmuscle cells. Invertebrates seem to use cytoplasmic actin in their muscles and have not evolved a separate muscle actin. Higher plant actins have properties which are in part muscle actinlike and in part cytoplasmic actinlike. All of the actins, including plant actins, are 75 to 95% homologous with each other. Each actin is encoded by at least one gene, so there are at least six genes encoding actins in most vertebrates. The chicken genome contains at least 11 actin genes and soybean contains a minimum of 2 actin genes. It appears that these actins evolved from the cytoplasmic actin found in protozoa and that they have been highly conserved in evolution (Vandekerckhove and Weber, 1984).

B. Higher Plant Actins and Actin Genes

Actin, isolated and partially purified from tomato (Vahey and Scordilis, 1980; Vahey *et al.*, 1982), is similar in its properties to vertebrate actins. Metcalf *et al.* (1980) showed that an antibody to mammalian cytoplasmic actin precipitates a 45-kDa protein from soybean. Nevertheless, most information on plant actins has come from analysis of the amino acid sequences derived from the actin genes which have been isolated and sequenced from soybean, maize, and *Arabidopsis* (Shah *et al.*, 1982, 1983; Hightower and Meagher, 1985, 1986; Nairn *et al.*, 1988). Comparisons of the derived amino acid sequences of these genes with each other and with other known actins revealed that they are 90–92% homologous with each other and approximately 86% homologous with animal actins. Animal cytoplasmic and muscle actins differ in amino acids in characteristic positions. Plant actins exhibit "cytoplasmic" amino acids in some of these positions, while in other positions they have "muscle" amino acids. Intron positions of the plant genes are also similar to both muscle and cytoplasmic actin genes of animals. All sequenced plant actin genes have introns in exactly the same positions. The first plant intron is also in the same position as an intron in the nematode *Caenorhabditis elegans* (cytoplasmic type) actin gene, while the second plant actin intron is in the same position as an intron in vertebrate skeletal muscle actin genes. This suggests that the plant actins may have arisen independently from the ancestral actin gene that gave rise to both the cytoplasmic and muscle actin isotypes. Other features of the plant actin genes suggest that they represent a third class of actins. The first nine amino acids in the plant actins are nearly identical with each other, yet this sequence is different from any other known actin. This raises a number of interesting possibilities and questions. Have plant actins acquired some unique functions in addition to those common to all actins? Do the different plant actins have the same function within the cell, and are they utilized to form the same MF arrays? Are the actin genes coordinately expressed, or is the expression of one or both of these genes restricted to some specific developmental stage?

Actins are encoded by small multigene families in most animals, fungi and protists. Similarly, there are at least three actin genes in *Arabidopsis*, eight in soybean (Hightower and Meagher, 1985), ten in tomato (Bernatzky and Tanksley, 1986), and six to ten in maize (Shah *et al.*, 1982). However, Baird and Meagher (1987) found evidence for over 200 actin genes in the petunia genome. At least some of these appear to be pseudogenes. Nevertheless, the actin multigene family in this higher plant species is both large and complex. Of the cloned actin sequences analyzed, the intron positions were conserved, but the petunia actin could be grouped into at least five subfamilies, representing genes which could encode actins with significantly different properties.

Although it is generally assumed that actin gene expression is constitutive, there is evidence that actin gene expression plays an important role in the growth and differentiation of animal cells (Farmer, 1986). There also is evidence that actin levels differ very substantially in plant tissues. Metcalf *et al.* (1984), using an antibody directed against calf thymus actin in a RIA for actin in plant homogenates, found a 15-fold higher level of actin in roots than in petioles or cotyledons. Actin levels in the radicle were 10-fold higher than those in the root tip.

C. Some Biochemical Properties of Actin

Muscle actin, as well as cytoplasmic actin, exists in two forms: G-actin (or globular actin) and F-actin (or fibrous actin). G-Actin is a single polypeptide chain with mass of about 42 kDa. As its name suggests, it has a roughly globular configuration. There is one high-affinity calcium-binding site/G-actin monomer which stabilizes the globular configuration of the molecule. G-Actin also has one ATP-binding site/monomer. F-Actin is a filamentous polymer, composed of G-actin monomers. The F-actin filaments consist of two helical aggregates of G-actin which are twisted around each other, with 13.5 subunits/turn. F-Actin filaments are the MFs seen in electron micrographs of cells. The actin self-assembly reaction can be inhibited by the fungal metabolite cytochalasin. Another fungal metabolite, phalloidin, binds to and stabilizes the actin filament.

D. Polymerization of G-Actin

Actin polymerization has been studied extensively and is the subject of several elegant theoretical treatments (Oosawa and Asakura, 1975; Hill and Kirschner, 1982; Frieden, 1985). G-Actin self-assembles. All the information necessary to form the F-actin filaments is in the G-actin monomers. The assembly reaction consists of four steps:

1. Activation (salt binds to the monomer, apparently inducing a conformational change in the protein which favors assembly).

2. Nucleation (the formation of oligomers having a higher probability of growing into filaments than decomposing into monomers. This is the slow step and is rate limiting for the whole polymerization process).
3. Elongation (the bidirectional growth of the polymers).
4. Annealing (the end-to-end joining of two filaments).

All of the steps are reversible. The polymerization reaction usually is followed by monitoring the viscosity of the solution, which increases dramatically as the G-actin assembles into F-actin filaments. The time course of spontaneous polymerization of monomers is sigmoidal with an initial lag period, which reflects the slow nucleation step.

F-Actin filaments are polar structures, with the polarity being a reflection of an inherent polarity of the G-actin monomer. The polarity of the filament can be demonstrated by the arrowhead pattern seen in the electron microscope when the actin filament binds heavy meromyosin (HMM, or S1, see below), a proteolytic cleavage fragment of myosin. The arrowheads point toward the "pointed" end and away from the "barbed" end of the actin filaments. The rate of addition of G-actin monomers is greater at the barbed end than at the pointed end of the filament. Furthermore, the rate of polymerization of ATP–actin is greater than that of ADP–actin. The overall rate of actin assembly is the sum of the individual assembly and disassembly reactions that occur at each end of the polymer. As was the case with tubulin, polymerization is sensitive to the monomer concentration, and there is a minimum concentration, the critical concentration (Cc), below which polymerization will not occur. The Cc may be the same at the two ends or it may differ, depending upon the reaction conditions. In the presence of calcium ions and KCl, the Cc is the same at the two ends, but it differs when the reaction occurs in the presence of magnesium ions. The difference in the critical concentration between the two ends can be substantial. When this is the case, actin subunits will flux through the filament (treadmilling). Hydrolysis of the bound ATP is not directly coupled to assembly. Rather, there is a delay after the ATP–actin subunit is incorporated into the filament, before hydrolysis occurs. As a result, the actin filament has ATP–actin "caps" on either end when it is growing rapidly. Since the rate of dissociation of the ADP–actin subunits from the filament is greater than that for ATP–actin subunits, the filament will shorten rapidly when the ADP–actin core is exposed at either end.

E. Actin-Binding Proteins

A variety of proteins have been shown to interact with actin monomers or actin filaments. The interaction of actin with these proteins can play an important role in determining the form and function of the actin filament and, as a result, the physical properties of the cytoplasm. These actin-binding proteins were reviewed (Pollard and Cooper, 1986). Although little is known about actin-binding proteins in plants, immunofluorescence studies recently have detected some putative actin-binding proteins in plant cells (Lim et al., 1986; Parke et al., 1986; and see below). These results indicate that higher plant MFs may prove to be biochemically analogous to those of animal cells. Investigations of the actin-binding proteins of higher plants could be expected to reveal not only the degree to which the evolution of the plant MF cytoskeleton has paralleled that of animals, but some novel proteins and mechanisms of interactions as well. The actin-binding proteins discussed in the rest of this section represent examples of the kinds of interactions that occur between actin and specific cytoplasmic proteins. They are presented not because we expect to find their homologs in plant cells, but because proteins with analogous function probably occur in plants. For example, filamin binds to actin and regulates its interaction with myosin in muscle. Vahey (1983) identified and partially characterized a tomato protein with a similar function, except that it bound to the actin-binding site on myosin instead of binding to actin. It is not known if this activation-inhibiting protein is part of the tomato MF cytoskeleton and plays a role in its function.

1. Myosin

There are both muscle and cytoplasmic myosins in animal cells. Vertebrate muscle and nonmuscle myosins consist of two heavy polypeptides and two different kinds of light chains. The two heavy chains, each near 200 kDa, are in an α-helical configuration throughout most of their length, with the α-helices of the two peptides twisted about each other to form a linear rod. The N-terminal ends of both chains are in a globular configuration. Each muscle myosin molecule also contains two pairs of light chains, one which consists of 20-kDa peptides while the other pair is 16 kDa. The four light chains are associated with the globular heads of the heavy chains. Thus, the vertebrate myosin molecule is divided into two different domains, a linear, helical tail that makes up about three-fourths of the length of the molecule, but only one-third of its mass, and a globular head that contains the light chains and most of the mass of the molecule. Digestion of the myosin monomers with the protease papain cleaves it at the junction of the heads with the tail to produce two head fragments/myosin. These are known as the S1 fragments, or heavy meromyosin (HMM). These S1 fragments exhibit a specific, high-affinity binding to F-actin filaments. The S1 or HMM particles are said to decorate the actin fibers, and this decoration has been used frequently as a diagnostic indicator of actin filaments seen in electron micrographs.

Muscle myosin monomers polymerize spontaneously when placed in salt solutions above about 140 mM. Thick filaments in muscle consist of polymers of approximately 500 myosin monomers. The myosin monomers assemble by noncovalent interactions to form a bipolar macromolecule in which the myosin heads are clustered at either end. The tails are packed together in a staggered array. Vertebrate nonmuscle myosin will not assemble spontaneously into myosin thick filaments. Instead, one of the myosin light chains must first be phosphorylated by a myosin kinase. This reaction is controlled by the calcium ion concentration. In the absence of Ca^{2+}, myosin isn't phosphorylated and myosin polymers do not form. The nonmuscle myosin polymers are more than an order of magnitude smaller than those forming the muscle thick filaments. Typically, 10–20 myosin monomers aggregate to form the cytoplasmic bipolar polymer, in contrast to the nearly 500 myosin subunits that form the sarcomere thick filament.

Myosin heads are ATPases which have a high affinity for F-actin. The isolated S1 fragments, obtained by proteolytic digestion of myosin with papain, retain their ATPase activity and their affinity for actin filaments. Both myosin and the S1 fraction of myosin are actin-dependent ATPases. The ATP-dependent binding of myosin to actin filaments, with the subsequent hydrolysis of the ATP and the release of myosin from actin, is the basis for muscle motility. The interaction of nonmuscle myosin with actin filaments is important in bringing about contraction, such as the contraction of the cleavage furrow in dividing animal cells, and there is evidence that the interaction of myosin with actin filaments is responsible for cytoplasmic streaming.

2. *Proteins That Bind to Actin Filament Ends*

Proteins that bind to the ends of actin filaments and prevent the removal or addition of actin monomers at that end are known as capping proteins. Usually, capping proteins bind to the barbed end of the actin filament, preventing further assembly or disassembly at that end. Some of the capping proteins not only cap actin filaments, but also may sever actin filaments and/or act as nucleation sites for actin assembly.

Simple capping proteins are proteins that may both cap filaments and act as nucleation sites, but they do not sever existing actin filaments. Typically they are dimers, with subunit molecular weights of about 30,000. Both capping proteins and the fungal metabolite cytochalasin inhibit the addition of monomers to the rapidly growing barbed ends of actin filaments. Capping proteins also prevent the disassembly of MFs from the barbed end. Paradoxically, capping proteins may also accelerate the overall rate of G-actin assembly. This is because nucleation is the slow, rate-limiting step in actin assembly. In the presence of capping proteins, the overall rate of assembly is faster since, although the individual capped actin filaments grow slowly from their pointed ends, there are many more growing MFs.

In addition to these simple capping proteins, there are capping proteins that also sever actin filaments. Two different groups of actin-binding proteins can be recognized with these properties: (1) the gelsolin–villin group consists of monomeric proteins with molecular weights between 90,000 and 95,000. Gelsolin is found in macrophages and platelets, as well as various other mammalian cell types. Villin is found in the microvilli of intestinal epithelial cells. (2) The fragmin and severin group is similar in function to those of the gelsolin–villin group, but fragmin and severin are smaller proteins, with molecular weights of 40,000 to 45,000, and they are found mostly in invertebrates. Both of these classes of proteins cap, nucleate, and sever. The severing activity is the most unique. They are believed to bind to the sides of actin filaments and to intercalate between the actin subunits, severing the microfilament. They remain associated with the barbed end of one of the fragments, leaving the pointed end free. As a result, they also cap one of the filaments. The result is shorter filaments which grow more slowly.

The function of capping proteins in cells is not known, but it seems likely that they are involved in local changes in cytoplasmic viscosity, such as in the sol–gel transformation. The capping and severing proteins reduce the filament length, thereby decreasing viscosity. Regions of the cytoplasm with many short MFs are more apt to be in a sol state than cytoplasm with the same amount of MFs, but organized into fewer longer MFs.

3. *Proteins That Bind to the Sides of Actin Filaments*

This category includes three different groups of proteins. First, there are those that bind to single actin filaments; that is, they have a single actin-binding site, such as myosin, the troponins, and tropomyosin. Second, there are those proteins that have two actin-binding sites and as a result can cross-bridge actin filaments. These include filamin (smooth muscle), α-actinin (many different tissues), and fodrin (brain). These proteins probably play an important part in determining whether the cytoplasm takes on a gel or a sol state. Third, there are bundling proteins, including fimbrin and villin (both found in the microvilli of intestinal epithelium) and fascin (mostly found in invertebrates). These are bivalent proteins that create bundles of actin filaments in which the filaments lie very close to each other (about 10 nm apart laterally).

4. *Actin-Binding Proteins in Plants*

Lim *et al.* (1986) identified a monoclonal antibody to muscle troponin T which decorated cytoskeletal arrays in animal, fungal, and higher plant cells in an immunofluorescence procedure. All four cytoskeletal arrays in onion root cells reacted with the antibody. The interpretation of these results, however, is not straightforward. In immunoblots of HeLa whole-cell proteins, the anti-troponin T antibody cross-reacted with other proteins, including a 35-kDa peptide associated with the MT cytoskeleton. Furthermore, immunoblots of brain MT proteins showed that the antibody also cross-reacted with MAP1A. As a result, the positive immunofluorescent images obtained in onion root cells may indicate the presence of a conserved epitope on a plant MAP, or, since MFs have been colocalized in plant MT arrays, it may mean that an actin-binding protein with an epitope similar to the reactive epitope of troponin T is present in the plant cytoskeleton. R. J. Cyr (personal communication) was unable to detect any reactive species in immunoblots of carrot whole-cell or MT proteins using this antibody.

Higher plant myosin has been isolated from tomato fruits (Vahey and Scordilis, 1980; Vahey et al., 1982) and from *Egeria densa* (Ohsuka and Inoué, 1979). In both cases the protein assembled into bipolar thick filaments in low-ionic-strength buffers, and exhibited an actin-stimulated ATPase activity. SDS-gel electrophoresis indicated that the tomato myosin had heavy chains of 200 kDa, similar in size to muscle myosin. Parke et al. (1986) used a monoclonal antibody to the myosin heavy chain of mouse to localize myosin in onion root tip cells. Immunoblotting of electrophoretically separated onion whole-cell proteins with the anti-myosin antibody revealed a single reactive 200-kDa peptide comigrating with authentic muscle myosin. Immunofluorescence procedures showed that this antigenic determinant colocalized with MFs in the onion cells. Staining was particularly intense in the phragmoplast.

Cytoplasmic streaming in the algae *Chara* and *Nitella* is dependent upon actin filaments that are stationary in, and possibly responsible for, the gel phase of the cytoplasm (Palevitz and Hepler, 1975). Myosin also is present in the alga *Nitella* (Kato and Tonomura, 1977). More recently, beads coated with skeletal muscle myosin were found to exhibit ATP-dependent movement on *Nitella* actin filaments *in vitro* (Sheetz and Spudich, 1983; Sheetz et al., 1984). Although these algae are only remotely related to higher plants, the ability of vertebrate myosin to recognize and utilize actin filaments in a correct functional manner suggests a high degree of conservation of the components of this motility system. Given these results and the preliminary data on the characterization of higher plant myosin, it would not be surprising to find that it was remarkably similar to vertebrate myosin.

VI. INTERMEDIATE FILAMENT PROTEINS

Plant biologists are less familiar with IFs than other components of the cytoskeleton, perhaps reflecting the lesser importance of these structures in plant cells. Nevertheless, recent work on these proteins suggests that they may be more widespread in their distribution than previously supposed. It is important that plant biologists be aware of this possibility and have a concept of the nature of these structures and their component proteins.

The tubulins and actins of plants differ in important respects from the tubulins and actins of vertebrates. Nevertheless, both tubulin and actin are highly conserved in amino acid sequence over the course of evolution. In contrast, the IF proteins are rather divergent, even in vertebrates. In vertebrates there are possibly seven different classes of IF proteins. Unlike other cytoskeletal proteins, these IF proteins are highly cell-type specific (Lazarides, 1982; Steinert and Parry, 1985).

A. Types of Intermediate Filament Proteins

1. Keratins

These IF proteins are found in epithelial cells of all vertebrates, and are similar structurally to the keratin that makes up hair (Fuchs et al., 1986; Roop and Steinert, 1986). Human epithelial tissues contain 20 different keratins, which range in size from 40 to 70 kDa. A given epithelial cell does not have all 20 keratin IF proteins, but rather different epithelial cells exhibit a characteristic group of keratins, and often the keratins expressed by a given epithelium change as the tissue develops (Sun and Green, 1978; Fuchs and Green, 1980).

Analysis by isoelectric focusing shows that the keratins are either acidic (type I) or basic (type II) (Dale et al., 1978; Fuchs et al., 1981). Isolated, purified epithelial keratins are not capable of self-assembly; however, if an acidic keratin is mixed with any basic keratin, the two spontaneously assemble to form IF filaments. They appear to be obligate heteropolymers (Franke et al., 1983). There is at least one gene for each of the 20 different epithelial IF proteins, and the expression of these genes is carefully regulated so that only a few are active in any particular epithelial cell, but they are expressed in pairs, one gene encoding a type I and one gene encoding a type II keratin (Moll et al., 1982; Fuchs and Hanukoglu, 1983; Roop et al., 1985).

2. Desmin

All types of muscle contain an IF protein known as desmin, an acidic protein of 53 kDa found in the Z disk of muscle where it binds and organizes myofilaments. Based on its immunological properties and also the fact that it is capable of self-assembly *in vitro* to form 10 nm fibers, it is an IF protein. It also will copolymerize with type II epithelial keratins *in vitro* to form IFs, although its expression is confined to muscle cells (Geisler and Weber, 1982; Quax et al., 1985).

3. Neurofilament (NF) Proteins

Mammalian neurons contain unique IFs known as neurofilaments. These IFs are composed of three different proteins with apparent M_r of 200,000 (NF-H), 160,000 (NF-M), and 68,000 (NF-L) (Geisler and Weber, 1981b; Liem and Hutchinson, 1982). The three proteins will coassemble *in vitro* to form 10 nm filaments. However, only the NF-L appears to be capable of efficient self-assembly (Geisler and Weber, 1981b).

4. Vimentin

Many cells of mesenchymal origin contain IFs constructed of the protein vimentin, which has a size of about 50 kDa. It is similar to desmin and NF-L in being able to form polymers spontaneously *in vitro*, with no other IF protein present. This is the most widely distributed IF protein in vertebrate cells (Quax *et al.*, 1983).

5. *Glial Cell IF Protein*

The central nervous system of vertebrates not only contains neurons, but also glial cells. Glial cells, however, do not contain NFs, but instead exhibit IFs that are similar to those of the mesenchymal cells. However, in addition to vimentin, glial cells contain a unique IF protein of 51 kDa which has been designated GFAP for glial fibrillary acidic protein. It has been shown to be capable of self-assembly *in vitro* to form 10 nm filaments (Geisler and Weber, 1983; Lewis *et al.*, 1984).

6. *Lamins*

The nuclear lamina is a network of filamentous proteins that line the nucleoplasmic surface of the inner membrane of the nuclear envelope. It appears to represent an anchoring site for chromatin during interphase. These filaments are composed of three immunologically related proteins in vertebrates, lamins A, B, and C. Recently, these fibrous proteins were shown to be IF proteins (Aebi *et al.*, 1986). They contain a 360-amino acid central domain with an α-helical conformation similar to that of other IF proteins (Fisher *et al.*, 1986; McKeon *et al.*, 1986). Furthermore, they coassemble with other IF proteins to form 10 nm diameter fibers (Goldman *et al.*, 1986).

7. *Tektins*

Tektins are relatively insoluble, filamentous proteins which associate with the A microtubules of the flagellar axoneme (Linck *et al.*, 1982). They have a high content of α-helical structure, similar to that of other IF proteins (Linck and Langevin, 1982). Recently, Chang and Piperno (1987) demonstrated that monoclonal antibodies specific for sea urchin flagellar axoneme tektins also react with vimentin, desmin, and nuclear lamins. Similarly, polyclonal antibodies to vimentin and desmin reacted with tektin. Although it has not been rigorously demonstrated that tektins are IF proteins, the evidence indicates that this is probable.

B. Structural Characteristics Common to Intermediate Filament Proteins

1. *Solubility*

The IF proteins exhibit a characteristic low solubility, even in buffers containing 1% Triton X-100, but they can be solubilized in 6 M urea. When the urea is removed by dialysis, many of the IF proteins will spontaneously reassemble to form the 10 nm filaments. This process has been used to purify the IF proteins by cyclic rounds of assembly–disassembly (Steinert *et al.*, 1976, 1981).

2. *Immunological Characteristics*

Antibodies have been raised to the different IF proteins and used to localize IF in cells by immunofluorescence. In some cases, these antibodies are IF class-specific (Lazarides and Hubbard, 1976; Franke *et al.*, 1978). However, monoclonal antibodies have been raised that will react with all of the IF proteins. This tells us that there are highly conserved regions present in all IF proteins which may account for the fact that they form morphologically similar structures. This common epitope is localized in the rod domain of the IF proteins (see below).

3. *Domain Organization*

Weber and co-workers demonstrated that mild digestion of desmin with chymotrypsin gives three fragments, each representing a different domain of the protein. The amino- and carboxy-terminal domains showed no ordered structure. However, the largest of the fragments was a peptide that would spontaneously assemble into a tetrameric rod, 50 nm long and 2–3 nm wide. These tetramers are not IFs, but they do have a rodlike structure. X-Ray diffraction analysis of the rods showed them to have an α-helical content of about 80% (Geisler *et al.*, 1982). A central rod domain with a similar structure is present in all IF proteins (Geisler and Weber, 1986). The α-helical rod domain is about 310 residues long and is flanked on either end by sequences that are neither conserved between different IFs nor helical. The amino-terminal region is called the headpiece while the carboxy-terminal domain is called the tailpiece. It is in the head- and tailpieces that the extreme variation in size and amino acid sequence among the IF proteins is found. The headpieces generally are basic. The tailpiece is highly variable between IF proteins. The tailpieces of the NF are highly charged due to many glutamic acid residues. In addition, they have many serine residues which may be phosphorylated to further increase the charge in this region (Geisler and Weber, 1986).

4. *Amino Acid Sequence*

Some isolated IF proteins have been directly sequenced (Geisler and Weber, 1981a, 1982, 1983; Geisler *et al.*, 1983, 1985). Additional sequence data have been deduced from the nucleotide sequence of cloned cDNAs encoding these proteins (Steinert *et al.*, 1983, 1984; Quax *et al.*, 1983, 1984). Overall, they display little homology. The two keratin subclasses are about 30% homologous, although members of each subgroup are 60–70% homologous. The keratins also are about 30% homologous to other IF proteins; however, the degree of homology of the other IF proteins with each other is between 50 and 65% (Steinert et al., 1981). Secondary structure predictions based on sequence data indicate that all the IF proteins contain a central α-helical rod domain of 311–314 amino acid residues. The rod regions of all the IF proteins exhibit a moderate to high degree of homology. For example, the rod region of vimentin is 73% homologous to that of desmin and 63% homologous to the rod region of GFAP, but only about 50% homologous to the rod region of NF-L and NF-M.

C. A Common Structure from Different Proteins

Despite the substantial divergence of the IF proteins, their predicted secondary structures and physical properties are similar. All IF proteins have a central α-helical domain of about 310 amino acid residues in which there is a regular heptad distribution of hydrophobic amino acids, characteristic of α-helices that interact to form coils (Geisler and Weber, 1982; Hanukoglu and Fuchs, 1982, 1983). A protein with a repeating heptad structure has the sequence $(a, b, c, d, e, f, g)_n$, where a and d are nonpolar amino acids. This repeating pattern of residues gives an inclined stripe of nonpolar amino acids running down the length of the α-helix in a spiral fashion. Since the nonpolar residues are unstable in water, two or more α-helices aggregate to form a stable structure in which their nonpolar residues interact with each other to form the coil, with the nonpolar residues buried inside the coil (McLachlan and Stewart, 1975, 1982). Interestingly, this type of protein structure was first determined by Crick (1953) in his work with wool, a keratin similar to the epithelial IF keratins.

D. Higher Plant Intermediate Filaments

There is evidence that higher plant cytoskeletal arrays contain elements that have some of the properties of IFs. Cytoskeletal preparations from various plant species exhibit bundles of 7 nm diameter fibers. Indirect immunofluorescence with the Pruss et al. (1981) monoclonal antibody, which recognizes the highly conserved part of IF proteins, decorates most of the cytoskeletal arrays of plant cells (Powell et al., 1982; Dawson et al., 1985). Immunoblots of SDS-PAGE separated whole cell or cytoskeletal proteins from onion root tips and cultured carrot cells, probed with the Pruss antibody, exhibited several (possibly five) reactive polypeptides, ranging from M_r 50–69 kDa (Dawson et al., 1985). Additional immunochemical data, a more complete characterization of the morphology of the reacting filaments, and biochemical data on the protein components of the putative plant IFs will be needed before the significance of these observations can be evaluated. Nevertheless, it appears highly likely that intermediate filaments contribute to the plant cytoskeleton.

The evidence for IFs outside the vertebrates is at present somewhat conflicting. There are reports of IF proteins in both *Drosophila* (Walter and Biessmann, 1984) and *Dictyostelium* (Koury and Eckert, 1984). However, genomes of several invertebrates failed to hybridize to cDNA probes for various IF proteins (Quax *et al.,* 1984; Fuchs and Marchuk, 1983). Since fairly stringent conditions were used for the hybridizations, this may mean only that invertebrate IF proteins are diverged. The fact that IF proteins (the nuclear lamins) are present in a ubiquitous eukaryotic structure such as the nuclear envelope suggests that these proteins are likely to play a role in the higher plant cytoskeleton as well.

ACKNOWLEDGMENTS

The research described in this review that was conducted in the author's laboratory was supported by grants from the Monsanto Company, the National Science Foundation, and the National Institutes of Health. The author is indebted to his former students, Drs. L. Morejohn, M. Bustos, R. Cyr, and M. Guiltinan for many stimulating discussions about the cytoskeleton. He is also grateful to Dr. Yasuyuki Yamada, Kyoto University, Kyoto, Japan, in whose laboratory this review was completed.

REFERENCES

Aebi, U., Fowler, W. E., Rew, P., and Sun, T.-T. (1983). *J. Cell Biol.* **97**, 113–143.
Aebi, U., Cohn, J., Buhle, L., and Gerace, L. (1986). *Nature (London)* **323**, 560–564.
Allen, R. D., Metuzals, J., Tasaki, I., Brady, S. T., and Gilbert, S. P. (1982). *Science* **218**, 1127–1128.
Allen, R. D., Weiss, D. G., Hayden, J. H., Brown, D. T., Fujiwake, H., and Simpson, M. (1985). *J. Cell Biol.* **100**, 1736–1752.
Amos, L. (1979). *In* "Microtubules" (K. Roberts and J. S. Hyams, eds.), pp. 1–64. Academic Press, New York.
Andreu, J. M., and Timasheff, S. N. (1982a). *Biochemistry* **21**, 534–543.
Andreu, J. M., and Timasheff, S. N. (1982b). *Biochemistry* **21**, 6465–6476.
Andreu, J. M., Gorbunoff, M. J., Lee, J. C., and Timasheff, S. N. (1984). *Biochemistry* **23**, 1742–1752.
Ashton, F. M., and Crafts, A. S. (1981). "Mode of Action of Herbicides." Wiley (Interscience), New York.
Baird, W. V., and Meagher, R. B. (1988). *EMBO J.* **6**, 3223–3231.
Bajer, A. S., and Molé-Bajer, J. (1986). *Ann. N.Y. Acad. Sci.* **466**, 767–784.
Baker, E. J., Schloss, J. A., and Rosenbaum, J. L. (1984). *J. Cell Biol.* **99**, 2074–2081.
Barra, H. S., Arce, C. A., Rodriguez, J. A., and Caputto, R. (1974). *Biochem. Biophys. Res. Commun.* **60**, 1384–1390.
Bartels, P. G., and Hilton, J. L. (1973). *Pestic. Biochem. Physiol.* **3**, 462–472.
Batra, J. K., Powers, L. J., Hess, F. D., and Hamel, E. (1986). *Cancer Res.* **46**, 1889–1893.
Ben Ze'ev, A., Farmer, S. R., and Penman, S. (1979). *Cell* **17**, 319–325.
Bergen, L. G., and Borisy, G. G. (1980). *J. Cell Biol.* **84**, 141–150.
Bergen, L. G., and Borisy, G. G. (1983). *J. Biol. Chem.* **258**, 4190–4194.
Bernatzky, R., and Tanksley, S. D. (1986). *Theor. Appl. Genet.* **72**, 414–421.

Bershadsky, A. D., and Vasiliev, J. M. (1987). "The Cytoskeleton." Plenum, New York.
Bhattacharyya, B., Sackett, D. L., and Wolff, J. (1985). *J. Biol. Chem.* **260,** 10208–10216.
Binder, L. I., Frankfurter, A., Kim, H., Caceres, A., Payne, M. R., and Rebhun, L. I. (1984). *Proc. Natl. Acad. Sci. U.S.A.* **81,** 5613–5617.
Bloom, G. S., Luca, F. C., and Vallee, R. B. (1985). *Biochemistry* **24,** 4185–4191.
Bolduc, C., Lee, V. D., and Huang, B. (1988). *Proc. Natl. Acad. Sci. U.S.A.* **85,** 131–135.
Bond, J. F., Fridovich-Keil, J. K., Pillus, L., Mulligan, R. C., and Solomon, F. (1986). *Cell* **44,** 461–468.
Borisy, G. G., Cleveland, D. W., and Murphy, D. B. (eds.) (1984). "The Molecular Biology of the Cytoskeleton." Cold Spring Harbor Lab., Cold Spring Harbor, New York.
Brenner, S. L., and Brinkley, B. R. (1982). *Cold Spring Harbor Symp. Quant. Biol.* **46,** 241–254.
Brietling, F., and Little, M. (1986). *J. Mol. Biol.* **189,** 367–370.
Brinkley, B. R. (1985). *Annu. Rev. Cell Biol.* **1,** 145–172.
Brunke, K. J., Young, E. E., Buchbinder, B. U., and Weeks, D. P. (1982). *Nucleic Acids Res.* **10,** 1295–1310.
Brunke, K., Anthony, J., Kalish, F., Sternberg, E., and Weeks, D. (1984). In "Molecular Biology of the Cytoskeleton" (G. G. Borisy, D. W. Cleveland, and D. B. Murphy, eds.), pp. 367–379. Cold Spring Harbor Lab., Cold Spring Harbor, New York.
Bryan, J. (1971). *Exp. Cell Res.* **66,** 129–136.
Bryan, J., and Wilson, L. (1971). *Proc. Natl. Acad. Sci. U.S.A.* **68,** 1762–1766.
Bulinski, J., and Borisy, G. G. (1979). *Proc. Natl. Acad. Sci. U.S.A.* **76,** 293–297.
Bulinski, J. C., Richards, J. E., and Piperno, G. (1988). *J. Cell Biol.* **106,** 1213–1220.
Burland, T. G., Gull, K., Schedl, T., Boston, R. S., and Dove, W. F. (1983). *J. Cell Biol.* **97,** 1852–1859.
Carlier, M.-F., Hill, T. L., and Chen, Y.-D. (1984). *Proc. Natl. Acad. Sci. U.S.A.* **82,** 771–775.
Caron, J., Jones, A. L., Rall, L. B., and Kirschner, M. W. (1985). *Nature* **317,** 648–651.
Chang, X., and Piperno, G. (1987). *J. Cell Biol.* **104,** 1563–1568.
Chen, Y.-D., and Hill, T. L. (1985). *Proc. Natl. Acad. Sci. U.S.A.* **82,** 4127–4131.
Chou, P. Y., and Fasman, G. D. (1978). *Adv. Enzymol.* **47,** 45–148.
Clayton, L., and Gull, K. (1982). In "Microtubules in Microorganisms" (P. Cappuccinelli and N. R. Morris, eds.), pp. 179–199. Dekker, New York.
Clayton, L., and Lloyd, C. W. (1984). *Exp. Cell Res.* **156,** 231–238.
Clayton, L., Black, C. M., and Lloyd, C. M. (1985). *J. Cell Biol.* **101,** 319–324.
Cleveland, D. W. (1986). In "Cell and Molecular Biology of the Cytoskeleton" (J. W. Shay, ed.), pp. 203–225. Plenum, New York.
Cleveland, D. W. (1987). *J. Cell Biol.* **104,** 381–383.
Cleveland, D. W., and Havercroft, J. C. (1983). *J. Cell Biol.* **97,** 919–924.
Cleveland, D. W., and Sullivan, K. F. (1985). *Annu. Rev. Biochem.* **54,** 331–365.
Cleveland, D. W., Hwo, S. Y., and Kirschner, M. W. (1977). *J. Mol. Biol.* **116,** 1582–1590.
Cleveland, D. W., Lopata, M. A., Sherline, P., and Kirschner, M. W. (1981). *Cell* **25,** 537–546.
Cleveland, D. W., Pittinger, M. F., and Feramisco, J. R. (1983). *Nature (London)* **305,** 738–740.
Cortese, F., Bhattacharyya, B., and Wolff, J. (1977). *J. Biol. Chem.* **252,** 1134–1140.
Crick, F. H. C. (1953). *Acta Crystallogr.* **6,** 685–688.
Cyr, R. J., Guiltinan, M. J., Bustos, M. M., Sotak, M., and Fosket, D. E. (1987a). *Biochim. Biophys. Acta* **914,** 28–34.
Cyr, R. J., Bustos, M. M., Guiltinan, M. J., and Fosket, D. E. (1987b). *Planta,* **171,** 365–376.
Dale, B. A., Holbrook, K. A., and Steinert, P. M. (1978). *Nature (London)* **276,** 729–731.
Davidse, L., and Flach, W. (1977). *J. Cell Biol.* **72,** 174–193.
Dawson, P. J., and Lloyd, C. W. (1985). *EMBO J.* **4,** 2451–2455.
Dawson, P. J., and Lloyd, C. W. (1987). *Biochem. Plants* **9.**
Dawson, P. J., Hulme, J. S., and Lloyd, C. W. (1985). *J. Cell Biol.* **100,** 1793–1798.
De Brabander, M. J., Van De Veire, R. M. L., Aerts, F. E. M., Borgers, M., and Janssen, P. A. J. (1976). *Cancer Res.* **36,** 905–916.

de la Torre, J., Carrascosa, J. L., and Avila, J. (1986). *Eur. J. Cell Biol.* **40**, 233–237.
Delmer, D. (1987). *Annu. Rev. Plant Physiol.* **38**, 259–290.
De Mey, J., Lambert, A. M., Bajer, A. S., Moeremans, M., and De Brabander, M. (1982). *Proc. Natl. Acad. Sci. U.S.A.* **79**, 1898–1902.
Dentler, W. L., Grannett, S., and Rosenbaum, J. L. (1975). *J. Cell Biol.* **65**, 237–241.
Detrich, H. W., III, and Overton, S. A. (1986). *J. Biol. Chem.* **261**, 10922–10930.
Do, C. V., Sears, E. B., Gilbert, S. P., and Sloboda, R. D. (1988). *Cell Motil. Cytoskel.* **10**, 246–254.
Drubin, D., Kirschner, M., and Feinstein, S. (1984). *In* "Molecular Biology of the Cytoskeleton" (G. G. Borisy, D. W. Cleveland, and D. B. Murphy, eds), pp. 343–356. Cold Spring Harbor Lab., Cold Spring Harbor, New York.
Drubin, D., Feinstein, S., Shooter, E., and Kirschner, M. (1985). *J. Cell Biol.* **101**, 1799–1807.
Duerr, A., Pallas, D., and Solomon, F. (1981). *Cell* **24**, 203–211.
Dustin, P. (1984). "Microtubules." Springer-Verlag, Berlin, Federal Republic of Germany.
Euteneurer, U., and McIntosh, J. R. (1980). *J. Cell Biol.* **70**, 509–515.
Falconer, M. M., and Seagull, R. W. (1985). *Protoplasma* **125**, 190–198.
Falconer, M. M., and Seagull, R. W. (1986). *Protoplasma* **133**, 140–148.
Falconer, M. M., and Seagull, R. W. (1987). *Protoplasma* **136**, 118–124.
Farmer, S. R. (1986). *In* "Cell and Molecular Biology of the Cytoskeleton" (J. W. Shay, ed.), pp. 131–149. Plenum, New York.
Farrell, K. W., Jordan, M. A., Miller, H. P., and Wilson, L. (1986). *J. Cell Biol.* **104**, 1035–1046.
Fedtke, C. (1982). "Biochemistry and Physiology of Herbicide Action." Springer-Verlag, Berlin, Federal Republic of Germany.
Field, D. J., Collins, R. A., and Lee, J. C. (1984). *Proc. Natl. Acad. Sci. U.S.A.* **81**, 4041–4045.
Fisher, D. Z., Chaudhary, N., and Blobel, G. (1986). *Proc. Natl. Acad. Sci. U.S.A.* **83**, 6450–6454.
Forer, A., and Jackson, W. T. (1979). *J. Cell Sci.* **37**, 323–347.
Franke, W. W., Seib, E., Osborn, M., Weber, K., Herth, W., and Falk, H. (1977). *Cytobiologie* **15**, 24–48.
Franke, W. W., Schmid, E., Osborn, M., and Weber, K. (1978). *Proc. Natl. Acad. Sci. U.S.A.* **75**, 5034–5038.
Franke, W. W., Schiller, D. L., Hatzfeld, M., and Winter, S. (1983). *Proc. Natl. Acad. Sci. U.S.A.* **80**, 7113–7117.
Frieden, C. (1985). *Annu. Rev. Biophys. Biophys. Chem.* **14**, 189–210.
Fuchs, E., and Green, H. (1980). *Cell* **19**, 1033–1042.
Fuchs, E., and Hanukoglu, I. (1983). *Cell* **34**, 332–334.
Fuchs, E., and Marchuk, D. (1983). *Proc. Natl. Acad. Sci. U.S.A.* **80**, 5857–5861.
Fuchs, E., Coppock, S. M., Green, H., and Cleveland, D. W. (1981). *Cell* **27**, 75–84.
Fuchs, E., Marchuk, D., and Tyner, A. (1986). *In* "Cell and Molecular Biology of the Cytoskeleton" (J. W. Shay, ed.), pp. 85–107. Plenum, New York.
Fukuda, H. (1987). *Plant Cell Physiol.* **28**, 517–528.
Fukuda, H., and Iwata, N. (1986). *Plant Cell Physiol.* **27**, 273–283.
Fukuda, H., and Komamine, A. (1985). *In* "Cell Culture and Somatic Cell Genetics of Plants" (I. K. Vasil, ed.), Vol. 2, pp. 149–212. Academic Press, New York.
Fulton, C., and Simpson, P. A. (1976). *In* "Cold Spring Harbor Conference on Cell Motility" (R. Goldman, T. Pollard, and J. Rosenbaum, eds.), pp. 987–1005. Cold Spring Harbor Lab., Cold Spring Harbor, New York.
Gay, D. A., Yen, T. J., Lau, J. T. Y., and Cleveland, D. W. (1987). *Cell* **50**, 671–679.
Geisler, N., and Weber, K. (1981a). *Proc. Natl. Acad. Sci. U.S.A.* **78**, 4120–4123.
Geisler, N., and Weber, K. (1981b). *J. Mol. Biol.* **151**, 565–571.
Geisler, N., and Weber, K. (1982). *EMBO J.* **1**, 1649–1656.
Geisler, N., and Weber, K. (1983). *EMBO J.* **2**, 2059–2063.

Geisler, N., and Weber, K. (1986). *In* "Cell and Molecular Biology of the Cytoskeleton" (J. W. Shay, ed.), pp. 41–68. Plenum, New York.
Geisler, N., Plessmann, U., and Weber, K. (1982). *Cell* **30**, 277–286.
Geisler, N., Kaufmann, E., Fischer, S., Plessmann, U., and Weber, K. (1983). *EMBO J.* **2**, 1295–1302.
Geisler, N., Fischer, S., Vandekerckhove, J., Van Damme, J., Plessmann, U., and Weber, K. (1985). *EMBO J.* **4**, 57–63.
Goldenring, J. R., Gonzalez, B., McGuire, Jr., S., and DeLorenzo, R. J. (1983). *J. Biol. Chem.* **258**, 12632–12640.
Goldman, A. R., Maul, G., Steinert, P. M., Yang, H.-Y., and Goldman, R. D. (1986). *Proc. Natl. Acad. Sci. U.S.A.* **83**, 3839–3843.
Goldstein, L. S. B., Laymon, R. A., and McIntosh, J. R. (1986). *J. Cell Biol.* **102**, 2076–2087.
Green, P. B. (1985). *J. Cell Sci. Suppl.* **2**, 181–201.
Guiltinan, M. J., Velton, J., Bustos, M. M., Cyr, R. J., Schell, J., and Fosket, D. E. (1987a). *Mol. Gen. Genet.* **207**, 328–334.
Guiltinan, M. J., Ma, D.-P., Yagegari, R., Barker, R., Bustos, M. M., Cyr, R. J., and Fosket, D. E. (1987b). *Plant Mol. Biol.* **10**, 171–184.
Gundersen, G. G., Kalnoski, M. H., and Bulinski, J. C. (1984). *Cell* **38**, 779–789.
Gunning, B. E. S., and Hardham, A. R. (1982). *Annu. Rev. Plant Physiol.* **33**, 651–698.
Gunning, B. E. S., and Wick, S. M. (1985). *J. Cell Sci. Suppl.* **2**, 157–179.
Gwo-Shu Lee, M., Loomis, C., and Cowan, N. J. (1984). *Nuc. Acids Res.* **12**, 5823–5836.
Hahne, G., and Hoffmann, F. (1984). *Proc. Natl. Acad. Sci. U.S.A.* **81**, 5449–5453.
Halperin, W. (1966). *Am. J. Bot.* **53**, 443–453.
Halperin, W., and Jensen, W. (1967). *J. Ultrastruct. Res.* **18**, 428–443.
Halperin, W., and Wetherell, D. F. (1965). *Science* **147**, 756–758.
Hanukoglu, I., and Fuchs, E. (1982). *Cell* **31**, 243–252.
Hanukoglu, I., and Fuchs, E. (1983). *Cell* **33**, 915–924.
Hart, J. W., and Sabnis, D. D. (1976). *Curr. Adv. Plant Sci.* **26**, 1095–1104.
Heath, I. B., and Seagull, R. W. (1982). *In* "The Cytoskeleton in Plant Growth and Development" (C. W. Lloyd, ed.), pp. 163–183. Academic Press, New York.
Henderson, D., Geisler, N., and Weber, K. (1982). *J. Mol. Biol.* **155**, 173–176.
Hepler, P. K., and Fosket, D. E. (1971). *Protoplasma* **72**, 213–236.
Hepler, P. K., and Jackson, W. T. (1968). *J. Cell Biol.* **38**, 437–446.
Hertel, C., and Marmé, D. (1983). *Pestic. Biochem. Physiol.* **19**, 282–290.
Hess, F. D. (1979). *Exp. Cell Res.* **119**, 99–109.
Hess, F. D., and Bayer, D. E. (1977). *J. Cell Sci.* **24**, 351–360.
Hightower, R. C., and Meagher, R. B. (1985). *EMBO J.* **4**, 1–8.
Hightower, R. C., and Meagher, R. B. (1986). *Genetics* **114**, 315–332.
Hill, T. L., and Chen, Y.-D. (1984). *Proc. Natl. Acad. Sci. U.S.A.* **81**, 5772–5776.
Hill, T. L., and Kirschner, M. (1982). *Int. Rev. Cytol.* **78**, 1–125.
Hiraoka, Y., Toda, T., and Yanagida, M. (1984). *Cell* **39**, 349–358.
Horio, T., and Hotani, H. (1986). *Nature (London)* **312**, 237–242.
Huber, G., Alaimo-Beuret, D., and Matus, A. (1985). *J. Cell Biol.* **100**, 496–507.
Hussey, P. J., and Gull, K. (1985). *FEBS Lett.* **181**, 113–118.
Hussey, P. J., Lloyd, C. W., and Gull, K. (1988). *J. Biol. Chem.* **263**, 5474–5479.
Ilker, R., Breidenbach, R. W., and Lyons, J. M. (1979). *In* "Low Temperature Stress in Crop Plants" (J. M. Lyons, D. Graham, and J. K. Raison, eds.), pp. 97–113. Academic Press, New York.
Jackson, W. T. (1982). *In* "The Cytoskeleton in Plant Growth and Development" (C. W. Lloyd, ed.), pp. 3–29. Academic Press, New York.
Job, D., Rauch, C. T., Fischer, E. H., and Margolis, R. L. (1982). *Biochemistry* **21**, 509–515.
Job, D., Pabion, M., and Margolis, R. (1985). *J. Cell Biol.* **101**, 1680–1689.
Juniper, B. E., and Lawton, J. R. (1979). *Planta* **145**, 411–416.

Kato, T., and Tonomura, Y. (1977). *J. Biochem.* (*Tokyo*) **82,** 777–782.
Kato, T., Kakiuchi, M., and Okamura, S. (1985). *J. Biochem.* (*Tokyo*) **98,** 371–377.
Keller, L. R., Schloss, J. A., Silflow, C. D., and Rosenbaum, J. R. (1984). *J. Cell Biol.* **98,** 1138–1143.
Kemphues, K. J., Raff, E. C., Raff, R. A., and Kaufman, T. C. (1980). *Cell* **21,** 445–451.
Kemphues, K. C., Kaufman, T. C., Raff, R. A., and Raff, E. C. (1982). *Cell* **31,** 655–670.
Kimmel, B. E., Samson, S., Wu, J., Hirschberg, R., and Yarbrough, L. R. (1985). *Gene* **35,** 237–248.
Kirchner, K., and Mandelkow, E.-M. (1985). *EMBO J.* **4,** 2397–2402.
Kirschner, M., and Mitchison, T. (1986). *Cell* **45,** 329–342.
Kobayashi, H., Fukuda, H., and Shibaoka, H. (1987). *Protoplasma* **138,** 69–71.
Korn, E. D. (1982). *Physiol. Rev.* **62,** 672–737.
Koury, S. T., and Eckert, B. S. (1984). *J. Cell Biol.* **99,** 320a.
Krauhs, E., Little, M., Kempf, T., Hofer-Warbinek, R., Ade, W., and Ponstingl, H. (1981). *Proc. Natl. Acad. Sci. U.S.A.* **78,** 4156–4160.
Kreis, T. E. (1987). *EMBO J.* **6,** 2597–2606.
Kumar, N. (1981). *J. Biol. Chem.* **256,** 10435–10441.
Kumar, N., and Flavin, M. (1981). *J. Biol. Chem.* **256,** 7678–7686.
Kuznetsov, S. A., and Gelfand, V. I. (1986). *Proc. Natl. Acad. Sci. U.S.A.* **83,** 8530–8534.
Laemmli, U. K. (1970). *Nature* (*London*) **227,** 680–685.
Lambert, A. M., and Bajer, A. S. (1977). *Cytobiologie* **15,** 1–23.
Lau, J. T. Y. L., Pittenger, M. F., and Cleveland, D. W. (1985). *Mol. Cell. Biol.* **5,** 1611–1620.
Lazarides, E. (1982). *Annu. Rev. Biochem.* **51,** 219–250.
Lazarides, E., and Hubbard, B. D. (1976). *Proc. Natl. Acad. Sci. U.S.A.* **73,** 4344–4348.
LeDizet, M., and Piperno, G. (1987). *Proc. Natl. Acad. Sci. U.S.A.* **84,** 5720–5724.
Lee, G., Cowan, N., and Kirschner, M. W. (1988). *Science* **239,** 285–288.
Lee, J. C., and Timasheff, S. N. (1977). *Biochemistry* **16,** 1754–1764.
Lee, Y. C., and Wolff, J. (1984). *J. Biol. Chem.* **259,** 8041–8044.
Lefebvre, P. A., and Rosenbaum, J. L. (1986). *Annu. Rev. Cell Biol.* **2,** 517–546.
Lefebvre, P. A., Silflow, C. D., Wieben, E. D., and Rosenbaum, J. L. (1980). *Cell* **20,** 469–477.
Leslie, R. J., Hird, R. B., Wilson, L., McIntosh, J. R., and Scholey, J. M. (1987). *Proc. Natl. Acad. Sci. U.S.A.* **84,** 2771–2775.
Lewis, S. A., Balcarek, J. M., Krek, V., Shelanski, M., and Cowan, N. J. (1984). *Proc. Natl. Acad. Sci. U.S.A.* **81,** 2743–2746.
Lewis, S. A., Gu, W., and Cowan, N. J. (1987). *Cell* **49,** 539–548.
L'Hernault, S. W., and Rosenbaum, J. L. (1985). *Biochemistry* **24,** 473–478.
Lichtenberg, B., Mandelkow, E.-M., Hagestedt, T., and Mandelkow, E. (1988). *Nature* **334,** 359–362.
Liem, S. A., and Hutchinson, S. B. (1982). *Biochemistry* **21,** 3221–3226.
Lim, S.-S., Hering, G. E., and Borisy, G. G. (1986). *J. Cell Sci.* **85,** 1–19.
Linck, R. W., and Langevin, G. L. (1982). *J. Cell Sci.* **58,** 1–22.
Linck, R. W., Albertini, D. F., Kenney, D. M., and Langevin, G. L. (1982). *Cell Motil.* (*Suppl.*) **1,** 127–132.
Littauer, U. Z., Giveon, D., Thierauf, M., Ginsburg, I., and Ponstingl, H. (1986). *Proc. Natl. Acad. Sci. U.S.A.* **83,** 7162–7166.
Lloyd, C. W. (ed.) (1982). "The Cytoskeleton in Plant Growth and Development." Academic Press, New York.
Lloyd, C. W. (1984). *Int. Rev. Cytol.* **86,** 1–51.
Lloyd, C. W. (1987). *Annu. Rev. Plant Physiol.* **38,** 119–139.
Lloyd, C. W., Slabas, A. R., Powell, A. J., and Lowe, S. B. (1980). *Planta* **147,** 500–506.
Ludueña, R. F., Myles, D. G., and Pfeffer, T. A. (1980). *Exp. Cell Res.* **130,** 455–459.
Ludwig, S. R., Oppenheimer, D. G., Silflow, C. D., and Snustad, D. P. (1987a). *Proc. Natl. Acad. Sci. U.S.A.* **84,** 5833–5837.

Ludwig, S. R., Oppenheimer, D. G., Silflow, C. D., and Snustad, D. P. (1987b). *Plant Molec. Biol.* **10,** 311–321.
McIntosh, J. R. (1983). *Mod. Cell Biol.* **2,** 115–142.
McKeon, F. D., Kirschner, M. W., and Caput, D. (1986). *Nature* **319,** 463–468.
McLachlan, A. D., and Stewart, M. (1975). *J. Mol. Biol.* **98,** 293–304.
McLachlan, A. D., and Stewart, M. (1982). *J. Mol. Biol.* **162,** 693–698.
Maccioni, R. B., Rivas, C. I., and Vera, J. C. (1988). *EMBO J.* **7,** 1957–1963.
Maekawa, S., and Sakai, H. (1978). *J. Biochem. (Tokyo)* **83,** 1065–1075.
Mandelkow, E.-M., Herrmann, M., and Ruhl, U. (1985). *J. Mol. Biol.* **185,** 311–327.
Margolis, R. L., and Rauch, C. T. (1981). *Biochemistry* **20,** 4451–4458.
Margolis, R. L., and Wilson, L. (1977). *Proc. Natl. Acad. Sci. U.S.A.* **74,** 3466–3470.
Margolis, R. L., and Wilson, L. (1978). *Cell* **13,** 1–8.
Margolis, R. L., and Wilson, L. (1981). *Nature (London)* **293,** 705–711.
Margolis, R. L., Rauch, C. T., and Job, D. (1986). *Proc. Natl. Acad. Sci. U.S.A.* **83,** 639–643.
Matus, A. (1988). *Annu. Rev. Neurosci.* **11,** 29–44.
Maurta, H., Greer, K., and Rosenbaum, J. L. (1986). *J. Cell Biol.* **103,** 571–579.
Metcalf, T. N., III, Szabo, L. J., Schubert, K. R., and Wang, J. H. L. (1980). *Nature (London)* **285,** 171–172.
Metcalf, T. N., III, Szabo, L. T., Schubert, K. R., and Wang, J. L. (1984). *Protoplasma* **120,** 91–99.
Mita, T., and Shibaoka, H. (1984). *Protoplasma* **119,** 100–109.
Mitchison, T., and Kirschner, M. (1984a). *Nature (London)* **312,** 232–237.
Mitchison, T., and Kirschner, M. (1984b). *Nature (London)* **312,** 237–242.
Mizuno, K. (1985). *Cell Biol. Int. Rep.* **9,** 13–21.
Mizuno, K., Koyama, M., and Shibaoka, H. (1981). *J. Biochem. (Tokyo)* **89,** 329–332.
Mizuno, K., Perkins, J., Sek, F., and Gunning, B. (1985). *Cell Biol. Int. Rep.* **9,** 5–12.
Moll, R., Franke, W. W., Schiller, D. L., Geiger, B., and Krepler, R. (1982). *Cell* **31,** 11–21.
Morejohn, L. C., and Fosket, D. E. (1982). *Nature (London)* **297,** 426–428.
Morejohn, L. C., and Fosket, D. E. (1984a). *Science* **224,** 874–876.
Morejohn, L. C., and Fosket, D. E. (1984b). *J. Cell Biol.* **99,** 141–147.
Morejohn, L. C., and Fosket, D. E. (1986). In "Cell and Molecular Biology of the Cytoskeleton" (J. W. Shay, ed.), pp. 257–329. Plenum, New York.
Morejohn, L. C., Bureau, T. E., Tocchi, L. P., and Fosket, D. E. (1984). *Proc. Natl. Acad. Sci. U.S.A.* **81,** 1440–1444.
Morejohn, L. C., Bureau, T. E., and Fosket, D. E. (1985). *Cell Biol. Int. Rep.* **9,** 849–857.
Morejohn, L. C., Bureau, T. E., Tocci, L. P., and Fosket, D. E. (1987a). *Planta* **170,** 230–241.
Morejohn, L. C., Bureau, T. E., Molé-Bajer, J., Bajer, A. S., and Fosket, D. E. (1987b). *Planta* **172,** 252–264.
Mudge, L. C., Gossett, B. J., and Murphy, T. R. (1984). *Weed Sci.* **32,** 591–594.
Murphy, D. B., and Wallis, K. T. (1983). *J. Biol. Chem.* **258,** 7870–7875.
Murphy, D. B., and Wallis, K. T. (1985). *J. Biol. Chem.* **260,** 12293–12301.
Murphy, D. B., Grasser, W. A., and Wallis, K. T. (1986). *J. Cell Biol.* **102,** 628–635.
Nairn, C. J., Winesett, L., and Ferl, R. J. (1988). *Gene* **65,** 247–257.
Nath, J. P., and Himes, R. H. (1986). *Biochem. Biophys. Res. Commun.* **135,** 1135–1143.
Neff, N. F., Thomas, J. H., Grisafi, P., and Botstein, D. (1983). *Cell* **33,** 211–219.
Ohsuka, K., and Inoué, A. (1979). *J. Biochem. (Tokyo)* **85,** 375–378.
Okamura, S. (1983). *Protoplasma* **118,** 199–205.
Okamura, S., Kato, T., and Nishi, A. (1984). *FEBS Lett.* **168,** 278–280.
Olmsted, J. B. (1981). *J. Cell Biol.* **89,** 418–423.
Olmsted, J. B. (1986). *Annu. Rev. Cell Biol.* **2,** 421–457.
Olmsted, J. B., Witman, G. B., Carlson, K., and Rosenbaum, J. L. (1971). *Proc. Natl. Acad. Sci. U.S.A.* **68,** 2273–2277.
Olmsted, J. B., Cox, J. V., Asnes, C. F., Parysek, L. M., and Lyon, H. D. (1984). *J. Cell Biol.* **99,** 28s–32s.

Oosawa, F., and Asakura, S. (1975). "Thermodynamics of Polymerization of Protein." Academic Press, New York.
Oppenheimer, D. G., Haas, N., Silflow, C. D., and Snustad, D. P. (1988). *Gene* **63**, 87–102.
Orbach, M. C., Porro, E. B., and Yanofsky, C. (1986). *Mol. Cell. Biol.* **6**, 2452–2461.
Palevitz, B. A. (1987). *J. Cell Biol.* **104**, 1515–1519.
Palevitz, B. A., and Hepler, P. K. (1975). *J. Cell Biol.* **65**, 29–38.
Parke, J., Miller, J., and Anderton, B. H. (1986). *Eur. J. Cell Biol.* **41**, 9–13.
Parthasarathy, M. V. (1985). *J. Cell Sci.* **39**, 1–12.
Parthasarathy, M. V., and Pesacreta, T. C. (1980). *Can. J. Bot.* **58**, 807–815.
Pashcal, B. M., Shpetner, H. S., and Vallee, R. B. (1987). *J. Cell Biol.* **105**, 1273–1282.
Pickett-Heaps, J. D. (1967). *Dev. Biol.* **15**, 206–236.
Pickett-Heaps, J. D. (1969). *Cytobios* **3**, 257–280.
Pickett-Heaps, J. D., and Northcote, D. H. (1966). *J. Cell Sci.* **1**, 121–128.
Pickett-Heaps, J. D., Tipit, D. H., and Porter, K. R. (1982). *Cell* **29**, 729–744.
Pittenger, M. F., and Cleveland, D. W. (1985). *J. Cell Biol.* **101**, 1941–1952.
Pollard, T. D., and Cooper, J. A. (1986). *Annu. Rev. Biochem.* **55**, 987–1035.
Ponstingl, H., Krauhs, E., Little, M., and Kempf, K. (1981). *Proc. Natl. Acad. Sci. U.S.A.* **78**, 2757–2761.
Pruss, R. M., Mirsky, R., Raff, M. C., Thorpe, R., Dowdling, A. J., Anderton, B. H. (1981). *Cell* **27**, 419–428.
Purich, D. L., and Kristofferson, D. (1984). *Adv. Protein Chem.* **36**, 133–212.
Quader, H., Deichgraber, G., and Schnepf, E. (1986). *Planta* **168**, 1–10.
Quax, W., Vree Egberts, W., Hendriks, W., Quax-Jeuken, Y., and Bloemendal, H. (1983). *Cell* **35**, 215–223.
Quax, W., van den Heuvel, R., Vree Egberts, W., Quax-Jeuken, Y., and Bloemendal, H. (1984). *Proc. Natl. Acad. Sci. U.S.A.* **81**, 5970–5974.
Quax, W., van den Brock, L., Vree Egberts, W., Ramaekers, F., and Bloemendal, H. (1985). *Cell* **43**, 327–338.
Quinlan, R. A., Pogson, I., and Gull, K. (1980). *J. Cell Sci.* **46**, 341–352.
Rinkin, A., Atsmon, D., and Gitler, C. (1980). *Plant Cell Physiol.* **21**, 829–837.
Rinkin, A., Atsmon, D., and Gitler, C. (1983). *Plant Physiol.* **71**, 747–748.
Robinson, D. G., and Herzog, W. (1977). *Cytobiologie* **15**, 463–474.
Robinson, D. G., and Quader, H. (1982). *In* "The Cytoskeleton in Plant Growth and Development" (C. W. Lloyd, ed.), pp. 109–126. Academic Press, New York.
Roop, D. R., and Steinert, P. M. (1986). *In* "Cell and Molecular Biology of the Cytoskeleton" (J. W. Shay, ed.), pp. 69–83. Plenum, New York.
Roop, D. R., Cheng, C. K., Toftgard, R., Stanley, J. R., Steinert, P. M., and Yuspa, S. H. (1985). *Ann. N.Y. Acad. Sci.* **455**, 426–435.
Rudolph, J. E., Kimble, M., Hoyle, H. D., Subler, M. A., and Raff, E. C. (1987). *Mol. Cell. Biol.* **7**, 2231–2242.
Sackett, D. L., and Wolff, J. (1986). *J. Biol. Chem.* **261**, 9070–9076.
Schedl, T., Burland, T. G., Gull, K., and Dove, W. F. (1984). *Genetics* **108**, 143–164.
Schiff, P. B., and Horowitz, S. B. (1980). *Proc. Natl. Acad. Sci. U.S.A.* **77**, 1561–1565.
Schliwa, M. (1986). "The Cytoskeleton." Springer-Verlag, New York.
Schloss, J. S., Silflow, C. D., and Rosenbaum, J. L. (1984). *Mol. Cell. Biol.* **4**, 424–434.
Schnapp, B. J., Vale, R. D., Scheetz, M. P., and Reese, T. S. (1985). *Cell* **40**, 455–462.
Schröder, H. C., Wehland, J., and Weber, K. (1985). *J. Cell Biol.* **100**, 276–281.
Schroeder, M., Wehland, J., and Weber, K. (1985). *Eur. J. Cell Biol.* **38**, 211–218.
Schulze, E., and Kirschner, M. (1986). *J. Cell Biol.* **102**, 1020–1031.
Seagull, R. W., Falconer, M. M., and Weerdenburg, C. A. (1987). *J. Cell Biol.* **104**, 995–1004.
Serrano, L., Avila, J., and Maccioni, R. B. (1984a). *Biochemistry* **23**, 4675–4681.
Serrano, L., de la Torre, J., Maccioni, R. B., and Avila, J. (1984b). *Proc. Natl. Acad. Sci. U.S.A.* **81**, 5989–5993.

Serrano, L., Montejo, E., Hernandez, M. A., and Avila, J. (1985). *Eur. J. Biochem.* **153,** 595–600.
Serrano, L., Valencia, A., Caballero, R., and Avila, J. (1986). *J. Biol. Chem.* **261,** 7076–7081.
Serrano, L., Wandosell, F., De La Torre, J., and Avila, J. (1988). *Biochem. J.* **252,** 683–691.
Shah, D. M., Hightower, R. C., and Meagher, R. B. (1982). *Proc. Natl. Acad. Sci. U.S.A.* **79,** 1022–1026.
Shah, D. M., Hightower, R. C., and Meagher, R. B. (1983). *J. Mol. Appl. Genet.* **2,** 111–126.
Shatz, P. J., Solomon, F., and Botstein, D. (1986). *Mol. Cell. Biol.* **6,** 3722–3733.
Sheetz, M. P., and Spudich, J. A. (1983). *Nature (London)* **303,** 31–35.
Sheetz, M. P., Chasan, R., and Spudich, J. A. (1984). *J. Cell Biol.* **99,** 1867–1871.
Sheir-Neiss, G., Lai, M., and Morris, N. (1978). *Cell* **15,** 639–647.
Shelanski, M. L., Gaskin, F., and Cantor, C. R. (1973). *Proc. Natl. Acad. Sci. U.S.A.* **70,** 765–768.
Sheldon, J. M., and Dickinson, H. G. (1986). *Planta* **168,** 11–23.
Silflow, C. D., and Rosenbaum, J. L. (1981). *Cell* **24,** 81–88.
Silflow, C. D., Chisholm, R. L., Conner, T. W., and Ranum, L. P. W. (1985). *Mol. Cell. Biol.* **5,** 2389–2398.
Silflow, C. D., Oppenheimer, D. G., Kopczak, S. D., Ploense, S. E., Ludwig, S. R., Haas, N., and Snustad, D. P. (1987). *Dev. Genet.* **8,** 435–460.
Simmonds, D. H. (1986). *Planta* **167,** 469–472.
Sloboda, R. D., and Rosenbaum, J. L. (1979a). *Methods Enzymol.* **85,** 409–416.
Sloboda, R. D., and Rosenbaum, J. L. (1979b). *Biochemistry* **18,** 48–55.
Sloboda, R. D., Rudolph, S. A., Rosenbaum, J. L., and Greengard, P. (1975). *Proc. Natl. Acad. Sci. U.S.A.* **72,** 177–181.
Solomon, F., Magendantz, M., and Salzman, A. (1979). *Cell* **18,** 431–438.
Soltys, B. J., and Borisy, G. G. (1985). *J. Cell Biol.* **100,** 1682–1689.
Staiger, C. J., and Schliwa, M. (1987). *Protoplasma* **141,** 1–12.
Steinert, P. M., and Parry, D. A. D. (1985). *Annu. Rev. Cell Biol.* **1,** 41–65.
Steinert, P. M., Idler, W. W., and Zimmermann, S. B. (1976). *J. Mol. Biol.* **108,** 547–567.
Steinert, P. M., Idler, W., Cabral, F., Gottesman, M. M., and Goldman, R. D. (1981). *Proc. Natl. Acad. Sci. U.S.A.* **78,** 3692–3696.
Steinert, P. M., Rice, R. H., Roop, D. B., Trus, B. L., and Steven, A. C. (1983). *Nature (London)* **302,** 794–800.
Steinert, P. M., Parry, P. A. D., Racoosin, E. L., Idler, W. W., Steven, A. C., Trus, B. L., and Roop, D. R. (1984). *Proc. Natl. Acad. Sci. U.S.A.* **81,** 5709–5713.
Steward, F. C., Mapes, M. O., Kent, A. E., and Holsten, R. D. (1964). *Science* **143,** 20–27.
Stossel, T. P., Chaponnier, C., Ezzell, R. M., Hartwig, J. H., Janmey, P. A., Kwiatkowski, D. J., Lind, S. E., Smith, D. B., Southwick, F. S., Yin, H. L., and Zaner, K. S. (1985). *Annu. Rev. Cell Biol.* **1,** 353–402.
Strachen, S. D., and Hess, F. D. (1983). *Pestic. Biochem. Physiol.* **20,** 141–150.
Studier, F. W. (1973). *J. Mol. Biol.* **79,** 237–248.
Sullivan, K. F., and Cleveland, D. W. (1984). *J. Cell Biol.* **99,** 1754–1760.
Sullivan, K. F., and Cleveland, D. W. (1986). *Proc. Natl. Acad. Sci. U.S.A.* **83,** 4327–4331.
Sullivan, K. F., Machlin, P. S., Ratrie, H., III, and Cleveland, D. W. (1986a). *J. Biol. Chem.* **261,** 13317–13322.
Sullivan, K. F., Havercroft, J. C., Machlin, P. S., and Cleveland, D. W. (1986b). *Mol. Cell. Biol.* **6,** 4409–4418.
Sun, T.-T., and Green, H. (1978). *Cell* **14,** 469–476.
Sung, Z. R., and Okimoto, R. (1981). *Proc. Natl. Acad. Sci. U.S.A.* **78,** 3683–3687.
Swan, J., and Solomon, F. (1984). *J. Cell Biol.* **99,** 164–174.
Szasz, J., Yaffe, M. B., Elzinga, M., Blank, G. S., and Sternlicht, H. (1986). *Biochemistry* **25,** 4572–4582.

Theurkauf, W. E., Baum, H., and Wensink, P. C. (1986). *Proc. Natl. Acad. Sci. U.S.A.* **83**, 8477–8481.
Thomas, J. H., Neff, N. F., and Botstein, D. (1985). *Genetics* **112**, 715–734.
Vahey, M. (1983). *J. Cell Biol.* **96**, 1761–1765.
Vahey, M., and Scordilis, S. (1980). *Can. J. Bot.* **58**, 797–801.
Vahey, M., Titus, M., Trautwein, R., and Scordilis, S. (1982). *Cell Motil.* **2**, 137–147.
Vale, R. D. (1987). *Annu. Rev. Cell Biol.* **3**, 347–378.
Vale, R. D., Schnapp, B. J., Reese, T. S., and Sheetz, M. P. (1985a). *Cell* **40**, 559–569.
Vale, R., Reese, T., and Sheetz, M. (1985b). *Cell* **42**, 39–50.
Vale, R., Schnapp, B., Mitchison, T., Steuer, E., Reese, T., and Sheetz, M. (1985c). *Cell* **43**, 623–632.
Vallee, P. (1980). *Proc. Natl. Acad. Sci. U.S.A.* **77**, 3206–3210.
Vallee, R. B. (1982). *J. Cell Biol.* **92**, 435–442.
Vallee, R. B., and Bloom, G. S. (1983). *Proc. Natl. Acad. Sci. U.S.A.* **80**, 6259–6263.
Vallee, R. B., Bloom, G. S., and Theurkauf, W. E. (1984). *J. Cell Biol.* **99**, 38s–44s.
Vandekerckhove, J., and Weber, K. (1984). *J. Mol. Biol.* **179**, 391–413.
Vandre, D. D., Davis, F. M., Rao, P. N., and Borisy, G. G. (1984). *Proc. Natl. Acad. Sci. U.S.A.* **81**, 4438–4443.
Vaughn, K. C. (1986a). *Pestic. Biochem. Physiol.* **26**, 66–74.
Vaughn, K. C. (1986b). *Abstr. Weed Sci. Soc. Am.* **26**, 77a.
Vaughn, K. C., Marks, M. D., and Weeks, D. P. (1987). *Plant Physiol.* **83**, 956–964.
Villasante, A., Wang, D., Dobner, P., Dolph, P., Lewis, S. A., and Cowan, N. J. (1986). *Mol. Cell. Biol.* **6**, 2409–2319.
Voter, W. A., and Erickson, H. P. (1982). *J. Ultrastruct. Res.* **80**, 374–382.
Walter, M. F., and Biesmann, H. (1984). *J. Cell Biol.* **99**, 1468–1477.
Wang, D., Villasante, A., Lewis, S. A., and Cowan, N. J. (1986). *J. Cell Biol.* **103**, 1903–1910.
Weatherbee, J. A., and Morris, N. R. (1984). *J. Biol. Chem.* **259**, 15452–15459.
Weatherbee, J. A., May, G. S., Gambino, J., and Morris, N. R. (1985). *J. Cell Biol.* **101**, 712–719.
Webster, D. R., Gundersen, G. G., Bulinski, J. C., and Borisy, G. G. (1987). *Proc. Natl. Acad. Sci. U.S.A.* **84**, 9040–9044.
Weeks, D. P., and Collis, P. S. (1976). *Cell* **9**, 15–27.
Wehland, J., Schroeder, M., and Weber, K. (1984). *Cell Biol. Int. Rep.* **8**, 147–150.
Weisenberg, R. C. (1972). *Science* **177**, 1104–1105.
Weisenberg, R. C., Borisy, G. G., and Taylor, E. W. (1968). *Biochemistry* **7**, 4466–4479.
White, E., Tolbert, E. M., and Katz, E. R. (1983). *J. Cell Biol.* **97**, 1011–1019.
Wiche, G. (1985). *Trends Biochem. Sci.* **10**, 67–70.
Wiche, G., Briones, E., Koszka, C., Artlieb, U., and Krepler, R. (1984). *EMBO J.* **3**, 4040–4043.
Wick, S. M., and Duniec, J. (1984). *Protoplasma* **122**, 45–55.
Wick, S. M., Seagull, R. W., Osborn, M., Weber, K., and Gunning, B. E. S. (1981). *J. Cell Biol.* **89**, 685–690.
Wilson, L., Anderson, K., and Chin, D. (1976). In "Cold Spring Harbor Conference on Cell Proliferation" (R. Goldman, T. Pollard and J. Rosenbaum, eds.), Book C, pp. 1051–1064. Cold Spring Harbor Lab., Cold Spring Harbor, New York.
Wilson, L., Miller, H. P., Farrell, K. W., Snyder, K. B., Thompson, W. C., and Purich, D. L. (1985). *Biochemistry* **24**, 5254–5262.
Wochok, Z.-S. (1973). *Cytobios* **7**, 87–95.
Yadav, N. S., and Filner, P. (1983). *Planta* **157**, 46–52.
Youngblom, J., Schloss, J. A., and Silflow, C. D. (1984). *Mol. Cell. Biol.* **4**, 2686–2696.
Zieve, G., and Solomon, F. (1982). *Cell* **28**, 233–242.
Zieve, G., and Solomon, F. (1984). *Mol. Cell. Biol.* **4**, 371–374.

Calmodulin and Calcium-Binding Proteins

11

ELIZABETH ALLAN
PETER K. HEPLER

I. Introduction
II. Calmodulin: Structure
III. The Function of Calmodulin in Plant Cells
   A. Control of Calmodulin-Regulated Processes
   B. Calmodulin-Binding Proteins
   C. Calmodulin-Dependent Processes
IV. Concluding Remarks
   References

I. INTRODUCTION

Calcium is a second messenger for a variety of fundamental and specialized cellular activities in plant and animal cells, mediating between primary signals and the metabolic response. Activation of a plant cell by a number of primary signals, including light, growth substances, and gravity, is thought to raise the intracellular free calcium from resting levels of 10^{-7} to up to 10^{-5} M. The response to elevated calcium in most instances appears to function through calcium-binding proteins that have affinities for calcium in the micromolar range.

The activator protein calmodulin is the most ubiquitous of the calcium-binding proteins in terms of species distribution, tissue distribution, and range of cellular activities regulated. It has been found in all eukaryotes and in all tissues so far examined. Though other calcium-binding proteins have been found in plants, including the regulatory subunit of quinate NAD:oxidoreductase, and a 13000–15000 MW protein from an oxygen-evolving photosystem II preparation (Sparrow and England, 1984), thus far, calmodulin is the only calcium-binding activator protein that has been identified in plant cells. Other calmodulin-like proteins have been found in plants (Grand et al., 1980; Bazari and Clarke, 1981; Van Eldik et al., 1980b), and are probably also calcium-binding proteins.

In this chapter, we concentrate on calmodulin and the processes it regulates. Ca^{2+} ATPase, known to be a calcium-binding protein in plant and animal cells (Stephens and Grisham, 1979), is discussed extensively elsewhere (Marmé and Dieter, 1983). Several other recent reviews discuss plant calmodulin (Charbonneau, 1980; Cormier et al., 1980; Roux and Slocum, 1982; Marmé and Dieter, 1983; Allan and Trewavas, 1987; Roberts et al., 1986b).

II. CALMODULIN: STRUCTURE

Calmodulin is a low-molecular-weight, acidic, calcium-binding protein highly conserved structurally throughout evolution (Watterson et al., 1984; Takagi et al., 1980; Yazawa et al., 1981; Sasagawa et al., 1982; Van Eldik et al., 1980a). It has 148 amino acids, with a molecular weight of about 16,700 (Watterson et al., 1980a), and an isoelectric point of 3.9–4.3. Several plant proteins and activities have been suggested to be dependent on it (Tables I and II).

TABLE I

Plant Proteins Suggested to Be Calmodulin Dependent

Protein	Reference
NAD kinase[a]	Dieter and Marmé (1980); Jarrett et al. (1983)
Ca^{2+}/calmodulin ATPase from plasma membrane[a]	Dieter and Marmé (1980)
Mg^{2+}/Ca^{2+} ATPase from spinach chloroplast envelope[a]	Nguyen and Seigenthaler (1985)
Plasma membrane protein kinases	Blowers et al. (1985)
Nuclear protein kinases	Datta et al. (1986)
Soluble leaf protein kinase	Polya and Micucci (1985)
Chromatin-associated protein kinase	Polya et al. (1983)
Glucan synthase kinase	Poovaiah and Veluthambi (1986)
Quinate NAD : oxidoreductase : phosphorylation of the catalytic subunit.	Ranjeva et al. (1983)
Chromatin-associated Ca^{2+} ATPase	Matsumoto et al. (1984)
Lipoxygenase	Leshem et al. (1982)
Superoxide dismutase	Leshem et al. (1982)
Lipolytic acyl hydrolase	Moreau and Isett (1985)

[a] Proteins demonstrated to bind directly to calmodulin through binding to a calmodulin-affinity column.

Calmodulin is highly conserved functionally. However, differences in the abilities of calmodulins from diverse sources to activate a variety of enzymes have been noted, both in the concentration required for half-maximal activation, and in the maximal level of activation achieved (Roberts *et al.*, 1984; Harmon *et al.*, 1984; Nakamura *et al.*, 1984; Marshak *et al.*, 1984; Lukas *et al.*, 1985). Slight amino acid variations are believed to account for these differences (Roberts *et al.*, 1985, 1986a).

Calcium appears to activate calmodulin by inducing a conformational change. On binding calcium, both plant (Dieter *et al.*, 1985) and animal calmodulins (Klee, 1977, 1980) undergo an increase in α-helix, and a hydrophobic domain is exposed (Tanaka and Hidaka, 1980; LaPorte *et al.*, 1980). Indirect evidence suggests that calcium-induced exposure of the hydrophobic site is important for its activating ability (LaPorte *et al.*, 1980; Hidaka *et al.*, 1980; Tanaka and Hidaka, 1980; Kauss, 1982; Hisanaga and Pratt, 1984).

There have been conflicting reports of the affinities of the four calcium-binding domains for calcium. However, estimates of K_d values are generally within $10^{-6}/10^{-5}$ M, with at least one site having a K_d of 10^{-6} M. Mg^{2+} at

TABLE II

Physiological Processes in Plants Suggested to Be Dependent on Calmodulin

Process	Reference
Cell division (anaphase A)	Vantard *et al.* (1985), Lambert and Vantard (1986)
Microtubule disassembly	Vantard *et al.* (1985)
NAD- and NADP-requiring pathways	Allan and Trewavas (1985, 1986b)
RNA transcription	Matsumoto *et al.* (1984), Datta *et al.* (1986)
Cytokinin-induced bud formation in *Funaria*	Saunders and Hepler (1983)
Cytokinin-induced tuberization in potato	Poovaiah and Veluthambi (1986)
Cytokinin-induced senescence retardation	Leshem *et al.* (1982)
Cytokinin-dependent coleoptile growth	Elliot *et al.* (1983)
Cytokinin-dependent betacyanin synthesis	Elliot (1983)
Phytochrome- and blue light-dependent chloroplast reorientation (*Mougeotia*)	Serlin and Roux (1984)
Phytochrome-dependent spore germination	Wayne and Hepler (1984)
Gibberellic acid-dependent α-amylase induction	Elliot *et al.* (1983)
Auxin-dependent coleoptile growth	Elliot *et al.* (1983), Raghothama *et al.* (1985)
Calcium transport	Dieter and Marmé (1983)
Geotropism	Biro *et al.* (1982)
(Auxin-stimulated) phosphatidylinositol turnover	Sandelius and Morré (1987)
Light-dependent reversal of phototaxis in *Chlamydomonas*	Hirschberg and Hutchinson (1980)

physiological concentrations (1×10^{-3} M) increases the K_d for Ca^{2+}, possibly through binding to one of the domains (Crouch and Klee, 1980). It is of interest that full Ca^{2+} occupancy is not required for either binding or activation of some calmodulin-dependent proteins, while binding of calmodulin to interacting proteins may occur at even lower calcium concentrations than those for activation. For example, in the case of calcineurin, binding occurs when two calcium sites are filled, and activation when three or four are filled (Kincaid and Vaughan, 1986). At least three sites are also required for activation of phosphodiesterase and myosin light chain kinase (Crouch and Klee, 1980; Stull *et al.*, 1980), while an effect on microtubule (MT) polymerization is observed when only one or two calcium ions are bound (Lee and Wolff, 1982). Some reports also suggest that binding of calmodulin to calmodulin-binding proteins increases its affinity for calcium (Olwin and Storm, 1985; Kincaid and Vaughan, 1986): an increase of an order of magnitude in calcium affinity is obtained in the presence of troponin I, myosin light chain kinase, and phosphodiesterase, respectively (Olwin and Storm, 1985). This positive cooperativity could result in activation of calmodulin during small transient increases in calcium.

III. THE FUNCTION OF CALMODULIN IN PLANT CELLS

A. Control of Calmodulin-Regulated Processes

Many calmodulin-dependent proteins may be present within a single cell. Concurrent activation of all of these proteins with an increase in intracellular calcium could lead to activation of potentially conflicting processes (e.g., phosphodiesterase and adenylate cyclase in animal cells). However, several mechanisms exist to avoid this situation. Selective activation of proteins may be achieved by compartmentation of calmodulin, calmodulin-binding proteins, or calcium fluxes; regulation of calmodulin or calmodulin-binding protein concentration; different requirements for calmodulin; regulation of the sensitivity of calmodulin-binding proteins for calmodulin; different calcium requirements; and the action of inhibitors and promoters. In the remainder of this chapter, we discuss the role of calmodulin with reference to these regulatory mechanisms.

The diversity of regulatory controls for calmodulin-dependent processes complicates the interpretation of calmodulin regulation *in vivo*. The presence of more than one calmodulin-binding protein in the same subcellular compartment and their involvement in the same physiological event also complicates analysis of calmodulin function, for example, in the mitotic spindle. To understand the *in vivo* role of calmodulin it is therefore essential to obtain detailed knowledge of the reactions and their requirements.

1. Concentration and Localization of Calmodulin

a. Quantitative Estimation of Calmodulin. Extraction of plant calmodulin suggests that the average cytoplasmic concentration of this protein is at least 1 μM if account is taken of losses during extraction and the space occupied by the cell wall and vacuole (Charbonneau and Cormier, 1979; Marmé and Dieter, 1983; Muto and Miyachi, 1984; Allan and Trewavas, 1985). This is sufficient to fully activate a number of plant enzymes (Table III). The concentration of calmodulin varies with tissue and stage of development. On a fresh weight basis, it increases during the first 24 hr of germination (Cocucci, 1984), and is in highest concentration in pea root in the cap, with lowest concentrations at the base of the apical meristem at the onset of rapid cell elongation (Allan and Trewavas, 1985). In the shoot, young tissues have the highest concentration on a fresh weight basis (Muto and Miyachi, 1984). It also increases during cell division in animal cells (Chafouleas *et al.*, 1982).

Although most calmodulin-dependent enzymes are half-maximally activated in the nanomolar calmodulin range, others require high, micromolar, levels. For example, the effect of calmodulin on MT dynamics, in the absence of protein kinase activity, requires micromolar levels of calmodulin (Marcum *et al.*, 1978; Job *et al.*, 1981). Two cytoskeletal proteins, microtubule-associated protein 2 (MAP2) and spectrin, in animal cells also have high calmodulin requirements for activation [K_d 7 μM for MAP2 (Lee and Wolff, 1984); K_d 2.8–3 μM for spectrin (Sobue *et al.*, 1981)]. High total concentrations of calmodulin may also be required if several calmodulin-binding proteins are present since the free concentration will be reduced. Thus, unusually high levels may be physiologically significant. For example, high calmodulin concentrations in apical meristems might reflect a requirement

TABLE III

Calcium- and Calmodulin-Dependence of Plant Enzymes

Enzyme	Concentration for half-maximal activation
Pea NAD kinase	1 nM pea calmodulin (Muto, 1983)
Zucchini NAD kinase	72 nM zucchini calmodulin (Marmé and Dieter, 1983)
Corn Ca^{2+} ATPase	160 nM bovine calmodulin (Marmé and Dieter, 1983)
Carrot quinate NAD : oxidoreductase	500 nM bovine calmodulin (Marmé and Dieter, 1983)
Wheat germ histone H1 kinase	1400 μM sheep brain calmodulin (Polya *et al.*, 1983)
Zucchini NAD kinase	Linear increase in activity 10^{-7}–10^{-3} M calcium (Dieter, 1986)
Pea NAD kinase	70 μM calcium (Muto, 1983)
Wheat germ histone H1 kinase	60–80 μM calcium (Polya *et al.*, 1983; Polya and Micucci, 1984).

for the rapid cytoskeletal changes occurring during cell division and early cell expansion, and/or activity of a large number of calmodulin-binding proteins such as Ca^{2+} ATPase, protein kinases, and NAD kinase.

b. Subcellular Localization of Calmodulin. Localization of calmodulin has been estimated by assay following subcellular fractionation and by immunocytochemical procedures, taking care to avoid artifactual redistribution of calmodulin. The study of Biro et al. (1984) indicates that, in oat coleoptiles, calmodulin or a calmodulin-like protein is present in nuclei, etioplasts, the cell wall, the outer mitochondrial membrane, and the space between the inner and outer mitochondrial membranes. Calmodulin has also been co-isolated with chromatin (Matsumoto et al., 1983), and with chloroplast stromal fractions (Jarrett et al., 1982; Muto, 1982). One fractionation study has estimated that the concentration in chloroplasts of wheat leaf cells is about 100 nM, a value too low to activate many calmodulin-binding proteins (Muto, 1982). The biological relevance of subcellular localization in plant cells is poorly understood at present, largely due to a lack of information on calmodulin-binding proteins and their compartmentation. However, localization of NAD kinase isozymes has helped to clarify the role of calmodulin in regulating NAD kinase activity (Section III,C,5).

Immunocytochemical examination of both plant and animal cells (Means and Dedman, 1980; Willingham et al., 1983; Wick et al., 1985; Vantard et al., 1985) and injection of fluorescently labeled calmodulin (Hamaguchi and Iwasa, 1980; Zavortink et al., 1983) have demonstrated that calmodulin appears to be localized in the mitotic apparatus (MA) during cell division. It is predominantly associated with the spindle poles, the chromosome-to-pole regions, and in nonpermeabilized cells, the interzone, and phragmoplast.

In the transition from interphase to prophase, MTs become redistributed from the cytoplasm to surround the nucleus, particularly at MT converging centers which tend to form at the poles of the future spindle (Wick and Duniec, 1983; Schmit et al., 1983). Tubular and lamellar ER also accumulate at the poles, as does anti-calmodulin staining (Wick et al., 1985; Vantard et al., 1985). A similar redistribution is noted in animal cells, where anti-calmodulin staining becomes associated with the centrioles and pericentriolar material (Willingham et al., 1983). The preprophase band of MTs in pea and onion root cells sometimes shows anti-calmodulin labeling (Wick et al., 1985).

Following breakdown of the nuclear envelope and formation of the spindle, chromosomes move individually to the metaphase plate. They are attached at the kinetochore to bundles of MTs which span the chromosome-to-pole region. The kinetochore MTs (KMTs) are connected by cross-bridges to tubular ER which ramifies along the KMTs from the poles (Hepler and Wolniak, 1984). These cross-bridges could consist of MAPs or dynein (Hepler and Wolniak, 1984). At these stages, anti-calmodulin staining is predominantly located at the poles and in the region between the chromosomes and the poles (Wick et al., 1985; Vantard et al., 1985).

Chromosome separation and movement to the poles at anaphase is generally a combination of two main events. One involves chromosome movement to the poles (anaphase A), which is associated with shortening of the KMTs (Cande, 1982), and possibly movement of KMTs along adjacent tubular ER (Hepler and Wolniak, 1984). The other involves spindle elongation (anaphase B), which is associated with rearranging and lengthening of non-KMTs (Cande, 1982). During late anaphase and telophase, the interzone becomes filled with large numbers of MTs which contribute to the phragmoplast at cytokinesis (Schmit et al., 1983; Wick et al., 1985; Vantard et al., 1985). Throughout anaphase, membranes retain their predominantly polar distribution, but progressively lose their KMT association. At telophase, KMTs disappear, and the nuclear envelope reforms around the two daughter sets of chromosomes.

Anti-calmodulin labeling remains associated with the poles and chromosome-to-pole regions through anaphase and early telophase. In plant (Wick et al., 1985) and animal cells (Willingham et al., 1983) that have not been permeabilized with detergent, anti-calmodulin staining is found also in the interzone during anaphase and telophase, and with the phragmoplast during cytokinesis in plant cells. Detergent-treated *Haemanthus* cells do not show staining in the interzone or phragmoplast (Vantard et al., 1985). Chromosome-to-pole labeling resembles, but is not identical to, that of cold-stable anti-tubulin staining in the chromosome-to-pole region, which is believed to represent KMT staining (Vantard et al., 1985). The staining appears to be more diffuse, and is similar to chlorotetracycline (CTC) fluorescence (Vantard et al., 1985), which is used as a membrane calcium probe, and to spindle membrane distribution (Hepler and Wolniak, 1984).

Despite the apparently varied distribution of calmodulin in the MA, interest has centered on the possible role of calmodulin in chromosome movement, and most particularly on the similarity of anti-calmodulin staining and the cold-stable KMTs since brain cold-stable MTs are known to be particularly sensitive to calmodulin *in vitro*. There is some evidence for this at the ultrastructural level. In cells that have been fixed in detergent, anti-calmodulin appears to be associated both with KMTs and with fibrous material associated with the MTs (Vantard et al., 1985). It has therefore been suggested that calmodulin may be associated with MAPs and regulate MTs through an effect on these proteins. However, in animal cells that are neither detergent-extracted nor alcohol-treated (Means and Dedman, 1980), immunoelectron microscopy shows anti-calmodulin staining associated with several membrane systems in mitotic cells. In rat cerebellar Purkinje cells it was associated with the smooth ER that ramifies along the KMTs, with membranous vesicles and mitochondrial membranes, as well as with the centrosomes and kinetochores (Means and Dedman, 1980).

Immunocytochemical colocalization of anti-calmodulin staining and KMTs may be misleading. First, association of calmodulin with MTs in the spindle should not be equated with a demonstration that it is active, since calmodulin may bind MTs in the presence of ethylene glycol bis(β-aminoethyl ether) N,N'-tetracetic acid (EGTA). Second, the procedures for immunolocalization of calmodulin generally involve permeabilization of membranes either by alcohols or by cofixation in Triton X-100. Several problems in interpreting immunolocalization of calmodulin arise from the use of permeabilizing agents. Calmodulin associated with membranes or high-MW MAPs may be released during fixation. The detergent Triton X-100 is known to disrupt a membrane-bound calcium-sequestering system in the MA (Silver *et al.*, 1980), to artifactually induce membrane-associated MT assembly even in the presence of glutaraldehyde (Mesland and Spiele, 1984), to result in loss of some integral membrane components (Goldenthal *et al.*, 1985), and in loss of anti-high-MW MAP staining (Albertini *et al.*, 1984; Wiche *et al.*, 1984; Connolly and Kalnins, 1980) and MT "side arms" (Connolly and Kalnins, 1980) during processing for immunocytochemistry. Since cells do not fix instantaneously, particularly in the low fixative concentrations required for retention of antigenicity; the detergent effects may cause relocalization and permit binding to nearby "sticky" molecules. These could include nonspecific binding to a highly basic protein, specific binding to MTs through tubulin (Kotani *et al.*, 1985) or through the low-MW MAP tau (Connolly and Kalnins, 1980), or specific binding to dynein (Hirokawa *et al.*, 1985), myosin light chain kinase (Guerriero *et al.*, 1981; Parke *et al.*, 1986), or actin-binding proteins (Sobue *et al.*, 1982). Alternatively, the calmodulin released from membranes or high-MW MAP sites could be lost from the cell, leaving behind the detergent-resistant pattern of distribution. Extraction of diffuse anti-calmodulin staining in the cytoplasm with fixation in methanol/ acetone (Willingham *et al.*, 1983) or Triton/EGTA (Deery *et al.*, 1984) is necessary for observing "distinct" calmodulin localization in interphase fibroblast cells. This reinforces the idea that localization of calmodulin with MTs and MT-associated material may be a misrepresentation of the pattern of calmodulin present in the spindle *in vivo*.

Calmodulin has not been found in association with cytoplasmic MTs of interphase plant cells (Wick *et al.*, 1985). This was suggested to indicate that MT breakdown or MT sliding regulated by calmodulin are not required for relocalization of MTs to the spindle, but it is also possible that calmodulin is not involved in regulating this subpopulation of MTs.

2. *Concentration and Compartmentation of Calmodulin-Binding Proteins*

Compartmentation of calmodulin-binding proteins may not only allow selective activation of several proteins within a single cell, it may also permit calmodulin-dependent and -independent isozymes to be involved in different functions. For example, the absence of calmodulin-dependent NAD kinase from the chloroplast stroma and thylakoids has provided evidence against a role for the calmodulin-dependent isozyme in regulating NADP levels in photosystem I. Its variable subcellular distribution (Sauer and Robinson, 1985; Muto and Miyachi, 1986) probably reflects different NADP requirements and usages. Other calmodulin-dependent proteins that are known to have specific subcellular distributions include the chloroplast Ca^{2+}/Mg^{2+} ATPase, protein kinases, and cytoskeletal proteins.

Little is known about concentrations of calmodulin-binding proteins in plant cells. It is also possible that some of these proteins are greatly underestimated or undetected since calmodulin may interfere with their extraction and assay. Removal of calmodulin from extracts has revealed calmodulin dependence of quinate NAD:oxidoreductase from carrot (Ranjeva et al., 1983) and protein kinases from pea membranes (Blowers et al., 1985) and wheat germ (Polya and Micucci, 1984).

3. The Concentration and Distribution of Calcium

Various primary signals are known to affect calcium fluxes or calcium transport mechanisms across the plasma membrane (Allan and Trewavas, 1987); however, little is known about movement of calcium between organelles and the cytoplasm in response to primary signals. Light increases calcium movement into chloroplasts (Muto et al., 1982; Kreimer et al., 1985) and mitochondria (Dieter and Marmé, 1981; Serlin and Roux, 1984), and gravity appears to affect movement of calcium into and out of the vacuole in corn root (Dauwalder et al., 1985). Localized sequestration and release of calcium from vesicles and ER are also believed to be important in the spindle during cell division (Wick and Hepler, 1980). Changes in cellular Ca^{2+} will affect both cytoplasmic and organellar Ca^{2+} according to the relative k_m values and the specific activities of calcium accumulation mechanisms.

Organellar free calcium levels are not known; however, transmembrane fluxes may be considered to involve free calcium, and may be important at least in regulating activity of membrane-associated calmodulin (Allan and Trewavas, 1987). Since the response of a cell to a calcium-releasing signal will depend largely on the subcellular redistribution of calcium in relation to compartmentation of calcium-response elements, the lack of information on free calcium levels in plant cells is a serious limitation in assessing the biological activity of calmodulin.

4. The Affinities of Calmodulin-Binding Proteins for Calmodulin and Calcium

The concentration of calmodulin required for half-maximal activation varies from nanomolar for solubilized NAD kinase, Ca^{2+} ATPase activation, phosphorylation of MT proteins, and depolymerization of MTs through phosphorylation, to micromolar for histone kinase, spectrin, MAP2, and MT depolymerization in the absence of phosphorylating conditions. The Ca^{2+} requirement is usually about 10^{-6} M, although it is as high as millimolar for some proteins. Proteins requiring unusually high levels of calcium include the membrane-bound form of zucchini NAD kinase, pea seedling NAD kinase, and wheat germ soluble chromatin-associated histone H1 kinase (Table III). It is not clear whether these reflect *in vivo* values. However, it should be noted that relatively high resting calcium levels may be associated with membranes and chromosomes/chromatin, so that higher calcium requirements of proteins associated with these structures are perhaps not unrealistic.

The sensitivity of calmodulin-binding proteins for calmodulin and/or calcium may be altered; for example, in animal tissues, calmodulin-dependent (Sharma and Wang, 1986) and -independent phosphorylation of cAMP phosphodiesterase (Sharma and Wang, 1985) and myosin light chain kinase (Conti and Adelstein, 1981) decreases the affinities of the enzymes for calmodulin. Calmodulin-dependent phosphorylation of (animal) cAMP phosphodiesterase increases the concentration of calcium required for half-maximal activation (Sharma and Wang, 1986).

Activation of different calmodulin-dependent proteins may therefore occur at different concentrations of calcium or calmodulin, permitting selective activation of proteins within a single compartment.

B. Calmodulin-Binding Proteins

Determination of Calmodulin Dependence

Many physiological activities have been suggested to be dependent on calmodulin. In some instances, these can be traced to specific proteins known to be calmodulin-dependent by a variety of methods, including direct binding to a calmodulin-affinity column (Table I). In most instances, calmodulin dependence relies on indirect evidence. Several procedures may be used to identify calmodulin-regulated proteins. These include (a) calcium dependence generally in the micromolar range, (b) inhibition by calcium chelators, (c) calmodulin dependence *in vivo* and *in vitro,* (d) inhibition by calmodulin antagonists, (e) affinity chromatography, (f) binding of ^{125}I-labeled calmodulin to protein gels, and (g) equilibrium binding.

(a–d) The bulk of the evidence for calmodulin regulation of processes in plant cells is based on *in vivo* inhibition of a calcium-dependent event by a calmodulin antagonist, usually a phenothiazine. This provides an initial indication of possible calmodulin-dependent events. However, calmodulin inhibitors are not completely specific. Early binding studies using phenothiazines demonstrated that many proteins appeared to have low-affinity calcium-independent sites. When the binding of bovine serum albumin (BSA) and calmodulin to phenothiazines was compared, it was found that the concentration of trifluoperazine (TFP) required for half-maximal binding to calmodulin in the presence of calcium was much lower than that for BSA, and that the extent of binding of TFP to calmodulin was much greater in the presence of calcium than in the presence of EGTA. Above 100 μM TFP, however, binding to calmodulin and BSA was the same, regardless of the presence of calcium or EGTA (Levin and Weiss, 1977). Binding of TFP therefore becomes nonspecific at concentrations near 100 μM.

Many plant developmental activities claimed to be calmodulin dependent on the basis of phenothiazine inhibition are inhibited only at very high levels of phenothiazines, for example, GA_3-dependent amylase induction [IC_{50} for TFP of 400 μM (Elliot *et al.,* 1983)], lipolytic acyl hydrolase [IC_{50} for TFP ~500 μM (Moreau and Isett, 1985)] and cytokinin-dependent betacyanin synthesis in *Amaranthus* seedlings [IC_{50} for TFP of 150 μM (Elliot, 1983)]. Shoot growth in lettuce achenes is almost completely inhibited at 750 μM chlorpromazine (CPZ), while gravitropism in light grown shoots is inhibited above 500 μM CPZ. Germination is 60% inhibited at concentrations of 1 mM (Kordan, 1980). Purified enzymes that are well-established as calmodulin dependent by a variety of criteria display far lower IC_{50} values for TFP, between 10 and 50 μM. Although higher extracellular concentrations of inhibitors may be required to affect physiological processes *in vivo* as a result of incomplete penetration, a study of radioactively labeled CPZ uptake in *Avena* by Biro *et al.* (1982) indicates that the internal concentration of phenothiazines could be higher than the external concentration.

Phenothiazines are amphipathic, and may disrupt enzyme activities nonspecifically through activation/inhibition of proteins which are affected by interactions with hydrophobic components of membranes or with nonpolar sections of proteins. Binding to phospholipids associated with membrane-bound glucan synthase has been suggested to be the cause of phenothiazine inhibition of this enzyme, since it occurs in the presence of calcium-independent trypsin activation, and requires high levels of drugs. Also, calmodulin dependence of glucan synthase has not been demonstrated (Kauss *et al.,* 1983). Phenothiazine inhibition of calcium sequestration by the ER Ca^{2+} ATPase has also been suggested to be due to the capacity of these drugs to bind nonspecifically to membranes (Buckhout, 1984), although it could occur through direct binding to the enzyme (Wulfroth and Petzelt, 1985). Calmidazolium and compound 48/80 are also known to inhibit calcium/phospholipid-dependent activation of erythrocyte Ca^{2+} ATPase (Gietzen, 1983). Inhibition of Ca^{2+} ATPase by anti-calmodulin drugs *in vivo* would potentially affect all intracellular calcium-dependent activities.

Several *in vivo* effects of phenothiazines are known not to be due to their anti-calmodulin effect. For example, inhibition of the light-dependent increase in NAD kinase activity is indirect, possibly through inhibition of photosynthetic electron transport (Muto and Miyachi, 1986). Inhibition of photosystem II is at the Mn^{2+}, not the Ca^{2+}, site. The W series of calmodulin inhibitors, and calmidazolium, are generally acknowledged to be more specific than the phenothiazines. The phenothiazine analog, chlorpromazine sulfoxide, and the analogs of W-7 and W-13 (W-5 and W-12, respectively) have similar structures and hydrophobicity indices but lower or negligible anti-calmodulin effects, thus providing controls for some activities. These include chloroplast movement in *Mougeotia* (Serlin and Roux, 1984) and auxin-induced coleoptile elongation (Raghothama *et al.*, 1985). Nevertheless, the proposed calmodulin dependence of physiological activities based solely on *in vivo* calmodulin inhibitor experiments can be due to an indirect effect, and it is important that calmodulin-dependent proteins be isolated and characterized.

(e–g) Calmodulin affinity chromatography, ^{125}I-labeled calmodulin binding to blots of protein gels, and equilibrium binding are very useful in demonstrating calmodulin dependence through direct binding to calmodulin. Proteins that have been found to bind to a calmodulin-affinity column in the presence of calcium are listed in Table I.

C. Calmodulin-Dependent Processes

In the following sections, we discuss processes that have been suggested to be calmodulin dependent, in most cases based on inhibitor experiments. Regulation of NAD kinase and cell division has been investigated more extensively; however, the involvement of calmodulin in these areas remains unclear. A discussion of the roles of calmodulin in regulating NAD kinase and cell division will highlight some of the problems of interpretation resulting from incomplete information.

1. Protein Phosphorylation

Phosphorylation/dephosphorylation reactions are widespread regulatory mechanisms in animal cells. The two major regulators of protein kinases in animals are cAMP and Ca^{2+}, and many specific kinases have been identified as calcium/calmodulin dependent. Calcium/calmodulin-dependent protein kinase activity has also been identified in plant cells. Using sodium dodecyl sulfate-polyacrylamide gel electrophoresis, calcium/calmodulin-dependent incorporation of ^{32}P into protein bands has been detected in membrane (Hetherington and Trewavas, 1982, 1984; Salimath and Marmé, 1983; Veluthambi and Poovaiah, 1984) and soluble fractions (Veluthambi and Poovaiah, 1984). However, the phosphorylated proteins remain largely uncharacterized. Glucan synthase kinase appears to be calmodulin dependent (Poovaiah and Veluthambi, 1986). The kinase phosphorylating the catalytic subunit of quinate NAD : oxidoreductase has also been demonstrated to be calmodulin dependent. Following calcium/calmodulin-dependent phosphorylation of the catalytic subunit (Ranjeva *et al.*, 1983), binding of the regulatory and phosphorylated catalytic subunits may take place independently of calcium under dark conditions (Ranjeva et al., 1986). The enzyme is then dependent on calcium for activity. Calcium/calmodulin-dependent protein kinases from several sources, including membrane (Hetherington and Trewavas, 1984), soluble (Polya and Micucci, 1984), and chromatin (Polya et al., 1983) fractions, can phosphorylate histone, but whether histone is an *in vivo* substrate is uncertain, since calcium levels required for activation are relatively high (Polya et al., 1983; Polya and Micucci, 1984).

Several MT proteins from animal cells have been demonstrated to be phosphorylated by calmodulin-dependent protein kinases, including MAP2 (Schulman et al., 1985; Goldenring et al., 1985; Larson et al., 1985; Yamamoto et al., 1985; Vallano et al., 1985), tau (Yamamoto et al., 1983, 1985), tubulin (Burke and DeLorenzo, 1981; Goldenring et al., 1985; Larson et al., 1985; Yamamoto et al., 1985), and stable-tubule-only-polypeptides (STOPs) (Larson et al., 1985). Calmodulin-dependent phosphorylation of MAP2, tau, or tubulin inhibits assembly of cold-labile MTs (Yamamoto et al., 1983, 1985), while MT disassembly is promoted by calmodulin-dependent phosphorylation in both cold-labile (Murthy and Flavin, 1983; Yamauchi and Fujisawa, 1983) and cold-stable MTs (Margolis and Job, 1984). Microtubule-associated proteins have still to be investigated in plant cells.

2. Intracellular Motility

a. Cytoplasmic Streaming. Cytoplasmic streaming in many plant cells is inhibited by elevated levels of Ca^{2+}. The possible involvement of calmodulin has been investigated by the use of inhibitors. TFP at 5 μM (Beilby and MacRobbie, 1984) does not affect the maintenance or cessation of streaming in *Chara*. However, TFP and CPZ both inhibit streaming in cultured tomato cells at 10 μM, although the more specific inhibitors W-7 and R24571 do not until 1 mM: orders of magnitude higher than their IC_{50} for calmodulin (Woods et al., 1984). Tominaga et al. (1987) have suggested that calmodulin might affect cessation or recovery of streaming through a dephosphorylation event.

b. Actin Structure and Function. Calmodulin could affect intracellular motility through modulating actin structure and function or through MAPs. It is known to affect actin–myosin interaction through myosin light chain kinase (Yagi et al., 1978), to increase the rate of polymerization of G-actin to F-actin (Piazza and Wallace, 1985), and to affect actin gelation (Sobue et al., 1982; Kotani et al., 1985). Calmodulin does not appear to modulate actin viscosity on its own, but acts through a variety of actin-binding proteins (Kotani et al., 1985). The calmodulin-binding proteins MAP2 and tau may interact with actin; MAP2 may promote gelation, while tau promotes bundling. Calmodulin inhibits their interaction with actin, preventing gelation and bundling, respectively (Kotani *et al.*, 1985). In the presence of calcium, calmodulin may also bind caldesmon, abolishing caldesmon/F-actin interaction. This permits interaction of F-actin and filamin, and hence gelation of F-actin (Sobue *et al.*, 1982). Calmodulin also binds spectrin, a protein whose actin cross-linking activity (Stromqvist *et al.*, 1985) is inhibited by calcium (Fishkind *et al.*, 1985). The calmodulin-binding protein tau may also bind to spectrin, inhibiting its F-actin cross-linking activity (Carlier *et al.*, 1984); however, it is not known if calcium inhibition and/or tau inhibition of spectrin-induced F-actin cross-linking are mediated via calmodulin.

Calmodulin therefore appears to be able to mediate opposing effects on actin structure and function. Although calmodulin-dependent actin-binding proteins have not yet been identified in plant cells, a protein with the same molecular weight as spectrin that cross-reacts with antibodies to red blood cell spectrin has recently been identified in maize roots (Wang and Yen, 1986).

3. Polar Tip Growth

Calcium appears to be required for tip-growing single cells, and has been found to display a tip-to-base gradient (Reiss *et al.*, 1986). Hausser *et al.* (1984) have investigated the possibility of calmodulin involvement through localization of the fluorescent anti-calmodulin drugs fluphenazine and CPZ in a variety of polar tip-growing cells. From their accumulation in the extreme tip of these cells in the early stage of growth, it has been suggested that calmodulin is involved in the manifestation of polarity, possibly regulating calcium fluxes through membranes. In later stages, phenothiazine fluorescence becomes uniformly distributed. Since cytochalasin B, which stops tip growth (Reiss *et al.*, 1986), alters the distribution of fluorescence, it was proposed that at this later stage calmodulin might be associated with microfilaments in the tip, regulating vesicle transport and cytoplasmic streaming (Hausser *et al.*, 1984). Another possible role for calmodulin in regulating tip growth is calcium-dependent vesicle fusion. However, membrane fusion may be mediated by calelectrin rather than calmodulin (Sudhof *et al.*, 1984).

4. Primary Messengers

Evidence from inhibitor experiments suggests that cytokinins and auxin may elicit some of their physiological responses through calmodulin. Raghothama *et al.* (1985) and Elliot *et al.* (1983) have found that auxin-induced coleoptile elongation is inhibited by low concentrations of phenothiazine inhibitors. Inactive analogs were ineffective (Raghothama *et al.*, 1985). Relatively low concentrations of phenothiazines also inhibit cytokinin-dependent processes, including budding in *Funaria* (Saunders and Hepler, 1983) and tuberization in potato (Poovaiah and Veluthambi, 1986). Several other activities may be inhibited by higher, probably nonspecific, concentrations of phenothiazines (Elliot, 1983; Elliot *et al.*, 1983).

Gravitropism has also been found to be inhibited by low concentrations of calmodulin inhibitors in *Avena* coleoptile (Biro *et al.*, 1982), at concentrations that do not inhibit growth.

Phytochrome has also been linked in some instances to calmodulin through inhibitor experiments. These include chloroplast reorientation in *Mougeotia* (Wagner *et al.*, 1984; Serlin and Roux, 1984); spore germination, including nuclear migration in *Onoclea* (Wayne and Hepler, 1984); and nuclear protein phosphorylation (Datta *et al.*, 1986).

Growth substances, gravity, and light have all been linked to induction of calcium fluxes (Hale and Roux, 1980; Roux, 1983). However, it is not clear whether their link with calmodulin is through changing calcium concentration, hence activating calmodulin, or/and whether they affect intracellular calcium concentration via calmodulin. For example, phytochrome appears to influence the calmodulin-dependent component of Ca^{2+} ATPase in corn coleoptile (Dieter and Marmé, 1981). In other instances, for example chloroplast reorientation and nuclear migration, calmodulin might directly affect proteins such as actin-binding proteins or myosin light chain kinase.

5. *Calmodulin and NAD Kinase*

a. **The Role of Calmodulin-Dependent NAD Kinase.** NAD kinase may play a central role in several metabolic pathways. Pyridine nucleotides are essential for a large number of enzymes, and affect a wide range of metabolic pathways. Since they are in rate-limiting concentrations in many types of plant cells (Yamamoto, 1963), and since conversion of NAD to NADP via NAD kinase activity is the only known mechanism of NADP formation, changes in NAD kinase activity would be expected to affect NADP-dependent, and possibly NAD-dependent, enzymes. Since pyridine nucleotides cannot cross the chloroplast envelope, compartmentation of NAD kinase isozymes will permit differential regulation of pyridine nucleotide levels in the cytoplasm and chloroplast.

There is indirect evidence suggesting that pyridine nucleotides may regulate the relative activities of the cytoplasmic NADP-requiring pentose phosphate pathway and the NAD-requiring glycolytic pathway in several systems, including bean hypocotyl (Yamamoto, 1966), *Vigna* (Yamamoto, 1963), carrot (ap Rees and Beevers, 1960), maize and pea roots (Gibbs and Beevers, 1955; Butt and Beevers, 1960; Black and Humphreys, 1962; Fowler and ap Rees, 1970). In pea roots, changes in the pathways coincide with a large increase in calcium/calmodulin-dependent NAD kinase (Allan and Trewavas, 1985). Thus in the root apex, an increase in activity of calmodulin-dependent NAD kinase could act as the regulatory switch for glycolysis and the pentose phosphate pathway. An increase in calmodulin-dependent NAD kinase activity has also been suggested to account for the relative increases in phosphorylated pyridine nucleotides and pentose phosphate pathway activity following fertilization of marine eggs and seed germination (Epel, 1982; Cocucci, 1984). The pentose phosphate pathway is a major supplier of carbon skeletons and reducing power in the form of NADPH for biosynthetic activities, so its regulation could modulate several other pathways. There is circumstantial evidence linking calmodulin-dependent NAD kinase in regulating some of these pathways, including amino acid and lignin biosynthesis (Allan and Trewavas, 1987). Cytoplasmic calmodulin-dependent NAD kinase could therefore play a pivotal role between NAD(H)-requiring, predominantly energy-producing pathways, and NADP(H)-requiring pathways, which are largely involved in biosynthetic activities.

b. Calmodulin Dependence of NAD Kinase. Calmodulin was initially discovered in plants as the activator of NAD kinase (Muto and Miyachi, 1977; Anderson and Cormier, 1978; Anderson et al., 1980). Originally, plant NAD kinase was proposed to be entirely dependent on calcium and calmodulin for activity (Jarrett et al., 1983). Several studies now indicate that there are both calcium/calmodulin-dependent and calcium/calmodulin-independent NAD kinase isozymes (Simon et al., 1982, 1984; Allan and Trewavas, 1985, 1986b; Muto and Miyachi, 1986). The proportions of the two forms of activity depend on the plant species and tissue (Allan and Trewavas, 1985; Muto and Miyachi, 1986).

Cell fractionation studies indicate that most calmodulin-independent NAD kinase is present in the chloroplast stroma of pea, wheat, and spinach (Muto and Miyachi, 1986) but also in the cytoplasm of pea, wheat (Muto and Miyachi, 1986), spinach (Simon et al., 1982; Muto and Miyachi, 1986), and zucchini (Dieter and Marmé, 1980). By contrast, most calmodulin-dependent NAD kinase appears to be cytoplasmic, or in the outer membranes of chloroplasts or mitochondria where it is accessible to cytoplasmic calcium. Calmodulin-dependent NAD kinase therefore appears to be potentially a major regulator of cytoplasmic NADP levels. There is also some calmodulin-dependent NAD kinase in the inner membrane of pea chloroplasts (Muto and Miyachi, 1986), and in the inner membrane of maize mitochondria, apparently facing the mitochondrial matrix (Sauer and Robinson, 1985).

Originally, reports indicated that NAD kinase was associated largely with chloroplasts (Ogren and Krogman, 1965; Heber and Santarius, 1965; Muto et al., 1981), and that its activity was entirely dependent on Ca^{2+} and calmodulin. Thus it seemed likely that light caused Ca^{2+} to enter the chloroplast (Nobel, 1969; Muto et al., 1982; Kreimer et al., 1986), accounting for the activation of NAD kinase (Jarrett et al., 1982, 1983). More recent assessment of this problem, however, indicates that virtually all of the calmodulin-dependent NAD kinase is not in the chloroplast stroma (Muto et al., 1981; Simon et al., 1982, 1984; Muto and Miyachi, 1986), and that the calmodulin-independent form rather than the calmodulin-dependent form is affected by light (Muto and Miyachi, 1986).

Light does not appear to be a general signal for activation of calmodulin-dependent NAD kinase. However, there may be some exceptions. Dieter (1986) has found a light-induced increase in NADP in corn coleoptile in which most NAD kinase is calmodulin dependent and associated with the outer mitochondrial membrane. Dieter and Marmé (1982, 1984) have proposed a mechanism for light-induced activation, again based on the calcium second messenger concept, but involving white (Dieter, 1984) or far red light (Marmé and Dieter, 1983) activation of cytoplasmic calmodulin-dependent NAD kinase. Activation is proposed to occur through increasing calcium as a result of inhibition of the calmodulin-dependent component of the plasma membrane Ca^{2+} ATPase and increased calcium efflux from mitochondria (Dieter and Marmé, 1981). This model awaits further examination of changes in the subcellular distribution of pyridine nucleotides, the possible contribution of calmodulin-independent NAD kinase activity, and measurement of changes in free cytosolic calcium.

Investigation of the light-dependent increase in NAD kinase activity has highlighted the necessity for examining the compartmentation of calmodulin-binding proteins in relation to subcellular calcium fluxes, and the contribution of calmodulin-independent isozymes.

6. Calmodulin and Cell Division

a. Introduction. There is a strong circumstantial evidence implicating calcium in cell division. Division is markedly affected by reducing calcium entry into cells by calcium channel blockers, chelation, or La^{3+}, or by reducing its availability in the spindle by microinjection of Ca^{2+} buffers (Kiehart, 1981; Izant, 1983; Hepler, 1985). Free (Keith *et al.*, 1985; Poenie *et al.*, 1985, 1986; Ratan *et al.*, 1986), and membrane-associated calcium (Wolniak *et al.*, 1983; Vantard *et al.*, 1985) also appears to undergo considerable changes *in vivo* during cell division. Evidence so far points to a role for calcium in triggering of cell division, nuclear envelope breakdown (Poenie *et al.*, 1985), the metaphase/anaphase transition (Hepler, 1985; Poenie *et al.*, 1986), and cytokinesis (Cande, 1980; Gunning, 1982; Poenie *et al.*, 1986). However, there is little agreement on the mode of action of calcium.

b. Calcium and Calcium/Calmodulin Regulation of MT Polymerization. Work with animal cells has shown that tubulin polymerization has an intrinsic sensitivity to calcium *in vitro* (Berkowitz and Wolff, 1981; Lee and Wolff, 1982) that is lost in the presence of MAPs or STOPS unless calmodulin is present (Rebhun *et al.*, 1980; Job *et al.*, 1982; Lee and Wolff, 1982, 1984). In the absence of MAPs calmodulin has no effect on MT polymerization (Rebhun *et al.*, 1980). The discovery that calcium regulates MT polymerization/depolymerization *in vitro* (Weisenberg, 1972) focused attention on the possibility that calcium acted on cell division by regulating MT dynamics in the spindle. In support of this hypothesis, perturbation of Ca^{2+} levels *in vivo* by microinjecting calcium into the spindle, excess Ca^{2+} in Ca/EGTA buffers, or treatment with calcium ionophore caused rapid, localized, and reversible loss of spindle birefringence in marine eggs (Salmon and Segall, 1980; Kiehart, 1981) and PtK_1 cells (Izant, 1983).

Calcium/calmodulin therefore appear to regulate MTs *in vivo* and *in vitro*. However, when the effects of calcium on spindle MTs are examined immunocytochemically it is found that, although cytoplasmic MTs are depolymerized by micromolar calcium, higher concentrations of calcium are required to depolymerize spindle MTs. Recent evidence suggests that millimolar levels could be required to depolymerize the cold-stable KMTs of *Haemanthus* endosperm (Lambert and Vantard, 1986; Vantard *et al.*, 1986). A cold-stable subpopulation of brain MTs is similarly resistant to millimolar calcium, although it becomes more sensitive in the presence of calmodulin (Job *et al.*, 1981; Pirollet *et al.*, 1983). The brain cold-stable MTs are also more sensitive to calmodulin-induced phosphorylation-independent MT depolymerization than are cold-labile MTs. As the KMTs in the spindle are also generally cold stable (Welsh *et al.*, 1979; Vantard *et al.*, 1985) [with the exception of sea urchin egg (Silver *et al.*, 1980)], it has been suggested that brain and spindle cold-stable subpopulations of MTs are both subject to the same regulatory mechanisms, including regulation by calmodulin.

c. Is Calmodulin Involved in Cell Division? Studies of the effects of calmodulin inhibitors or antibodies to calmodulin on mitosis are conflicting. In some studies of both plant (Wick *et al.*, 1985) and animal cells (Zavortink *et al.*, 1983), it has been found that mitosis may continue in the presence of calmodulin inhibitors. In pea and onion root cells, following inhibitor treatment, cells subsequently treated with anti-calmodulin revealed mitotic spindles with reduced or undetectable levels of antibody fluorescence (Wick *et al.*, 1985), indicating that cell division can proceed in the absence of calmodulin in the spindle. However, it is not known whether mitosis was normal in these cells, and it is also not certain that the inhibitors penetrated to all mitotic cells. Inhibition of calmodulin activity by antibodies (Zavortink *et al.*, 1983) was not demonstrated, thus negative results are inconclusive. It should also be noted that the antagonist TFP does not inhibit calmodulin-induced phosphorylation-independent depolymerization of cold-stable MTs (Job *et al.*, 1981), further complicating interpretation of inhibitor experiments.

A more detailed study of the effect of inhibitors on the progress of mitosis carried out on cultured animal (PtK_2) cells (Keith *et al.*, 1983a) demonstrated that W-7, R24571, or TFP, added at prophase, prolonged or arrested metaphase. If cells progressed through metaphase, subsequent anaphase and telophase were normal. Chromosome separation in single wall-less cells of

Haemanthus endosperm was also inhibited by calmidazolium (R24571) providing the drug was added at early metaphase (Lambert and Vantard, 1986), again indicating that calmodulin may regulate events in metaphase, the metaphase/anaphase transition, or anaphase A plus anaphase B.

Studies on earlier stages of cell division show that low concentrations of W-13 (Chafouleas *et al.*, 1982) and W-7 (Hidaka and Sasaki, 1985) block the transition from G_1 to S (Chafouleas *et al.*, 1982) and G_2 to M (Hidaka and Sasaki, 1985) in CHO-K1 cells.

The results of inhibitor experiments are therefore incomplete and inconclusive. Negative results could arise from slow penetration, metabolism, sequestration, or genuine lack of inhibitory effect, while a role in MT polymerization/depolymerization is not ruled out by negative results. Furthermore, studies of the effects on anaphase have not distinguished between anaphase A and anaphase B. It remains possible that calmodulin has a role at the G_1/S and G_2/M transition and at metaphase/anaphase.

d. Is Calmodulin Involved in Regulating Kinetochore-to-Pole MT Depolymerization? Interest in calmodulin has centered on the possibility that it is involved in regulating anaphase A (chromosome-to-pole movement) through preferential depolymerization of the cold-stable KMTs. Evidence for this attractive model rests largely on the effects of calcium and calmodulin on brain cold-stable MTs *in vitro*, and on immunocytochemical localization of calmodulin in the half spindles (Section III,A,1). Unfortunately, the experimental approach to this hypothesis has not yet been sufficiently rigorous to provide further evidence in support of it, and indeed much evidence is inconsistent with the model, indicating that the KMTs may be unusually stable to Ca^{2+} and calmodulin. Evidence concerning the possible role of calmodulin in regulating KMT depolymerization is discussed below.

First, MTs polymerized from tubulin isolated from sea urchin egg spindles by calcium or cold treatment are not sensitive to calmodulin (Keller *et al.*, 1982). However, it should be noted that it is not certain that proteins from KMTs were present. It is also possible that loss or modification of calmodulin-sensitizing factors occurred during purification.

Second, it is widely assumed that KMTs are very similar to brain cold-stable MTs, are stabilized by STOPS, and are therefore highly sensitive to calmodulin. However, regulation of MT depolymerization is complex. For example, MAP_1 and MAP_2 can also stabilize MTs to cold (Sloboda and Rosenbaum, 1979); MAP_2 and tau also appear to stabilize MTs to calcium (Job *et al.*, 1985), while tau (Kirschner *et al.*, 1986) and other MAPs (Horio and Hotani, 1986) appear to suppress dynamic instability of MTs. Calmodulin, even in large excess, does not affect MAP-induced MT stabilization (Job *et al.*, 1985). Since high-MW MAPs have been localized in the region of the cold-stable KMTs (Izant *et al.*, 1983; Wiche *et al.*, 1984), KMTs could conceivably be stabilized by factors other than STOPs and be unaffected by calmodulin. Isolation and characterization of spindle cold-stable and cold-labile MTs would help to clarify this.

Third, the role of calcium in regulating KMTs is in question since most studies show that very high concentrations of the ion are required. Up to millimolar levels of calcium as estimated by Ca^{2+}/EGTA buffers in detergent-extracted cells and perfusion of calcium in the presence of A23187 in living cells are required to depolymerize KMTs during various stages of chromosome migration *in vivo*, although pole-to-pole and cytoplasmic MTs were far more calcium sensitive (Vantard *et al.*, 1986; Lambert and Vantard, 1986). Millimolar calcium is also required for complete depolymerization of MTs polymerized from spindle tubulin from sea urchin eggs (Rebhun *et al.*, 1980). These high calcium levels are not entirely consistent with the proposed role of either Ca^{2+} or Ca^{2+}/calmodulin in regulating MT depolymerization during anaphase A. Despite the apparent requirement for high levels of calcium to cause depolymerization, it is possible that KMTs may be highly sensitive to the ion for a very brief period that has so far gone undetected, or that calcium may be released in very high concentration in the vicinity of MTs from adjacent membranes.

Fourth, it would appear from current data that there are two mechanisms of Ca^{2+}/calmodulin-dependent MT depolymerization, neither of which is likely to be operating during anaphase. One mechanism which may involve direct binding to MAPs or STOPS (Job *et al.*, 1981) requires relatively high levels of calcium and calmodulin [e.g., 100 μM calcium at 5 μM calmodulin, or 800 μM calcium at 1 μM calmodulin (Job *et al.*, 1981)]. The other mechanism, involving Ca^{2+}/calmodulin-dependent phosphorylation of MT protein, is active at low micromolar levels of Ca^{2+} and submicromolar calmodulin [cold-labile MTs (Fukunaga *et al.*, 1982; Yamamoto *et al.*, 1983, 1985), cold-stable MTs (Margolis and Job, 1984; Larson *et al.*, 1985)]. The low *in vivo* Ca^{2+} concentration is sufficient to activate this mechanism. However, as described in the previous section, there is no evidence as yet that KMTs can be depolymerized by low micromolar concentrations of calcium during anaphase.

Fifth, calmodulin inhibitor experiments generally do not support a role for calmodulin in regulating KMT depolymerization in anaphase A, but rather point to a role for calmodulin in the G_1/S or G_2/M transitions, metaphase, the metaphase/anaphase transition, or both anaphase A and B.

Sixth, immunocytochemical localization of calmodulin within the spindle places calmodulin in association with MTs (DeMey *et al.*, 1980; Vantard *et al.*, 1986), but it has not ruled out the possibility of association with MAPs and other structures. A major site of anti-calmodulin staining in nonpermeabilized cells is the smooth ER adjacent to the KMTs (Means and Dedman, 1980). Anti-calmodulin staining appears to be most concentrated near the poles of KMTs, with little staining near the chromosomes (DeMey *et al.*,

1980; Means and Dedman, 1980) and no staining at the kinetochore itself (DeMey *et al.*, 1980). Since the slow growth end of KMTs is at the pole, it has been suggested that calmodulin is in the correct position to regulate MT dynamics (Margolis, 1983). However, if MT depolymerization occurs at the kinetochore end as recently suggested (Mitchison, 1986), then calmodulin does not appear to be in the appropriate position for inducing depolymerization. On the other hand, it has been suggested that it could function as an attachment point for MTs at the pole by molecular linking to STOPS and tubulin (Margolis and Job, 1984).

Thus, although calmodulin regulation of MT depolymerization during anaphase A is a very attractive idea, further evidence is needed to establish the point. Major limitations with the hypothesis at present are the lack of studies on calcium and calmodulin dependence of spindle MTs, the apparently low Ca^{2+} levels in the spindle, and the apparently high calcium resistance of KMTs.

e. Calmodulin: Potential Involvement in MT–Membrane and MT–MT Interactions. There are a number of potential alternative or additional roles for calmodulin during cell division that tend to be overlooked. Some of these are discussed below.

Cross-bridging between MTs and membranes (Hepler *et al.*, 1970) and between MTs and MTs (Hepler *et al.*, 1970) has been reported in the MA. Cross-bridging side-arms of MTs may consist of proteins such as high-MW MAPs (Connolly and Kalnins, 1980) or dynein ATPase (Dentler *et al.*, 1980; Haimo *et al.*, 1979; Hirokawa *et al.*, 1985). Since calmodulin may bind both to MAPs and to the membrane cytoskeletal protein spectrin, calmodulin could be a component of these links.

When immunofluorescence of animal (Welsh *et al.*, 1979) and plant root tip cells (Wick *et al.*, 1985) is carried out in the absence of detergent or alcohol extraction, anti-calmodulin staining is observed in the interzone as well as in the half-spindles. In detergent- or alcohol-treated cells it is observed only in the half-spindles. This suggests that calmodulin may be associated with detergent-labile compartments such as membranes or high-MW MAPs in the interzone. This is supported by the observation that, during cell plate formation, the entire phragmoplast is stained with anti-calmodulin, even at late stages when MTs are confined to the outer areas of the growing cell plate, indicating that calmodulin is not exclusively associated with MTs or areas of MT disassembly (Wick *et al.*, 1985). In addition to containing MTs, the phragmoplast is rich in membranes: Golgi vesicles and a network of smooth tubular ER. Calmodulin might be involved, for example, in regulating Ca^{2+} ATPase activity, Golgi vesicle fusion, or movement of membranes along MTs in this region. The presence of anti-calmodulin in the kinetochore-to-pole region of detergent-treated cells does not, of course, preclude its association with detergent-labile compartments in this zone, as this binding may be relocalized to MT-associated components not present in MTs of the interzone, or may be lost, leaving behind detergent-resistant binding. Membrane-associated calmodulin in the kinetochore-to-pole region could be involved in Ca^{2+} ATPase activity, chromosome attachment to membranes, and chromosome movement.

f. Dynein ATPase. Immunocytochemistry (Mohri *et al.*, 1976; Hirokawa *et al.*, 1985) localizes dynein in both the kinetochore-to-pole and pole-to-pole MTs of the interzone region, while both anti-dynein (Sakai *et al.*, 1976) and dynein inhibitors (Cande and Wolniak, 1978; Cande, 1982) inhibit ATP-induced chromosome movement. Cande (1982) observed that, in detergent-extracted spindles, anaphase B but not anaphase A was inhibited by dynein inhibitors, thus pole separation could be mediated through dynein-induced sliding of antiparallel MTs that are present in the interzone (Haimo, 1985; McIntosh *et al.*, 1985). However, Cande's results do not preclude an additional involvement of dynein in anaphase A (Pratt *et al.*, 1980; Pratt, 1984); membranes will have been destroyed, thus MT–dynein–membrane interactions, which might constitute part of the force generating mechanism, would not have been operating. Dynein has also been proposed to be involved in prometaphase antipolar chromosome movement (Pickett-Heaps *et al.*, 1982). Although spindle dynein has not yet been analyzed for calmodulin dependence, the possibility exists, especially since isoforms of this protein from ciliary axonemes (Blum *et al.*, 1980) and sea urchin eggs (Hisanaga and Pratt, 1983) are regulated by calmodulin.

g. Regulation of Calcium Levels: Ca^{2+} ATPase and Calcium Chelation. Calmodulin could be involved in sequestering calcium at the pole or/and in the half-spindles. It has been shown that calmodulin can protect MTs in the absence of MAPs or STOPS against calcium through chelation (Lee and Wolff, 1982). It could also affect calcium sequestration through regulating Ca^{2+} ATPase activity. Localization in the mitotic apparatus closely resembles membrane-associated calcium as demonstrated by CTC fluorescence. Immunocytochemistry of nonpermeabilized cells also places calmodulin in association with ER and vesicles in the half-spindles, which are the organelles believed to be involved in calcium sequestration and release in the spindle. Spindle Ca^{2+} ATPase activity has been investigated, and has not been demonstrated to be calmodulin dependent (Nagle and Egrie, 1981). However, the authors point out that they had not investigated possible indirect activation via protein phosphorylation, and that the ATPase observed was not demonstrated to be the calcium-sequestering Ca^{2+} ATPase of the spindle. The possibility of calmodulin regulation of an ER Ca^{2+} ATPase or calcium-release mechanism is therefore still open.

IV. CONCLUDING REMARKS

There is considerable evidence that calmodulin regulates a wide range of basic and specialized metabolic activities in animal cells. Although the evidence in many instances is not so conclusive in plant cells, many areas of metabolism have been indicated as calmodulin regulated. It is also clear that the regulation of calmodulin activity is complex and involves numerous controlling mechanisms. Thus, *in vivo* roles for calmodulin are not always as obvious as they might first appear. This complexity is both exciting and challenging, and the elucidation of the undoubtedly important roles of calmodulin in plant cell metabolism remains as a future challenge.

ACKNOWLEDGMENTS

This work was supported by a NATO postdoctoral fellowship to E.A. and by NSF grants PCM-8402414 and DCB-8502723 to P.K.H.

REFERENCES

Albertini, D. F., Herman, B., and Sherline, P. (1984). *Eur. J. Cell Biol.* **33**, 134–143.
Allan, E. F. (1984). Ph.D. thesis. University of Edinburgh, Edinburgh, Scotland.
Allan, E. F., and Trewavas, A. J. (1985). *Planta* **165**, 493–501.
Allan, E. F., and Trewavas, A. J. (1986a). In "Molecular and Cellular Aspects of Calcium in Plant Development" (A. J. Trewavas, ed.), pp. 311–312. Plenum, New York.
Allan, E. F., and Trewavas, A. J. (1986b). In "Molecular and Cellular Aspects of Calcium in Plant Development" (A. J. Trewavas, ed.), pp. 357–358. Plenum, New York.
Allan, E. F., and Trewavas, A. J. (1987). In "The Biochemistry of Plants" (P. K. Stumpf and E. E. Conn, eds.), Vol. 12, pp. 117–144. Academic Press, New York.
Andersen, J. P., Lassen, K., and Moller, J. V. (1985). *J. Biol. Chem.* **260**, 371–380.
Anderson, J. M., and Cormier, M. J. (1978). *Biochem. Biophys. Res. Commun.* **84**, 595–602.
Anderson, J. M., Charbonneau, H., Jones, H. P., McCann, R. O., and Cormier, M. J. (1980). *Biochemistry* **19**, 3113–3120.
Anderson, L. E. (1974). In "Proceedings of the Third International Congress on Photosynthesis" (M. Avron, ed.), pp. 1393–1405. Elsevier, Amsterdam.
ap Rees, T., and Beevers, H. (1960). *Plant Physiol.* **35**, 830–838.
Bazari, W. L., and Clarke, M. (1981). *J. Biol. Chem.* **256**, 3598–3603.
Beilby, M. J., and MacRobbie, E. A. (1984). *J. Exp. Bot.* **35**, 568–580.
Berkowitz, S. A., and Wolff, J. (1981). *J. Biol. Chem.* **256**, 11216–11223.
Biro, R. L., Hale, C. C., II, Wiegland, O. F., and Barlow, S. J. (1982). *Ann. Bot.* **50**, 737–745.
Biro, R. L., Daye, S., Serlin, B. S., Terry, M. E., Datta, N., Sopory, S. K., and Roux, S. J. (1984). *Plant Physiol.* **75**, 382–386.
Black, C. C., and Humphreys, T. E. (1962). *Plant Physiol.* **37**, 66–73.
Blowers, D. P., Hetherington, A., and Trewavas, A. (1985). *Planta* **166**, 208–215.
Blum, J. J., Hayes, A., Jamieson, G. A., and Vanaman, T. C. (1980). *J. Cell Biol.* **87**, 386–397.
Boynton, A. L., Whitfield, J. F., and MacManus, J. P. (1980). *Biochem. Biophys. Res. Commun.* **95**, 745–749.

Bradbury, E. M., Inglis, R. J., and Matthews, H. R. (1974). *Nature (London)* **249**, 553–555.
Brandl, C. J., Green, N. M., Korczak, B., and MacLennan, D. H. (1986). *Cell* **44**, 597–607.
Brostrom, C. O., and Wolff, D. J. (1981). *Biochem. Pharmacol.* **30**, 1395–1405.
Buckhout, T. J. (1984). *Plant Physiol.* **76**, 962–967.
Burgess, W. H., Jemiolo, D. K., and Kretsinger, R. H. (1980). *Biochim. Biophys. Acta* **623**, 257–270.
Burke, B. E., and DeLorenzo, R. J. (1981). *Proc. Natl. Acad. Sci. U.S.A.* **78**, 991–995.
Butt, V. S., and Beevers, H. (1960). *Biochem. J.* **76**, 51.
Cande, W. Z. (1980). *J. Cell Biol.* **87**, 326–335.
Cande, W. Z. (1982). *Nature (London)* **295**, 700–701.
Cande, W. Z., and Wolniak, S. M. (1978). *J. Cell Biol.* **79**, 573–580.
Carlier, M.-F., Simon, C., Cassoly, R., and Prakel, L.-A. M. (1984). *Biochimie* **66**, 305–311.
Chafouleas, J. G., Bolton, W. E., Boyd, A. E., III, and Means, A. R. (1982). *Cell* **28**, 41–50.
Charbonneau, H. (1980). In "Calcium Binding Proteins: Structure and Function" (F. L. Siegel, E. Carafoli, R. H. Kretsinger, D. H. MacLennan, and R. H. Wasserman, eds.), pp. 155–164. Elsevier/North-Holland, Amsterdam.
Charbonneau, H., and Cormier, M. J. (1979). *Biochem. Biophys. Res. Commun.* **90**, 1039–1047.
Cheung, W. Y. (1967). *Biochem. Biophys. Res. Commun.* **29**, 478–482.
Cheung, W. Y. (1970). *Biochem. Biophys. Res. Commun.* **38**, 533–538.
Cocucci, M. (1984). *Plant Cell Environ.* **7**, 215–221.
Cohen, P. (1980). In "Calcium and Cell Function" (W. Y. Cheung, ed.), Vol. 1, pp. 183–189. Academic Press, New York.
Connolly, J. A., and Kalnins, V. I. (1980). *Eur. J. Cell Biol.* **21**, 296–300.
Conti, M. A., and Adelstein, R. S. (1981). *J. Biol. Chem.* **256**, 3178–3181.
Cormier, M. J., Anderson, J. M., Charbonneau, H., Jones, H. P., and McCann, R. O. (1980). In "Calcium and Cell Function" (W. Y. Cheung, ed.), Vol. 1, pp. 201–218. Academic Press, New York.
Crouch, T. H., and Klee, C. B. (1980). *Biochemistry* **19**, 3692–3698.
Das, R., and Sopory, S. K. (1985). *Biochem. Biophys. Res. Commun.* **128**, 1445–1460.
Datta, N., Chen, Y.-R., and Roux, S. J. (1986). In "Molecular and Cellular Aspects of Calcium in Plant Development" (A. J. Trewavas, ed.), pp. 115–122. Plenum, New York.
Dauwalder, M., Roux, S. J., and Rabenberg, L. K. (1985). *Protoplasma* **129**, 137–148.
Deery, W. J., Means, A. R., and Brinkley, B. R. (1984). *J. Cell Biol.* **98**, 904–910.
DeMey, J., Moermans, M., Geuens, G., Nuydens, R., VanBelle, H., and DeBrabander, M. (1980). In "Microtubules and Microtubule Inhibitors 1980" (M. DeBrabander and J. DeMey, eds.), pp. 227–241. Elsevier, Amsterdam.
Dentler, W. L., Pratt, M. M., and Stephens, R. E. (1980). *J. Cell Biol.* **84**, 381–403.
Dieter, P. (1984). *Plant Cell Environ.* **7**, 371–380.
Dieter, P. (1986). In "Molecular and Cellular Aspects of Calcium in Plant Development" (A. J. Trewavas, ed.), pp. 91–98. Plenum, New York.
Dieter, P., and Marmé, D. (1980). *Cell Calcium* **1**, 279–286.
Dieter, P., and Marmé, D. (1981). *Biochem. Biophys. Res. Commun.* **101**, 749–755.
Dieter, P., and Marmé, D. (1982). *Eur. J. Cell Biol.* **27**, 110.
Dieter, P., and Marmé, D. (1983). *Planta* **159**, 277–281.
Dieter, P., and Marmé, D. (1984). *Plant, Cell and Environ.* **7**, 371–380.
Dieter, P., Cox, J. A., and Marmé, D. (1985). *Planta* **166**, 216–218.
Elliot, D. C. (1983). *Plant Physiol.* **72**, 215–218.
Elliot, D. C., Batchelor, S. M., Cassar, R. A., and Marinos, N. G. (1983). *Plant Physiol.* **72**, 219–224.
Epel, D. (1982). In "Calcium and Cell Function" (W. Y. Cheung, ed.), Vol. 2, pp. 355–383. Academic Press, New York.
Fischer, E. H., Sumerwell, W. N., Junge, J., and Stein, E. A. (1964). *Proc. Int. Biochem. Congr., 4th* **8**, 124–137.

Fishkind, D. J., Bouder, I. M., and Brigg, D. A. (1985). *J. Cell Biol.* **101,** 1071 (Abstr.).
Fowler, M. W., and ap Rees, T. (1970). *Biochim. Biophys. Acta* **201,** 33–44.
Fukunaga, K., Yamamoto, H., Matsui, K., Higashi, K., and Miyamoto, E. (1982). *J. Neurochem.* **39,** 1607–1617.
Gariépy, J., and Hodges, R. S. (1983). *Biochemistry* **22,** 1586–1594.
Gibbs, M., and Beevers, H. (1955). *Plant Physiol.* **30,** 343–347.
Gietzen, K. (1983). *Biochem. J.* **216,** 611–616.
Goldenring, J. R., Vallano, M. L., and DeLorenzo, R. J. (1985). *J. Neurochem.* **45,** 900–905.
Goldenthal, K. L., Hedman, K., Chen, J. W., August, J. T., and Willingham, M. C. (1985). *J. Histochem. Cytochem.* **33,** 813–820.
Grand, R. J. A., Nairn, A. C., and Perry, S. V. (1980). *Biochem. J.* **185,** 755–760.
Guerriero, V., Rowley, D. R., and Means, A. E. (1981). *Cell* **27,** 449–458.
Gunning, B. E. S. (1982). *In* "The Cytoskeleton in Plant Growth and Development" (C. W. Lloyd, ed.), pp. 229–292. Academic Press, London.
Haimo, L. T. (1985). *Can. J. Biochem. Cell Biol.* **63,** 519–532.
Haimo, L. T., Telzer, B. R., and Rosenbaum, J. L. (1979). *Proc. Natl. Acad. Sci. U.S.A.* **76,** 5759–5763.
Hale, C. C., II, and Roux, S. J. (1980). *Plant Physiol.* **65,** 658–662.
Hamaguchi, Y., and Iwasa, F. (1980). *Biomed. Res.* **1,** 502–509.
Harmon, A. C., Jarrett, H. W., and Cormier, M. J. (1984). *Anal. Biochem.* **141,** 168–178.
Haug, A., and Weis, C. (1986a). *In* "Molecular and Cellular Aspects of Calcium in Plant Development" (A. J. Trewavas, ed.), pp. 19–26. Plenum, New York.
Haug, A., and Weis, C. (1986b). *In* "Molecular and Cellular Aspects of Calcium in Plant Development" (A. J. Trewavas, ed.), pp. 323–324. Plenum, New York.
Hausser, I., Herth, W., and Reiss, H.-D. (1984). *Planta* **162,** 33–39.
Heber, U. W., and Santarius, K. A. (1965). *Biochim. Biophys. Acta* **109,** 390–408.
Heldt, H. W., and Rapley, L. (1970). *FEBS Lett.* **10,** 143–148.
Hepler, P. K. (1985). *J. Cell Biol.* **100,** 1363–1368.
Hepler, P. K., and Wayne, R. (1985). *Annu. Rev. Plant Physiol.* **36,** 397–439.
Hepler, P. K., and Wolniak, S. M. (1984). *Int. Rev. Cytol.* **90,** 169–238.
Hepler, P. K., McIntosh, J. R., and Cleland, S. (1970). *J. Cell Biol.* **45,** 438–444.
Hepler, P. K., Wick, S. M., and Wolniak, S. M. (1981). *Int. Cell Biol.* 673–686.
Hetherington, A., and Trewavas, A. (1982). *FEBS Lett.* **145,** 67–71.
Hetherington, A., and Trewavas, A. (1984). *Planta* **161,** 409–417.
Hidaka, H., and Sasaki, Y. (1985). *In* "Calmodulin Antagonists and Cellular Physiology" (H. Hidaka and D. J. Hartshorne, eds.), pp. 91–97. Academic Press, Orlando, Florida.
Hidaka, H., Yamaki, M., Naka, T., Tanaka, H., Hiyashi, H., and Kobayashi, R. (1980). *Mol. Pharmacol.* **17,** 66–72.
Hidaka, H., Endo, T., Kawamoto, S., Yamada, E., Umekawa, H., Tanabe, K., and Hara, K. (1983). *J. Biol. Chem.* **258,** 2705–2709.
Higuchi, T., and Shimada, M. (1967). *Plant Cell Physiol.* **8,** 71–78.
Hirokawa, N., Takemura, R., and Hisanaga, S. (1985). *J. Cell Biol.* **101,** 1858–1870.
Hirschberg, R., and Hutchinson, W. (1980). *Can. J. Microbiol.* **26,** 265–267.
Hisanaga, S., and Pratt, M. M. (1984). *Biochemistry* **23,** 3032–3037.
Hopkins, D. W., and Briggs, W. R. (1973). *Plant Physiol.* **51,** 284 (Abstr.).
Horio, T., and Hotani, H. (1986). *Nature* **321,** 605–607.
Izant, J. G. (1983). *Chromosoma* **88,** 1–10.
Izant, J. G., Weatherbee, J. A., and McIntosh, J. R. (1983). *J. Cell Biol.* **96,** 424–434.
Jameson, L., Frey, T., Zeeberg, B., Dalldorf, F., and Caplow, M. (1980). *Biochemistry* **19,** 2472–2479.
Jarrett, H. W., Brown, C. J., Black, C. C., and Cormier, M. J. (1982). *J. Biol. Chem.* **257,** 13795–13804.
Jarrett, H. W., DaSilva, T., and Cormier, M. J. (1983). *In* "Metals and Micronutrients: Their

Uptake and Utilization by Plants'' (D. A. Robb and D. S. Pierpoint, eds.), pp. 205–218. Academic Press, London.
Job, D., Fischer, E. H., and Margolis, R. L. (1981). *Proc. Natl. Acad. Sci. U.S.A.* **78,** 4679–4682.
Job, D., Rauch, C. T., Fischer, E. H., and Margolis, R. L. (1982). *Biochemistry* **21,** 509–515.
Job, D., Pabion, M., and Margolis, R. L. (1985). *J. Cell Biol.* **101,** 1680–1689.
Kakiuchi, S., and Yamazaki, R. (1970). *Biochem. Biophys. Res. Commun.* **41,** 1104–1110.
Kauss, H. (1982). *Plant Sci. Lett.* **26,** 103–109.
Kauss, H., Kohle, H., and Jeblick, W. (1983). *FEBS Lett.* **158,** 84–88.
Keith, C., DiPaola, M., Maxfield, F. R., and Shelanski, M. L. (1983a). *J. Cell Biol.* **97,** 42a (Abstr.) 161.
Keith, C., DiPaola, M., Maxfield, F. R., and Shelanski, M. L. (1983b). *J. Cell Biol.* **97,** 1918–1924.
Keith, C. H., Ratan, R., Maxfield, F. R., Bajer, A., and Shelanski, M. L. (1985). *Nature (London)* **316,** 848–850.
Keller, T. C. S., III, Jemiolo, D. K., Burgess, W. H., and Rebhun, L. I. (1982). *J. Cell Biol.* **93,** 797–803.
Kiehart, D. P. (1981). *J. Cell Biol.* **88,** 604–617.
Kincaid, R. L., and Vaughan, M. (1986). *Proc. Natl. Acad. Sci. U.S.A.* **83,** 1193–1197.
Kirschner, M., Schulze, E., Kristofferson, D., Drubin, D., and Mitchison, T. (1986). *In* "The Molecular Biology of the Cytoskeleton," ASCB Summer Res. Conf.
Klee, C. B. (1977). *Biochemistry* **16,** 1017–1024.
Klee, C. B. (1980). *In* "Calcium and Cell Function" (W. Y. Cheung, ed.), Vol. 1, pp. 59–77. Academic Press, New York.
Klevit, R. E., Blumenthal, D. K., Wemmer, D. E., and Krebs, E. G. (1985). *Biochemistry* **24,** 8152–8157.
Kordan, H. A. (1980). *Z. Pflanzenphysiol.* **100,** 273–278.
Kotani, S., Nishida, E., Kumagai, H., and Sakai, H. (1985). *J. Biol. Chem.* **260,** 10779–10783.
Kreimer, G., Melkonian, M., Holtum, J. A. M., and Latzko, E. (1985). *Planta* **166,** 515–523.
Lambert, A. M., and Vantard, M. (1986). *In* "Molecular and Cellular Aspects of Calcium in Plant Development" (A. J. Trewavas, ed.), pp. 175–183. Plenum, New York.
LaPorte, D. C., Wierman, B. M., and Storm, D. R. (1980). *Biochemistry* **19,** 3814–3819.
Larson, R. E., Goldenring, J. R., Vallano, M. L., and DeLorenzo, R. J. (1985). *J. Neurochem.* **44,** 1566–1574.
Lee, Y. C., and Wolff, J. (1982). *J. Biol. Chem.* **257,** 6306–6310.
Lee, Y. C., and Wolff, J. (1984). *J. Biol. Chem.* **259,** 1226–1230.
Leshem, Y., Ferguson, I. B., Wurzburger, Y., and Frier, A. A. (1982). *In* "Eleventh International Conference on Plant Growth Substances" (P. F. Wareing, ed.), pp. 569–578. Plenum, New York.
Levin, R. M., and Weiss, B. (1977). *Mol. Pharmacol.* **13,** 690–697.
Levin, R. M., and Weiss, B. (1978). *Biochim. Biophys. Acta* **540,** 197–204.
Lin, C. T., Sun, D., Song, G.-X., and Wu, J.-Y. (1986). *J. Histochem. Cytochem.* **34,** 561–567.
Lin, Y. M., Liu, Y. P., and Cheung, W. Y. (1974). *J. Biol. Chem.* **249,** 4943–4954.
Lukas, T. J., Wiggins, M. E., and Watterson, D. M. (1985). *Plant Physiol.* **78,** 477–483.
McIntosh, J. R., Roos, U.-P., Neighbors, B., and McDonald, K. L. (1985). *J. Cell Sci.* **75,** 93–129.
Marcum, J. M., Dedman, J. R., Brinkley, B. R., and Means, A. R. (1978). *Proc. Natl. Acad. Sci. U.S.A.* **75,** 3771–3775.
Margolis, R. L. (1983). *In* "Calcium and Cell Function" (W. Y. Cheung, ed.), Vol. 4, pp. 313–335. Academic Press, New York.
Margolis, R. L., and Job, D. (1984). *Adv. Cyclic Nucleotide Protein Phosphorylation Res.* **17,** 417–425.

Marmé, D., and Dieter, P. (1983). *In* "Calcium and Cell Function" (W. Y. Cheung, ed.), Vol. 4, pp. 263–311. Academic Press, New York.
Marshak, D. R., Clarke, M., Roberts, D. M., and Watterson, D. M. (1984). *Biochemistry* **23**, 2891–2899.
Matsumoto, H., Tanigawa, M., and Yamaya, T. (1983). *Plant Cell Physiol.* **24**, 593–602.
Matsumoto, M., Yamaya, T., and Tanigawa, M. (1984). *Plant Cell Physiol.* **25**, 191–195.
Means, A. R., and Dedman, J. R. (1980). *Nature (London)* **285**, 73–77.
Mesland, D. A. M., and Spiele, H. (1984). *J. Cell Sci.* **68**, 113–137.
Minowa, O., and Yagi, K. (1984). *J. Biochem. (Tokyo)* **96**, 1175–1182.
Mitchison, T. J. (1986). *In* "The Molecular Biology of the Cytoskeleton," ASCB Summer Res. Conf.
Mitchison, T. J., and Kirschner, M. W. (1985). *J. Cell Biol.* **101**, 766–777.
Mohri, H., Mohri, T., Mabuchi, I., Yazaki, I., Sakai, H., and Ogawa, K. (1976). *Dev. Growth Differ.* **18**, 391–398.
Moreau, R. A., and Isett, T. F. (1985). *Plant Sci.* **40**, 95–98.
Murthy, A. S. N., and Flavin, M. (1983). *Eur. J. Biochem.* **137**, 37–46.
Muto, S. (1982). *FEBS Lett.* **147**, 161–164.
Muto, S. (1983). *Z. Pflanzenphysiol.* **109**, 385–393.
Muto, S., and Miyachi, S. (1977). *Plant Physiol.* **59**, 55–60.
Muto, S., and Miyachi, S. (1984). *Z. Pflanzenphysiol.* **114**, 421–431.
Muto, S., and Miyachi, S. (1986). *In* "Molecular and Cellular Aspects of Calcium in Plant Development" (A. J. Trewavas, ed.), pp. 107–114. Plenum, New York.
Muto, S., and Shimogawara, K. (1985). *FEBS Lett.* **193**, 88–92.
Muto, S., Miyachi, S., Usada, H., Edwards, G. E., and Bassham, J. A. (1981). *Plant Physiol.* **68**, 324–328.
Muto, S., Izawa, S., and Miyachi, S. (1982). *FEBS Lett.* **139**, 250–254.
Nagle, B. W., and Egrie, J. C. (1981). *In* "Mitosis and Cytokinesis" (A. M. Zimmerman and A. Forer, eds.), pp. 337–361. Raven, New York.
Nakamura, T., Fujita, K., Eguchi, Y., and Yazawa, M. (1984). *J. Biochem. (Tokyo)* **95**, 1551–1557.
Nguyen, T. D., and Siegenthaler, P. A. (1985). *Biochim. Biophys. Acta* **840**, 99–106.
Nobel, P. S. (1969). *Biochim. Biophys. Acta* **172**, 134–143.
Ogren, W. L., and Krogman, D. W. (1965). *J. Biol. Chem.* **240**, 4603–4608.
Oláh, Z., and Kiss, Z. (1982). *FEBS Lett.* **195**, 33–37.
Olwin, B. B., and Storm, D. R. (1985). *Biochemistry* **24**, 8081–8086.
Paliyath, G., and Poovaiah, B. W. (1984). *Proc. Natl. Acad. Sci. U.S.A.* **81**, 2065–2069.
Parke, J., Miller, C., and Anderton, B. H. (1986). *Eur. J. Cell Biol.* **41**, 9–13.
Piazza, G. A., and Wallace, R. W. (1985). *Proc. Natl. Acad. Sci. U.S.A.* **82**, 1683–1687.
Pickett-Heaps, J. D., Tippit, D. H., and Porter, K. R. (1982). *Cell* **29**, 729–744.
Pirollet, F., Job, D., Fischer, E. H., and Margolis, R. L. (1983). *Proc. Natl. Acad. Sci. U.S.A.* **80**, 1560–1564.
Poenie, M., Alderton, J., Tsien, R. Y., and Steinhardt, R. A. (1985). *Nature (London)* **315**, 147–149.
Poenie, M., Alderton, J., Steinhardt, R. A., and Tsien, R. Y. (1986). *Science* **233**, 886–889.
Polya, G. M., and Micucci, V. (1984). *Biochim. Biophys. Acta* **785**, 68–74.
Polya, G. M., and Micucci, V. (1985). *Plant Physiol.* **79**, 968–972.
Polya, G. M., Davies, J. R., and Micucci, V. (1983). *Biochim. Biophys. Acta* **761**, 1–12.
Polya, G. M., Micucci, V., Basiliadis, S., Lithgow, T., and Lucantoni, H. (1986). *In* "Molecular and Cellular Aspects of Calcium in Plant Development" (A. J. Trewavas, ed.), pp. 75–82. Plenum, New York.
Poovaiah, B. W., and Veluthambi, K. (1986). *In* "Molecular and Cellular Aspects of Calcium in Plant Development" (A. J. Trewavas, ed.), pp. 83–90. Plenum, New York.

Pratt, M. M., Otter, T., and Salmon, E. D. (1980). *J. Cell Biol.* **86,** 738–745.
Pratt, M. M. (1984). *Int. Rev. Cytol.* **87,** 83–105.
Pryke, J. A., and ap Rees, T. (1976). *Planta* **132,** 279–284.
Raghothama, K. G., Mizrah, Y., and Poovaiah, B. W. (1985). *Plant Physiol.* **79,** 28–33.
Ranjeva, R., Refeno, G., Boudet, A. M., and Marmé, D. (1983). *Proc. Natl. Acad. Sci. U.S.A.* **80,** 5222–5224.
Ranjeva, R., Graziana, R., Dillenschneider, M., Charpenteau, M., and Boudet, A. M. (1986). *In* "Molecular and Cellular Aspects of Calcium in Plant Development" (A. J. Trewavas, ed.), pp. 41–48. Plenum, New York.
Ratan, R. R., Shelanski, M. L., and Maxfield, F. R. (1986). *Proc. Natl. Acad. Sci. U.S.A.* **83,** 5136–5140.
Rebhun, L. I., Jemiolo, D., Keller, T., Burgess, W., and Kretsinger, R. I. (1980). *In* "Microtubules and Microtubule Inhibitors 1980" (M. DeBrabander and J. DeMey, eds.), pp. 243–252. Elsevier, Amsterdam.
Reiss, H.-D., Herth, W., and Schneff, E. (1986). *In* "Molecular and Cellular Aspects of Calcium in Plant Development" (A. J. Trewavas, ed.), pp. 211–217. Plenum, New York.
Roberts, D. M., Burgess, W. H., and Watterson, D. M. (1984). *Plant Physiol.* **75,** 796–798.
Roberts, D. M., Crea, R., Malecha, M., Alvarado-Urbina, G., Chiarello, R. H., and Watterson, D. M. (1985). *Biochemistry* **24,** 5090–5098.
Roberts, D. M., Rowe, P. M., Siegel, F. L., Lukas, T. J., and Watterson, D. M. (1986a). *J. Biol. Chem.* **261,** 1491–1494.
Roberts, D. M., Lukas, T. J., and Watterson, D. M. (1986b). *CRC Crit. Rev. Plant Sci.,* submitted for publication.
Roux, S. J. (1983). *Symp. Soc. Exp. Biol.* **36,** 561–580.
Roux, S. J., and Slocum, R. D. (1982). *In* "Calcium and Cell Function" (W. Y. Cheung, ed.), Vol. 3, pp. 409–453. Academic Press, New York.
Saitoh, Y., Wells, J. N., and Hardham, J. G. (1982). *Fed. Proc., Fed. Am. Soc. Exp. Biol.* **41,** 1728.
Sakai, H., Mabuchi, I., Shimoda, S., Kuriyama, R., Ogawa, K., and Mohri, H. (1976). *Dev. Growth Differ.* **18,** 211–219.
Salimath, B. P., and Marmé, D. (1983). *Planta* **158,** 560–568.
Salmon, E. D., and Segall, R. R. (1980). *J. Cell Biol.* **86,** 355–365.
Sandelius, A. S., and Morré, D. J. (1987). *Plant Physiol.* **84,** 1022–1027.
Sasagawa, T., Ericsson, L. H., Walsh, K. A., Schreiber, W. E., Fischer, E. H., and Titani, K. A. (1982). *Biochemistry* **21,** 2565–2569.
Sauer, A., and Robinson, D. G. (1985). *Planta* **166,** 227–233.
Saunders, M. J., and Hepler, P. K. (1983). *Dev. Biol.* **99,** 41–49.
Schafer, A., Bygrave, F., Matzenauer, S., and Marmé, D. (1985). *FEBS Lett.* **187,** 25–28.
Schleicher, M., Lukas, T. J., and Watterson, D. M. (1984). *Arch. Biochem. Biophys.* **229,** 33–42.
Schmit, A.-C., Vantard, M., DeMey, J., and Lambert, A.-M. (1983). *Plant Cell Rep.* **2,** 285–288.
Schulman, H., Kuret, J., Jefferson, A. B., Nose, P. S., and Spitzer, K. H. (1985). *Biochemistry* **24,** 5320–5327.
Schumaker, K. S., and Sze, H. (1985). *Plant Physiol.* **79,** 1111–1117.
Serlin, B. S., and Roux, S. J. (1984). *Plant Physiol.* **81,** 6368–6372.
Serlin, B. S., Dauwalder, M., and Roux, S. J. (1986). *In* "Molecular and Cellular Aspects of Calcium in Plant Development" (A. J. Trewavas, ed.), pp. 321–322. Plenum, New York.
Sharma, R. K., and Wang, J. H. (1985). *Proc. Natl. Acad. Sci. U.S.A.* **82,** 2603–2607.
Sharma, R. K., and Wang, J. H. (1986). *J. Biol. Chem.* **261,** 1322–1328.
Silver, R. B., Cole, R. D., and Cande, W. Z. (1980). *Cell* **19,** 505–516.
Simon, P., Dieter, P., Bonzon, M., Greppin, H., and Marmé, D. (1982). *Plant Cell Rep.* **1,** 119–122.

Simon, P., Bonzon, M., Greppin, H., and Marmé, D. (1984). *FEBS Lett.* **167,** 332–338.
Sloboda, R. D., and Rosenbaum, J. L. (1979). *Biochemistry* **18,** 48–55.
Sobue, K., Muramoto, Y., Fujita, M., and Kakiuchi, S. (1981). *Biochem. Biophys. Res. Commun.* **100,** 1063–1070.
Sobue, K., Morimoto, K., Kanda, K., Maruyama, K., and Kakiuchi, S. (1982). *FEBS Lett.* **138,** 289–292.
Solomon, F. (1977). *Biochemistry* **16,** 358–363.
Sparrow, R. W., and England, R. R. (1984). *FEBS Lett.* **177,** 95–98.
Stephens, E. M., and Grisham, C. M. (1979). *Biochemistry* **18,** 4876–4885.
Stratton, B. R., and Trewavas, A. J. (1982). *Plant Cell Environ.* **4,** 419–426.
Stromqvist, M., Backman, C., and Shambhag, V. P. (1985). *FEBS Lett.* **190,** 15–20.
Stull, J. T., Manning, D. R., High, C. W., and Blumenthal, D. K. (1980). *Fed. Proc., Fed. Am. Soc. Exp. Biol.* **39,** 1552–1557.
Sudhof, T. C., Ebbecke, M., Walker, J. H., Fritsche, U., and Boustead, C. (1984). *Biochem. J.* **23,** 1103–1109.
Takagi, T., Nemoto, T., Konishi, K., Yazawa, M., and Yagi, K. (1980). *Biochem. Biophys. Res. Commun.* **96,** 377–381.
Tanaka, T., and Hidaka, H. (1980). *J. Biol. Chem.* **255,** 11078–11080.
Teshima, Y., and Kakiuchi, S. (1978). *J. Cyclic Nucleotide Res.* **4,** 219–231.
Tezuka, T., and Yamamoto, Y. (1972). *Plant Physiol.* **50,** 458–462.
Tezuka, T., and Yamamoto, Y. (1974). *Plant Physiol.* **53,** 717–722.
Tominaga, Y., Wayne, R., Tung, H. Y. L., and Tazawa, M. (1987). *Protoplasma* **136,** 161–169.
Vallano, M. L., Goldenring, J. R., Buckholz, T. M., Larson, R. E., and DeLorenzo, R. J. (1985). *Proc. Natl. Acad. Sci. U.S.A.* **82,** 3202–3206.
Vanaman, T. C. (1980). In "Calcium and Cell Function" (W. Y. Cheung, ed.), Vol. 1, pp. 41–58. Academic Press, New York.
Van Eldik, L. J., Grossman, A. R., Iverson, D. B., and Watterson, D. M. (1980a). *Proc. Natl. Acad. Sci. U.S.A.* **77,** 1912–1916.
Van Eldik, L. J., Piperno, G., and Watterson, D. M. (1980b). *Proc. Natl. Acad. Sci. U.S.A.* **77,** 4779–4783.
Vantard, M., Lambert, A. M., DeMey, J., Picquot, P., and Van Eldik, L. J. (1985). *J. Cell Biol.* **101,** 488–499.
Vantard, M., Stoeckel, H., Picquot, P., Van Eldik, L. J., and Lambert, A. M. (1986). In "Molecular and Cellular Aspects of Calcium in Plant Development" (A. J. Trewavas, ed.), pp. 375–377. Plenum, New York.
Veluthambi, K., and Poovaiah, B. W. (1984). *Plant Physiol.* **76,** 359–365.
Wang, Y., and Yen, L. (1986). *Chin. Biochem. J.,* in press.
Watterson, D. M., Harrelson, W. G., Keller, P. M., Sharief, F., and Vanaman, T. C. (1976). *J. Biol. Chem.* **251,** 4501–4513.
Watterson, D. M., Sharief, F., and Vanaman, T. C. (1980a). *J. Biol. Chem.* **255,** 962–975.
Watterson, D. M., Mendel, P. A., and Vanaman, T. C. (1980b). *Biochemistry* **19,** 2672–2676.
Watterson, D. M., Burgess, W. H., Lukas, T. J., Iverson, D., Marshak, D. R., Schleicher, M., Erickson, B. W., Fok, K.-F., and Van Eldik, L. J. (1984). *Adv. Cyclic Nucleotide Protein Phoshorylation Res.* **16,** 205–226.
Wayne, R., and Hepler, P. K. (1984). *Planta* **160,** 12–20.
Weisenberg, R. C. (1972). *Science* **177,** 1104–1105.
Welsh, M. J., Dedman, J. R., Brinkley, B. R., and Means, A. R. (1979). *J. Cell Biol.* **81,** 624–634.
Wiche, G., Briones, E., Koszka, C., Artlieb, U., and Krepler, R. (1984). *EMBO J.* **3,** 991–998.
Wick, S. M., and Duniec, J. (1983). *J. Cell Biol.* **97,** 235–243.
Wick, S. M., and Hepler, P. K. (1980). *J. Cell Biol.* **86,** 500–513.
Wick, S. M., Muto, S., and Duniec, J. (1985). *Protoplasma* **126,** 198–206.

Willingham, M. C., Wehland, J., Klee, C. B., Richert, N. D., Rutherford, A. V., and Pastan, I. H. (1983). *J. Histochem. Cytochem.* **31,** 445–461.
Wolniak, S. M., Hepler, P. K., and Jackson, W. T. (1983). *J. Cell Biol.* **96,** 598–605.
Woodford, T. A., and Pardee, A. B. (1986). *J. Biol. Chem.* **261,** 4669–4676.
Woods, C. M., Polito, V. S., and Reid, M. S. (1984). *Protoplasma* **121,** 17–24.
Wulfroth, P., and Petzelt, C. (1985). *Cell Calcium* **6,** 295–310.
Yagi, K., Yazawa, M., Kakiuchi, S., Ohshima, M., and Venishi, K. (1978). *J. Biol. Chem.* **253,** 1338–1340.
Yamamoto, H., Fukunaga, K., Tanaka, E., and Miyamoto, E. (1983). *J. Neurochem.* **41,** 1119–1125.
Yamamoto, H., Fukunaga, K., Goto, S., Tanaka, E., and Miyamoto, E. (1985). *J. Neurochem.* **44,** 759–768.
Yamamoto, Y. (1963). *Plant Physiol.* **38,** 45–54.
Yamamoto, Y. (1966). *Plant Physiol.* **41,** 519–522.
Yamauchi, T., and Fujisawa, H. (1983). *Biochem. Biophys. Res. Commun.* **110,** 287–291.
Yazawa, M., Yagi, K., Toda, H., Kondo, K., Narita, K., Yamazaki, R., Sobue, K., Kakiuchi, S., Nagao, S., and Nazawa, Y. (1981). *Biochem. Biophys. Res. Commun.* **99,** 1051–1057.
Zavortink, M., Welsh, M. J., and McIntosh, J. R. (1983). *Exp. Cell Res.* **149,** 375–385.

Plant Hydroxyproline-Rich Glycoproteins

12

ALLAN M. SHOWALTER
JOSEPH E. VARNER

I. Introduction
II. Cell Wall Hydroxyproline-Rich Glycoproteins or "Extensins"
   A. Cellular and Tissue Distribution
   B. Structure
   C. Biosynthesis and Degradation
   D. Function
   E. Response to Stress
   F. Hydroxyproline-Rich Glycoproteins in Nondicotyledonous Plant Species
III. Arabinogalactan Proteins
   A. Cellular and Tissue Distribution
   B. Structure
   C. Biosynthesis and Degradation
   D. Function
   E. Distribution of Arabinogalactan Proteins in the Plant Kingdom
IV. Solanaceous Lectins
   A. Cellular and Tissue Distribution
   B. Structure
   C. Biosynthesis and Degradation
   D. Function
   E. Distribution of the Solanaceous Lectins in the Plant Kingdom
V. Summary and Insights into Future Plant Hydroxyproline-Rich Glycoprotein Research
   References

I. INTRODUCTION

Hydroxyproline (Hyp) is an unusual and abundant amino acid in nature. It is synthesized by the posttranslational hydroxylation of peptidyl proline residues and is a constituent of only a handful of proteins, including collagen in animals and cell wall hydroxyproline-rich glycoproteins (HRGPs) and arabinogalactan proteins (AGPs) in plants. These three proteins, in which hydroxyproline content is extremely high (10–50%), represent three of the most abundant proteins on our planet! Hydroxyproline is thus one of the most abundant amino acids in nature despite its absence from almost all other proteins.

There are at least three classes of HRGPs in plants: (1) the cell wall HRGPs or "extensins" (reviewed previously in Lamport, 1980; Lamport and Catt, 1981), (2) the AGPs (reviewed in Clarke et al., 1979; Fincher et al., 1983), and (3) certain lectins from the Solanaceae family (see Allen, 1983; Kilpatrick, 1983). These three classes of HRGPs are readily separated and distinguished from one another by virtue of their unique biochemical and physical properties. In this chapter, we examine each of the plant HRGPs with respect to these properties and explore how these properties are related to their proposed biological functions.

II. CELL WALL HYDROXYPROLINE-RICH GLYCOPROTEINS OR "EXTENSINS"

A. Cellular and Tissue Distribution

In the 1950s, Steward et al. (1954) discovered hydroxyproline in plants and elucidated its occurrence in protein. Shortly thereafter, Dougall and Shimbayashi (1960) and Lamport and Northcote (1960) independently determined that the bulk of the peptidyl hydroxyproline in plants exists in the cell wall. These studies and a number of subsequent investigations, including the recent immunoelectron microscopic investigation of the cell wall HRGPs in carrot root tissue (Stafstrom and Staehlin, 1988), have made it clear that cell wall HRGPs constitute a major protein component of plant cell walls, most notably those of dicotyledonous plants and certain green algae (Tables I–III). These HRGPs also have a characteristic tissue distribution in plants, being especially abundant in roots, stems, callus cultures, cell cultures, and seed coats and much less abundant in leaves (Tables II and III).

B. Structure

Initially, characterization of these abundant cell wall HRGPs proved to be difficult because of the inability to extract these proteins from the cell wall. Lamport (1977) circumvented this obstacle by hydrolyzing plant cell walls with acid/protease treatment and characterized the resulting glycopeptide fragments that went into solution. The amino acid sequences which Lamport elucidated using this approach (Fig. 1) are most remarkable in their repetition of a unique pentapeptide sequence, Ser-(Hyp)$_4$, their abundance of

TABLE I

Hydroxyproline Content of Various Green Algae[a]

Alga	Percentage dry weight of hydroxyproline in whole cells
Chlamydomonas reinhardtii	0.61
Tetraspora sp.	0.43
Chlorella pyrenoidosa	0.055
Chlorella ellipsoidea	0.12
Chlorella vulgaris	0.35
Scenedesmus obliquus	0.06
Hydrodictyon reticulatum	0.03
Bryopsis plumosa	0.23
Codium fragile	0.12
Ulva linza	0.16
Monostroma sp.	0.20
Oedogonium faveolatum	0.04
Spirogyra sp.	0.04
Cladophora sp.	0.04
Chaetomorpha aerea	0.12
Nitella sp.	0.004

[a] Adapted from Gotelli and Cleland (1968).

tyrosine and lysine residues, and the occurrence of an unusual tyrosine derivative which was later identified as isodityrosine (IDT; Fig. 2).

Subsequent investigations using different plant material and/or extraction methods resulted in the isolation and biochemical characterization of several soluble HRGPs (Tables IV and V). These studies established that these cell wall proteins are particularly rich in hydroxyproline and that most of the remaining amino acids are serine, histidine, lysine, tyrosine, and valine. The relatively high content of lysine in these proteins is undoubtedly responsible for making them extremely basic, with isoelectric points in the range of 10–12. Studies of tomato HRGPs (Smith *et al.*, 1984, 1986) have documented the presence of more than one type of cell wall HRGP in a given tissue, indicating the probable existence of a multigene family for these proteins. The amino acid sequence repeats of three of these soluble HRGPs are known and presented in Fig. 3. Again, these sequences show the presence of Ser-(Hyp)$_4$ pentapeptide repeats and of numerous tyrosine and lysine residues. Note that the tomato sequences shown here are distinct, not only from each other but also from the glycopeptide sequences elucidated by Lamport following acid/protease treatment of tomato cell walls, with two exceptions, Ser-(Pro)$_4$-Thr-Hyp-Val-Tyr-Lys and Ser-(Pro)$_4$-Val-1/2IDT-Lys-1/2IDT-Lys (Fig. 1). Hence, it appears that there are at least four different tomato cell

TABLE II

Hydroxyproline Content of Various Tissue Cultures[a]

Species	Percentage dry weight of hydroxyproline in cell walls
Ginko biloba (2n)[b]	1.1
Ginko biloba (n)[b]	0.07
Oryza sativa[c]	0.2
Phaseolus vulgaris[d]	2.7
Lycopersicon esculentum[d]	2.6
Solanum tuberosum[d]	2.4
Galega officinalis[d]	1.6
Apium graveolens[d]	1.5
Nicotiana tabacum var. Turkish[d]	1.5
Nicotiana tabacum var. Xanthi[d]	1.3
Acer pseudoplatanus[d]	1.2
Daucus carota[d]	1.1
Lepidium sativum[d]	0.95
Camellia sinensis[d]	0.8
Centaurea cyanus[d]	0.22–0.5
Rosa (Paul's Scarlet)[d]	0.28

[a] Adapted from Lamport (1965).
[b] Gymnosperm.
[c] Angiosperm: monocot.
[d] Angiosperm: dicot.

wall HRGPs (i.e., P1a, P1b, P2, and another, tentatively named P3), one of which (P3) is particularly insoluble and can at present be analyzed only in glycopeptide form following acid/protease treatment of cell walls.

Chen and Varner (1985a,b) have recently isolated and characterized cDNA and genomic clones encoding a carrot HRGP; the complete HRGP

```
A. Ser-Hyp-Hyp-Hyp-Hyp-Ser-Hyp-Ser-Hyp-Hyp-Hyp-Hyp-("Tyr"-Tyr)-Lys
   [Ser-Hyp-Hyp-Hyp-Hyp-Ser-Hyp-Ser-Hyp-Hyp-Hyp-Hyp-1/2IDT-Tyr-1/2IDT-Lys]

B. Ser-Hyp-Hyp-Hyp-Hyp-Ser-Hyp-Lys

C. Ser-Hyp-Hyp-Hyp-Hyp-Thr-Hyp-Val-Tyr-Lys

D. Ser-Hyp-Hyp-Hyp-Hyp-Lys

E. Ser-Hyp-Hyp-Hyp-Hyp-Val-("Tyr"-Lys-Lys)
   [Ser-Hyp-Hyp-Hyp-Hyp-Val-1/2IDT-Lys-1/2IDT-Lys]
```

Fig. 1. Amino acid sequences of five HRGP glycopeptides solubilized from tomato cell walls by acid/protease treatment (from Lamport, 1977). The sequences of two of these peptides following the identification of "Tyr" as isodityrosine [IDT] (see Fig. 2) are shown in brackets (from Epstein and Lamport, 1984).

TABLE III

Hydroxyproline Content of Various Plants and Tissues[a]

Species	Tissue	Percentage dry weight of hydroxyproline in cell walls
Avena sativa[b]	Coleoptile	0.05
Zea mays[b]	Coleoptile and leaf	0.05
Festuca sp.[b]	Leaf	0.05
Pisum sativum[c]	Root	1.1
	Epicotyl	0.58
Phaseolus vulgaris[c]	Hypocotyl	0.88
Eruca sativa[c]	Hypocotyl	0.17
Citrus paradisi[c]	Pericarp	0.14
Helianthus annuus[c]	Hypocotyl	0.1
Pyrus communis[c]	Pericarp	0.04
Glycine max[c,d]	Seed coat (outer layer)	1.6
	Seed coat (inner layer)	0.45
	Root nodule (cortex)	1.8
	Root nodule (medulla)	0.38

[a] Adapted from Lamport (1965) with the exception of *Glycine max* data.
[b] Angiosperm: monocot.
[c] Angiosperm: dicot.
[d] Calculated from Cassab *et al.* (1985) and Cassab (1986).

amino acid sequence encoded by the carrot genomic clone pDC5A1 is presented in Fig. 4. The protein encoded by this sequence is 306 amino acids in length and contains a 32-amino acid putative signal peptide sequence at the amino terminus, presumably essential for the delivery of this protein to the

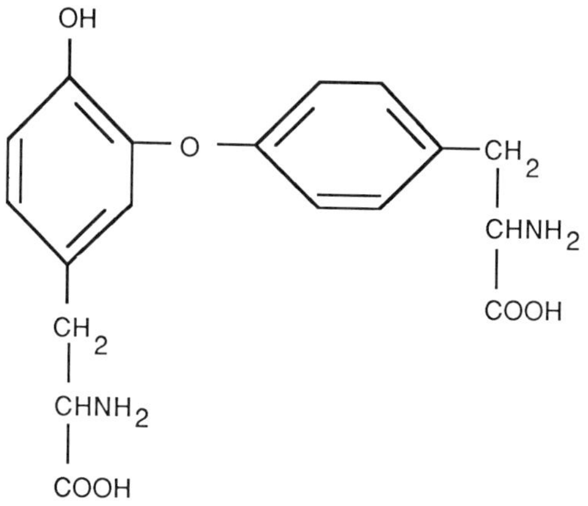

Fig. 2. Structure of isodityrosine. From Fry (1982).

TABLE IV
Biochemical Properties of Various Soluble Cell Wall HRGPs

Biochemical properties	Carrot[a]	Tomato P1[b]	Tomato P2[c]	Tobacco[d]	Potato[e]	Melon[f]	Soybean[g]
Tissue source	Roots	Tissue culture	Tissue culture	Callus tissue	Tubers	Callus culture	Seed coats
Estimated mass of glycoprotein (kDa)	86	nd[h]	nd	90–120	91	nd	100
Isoelectric point	~11	i	i	i	>11	i	i
Protein (%)	35	40–50	40–50	74	39	36	nd
Abundant amino acids	Hyp,Ser,Lys, Tyr,Val,His	Hys,Ser,Lys, Tyr,Val,His, Thr,Pro	Hyp,Ser,Lys, Val,Tyr	Hyp,Ser,Lys, Val,Tyr,His, Pro	Hyp,Ser,Lys, Val,Tyr,His, Pro	Hyp,Ser,Lys, Val,Tyr,Pro	Hyp,Ser,Lys, Pro,Tyr,His
Carbohydrate (%)	65	50–60	50–60	26	61	64	nd
Carbohydrates present (% of total CHO)	97% Ara 3% Gal	90% Ara 6–7% Gal 2% Glc	90% Ara 6–7% Gal 2% Glc	87% Ara 8% Gal 5% Glc	91% Ara 5% Gal 3% Glc 1% GlcN	77% Ara 22% Gal 1% Glc	90% Ara 9% Gal
Glycopeptide linkages	Hyp-Ara Ser-Gal	Hyp-Ara	Hyp-Ara	nd	nd	Hyp-Ara Ser-Gal	Hyp-Ara

[a] From Stuart and Varner (1980) and van Holst and Varner (1984).
[b] From Smith et al. (1984). Tomato P1 has now been separated into two HRGP components, P1a and P1b, with almost identical properties (Smith et al., 1986).
[c] From Smith et al. (1984).
[d] From Mellon and Helgeson (1982).
[e] From Leach et al. (1982).
[f] From Esquerré-Tugayé et al. (1985).
[g] From Cassab et al. (1985).
[h] nd, Not determined.
[i] The isoelectric point was estimated based on amino acid composition and found to be in the range of 10–10.5.

TABLE V
Amino Acid Compositions[a] of Various Soluble Cell Wall HRGPs

Amino acid	Carrot[b]	Tomato P1a[c]	Tomato P1b[c]	Tomato P2[d]	Tobacco[e]	Potato[f]	Melon[g]	Soybean[h]
Hyp	45.5	32.7	30.9	41.8	38.0	41.7	35.0	36.2
Asx	0.3	1.4	1.9	0.7	5.5	0.7	2.0	2.1
Thr	1.2	6.2	6.5	1.0	5.6	3.2	1.0	1.3
Ser	14.0	9.8	9.5	12.1	8.9	9.4	9.8	8.2
Glx	0.3	1.5	1.6	0.3	0.43	1.1	1.2	2.4
Pro	0.9	9.6	10.2	0.8	8.4	9.2	6.5	9.9
Gly	0.4	1.7	2.0	0.3	0.44	1.1	3.8	4.0
Ala	0.4	2.9	1.9	0.5	1.2	0.9	2.1	1.9
Val	5.9	8.3	7.1	5.1	5.2	3.8	11.1	2.5
Cys	0.0	0.0	0.0	0.0	0.22	0.1	0.0	1.0
Met	0.0	0.0	0.0	0.0	0.20	0.0	0.0	2.2
Ile	0.3	1.0	1.1	0.9	0.27	0.3	1.2	0.9
Leu	0.3	1.0	0.8	0.2	0.75	0.2	1.3	1.3
Tyr	11.0	7.7	8.0	14.9	12.2	6.2	7.1	8.5
Phe	0.0	0.0	0.0	0.2	0.0	0.1	0.7	0.5
His	11.8	6.1	7.1	1.0	5.0	5.1	2.0	8.8
Lys	6.5	9.5	10.3	20.1	12.4	15.9	15.2	10.8
Arg	0.0	0.7	1.2	0.1	0.0	0.2	Trace	1.0
Trp	1.2	nd[i]	nd	nd	nd	1.9	nd	nd

[a] Expressed in mole %.
[b] From van Holst and Varner (1984).
[c] From Smith et al. (1986).
[d] From Smith et al. (1984).
[e] From Mellon and Helgeson (1982).
[f] From Leach et al. (1982).
[g] From M. T. Esquerré-Tugayé (personal communication).
[h] From Cassab et al. (1985).
[i] nd, Not determined.

```
P1(H5).     Ser-Hyp-Hyp-Hyp-Hyp-Thr-Hyp-Val-Tyr-Lys

P1(H20).    Ser-Hyp-Hyp-Hyp-Hyp-Val-Lys-Pro-Tyr-His-Pro-Thr-Hyp-Val-Tyr-Lys

P2(H3).     Tyr-Lys

P2(H4).     Ser-Hyp-Hyp-Hyp-Hyp-Val-Tyr-Lys

P2(H11).    Ser-Hyp-Hyp-Hyp-Hyp-Val-1/2IDT-Lys-1/2IDT-Lys

P2(H5/6).   Ser-Hyp-Hyp-Hyp-Hyp-Ile-Tyr-Lys (minor component)

P2(H12).    Ser-Hyp-Hyp-Hyp-Hyp-Ile-1/2IDT-Lys-1/2IDT-Lys (minor component)
```

Fig. 3. Amino acid sequences of tryptic peptides from two tomato HRGPs (P1 and P2). The "H" designation for each tryptic peptide follows Smith's nomenclature. It should be noted here that P1 refers to two different HRGPs named P1a and P1b. P1a and P1b have remarkably similar amino acid compositions and tryptic peptide maps, and moreover, as shown here, the two major tryptic peptides from P1a (i.e., H5 and H20) have identical sequences to those from P1b. From Smith *et al.* (1986).

plant cell wall. Distributed throughout the remaining protein sequence are 25 Ser-(Pro)$_4$ pentapeptide repeats which are presumably all posttranslationally modified to the Ser-(Hyp)$_4$ repeats characteristic of these proteins. In addition to serine and (hydroxy)proline, this carrot HRGP sequence is especially rich in tyrosine, lysine, and histidine. A number of other less frequently repeated amino acid sequences are also present in this sequence, including Tyr-Lys-Tyr-Lys (11 times) and Thr-Pro-Val (8 times). A comparison of amino acid composition and types of repeating sequences present in this carrot sequence and in the tomato sequences presented above (Figs. 1 and 3) shows a high degree of homology, especially between the carrot sequence and the tomato P1a and P1b sequences. The carrot HRGP cDNA and genomic clones have also been used as probes in Southern blot hybridizations with various plant DNAs (A. M. Showalter, unpublished observations). In these experiments, multiple bands of hybridization appear in each plant DNA lane, consistent with the hypothesis that the cell wall HRGPs are members of a multigene family in several plant species.

Showalter *et al.* (1985) and Showalter and Varner (1987) have also isolated and characterized tomato cDNA and genomic clones for HRGPs using the carrot clones as probes, and these sequences are presented in Fig. 5. The gene sequence contains numerous Ser-(Pro)$_4$-Val-His and Ser-(Pro)$_4$-Val-Ala repeats. These repeats, although they have not yet been observed on a protein level, have also been observed to be encoded by an *Arabidopsis* HRGP gene clone (L. Herrera-Estrella, personal communication). The tomato cDNA sequence, however, shows perfect agreement (excluding the posttranslational modifications) with a known protein sequence, a tomato HRGP glycopeptide sequence originally characterized by Lamport. Both the cDNA and peptide sequences specify Ser-(Pro)$_4$-Ser-Pro-Ser-(Pro)$_4$-(Tyr)$_3$-Lys hexadecapeptide repeat units (compare Fig. 1A with Fig. 5B).

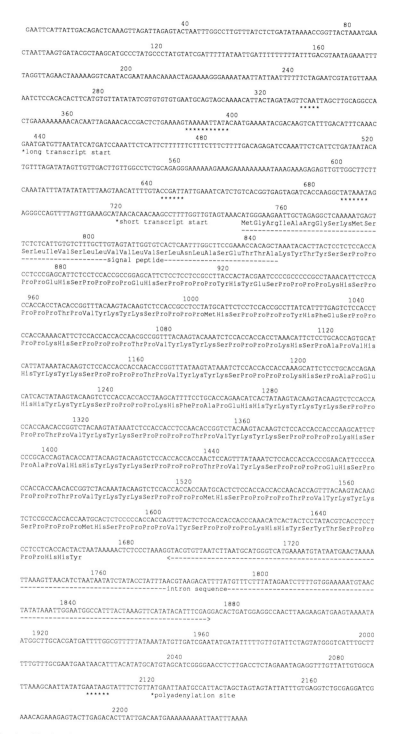

Fig. 4. Nucleotide and amino acid sequence of the carrot HRGP genomic clone pDC5A1. From Chen and Varner (1985b).

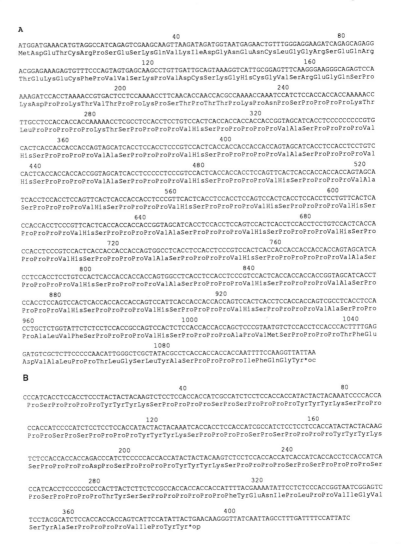

Fig. 5. Nucleotide and amino acid sequence of the tomato HRGP genomic clone Tom 5 (A) and of the tomato HRGP cDNA clone pTom 17-1 (B). *oc and *op represent ochre and opal termination codons, respectively. From Showalter et al. (1985) and Showalter and Varner (1987).

Hironaka et al. (1985, personal communication) have also used carrot HRGP sequences to isolate petunia cDNA clones coding for cell wall HRGPs. These petunia clones encode Ser-(Pro)$_4$-Ser-Pro-Ser-(Pro)$_4$-(Tyr)$_3$-Lys and Ser-(Pro)$_4$-Thr-Pro-Val-Tyr-Lys repeats; such sequences have been elucidated previously in acid/protease-generated tomato cell wall glycopeptides and tomato P1a and P1b sequences. Corbin et al. (1987), using the tomato HRGP gene sequence, have isolated three HRGP cDNA clones from bean cell cultures and observed the presence of numerous Ser-(Pro)$_4$-Ser-Pro-Ser-(Pro)$_4$-(Tyr)$_3$-Lys repeat units in two of the cDNAs and numerous Ser-(Pro)$_4$-Lys-His-Ser-(Pro)$_4$-(Tyr)$_3$-His repeats in the third cDNA.

All of the HRGP cDNA and genomic clones studied to date demonstrate a codon usage bias, particularly with respect to proline, serine, and tyrosine. These preferences are generally the same in all of the clones, being CCA for proline, UCU or UCA for serine, and UAC for tyrosine. The other amino acids have codon usage biases which vary depending upon which clone is inspected.

A comparison of all known HRGP repeated amino acid sequences to date from either protein or DNA sequence determinations is shown in Table VI. This table clearly shows the conservation of the Ser-(Hyp/Pro)$_4$ repeats but also reveals substantial variability in the flanking amino acid sequences. These flanking sequences are presumably of great importance in refining and elaborating the biophysical and structural information of the Ser-(Pro)$_4$ repeats, and consequently in functionally distinguishing a plant's set of cell wall HRGPs.

Carbohydrate accounts for approximately 40–60% of the weight of all cell wall HRGPs studied to date (see Table IV). Lamport (1967, 1969) has demonstrated that alkali-stable O-glycosidic linkages between arabinose and hydroxyproline constitute the basis for most of the glycosylation of these HRGPs. Specifically, one to four arabinose residues are attached to most but not all hydroxyproline residues in the form of short arabinoside chains. The glycosidic linkages which occur in these chains are very specific (Fig. 6). The hydroxyproline O-glycosidic linkage is widely distributed in the plant kingdom (Table VII). In dicot HRGPs, hydroxyproline residues are most frequently modified with tri- and tetraarabinosides; in general, the hydroxyproline arabinoside compositions are species-specific and show a lesser degree of substituted hydroxyproline in less advanced plant species (Table VII). Lamport *et al.* (1973) have also found single galactose units O-glycosidically attached to some but not all serine residues in several HRGP glycopeptides (Fig. 6). Consistent with the above findings, others have found that arabinose and galactose are the major carbohydrate constituents of various soluble HRGPs and account for approximately 90 and 5%, respectively, of the total carbohydrate moieties (see Table IV). Other carbohydrates such as glucose and *N*-acetylglucosamine may be attached to cell wall HRGPs, although such linkages have never been found. Alternatively, these carbohydrates may represent contamination from other cell wall components.

Under the electron microscope, soluble carrot and tomato cell wall HRGPs appear as thin, kinked rods of approximately 80 nm in length (Stafstrom and Staehelin, 1986; Heckman *et al.*, 1988) (Fig. 7). Consistent with these microscopic observations, circular dichroism analysis indicates that cell wall HRGPs exist in a polyproline II helix (i.e., a left-handed helix with

TABLE VI

Amino Acid Sequence Repeats in Cell Wall HRGPs

Sequence repeat[a]	Derived from	Plant	Reference
SOOOOSOSOOOYYYK	Acid/protease-treated cell walls	Tomato cell cultures	Lamport (1977), Epstein and Lamport (1984)
SPPPPSPSPPPPYYYK	cDNA clone	Tomato stem (unwounded)	Showalter and Varner (1987)
SPPPPSPSPPPPYYYK	cDNA clone	Petunia callus	C. M. Hironaka (unpublished observations)
SPPPPSPSPPPPYYYK	cDNA clone	Bean cell cultures	Corbin et al. (1987)
SOOOOK	Acid/protease-treated cell walls	Tomato cell cultures	Lamport (1977)
SOOOOK	Acid/protease-treated cell walls	Melon seedlings	Esquerré-Tugayé and Lamport (1979)
SPPPK	Genomic clone (pDC5A1)	Carrot	Chen and Varner (1985b)
SOOOOTOVYK	Acid/protease-treated cell walls	Tomato cell cultures	Lamport (1977)
SOOOOTOVYK	Soluble HRGP tryptic peptide	Tomato cell cultures	Smith et al. (1986)
SPPPPTPVYK	Genomic clone (pDC5A1)	Carrot	Chen and Varner (1985b)
SPPPPTPVYK	cDNA clone	Petunia callus	C. M. Hironaka (unpublished observations)
SOOOOVYKYK	Acid/protease-treated cell walls	Tomato cell cultures	Lamport (1977), Epstein and Lamport (1984)
SOOOOVYKYK	Soluble HRGP tryptic peptide	Tomato cell cultures	Smith et al. (1986)
SOOOOSOK	Acid/protease-treated cell walls	Tomato cell cultures	Lamport (1977)
SOOOOVKPYHPTOVYK	Soluble HRGP tryptic peptide	Tomato cell cultures	Smith et al. (1986)
SPPPPVA	Genomic clone (Tom 5)	Tomato	Showalter and Varner (1987)
SPPPPVA	Genomic clone	Arabidopsis	L. Herrera-Estrella (unpublished observations)
SPPPPVH	Genomic clone (Tom 5)	Tomato	Showalter and Varner (1987)
SPPPPVH	Genomic clone	Arabidopsis	L. Herrera-Estrella (unpublished observations)
SPPPPKHSPPPPYYYH	cDNA clone	Bean cell cultures	Corbin et al. (1987)

[a] S, Serine; O, hydroxyproline; P, proline; Y, tyrosine; K, lysine; T, threonine; V, valine; H, histidine; and A, alanine. Note that hydroxyproline residues cannot be distinguished from proline residues based on DNA sequence analysis, and therefore the DNA-derived amino acid sequences are recorded here using proline residues.

L-Ara$_f$ $\xrightarrow{1\ \alpha\ 3}$ L-Ara$_f$ $\xrightarrow{1\ \beta\ 2}$ L-Ara$_f$ $\xrightarrow{1\ \beta\ 2}$ L-Ara$_f$ $\xrightarrow{1\ \beta\ 4}$ Hyp

L-Ara$_f$ $\xrightarrow{1\ \beta\ 2}$ L-Ara$_f$ $\xrightarrow{1\ \beta\ 2}$ L-Ara$_f$ $\xrightarrow{1\ \beta\ 4}$ Hyp

L-Ara$_f$ $\xrightarrow{1\ \beta\ 2}$ L-Ara$_f$ $\xrightarrow{1\ \beta\ 4}$ Hyp

L-Ara$_f$ $\xrightarrow{1\ \beta\ 4}$ Hyp

D-Gal$_p$ $\xrightarrow{1\ \alpha\ 3}$ Ser

Fig. 6. O-Glycosidic linkages found in cell wall HRGPs. From Akiyama et al. (1980) and Lamport et al. (1973).

three residues per turn and a pitch of 9.36 Å) (van Holst and Varner, 1984). Moreover, it appears that the carbohydrate moiety of this glycoprotein serves to stabilize this helical conformation, presumably by intramolecular hydrogen bonding, since deglycosylation with anhydrous hydrogen fluoride results in a distortion or unwinding of the helix.

TABLE VII

Hydroxyproline Arabinoside Composition of Various Plants[a]

Group	Species	Percentage hydroxyproline arabinosides				
		Hyp-Ara$_4$	Hyp-Ara$_3$	Hyp-Ara$_2$	Hyp-Ara$_1$	Hyp
Algae	Chlorella vulgaris	0	4	16	6	74
Liverworts	Sphaerocarpos	1	30	26	3	40
Mosses	Funaria hygrometrica	3	8	18	8	63
Ferns	Onoclea sensibilis (leaves)	21	32	6	8	33
Horsetails	Equisetum sp. (sporophyte)	19	19	5	5	52
Gymnosperms	Ginko biloba (culture)	33	44	6	4	13
	Cupressus sp. (culture)	26	34	8	6	26
	Ephedra sp. (culture)	27	37	4	6	26
Angiosperms (monocots)	Zea mays (pericarp)	4	13	2	15	66
	Avena sativa (coleoptile)	6	11	3	5	75
	Iris kaempferi (pericarp)	10	11	3	5	71
	Allium porrum (pericarp)	6	13	3	5	73
Angiosperms (dicots)	Acer pseudoplatanus (culture)	75	17	3	2	3
	Lycopersicon esculentum (culture)	52	28	4	6	10
	Convolvulus arvensis (culture)	63	22	6	4	5
	Vicia tetrasperma (culture)	52	31	5	4	8
	Pisum sativum (root)	33	41	7	6	13

[a] From Lamport and Miller (1971).

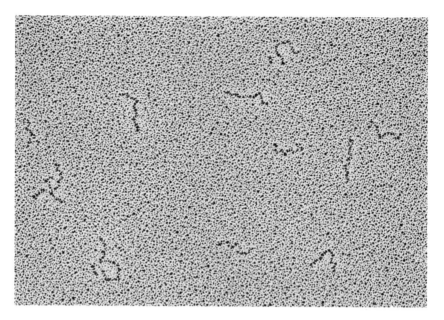

Fig. 7. Electron micrograph of carrot cell wall HRGP molecules. These molecules have an average length of 84 nm. Courtesy of J. P. Stafstrom and L. A. Staehelin; see also Stafstrom and Staehelin (1986).

C. Biosynthesis and Degradation

The general biosynthetic pathway for the cell wall HRGPs is well established [see Robinson *et al.* (1985) and Sadava and Chrispeels (1978) for reviews], although the subcellular localization of some of the reactions is still controversial. Following transcription, cell wall HRGP mRNA is translated in the cytoplasm into a proline-rich polypeptide. The existence of a putative signal peptide, identified at both a protein (Smith, 1978) and a gene level (Chen and Varner, 1985b) makes it likely that translation occurs on membrane-bound ribosomes. The proline-rich precursor is then posttranslationally modified by the action of prolyl hydroxylase to yield an hydroxyproline-rich polypeptide (Sadava and Chrispeels, 1978). Several reports have localized this enzyme in various membrane-bound fractions (Bolwell *et al.,* 1985a; Tanaka *et al.,* 1980; Wienecke *et al.,* 1982), while one report has indicated a cytoplasmic localization (Sadava and Chrispeels, 1978). These apparently conflicting data may reflect cross-contamination of subcellular fractions due to weak associations with membranes and/or heterogeneity of the enzyme. Prolyl hydroxylase requires molecular oxygen, Fe^{2+}, ascorbate, and α-ketoglutarate like the animal counterpart; unlike animal prolyl hydroxylase, however, the plant enzyme can apparently utilize polyproline as a synthetic substrate. This observation has led to the idea that the plant enzyme requires a polyproline II helix for activity (Tanaka et al., 1981). Not all of the prolines in these glycoproteins are hydroxylated (see Fig. 3), and the question of substrate specificity remains both intriguing and unanswered. One possibility is that the enzyme can read a primary sequence and sequentially hydroxylates only those prolines preceded by amino acids containing hydroxy groups; this idea is consistent with all presently known cell wall HRGP amino acid sequence data (see Table VI) but clearly requires further verification.

After hydroxylation is complete, another important posttranslation modification occurs: the attachment of arabinose residues to hydroxyproline. This modification occurs in the Golgi and presumably involves at least three different arabinosyl transferase activities responsible for the specific linkages of arabinose residues to hydroxyproline and arabinosylated hydroxyproline (see Fig. 6) (Karr, 1972; Owens and Northcote, 1981; Sadava and Chrispeels, 1978). Little is known about the galactosyl transferase responsible for attaching single galactose residues to some but not all serine residues. As in the case of proline hydroxylation, the substrate specificity is unknown for these glycosyl transferases; it will be interesting to learn how particular hydroxyproline and serine residues in the glycoprotein are selected for glycosylation and what determines the extent of glycosylation in the case of the former.

Finally, the glycoprotein is secreted into the cell wall. Neither continued protein synthesis nor posttranslational modifications (hydroxylation and subsequent arabinosylation) are required for this to occur since neither cycloheximide treatment of pulse-labeled HRGP nor addition of α,α'-dipyridyl, a ferrous ion chelator and prolyl hydroxylase inhibitor, blocks secretion (Sadava and Chrispeels, 1978; Smith, 1981). This secretory process, however, does require energy since metabolic uncouplers of oxidative phosphorylation and inhibitors of electron transport block cell wall HRGP secretion (Sadava and Chrispeels, 1978). Soon after its secretion into the wall, the glycoprotein becomes insoluble. This insolubilization is correlated with the posttranslational formation of isodityrosine cross-links (Cooper and Varner, 1983; Fry, 1982) (see Fig. 2). If such cross-links were intermolecular, they could account for the insolubilization; however, only intramolecular isodityrosine cross-links have been found to date (Epstein and Lamport, 1984). A wall-bound peroxidase/ascorbate system is thought to control the formation of isodityrosine and thereby bring about insolubilization, since insolubilization can be inhibited by various peroxidase inhibitors and free radical scavengers (Cooper and Varner, 1983). Very recently, Everdeen et al. (1988) have reportedly isolated an "extensin peroxidase" capable of cross-linking HRGP monomers in vitro; the nature of the cross-links, however, remains to be determined.

The rate at which the cell wall HRGPs turn over in the wall is unknown, and, in general, there is little information on the degradation or turn over of any of the three classes of HRGPs in plants. It is thought that one or more HRGPs must turn over because there is an effective salvaging pathway for converting free hydroxyproline to proline (Varner, 1980); such a pathway is of evolutionary interest since it has not been reported to occur in animals. The cell wall HRGPs, however, appear to be especially stable since they are so difficult to solubilize and digest proteolytically.

D. Function

The biochemical and physical properties of the cell wall HRGPs afford clues as to their functions. The abundance of these proteins in the cell walls of higher plants, particular dicots, along with their kinked rodlike structure, is compatible with a structural role in the plant cell wall. Moreover, the digestion of cell wall polysaccharide components with anhydrous hydrogen fluoride leaves behind an architectural framework composed largely of HRGPs, also supporting a structural role for these cell wall proteins (Mort and Lamport, 1977). Associated with this structural role, these glycoproteins probably play active roles in cell growth and development, since HRGP levels vary depending upon tissue type and developmental stage (Cassab et al., 1985; Steward et al., 1954) and since the secretion of underhydroxylated HRGP results in the abnormal regeneration of cell walls in tobacco protoplasts (Cooper et al., 1986). High levels of cell wall HRGPs have also been correlated with the cessation of cell elongation (i.e., extension) (Cleland and Karlsnes, 1967; Monro et al., 1974; Sadava et al., 1973). Lamport (1963) coined the word "extensin" to refer to this protein and its presumed, but as yet unconfirmed, role in cellular extension growth. These glycoproteins are not known to interact covalently with any other cell wall components except themselves but probably do interact ionically with other wall components. In this regard, the negatively charged uronic acid residues of pectin are likely sites for interaction with the positively charged cell wall HRGPs. Others have elaborated on this idea by proposing models in which these glycoproteins are organized around cellulose microfibrils to greater (Lamport and Epstein, 1983) or lesser (Cooper et al., 1984) extents. Evidence for (or against) such models, however, is lacking and difficult to obtain.

The hydroxyproline content and extracellular structural role of cell wall HRGPs have inspired some comparison to animal collagen. Aside from these two factors, the two protein classes are distinct from one another and probably have little if any evolutionary relationship; the proteins differ in the amounts of hydroxyproline, the degree and type of glycosylation, the primary sequence repeating units, and the structural organization of their genes. Apparently, plants and animals have found two different solutions to the problem of the production of a structural (i.e., rodlike) extracellular matrix protein.

The observations that these glycoproteins accumulate in the cell wall in response to wounding and infection (see Section II,E) support a role in plant defense. These proteins may function in defense simply by forming a more dense, impenetrable cell wall barrier or by providing nucleation sites for the deposition of lignin (Whitmore, 1978), again resulting in a more protective wall barrier to potential pathogens. In addition, purified HRGPs possess the ability to agglutinate bacteria (Leach *et al.*, 1982; Mellon and Helgeson, 1982; van Holst and Varner, 1984); this ability is evidently derived from their positively charged nature (polylysine also can agglutinate bacteria). Consequently, these glycoproteins may also function in defense by immobilizing pathogens in the wall.

E. Response to Stress

A number of different conditions affect the expression of cell wall HRGPs, including mechanical wounding, ethylene treatment, elicitor treatment, infection, heat treatment, and red light illumination. There is an accumulation of cell wall HRGPs in carrot disks (Chrispeels *et al.*, 1974; Stuart and Varner, 1980) and in bean hypocotyls (Klis *et al.*, 1983) following slicing (wounding). HRGP mRNA levels also increase following wounding of carrot disks (Chen and Varner, 1985a,b). Similarly, wounded tomato stems accumulate cell wall HRGPs and HRGP mRNA (Fig. 8), although the pattern of RNA accumulation here is different from that of the carrot (Showalter and Varner, 1987). Furthermore, cells in callus or tissue culture show much higher levels of cell wall hydroxyproline than their uncultured counterparts (see Tables II and III), indicating that these tissues behave much like wounded tissue. Ethylene, a plant hormone whose level is known to increase following wounding, stimulates and may mediate wound-induced accumulation of cell wall HRGPs (Esquerré-Tugayé *et al.*, 1979; Ridge and Osborne, 1970). Recent studies also indicate that ethylene-treated carrot roots accumulate HRGP mRNA (Ecker and Davis, 1987; Tierney *et al.*, 1988).

Several groups have observed that HRGPs accumulate in response to infection and that this accumulation is correlated with the expression of disease resistance. Specifically, Esquerré-Tugayé and Lamport (1979) and Esquerré-Tugayé *et al.* (1985) have found an approximately 10-fold increase in cell wall HRGP in melon plants in response to infection with the anthracnose fungus *Colletotrichum lagenarium* (Table VIII). This same group has also found that artificial enhancement or suppression of HRGP levels in melon plants results respectively in increased or decreased resistance to *Colletotrichum lagenarium* (Esquerré-Tugayé *et al.*, 1979). Hammerschmidt *et al.* (1984) have observed that cell wall hydroxyproline levels, indicative of HRGP levels, increase more rapidly in resistant cultivars than in susceptible

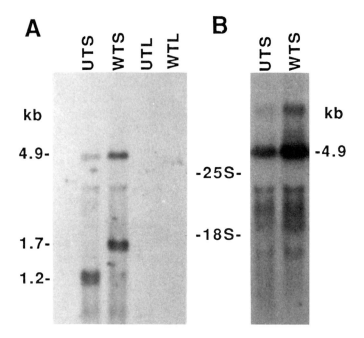

Fig. 8. RNA blot of unwounded and wounded tomato stem and leaf RNA probed with the tomato HRGP gene clone Tom 5 (A) and the tomato HRGP cDNA clone pTom 17-1 (B). Lanes UTS, WTS, UTL, and WTL contain 10 μg of total RNA isolated from unwounded tomato stems, wounded tomato stems, unwounded tomato leaves, and wounded tomato leaves, respectively. The migration positions of tomato 25S and 18S rRNA and the sizes of the major hybridizing transcripts are indicated. The RNA blot hybridization patterns seen here with the genomic and cDNA HRGP probes are similar (but not identical), as are their respective DNA sequences (see Fig. 5). Presumably, the repeated Ser-(Pro)$_4$ encoding sequences are largely responsible for the similarity observed in these two blot hybridizations, while the differences may be attributable to the repeated Val-His and Val-Ala encoding sequences characteristic of Tom 5 and the repeated (Tyr)$_3$-Lys encoding sequences characteristic of pTom 17-1. From Showalter and Varner (1987).

cultivars of cucumber plants infected with the fungus *Cladosporium cucumerinum*. Treatment of melon and soybean hypocotyls with components of fungal cell walls (i.e., elicitors) or plant cell walls (i.e., endogenous elicitors) also results in the stimulation of HRGP synthesis (Roby *et al.*, 1985). Similarly, elicitor treatment of bean cell cultures results in the accumulation of cell wall hydroxyproline (Bolwell *et al.*, 1985b) and in the accumulation of a cell wall HRGP which apparently is analogous to the solanaceous lectins to be discussed later in the chapter (Bolwell, 1987). HRGP mRNA has also been shown to accumulate in elicitor-treated bean cell suspension cultures and in race : cultivar-specific interactions between bean hypocotyls and the partially biotrophic fungus *Colletotrichum lindemuthianum* (Showalter *et al.*, 1985). Moreover, HRGP mRNA accumulated earlier in an incompatible

TABLE VIII

Time Course of Hydroxyproline Accumulation in Melon Seedlings Inoculated with *Colletotrichum lagenarium*[a]

Time (in days) after inoculation	Hydroxyproline content (μg/stem)	
	Inoculated	Control
1	3.0	3.0
2	2.0	4.5
3	5.5	4.0
4	12	4.5
5	33	5.5
6	64	6.0
7	57	5.0

[a] Adapted from Esquerré-Tugayé *et al.* (1985).

interaction (host resistant) than in a compatible interaction (host susceptible), and this accumulation was correlated with the expression of hypersensitive resistance in the case of the former. Recent work also indicates that an endogenous elicitor from carrot roots stimulates carrot cell cultures to accumulate HRGP mRNA (Tierney *et al.*, 1988).

Precisely how HRGP accumulation is brought about by wounding and infection is unknown, although some progress is being made toward answering this question. Not only does HRGP mRNA accumulation occur, but this accumulation, as shown by nuclear run-off experiments, is controlled at a transcriptional level at least in wounded bean hypocotyls and elicitor-treated bean cell cultures (Lawton and Lamb, 1987). The precise signal transduction pathways for wound- and infection-regulated HRGP accumulation remain to be elucidated despite the indication that ethylene and elicitors respectively may be involved at relatively early stages.

Two groups have suggested that cell wall HRGPs are heat-shock proteins. Stermer and Hammerschmidt (1985) have shown that cell wall hydroxyproline levels in cucumber plants increase in response to heat treatment and that such heat-shocked plants are more resistant to fungal infection. J. Zimmerman and co-workers (personal communication) have provided evidence for the accumulation of HRGP mRNA in heat-shocked carrot cell cultures.

Finally, there is some indication that cell wall HRGP levels are regulated by phytochrome, since red-light treatment of etiolated pea epicotyls increases the level of wall-bound hydroxyproline and this effect can be reversed by far-red light (Table IX). Such regulation may represent a level of developmental control for the synthesis of HRGPs.

TABLE IX

Effect of Light on *in Situ* Elongation and Hydroxyproline Content[a]

Time (hr)	Treatment	Final length (cm)	Hydroxyproline content (μg/g fresh weight)
1	Dark		54
	Red		58
	Far red		53
	Red + far red		54
3	Dark	1.16	54
	Red	1.08	61
	Far red	1.15	52
	Red + far red	1.14	48
6	Dark	1.32	64
	Red	1.14	88
	Far red	1.29	68
	Red + far red	1.28	72
12	Dark	1.39	72
	Red	1.11	86
	Far red	1.43	73
	Red + far red	1.34	79

[a] A 1-cm region was marked on the third internode of intact pea plants. Light treatments (5 min) were given and the plants returned to the darkness for the indicated period. The region between the marks was measured with a vernier caliper and then excised for hydroxyproline determinations. Each datum is the average of at least two separate analyses; there were at least 15 plants in each treatment group in each analysis. From Pike *et al.* (1979).

F. Hydroxyproline-Rich Glycoproteins in Nondicotyledonous Plant Species

Thus far we have only considered dicotyledonous plant cell wall HRGPs in our discussion since these are the best characterized in the plant kingdom. Other plant species, however, do contain cell wall HRGPs, although considerable variability exists with respect to their hydroxyproline richness and abundance in the cell wall. For example, certain green algae contain a wall matrix in which hydroxyproline may be the only constituent while maize and several other monocots contain HRGPs which are apparently only minor cell wall components.

In *Chlamydomonas,* HRGPs are the major constituents of cell wall structure and consist of a set of proteins which are characteristically distributed in this organism's numerous cell wall layers (Goodenough *et al.,* 1986). The amino acid composition of some of these cell wall HRGPs is shown in Table X. As a group, these HRGPs contain arabinose and galactose as major

TABLE X

Amino Acid Compositions[a] of Various *Chlamydomonas* HRGPs

Amino acid	GP1[b]	GP2[b]	GP3[b]	Plus[c] agglutinin	Minus[c] agglutinin
Hyp	32.3	14.7	5.7	12.3	12.0
Asx	4.7	8.1	10.6	9.5	10.4
Thr	4.2	6.6	7.4	5.9	6.7
Ser	15.8	8.3	9.3	10.3	11.3
Glx	3.2	7.3	7.4	8.8	8.3
Pro	2.7	7.7	4.5	4.3	5.6
Gly	6.6	8.5	11.0	8.8	8.8
Ala	10.1	8.5	10.2	7.7	7.8
Val	4.4	5.7	5.9	4.6	6.2
Cys	0.9	1.6	4.7	5.1	nd[d]
Met	0.8	1.0	0.8	1.0	0.8
Ile	3.3	1.3	3.2	2.7	3.5
Leu	2.9	7.0	7.0	5.1	3.6
Tyr	0.9	3.6	1.2	1.5	1.1
Phe	1.3	3.8	3.6	2.8	3.1
His	0.8	0.2	1.0	3.1	2.0
Lys	3.6	3.0	4.5	2.2	4.1
Arg	1.7	3.0	2.1	3.6	2.2
Trp	nd	nd	nd	nd	nd

[a] Expressed per 100 residues.
[b] Adapted from Goodenough *et al.* (1986).
[c] Adapted from Adair (1985).
[d] nd, Not determined.

carbohydrate constituents; these sugars are glycosidically attached to hydroxyproline residues in the form of short heterooligosaccharides which vary with respect to the number and sequence of arabinose and galactose residues present (see Roberts *et al.*, 1985). These wall proteins possess the remarkable ability of *in vitro* self-assembly (Catt *et al.*, 1978; Goodenough *et al.*, 1986; Adair *et al.*, 1987) and thus offer an excellent opportunity to explore the molecular interactions between cell wall HRGPs and their contribution to cell wall morphology since, unlike higher plants, *Chlamydomonas* lacks cellulose and other matrix polysaccharides. *Chlamydomonas* also possesses a fascinating class of HRGPs which function as sexual agglutinins (Adair, 1985); these HRGPs bring together gametes of opposing mating types (see Table X for the amino acid compositions of the sexual agglutinins isolated from *plus* and *minus* mating types). Both amino acid composition and electron microscopic appearance of the various *Chlamydomonas* HRGPs support the notion that the sexual agglutinins arose through evolutionary adaptation of preexisting cell wall HRGPs (Goodenough, 1985). An analogous evolutionary process may have occurred in the dicot HRGPs, allowing them to function in defense as pathogen agglutinins. *Volvox,* a multicellular green algae, also contains an extremely hydroxyproline-rich extracellular matrix, and one study has reported a "polyhydroxyproline" component with fewer than 5% nonhydroxyproline residues as a major constituent of this matrix (Mitchell, 1980).

In monocots, evidence exists for cell wall HRGPs or analogs thereof. In corn, iris, and yucca seed coats, hydroxyproline is relatively abundant in cell walls in comparison to other monocots which have been surveyed (VanEtten et al., 1961, 1963). Hydroxyproline arabinoside compositions of HRGPs from numerous monocots show that most of the hydroxyproline is not arabinosylated, unlike that of the dicot HRGPs (Table VII). Preliminary results show that carrot HRGP gene sequences cross-hybridize to corn (D. Shah, personal communication), wheat, oat (S. Dhawale, personal communication), and rice (H. Kende, personal communication) DNA or RNA, indicating some regions of conserved sequence and perhaps a common evolutionary origin. The contribution of these HRGPs to cell wall structure and plant defense is largely unknown, although one report has appeared supporting the idea that wheat cell wall HRGPs are not involved in resistance to *Erysiphe graminis* (Clarke et al., 1983).

Recently, Kieliszewski and Lamport (1987) and Hood et al. (1988) have isolated and characterized cell wall HRGPs from corn cell cultures and pericarp tissue respectively. Both groups found these corn HRGPs to be rich in hydroxyproline (22–25 mole %), threonine (18–25 mole %), proline (14–15 mole %), and lysine (11–14 mole %).

III. ARABINOGALACTAN PROTEINS

A. Cellular and Tissue Distribution

Arabinogalactan proteins (AGPs) are found in almost all tissues of higher plants; they occur in leaves, stems, roots, root nodules, cell cultures, calli, floral parts, seeds, and in the trunks of some angiosperms and gymnosperms (Fincher et al., 1983; Jermyn and Yeow, 1975). Conclusive cellular localization of the AGPs is difficult because of their extreme solubility. It is clear, however, that AGPs are most often found as constituents of the extracellular milieu. Cells in suspension culture from various tissue sources secrete AGPs into the culture medium; also, certain specialized cells such as the stylar canal cells and some secretory cells which produce gummy exudates secrete especially large quantities of AGPs into the cell wall compartment (Clarke et al., 1979). AGPs are also reportedly associated with the plasma membrane, and their putative precursors are found intracellularly (van Holst et al., 1981). Yariv's antigen (Fig. 9) specifically precipitates most AGPs as a red–orange complex and has greatly aided the cellular and tissue localization studies, even though the interaction between Yariv's antigen and AGPs is not well understood.

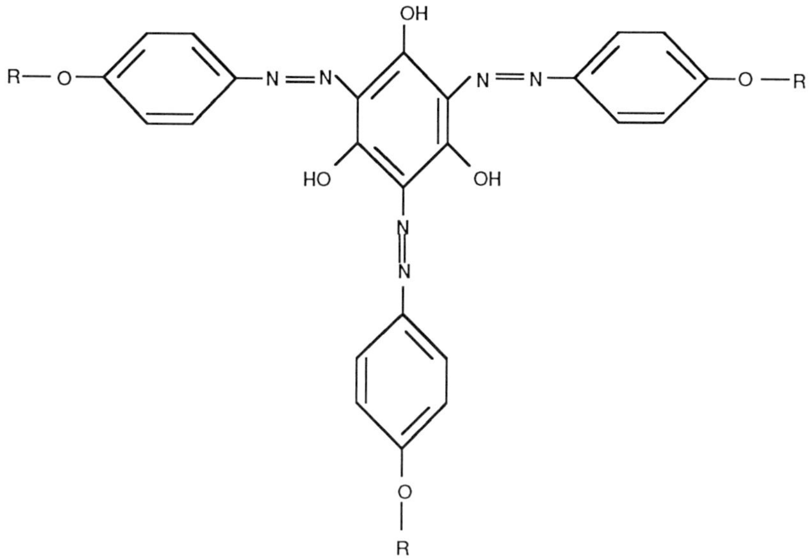

Fig. 9. Structure of Yariv's antigen, a β-glucosyl artificial carbohydrate antigen (R, glucosyl residue). From Yariv et al. (1962).

B. Structure

AGPS are proteoglycans in which protein typically accounts for only 2–10% of the weight; their molecular weights are extremely heterogeneous, presumably reflecting different extents of glycosylation. AGPs are readily solubilized during tissue extraction with low-ionic-strength aqueous solutions. In addition, most AGPs are soluble in saturated ammonium sulfate. These solubility properties have greatly facilitated their isolation and characterization; some of the biochemical properties for several AGPs are summarized in Table XI. The protein moiety of AGPs is typically rich in hydroxyproline, serine, alanine, threonine, and glycine (Table XII) and is resistant to proteolysis in its native state, presumably a property conferred by extensive glycosylation. AGPs have isoelectric points in the range of pH 2–5. The N-terminal sequences of four different AGPs, three from carrot and one from ryegrass, have recently been determined (Fig. 10); all four sequences contain Ala-Hyp repeats. Pending further investigation, such repeats may become a diagnostic characteristic of AGPs, since they do not occur in any of the known cell wall HRGPs and seem unlikely to occur in the Solanaceae lectins.

Carbohydrates account for most of the weight of AGPs, and, as their name implies, AGPs contain D-galactose and L-arabinose as the major carbohydrate constituents. However, D-mannose, D-xylose, D-glucose, D-glucuronate (and its 4-O-methyl derivative), D-galacturonate (and its 4-O-methyl

TABLE XI
Biochemical Properties of Various AGPs

Biochemical properties	Lolium[a]	Phaseolus[b]	Radish[c]	Rape[c]	Acacia[d] (gum arabic)
Tissue source	Tissue culture	Hypocotyls	Leaves	Leaves	nd[e]
Estimated mass of glycoprotein (kDa)	220–280	140	75	50	
Isoelectric point	nd	2.3	nd	nd	nd
Protein (%)	7	10	2.5	9.7	2
Abundant amino acids	Hyp,Ala,Ser	Hyp,Ala,Ser	Hyp,Ala,Ser	Hyp,Ala,Ser	Hyp,Ser
Carbohydrate (%)	93	90	97.5	90.3	98
Carbohydrates present (% of total CHO)	64% Gal 36% Ara	61% Gal 22% Ara 14% UA[e]	59.6% Gal 28.7% Ara	52.6% Gal 38.3% Ara	37% Gal 30% Ara 20% UA[e] 13% Rha
Glycopeptide linkages Hyp-Gal?	nd	Hyp-Ara Hyp-Glc?	nd	nd	nd

[a] From Anderson et al. (1977).
[b] From van Holst and Klis (1981) and van Holst et al. (1981).
[c] From Nakamura et al. (1984).
[d] From Akiyama et al. (1984).
[e] nd, Not determined; UA, uronic acid.

TABLE XII

Amino Acid Compositions[a] of Various AGPs

Amino acid	Lolium[b]	Phaseolus[c]	Radish[d]	Rape[d]	Acacia[e]
Hyp	14.8	26.1	22.1	25.7	30.3
Asx	5.5	3.9	2.9	3.5	4.5
Thr	6.5	7.2	11.0	9.1	7.9
Ser	9.8	18.3	14.4	14.6	15.1
Glx	4.9	4.6	9.4	8.3	3.6
Pro	5.6	2.6	4.8	4.8	9.1
Gly	6.7	6.5	5.5	5.4	5.1
Ala	22.5	16.3	17.7	13.4	2.1
Val	5.1	3.3	5.7	4.0	3.3
Cys	1.0	0.7	nd[f]	nd	nd
Met	1.6	0.7	1.6	6.7	0.3
Ile	1.3	1.3	1.2	0.6	1.5
Leu	5.0	3.3	2.4	2.3	7.0
Tyr	1.2	0.7	nd	nd	0.9
Phe	1.5	0.7	nd	nd	2.7
His	0.6	0.7	nd	nd	4.8
Lys	3.3	2.6	1.3	1.6	3.0
Arg	2.2	0.7	nd	nd	0.6
Trp	nd	nd	nd	nd	nd

[a] Expressed per 100 residues.
[b] Adapted from Anderson et al. (1977).
[c] Adapted from van Holst et al. (1981).
[d] Adapted from Nakamura et al. (1984).
[e] Adapted from Akiyama et al. (1984).
[f] nd, Not determined.

derivative), and L-fucose can also be present. The precise arrangement of carbohydrates in these proteoglycans is largely unknown; however, some structures such as the ones shown in Figs. 11 and 13 have been deduced. It is also not clear how the carbohydrate moiety is attached to the protein back-

```
Carrot(A)  Ala-Asp-Ala-Hyp-Ala-Hyp-Ser-Hyp-Ala-Hyp/Ser-Hyp
               (Asn)                (Thr)           (?)

Carrot(B)  Asp-Glu-Ala-Hyp-Ala-Hyp-Ala-Hyp-Ser-Hyp-Met

Carrot(C)  Gly-Hyp-Ala-Hyp-Ala-Hyp-Ala-Hyp-Gln-Val
               (Glu)                    (?)  (?)


Lolium     Ala-Glu-Ala-Hyp-Ala-Hyp-Ala-Hyp-Ala-Ser
```

Fig. 10. N-terminal amino acid sequence of three carrot AGPs (from Jermyn and Guthrie, 1985) and one *Lolium* AGP (from Gleeson et al., 1985). Uncertainties in these sequences are as indicated with parentheses.

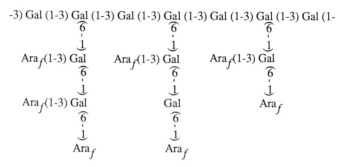

Fig. 11. Tentative structure for the arabinogalactan portion of the *Lolium* AGP. Ara_f represents arabinofuranose. From Anderson et al. (1977).

bone. In various AGPs, galactose has been found attached to serine, threonine, and hydroxyproline, and arabinose has been found linked to hydroxyproline. It is not known whether these linkages reflect attachments of mono-, oligo-, or polysaccharide substituents. It is important to note that carbohydrate moieties similar, if not identical, to those present in AGPs exist in the cell wall unattached to protein; these are the arabinogalactans (AGs) and are reviewed elsewhere (Clarke et al., 1979).

Electrophoretic separation of AGPs in the presence of Yariv's antigen

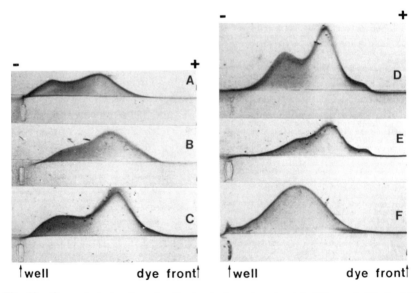

Fig. 12. Crossed electrophoresis of extracts from tomato style (A), petal (B), leaf (C), petiole (D), stem (E), and callus derived from stem (F) in the presence of Yariv's antigen. From van Holst and Clarke (1986).

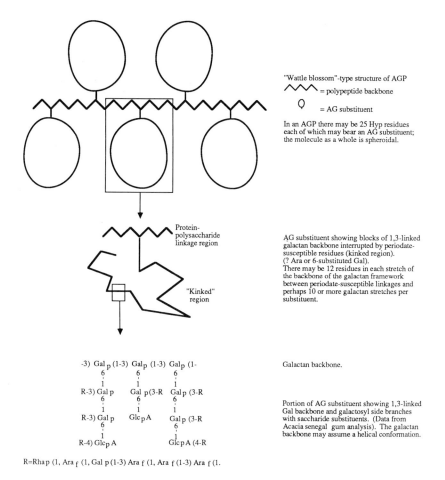

Fig. 13. Hypothetical structure of an AGP. From Fincher *et al.* (1983).

shows that AGPs are expressed in a tissue-specific manner and that a given plant tissue can contain more than one kind of AGP (Fig. 12). Whether these AGPs differ with respect to their protein moieties, carbohydrate moieties, or both is not yet known.

According to circular dichrometry, about 30% of the protein moiety of an AGP is in a polyproline II helix (van Holst and Fincher, 1984). Little is known about the conformation of the rest of the protein nor is much known about the conformations of the carbohydrate chains. One model, based on the above information and the observation that some AGPs have low viscosities, predicts that carbohydrate moieties with ovoid or spheroidal shapes are attached at several positions on a core protein, resulting in a "wattle blossom" structure (Fig. 13).

C. Biosynthesis and Degradation

Details of the biosynthesis and degradation of AGPs are scant (Fincher et al., 1983). Translation of the mRNA encoding the protein moiety most likely occurs on the rough endoplasmic reticulum. Hydroxylation of peptidyl prolines probably occurs in the endoplasmic reticulum, but this is not without debate; whether this prolyl hydroxylase is the same or different from the one(s) which hydroxylate the cell wall HRGPs is unknown. It will be interesting to test whether synthetic Ala-Pro oligomers (based on the recent amino acid sequence data of AGPs) can serve as substrates for the hydroxylase, particularly with respect to the hypothesis that an hydroxy amino acid must precede a proline residue to signal its hydroxylation in the cell wall HRGPs. Carbohydrates are apparently added to the protein moiety in the Golgi, although whether the carbohydrates are attached one at a time or in blocks is unknown. Whether underhydroxylated and underglycosylated AGPs are secreted, as in the case of cell wall HRGPs, is not certain, although indirect evidence indicates that this may be so (Pollard and Fincher, 1981).

D. Function

No function(s) for AGPs is unequivocally established. Based on their predominately extracellular location and their various biochemical and physical properties, AGPs have been proposed to act as glues, lubricants, and humectants (Fincher et al., 1983; Lamport and Catt, 1981). Their general abundance in the middle lamellae of the wall, and particularly in the styles of angiosperms and the medulla of root nodules (Cassab, 1986), makes them likely candidates for functioning in cell–cell recognition.

There is some indication that AGPs accumulate in response to wounding (Fincher et al., 1983) and pathogen infection (G. J. van Holst, personal communication). Consequently, a role for AGPs in plant defense is a possibility that requires further investigation.

E. Distribution of Arabinogalactan Proteins in the Plant Kingdom

AGPs have been found in flowering plants (both monocots and dicots) from every taxonomic group tested (Clarke et al., 1979; Fincher et al., 1983). In lower plants, AGPs have been reported in seven ferns, two mosses, and one liverwort (Clarke et al., 1978). Recently, AGPs have been found in the bryophyte *Plagiochila arctica* (D. Basile and J. Varner, personal communication). In addition, the colonial green alga *Eudorina californica* secretes a mucilage that has the general composition and properties of the AGPs (Tautvydas, 1978).

IV. SOLANACEOUS LECTINS

A. Cellular and Tissue Distribution

Although plant lectins were discovered nearly a century ago (Rüdiger, 1982), it was only in the 1970s that the solanaceous lectins were recognized as a unique group (Lord, 1985). The solanaceous lectins are related to one another by their immunological and hemagglutinating properties. They have been isolated from *Datura stramonium* seeds, potato tubers and fruit, and from tomato fruit. The *Datura* lectin, while most abundant in mature seeds, can also be detected immunologically in immature seeds, petals, stamens, anthers, and ovaries, and in trace levels in emerging cotyledons and roots (Kilpatrick *et al.*, 1979).

The cellular localization of the solanaceous lectins is unclear. In *Datura*, immunofluorescence data indicate that the lectin is associated primarily with the plasmalemma and with intracellular organelle membranes (Jeffree and Yeoman, 1981). Studying the cellular distribution of potato tuber lectin by fractionation experiments, Muray and Northcote (1978) found the lectin associated with root membranes, but they apparently did not examine the cell wall for the presence of the lectin. A recent subcellular fractionation study of potato tuber disks by Casalongué and Pont Lezica (1985) indicated that most of the lectin is tightly, but noncovalently, bound to the cell wall. Thus, these lectins may have both an intra- and extracellular distribution. It is also possible that these lectins are not a homogeneous group, and consequently their cellular distribution may depend upon the particular solanaceous lectin in question.

B. Structure

The solanaceous lectins are glycoproteins which can be distinguished from other plant lectins by their high hydroxyproline and arabinose content. Highly purified solanaceous lectins have been isolated (on tri-*N*-acetyl-chitotriose-Sepharose 6B or chitin affinity columns) and biochemically characterized from *Datura stramonium* seeds and potato tubers; some of their properties are shown in Table XIII. The *Datura* and potato lectins have monomeric sizes of approximately 30 and 50 kDa, respectively, and probably exist *in vivo* in equilibrium with their dimeric forms. These lectins contain about 50–60% protein and have similar amino acid compositions (Table XIV), especially with respect to the abundance of the four amino acids, hydroxyproline, serine, cysteine, and glycine. Exhaustive pronase digestion of the *Datura* and potato lectins following reductive alkylation release glycopeptides of about 18 and 33 kDa, respectively, which are rich in serine and hydroxyproline but not glycine and cysteine. Hence, there are at least two distinct domains in these lectins; one is rich in serine and hydroxyproline,

TABLE XIII

Biochemical Properties of *Datura* and Potato Lectins

Biochemical properties	*Datura*[a]	Potato[b]
Tissue source	Seeds	Tubers
Estimated mass of glycoprotein (kDa)	30	50
Isoelectric point	nd[c]	~9.5
Protein (%)	60	50
Abundant amino acids	Hyp,Ser Gly,Cys	Hyp,Ser Gly,Cys
Carbohydrate (%)	40	50
Carbohydrates present (% of total CHO)	84% Gal 16% Gal	95% Ara 5% Gal
Glycopeptide linkages	Hyp-Ara Ser-Gal	Hyp-Ara Ser-Gal

[a] From Desai *et al.* (1981).
[b] From Allen *et al.* (1978).
[c] nd, Not determined.

and the other(s) is rich in glycine and cysteine and is extensively cross-linked by disulfide bridges. Amino acid sequences from neither the *Datura* nor the potato lectins are yet known.

The carbohydrate moieties of these lectins contain only two sugars, L-arabinose (~90%) and D-galactose (~10%). These sugars are present exclusively in the serine/hydroxyproline-rich domain released by pronase treatment (Allen *et al.*, 1978; Desai *et al.*, 1981). The linkages of these sugars are identical to those found in the cell wall HRGPs (Fig. 6), and hydroxyproline tri- and tetraarabinosides are the most abundant hydroxyproline arabinosides present (Ashford *et al.*, 1982b). Because the amino acid and sugar compositions and the sugar linkages of the serine/hydroxyproline-rich domain of the lectins are remarkably similar to those of the cell wall protein, a close evolutionary relationship between these two glycoproteins probably exists. One intriguing hypothesis is that the serine/hydroxyproline-rich domain of these lectins is derived from a portion of the cell wall HRGP gene which fused with a gene encoding a cysteine/glycine-rich protein.

Circular dichroism studies indicate that the potato lectin, because of the presence of the serine/hydroxyproline-rich domain, is partially in a polyproline II helix (van Holst *et al.*, 1986). As with the carrot cell wall HRGP, the carbohydrate moiety serves to stabilize the polyproline II conformation, since deglycosylation of the serine/hydroxyproline-rich domain of the lectin destroys the helix.

TABLE XIV

Amino Acid Compositions[a] of Native *Datura* and Potato Lectins and Their Pronase-Resistant Glycopeptides

Amino acid	*Datura* lectin[b]		Potato lectin[c]	
	Native	+Pronase	Native	+Pronase
Hyp	13.5	49.6	20.4	66.2
Asx	6.7	3.5	4.9	1.6
Thr	5.6	0.8	5.7	0.8
Ser	12.9	19.0	12.7	19.2
Glx	6.7	6.0	7.0	1.2
Pro	6.2	5.4	7.0	0.7
Gly	12.9	6.0	12.3	2.9
Ala	3.9	1.2	4.1	3.6
Val	5.6	<0.4	0.4	<0.1
Cys	12.4	0.0	10.6	<0.1
Met	0.6	0.6	0.4	0.0
Ile	1.1	<0.4	1.6	<0.3
Leu	1.7	2.5	1.2	2.8
Tyr	2.2	2.5	3.3	<0.3
Phe	1.1	<0.4	0.2	<0.3
His	0.6	<0.4	0.0	nd[d]
Lys	1.7	<0.4	3.7	nd
Arg	2.8	<0.4	1.2	nd
Trp	1.7	<0.4	3.3	nd

[a] Expressed per 100 residues.
[b] Adapted from Desai *et al.* (1981).
[c] Adapted from Allen *et al.* (1978).
[d] nd, Not determined.

C. Biosynthesis and Degradation

Since the solanaceous lectins have a glycosylated serine/hydroxyproline-rich domain similar to the cell wall HRGP, it is reasonable to assume that the lectin is synthesized in a manner similar to that of the cell wall HRGP. Owens and Northcote (1981) have indeed shown that Golgi membranes from potato cell suspension cultures, which synthesize cell wall HRGP but not potato lectin, are capable of arabinosylating exogenously added deglycosylated potato lectin in the presence of UDParabinose; the product is indistinguishable from the native lectin by sodium dodecyl sulfate-polyacrylamide gel electrophoresis. Broekaert *et al.* (1986) have proposed a more radical biosynthetic scheme for the *Datura* lectin involving several lectin precursors which are posttranslationally processed in a series of steps involving proteolytic cleavages and ligations of these precursors. This complex and novel model, however, requires verification.

D. Function

The solanaceous lectins, like other plant lectins, bind sugars and agglutinate red blood cells. The agglutinating activity of the solanaceous lectins is specifically inhibited by oligomers of N-acetylglucosamine (Allen, 1983). Interestingly, the agglutinating activity of at least the potato lectin is lost following reduction but not after deglycosylation (Allen, 1983). Moreover, tyrosyl and tryptophyl groups are apparently involved with this agglutinating activity, since chemical modifications of these groups abolish it (Ashford *et al.*, 1981).

Although the physiological role(s) of the solanaceous lectins (and the other plant lectins as well) is unknown, it is thought that their ability to bind sugars is somehow involved with their function. Consequently, one of the most frequently suggested roles for the lectins involves various forms of cell–cell interaction. Other proposed roles include sugar transport, stabilization of seed storage proteins, control of cell division, and an antibody-like action.

E. Distribution of the Solanaceous Lectins in the Plant Kingdom

The solanaceous lectins are related serologically. Antibodies against *Datura stramonium* seed lectin cross-react with tomato fruit, with potato fruit and tuber lectins, and with *Nicandra* and *Capsicum* seed lectins (Kilpatrick *et al.*, 1980). Similarly, antibodies against potato lectin cross-react with tomato and *Datura* lectins (Ashford *et al.*, 1982a). It is unknown whether these other solanaceous lectins have similar amino acid and sugar compositions to the *Datura* and potato lectins. It is clear, however, that the solanaceous lectins have no immunological relationship to the lectins from nonsolanaceous plants. Why the occurrence of hydroxyproline-rich, arabinose-rich lectins is apparently confined to the Solanaceae remains a mystery.

V. SUMMARY AND INSIGHTS INTO FUTURE PLANT HYDROXYPROLINE-RICH GLYCOPROTEIN RESEARCH

The biochemical and physical properties of several cell wall HRGPs, AGPs, and solanaceous lectins are now known. These studies have defined and delineated the differences among these three classes of plant HRGPs. The most recent progress in this area has come with the application of molecular cloning to the cell wall HRGP field and has resulted in the elucidation of a complete carrot HRGP sequence. Similar molecular data are likely to follow soon in the AGP and solanaceous lectin fields, allowing for further refinement of the structural organization of these two HRGP classes. Unfortunately, we still lack information on how these glycoproteins interact with their environments (i.e., what other molecules does each HRGP interact with and what is the chemistry of such interactions?).

Despite the wealth of structural information on the plant HRGPs, relatively little is known about their functions. In many instances, correlations exist between certain physiological or stress conditions in plants and the levels of the HRGPs (particularly the cell wall HRGPs); however, these correlations (as well as the structural information) only provide clues to the functions. In order to assign definitive functions, more conclusive experimentation will have to be performed. Perhaps, one of the best (but not necessarily the easiest) approaches will be to isolate and characterize mutants of the various HRGP molecules and observe their phenotypes under normal and stress conditions. *Chlamydomonas* already has several cell wall mutants which could be used as a starting point for such experimentation. *Arabidopsis,* with its relatively small and simple genome, may prove to be a useful model system for generating and studying such mutants in higher plants. "Artificial" HRGP mutants could also be produced by chemically inhibiting proline hydroxylation (and subsequent glycosylation), and the resulting "mutants" studied as above. Granted identifying and characterizing HRGP mutants is risky and labor intensive; however, such an approach may be the most convincing way to assign unambiguous biological functions to these extraordinary glycoproteins.

ACKNOWLEDGMENTS

The preparation of this chapter was supported by a research grant from the Department of Energy (DE-FG 02-84ER13255) and by an unrestricted grant from the Monsanto Co. We are particularly grateful to our many colleagues who have provided us with information prior to publication and to Susan Worst for her critical review of this manuscript.

REFERENCES

Adair, W. S. (1985). *J. Cell Sci. Suppl.* **2,** 233–260.
Adair, W. S., Steinmetz, S. A., Mattson, D. M., Goodenough, U. W., and Heuser, J. E. (1987). *J. Cell Biol.* **105,** 2373–2382.
Akiyama, Y., Mori, M., and Kato, K. (1980). *Agric. Biol. Chem.* **44,** 2487–2489.
Akiyama, Y., Eda, S., and Kato, K. (1984). *Agric. Biol. Chem.* **48,** 235–237.
Allen, A. K. (1983). *In* "Chemical Taxonomy, Molecular Biology, and Function of Plant Lectins" (I. J. Goldstein and M. E. Etzler, eds.), pp. 71–85. Liss, New York.
Allen, A. K., Desai, A. N., Neuberger, A., and Creeth, J. M. (1978). *Biochem. J.* **171,** 665–674.
Anderson, R. L., Clarke, A. E., Jermyn, M. A., Knox, R. B., and Stone, B. A. (1977). *Aust. J. Plant Physiol.* **4,** 143–158.
Ashford, D., Menon, R., Allen, A. K., and Neuberger, A. (1981). *Biochem. J.* **199,** 399–408.
Ashford, D., Allen, A. K., and Neuberger, A. (1982a). *Biochem. J.* **201,** 641–645.

Ashford, D., Desai, N. N., Allen, A. K., Neuberger, A., O'Neill, M. A., and Selvendran, R. R. (1982b). *Biochem. J.* **201**, 199–208.
Bolwell, G. P. (1987). *Planta* **172**, 184–191.
Bolwell, G. P., Robbins, M. P., and Dixon, R. A. (1985a). *Biochem. J.* **229**, 693–699.
Bolwell, G. P., Robbins, M. P., and Dixon, R. A. (1985b). *Eur. J. Biochem.* **148**, 571–578.
Broekaert, W. F., Peumans, W. J., and Allen, A. K. (1986). Submitted for publication.
Casalongué, C., and Pont Lezica, R. (1985). *Plant Cell Physiol.* **26**, 1533–1539.
Cassab, G. I. (1986). *Planta* **168**, 441–446.
Cassab, G. I., Nieto-Sotelo, J., Cooper, J. B., van Holst, G. J., and Varner, J. E. (1985). *Plant Physiol.* **77**, 532–535.
Catt, J. W., Hills, G. J., and Roberts, K. (1978). *Planta* **138**, 91–98.
Chen, J., and Varner, J. E. (1985a). *Proc. Natl. Acad. Sci. U.S.A.* **82**, 4399–4403.
Chen, J., and Varner, J. E. (1985b). *EMBO J.* **4**, 2145–2151.
Chrispeels, M. J., Sadava, D., and Cho, Y. P. (1974). *J. Exp. Bot.* **25**, 1157–1166.
Clarke, A. E., Gleeson, P. A., Jermyn, M. A., and Knox, R. B. (1978). *Aust. J. Plant Physiol.* **5**, 707–722.
Clarke, A. E., Anderson, R. L., and Stone, B. A. (1979). *Phytochemistry* **18**, 521–540.
Clarke, J. A., Lisker, N., Ellingboe, A. H., and Lamport, D. T. A. (1983). *Plant Sci. Lett.* **30**, 339–346.
Cleland, R., and Karlsnes, A. M. (1967). *Plant Physiol.* **42**, 669–671.
Cooper, J. B., and Varner, J. E. (1983). *Biochem. Biophys. Res. Commun.* **112**, 161–167.
Cooper, J. B., Chen, J. A., and Varner, J. E. (1984). In "Structure, Function, and Biosynthesis of Plant Cell Walls" (W. M. Dugger and S. Bartnicki-Garcia, eds.), pp. 75–88. Am. Soc. Plant Physiol., Rockville, Maryland.
Cooper, J. B., Heuser, J. E., and Varner, J. E. (1986). Submitted for publication.
Corbin, D. R., Sauer, N., and Lamb, C. J. (1987). *Mol. Cell Biol.* **7**, 4337–4344.
Desai, N. N., Allen, A. K., and Neuberger, A. (1981). *Biochem. J.* **197**, 345–353.
Dougall, D. K., and Shimbayashi, K. (1960). *Plant Physiol.* **35**, 396–404.
Ecker, J. R., and Davis, R. W. (1987). *Proc. Natl. Acad. Sci. U.S.A.* **84**, 5202–5206.
Epstein, L., and Lamport, D. T. A. (1984). *Phytochemistry* **23**, 1241–1246.
Esquerré-Tugayé, M. T., and Lamport, D. T. A. (1979). *Plant Physiol.* **64**, 314–319.
Esquerré-Tugayé, M. T., Lafitte, C., Mazau, D., Toppan, A., and Touzé, A. (1979). *Plant Physiol.* **64**, 320–326.
Esquerré-Tugayé, M. T., Mazau, D., Pélissier, B., Roby, D., Rumeau, D., and Toppan, A. (1985). In "Cellular and Molecular Biology of Plant Stress" (J. L. Key and T. Kosuge, eds.), pp. 459–473. Liss, New York.
Everdeen, D. S., Kiefer, S., Willard, J. J., Muldoon, E. P., Dey, P. M., Li, X., and Lamport, D. T. A. (1988). *Plant Physiol.* **87**, 616–621.
Fincher, G. B., Stone, B. A., and Clarke, A. E. (1983). *Annu. Rev. Plant Physiol.* **34**, 47–70.
Fry, S. C. (1982). *Biochem. J.* **204**, 449–455.
Gleeson, P. A., Stone, B. A., and Fincher, G. B. (1985). *AGP News* **5**, 30–36.
Goodenough, U. W. (1985). In "The Origin and Evolution of Sex" (H. O. Halvorson and A. Monroy, eds.), pp. 123–140. Liss, New York.
Goodenough, U. W., Gebhart, B., Mecham, R. P., and Heuser, J. E. (1986). *J. Cell Biol.* **103**, 405–417.
Gotelli, I. B., and Cleland, R. (1968). *Am. J. Bot.* **55**, 907–914.
Hammerschmidt, R., Lamport, D. T. A., and Muldoon, E. P. (1984). *Physiol. Plant Pathol.* **24**, 43–47.
Heckman, J. W., Jr., Terhune, B. T., and Lamport, D. T. A. (1988). *Plant Physiol.* **86**, 848–856.
Hironaka, C. M., Fraley, R. T., Chen, J., Varner, J. E., and Shah, D. M. (1985). *Proc. Int. Congr. Plant Mol. Biol., 1st* p. 152 (Abstr.).
Hood, E. E., Shen, Q. X., and Varner, J. E. (1988). *Plant Physiol.* **87**, 138–142.
Jeffree, C. E., and Yeoman, M. M. (1981). *New Phytol.* **87**, 463–471.

Jermyn, M., and Guthrie, R. (1985). *AGP News* **5**, 4–25.
Jermyn, M. A., and Yeow, Y. M. (1975). *Aust. J. Plant Physiol.* **2**, 501–531.
Karr, A. L., Jr. (1972). *Plant Physiol.* **50**, 275–282.
Kieliszewski, M., and Lamport, D. T. A. (1987). *Plant Physiol.* **85**, 823–827.
Kilpatrick, D. C. (1983). *In* "Chemical Taxonomy, Molecular Biology, and Function of Plant Lectins" (I. J. Goldstein and M. E. Etzler, eds.), pp. 63–70. Liss, New York.
Kilpatrick, D. C., Yeoman, M. M., and Gould, A. R. (1979). *Biochem. J.* **184**, 215–219.
Kilpatrick, D. C., Jeffree, C. E., Lockhart, C. M., and Yeoman, M. M. (1980). *FEBS Lett.* **113**, 129–133.
Klis, F. M., Rootjes, M., Groen, S., and Stegwee, D. (1983). *Z. Pflanzenphysiol.* **110**, 301–307.
Lamport, D. T. A. (1963). *J. Biol. Chem.* **238**, 1438–1440.
Lamport, D. T. A. (1965). *Adv. Bot. Res.* **2**, 151–218.
Lamport, D. T. A. (1967). *Nature (London)* **216**, 1322–1324.
Lamport, D. T. A. (1969). *Biochemistry* **8**, 1155–1163.
Lamport, D. T. A. (1977). *Recent Adv. Phytochem.* **11**, 79–115.
Lamport, D. T. A. (1980). *In* "The Biochemistry of Plants" (J. Preiss, ed.), Vol. 3, pp. 501–541. Academic Press, New York.
Lamport, D. T. A., and Catt, J. W. (1981). *In* "Plant Carbohydrates II" (W. Tanner and F. A. Loewus, eds.), pp. 133–165. Springer-Verlag, New York.
Lamport, D. T. A., and Epstein, L. (1983). *In* "Current Topics in Plant Biochemistry and Physiology" (D. D. Randall, D. G. Blevins, R. L. Larson, and B. J. Rapp, eds.), Vol. 2, pp. 73–83. University of Missouri–Columbia, Columbia, Missouri.
Lamport, D. T. A., and Miller, D. H. (1971). *Plant Physiol.* **48**, 454–456.
Lamport, D. T. A., and Northcote, D. H. (1960). *Nature (London)* **188**, 665–666.
Lamport, D. T. A., Katona, L., and Roerig, S. (1973). *Biochem. J.* **133**, 125–131.
Lawton, M. A., and Lamb, C. J. (1987). *Mol. Cell. Biol.* **7**, 335–341.
Leach, J. E., Cantrell, M. A., and Sequeira, L. (1982). *Plant Physiol.* **70**, 1353–1358.
Lord, J. M. (1985). *New Phytol.* **101**, 351–366.
Mellon, J. E., and Helgeson, J. P. (1982). *Plant Physiol.* **70**, 401–405.
Mitchell, B. A. (1980). M.S. thesis. Michigan State University, East Lansing, Michigan.
Monro, J. A., Bailey, R. W., and Penny, D. (1974). *Phytochemistry* **13**, 375–382.
Mort, A. J., and Lamport, D. T. A. (1977). *Anal. Biochem.* **82**, 289–309.
Muray, R. H. A., and Northcote, D. H. (1978). *Phytochemistry* **17**, 623–629.
Nakamura, K., Tsumuraya, Y., Hashimoto, Y., and Yamamoto, S. (1984). *Agric. Biol. Chem.* **48**, 753–760.
Owens, R. J., and Northcote, D. H. (1981). *Biochem. J.* **195**, 661–667.
Pike, C. S., Un, H., Lystash, J. C., and Showalter, A. M. (1979). *Plant Physiol.* **63**, 444–449.
Pollard, P. C., and Fincher, G. B. (1981). *Aust. J. Plant Physiol.* **8**, 121–132.
Ridge, I., and Osborne, D. J. (1970). *J. Exp. Bot.* **21**, 843–856.
Roberts, K., Greif, C., Hills, G. J., and Shaw, P. J. (1985). *J. Cell Sci. Suppl.* **2**, 105–127.
Robinson, D. G., Andreae, M., and Sauer, A. (1985). *In* "Biochemistry of Plant Cell Walls" (C. T. Brett and J. R. Hillman, eds.), pp. 155–176. Cambridge University Press, New York.
Roby, D., Toppan, A., and Esquerré-Tugayé, M. T. (1985). *Plant Physiol.* **77**, 700–704.
Rüdiger, H. (1982). *Planta Med.* **46**, 3–9.
Sadava, D., and Chrispeels, M. J. (1978). *In* "Biochemistry of Wounded Plant Tissues" (G. Kahl, ed.), pp. 85–102. de Gruyter, New York.
Sadava, D., Walker, F., and Chrispeels, M. J. (1973). *Dev. Biol.* **30**, 42–48.
Showalter, A. M., and Varner, J. E. (1987). *In* "Molecular Strategies for Crop Protection" (C. Arntzen and C. Ryan, eds.), pp. 375–392. Liss, New York.
Showalter, A. M., Bell, J. N., Cramer, C. L., Bailey, J. A., Varner, J. E., and Lamb, C. J. (1985). *Proc. Natl. Acad. Sci. U.S.A.* **82**, 6551–6555.
Smith, J. J., Muldoon, E. P., and Lamport, D. T. A. (1984). *Phytochemistry* **23**, 1233–1239.

Smith, J. J., Muldoon, E. P., Willard, J. J., and Lamport, D. T. A. (1986). *Phytochemistry* **25,** 1021–1030.
Smith, M. A. (1978). Ph.D. dissertation. Washington University, St. Louis, Missouri.
Smith, M. A. (1981). *Plant Physiol.* **68,** 956–963.
Stafstrom, J. P., and Staehelin, L. A. (1986). *Plant Physiol.* **81,** 234–241.
Stafstrom, J. P., and Staehelin, L. A. (1988). *Planta* **174,** 321–332.
Stermer, B. A., and Hammerschmidt, R. (1985). *In* "Cellular and Molecular Biology of Plant Stress" (J. L. Key and T. Kosuge, eds.), pp. 291–302. Liss, New York.
Steward, F. C., Wetmore, R. H., Thompson, J. F., and Nitsch, J. P. (1954). *Am. J. Bot.* **41,** 123–134.
Stuart, D. A., and Varner, J. E. (1980). *Plant Physiol.* **66,** 787–792.
Tanaka, M., Shibata, H., and Uchida, T. (1980). *Biochim. Biophys. Acta* **616,** 188–198.
Tanaka, M., Sato, K., and Uchida, T. (1981). *J. Biol. Chem.* **256,** 11397–11400.
Tautvydas, K. J. (1978). *Planta* **140,** 213–220.
Tierney, M. L., Wiechert, J., and Pluymers, D. (1988). *Mol. Gen. Genet.* **211,** 393–399.
VanEtten, C. H., Miller, R. W., Earle, F. R., Wolff, I. A., and Jones, Q. (1961). *J. Agric. Food Chem.* **9,** 433–435.
VanEtten, C. H., Miller, R. W., Wolff, I. A., and Jones, Q. (1963). *J. Agric. Food Chem.* **11,** 399–410.
van Holst, G. J., and Clarke, A. E. (1986). *Plant Physiol.* **80,** 786–789.
van Holst, G. J., and Fincher, G. B. (1984). *Plant Physiol.* **75,** 1163–1164.
van Holst, G. J., and Klis, F. M. (1981). *Plant Physiol.* **68,** 979–980.
van Holst, G. J., and Varner, J. E. (1984). *Plant Physiol.* **74,** 247–251.
van Holst, G. J., Klis, F. M., De Wildt, P. J. M., Hazenberg, C. A. M., Buijs, J., and Stegwee, D. (1981). *Plant Physiol.* **68,** 910–913.
van Holst, G. J., Martin, S. R., Allen, A. K., Ashford, D., Desai, N. N., and Neuberger, A. (1986). *Biochem. J.* **223,** 731–736.
Varner, J. E. (1980). *Biochem. Biophys. Res. Commun.* **96,** 692–696.
Whitmore, F. W. (1978). *Plant Sci. Lett.* **13,** 241–245.
Wienecke, K., Glas, R., and Robinson, D. G. (1982). *Planta* **155,** 58–63.
Yariv, J., Rapport, M. M., and Graf, L. (1962). *Biochem. J.* **85,** 383–388.

Protein Degradation 13

RICHARD D. VIERSTRA

I. Introduction
II. Functions of Protein Degradation
III. Mechanisms for Degrading Proteins
 A. Ubiquitin-Dependent Proteolytic Pathway
 B. Pathways for Degrading Proteins in Organelles
IV. Conclusions
 References

I. INTRODUCTION

Plant proteins, like those in bacteria and animal cells, are continuously turned over (Mothes, 1933; Gregory and Sen, 1937). This "protein cycle" is accomplished by two separate sets of reactions: those involving the synthesis of the protein and processing into the mature form, and those responsible for degrading the mature protein back to amino acids. While protein turnover is commonly equated with degradation, it should be emphasized that turnover refers to the net flux of amino acids through a protein and not just the catabolic reactions. The purpose of this chapter is to review the current understanding of why plant proteins are degraded and the possible mechanisms used to accomplish this goal. With such knowledge, one can appreciate the importance and complexity of this process in plant cell physiology.

Although synthesis and degradation are equally important in determining the concentration of any plant protein, our understanding of mechanisms responsible for protein degradation has lagged considerably behind our understanding of the synthetic processes. This lack of understanding has been the result of several factors, including (1) the attention given to protein synthetic processes following the elucidation of the genetic code; (2) evidence from studies on rapidly dividing bacteria suggesting that protein degradation is inconsequential as compared with cell division for diluting enzyme levels; and (3) the many technical barriers that hamper measuring protein breakdown rates and studying how specific proteins are degraded *in vivo*. One of the main technical problems is the difficulty of distinguishing proteolytic events that occur *in vivo* from those that occur artifactually *in vitro* following tissue homogenization. As a result, we have only recently begun to understand in a rudimentary way the physiological relevance of plant protein breakdown as well as the mechanisms responsible for catabolism. This lack of understanding also applies to bacterial and animal systems where the same biases and technical barriers are evident.

II. FUNCTIONS OF PROTEIN DEGRADATION

Superficially, protein degradation appears to be wasteful. Why expend enormous amounts of energy synthesizing a protein only to degrade it as soon as 10 min later? The tremendous potential of protein breakdown for regulating cell physiology becomes evident, however, when one considers the difficulties inherent in modulating the level of an intracellular protein solely by regulating the expression of its gene alone (Paskin and Mayer, 1977). Clearly, the level of any protein can be controlled only through the concerted effort of both the synthetic and catabolic processes. The ability to modulate protein levels is required by organisms for adapting their metabolism to changes in environment and to internal developmental signals. A survey of enzymes representing either the first or rate-limiting steps in key metabolic reactions shows that these enzymes *all* have fast degradation rates (Goldberg and St. John, 1976). This short half-life allows cells to alter the flow of substrates through particular metabolic pathways by altering the steady-state level of these crucial enzymes. Regulation of plant nitrogen assimilation from nitrate by nitrate reductase is a case in point. This enzyme catalyzes the first step in the pathway and is rapidly degraded *in vivo* regardless of growth conditions [$t_{1/2}$ ~4.3 hr in tobacco cells (Zielke and Filner, 1971)]. Synthesis of nitrate reductase is induced substantially when plants are fed nitrate, allowing the protein to accumulate to high levels in the presence of the substrate (Crawford *et al.*, 1986). When the concentration of nitrate is low or ammonia is high, synthesis is repressed and the level of the enzyme drops rapidly as a result of ongoing protein breakdown (Zielke and Filner, 1971; Somers *et al.*, 1983).

This same type of regulation also may be utilized for key regulatory proteins involved in plant development, with the presence or absence of particular proteins determining particular developmental states. A good example of this in plants is the morphogenic photoreceptor, phytochrome (Pratt, 1979). Phytochrome exists in two forms, the red light-absorbing form, Pr, and the far-red light-absorbing form, Pfr. It is synthesized as Pr which is both stable ($t_{1/2}$ >100 hr) and biologically inactive. Upon photoconversion to Pfr by red light, the molecule becomes biologically active, stimulating a diverse array of developmental responses from seed germination to flowering. Unlike Pr, Pfr is rapidly broken down [$t_{1/2}$ 1–2 hr (Pratt et al., 1974; Shanklin et al., 1987)]. Thus, the plant cell has developed a method for accumulating this developmental regulator in an inactive form and rapidly degrading it once activated. The rapid removal of Pfr in conjunction with the synthesis of new Pr allows plants to recognize more easily changing light conditions and alter their morphology accordingly.

Protein degradation serves important housekeeping functions (Goldberg and St. John, 1976). Abnormal proteins continually arise in cells by a variety of methods, including mutations, biosynthetic errors, spontaneous denaturation, free radical-induced damage, and environmental conditions (e.g., heat stress and desiccation). Such proteins would reach toxic levels within the cell without mechanisms for their removal. This would be especially true for slow growing plant cells where cell division is not rapid enough to dilute out such abnormal proteins. Thus, it is not surprising that all cells, both prokaryotic and eukaryotic, have mechanisms that rapidly remove abnormal proteins (Goldberg, 1972; Etlinger and Goldberg, 1977; Canut et al., 1986).

For plants in particular, protein breakdown is involved in several important aspects of growth and development. Seed germination requires the selective catabolism of seed storage proteins (Wilson, 1986). The derived amino acids are then used to synthesize new proteins needed for growth of the seedling prior to becoming auxotrophic. Seed storage proteins are synthesized during seed maturation with the majority of them localized in specialized organelles, protein bodies. During seed germination, these proteins are at least partially degraded by specific proteases that are synthesized *de novo* and transported into the protein body (Jacobsen and Varner, 1967; Chrispeels et al., 1976; Baumgartner et al., 1978; Rogers et al., 1985). The synthesis of several of these proteases is hormonally induced. It is not completely clear whether the specificity of seed storage protein catabolism is a function of the substrate specificity of the proteases or the compartmentation of the protein. Following the germination of dicotyledonous seeds, the transition of microbodies in the cotyledons from glyoxysomes to peroxisomes requires the specific degradation of glyoxysomal enzymes within the organelle (Titus and Becker, 1985; Eising and Gerhardt, 1987).

Proteolysis is important for chloroplast development. Here, proteases are involved not only in protein degradation but also in the posttranslational processing of nuclear-encoded chloroplast proteins (Schmidt and Mishkind, 1986). This processing involves the proteolytic removal of the N-terminal transit sequence after the import of nuclear-encoded proteins into the chloroplast. It has been proposed that degradation of specific structural proteins is required for converting etioplasts to functional chloroplasts (Hampp and De Filippis, 1980). One etioplast enzyme involved in chlorophyll synthesis,

NADPH-protochlorophyllide oxidoreductase, is rapidly degraded during chloroplast biogenesis (Santel and Apel, 1981; Kay and Griffiths, 1983). Proteolysis also maintains the stoichiometric accumulation of enzyme subunits in the chloroplast. Schmidt and Mishkind (1983) showed that, in the absence of the chloroplast-encoded large subunit of RuBisCO, the nuclear-encoded small subunit of *Chlamydomonas* is rapidly degraded ($t_{1/2}$ ~10 min). In a similar fashion, chlorophyll a/b-binding protein and plastocyanin have been shown to be rapidly degraded in the absence of their respective cofactors, chlorophyll and Cu^{2+} (Bennett, 1981; Merchant and Bogorad, 1986).

Protein degradation also plays a role in chloroplast repair. Because of the nature of the photosynthetic light reactions, accidental oxidation of chloroplast proteins is inevitable, particularly when plants are exposed to supraoptimal light intensities. The chloroplast-encoded 32-kDa (or Q_B) protein is especially sensitive to light conditions, slowly degraded under low light, but rapidly degraded under intense light (Bennett, 1984). This membrane-bound protein is an integral part of the photosynthetic electron transport chain, situated between photosystem II and plastocyanin. It has been proposed that, under intense light, photosynthetic electron transport becomes saturated, resulting in the conversion of excess energy into oxygen radicals (Kyle *et al.*, 1984; Matoo *et al.*, 1984). These radicals oxidatively damage the 32-kDa protein, leading to photoinhibition of photosynthesis. Such damaged 32-kDa proteins then are rapidly degraded and replaced to restore electron flow.

During plant growth, a variety of plant proteins is continuously synthesized and degraded. Measuring the actual turnover rates is problematic because plants contain large pools of free amino acids, and the presence of the cell wall prohibits short pulse/chase studies (Huffaker and Peterson, 1974). Taking these problems into account, Trewavas (1972) and Boudet *et al.* (1975) determined the turnover rate of total protein in expanding *Lemna minor* leaves by two different methods. From such measurements, the calculated half-lifes of total protein were approximately 8 and 5 days, respectively. Thus, many proteins must be degraded very slowly or not at all in growing leaves. However, several short-lived proteins have been identified and shown to be altered by environmental signals. Nitrate and nitrite reductases are rapidly broken down with their steady-state level modulated by available nitrate (Somers *et al.*, 1983; Gupta and Beevers, 1984). ATP-sulfurylase, an enzyme involved in sulfur assimilation from sulfate, is degraded when tobacco cells are fed methionine or cysteine (Reuveny and Filner, 1977). Several of the enzymes responsible for light- or wound-induced flavonoid biosynthesis (including phenylalanine ammonia-lyase) display rapid breakdown rates (Hahlbrock *et al.*, 1976; Lamb *et al.*, 1979). This rapid breakdown, along with a transient induction of gene expression, accounts for the sigmoidal accumulation of these enzymes in light or after wounding (Dixon and Lamb, 1979; Cramer *et al.*, 1985).

Protein degradation is intimately involved in plant senescence. This process is not an uncontrolled cell autolysis but is actually an extremely complex process initiated by a series of specific signals (Thomas and Stoddart, 1980; Nooden, 1984). Interestingly, senescence can be effectively retarded by cytokinins (Martin and Thimann, 1972; Peterson and Huffaker, 1975) and requires the synthesis of new proteins (Martin and Thimann, 1972; Cooke et al., 1980). Degradation proceeds through a series of predetermined steps until much of the soluble protein has been degraded and transported to areas of growth or storage. Up to 70% of the leaf protein can be mobilized out of senescencing leaves before death (Dalling and Nettleton, 1986). One of the first proteins degraded is RuBisCO, located in the chloroplast (Peterson and Huffaker, 1975; Wittenbach, 1978). Because it accounts for up to 50% of total leaf protein, it represents the major source of mobilized nitrogen and carbon in senescencing leaves (Wittenbach, 1978). A similar temporal pattern of degradation is observed during plant starvation (Cooke et al., 1980), suggesting that starvation is, in fact, premature senescence. In this case, the derived amino acids are reincorporated into new protein. In certain plants, the number of chloroplasts does not significantly decrease during senescence, implying that chloroplast proteins are degraded by proteases within the chloroplast (Wardley et al., 1984). It was proposed earlier that chloroplast proteins are degraded by vacuolar proteases (Matile, 1982), based on the observation that the vacuole contains a wide variety of hydrolytic activities including proteases (Boller and Kende, 1979), and by analogy with the animal lysozome (Hershko and Ciechanover, 1982). As yet, there is no evidence to support a mechanism whereby chloroplast enzymes traverse at least three membranes to reach these vacuolar proteases. Instead, more recent studies indicate that chloroplasts have their own proteolytic activities that may perform this role (Malek et al., 1984; Liu and Jagendorf, 1984).

Clearly, protein degradation has important functions in plant physiology. In the future, however, protein degradation actually may become a hindrance to plant scientists in efforts to genetically engineer crop plants. A major goal of such engineering is to introduce genes for "foreign" proteins that confer beneficial traits into plants. Obviously, plants, like bacteria, can recognize "foreign" proteins and degrade them (Murai et al., 1983), in many cases preventing adequate expression. This response was appreciated in the development of bacterial expression vectors and was overcome by the use of bacterial mutants missing the responsible protease (Larimore et al., 1982; Huynh et al., 1985). Comparable mutants in plants are not yet available.

III. MECHANISMS FOR DEGRADING PROTEINS

At first glance, protein degradation would appear to be simply a protease (or set of proteases) digesting a protein. However, the process cannot be that simple. Almost all proteases isolated to date, while in many cases specific for certain amino acids (e.g., trypsin specifically cleaves after Arg and Lys residues), are not protein specific. As a result, such proteases must be regulated or sequestered to prevent random digestion of all intracellular proteins. In addition, protein degradation is highly selective, with proteins having half-lives, as short as 10 min coexisting in the same cellular mileu with proteins that are essentially not degraded at all (Goldberg and St. John, 1976). This selectivity implies that each protein is individually recognized by the proteolytic machinery. The correlation of degradation rates with various physicochemical parameters (e.g., molecular mass, pI, thermal stability, amino terminus, specific sequences) suggests that the structure of the protein plays a role in this recognition (Goldberg and Dice, 1974; Davies, 1982; Coates and Davies, 1983; Bachmair et al., 1986; Rogers et al., 1986). Such recognition could be accomplished by either having a specific protease that initiates the catabolism of each protein or having a mechanism(s) that specifically identifies and/or marks proteins for degradation by a few nonspecific proteases. The rate of recognition or marking would then determine the protein's half-life. The situation becomes even more complex with the observation that turnover rates of proteins can change during a cell's life cycle. If each protein had its own protease, then the amount or activity of this protease would have to change during these transitions. For a few important proteins (e.g., seed storage proteins), having a specific protease responsible for initiating breakdown may be feasible. However, for the bulk of intracellular proteins, the use of an identification/marking system in conjunction with a few proteases is more amenable to control.

It has been known for some time that plants contain a wide variety of proteolytic activities (see Ryan and Walker-Simmons, 1981; Dalling, 1986), but the specific function(s) of most of these activities remains unknown. It is apparent that the complete disassembly of a protein is very rapid *in vivo,* so rapid in fact that only a few proteolytic intermediates have been captured and identified in plants (Wilson and Chen, 1983; Shanklin et al., 1987; Greenberg et al., 1987). This failure to detect intermediates is one of the major barriers to establishing how specific proteins are broken down and identifying the protease(s) involved. Like animal and bacterial cells (Goldberg and St. John, 1976), *in vivo* protein degradation in plants appears to require energy, usually in the form of ATP (Liu and Jagendorf, 1984; Malek et al., 1984; Vierstra, 1987b). Initially, this requirement was unexpected because peptide bond hydrolysis is an exergonic reaction and most if not all purified proteases are ATP-independent. In two cases, ATP hydrolysis has been shown to activate the proteolytic machinery toward specific proteins (Hershko and Ciechanover, 1982; Larimore et al., 1982; Menon and Goldberg, 1987). Another unusual feature is that proteins destined for degradation, especially abnormal proteins, aggregate within the cell prior to degradation (Prouty et al., 1975; Klemes et al., 1981; Mackenzie et al., 1975).

Whether this aggregation is a spontaneous precipitation or the result of an active process designed to sequester such proteins is unknown.

The central questions in studies on protein breakdown are what mechanisms are used by plants to degrade proteins *in vivo,* and how are such mechanisms regulated? Evidence from a variety of studies in plants (and analogy to animal systems) provides a number of clues and indicates that each cell compartment may have its own proteolytic system(s).

A. Ubiquitin-Dependent Proteolytic Pathway

The best characterized proteolytic pathway involves the small (76 amino acids) heat-stable protein, ubiquitin. The pathway was first discovered in rabbit reticulocytes by Hershko and co-workers and has since been identified in other mammals, fungi, and higher plants (for reviews see Hershko and Ciechanover, 1982, 1986; Finley and Varshavsky, 1985; Vierstra, 1987a). As the name implies, ubiquitin is indeed ubiquitous among eukaryotes, being present in all animal, plant, and fungal species examined. It is arguably the most conserved protein yet identified (Sharp and Li, 1987), its amino acid sequence being invariant among mammals (including humans), fish, amphibians, birds, and insects. This sequence differs by only two residues out of 76 to that found in *Dictyostelium* and by three residues to that found in yeast and several higher plants [oats, *Arabidopsis,* and barley (Vierstra *et al.,* 1986; Burke *et al.,* 1988; Gausing and Barkardottir, 1986)]. X-Ray crystallographic structures of animal, fungal, and plant ubiquitins are identical (Vijay-Kumar *et al.,* 1987), showing a compact globular structure with a flexible protruding C-terminus. Ubiquitin has been found in the nucleus and cytoplasm but not in chloroplasts or mitochondria (Goldknopf and Busch, 1977; Hershko *et al.,* 1982; K. J. Reinke, G. A. De Zoeten, and R. D. Vierstra, unpublished observations). As a result of this, it has been concluded, although not unequivocally, that the pathway is present only in the cytoplasm and nucleus.

In the ubiquitin-dependent proteolytic pathway, the main function of ubiquitin is to become *covalently* attached to proteins destined for catabolism (Hershko and Ciechanover, 1982, 1986). This ubiquitin–protein conjugate then serves as a committed intermediate for proteolysis. Ligation occurs via an unusual peptide linkage between the C-terminal glycine carboxyl group of ubiquitin and free N-terminal and internal lysyl amino groups on the target protein. Linkage requires ATP and a multienzyme system which in reticulocytes contains at least seven distinct proteins [see Fig. 1 (Hershko and Ciechanover, 1986)]. First, ubiquitin is activated via adenylation of its C-terminus by ubiquitin-activating enzyme (E1) and ATP. The activated molecule is then covalently attached to E1 by an energy-rich thiol ester linkage with the concomitant release of AMP. The ubiquitin moiety is transferred to a family of ubiquitin carrier proteins (E2s) through transesterification, and

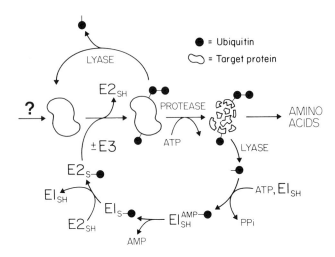

Fig. 1. Proposed pathway for ubiquitin-dependent proteolysis. The cycle begins with the activation of free ubiquitin by the enzyme, E1, followed by transfer of activated ubiquitin to E2. E2-bound ubiquitin is ligated to the target protein with or without the help of E3. Once the ubiquitin–protein conjugate is formed, either the target protein is degraded by an ATP-dependent protease(s) with the release of free ubiquitin or the conjugate is disassembled by ubiquitin–protein lyase(s) with the regeneration of both proteins intact.

finally ligated to the target protein with or without the help of a family of ubiquitin protein ligases (E3s). Similar pathways for ubiquitin conjugation have also been detected in yeast and a variety of higher plants (Ozkaynak *et al.*, 1984; Vierstra, 1987b; Hatfield and Vierstra, 1989). With plants, conjugation reactions are easily observed *in vitro* by adding radiolabeled ubiquitin and ATP to crude plant extracts and visualizing newly synthesized conjugates by SDS-PAGE and autoradiography (Fig. 2). Characterization of the pathways in yeast and higher plants indicates that not only do they share the same mechanism for conjugation as reticulocytes but that, like ubiquitin, the various proteins involved in conjugation have been highly conserved through evolution (Jentsch *et al.*, 1987; Vierstra, 1987b; Hatfield and Vierstra, 1989).

Once a protein is tagged with one or more ubiquitins, it has two possible fates (Fig. 1). The conjugate can be degraded by an ATP-dependent protease(s) specific for ubiquitin–protein conjugates with the release of free, undigested, ubiquitin. Recently, a large (1000–1500 kDa) complex containing at least nine different subunits has been partially purified from reticulocytes that may serve this role (Hough *et al.*, 1987; Waxman *et al.*, 1987). The protease complex is effectively inhibited by micromolar concentrations of hemin. In higher plants and yeast, proteolytic activities that degrade ubiquitin–protein conjugates have also been detected and, like their reticulocyte counterpart(s), they require ATP and are sensitive to hemin (Vierstra

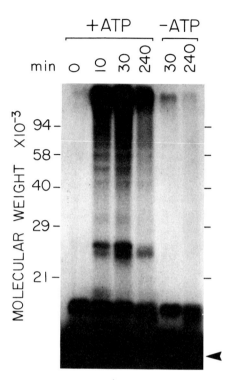

Fig. 2. Detection of ATP-dependent ubiquitin conjugation in a crude extract from etiolated oats. ^{125}I-Labeled human ubiquitin was added to a crude oat extract along with either ATP and an ATP-regenerating system [phosphocreatine and phosphocreatine kinase (+ATP)] or an ATP-depleting system [deoxyglucose and hexokinase (−ATP)]. The reaction mixture was incubated at 30°C and aliquots were removed at various times and analyzed by SDS-PAGE and autoradiography. Arrowhead to the right indicates the position of free ubiquitin. From Vierstra (1987b).

and Sullivan, 1988). As a result of this ligation/degradation cycle, the ubiquitin system fits one model of protein degradation where proteins are selectively marked for breakdown by a relatively nonspecific protease (see above). Recent evidence that ubiquitin itself has limited proteolytic activity suggests that the molecule may be more directly involved in protein catabolism in addition to serving as a tag for other proteases (Fried et al., 1987).

The second fate is that ubiquitin–protein conjugates can be disassembled by ubiquitin–protein lyase (or isopeptidase), which cleaves only the peptide bond between ubiquitin and the target protein liberating the target protein and ubiquitin intact (Matsui et al., 1982; Hershko et al., 1984a). This lyase(s) may function to (1) release incorrectly conjugated proteins, (2) regulate the ubiquitination level of specific proteins, and/or (3) remove proteolytic fragments generated during conjugate digestion from ubiquitin's C-terminus. Lyase activity is present in mammals, yeast, and higher plants, does not require ATP, and like the conjugate-specific protease, is sensitive to micromolar concentrations of hemin (Hershko *et al.*, 1984a; Vierstra and Sullivan, 1988).

While the reactions directly involved in conjugate formation and degradation/disassembly are at least partially understood, the important reactions that identify which proteins become targets for ubiquitination, and as a result determine the selectivity of the pathway, are unknown. Clearly, structural features on the target proteins play a role in the recognition process (Hershko *et al.*, 1984b; Bachmair *et al.*, 1986; Gregori *et al.*, 1987; Shanklin *et al.*, 1987). The following evidence has implicated the N-terminus more specifically: (1) The *in vivo* degradation rate of a protein can be correlated with the nature of the N-terminal amino acid [N-end rule (Bachmair *et al.*, 1986)]; (2) Proteins with blocked N-termini are poor substrates for conjugation (Hershko *et al.*, 1984b); (3) modifying the N-terminus by the posttranslational addition of a new amino acid residue alters the susceptibility of proteins to degradation by the ubiquitin pathway (Ferber and Ciechanover, 1987); and (4) E3 displays preferential binding to proteins with free N-termini (Hershko *et al.*, 1986). How the ubiquitin system recognizes various features of the N-terminus or other domains on the target proteins is unclear. That individual members of the E2 and E3 families show preferences for particular target proteins may provide one such mechanism (Pickart and Rose, 1985; Lee *et al.*, 1986). Here, unique features on a particular protein or group of proteins could be recognized and ubiquitinated by specific isoforms of E2 and E3. This mechanism is supported by recent evidence demonstrating that one particular E2 in yeast (RAD6) prefers attaching ubiquitin to histones H2A and H2B *in vitro* (Jentsch *et al.*, 1987). It is also possible that selectivity is controlled by the disassembly and degradation reactions. Under this scenario, ubiquitin conjugation would be nonselective, with specificity residing in the decision whether a conjugate is either degraded or disassembled.

A wide variety of studies now indicate that the ubiquitin-dependent proteolytic pathway has several important functions in eukaryotic cell physiology. While most of these studies were conducted with animal cells and yeast, it is likely, considering the conservation of the pathway, that the results will also pertain to higher plants. The ubiquitin pathway appears to be the major route for degrading abnormal proteins, indicating its role as the cell's "housekeeper" (Hershko *et al.*, 1982; Chin *et al.*, 1982; Ciechanover *et al.*, 1984; Finley *et al.*, 1987). Using a mouse carcinoma cell line containing a temperature-sensitive E1, Varshavsky and co-workers demonstrated that >90% of short-lived proteins ($t_{1/2}$ <5 hr) are degraded via ubiquitinated intermediates (Ciechanover *et al.*, 1984; Finley *et al.*, 1984). Because many important regulatory proteins have short half-lives (Goldberg and St. John, 1976), it follows that this pathway could have a major role in controlling cell physiology and development. This role is exemplified in plants by the finding that the levels of the active form of phytochrome may be controlled by the ubiquitin pathway (Shanklin *et al.*, 1987).

Ubiquitin-dependent proteolysis also may help cells cope with stress. In cultured chicken cells, yeast, and *Arabidopsis*, specific ubiquitin genes are heat inducible (Bond and Schlesinger, 1985; Finley *et al.*, 1987; Burke *et al.*, 1988). The levels of ubiquitin conjugates change during heat shock (Carlson *et al.*, 1987; Parag *et al.*, 1987), and mutants in the ubiquitin pathway are hypersensitive to heat (Ciechanover *et al.*, 1984; Finley *et al.*, 1987). In yeast, ubiquitin mutants are also hypersensitive to desiccation and starvation (Finley *et al.*, 1987). The association of plant ubiquitin with viral coat proteins (Dunigan *et al.*, 1988) and viral-induced inclusion bodies (K. J. Reinke, G. A. De Zoeten, and R. D. Vierstra, unpublished observations) suggests that the pathway may be connected to the defense against viral infection in plants. Taken together, it has been proposed that ubiquitin-dependent proteolysis helps ameliorate stress-induced damage by removing unwanted proteins (denatured or viral) that would otherwise accumulate to toxic levels (Finley *et al.*, 1984).

In addition to its functions in protein degradation, ubiquitin conjugation may have several other functions as well. These include regulating chromatin structure (Goldknopf and Busch, 1977; Goldknopf *et al.*, 1980; Wu *et al.*, 1981), cell cycle (Matsumoto *et al.*, 1983), DNA repair (Jentsch *et al.*, 1987), cell–cell interactions (Siegelman *et al.*, 1986), and signal transduction (Yarden *et al.*, 1986).

For plants in particular, the functions of the ubiquitin-dependent proteolytic pathway are still unclear. Plants contain a diverse array of ubiquitin–protein conjugates *in vivo* as observed by immunoblot analysis (Vierstra *et al.*, 1985; Burke *et al.*, 1988), suggesting that a variety of proteins are degraded via ubiquitinated intermediates. In fact, most plant ubiquitin actually exists in the conjugated form. With the exception of phytochrome (Shanklin *et al.*, 1987), the identity of these target proteins remains to be elucidated.

While the ubiquitin-dependent proteolytic pathway is likely to be one of the major catabolic pathways in the cytoplasm of plants, additional pathways are also likely. Goldberg and co-workers have identified several pathways for degrading proteins in reticulocytes that are independent of ubiquitin (Tanaka *et al.*, 1983; Fagan *et al.*, 1986). However, alternative cytoplasmic pathways in plants have not yet been detected.

B. Pathways for Degrading Proteins in Organelles

The most conspicuous organelle in plants is the vacuole, occupying as much as 90% of the total cell volume. It contains most of the proteolytic activities measured *in vitro* (Boller and Kende, 1979; Matile, 1982; Waters *et al.*, 1982). In barley leaves, for example, two vacuolar proteases are responsible for >95% of the activity capable of degrading purified RuBisCO *in vitro* (Thayer and Huffaker, 1984). Given the high activity of these vacuolar proteases, it is likely that most of the problems associated with purifying intact proteins from plants derive from the release of these proteases from the vacuole during tissue homogenization (Grey, 1982). A wide range of proteases exists, including endo- and exopeptidases, carboxy- and aminopeptidases, and serine-, sulfhydryl-, and metalloproteases (Matile, 1982). The best characterized plant proteases, papain, ficin, and bromelain, are located in the vacuoles of papaya, figs, and pineapple, respectively (Ryan and Walker-Simmons, 1981). No ATP-dependent proteases have been identified. As a result, the vacuole has been likened to the animal lysozome, and described as the "lytic compartment of the plant cell" (Matile, 1982). However, it has not yet been determined that these proteases are actually involved in protein degradation *in vivo*. The failure of yeast to exhibit noticeable phenotypes when missing major vacuolar proteases through mutation suggests that these proteases are not involved in breaking down important intracellular proteins (Wolf, 1984). Alternatively, it has been proposed that these proteases, along with other hydrolytic activities in the vacuole, serve as a defense against herbivores and fungi (Boller, 1986). Consistent with this proposal is the fact that several proteinaceous inhibitors of animal and fungal proteases are also localized in the vacuole (Walker-Simmons and Ryan, 1977).

The second largest organelle in plants is the chloroplast. While many chloroplast proteins are subjected to rapid turnover during this organelle's life cycle (see above), the degradative mechanisms used to accomplish this process are unknown. Given the fact that chloroplasts are autonomous with respect to functions required for gene transcription and protein synthesis (with the help of nuclear-encoded proteins), it seems reasonable that chloroplasts would also have their own mechanism(s) for protein degradation. Thus, as Dalling and Nettleton (1986) point out, it appears unnecessarily complicated to provide models where other cell compartments, such as the vacuole, are responsible for chloroplast protein breakdown.

Evidence now indicates that chloroplasts have their own proteases, both ATP-dependent and -independent (Hampp and De Filippis, 1980; Liu and Jagendorf, 1984, 1986; Malek *et al.,* 1984). Whether they are encoded in the chloroplast genome or the nuclear genome and posttranslationally transported into the chloroplast is not yet known. Indirect evidence suggests that important activities may indeed be transported into the chloroplast. For example, cycloheximide (an inhibitor of protein synthesis on 80S cytoplasmic ribosomes), but not chloramphenicol (an inhibitor of protein synthesis of chloroplast 70S ribosomes), retards chloroplast senescence and prevents the accumulation of two chloroplast proteases (Martin and Thimann, 1972; Cook *et al.,* 1980). Yoshida (1961) demonstrated in *Elodea* that, while chloroplasts in nucleate cells senesced, their counterparts in enucleate cells did not. Nuclear mutations exist that accelerate the turnover of chloroplast proteins (Leto *et al.,* 1985). Finally, in tobacco, where the entire chloroplast genome has been sequenced, only 11 unidentified open reading frames remain, with none showing homology with known proteases (Shinozaki et al., 1986). Thus it seems unlikely that chloroplasts could have a complex proteolytic pathway like that of the cytoplasmic ubiquitin system (Hershko and Ciechanover, 1986) without involving the nuclear genome.

Although no evidence currently exists for the presence of proteases within these organelles, it is likely that both plant mitochondria and microbodies are also capable of degrading proteins. In animal cells, mitochondrial proteins are turned over at distinct rates and mitochondria contain proteases (Goldberg and St. John, 1976). An ATP-dependent proteolytic activity responsible for degrading abnormal proteins in liver mitochondria has been characterized (Desautels and Goldberg, 1982). As mentioned above, selective degradation of glyoxysomal enzymes occurs during the transition of microbodies from glyoxysomes to peroxisomes (Titus and Becker, 1985; Eising and Gerhardt, 1987). This suggests that either proteases are imported in to perform this function or the glyoxysomal enzymes are exported and degraded outside the microbody.

IV. CONCLUSIONS

It is clear that protein degradation is an integral part of plant cell physiology and thus, quite likely an important regulatory feature in plant development. It is also apparent that protein degradation is an extremely complex process. Attempting to assign a function to *any* plant protease from *in vitro* studies using purified protein as substrates must be treated with caution if the intracellular locations of both the substrate and protease are not known. Given the wealth of information derived from mutational studies with bacterial and animal proteolytic pathways (Wolf, 1984; Ciechanover *et al.*, 1984; Finley *et al.*, 1984, 1987; Goff and Goldberg, 1987), more emphasis should be placed on this experimental approach. It was relatively easy to demonstrate that plant cells contain a variety of proteolytic activities. The next, more difficult, questions are what are the natural substrates for these proteases, and how is proteolysis regulated *in vivo*? A combined physiological, molecular, and genetic approach will be necessary to more fully understand protein degradation in plants.

REFERENCES

Bachmair, A., Finley, D., and Varshavsky, A. (1986). *Science* **234,** 179.
Baumgartner, B., Tokuyasu, K. T., and Chrispeels, M. J. (1978). *J. Cell Biol.* **79,** 10.
Bennett, J. (1981). *Eur. J. Biochem.* **118,** 61.
Bennett, J. (1984). *Nature (London)* **310,** 547.
Boller, T. (1986). In "Plant Proteolytic Enzymes" (M. J. Dalling, ed.), Vol. 1, pp. 67–96. CRC Press, New York.

Boller, T., and Kende, H. (1979). *Plant Physiol.* **663,** 1123.
Bond, U., and Schlesinger, M. J. (1985). *Mol. Cell. Biol.* **5,** 949.
Boudet, A., Humphrey, T. J., and Davies, D. D. (1975). *Biochem. J.* **152,** 409.
Burke, T. R., Callis, J., and Vierstra, R. D. (1988). *Molec. Gen. Genetics* **213,** 435–443.
Canut, H., Ailbert, G., Carrasco, A., and Boudet, A. M. (1986). *Plant Physiol.* **81,** 460.
Carlson, N., Rogers, S., and Rechsteiner, M. (1987). *J. Cell Biol.* **104,** 547.
Chin, D. T., Kuehl, L., and Rechsteiner, M. (1982). *Proc. Natl. Acad. Sci. U.S.A.* **79,** 5857.
Chrispeels, M. J., Baumgartner, B., and Harris, N. (1976). *Proc. Natl. Acad. Sci. U.S.A.* **73,** 3168.
Ciechanover, A., Finley, D., and Varshavsky, A. (1984). *Cell* **37,** 57.
Coates, J. B., and Davies, D. D. (1983). *Planta* **158,** 550.
Cook, R. J., Roberts, K., and Davies, D. D. (1980). *Plant Physiol.* **66,** 1119.
Cramer, C. L., Bell, J. N., Ryder, T. B., Bailey, J. A., Schuch, W., Bolwell, G. P., Robbins, M. P., Dixon, R. A., and Lamb, C. J. (1985). *EMBO J.* **4,** 285.
Crawford, N. M., Campbell, W. H., and Davis, R. W. (1986). *Proc. Natl. Acad. Sci. U.S.A.* **83,** 8073.
Dalling, M. J. (ed.) (1986). "Plant Proteolytic Enzymes," Vols. 1 and 2. CRC Press, New York.
Dalling, M. J., and Nettleton, A. M. (1986). *In* "Plant Proteolytic Enzymes" (M. J. Dalling, ed.), Vol. 2, pp. 125–153. CRC Press, New York.
Davies, D. D. (1982). *Encycl. Plant Physiol., New Ser.* **14A,** 189–228.
Desautels, M., and Goldberg, A. L. (1982). *Proc. Natl. Acad. Sci. U.S.A.* **79,** 1869.
Dixon, R. A., and Lamb, C. J. (1979). *Biochim. Biophys. Acta* **586,** 453.
Dunigan, D. D., Dietzgen, R. G., Schoelz, J. E., and Zaitlin, M. (1988). *Virology* **165,** 310–312.
Eising, R., and Gerhardt, B. (1987). *Plant Physiol.* **84,** 225.
Etlinger, J. D., and Goldberg, A. L. (1977). *Proc. Natl. Acad. Sci. U.S.A.* **74,** 54.
Fagan, J. M., Waxman, L., and Goldberg, A. L. (1986). *J. Biol. Chem.* **261,** 5705.
Ferber, S., and Ciechanover, A. (1987). *Nature (London)* **326,** 808.
Finley, D., and Varshavsky, A. (1985). *Trends Biochem. Sci.* **10,** 343.
Finley, A., Chiechanover, A., and Varshavsky, A. (1984). *Cell* **37,** 43.
Finley, D., Ozkaynak, E., and Varshavsky, A. (1987). *Cell* **48,** 1035.
Fried, V. A., Smith, H. A., Hildebrandt, E., and Weiner, K. (1987). *Proc. Natl. Acad. Sci. U.S.A.* **84,** 3685.
Gausing, K., and Barkardottir, R. (1986). *Eur. J. Biochem.* **158,** 57.
Goff, S. A., and Goldberg, A. L. (1987). *J. Biol. Chem.* **262,** 4508.
Goldberg, A. L. (1972). *Proc. Natl. Acad. Sci. U.S.A.* **69,** 422.
Goldberg, A. L., and Dice, J. F. (1974). *Annu. Rev. Biochem.* **43,** 835.
Goldberg, A. L., and St. John, A. (1976). *Annu. Rev. Biochem.* **45,** 747.
Goldknopf, I. L., and Busch, H. (1977). *Proc. Natl. Acad. Sci. U.S.A.* **74,** 864.
Goldknopf, I. L., Wilson, G., Ballal, N. R., and Busch, H. (1980). *J. Biol. Chem.* **255,** 10555.
Greenberg, B. M., Gaba, V., Mattoo, A. K., and Edelman, M. (1987). *EMBO J.* **6,** 2865.
Gregori, L., Marriott, D., Putkey, J. A., Means, A. R., and Chau, V. (1987). *J. Biol. Chem.* **262,** 2562.
Gregory, F. G., and Sen, G. K. (1937). *Ann. Bot.* **1,** 521.
Grey, J. C. (1982). *In* "Methods in Chloroplast Molecular Biology" (M. Edelman, R. B. Hallick, and N.-H. Chua, eds.), pp. 1093–1102. Elsevier/North-Holland, Amsterdam.
Gupta, S. C., and Beevers, L. (1984). *Plant Physiol.* **75,** 251.
Hahlbrock, K., Knobloch, K.-H., Kreuzaler, F., Potts, J. R. M., and Wellman, E. (1976). *Eur. J. Biochem.* **61,** 199.
Hampp, R., and De Filippis, L. F. (1980). *Plant Physiol.* **65,** 663.
Hatfield, P. H., and Vierstra, R. D. (1989). *Biochemistry* (in press).
Hershko, A., and Ciechanover, A. (1982). *Annu. Rev. Biochem.* **51,** 335.
Hershko, A., and Ciechanover, A. (1986). *Prog. Nucleic Acid Res. Mol. Biol.* **33,** 19.
Hershko, A., Eytan, E., Ciechanover, A., and Haas. A. L. (1982). *J. Biol. Chem.* **257,** 13964.

Hershko, A., Leshinsky, E., Ganoth, D., and Heller, H. (1984a). *Proc. Natl. Acad. Sci. U.S.A.* **81,** 1619.
Hershko, A., Heller, H., Eytan, E., Kaklij, G., and Rose, I. A. (1984b). *Proc. Natl. Acad. Sci. U.S.A.* **81,** 7021.
Hershko, A., Heller, H., Eytan, E., and Reiss, Y. (1986). *J. Biol. Chem.* **261,** 11992.
Hough, R., Pratt, G., and Rechsteiner, M. (1987). *J. Biol. Chem.* **262,** 8303.
Huffaker, R. C., and Peterson, L. W. (1974). *Annu. Rev. Plant Physiol.* **25,** 363.
Huynh, T. V., Young, R. A., and Davis, R. W. (1985). *In* "DNA Cloning: A Practical Approach" (D. M. Glover, ed.), Vol. 1, pp. 49–78. IRL Press, Washington, D.C.
Jacobsen, J. V., and Varner, J. E. (1967). *Plant Physiol.* **42,** 1596.
Jentsch, S., McGrath, J. P., and Varshavsky, A. (1987). *Nature (London)* **329,** 131.
Kay, S. A., and Griffiths, W. T. (1983). *Plant Physiol.* **72,** 229.
Klemes, Y., Etlinger, J. D., and Goldberg, A. L. (1981). *J. Biol. Chem.* **256,** 8436.
Kyle, D. J., Ohad, I., and Arntzen, C. J. (1984). *Proc. Natl. Acad. Sci. U.S.A.* **81,** 4070.
Lamb, C. J., Merritt, T. K., and Butt, V. S. (1979). *Biochim. Biophys. Acta* **582,** 196.
Larimore, F. S., Waxman, L., and Goldberg, A. L. (1982). *J. Biol. Chem.* **257,** 4187.
Lee, P. L., Merritt, T. K., Murakami, K., and Hatcher, V. B. (1986). *Biochemistry* **25,** 3134.
Leto, K. J., Bell, E., and McIntosh, L. (1985). *EMBO J.* **4,** 1645.
Liu, X.-Q., and Jagendorf, A. T. (1984). *FEBS Lett.* **166,** 248.
Liu, X.-Q., and Jagendorf, A. T. (1986). *Plant Physiol.* **81,** 603.
Mackenzie, J. M., Coleman, R. A., Briggs, W. R., and Pratt, L. H. (1975). *Proc. Natl. Acad. Sci. U.S.A.* **72,** 799.
Malek, L., Bogorad, L., Ayers, A. R., and Goldberg, A. L. (1984). *FEBS Lett.* **166,** 253.
Martin, C., and Thimann, K. V. (1972). *Plant Physiol.* **49,** 64.
Matile, P. H. (1982). *Encycl. Plant Physiol., New Ser.* **14A,** 169–188.
Matsui, S.-I., Sandberg, A. A., Negoro, S., Seon, B. K., and Goldstein, G. (1982). *Proc. Natl. Acad. Sci. U.S.A.* **79,** 1535.
Matsumoto, Y., Yasuda, H., Marunouchi, T., and Yamada, M. (1983). *FEBS Lett.* **151,** 139.
Mattoo, A. K., Hoffman-Falk, H., Marder, J. B., and Edelman, M. (1984). *Proc. Natl. Acad. Sci. U.S.A.* **81,** 1380.
Menon, A. S., and Goldberg, A. L. (1987). *J. Biol. Chem.* **262,** 14929.
Merchant, S., and Bogorad, L. (1986). *J. Biol. Chem.* **261,** 15850.
Mothes, K. (1933). *Planta* **19,** 117.
Murai, N., Sutton, D. W., Murray, M. G., Slighton, J. L., Merlo, D. J., Reichart, N. A., Sengupta-Gopalan, C., Stock, C. A., Barker, R. F., Kemp, J. D., and Hall, T. C. (1983). *Science* **222,** 476.
Nooden, L. D. (1984). *Physiol. Plant.* **62,** 273.
Ozkaynak, E., Finley, D., and Varshavsky, A. (1984). *Nature (London)* **312,** 663.
Parag, H. A., Raboy, B., and Kulka, R. G. (1987). *EMBO J.* **6,** 55.
Paskin, N., and Mayer, R. J. (1977). *Biochim. Biophys. Acta* **474,** 1.
Peterson, L. W., and Huffaker, R. C. (1975). *Plant Physiol.* **55,** 1009.
Pickart, C. M., and Rose, I. A. (1985). *J. Biol. Chem.* **260,** 1573.
Pratt, L. H. (1979). *Photochem. Photobiol. Rev.* **4,** 59.
Pratt, L. H., Kidd, G. H., and Coleman, R. A. (1974). *Biochim. Biophys. Acta* **365,** 93.
Prouty, W. F., Karnovsky, M. J., and Goldberg, A. L. (1975). *J. Biol. Chem.* **250,** 1112.
Reuveny, Z., and Filner, P. (1977). *J. Biol. Chem.* **252,** 1858.
Rogers, J. C., Dean, D., and Heck, G. R. (1985). *Proc. Natl. Acad. Sci. U.S.A.* **82,** 6512.
Rogers, S., Wells, R., and Rechsteiner, M. (1986). *Science* **234,** 364.
Ryan, C. A., and Walker-Simmons, M. (1981). *In* "The Biochemistry of Plants" (P. K. Stumpf and E. E. Conn, eds.), Vol. 6, pp. 321–350. Academic Press, New York.
Santel, H. J., and Apel, K. (1981). *Eur. J. Biochem.* **120,** 95.
Schmidt, G. W., and Mishkind, M. L. (1983). *Proc. Natl. Acad. Sci. U.S.A.* **80,** 2632.
Schmidt, G. W., and Mishkind, M. L. (1986). *Annu. Rev. Biochem.* **55,** 879.

Shanklin, J., Jabben, M., and Vierstra, R. D. (1987). *Proc. Natl. Acad. Sci. U.S.A.* **84,** 359.
Sharp, P. M., and Li, W.-H. (1987). *J. Mol. Evol.* **25,** 58.
Shinozaki, K., Ohme, M., Tanaka, M., Wakasugi, T., Hayashida, N., Matsubayashi, T., Zaita, N., Chunwongse, J., Obokata, J., Yamaguchi-Shinozaki, K., Ohto, C., Torazawa, K., Meng, B. Y., Sugita, M., Deno, H., Kamogashira, T., Yamada, K., Kusuda, J., Takaiwa, F., Kato, A., Tohdoh, N., Shimada, H., and Sugiura, M. (1986). *EMBO J.* **5,** 2043.
Siegelman, M., Bond, M. W., Gallatin, W. M., St John, T., Smith, H. L., Fried, V. A., and Weissman, I. L. (1986). *Science* **231,** 823.
Somers, D. A., Kuo, T.-M., Kleinhofs, A., Warner, R. L., and Oaks, A. (1983). *Plant Physiol.* **72,** 949.
Tanaka, K., Waxman, L., and Goldberg, A. L. (1983). *J. Cell Biol.* **96,** 1580.
Thayer, S. S., and Huffaker, R. C. (1984). *Plant Physiol.* **75,** 70.
Thomas, H., and Stoddart, J. L. (1980). *Annu. Rev. Plant Physiol.* **31,** 83.
Titus, D. E., and Becker, W. M. (1985). *J. Cell Biol.* **101,** 1288.
Trewavas, A. (1972). *Plant Physiol.* **49,** 40.
Vierstra, R. D. (1987a). *Physiol. Plant.* **70,** 103.
Vierstra, R. D. (1987b). *Plant Physiol.* **84,** 332.
Vierstra, R. D., and Sullivan, M. L. (1988). *Biochemistry* **27,** 3290.
Vierstra, R. D., Langan, S. M., and Haas, A. L. (1985). *J. Biol. Chem.* **260,** 12015.
Vierstra, R. D., Langan, S. M., and Schaller, G. E. (1986). *Biochemistry* **25,** 3105.
Vijay-Kumar, S., Buggs, C. E., Wilkinson, K. E., Vierstra, R. D., Hatfield, P. H., and Cook, W. J. (1987). *J. Biol. Chem.* **262,** 6396.
Walker-Simmons, M., and Ryan, C. A. (1977). *Plant Physiol.* **60,** 61.
Wardley, T. M., Bhalla, P. L., and Dalling, M. J. (1984). *Plant Physiol.* **75,** 421.
Waters, S. P., Noble, E. R., and Dalling, M. J. (1982). *Plant Physiol.* **69,** 575.
Waxman, L., Fagan, J. M., and Goldberg, A. L. (1987). *J. Biol. Chem.* **262,** 2451.
Wilson, K. A. (1986). *In* "Plant Proteolytic Enzymes" (M. J. Dalling, ed.), Vol. 2, pp. 19–46. CRC Press, New York.
Wilson, K. A., and Chen, J. C. (1983). *Plant Physiol.* **71,** 341.
Wittenbach, V. A. (1978). *Plant Physiol.* **62,** 604.
Wolf, D. H. (1984). *Biochem. Soc. Trans.* **13,** 279.
Wu, R. S., Kohn, K. W., and Bonner, W. M. (1981). *J. Biol. Chem.* **256,** 5916.
Yarden, Y., Escobedo, J. A., Kuang, W.-J., Yang-Feng, T. L., Daniel, T. O., Tremble, P. M., Chen, E. Y., Ando, M. E., Harkins, R. N., Francke, U., Fried, V. A., Ullrich, A., and Williams, L. T. (1986). *Nature (London)* **323,** 226.
Yoshida, Y. (1961). *Protoplasma* **54,** 476.
Zielke, H. R., and Filner, P. (1971). *J. Biol. Chem.* **246,** 1772.

Viroids 14

T. O. DIENER
R. A. OWENS

I. Introduction
II. The Biochemical Uniqueness of Viroids
III. The Biochemical Significance of Viroids
IV. Molecular Structure
    A. Early Studies
    B. Primary and Secondary Structures
    C. Nucleotide Sequences and Viroid Domains
V. Viroid Function
    A. Subcellular Location
    B. *In Vivo* Structure
    C. Question of Translation
    D. Mechanism of Replication
VI. Analysis of Structure/Function Relationships
    A. Requirements for Infectivity of Viroid cDNAs
    B. Mutational Analysis of Biological Function
VII. Mechanisms of Pathogenicity
VIII. Possible Viroid Origins
    References

I. INTRODUCTION

With the discovery of the viroids in the early 1970s, a group of replicating entities far simpler than viruses has become available to the molecular biologist. It is likely that their study will contribute, as has the study of viruses, to our knowledge of basic biological phenomena.

Viroids are low-molecular-weight RNAs ($1.1-1.3 \times 10^5$) of a unique, previously unknown structure that have been isolated from certain higher plant species afflicted with specific maladies. They are not detectable in healthy individuals of the same species but, when introduced into such individuals, they are replicated autonomously in spite of their small size and cause the appearance of the characteristic disease syndrome. Thus, viroids are the causative agents of the diseases in question. Unlike viral nucleic acids, viroids are not encapsidated, and no virion-like particles can be isolated from infected tissue.

Although viroids have been discovered because of their propensity to cause readily recognizable disease symptoms in certain hosts, they are sometimes replicated in other species without causing obvious damage to the host. Recent results, in fact, suggest that viroids may be more common in nature than previously believed and that they are not restricted to diseased plants.

Originally, the term "viroid" was introduced on the basis of the properties of the infectious agent responsible for the potato spindle tuber disease (Diener, 1971b). These properties were found to differ fundamentally from those of conventional viruses in at least five important respects:

1. The pathogen exists *in vivo* as an unencapsidated RNA.
2. Virion-like particles are not detectable in infected tissue.
3. The infectious RNA is of low molecular weight.
4. Despite its small size, the infectious RNA is replicated autonomously in susceptible cells; that is, no helper virus is required.
5. The infectious RNA consists of one molecular species only.

Work from several laboratories has subsequently confirmed these postulates and has led to a vast increase in our knowledge of the structural and functional properties of these unusual pathogens.

Although the above five criteria constituted ample justification to classify these pathogens as a taxon separate from viruses, recent studies have shown that this disparity is far greater than initially envisioned. In this chapter, the basic distinctions between viruses and viroids are first examined, known structural and functional properties of viroids are then summarized, and finally the question of viroid origin is considered. No attempt is made to cover aspects of viroids and viroid diseases that are not directly of biochemical or molecular biological concern. For more general treatments, the reader is referred to a number of reviews (Diener, 1979a,b, 1983; Sänger, 1982, 1984; see also Maramorosch and McKelvey, 1985; Diener, 1987).

II. THE BIOCHEMICAL UNIQUENESS OF VIROIDS

Aside from the five parameters listed above, viroids differ from viruses in a still more fundamental fashion. Whereas all viruses contain genetic information that is translated into one or more virus-specific proteins at some stage during their reproductive cycle, viroids do not contain such information. Viruses are entirely dependent upon their hosts' translational capacities, utilizing the latter to code for enzymes (or at least subunits thereof) and for structural proteins. Viruses may, therefore, be regarded as obligate parasites of their hosts' translational machinery. At the same time they are either independent or only partially dependent on their hosts' transcriptional systems. Viroids, on the other hand, are completely dependent on the host transcriptional systems and have altogether dispensed with the need for translation. Undoubtedly, this complete dependence on host biochemical systems permits the minimal information content of viroids. All they require are appropriate signal sequences to trigger their replication in susceptible cells. Viroids, therefore, represent an extreme form of obligate parasitism.

III. THE BIOCHEMICAL SIGNIFICANCE OF VIROIDS

Because of their unique properties, viroids are of considerable interest and importance to molecular biology. The primary reasons for this can be stated as follows:

1. What are the molecular signals which viroids possess (and cellular RNAs evidently lack) that induce so far unspecified host enzyme(s) to accept them as templates and to transcribe viroids into RNA strands of opposite polarity?

2. What are the molecular mechanisms of viroid replications? Are these mechanisms also operative in uninfected cells?

3. How do viroids induce disease in infected organisms? In the absence of viroid-specified proteins, one is forced to assume that this process is due to a direct interaction between the viroid (or its complement) and certain host constituents. The nuclear location of viroids suggests a direct interaction with the host genome—with potentially important lessons for the understanding of gene regulation in eukaryotic cells.

4. Why are viroids apparently restricted to higher plants? Or do viroids, after all, have animal counterparts?

5. How did viroids originate?

It is with these questions in mind that the present status of viroid research is summarized below.

IV. MOLECULAR STRUCTURE

A. Early Studies

Early electron microscopy of purified potato spindle tuber viroid (PSTV) revealed a uniform population of rods (~50 nm long) with widths similar to double-stranded (ds) DNA (Sogo *et al.*, 1973), suggesting that the viroid is a dsRNA. The thermal denaturation properties of PSTV, however, were not compatible with this concept (Diener, 1972). T_m values of 50° to 58°C [0.01–0.1 × SSC] for various viroids suggested that native viroids are single-stranded RNA molecules with hairpin-like configurations and extensive regions of intramolecular base pairing (Diener, 1972; Sogo et al., 1973).

Visualization of fully denatured viroid molecules led to the discovery that most are covalently closed circular molecules with circumferences approximately twice the length of native molecules (McClements, 1975; Sänger et al., 1976; McClements and Kaesberg, 1977). A variable proportion of linear molecules, about twice the length of native viroids, could also be discerned in purified viroid preparations. Although some of the linear viroid molecules result from nicking of circular molecules during isolation and purification (Sänger et al., 1979), linear viroid molecules are as infectious as circular ones (Owens et al., 1977; Palukaitis and Symons, 1980) and apparently exist as such in infected cells (Hadidi and Diener, 1978). Quantitative thermodynamic and kinetic studies of viroid thermal denaturation (Henco et al., 1977) suggested that viroids contain an uninterrupted double helix of 52 bp as well as several short, double-helical stretches, and Henco and co-workers proposed a tentative model for the secondary structure of viroids. Subsequent refinement of this model (Langowski et al., 1978) indicated that viroids exist in their native conformation as extended rodlike structures characterized by a series of double-helical sections and internal loops. Between 250 and 300 nucleotides are needed to account for the thermodynamic properties of the molecules. On the average, each helical sequence of 4 to 5 bp is followed by a defect in the form of an internal loop of two bases. Thus, the rigid, rodlike structure of the native viroid is based on a defective rather than a homogeneous RNA helix (Domdey et al., 1978).

Detailed analysis of the fine structure melting of several viroids by fast and slow temperature jump methods disclosed intermediate (T_m ~57°C) and high temperature transitions (T_m = 68°C) in addition to the highly cooperative main transition at 46.5° to 49°C (Henco et al., 1979; Gross, 1980). The intermediate transition corresponds to the dissociation of two hairpins with 5–10 bp each and 10–20 nucleotides in the loop, whereas the high transition corresponds to a hairpin of 10 bp and more than 40 bases in the loop (Henco et al., 1979).

For a more detailed discussion of viroid structure, the reader is referred to three excellent reviews (Riesner et al., 1979; Gross and Riesner, 1980; Riesner and Gross, 1985). Here we shall examine in some detail only the known primary and secondary structures of viroids.

B. Primary and Secondary Structures

The first complete nucleotide sequence of a viroid was that of the type strain of PSTV (Gross et al., 1978). This sequence was arranged in a secondary structure that maximized the number of base pairs in the molecule, and the model was then refined by considering (1) the location of cleavage sites resulting from controlled enzymatic digestions of PSTV, and (2) modification of reactive cytidines by bisulfite (Gross *et al.*, 1978). Studies of the binding of specific tRNA anticodons to complementary sequences in certain loop regions of PSTV confirmed this model (Wild *et al.*, 1980).

Features of particular interest include a stretch of 18 purines, mostly adenosines, in positions 48–65 and the absence of modified nucleotides and AUG translation initiator triplets (Gross *et al.*, 1978). These studies also indicated that tertiary folding of PSTV does not occur *in vitro*. The growing library of viroid nucleotide sequences allows features that are common to all viroids to be distinguished from those that are unique. Figure 1 presents nucleotide sequences and most probable secondary structures for eight viroids. Although their lengths vary from about 250 to 370 nucleotides, all viroids contain extensive regions of intramolecular complementarity which result in the formation of many short base-paired regions interrupted by small mismatched loops. Thermodynamically, unbranched structures are preferred over branched structures, thus the molecules assume rodlike, quasi-double-stranded configurations.

While overall sequence homologies between two different viroids vary from 35 to 76% (Keese and Symons, 1985), extensive homologies among all viroids exist in the upper and lower central portion of the structure. This region, known as the *central conserved region,* is virtually identical in all viroids except the avocado sunblotch viroid (ASBV) and is assumed to be essential to viroid function.

Another general feature of viroids is the presence of 1–3 pairs of inverted repeat sequences that each extend over 9 to 10 nucleotides and are capable of forming hairpin-like structures. These hairpins do not exist in the thermodynamically preferred configuration of the viroid in solution at room temperature, but are transiently formed during thermal denaturation of the viroid (Henco *et al.*, 1979; Steger *et al.*, 1984). The universal presence of at least one pair of these inverted repeats suggests that they are essential to viroid function.

C. Nucleotide Sequences and Viroid Domains

Sequence analysis of naturally occurring PSTV isolates that vary in the severity of symptoms they produce in tomato has shown that what appear to be minor sequence variations can have profound biological effects (Sänger, 1982; Schnölzer *et al.*, 1985). As shown in Fig. 2, most of the nucleotide exchanges, insertions, or deletions are clustered within a small region about 50 nucleotides from the left-hand terminal loop. In some isolates, an additional A is also found in the upper portion of the right half of the structure (Fig. 2). Although this insertion is believed by some investigators (Sänger, 1984) to be required to maintain a nucleotide number of 359, a requirement

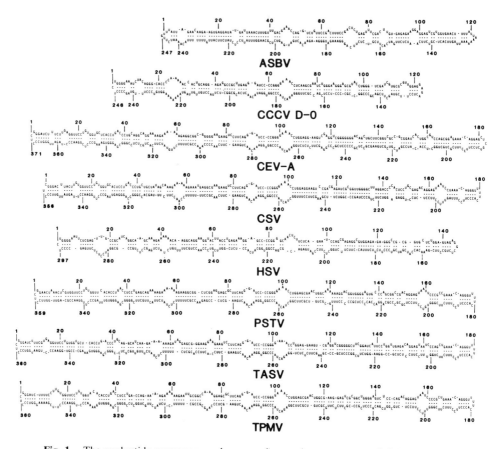

Fig. 1. The nucleotide sequences and proposed secondary structures of eight viroid "species." In the PSTV sequence, the central conserved region (C domain) extends from nucleotides 74 to 120 and 241 to 286; the pathogenicity-modulating region (P domain) from 47 to 73 and 287 to 314, including the polypurine stretch (nucleotides 47 to 73); the variable region (V domain) from 121 to 148 and 213 to 240; and the terminal regions (T domains) from 315 to 46 and 149 to 212. Abbreviations: ASBV, avocado sunblotch viroid; CCCVD-O, the D-O strain of the coconut cadang-cadang viroid; CEV-A, the Australian isolate of the citrus exocortis viroid; CSV, chrysanthemum stunt viroid; HSV, hop stunt viroid; PSTV, potato spindle tuber viroid; TASV, tomato apical stunt viroid; and TPMV, tomato planta macho viroid. Reproduced with permission from Keese and Symons (1987).

for a constant nucleotide number is difficult to understand. Characterized citrus exocortis viroid (CEV) strains contain 371–375 nucleotides (Visvader and Symons, 1985), and viroids such as tomato planta macho viroid and tomato apical stunt viroid that are closely related to PSTV contain 360 nucleotides (Kiefer et al., 1983).

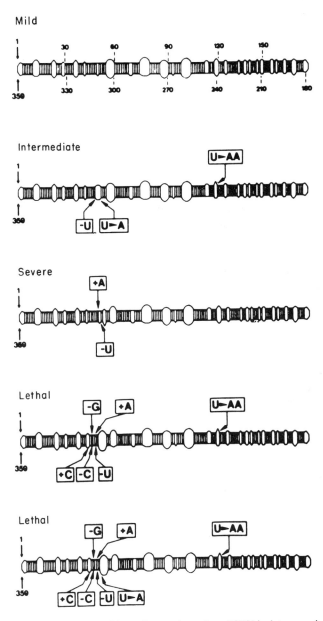

Fig. 2. Location of the nucleotide exchanges in various PSTV isolates causing mild, intermediate, severe, and lethal symptoms in tomato. The exchanges are related to the mild isolate as standard and indicated by boxes. It should be noted that the resulting effects on the secondary structure are not shown. Reproduced with permission from Sänger (1984).

An interesting observation connecting viroid structure with symptom severity was made by Schnölzer et al. (1985), who recognized that increasing symptom severity of PSTV could be correlated with the decreasing thermodynamic stability of a portion of the secondary structure. This region is often referred to as the "virulence-modulating" or "pathogenicity" region. Although the majority of sequence differences among various CEV isolates occur within an analogous region, no such relationship between symptom severity and structural stability can be observed (Visvader and Symons, 1985).

Comparative analysis of viroid sequences has also led to the recognition of other regularities. Thus, Keese and Symons (1985) divide viroids of the PSTV group into five domains (see legend to Fig. 1): (1) the central conserved region, which is the most highly conserved portion of the viroid structure (C domain); (2) the above-mentioned pathogenicity-modulating region, including the polypurine stretch (P domain); (3) a portion of the right half viroid molecule which is the most variable portion among the several viroids analyzed (V domain); and (4, 5) the two terminal loops, which show a high degree of sequence homology (T domains).

V. VIROID FUNCTION

When viroids are introduced into susceptible cells, they replicate without the assistance of a helper virus (Diener, 1971b). This basic biological fact raises a number of intriguing questions.

1. By what mechanisms are viroids replicated? Because viroids introduce only a very limited amount of genetic information into host cells, it appears that preexisting host enzymes are largely or entirely responsible for viroid replication.

2. How do viroids incite diseases in certain hosts, yet replicate in the majority of susceptible plant species without discernible damage to the host?

3. How did viroids originate?

A. Subcellular Location

Bioassays of subcellular fractions from PSTV-infected tomato leaves demonstrated that only the tissue debris and nuclear fractions contain appreciable infectivity; chloroplasts, mitochondrial, ribosomal, and "soluble" fractions contain only traces of infectivity (Diener, 1971a; Schumacher et al., 1983). CEV is also located primarily in the nuclear fraction (Sänger, 1972), but a significant portion has also been reported to be associated with a plasma-membrane-like component of the endomembrane system of infected *Gynura aurantiaca* (Semancik et al., 1976).

The fact that infectious PSTV is located primarily in the nuclei of infected cells does not prove that it is synthesized there. However, experiments with an *in vitro* RNA-synthesizing system, in which purified cell nuclei from infected tomato leaves were used as an enzyme source, suggested that this is the case (Takahashi and Diener, 1975). It appears, therefore, that the infecting viroid migrates to the nucleus (by a mechanism as yet unknown) and is replicated there. The absence of significant amounts of PSTV in the cytoplasmic fraction of infected cells suggests that most of the progeny viroid remains in the nucleus. Recently, a more extensive fractionation of the nuclei revealed that the viroid is associated exclusively with nucleoli (Schumacher *et al.*, 1983).

B. *In Vivo* Structure

Although purified viroids are composed solely of RNA (and yet have all the biological properties of the pathogen present in crude extracts), and no virion-like nucleoprotein particles have ever been detected in infected plants, the possibility nevertheless exists that viroids occur in the form of complexes with cellular constituents *in vivo*. This possibility has been discussed (Diener, 1979a,b; Sänger, 1982), but only recently has evidence been obtained for the existence of such complexes. By *in vitro* reconstitution experiments and isolation/characterization of *in vivo* complexes, Wolff *et al.* (1985) have shown that PSTV is associated with histones, as well as with M_r 41,000 and 31,000 proteins. The predominant viroid-containing species present in a nucleosomal fraction obtained from nucleoli was a 12–15S complex (Wolff *et al.*, 1985).

C. Question of Translation

Viroids are of sufficient chain length to code for a polypeptide of about 10,000 Da, although the uneven nucleotide number of circular PSTV theoretically permits three rounds of translation with a frame shift each time. Neither PSTV nor CEV, however, functions as a messenger RNA in a variety of cell-free protein-synthesizing systems (Davies *et al.*, 1974; Hall *et al.*, 1974). CEV is also not translated in *Xenopus laevis* oocytes (even after polyadenylation *in vitro*) and does not interfere with the translation of endogenous messenger RNAs (Semancik *et al.*, 1977).

A second approach to the question of viroid translation has involved a search for viroid specific proteins in infected host tissue. Comparisons of protein species in healthy and PSTV-infected tomato (Zaitlin and Hariharasubramanian, 1972) and healthy and CEV-infected *G. aurantiaca* (Conejero and Semancik, 1977) did not reveal qualitative differences between healthy and infected plants. In both studies, synthesis of at least two proteins was enhanced in infected as compared with healthy tissue, but subsequent studies indicated that these proteins are host- and not viroid-specific (Flores *et al.*, 1978; Conejero *et al.*, 1979). Two-dimensional gel electrophoretic analyses of proteins synthesized in uninfected or PSTV-infected tomato cells derived from suspension cultures revealed neither quantitative nor qualitative changes resulting from maintenance of the viroid in the cell line (Zelcer *et al.*, 1981).

The apparent lack of mRNA activity for PSTV and CEV is not surprising because both viroids lack an AUG initiation codon (see Fig. 1). Although all sequenced viroids contain one or more GUG codons, it is unlikely that these function as translation initiators, because AUG has been shown to be the only initiation codon in eukaryotic cells (Sherman *et al.*, 1980). On the other hand, both characterized CSV strains contain one AUG codon (Haseloff and Symons, 1981), and ASBV contains three AUG codons (Symons, 1981). Potential polypeptide products of CSV and ASBV, ranging in length from 24 to 63 amino acids, could be translated from the nucleotide sequences, starting with the various AUG codons, but major differences between the possible translation products of CSV and PSTV led Haseloff and Symons (1981) to conclude that it is unlikely that either viroid codes for proteins involved in their replication.

Viroids might, however, be translated *in vivo* from a complementary RNA strand synthesized by preexisting host enzymes using the infecting viroid as a template. RNA sequences complementary to viroids have been identified in infected tissue (see Section V), and it has been suggested that these might act as mRNAs (Matthews, 1978). cCEV and cASBV contain single AUG codons which could permit synthesis of a polypeptide with 42 and 5 amino acid residues, respectively (Visvader *et al.*, 1982; Symons, 1981), but the complementary strands of PSTV and CSV (cPSTV or cCSV) lack AUG initiation codons. Visvader *et al.* (1985) have pointed out that conservation of only a single polypeptide containing 15 amino acids and initiated by a GUG among CEV variants suggests that CEV does not encode any functional polypeptides.

Although more sensitive methods of analysis might conceivably yet disclose the presence of viroid-specified polypeptides in infected cells, this appears unlikely, and one must conclude that viroids do not act as mRNAs. If so, the complementary RNA sequences found in infected tissues must be synthesized entirely by preexisting (but possibly activated) host enzymes. Hence, viroids are the only autonomously replicating pathogens that do not code for pathogen-specific proteins such as replicases or subunits thereof.

D. Mechanism of Replication

1. RNA- or DNA-Dependent Replication?

Theoretically, viroid replication could involve transcription from either RNA or DNA templates. An RNA-directed mechanism requires, as an intermediate, RNA sequences complementary to the entire viroid in infected tissue, as well as a preexisting host enzyme with the specificity of an RNA-directed RNA polymerase. A DNA-directed mechanism would require the presence of DNA sequences complementary to the entire viroid. Although, in the past, data favoring both DNA- and RNA-directed replication have been reported, it is now evident that viroid replication occurs from RNA templates.

To distinguish between RNA- and DNA-directed replication, the effects of certain antibiotic compounds on viroid replication have been investigated. Three such studies (Diener and Smith, 1975; Takahashi and Diener, 1975; Mühlbach and Sänger, 1979) reporting inhibition by actinomycin D seem to suggest that DNA-directed synthesis is involved in viroid replication. Contradictory results have been reported by Grill and Semancik (1980), who concluded that actinomycin D had no specific inhibitory effect on viroid replication and that the inhibitory effects previously reported were due to a general toxic effect of actinomycin D on cell metabolism.

The inhibitory effect of α-amanitin reported by Mühlbach and Sänger (1979) is less likely to be due to nonspecific, secondary effects of the compound. This conclusion is strengthened by the demonstration that intracellular α-amanitin concentrations sufficient to inhibit viroid replication by about 75% did not affect the biosynthesis of either tobacco mosaic virus RNA or prominent cellular RNA species (tRNA, 5S RNA, 7S RNA, and ribosomal RNA).

2. Enzymes Involved

Rackwitz *et al.* (1981) have shown that purified DNA-directed RNA polymerase II isolated from plant cells can transcribe several synthetic and natural RNA templates *in vitro,* albeit at an efficiency about two orders of magnitude lower than with DNA templates. Of all natural RNA templates tested, purified viroids were transcribed with the highest efficiency. Gel electrophoretic analysis of the *in vitro* transcription products under denaturing conditions revealed both full-length linear molecules and several smaller viroid-complementary molecules (Rackwitz *et al.,* 1981).

These findings suggest that viroids are replicated by a novel mechanism in which the infectious RNA molecules are copied entirely by a preexisting (or activated) host enzyme, and that this enzyme may be normally DNA-directed RNA polymerase II. Possibly, it is the quasi-double-stranded DNA-like native structure of viroids which permits the enzyme to function in this capacity. Indeed, it has been shown that viroids readily form binary complexes with RNA polymerase II, that they compete with DNA for the template binding sites on the enzyme, and that they strongly inhibit DNA-directed RNA synthesis (Rackwitz *et al.,* 1981). In this view, the infecting viroid molecule commandeers nuclear RNA polymerase II for its own replication, and the viroids may be regarded, as has been suggested previously

(Diener, 1980), as "selfish" RNAs in the sense introduced for noncoding DNA (for a review, see Lewin, 1981).

In a recent study on the synthesis of PSTV "plus" and "minus" strands in isolated nuclei, Spiesmacher et al. (1985) evaluated the synthesis of PSTV-related RNA by "transcription–hybridization" analysis. They confirmed that the nucleus is the site of PSTV synthesis and concluded that viroid RNAs of both polarities are synthesized in the nucleus. Inhibition experiments with actinomycin D and α-amanitin suggested that infectious PSTV (the plus strand) is synthesized by DNA-dependent RNA polymerase I and the minus strand by RNA polymerase II (Spiesmacher et al., 1985). However, involvement of either of these enzymes in the *in vivo* replication of viroids remains to be unequivocally demonstrated.

Nuclear RNA polymerase II, however, is not the only host enzyme capable of transcribing viroid templates. Uninfected plants of several species contain enzymes with the specificity of RNA-dependent RNA polymerases (Fraenkel-Conrat, 1979). Boege et al. (1982) have shown that such an enzyme, isolated from healthy tomato, will accept PSTV as template for the *in vitro* synthesis of full-length linear minus strands.

3. Replication Intermediates

Another approach to the study of viroid replication involves the identification of viroid-related RNA or DNA sequences in nucleic acid extracts by molecular hybridization. Three types of molecular probes have been used in such studies: (1) purified viroids labeled *in vitro* with ^{125}I; (2) *in vitro*-prepared, single-stranded, viroid-complementary DNA (cDNA); and (3) double-stranded, viroid-related DNA obtained by recombinant DNA technology.

Initially, in experiments using ^{125}I-labeled viroids as probes, two groups reported viroid-complementary sequences in the DNA of viroid-infected (Semancik and Geelen, 1975) and even uninfected host plants (Hadidi et al., 1976). Later work, however, showed these reports to be in error. In the former case, the reported DNA sequences were subsequently identified as viroid-complementary RNA (Grill and Semancik, 1978) and, in the latter case, it is clear that neither DNA from uninfected nor DNA from PSTV-infected tomato plants contains sequences complementary to major portions of the viroid. Thus, no viroid-complementary regions could be identified in DNA from either uninfected or viroid-infected tomato plants by conventional solution and filter hybridization techniques (Zaitlin et al., 1980) or by Southern blot hybridization using either ^{125}I-labeled PSTV (Branch and Dickson, 1980) or fully defined ^{32}P-labeled, cloned ds PSTV cDNA (Hadidi et al., 1981) as probes. Because the experimental sensitivity was adequate to detect less than one copy of viroid complement per haploid genome, the latter two studies rule out the presence of even a single complete and contiguous complement of PSTV in host DNA. These experiments, however, do not rule out the possibility that viroid-related sequences might be randomly located on host chromosomes or that host DNA contains only a small portion of the PSTV genome. Short, viroid-complementary DNA sequences could conceivably serve as recognition sites and be involved in viroid pathogenesis, but any such viroid-related DNA sequences could not act as templates for the synthesis of progeny viroids.

Further evidence that viroids are replicated from RNA intermediates and not from DNA integrated into the host genome comes from the observation that the primary structure of a viroid is faithfully maintained, regardless of the host in which the viroid is replicated (Dickson et al., 1978; Niblett et al., 1978; Owens et al., 1978). This result would be expected if the incoming viroid (and not the host DNA) serves as the template for progeny synthesis. Finally, the most convincing evidence for an RNA-directed mechanism of viroid replication is the presence of viroid-complementary RNA molecules in nucleic acid extracts from infected plants. Such molecules are not found in extracts from uninfected plants and presumably represent intermediates in the process of viroid replication.

Viroid-complementary RNA sequences (minus strands) were first detected in extracts from CEV-infected tomato and G. aurantiaca leaves by solution hybridization with a ^{125}I-labeled viroid probe (Grill and Semancik, 1978). Grill et al. (1980) could draw no firm conclusions about the size(s) of the detected viroid-complementary molecules because the RNAs were not denatured before analysis and the gel electrophoresis was conducted under nondenaturing conditions. Viroid-complementary RNA templates from which progeny viroids are transcribed must contain a full complement of the viroid sequence, i.e., they must be of equal or greater length than the viroid.

The first convincing evidence for the existence in infected plants of full-length RNA molecules complementary to a viroid was obtained by Owens and Cress (1980) in blot hybridization experiments using viroid-specific recombinant dsDNA probes. Hybridization experiments demonstrated the presence of RNA molecules of the same mobility (and presumably molecular weight) as linear PSTV, but of opposite polarity. PSTV-complementary RNA molecules of this size were found after treatment of the nucleic acid extracts with RNase in the presence of 0.3 M NaCl, denaturation of ribonuclease-resistant RNAs, and gel electrophoresis at 55°C in the presence of 8 M urea.

In addition to full-length minus strands, viroid-infected cells also contain viroid-specific molecules longer than unit length. The first suggestion that such viroid-related RNAs may exist was obtained in blot hybridization experiments with nucleic acid extracts from PSTV-infected plants, in which two minus strand RNA species that migrated more slowly than PSTV were observed (Hadidi and Hashimoto, 1981). PSTV-specific RNA molecules with electrophoretic mobilities slower than those of circular or linear PSTV were also observed in another study in which nucleic acid extracts from PSTV-infected plants were separated by gel electrophoresis and in which viroid-specific molecules were identified by Northern blot hybridization

(Rohde and Sänger, 1981). Seven RNA species complementary to PSTV were observed, six of which migrated more slowly than circular PSTV and one with about the same mobility as linear PSTV (Rohde and Sänger, 1981).

Convincing evidence for longer-than-unit-length PSTV minus strands was obtained in similar blot hybridization studies with fully denaturing gel electrophoresis systems (Branch *et al.*, 1981). The four discrete bands identified contained approximately 700, 1050, 1500, and 1800 nucleotides, suggesting that they correspond to multimers of PSTV expected to contain 718 (dimer), 1077 (trimer), 1436 (tetramer), and 1795 (pentamer) bases. No unit-length minus strands were detected, probably because of interference by the significant quantities of unlabeled PSTV moving to the same position in the gel as minus-strand monomers (Hutchins *et al.*, 1985). Enzymatic studies indicated that the PSTV minus strands are composed exclusively of RNA and are extracted as complexes containing extensive ds regions.

Branch *et al.* (1981) hypothesized that the longer-than-unit-length PSTV minus strands play a role in viroid replication and that complexes containing ds regions of the length of PSTV represent replication intermediates. Viroid replication was suggested to occur by a rolling circle-type mechanism.

The molecular structure of the PSTV-specific RNA molecules has been further clarified by Owens and Diener (1982), who presented evidence that these molecules may represent intermediates in viroid replication. Hybridization probes specific for either PSTV or its minus strand were prepared from a cloned dsDNA representing the complete 359-nucleotide sequence of PSTV. Blot hybridization experiments revealed the presence of viroid-related, mostly dsRNA species that migrate more slowly than unit-length PSTV in RNA extracts from infected tissue. The more slowly migrating dsRNA zones would contain minus-strand dimers and higher multimers, whereas the more rapidly migrating dsRNA zone would contain circular ds molecules of unit length with ss minus-strand tails of varying lengths. These structures are probably fragments of a larger PSTV replicative intermediate complex because semisynchronous synthesis of PSTV is accompanied by simultaneous synthesis of dsPSTV (Owens and Diener, 1982).

The formation of longer-than-unit-length structures can most readily be explained if one assumes that minus strands are synthesized on a circular PSTV template and that this synthesis continues past the origin of replication, leading to the synthesis of linear dimers and higher multimers of cPSTV. Such a scheme resembles the rolling circle model previously advanced to explain replication of certain viral RNAs (Brown and Martin, 1965). Additional experiments suggested that unit-length linear PSTV synthesized from a minus strand template may be circularized while still complexed to the template.

The study of viroid replication intermediates has been extended to CEV by Branch and Robertson (1984) and to ASBV by Bruening *et al.* (1982). Multimeric strands of either polarity could be identified with ASBV, and large amounts of plus-strand oligomers up to octamers could be identified (Bruening et al., 1982). Based on these results, theoretical schemes for viroid replication that are based on a rolling circle-type mechanism have been proposed (Owens and Diener, 1982; Branch and Robertson, 1984; Ishikawa et al., 1984; Hutchins et al., 1985) (Fig. 3).

In summary, much of the mystery that has surrounded the mechanism of viroid replication in the past appears to have been dispelled. Although results of the studies described above differ in detail, they nevertheless converge toward a unified concept of the molecular mechanisms involved. This concept includes the following postulates:

1. Viroids are transcribed from complementary RNA (and not DNA) templates.

2. These templates, as well as progeny viroids, are synthesized by preexisting (or activated) host enzyme(s)—possibly RNA polymerase I and/or RNA polymerase II functioning as RNA-dependent RNA polymerases (replicases).

3. Replication intermediates are synthesized from infecting circular viroids by a rolling circle-type mechanism that results in the formation of oligomers of opposite polarity from which progeny viroids are in turn transcribed.

With ASBV, a somewhat different mechanism may be operative, as evidenced by the high level of plus-strand oligomers and the low level of minus-

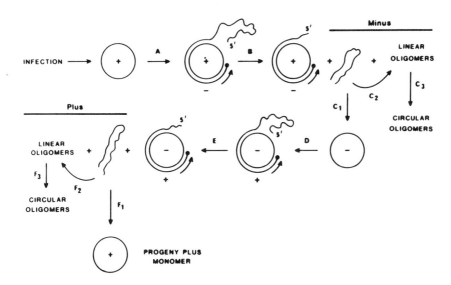

Fig. 3. Rolling circle model for the replication of viroids and virusoids. The small filled circle represents the RNA polymerase and the associated arrow the direction of RNA synthesis. A, B, C, D, E, F represent sequential steps in the replication scheme, with alternative paths labeled 1, 2, 3, etc. Reproduced with permission from Hutchins et al. (1985).

strand molecules (Bruening *et al.*, 1982). These authors have suggested that infecting monomeric ASBV serves as template for the synthesis of a circular minus strand, from which the oligomeric plus-strand viroids are transcribed by the rolling circle mechanism (Bruening *et al.*, 1982).

4. The presence in infected plants of oligomeric viroid replication intermediates requires that a specific cleavage–ligation mechanism exists for their precise cleavage into monomeric viroid lengths and ligation into circular progeny viroids.

4. Viroid Processing

Conceptually, the process of viroid cleavage and ligation (to form the viroid circle) resembles the cleavage–ligation reaction by which introns are spliced out of precursor RNAs and exons are joined to form functional RNA. Possible connections between viroids and introns have been postulated by several authors (Roberts, 1978; Crick, 1979; Diener, 1979a), and the presence of a nucleotide sequence in the PSTV complement that exhibits complementarity with the 5'-end of small nuclear RNA U1 (as do nuclear-encoded mRNA introns) (Diener, 1981; Gross *et al.*, 1982) has added plausibility to such speculations. Recent evidence indicates, however, that nuclear mRNA introns are excised in the form of lariat RNAs containing, at the branch site, an unusual nuclease-resistant structure with 2'-5' and 3'-5' phosphodiester bonds joined to a single residue (Konarska *et al.*, 1985). No such structures have been detected in viroids.

Analysis of viroid nucleotide sequences has disclosed the presence of features characteristic of the group I introns found in nuclear rRNA and certain mitochondrial mRNA and rRNA introns (Dinter-Gottlieb, 1986). These features include a 16-nucleotide consensus sequence and three pairs of short complementary sequences (boxes 9L and 2; 9R and 9R'; A and B, see Fig. 1 in Diener, 1986).

VI. ANALYSIS OF STRUCTURE/FUNCTION RELATIONSHIPS

A. Requirements for Infectivity of Viroid cDNAs

Evidence from several laboratories has demonstrated that appropriately constructed, viroid-specific cDNAs obtained by recombinant DNA technology, as well as *in vitro* RNA transcripts derived therefrom, are infectious when introduced into viroid-susceptible plants (Cress *et al.*, 1983; Ohno *et al.*, 1983; Tabler and Sänger, 1984, 1985; Meshi *et al.*, 1984; Ishikawa *et al.*, 1984; Owens *et al.*, 1986). Such plants develop the characteristic symptoms of viroid infection and progeny viroids of the predicted sequence are synthesized. Evidently, the infecting cDNAs must be transcribed into viroid-specific RNAs, from which progeny viroids are then synthesized. Although the exact mechanisms of this process are unknown, experimental results point to specific requirements that must be met in order for the cDNAs or their RNA transcripts to be infectious (Table I).

Most conspicuous are the high levels of infectivity with all multimeric viroid cDNAs tested, regardless of polarity, and the low levels or absence of infectivity with DNAs containing monomeric inserts. Excision of monomeric DNA inserts from their vectors results in greatly increased levels of infectivity (Table I). As has been suggested (Tabler and Sänger, 1984; Meshi et al., 1984), it is likely that these restriction fragments undergo ligation *in vivo* to form multimers, particularly when (as is the case with all constructs reported) excision results in the formation of cohesive ends, thus facilitating ligation. *In vivo* ligation by a host enzyme cannot, however, explain the trace levels of infectivity often observed with plasmids containing monomeric viroid cDNA inserts (Table I).

TABLE I

Infectivities of Cloned, Viroid-Specific cDNAs and RNA Transcripts

Size of viroid-specific insert	Type of nucleic acid inoculated	Polarity of transcript[a]	Infectivity[b]	References[c]
Monomer[d]	ssDNA	+	− or ±	1
		−	−	1
	dsDNA in vector	+	− or ±	1, 2
		−	− or ±	1, 2
	excised	+	+ or ++	1, 3
	RNA (*E. coli*)	+	±	2
	RNA (*in vitro* transcripts)	+	−	4, 5
		−	−	4, 5
Multimer	ssDNA	+	+++	1
		−	+++	1
	dsDNA in vector	+	+++	1, 2
		−	+++	1, 2
	excised	+	+++	1, 2
	RNA (*in vitro* transcripts)	+	+++	4, 5
		−	−	4, 5

[a] +, polarity of infectious viroid; −, polarity of viroid complement.

[b] −, no infectivity; ±, trace; +, low; ++, medium; +++, high levels of infectivity.

[c] (1) Tabler and Sänger (1984); (2) Cress et al. (1983); (3) Meshi et al. (1984); (4) Ishikawa et al. (1984); (5) Tabler and Sänger (1985).

[d] With six or fewer terminal nucleotide duplications.

From Diener (1986).

The circularity of viroids and the presumed rolling circle type of replication suggest that repeat units of the viroid sequence may be needed to permit transcription of the entire sequence from a linear template with a specific initiation site (which is unlikely to be at the exact beginning of a monomeric insert). Naturally occurring linear viroid monomers (Owens et al., 1977) and artificially nicked viroids with 2′,3′-cyclic phosphate termini (Hashimoto et al., 1985) are as infectious as circular molecules, but such linear molecules are presumably ligated in vivo by an RNA ligase before replication takes place, as usual, by transcription from the resulting circular template. RNAs transcribed in vitro from DNAs containing monomeric inserts of viroid sequences are noninfectious, whereas those from multimeric inserts are highly infectious, provided that the transcripts are of the polarity of the infectious viroid (Table I).

All of these results indicate that more than a monomeric, viroid-specific DNA or RNA is required for expression of infectivity exceeding trace levels. Infectivity levels of dimeric inserts are as high as those of trimeric or tetrameric inserts, but whether less than a complete dimer would suffice is not evident from these experiments.

An important clue regarding the required extent of sequence duplication has come from experiments with monomeric, ds PSTV cDNAs inserted into different vectors. Whereas plasmid pBR322 containing a monomeric, BamHI-derived ds PSTV cDNA unit, inserted into the BamHI site of the plasmid, is either noninfectious (Cress et al., 1983) or only marginally infectious (Tabler and Sänger, 1984), the same cDNA inserted into either bacteriophage M13 DNA (Tabler and Sänger, 1984) or plasmids pUC9 or pSP64 (Owens et al., 1986) results in constructs with relatively high levels of infectivity (but not as high as those of plasmids with dimeric inserts) (not shown in Table I).

Examination of the plasmid sequences adjacent to the junction between the vector DNA and the viroid-specific insert reveals that insertion of the BamHI PSTV unit into plasmid pBR322 leads to a clone consisting of the 359 nucleotides of the monomeric PSTV sequence plus 6 PSTV-specific nucleotides originating from the vector, whereas insertion into M13, pUC9, or pSP64 vectors results in clones consisting of the 359-nucleotide monomeric PSTV sequence plus, coincidentally, 11 PSTV-specific nucleotides (GGATCCCCGGG). Tabler and Sänger (1984) have pointed out that this difference of five nucleotides seems to be essential for the infectivity of the cloned PSTV cDNA and that, interestingly, the viroid region in question is part of the central region that is strictly conserved in all viroids.

Clearly, far less than a complete dimer is required for infectivity, and there seems to be a direct relationship between the extent of sequence duplication and the level of infectivity. Whereas a 6-nucleotide duplication results in a clone with trace amounts or no infectivity, an 11-nucleotide duplication results in a clone with a substantially higher infectivity. The infectivity of clones containing an 11-nucleotide duplication, however, is still below that of a clone containing a complete duplication of the viroid sequence. Diener (1986) has proposed a hypothetical model for viroid processing that may explain this requirement for sequence duplication.

The model identifies a thermodynamically extremely stable base-paired configuration that partially or completely dimeric, as well as higher, viroid oligomers can assume. It postulates that this structure, which involves structural features common to all viroids (the central conserved region and secondary hairpin I), is essential for precise cleavage and ligation (Fig. 4). The model explains the strong infectivity of recombinant plasmids containing tandem repeats of two or more viroid sequences, the lesser infectivity of certain plasmids containing partially duplicated viroid-specific inserts, and the marginal infectivity of plasmids containing monomeric inserts. The model also accounts for the fact that vector-derived sequences on either or both sides of the viroid sequence(s) of a restriction fragment are precisely excised and do not appear in progeny viroids. It is also compatible with recent evidence clearly implicating the upper conserved region of viroids as the cleavage–ligation site (Meshi *et al.*, 1985; Visvader *et al.*, 1985), as well as with results from site-specific mutagenesis of infectious viroid cDNAs (Visvader *et al.*, 1985; Owens *et al.*, 1986).

B. Mutational Analysis of Biological Function

To date, it has proven more difficult than expected to bring the full power of site-directed DNA mutagenesis techniques to bear upon the question of how viroid conformation is related to biological function.

Mutational analysis of PSTV and HSV has revealed that the viroid molecule is exceedingly sensitive to nucleotide exchanges, deletions, or additions. Even nucleotide exchanges that seem unlikely to affect the thermodynamic stability of the native configuration, such as the conversion of a G-C to a G-U base pair, often are lethal (Ishikawa *et al.*, 1985, Hammond and Owens 1987; Owens *et al.*, 1987).

A major problem has been the lack of *in vitro* assays for defined biological functions similar to those used by workers studying the role(s) of brome mosaic virus RNA 3 in viral replication (Dreher *et al.*, 1984). In this regard, a recent report of the conversion of multimeric PSTV RNAs synthesized *in vitro* into unit-length, circular molecules by plant nuclear extracts (Tsagris *et al.*, 1987) is encouraging.

Recent infectivity studies with PSTV (Hammond *et al.*, 1989) and TASV (Candresse, Owens, and Diener, unpublished) cDNAs containing site-

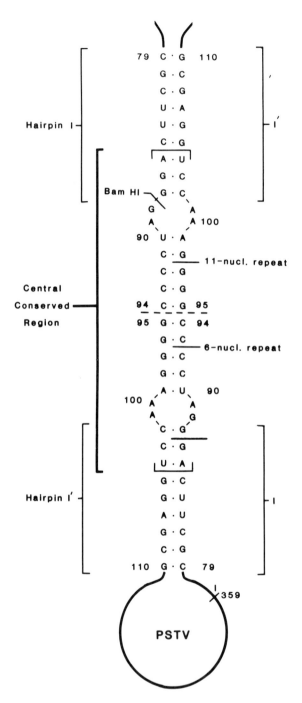

Fig. 4. The highly base-paired, thermodynamically stable configuration that dimeric or higher oligomeric viroids can assume. Contributions to this structure by the central conserved region and secondary hairpin I (I, I′) are indicated. The extent of sequence duplication of some recombinant clones containing *Bam*HI-bordered, viroid-specific inserts is shown. From Diener (1986).

directed alterations are compatible with processing of longer-than-unit-length molecules at the putative splice site discussed in the previous section, but other experiments suggest that alternative, albeit less preferred, sites can be utilized (Hammond *et al.*, 1989; Sänger *et al.*, 1988). As pointed out before (Diener, 1986), the lower central conserved region in longer-than-unit-length molecules of PSTV-group viroids can assume a palindromic structure similar to that of the upper central conserved region. Although the structure possible with the lower strand is thermodynamically less stable than that possible with the upper strand, the lower-strand processing site may be used with constructs in which, because of site-directed mutagenesis, the upper site is not available (Hammond *et al.*, 1989).

Undoubtedly, this model represents an oversimplification of reality. For example, it does not consider the likely possibility that viroid replication takes place in specific viroid–host protein complexes. As discussed above, association of viroids with specific proteins has been demonstrated (Wolff *et al.*, 1985) and, in such complexes, the viroid RNA may assume a secondary structure that greatly differs from that believed to exist in solution.

VII. MECHANISMS OF PATHOGENICITY

Another intriguing question concerns viroid pathogenicity: By what mechanisms do viroids incite diseases in certain hosts, yet replicate in other susceptible species without inflicting discernible damage? The nuclear location of viroids and their apparent inability to act as mRNAs suggest that viroid-induced disease symptoms may be caused by direct interaction of the viroid with the host genome; that is, by interference with gene regulation in the infected cells. If so, viroids might be regarded as abnormal regulatory molecules (Diener, 1971b).

Many of the symptoms induced by viroid infection, such as stunting of plants, epinasty, curling, and deformation of leaves, suggest viroid-induced disturbances in the metabolism of growth substances. Indeed, in a comparison of the concentrations of some plant growth substances in healthy and CEV-infected *G. aurantiaca* plants, an auxinlike substance of unknown chemical nature was found to be formed as a consequence of viroid infection (Rodriguez *et al.*, 1978). Also, a significant decrease in the levels of endogenous gibberellins (probably GS_3 and/or GA_1) was observed in viroid-infected, as compared with healthy plants, but no changes in the levels of abscisic or indoleacetic acids were detectable.

Certain alterations of protein synthesis have been noted in both PSTV- and CEV-infected plants. These aberrations may be related to the pathogenic properties of viroids and may constitute intermediate points in the causal chain leading from viroid infection to the appearance of macroscopic symptoms in infected plants. With CEV, two low molecular weight proteins, P_1 of 15,000 and P_2 of 18,000, accumulate in infected plants (Flores et al., 1978). Convincing evidence that P_1 and P_2 are host proteins and not translation products of the viroid or its complementary strand has been reported (Conejero et al., 1979). These authors demonstrated that the molecular weights of P_1 and P_2 depend upon the particular CEV-infected host species from which they are isolated and that senescence in healthy G. aurantiaca plants induces the same low molecular weight proteins as does infection with CEV.

In addition to the macroscopic symptoms characteristic of certain viroid-infected plants, cytopathic effects of viroid infection have also been observed. Thus, CEV infection of G. aurantiaca plants has been reported to result in the appearance of membranous structures derived from the plasma membrane, so-called "plasmalemmasomes" or "paramural bodies" (Semancik and Vanderwoude, 1976). These bodies vary in size, internal structure, and shape, and their origin and function are not yet entirely understood. Appearance of these paramural bodies has been regarded as the primary cytopathic effect of viroid infection and, in view of the claimed association of CEV with plasma membranes (Semancik et al., 1976), their appearance has been regarded as suggesting a direct causal relationship with the pathogenic RNA.

In another study of cytopathic effects of viroid infection, however, plasmalemmasomes were found to be present at equal frequency in CEV-infected and healthy G. aurantiaca plants (Sänger, 1982), indicating that they could not be the primary cytopathic effect of viroid infection. Viroid infection, nevertheless, was shown to affect the structure of plasmalemmasomes.

Hari (1980), on the other hand, reported that PSTV-infected tomato leaf cells develop paramural bodies, as has been reported earlier for CEV-infected G. aurantiaca cells. Another, previously unreported, cytopathic effect involves chloroplasts of infected cells, in which aberrations of the thylakoid membrane systems and lack of development of grana were observed (Hari, 1980).

A possible clue to explain the mechanism of viroid pathogenesis may be the recent discovery that an $M_r 68K$ host protein is differentially phosphorylated in extracts from viroid-infected and mock-inoculated tissues. This phosphoprotein is immunologically related to a double-stranded RNA-dependent protein kinase from virus-infected, interferon-treated human cells (Hiddinga et al., 1988). These findings suggest, but do not prove, that this protein, which is similar to double-stranded RNA-dependent protein kinases implicated in mammalian systems in the regulation of protein synthesis and virus replication, is involved in both plant virus and viroid pathogenesis. If so, plant viruses and viroids would affect similar host systems and thus explain the similarity of viroid- and virus-induced symptoms in plants.

VIII. POSSIBLE VIROID ORIGINS

At the time when the viroid concept was initially advanced (Diener, 1971b), viroids could be regarded as either very primitive or degenerate relatives of conventional viruses. Knowledge accumulated since then and summarized in this chapter has rendered this concept increasingly less likely. The lack of mRNA activity and novel molecular structure (see Sections IV,B and V,C) imply a far greater phylogenetic distance from viruses than could be previously imagined.

Comparative sequence analysis of five related viroids (PSTV, CEV, CSV, TASV, and TPMV) has revealed striking similarities with the ends of transposable genetic elements (Kiefer *et al.*, 1983). The presence of inverted repeats often ending with U-G and C-A and flanking imperfect direct repeats suggests that viroids may have originated from transposable elements or retroviral proviruses by deletion of interior sequences. Alternatively, these similarities between viroids and transposable genetic elements could be a consequence of convergent evolution. In the former view, viroids would represent (be derived from?) RNA elements with the capacity to integrate into cellular RNA. No evidence for RNA integration mechanisms exists, but Zimmern (1982) has speculated that viroids and RNA viruses may have evolved comparatively recently from two interacting classes of RNA molecules—mobile, circular, viroid-like "signal RNAs" and linear "antenna RNAs." These two classes of RNA are assumed to be normally involved in genetic exchange between cells via RNA recombination and amplification.

On the other hand, it is also possible that viroids originated quite early in precellular (or early cellular) evolution when the primary genetic material was RNA. Two lines of evidence, recent studies of protein-free RNA processing reactions and eukaryotic gene structure as well as comparative sequence analysis of prokaryotic and eukaryotic ribosomal RNAs, have led Darnell and Doolittle (1986) to propose an hypothesis in which the precursor of the eukaryotic nuclear genome is traced back to the earliest stages in evolution. In this scheme, protein-free RNA synthesis as well as site-specific RNA cleavage and splicing could all have been available for use in primitive RNA-dominated genetic systems, and fairly sophisticated RNA molecules are envisioned as early participants in evolution. In this view, viroids may be *living fossils,* RNA molecules that have survived (and evolved) since their origin during the very early prebiotic stages of evolution.

REFERENCES

Boege, F., Rohde, W., and Sänger, H. L. (1982). *Biosci. Rep.* **2,** 185–194.
Branch, A. D., and Dickson, E. (1980). *Virology* **104,** 10–26.
Branch, A. D., and Robertson, H. D. (1984). *Science* **223,** 450–455.

Branch, A. D., Robertson, H. D., and Dickson, E. (1981). *Proc. Natl. Acad. Sci. U.S.A.* **78**, 6381–6385.
Brown, F., and Martin, S. J. (1965). *Nature (London)* **208**, 861–863.
Bruening, G., Gould, A. R., Murphy, P. J., and Symons, R. H. (1982). *FEBS Lett.* **148**, 71–78.
Conejero, V., and Semancik, J. S. (1977). *Virology* **77**, 221–232.
Conejero, V., Picazo, I., and Segado, P. (1979). *Virology* **97**, 454–456.
Cress, D. E., Kiefer, M. C., and Owens, R. A. (1983). *Nucleic Acids Res.* **11**, 6821–6835.
Crick, F. (1979). *Science* **204**, 264–271.
Darnell, J. E., and Doolittle, W. F. (1986). *Proc. Natl. Acad. Sci. U.S.A.* **83**, 1271–1275.
Davies, J. W., Kaesberg, P., and Diener, T. O. (1974). *Virology* **61**, 281–286.
Dickson, E., Diener, T. O., and Robertson, H. D. (1978). *Proc. Natl. Acad. Sci. U.S.A.* **75**, 951–954.
Diener, T. O. (1971a). *Virology* **43**, 75–89.
Diener, T. O. (1971b). *Virology* **45**, 411–428.
Diener, T. O. (1972). *Virology* **50**, 606–609.
Diener, T. O. (1979a). *Science* **205**, 859–866.
Diener, T. O. (1979b). "Viroids and Viroid Diseases." Wiley (Interscience), New York.
Diener, T. O. (1980). *Plant Dis. Etiol., Meet. Fed. Br. Plant Pathol. Soc. Gen. Microbiol., 1980* p. 8 (Abstr.).
Diener, T. O. (1981). *Proc. Natl. Acad. Sci. U.S.A.* **78**, 5014–5015.
Diener, T. O. (1983). *Adv. Virus Res.* **28**, 241–283.
Diener, T. O. (1986). *Proc. Natl. Acad. Sci. U.S.A.* **83**, 58–62.
Diener, T. O. (ed.) (1987). "The Viroids." Plenum, New York.
Diener, T. O., and Smith, D. R. (1975). *Virology* **63**, 421–427.
Dinter-Gottlieb, G. (1986). *Proc. Natl. Acad. Sci. U.S.A.* **83**, 6250–6254.
Domdey, H., Jank, P., Sänger, H. L., and Gross, H. J. (1978). *Nucleic Acids Res.* **5**, 1221–1236.
Dreher, T. W., Bujarski, J. J., and Hall, T. C. (1984). *Nature (London)* **311**, 171–175.
Flores, R., Chroboczek, J., and Semancik, J. S. (1978). *Physiol. Plant Pathol.* **13**, 193–201.
Fraenkel-Conrat, H. (1979). *Trends Biochem. Sci.* **4**, 184–186.
Grill, L. K., and Semancik, J. S. (1978). *Proc. Natl. Acad. Sci. U.S.A.* **75**, 896–900.
Grill, L. K., and Semancik, J. S. (1980). *Nature (London)* **283**, 399–400.
Grill, L. K., Negruk, V. I., and Semancik, J. S. (1980). *Virology* **107**, 24–33.
Gross, H. J. (1980). *Hoppe-Seyler's Z. Physiol. Chem.* **361**, 477–492.
Gross, H. J., and Riesner, D. (1980). *Angew. Chem., Int. Ed. Engl.* **19**, 231–243.
Gross, H. J., Domdey, H., Lossow, C., Jank, P., Raba, M., Alberty, H., and Sänger, H. L. (1978). *Nature (London)* **273**, 203–208.
Gross, H. J., Krupp, G., Domdey, H., Raba, M., Alberty, H., Lossow, C. H., Ramm, K., and Sänger, H. L. (1982). *Eur. J. Biochem.* **121**, 249–257.
Hadidi, A., and Diener, T. O. (1978). *Virology* **86**, 57–65.
Hadidi, A., and Hashimoto, J. (1981). *Phytopathology* **71**, 222.
Hadidi, A., Jones, D. M., Gillespie, D. H., Wong-Staal, F., and Diener, T. O. (1976). *Proc. Natl. Acad. Sci. U.S.A.* **73**, 2453–2457.
Hadidi, A., Cress, D. E., and Diener, T. O. (1981). *Proc. Natl. Acad. Sci. U.S.A.* **78**, 6932–6935.
Hall, T. C., Wepprich, R. K., Davies, J. W., Weathers, L. G., and Semancik, J. S. (1974). *Virology* **61**, 486–492.
Hammond, R. H., and Owens, R. A. (1987). *Proc. Natl. Acad. Sci.* **84**, 3967–3971.
Hammond, R. H., Owens, R. A., and Diener, T. O. (1989). *Virology* (in press).
Hari, V. (1980). *Phytopathology* **70**, 385–387.
Haseloff, J., and Symons, R. H. (1981). *Nucleic Acids Res.* **9**, 2741–2752.
Hashimoto, J., Suzuki, K., and Uchida, T. (1985). *J. Gen. Virol.* **66**, 1545–1551.
Henco, K., Riesner, D., and Sänger, H. L. (1977). *Nucleic Acids Res.* **4**, 177–194.

Henco, K., Sänger, H. L., and Riesner, D. (1979). *Nucleic Acids Res.* **6**, 3041–3059.
Hiddinga, H. J., Crum, C. J., and Roth, D. A. (1988). *Science* **241**, 451–453.
Hutchins, C. J., Keese, P., Visvader, J. E., Rathjen, P. D., McInnes, J. L., and Symons, R. H. (1985). *Plant Mol. Biol.* **4**, 293–304.
Ishikawa, M., Meshi, T., Ohno, T., Okada, Y., Sano, T., Ueda, I., and Shikata, E. (1984). *Mol. Gen. Genet.* **196**, 421–428.
Ishikawa, M., Meshi, T., Okada, Y., Sano, T., and Shikata, E. (1985). *J. Biochem. (Tokyo)* **98**, 1615–1620.
Keese, P., and Symons, R. H. (1985). *Proc. Natl. Acad. Sci. U.S.A.* **82**, 4582–4586.
Keese, P., and Symons, R. H. (1987). *In* "The Viroids" (T. O. Diener, ed.). Plenum, New York, pp. 37–62.
Kiefer, M. C., Owens, R. A., and Diener, T. O. (1983). *Proc. Natl. Acad. Sci. U.S.A.* **80**, 6234–6238.
Konarska, M. M., Grabowski, P. J., Padgett, R. A., and Sharp, P. A. (1985). *Nature (London)* **313**, 552–557.
Langowski, J., Henco, K., Riesner, D., and Sänger, H. L. (1978). *Nucleic Acids Res.* **5**, 1589–1610.
Lewin, R. (1981). *Science* **213**, 634.
McClements, W. L. (1975). Ph.D. thesis. University of Wisconsin, Madison, Wisconsin.
McClements, W. L., and Kaesberg, P. (1977). *Virology* **76**, 477–484.
Maramorosch, K., and McKelvey, J. J. (eds.) (1985). "Subviral Pathogens of Plants and Animals: Viroids and Prions." Academic Press, New York.
Matthews, R. E. F. (1978). *Nature (London)* **276**, 850.
Meshi, T., Ishikawa, M., Ohno, T., Okada, Y., Sano, T., Ueda, I., and Shikata, E. (1984). *J. Biochem. (Tokyo)* **95**, 1521–1524.
Meshi, T., Ishikawa, M., Watanabe, Y., Yamaya, J., Okada, Y., Sano, T., and Shikata, E. (1985). *Mol. Gen. Genet.* **200**, 199–206.
Mühlbach, H.-P., and Sänger, H. L. (1979). *Nature (London)* **278**, 185–188.
Niblett, C. L., Dickson, E., Fernow, K. H., Horst, R. K., and Zaitlin, M. (1978). *Virology* **91**, 198–203.
Ohno, T., Ishikawa, M., Takamatsu, N., Meshi, T., Okada, Y., Sano, T., and Shikata, E. (1983). *Proc. Jpn. Acad., Ser. B* **59**, 251–254.
Owens, R. A., and Cress, D. E. (1980). *Proc. Natl. Acad. Sci. U.S.A.* **77**, 5302–5306.
Owens, R. A., and Diener, T. O. (1982). *Proc. Natl. Acad. Sci. U.S.A.* **79**, 113–117.
Owens, R. A., Erbe, E., Hadidi, A., Steere, R. L., and Diener, T. O. (1977). *Proc. Natl. Acad. Sci. U.S.A.* **74**, 3859–3863.
Owens, R. A., Smith, D. R., and Diener, T. O. (1978). *Virology* **89**, 388–394.
Owens, R. A., Hammond, R. W., Gardner, R. C., Kiefer, M. C., Thompson, S. M., and Cress, D. E. (1986). *Plant Mol. Biol.* **6**, 179–192.
Owens, R. A., Hammond, R. W., and Diener, T. O. (1987). *In* "Plant Molecular Biology" (D. Von Wettstein and N.-H. Chua, eds.). Plenum, New York, pp. 483–494.
Palukaitis, P., and Symons, R. H. (1980). *J. Gen. Virol.* **46**, 477–489.
Rackwitz, H. R., Rohde, W., and Sänger, H. L. (1981). *Nature (London)* **291**, 297–301.
Riesner, D., and Gross, H. J. (1985). *Annu. Rev. Biochem.* **54**, 531–564.
Riesner, D., Henco, K., Rokohl, U., Klotz, G., Kleinschmidt, A. K., Domdey, H., Jank, P., Gross, H. J., and Sänger, H. L. (1979). *J. Mol. Biol.* **133**, 85–115.
Riesner, D., Steger, G., Schumacher, J., Gross, H.-J., Randles, J. W., and Sänger, H. L. (1983). *Biophys. Struct. Mech.* **9**, 145–170.
Roberts, R. J. (1978). *Nature (London)* **274**, 530.
Rodriguez, J. L., Garcia-Martinez, J. L., and Flores, R. (1978). *Physiol. Plant Pathol.* **13**, 355–363.
Rohde, W., and Sänger, H. L. (1981). *Biosci. Rep.* **1**, 327–336.

Sänger, H. L. (1972). *Adv. Biosci.* **8,** 103–116.
Sänger, H. L. (1982). *Encycl. Plant Physiol., New Ser.* **14B,** 368–454.
Sänger, H. L. (1984). *In* "The Microbe 1984: Part I. Viruses" (B. W. J. Mahy and J. R. Pattison, eds.), Soc. Gen. Microbiol. Symp. 36, pp. 281–334. Cambridge University Press, London.
Sänger, H. L., Klotz, G., Riesner, D., Gross, H. J., and Kleinschmidt, A. K. (1976). *Proc. Natl. Acad. Sci. U.S.A.* **73,** 3852–3856.
Sänger, H. L., Ramm, K., Domdey, H., Gross, H. J., Henco, K., and Riesner, D. (1979). *FEBS Lett.* **99,** 117–122.
Sänger, H. L., Tabler, M., and Tsagris, M. (1988). Fifth Intntl. Congress Plant Pathology, Kyoto, Japan, Abstracts, p. 40.
Schnölzer, M., Haas, B., Ramm, K., Hofmann, H., and Sänger, H. L. (1985). *EMBO J.* **4,** 2181–2190.
Schumacher, J., Sänger, H. L., and Riesner, D. (1983). *EMBO J.* **2,** 1549–1555.
Semancik, J. S., and Geelen, J. L. M. C. (1975). *Nature (London)* **256,** 753–756.
Semancik, J. S., and Vanderwoude, W. J. (1976). *Virology* **69,** 719–726.
Semancik, J. S., Tsuruda, D., Zaner, L., Geelen, J. L. M. C., and Weathers, L. G. (1976). *Virology* **69,** 669–676.
Semancik, J. S., Conejero, V., and Gerhart, J. (1977). *Virology* **80,** 218–221.
Sherman, F., McKnight, G., and Stewart, J. W. (1980). *Biochim. Biophys. Acta* **609,** 343–346.
Sogo, J. M., Koller, T., and Diener, T. O. (1973). *Virology* **55,** 70–80.
Spiesmacher, E., Mühlbach, H.-P., Tabler, M., and Sänger, H. L. (1985). *Biosci. Rep.* **5,** 251–265.
Steger, G., Hofmann, H., Förtsch, J., Gross, H. J., Randles, J. W., Sänger, H. L., and Riesner, D. (1984). *J. Biomol. Struct. Dyn.* **2,** 543–571.
Symons, R. H. (1981). *Nucleic Acids Res.* **9,** 6527–6537.
Tabler, M., and Sänger, H. L. (1984). *EMBO J.* **3,** 3055–3062.
Tabler, M., and Sänger, H. L. (1985). *EMBO J.* **4,** 2191–2199.
Takahashi, T., and Diener, T. O. (1975). *Virology* **64,** 106–114.
Tsagris, M., Tabler, M., Mühlbach, H.-P., and Sänger, H. L. (1987). *EMBO J.* **6,** 2173–2183.
Visvader, J. E., and Symons, R. H. (1985). *Nucleic Acids Res.* **13,** 2907–2920.
Visvader, J. E., Gould, A. R., Bruening, G. E., and Symons, R. H. (1982). *FEBS Lett.* **137,** 288–292.
Visvader, J. E., Forster, A. C., and Symons, R. H. (1985). *Nucleic Acids Res.* **13,** 5843–5856.
Wild, U., Ramm, K., Sänger, H. L., and Riesner, D. (1980). *Eur. J. Biochem.* **103,** 227–235.
Wolff, P., Gilz, R., Schumacher, J., and Riesner, D. (1985). *Nucleic Acids Res.* **13,** 355–367.
Zaitlin, M., and Hariharasubramanian, V. (1972). *Virology* **47,** 296–305.
Zaitlin, M., Niblett, C. L., Dickson, E., and Goldberg, R. B. (1980). *Virology* **104,** 1–9.
Zelcer, A., Van Adelsberg, J., Leonard, D. A., and Zaitlin, M. (1981). *Virology* **109,** 314–322.
Zimmern, D. (1982). *Trends Biochem. Sci.* **7,** 205–207.

Biochemistry of DNA Plant Viruses

15

ROBERT J. SHEPHERD

I. Introduction
II. Caulimoviruses (Double-Stranded DNA Viruses)
   A. Biology and Special Features
   B. Intrinsic Properties of Viruses and Viral DNA
   C. Genetic Organization
   D. Transcription and Translation
   E. Mapping of Polypeptide Products and Gene Functions of CaMV
   F. Replication of Caulimoviruses
   G. The Reverse Transcriptase of CaMV
   H. Insect Transmission of Caulimoviruses
III. Geminiviruses (Single-Stranded DNA Viruses)
   A. Biology and Intrinsic Features of Geminiviruses
   B. Properties of Virions and Viral DNA
   C. Genetic Organization of Geminiviruses
IV. Prospects for Using DNA Viruses as Gene Vectors
   References

I. INTRODUCTION

Two groups of DNA viruses, the caulimoviruses and geminiviruses, infect higher plants. The biochemistry of these viruses has been the subject of much recent activity and several notable discoveries will be reviewed here. The DNA viruses of lower plants, such as those of green algae, are not discussed.

The numbers and types of plant DNA viruses are relatively few in comparison with RNA viruses. Although the reason for this disparity is not known, it has been suggested that DNA viruses have had difficulty in adapting to higher plants and that the few that have adapted have done so by developing unusual biological properties (Howell, 1985b). Perhaps one factor to account for this is that DNA replication is more stringently regulated in plants than in other organisms.

Several reviews of the caulimoviruses have appeared recently; especially of note are those by Hohn et al. (1982), Howell (1982, 1985b), Dixon and Hohn (1985), Covey (1985), and Shepherd (1979, 1985). The properties of geminiviruses have also been reviewed, see those of Goodman (1981a,b), Howell (1982), Howarth (1985), Harrison (1985), Lazarowitz (1987), and Stanley (1985). In addition, Hohn et al. (1985) have reviewed evidence for reverse transcription of the cauliflower mosaic virus (CaMV) genome.

The DNA genomes of these viruses are easy to manipulate genetically. The naked DNAs of both caulimoviruses and some geminiviruses are infectious when rubbed over leaves of healthy plants with an abrasive, and the DNA of both groups can be cloned in an infectious form in bacteria. Cloning has facilitated the preparation of mutant genomes, propagation of various partial genomes, and preparation of recombinant genomes *in vitro* to test the biological activity of particular segments of DNA. Ease of DNA restructuring has greatly simplified the use of these viruses as model systems to study gene expression and pathogenesis in plants.

When extracts containing virus or their DNA are rubbed over the leaves of healthy plants with a mild abrasive, spontaneous infection occurs at a few sites on the leaves. From these initial infection foci the viruses multiply and move from cell to cell to establish systemic infections. The multiplication and movement of this foreign DNA in the plant is equivalent to causing a genetic transformation of the whole plant. This has led some investigators to consider these viruses as potential agents for transducing foreign genes into plants. However, the viruses are not carried through germ line cells and progeny of infected parents are free of infection.

Various investigators have attempted to use the DNA viruses as gene vectors for higher plants. Foreign DNA spliced into these viral genomes in the proper manner can be inoculated to plants to cause infection followed by subsequent movement of the foreign DNA throughout the plant. However, it is doubtful that any of the DNA plant viruses have conventional origins for DNA replication and the usual approaches cannot be applied using these viruses as gene vectors. For example, the reverse transcriptional mode for replication of caulimoviruses complicates their use as genetic vehicles. These and other problems in vector construction are addressed in this chapter.

Of the 25 taxa of viruses that infect plants, only two consist of viruses with DNA as their genomic material (Matthews, 1979). Most other plant viruses have single-stranded RNA genomes, with the great majority consisting of positive sense, single-stranded RNA viruses in which the genomic material also functions as messenger RNA. Two groups that are exceptions are the reoviridae, which have double-stranded RNA genomes, and the rhabdoviridae which have negative sense, single-stranded RNA genomes (Matthews, 1979).

The name "caulimovirus" was proposed for the double-stranded DNA viruses related to cauliflower mosaic, dahlia mosaic, and carnation etched ring viruses by the International Committee on Taxonomy of Viruses (Harrison et al., 1971). *Caulimo* is a contraction from *cauli*flower *mo*saic with cauliflower mosaic virus having been chosen as the type member for the group. Over a dozen viruses are now known to be in this group (Table I) and new caulimoviruses are being described continually. The viruses have roughly spherical particles of about 50 nm with a single piece of relaxed, circular, double-stranded DNA of about 8000 bp. Several members have aphids (order Homoptera, family Aphididae) as arthropod vectors.

The name "geminivirus" (from the Latin *gemini* meaning "twins") was proposed by Harrison et al. (1977) for the small viruses with paired particles that have single-stranded DNA genomes. The maize streak virus, which has

TABLE I

Caulimoviruses and Their Host Ranges

Virus[a]	Susceptible species and family
Blueberry red ringspot virus (BRRV)	*Vaccinium corymbosum* (Ericareae)
Carnation etched ring virus (CERV)	*Dianthus caryophyllus, D. barbatus, Saponaria* sp., and *Silene* sp. (in Caryophyllaceae)
Cassava vein mosaic virus (CVMV)	*Manihot esculenta* (Euphorbiaceae)
Cauliflower mosaic virus (CaMV)	Many *Brassica* sp. and other Cruciferae: some strains infect *Nicotiana* and *Datura* sp. (Solanaceae)
Dahlia mosaic virus (DaMV)	*Dahlia variabilis, Verbesina encelioides, Zinnia elegans,* (Compositae); some strains infect species of Amaranthaceae, Chenopodiaceae, and Solanaceae
Figwort mosaic virus (FMV)	*Scrophularia californica* (Scrophulariaceae) and various species of Chenopodiaceae and Solanaceae
Horseradish latent virus (HRLV)	*Armoracia rusticana, Brassica* sp. and other Cruciferae
Mirabilis mosaic virus (MMV)	*Mirabilis* sp. (Nyctaginaceae)
Peanut chlorotic streak virus (PCLSV)	*Arachis hypogaea* and other Leguminosae, *Datura stramonium, Nicotiana* sp., and other Solanaceae
Soybean chlorotic mottle virus (SoyCMV)	*Glycine max, Dolichos lablab, Phaseolus vulgaris*, and other Leguminosae
Strawberry vein banding virus (SVBV)	*Fragaria* sp. (Rosaceae)
Thistle mottle virus (ThMV)	*Cirsium arvense* (Compositae)

[a] CERV, CaMV, DaMV, FMV, HRLV, MMV, and SVBV have been transmitted by aphids. Hull (1984) has summarized biological and biophysical properties of individual caulimoviruses.

a genome consisting of one molecule of circular single-stranded DNA of about 2600 nucleotides, is the type member of the group. This virus and certain others have leafhoppers (order Homoptera, family Cicadellidae) as their arthropod vectors. Other members have genomes consisting of two single-stranded DNA molecules, each of about 2500–2700 nucleotides. The latter have whiteflies (order Homoptera, family Aleyrodidae) as insect vectors. About a dozen viruses are known in this group.

II. CAULIMOVIRUSES (DOUBLE-STRANDED DNA VIRUSES)

A. Biology and Special Features

Caulimoviruses cause a variety of diseases on cultivated and noncultivated agricultural species in temperate climates, with some members of the group occurring in tropical regions. Examples of tropical viruses are cassava vein mosaic and peanut chlorotic streak viruses (Table I) which occur in Brazil and subtropical India, respectively. The diseases are characterized by leaf discolorations (mosaics and mottles), and may be accompanied by vein-banding, vein-clearing, necrosis, puckering, and other types of distortion, and varying degrees of stunting. The severity of the disease differs with virus strain and host species and variety.

A single gene of CaMV may be largely responsible for disease induction (Daubert et al., 1984) and host range (Schoelz et al., 1986). Small in-frame insertions in this region of the CaMV genome attenuate disease (Daubert et al., 1983), while other portions of the viral genome seem to affect the host reaction to a lesser degree.

The caulimoviruses have moderately restricted host ranges. Generally, a virus in the group will infect many species in one or two families, but few plants in other families (Shepherd and Lawson, 1981) (Table I).

In nature, caulimoviruses are transmitted mainly by small, soft-bodied, sucking insects of the family Aphididae, commonly called aphids. These insects reproduce in prodigious numbers at certain times of the year, producing both winged and nonwinged forms which carry virus on their mouthparts. Transmission occurs as the result of feeding activity. Complementation experiments using transmissible and nontransmissible strains of CaMV have shown that an accessory factor produced in infected plants is required for insect transmission (Lung and Pirone, 1973, 1974). The accessory factor is probably a virus-specified protein produced in infected plants in addition to the virus itself.

The caulimoviruses induce in the cytoplasm of infected plants conspicuous inclusion bodies that appear as round or elongated bodies of one to several microns that enlarge slowly throughout the course of infection. Occurring in virtually every cell of systemically infected plants, they consist of an electron-dense, granular matrix interspersed with less dense vacuole-like regions. Virus particles of about 50 nm diameter are packed into the matrix and the vacuoles. No external membrane surrounds the body (Fig. 1). Most of the virus in the cell occurs as occluded material in the inclusion body. Only occasional virions are found in the cytoplasm or in abnormally enlarged plasmodesmata.

Incipient inclusion bodies appear as tiny masses of electron-dense matrix material surrounded by ribosomes (Kitajima and Lauritis, 1969) (Fig. 1), suggesting that the inclusion body is the major site in the cell for virus assembly. Autoradiographic evidence also indicates that the inclusion bodies may be sites active in viral DNA synthesis (Kamei et al., 1969; Fujisawa et al., 1971, 1972). Recent experiments, to be discussed later in this chapter, suggest that inclusion bodies are the site for reverse transcriptional synthesis of viral DNA.

The caulimoviruses also cause a conspicuous enlargement of the plasmodesmata (protoplasmic channels between cells) and virions are frequently found in these enlarged structures (Kitajima and Lauritis, 1969). This cytological effect may be a reflection of a virus function involved in cell-to-cell movement of virus.

B. Intrinsic Properties of Viruses and Viral DNA

Most features of virions and DNA of the caulimoviruses are based on work with CaMV since no other virus has been as well characterized. The virions are isometric particles about 50 nm in diameter with little substructure. With neutron scattering, Chauvin et al. (1979) found that the particle was 49 nm in diameter with a protein shell 6.5 nm thick with the DNA lining the interior of the shell. Kruse et al. (1987) have repeated these experiments using virus in buffers containing various amounts of D_2O to calculate a more detailed distribution of protein and DNA in the virion (Table II). An outer protein shell of low density containing approximately 60% of the total protein and no nucleic acid extends from 215 to 250 Å. Most of the DNA is distributed in shells II and III, which lie between 150 and 215 Å. Shell III contains about 42% of the DNA and 26% of the protein. The innermost region contains 11% of the protein (Table II).

Purified CaMV contains 16% DNA based on its phosphorus content (1.63%) and quantitative diphenylamine tests. The sedimentation coefficient is 208S in 0.1 M NaCl, 0.01 M phosphate, pH 7.2, at infinite dilution. The measured partial specific volume is 0.704 g/cm and the diffusion coefficient is 0.75 (± 0.04) $\times 10^{-7}$ cm^2/sec (Hull et al., 1976). From these features the molecular weight equivalent for the particle can be calculated as 22.8×10^6 (Hull et al., 1976). A value of $19.3–20.0 \times 10^6$ has been estimated by Kruse et al. (1987) from neutron small angle scattering.

Virus subjected to sodium dodecyl sulfate degradation gives several protein components during gel electrophoresis. Several of these arise from ag-

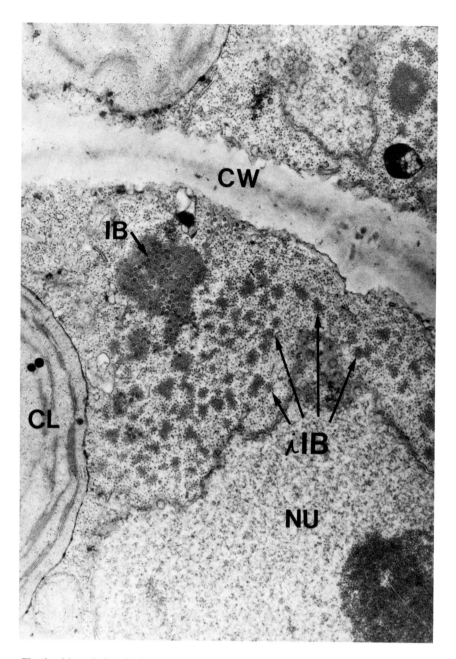

Fig. 1. Mesophyll cell of *Brassica campestris* infected with cauliflower mosaic virus. One well-developed inclusion body (IB) containing spherical virus particles occurs in the cytoplasm near the cell wall (CW). A few scattered virus particles occur free in neighboring regions of the cytoplasm. Note the numerous incipient inclusions (iIB), consisting mainly of electron-dense matrix material, that occur in the cytoplasm near the well-developed inclusion body. Prominent arrays of ribosomes cluster around these incipient bodies. CL, chloroplast. Courtesy of Kim Reinke (Department of Plant Pathology, University of Wisconsin).

TABLE II

Volume Fractions Occupied by Protein, DNA, and Water in
Each Shell of the Spherically Averaged CaMV Model[a]

Shell (Å)	Volume fraction of indicated component:		
	Protein	DNA	Water
I 120–150	0.27 (11.0)[b]	0.04 (6.4)[b]	0.74
II 150–185	0.04 (2.9)	0.18 (51.6)	0.78
III 185–215	0.29 (25.8)	0.12 (42.0)	0.59
IV 215–250	0.43 (60.3)	0.00 (0)	0.57

[a] As estimated by Kruse et al. (1987). Volume fractions are accurate to 0.05 for large and 0.02 for small values.

[b] Number in parentheses indicates the percentage of total protein or DNA present in the corresponding shell.

gregation or proteolytic degradation of a single protein of about 44 kDa (Al Ani et al., 1979). The protein is glycosylated (Hull and Shepherd, 1976; duPlessis and Smith, 1981) and phosphorylated (Hahn and Shepherd, 1980), and is probably processed from a higher-molecular-weight precursor (Hahn and Shepherd, 1982). As virions age *in vitro* after purification and to a lesser extent *in vivo*, the 44-kDa protein is reduced in size to 37 kDa (Hull and Shepherd, 1976).

Native DNA freed of protein at a concentration of about 0.5 μg/ml infects about half the seedlings (*Brassica campestris*) to which it is applied using an abrasive (Shepherd et al., 1970). Manual inoculation with an abrasive in this manner probably causes breaks in the cuticle and outer cell wall to expose the plasma membrane through which viral DNA is absorbed. Howell et al. (1980) and Lebeurier et al. (1980) found that cloned DNA of CaMV is infectious. The linear molecule when released from the plasmid vector by restriction at the cloning site is infectious to turnip seedlings to about the same extent as native DNA prepared from virions. Cloned DNA in sterile water at a concentration of about 10 μg/ml (10^{12} DNA molecules/ml) infects a high percentage of plants (Geldreich et al., 1986).

CaMV DNA from purified virions contains two sedimenting species of 17.1S and 19.0S which under the electron microscope consist of linear and relaxed circular molecules of 2.31 μm contour length (Shepherd and Wakeman, 1971). Only the circular form is infectious (Hull and Shepherd, 1977). DNA of CaMV has a T_m of 87.2°C in 0.15 M NaCl, 0.015 M sodium citrate and a buoyant density in CsCl of 1.70/g/cm^3 (Shepherd et al., 1970).

Several topological forms of CaMV DNA are released from degraded virions. These occur as simple relaxed circles or circles with one or more knots in the molecule as revealed by high-resolution electron microscopy (Menissier-deMurcia et al., 1983) (Fig. 2). These forms separate as a series of closely spaced bands when CaMV DNA is migrated electrophoretically into agarose gels (Fig. 2). The knotted forms of CaMV DNA probably arise from copying of the linear full-length RNA molecule which serves as a replicative intermediate in the viral replication cycle. Being linear, the ends

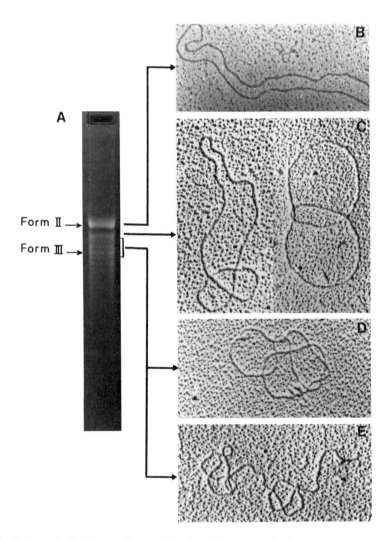

Fig. 2. Topological forms of encapsidated cauliflower mosaic virus DNA. The heterogeneity exhibited during gel electrophoresis of viral DNA is related to knotting of the molecules. Supercoiled DNA (form I) is not found in virions. Open circular (form II) is the most slowly migrating form in 1% agarose gels shown here (A). Linear (form III), which may be knotted, and relaxed circular knotted forms migrate more rapidly in gels. The panel on the right shows electron micrographs of unknotted circular (B), knotted circular (C and D), and knotted linear (E) DNA taken from 1% agarose gels. From Menissier et al. (1983). Courtesy of J. Menissier.

of the RNA are free to become knotted during thermal agitation before being reverse transcribed to produce circular DNA (see Section II,F).

A supercoiled form of the DNA active in transcription occurs in the nucleus (Olszewski et al., 1982; Menissier et al., 1984). It probably occurs as a minichromosome similar to that of simian virus 40 in nuclei of mammalian cells (Olszewski et al., 1982; Olszewski and Guilfoyle, 1983). The transcriptionally active minichromosome contains DNA associated with histones with a nucleosomal structure (Menissier-deMurcia et al., 1983) (Fig. 3). The repeating subunit is similar to that of plant chromatin (Phillips and Gigot, 1977).

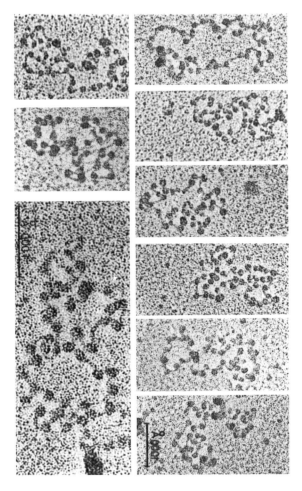

Fig. 3. Cauliflower mosaic virus minichromosomal DNA isolated from nuclei of infected turnip (*Brassica campestris*). Each panel shows a single circular DNA molecule. These exhibit an average of 41 globular nucleosomes per DNA molecule. From Menissier et al. (1983). Courtesy of J. Menissier.

Native DNA of CaMV virions has a relaxed nature because three single-stranded interruptions occur in the molecule at specific places (Volovitch *et al.*, 1976, 1978), two in the plus strand (also called the β-strand) and one in the minus strand (the α-strand) (Fig. 4). Some other caulimoviral DNAs, such as those of dahlia mosaic and figwort mosaic viruses, have three interruptions in the β-strand (Hull and Donson, 1982; Richins and Shepherd, 1983). The DNA of some deletion strains of CaMV, such as CM4-184 and some of the other caulimoviruses, e.g., those of strawberry vein banding and peanut chlorotic streak viruses, contain only two interruptions, one in each strand. By convention, the single interruption in the minus strand that is

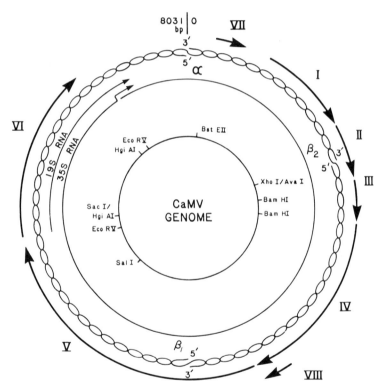

Fig. 4. A physical map of the cauliflower mosaic virus genome. The circular double-stranded DNA of about 8000 bp is indicated by the interwoven line. Encapsidated DNA has three specifically located interruptions, one (designated α) in the transcribed or minus strand, and two (designated β_1 and β_2) in the complementary or plus strand. The heavy peripheral arrows indicate the location of the eight translational open reading regions (tentative genes) which occur in the 35S RNA transcript. The two main RNA transcripts, 19S and 35S RNA, are indicated by the lighter inner lines. A restriction map showing a few salient restriction sites is shown in the center of the diagram. The polarity of the CaMV genome was established by Hull *et al.* (1979).

transcribed into RNA has been chosen as the zero point of the physical maps of caulimoviral genomes (Fig. 4).

DNA sequencing has shown that the interruptions are actually triple-stranded regions (overlaps) in which a short, redundant, non-base-paired tail protrudes over the double helix (Frank *et al.*, 1980). The 5′-OH terminus is probably the one which is displaced since it is most reactive to labeling with polynucleotide kinase (Frank *et al.*, 1980). Ribonucleotides are found covalently attached to the 5′-termini at interruption sites (Richards *et al.*, 1981). As will be discussed later, these are probably residual RNA primers left *in situ* from replication of the DNA genome by reverse transcription.

C. Genetic Organization

The genetic constitution of the caulimoviruses is based largely on sequencing of the DNA of three strains of CaMV and two other viruses in the group, carnation etched ring virus (CERV) and figwort mosaic virus (FMV). All three viruses have a similar genetic structure. The CaMV genome is a circular molecule of about 8000 bp (Fig. 4) with eight open reading frames (tentative genes) of sufficient size to code for polypeptides of greater than 11 kDa. These are indicated by the peripheral arrows in Fig. 4. The small region VIII that overlaps the C-terminal portion of region IV appears in all three strains of CaMV that have been sequenced, suggesting it may be of biological significance. However, this region is missing from CERV and FMV (Hull *et al.*, 1986; Richins *et al.*, 1987).

An unusual feature of the arrangement of the different coding regions, with the exception of region VI, is that they are very closely spaced, with generally only one or two nucleotides between the termination of one region and the start codon of the next. In several cases, short overlaps occur between successive regions, and in every case, the reading frame changes between consecutive genes. Gene VI of CaMV is separated from other coding regions by a short intergenic region of about 100 bp between region V and region VI and a large intergenic region of about 700 bp separates the end of gene VI and the initiation codon of gene VII (Fig. 4).

The various genes of CaMV code for polypeptides of 11 to 79 kDa (regions VIII and V, respectively) (Table III). Gene V, the reverse transcriptase gene, is the most highly conserved portion of the caulimoviral genome (Richins *et al.*, 1987). Genes VII and VIII are the least conserved regions, placing doubt on their biological significance. Gene VI is the least conserved coding region of genes common to all caulimoviruses (Richins *et al.*, 1987).

The analysis of mutants has provided evidence for the functional importance of genes I through VI (Howell *et al.*, 1982; Dixon *et al.*, 1983; Daubert *et al.*, 1983). Large insertions in gene II (Gronenborn *et al.*, 1981), or deletion of nearly the entire gene does not affect virus activity (Howarth *et al.*, 1981). This gene has been implicated in insect transmission of CaMV. Gene

TABLE III

Protein Coding Regions of the CaMV, CERV, and FMV Genomes

Coding region	Protein molecular weights and amino acid homologies		
	CaMV	CERV	FMV
I	328 (36,825)[a]	319 (36,531)[a] 48.3%[b]	323 (36,957) 54%, 49%[c]
II	159 (17,878)	168 (18,799) 53.6%	164 (18,934) 42%, 38%
III	129 (14,136)	128 (14,232) 41.4%	115 (12,659) 37%, 31%
IV	488 (56,614)	494 (56,888) 41.3%	489 (57,346) 38%, 33%
V	678 (78,698)	659 (76,518) 66.9%	666 (76,938) 64%, 64%
VI	520 (57,833)	496 (56,423) 33.1%	512 (58,133) 26%, 21%
VII	11,398[d]	10,301[d]	7,447
VIII	12,430	—	—
IX	—	—	11,924

[a] Assuming that translation begins with the first in frame ATG and ends with the first in frame stop codon within each open reading frame. The first figure gives the number of amino acid residues followed by the molecular weight in parentheses.

[b] These numbers are the percent direct homology of the amino acid sequences in each putative coding region.

[c] The first percentage given is the homology of FMV versus CaMV, the second gives the homology of FMV versus CERV.

[d] The homologies between region VII were so low that a reliable degree of homology could not be determined.

VII can also be interrupted by large inserts (Gardner, 1982; Daubert et al., 1983) or eliminated completely (Dixon and Hohn, 1984) without destroying virus infectivity. This region may be of some significance not related to its coding qualities since all caulimoviruses that have been sequenced have a small open reading frame in the large intergenic region upstream of the first essential coding region (gene I).

D. Transcription and Translation

Only one strand of the CaMV genome is transcribed to RNA (Howell and Hull, 1978; Odell and Howell, 1980; Guilfoyle, 1980). Two major polyadenylated transcripts, conspicuous in infected plants (Covey et al., 1981; Odell et al., 1981) and protoplasts (Howell and Hull, 1978), hybridize specifically to the DNA strand with a single interruption (minus strand). Up to 5% of the poly(A) RNA of infected plants is virus specific (Covey and Hull, 1981).

Of the two major species of polyadenylated RNA, the 19S transcript spans the gene VI region of the genome (Odell and Howell, 1980; Covey and Hull, 1981; Odell et al., 1981). Fine-scale mapping by S_1 nuclease and primer extension has positioned the 5'-end at nucleotide 5765 (Covey et al., 1981;

Dudley et al., 1982; Guilley et al., 1982). The other major virus-specific transcript, the 35S RNA, hybridizes to all EcoRI fragments on Northern blots and has a size similar to that of the minus strand of viral DNA. UV mapping experiments suggest that the synthesis of this transcript begins in the large intergenic region to the left of the α interruption (Fig. 4). Low-resolution S_1 nuclease mapping has confirmed this position and its 5'-extremity has been mapped to nucleotide 7435 (Covey et al., 1981; Dudley et al., 1982; Guilley et al., 1982). The 3'-end of the 35S transcript has been mapped to nucleotide 7615. Thus the 3'-end of 35S RNA is 180 nucleotides downstream of its 5'-end, leaving a direct repeat of 180 nucleotides on each end of the RNA molecule (Covey et al., 1981; Guilley et al., 1982). The significance of other species of 35S RNA that initiate and terminate at other locations on the CaMV genome is not clear (Condit and Meagher, 1983).

The 19S transcript terminates at the same point as the 35S transcript, i.e., position 7615. Hence the transcriptional terminator is probably the same for both RNAs. The lack of termination during initial transcription past the terminator to produce 35S RNA is not understood, and elements important to transcriptional termination have not been defined.

CaMV genomes integrated into plant chromosomes using an *Agrobacterium*-Ti plasmid exhibit a pattern of transcription similar to virus-infected plants, though the level of transcription of region VI (19S RNA) in *Nicotiana* was only 10% of that in *Brassica*, the native host of CaMV (Gardner et al., 1984; Shewmaker et al., 1985).

Mapping of RNA transcripts indicates that promoter elements exist in each intergenic region of the CaMV genome. TATA-box-like sequences occur about 30 bp upstream of each 35S RNA transcription initiation site, TATATAA for Cabb S, CM1841, and D/H strains. About 30 bp upstream of the cap site of the 19S RNA the sequence is TATTTAA for Cabb S and D/H strains and TATATAA for CM1841 (Frank et al., 1980; Gardner et al., 1981; Balazs et al., 1982). These are consistent with eukaryotic consensus sequences for the initiation of transcription (Breathnach and Chambon, 1981).

The upstream region of the 35S transcript has been used to promote the expression of a human growth hormone gene in plants, and effects of 5' deletions on its activity have been analyzed (Odell et al., 1985). The polyadenylation signal at +180 (downstream position with respect to the start site for the initiation of transcription) was removed by 3' deletion. The segment from +9 to -46 was adequate for accurate initiation of transcription but upstream sequences from -46 to -105 greatly increased the level of transcription (Odell et al., 1985). A CAAT sequence (CAT-box) and an inverted repeat in this region plus a sequence similar to the core enhancer of animals may be elements important in promoting transcription. The 35S promoter behaved as a constitutive promoter in all parts of the tobacco plant (leaves, stems, flowers, and roots) (Odell et al., 1985). CaMV region VI 5'/

3' expression signals are active in tobacco and petunia and a variety of other plants which are not hosts of CaMV (Paszkowski *et al.*, 1984; Koziel *et al.*, 1984; Potrykus *et al.*, 1985; Sanders *et al.*, 1987).

The report of Okada *et al.* (1986) that the 35S promoter of CaMV is active only during the S phase of dividing cells is difficult to reconcile with other observations on the biology of this virus. These investigators compared gene expression of a chloramphenicol acetyltransferase (CAT) gene linked to either a nopaline synthase promoter or a CaMV 35S promoter. Those DNA constructs were then electroporated into synchronized tobacco suspension cell protoplasts. In the case of the 35S viral promoter, CAT activity varied markedly with the cell cycle and the peak of CAT activity was found in cells at S phase. With the nopaline synthase promoter, no periodic activity associated with the cell cycle was observed. The S phase-specific activity associated with the CaMV 35S promoter in tobacco cells is unlikely to occur in natural hosts of the virus such as *Brassica campestris* (turnip), where the virus multiplies for a prolonged period after cell division ceases. Furthermore, prodigious levels of CaMV-specific transcription have been obtained in isolated nuclei from infected turnip leaves that have few if any cells in S phase (Guilfoyle, 1980). Thus the activity of the 35S promoter may differ in host and nonhost cells, or activity in rapidly dividing suspension cells cultures may be different from that in leaf parenchyma.

The 35S promoter of CaMV has been widely used in Ti-plasmid gene vectors to transcribe heterologous genes in transformed plants. For example, the popular and effective pMON series of gene vectors uses the 35S CaMV promoter to obtain foreign gene expression (Shah *et al.*, 1986). Promoter strength is an important consideration in transformation experiments where a high level of transcription of selected coding sequences is frequently desired. In tests where the 35S CaMV promoter has been compared with the nopaline synthase promoter, another commonly used promoter for chimeric gene constructs in Ti-plasmid gene vectors, the CaMV promoter gave transcription levels 30-fold higher. The corresponding levels of expression of neomycin phosphotransferase activity in a chimeric construct were 110-fold higher for the viral sequence versus the nopaline synthase promoter (Sanders *et al.*, 1987).

The specific sequences contributing to the transcriptional activity of the 35S promoter of CaMV have not been defined. However, the upstream sequences appear to act as a transcriptional enhancer that is active with heterologous as well as natural promoters. Insertion of a single copy of the upstream sequences of the 35S promoter (nucleotides -90 to -343 from the transcription start site), coupled to a nopaline synthase promoter directing a neomycin phosphotransferase II gene, increased transcription about 40-fold in transgenic tobacco plants. When the 35S promoter upstream sequences were used in tandem repeats with the nopaline synthase promoter, a further 10-fold increase in transcription occurred (Kay *et al.*, 1987).

A high level of virus-specific transcription has been found in nuclei isolated from CaMV-infected *Brassica*. Based on its sensitivity to α-amanitin and ionic strength, transcription is probably catalyzed by endogenous DNA-dependent RNA polymerase II (Guilfoyle, 1980). As mentioned earlier, transcriptionally active complexes in turnip nuclei have been shown to contain covalently closed, circular (supercoiled) forms of CaMV DNA, with a nucleosome structure similar to chromatin (Olszewski *et al.*, 1982; Menissier *et al.*, 1983).

Analyses of the translational products of total cellular RNA from CaMV-infected plants as compared with that from healthy plants revealed a major new polypeptide product of about 62 kDa, termed P62 by some investigators (Odell and Howell, 1980; Al Ani *et al.*, 1980; Covey and Hull, 1981). By hybrid-arrested translation this protein has been associated with *Eco*RI fragment *b*, corresponding to region VI of the CaMV genome (Odell and Howell, 1980; Ali Ani *et al.*, 1980), a conclusion confirmed by hybrid selection translation experiments (Covey and Hull, 1981). Based on these observations, the 19S RNA transcript has been implicated as the mRNA for P62.

Plant *et al.* (1985) have characterized a second subgenomic CaMV-specific RNA which is of somewhat dubious origin. It can be translated in a cell-free system to yield a 75-kDa product. The RNA was purified from total poly(A) RNA from infected leaves by hybrid selection on CaMV DNA attached to diazobenzyloxymethyl (DBM) paper. The inclusion body protein, P62, was a major translation product of total selected RNA in rabbit reticulocyte lysates. In addition, polypeptides of 80, 75, 57, 54, and 38 kDa were obtained. When various restriction fragments of CaMV DNA were attached to DBM paper for hybrid selection in the same manner, those containing region V, the probable reverse transcriptase of CaMV, selected RNA that gave a 75-kDa translation product. This RNA was found to sediment in sucrose density gradients at about 22S. The precise sequences represented by this RNA have not yet been established by nucleotide mapping, nor is it known how this transcript originates.

Other minor transcripts of CaMV have been described (Guilley *et al.*, 1982; Condit *et al.*, 1983) but none have been associated with a particular biological role.

The mechanism of translation of genes I through V of CaMV has been the subject of much speculation. No subgenomic forms of genes I–IV have been found as RNA transcripts. If readily translatable forms of these regions occurred, one might expect to find RNA transcripts with these genes in 5'-proximal positions. With the exception of 22S RNA, which has been translated to give the gene V product (Plant *et al.*, 1985), no subgenomic mRNA of the closely packed genes I–V has been reported. Thus it seems likely that there are no promoters leading to the production of subgenomic transcripts of CaMV other than the 19S promoter. Processing of 35S RNA also seems unlikely because of the lack of RNA splicing signals in suitable positions in

CaMV DNA. Furthermore, when intervening sequences of other genes are inserted into the nonessential gene II position of the genome, these are spliced out of the viral genome during subsequent virus multiplication in plants (Hohn *et al.*, 1986). Consequently, it seems plausible that 35S RNA serves as mRNA for genes I through V, and that it is translated as a polycistronic messenger.

Most eukaryotic mRNAs are functionally monocistronic with ribosomes recognizing the 5′-end and scanning inward to translate the 5′-proximal cistron most efficiently (Kozak, 1986a). The efficiency of translation in eukaryotes is also influenced by the sequence surrounding the AUG initiation codon (Kozak, 1986b). Furthermore, if the 5′-proximal AUG occurs in a poor context, ribosomes may scan past the first AUG to downstream AUGs that occurs in favorable contexts, thus allowing relatively efficient translation of internal cistrons (Kozak, 1986b). With respect to translation of CaMV, it has been noted that each of the first AUG codons that appear in the open reading regions of the CaMV genome is in a favorable context (Covey, 1985). However, the significance of context may be greater in animal cells than in plant translation systems (Lutcke *et al.*, 1987). Consequently, leaky scanning is not an attractive hypothesis to account for the translation of internal cistrons of CaMV 35S RNA.

The suggestion that 35S RNA is translated as a polycistronic mRNA *in vivo* by reinitiation, i.e., by ribosomes moving directly from the stop codon of one open reading region to a nearby start codon of another open reading region without detaching from the mRNA, has been proposed as a mechanism for the translation of CaMV genes VII and I–V (Dixon *et al.*, 1983). This notion is supported by the successful translation *in vivo* of foreign genes inserted into the CaMV genome between genes I and III if the foreign gene is positioned so that its ATG codon is only a few nucleotides downstream from the termination codon of region I (Brisson *et al.*, 1984). There are a number of precedents for translation initiation on internal AUGs in eukaryotic cells (Kozak, 1986c; Peabody and Berg, 1986a,b). Generally, internal genes are expressed more efficiently if translation that has been initiated on an upstream region is terminated in the vicinity of the internal AUG.

Unlike the 19S RNA transcript, 35S RNA of CaMV cannot be translated in a reticulocyte system *in vitro* (Guilley *et al.*, 1982). Thus, this system cannot be used to study the requirements for internal initiation of the RNA. However, positive evidence for such a mechanism has been obtained by analysis of CaMV mutants with frame shifts in a small, nonessential, open reading region (gene VII) between genes VI and I (Fig. 4) (Dixon and Hohn, 1984). Virus infectivity (and by inference the translation of regions I through IV) is dependent on translation beginning and ending with gene VII. Mutation of cloned DNA has shown that if the start codon of gene VII is present, an in-frame termination codon before the ATG of gene I (see Fig. 4) is necessary for the virus to be biologically active. Deletion of the stop codon of region VII leads to loss of virus activity; restoring this codon to an inactive mutant leads to its reactivation. Infectivity can also be restored to inactive mutants of this sort by removal of the intergenic region between the α interruption and the region I start codon (Dixon and Hohn, 1984).

E. Mapping of Polypeptide Products and Gene Functions of CaMV

Several biological functions have been mapped on the CaMV genome by constructing artificial recombinants between strains with different phenotypes and by selective inactivation of functions by insertional mutagenesis. Antibodies prepared to synthetic peptides whose amino acid sequence has been inferred from the DNA sequence, or alternatively, prepared to products of gene fusions between viral genes and β-galactosidase (which can be induced to high levels in bacterial cells), have been useful for these experiments. Gene products have been identified for the six largest open reading regions of the CaMV genome, showing that each of these genes is expressed in infected plants.

A tentative assignment of region IV as the coat protein gene was made as a result of sequencing of CaMV DNA (Frank *et al.*, 1980). Only open reading region IV coded for a protein with an unusually high content of basic amino acids similar to that of CaMV coat protein (23% lysine plus arginine) (Brunt *et al.*, 1975). Confirmation came from expression of the coat protein in *Escherichia coli* (Daubert *et al.*, 1982), where only the gene IV sequence led to the production of virus antigen.

Polypeptides of CaMV particles produce an array of proteins in gels (Kelley *et al.*, 1974; Hull and Shepherd, 1976; Brunt *et al.*, 1975), suggesting that several polypeptides exist in the virion. However, several of these components may arise from a single protein by proteolysis or aggregation (Al Ani *et al.*, 1979). The major components of mature virions are proteins of 44 and 37 kDa. These are probably cleaved from a higher-molecular-weight precursor after or concomitant with encapsidation of DNA (Hahn and Shepherd, 1982). The precursor and to a lesser extent the proteolytically reduced forms are phosphorylated (Hahn and Shepherd, 1980) and glycosylated (duPlessis and Smith, 1981). A kinase capable of phosphorylating endogenous virus capsid proteins is associated with CaMV particles (Martinez-Izquierdo and Hohn, 1987; Menissier-deMurcia *et al.*, 1986).

In CaMV-infected plants, a protein of about 62 kDa accumulates to a level approaching that of virion structural proteins. When total RNA or poly(A) RNA from CaMV-infected leaves is translated in a cell-free system, this protein appears as a conspicuous product. However, translation of the mRNA of this protein is arrested in a cell-free system when an *Eco*RI fragment corresponding to region VI of the CaMV genome is hybridized to the RNA (Odell and Howell, 1980; Al Ani *et al.*, 1980). In addition, when infected leaf poly(A) RNA is hybrid-selected on CaMV DNA immobilized on DBM paper, followed by its translation in a cell-free system, the same 62-kDa protein is the major translation product. Partial protease digestion of the major protein produced in these hybrid selection experiments gives an array of peptides similar to that obtained from one of the major proteins of viral inclusion bodies (Covey and Hull, 1981). In addition, purified inclusion bodies contain a major constituent of about 62 kDa (Shepherd et al., 1980; Shockey et al., 1980). Antiserum prepared to this protein reacted with CaMV-specific, in vitro translation products, and the N-terminal amino acid sequence corresponded to that predicted from the region VI CaMV DNA sequence (Xiong et al., 1982). These observations strongly suggest that P62 is an inclusion body protein, coded for by gene VI, that accumulates in infected cells. The amount of P62 in infected cells appears to be far greater than other viral proteins, leading to speculation that the protein may have a structural rather than an enzymatic role in the biology of CaMV. It may be the electron-dense matrix material seen in electron micrographs of viral inclusion bodies (Fig. 1).

Some of the biological consequences of region VI have been identified. Small noninactivating insertions in region VI have a dramatic effect on disease symptoms (Daubert et al., 1983) and insertion of the gene into tobacco chromosomes by *Agrobacterium* Ti plasmid transformation results in a chlorotic mottling disease similar to that of virus-infected plants (Goldberg et al., 1987; Baughman et al., 1988). Other experiments have shown that region VI determines whether the virus elicits a hypersensitive reaction (local necrotic lesions) or a compatible reaction in the host plant conducive to systemic development of the virus (Schoelz et al., 1986; Schoelz and Shepherd, 1988). The elicitation of the host defense reaction as indicated by hypersensitivity may be related to the role of region VI as the major host range determinant of CaMV (Schoelz et al., 1987).

Recent experiments have shown that major changes occur in gene VI of caulimoviruses when these viruses are adapted to new host plants. In an experiment in which a cloned strain of FMV (Table I) was propagated in *Datura innoxia* (a new host) for 2 years and the virus then recloned, there were a number of restriction site changes clustered in region VI. When the DNAs of original and adapted strains were sequenced, stretches of gene VI showed extensive changes at the amino acid level. These amino acid exchanges suggest that gene VI is probably under a host-imposed selection that leads to its rapid evolution during virus adaptation to the new host.

The expression of gene I of CaMV has been studied by use of a specific antibody. Gene I DNA sequences have been fused with the bacterial β-galactosidase gene followed by expression of the large chimeric fusion protein in *E. coli*. The protein, readily purified from bacteria because of its insolubility, was used to prepare immune serum. Proteins of two sizes with gene I specificity have been detected in infected plants. The larger (45–46 kDa) is found only in the insoluble fraction of tissue homogenates, and may be a constituent of viral inclusion bodies (Harker et al., 1987; Young et al., 1987; Martinez-Izquierdo et al., 1987). The smaller (36–38 kDa) is associated with virus particles, including those passed through multiple sucrose density gradients. However, it is diminished when virus is banded out isopycnally in CsCl (Young et al., 1987), probably because of the chaotropic effects of the salt.

Gene I of CaMV and CERV has been reported to possess a low level of amino acid sequence homology to the 30K cistron of tobacco mosaic virus (TMV) (Hull et al., 1986). The 30K cistron of TMV is involved in cell-to-cell movement of virus (Nishiguchi et al., 1978, 1980; Ohno et al., 1983; Meshi et al., 1987).

The insect transmission function of CaMV has been associated with gene II by recombination of aphid-transmissible and aphid-nontransmissible strains of CaMV and by insertional inactivation (Armour et al., 1983; Daubert et al., 1983, 1984; Woolston et al., 1983). An 18-kDa polypeptide may be the product of this gene (Woolston et al., 1983). Insect transmission is discussed in greater detail below.

Proteins which are probably the gene products for other regions of the CaMV genome have been detected in infected plants. Xiong et al. (1984) found that antibodies to a synthetic 19-amino acid peptide corresponding to the N-terminal sequence of CaMV region III reacted with a 15-kDa protein from infected leaves. The same antibody was used to show that the reactive polypeptide was localized in virions and that the virion-associated protein was an 11-kDa form, in contrast to the 15-kDa form of semipurified inclusions (Giband et al., 1986). In addition, the latter was discovered to have a strong affinity for double-stranded DNA (Giband et al., 1986), the significance of which is not clear.

Several lines of evidence indicate that gene V of CaMV specifies the reverse transcriptase that participates in replication of viral DNA. The gene product inferred from the nucleotide sequence of gene V has a high level of amino acid homology with known reverse transcriptases (Toh et al., 1985; Volovitch et al., 1984; Saigo et al., 1984). Moreover, when CaMV is cloned and expressed in yeast (*Saccharomyces cerevisiae*), gene V specifies a significant level of reverse transcriptase activity (Takatsuji et al., 1986).

An 80-kDa protein detected in CaMV-infected plants or isolated replication complexes may be the product of gene V. Antibodies prepared to a synthetic peptide with an amino acid sequence corresponding to the C-terminal of gene V (as predicted from the DNA sequence) detect proteins from CaMV-infected plants on Western blots. A polypeptide of 80 kDa plus a variety of lower-molecular-weight forms react with this antibody (Ziegler et al., 1985; Laquel et al., 1986). The immune serum specifically inhibits the unique polymerase associated with CaMV replication complexes (Laquel et al., 1986), indicating that these activities are specified by the same enzyme

(Menissier *et al.*, 1984). Additional properties of the reverse transcriptase of CaMV are given below.

Proteins corresponding to CaMV genes VII or VIII have not been reported, and it is not known whether these open reading regions are expressed in infected plants.

F. Replication of Caulimoviruses

Much information has accumulated recently suggesting that the DNA of caulimoviruses is replicated by reverse transcription. Though this process may illustrate the overall course of replication, the particulars for this less conventional mode of replication are not well understood.

Investigations on the replicative intermediates of CaMV DNA (partially assembled molecules) suggest the two DNA strands are synthesized by separate processes, instead of by a conventional replication fork moving around the circular DNA molecule (Guilley *et al.*, 1983; Pfeiffer and Hohn, 1983; Hull and Covey, 1983a,b; Guilfoyle *et al.*, 1983; Marco and Howell, 1984). The most common mode of replication, well documented with bacterial plasmids and various bacterial and animal viruses, in which DNA is copied directly to DNA, requires origins of DNA replication in addition to various proteins (Kornberg, 1979). In contrast, duck hepatitis virus has been found to replicate its genome by reverse transcription (Summers and Mason, 1982). Replication of this virus has served as a model for CaMV, although there are definite differences in the replication of the two viral DNAs.

Early autoradiographic observations indicated that the inclusion bodies of CaMV have a role in viral DNA replication (Kamei *et al.*, 1969; Favali *et al.*, 1973). Recently, these abnormal cytological structures, which occur exclusively in the cytoplasm, have been further implicated in viral DNA replication (Modjtahedi *et al.*, 1984; Bonneville *et al.*, 1984; Mazzolini *et al.*, 1985). Autoradiographic analysis of isolated cellular components has demonstrated that virus-specific DNA synthesis occurs most abundantly in these structures. Inclusion bodies have been prepared essentially free of cellular DNA but with retention of a high level of CaMV-specific replication activity, particularly minus-strand synthesis (Mazzolini *et al.*, 1985). As mentioned later, the DNA polymerase activity associated with these isolated fractions shows template preferences characteristic of CaMV reverse transcriptase.

Three unusual features of CaMV biology stimulated the search for reverse transcription of its genome. The first was the observation that the 3'-terminus (nucleotide 7615) of the 35S RNA transcript mapped downstream of its own 5' initiation site (at nucleotide 7435) (Covey *et al.*, 1981; Guilley *et al.*, 1982). This meant the 35S transcript was more than a full-length copy of the DNA genome and that it had a 180-nucleotide direct repeat at each end of the

RNA molecule. The second observation was the presence of 14 nucleotides adjacent to the α discontinuity in the plus strand (see Fig. 4) of the DNA sequence that was complementary to the 3′-end of methionine initiator tRNA (Guilley *et al.*, 1983; Hull and Covey, 1983b; Pfeiffer and Hohn, 1983; Howell *et al.*, 1983; Guilfoyle *et al.*, 1983). This suggested that tRNA^{met} binds to 35S RNA at this point and primes DNA synthesis at its free 3′-OH. The third observation related to some unusual small single-stranded DNA molecules of viral specificity that had ribonucleotides covalently attached to their 5′-end. Covey *et al.* (1983) described a small DNA (725 nucleotides) with minus-stranded polarity that had covalently attached RNA. (It could be reduced in length by about 100 nucleotides by treatment with alkali or ribonuclease A.) This molecule mapped to the large intergenic region to the left of the α interruption (see Fig. 7). Guilley *et al.* (1983) found a similar species of single-stranded DNA of about 600 nucleotides in length that occurred in denatured viral DNA preparations. This DNA had the same polarity and 5′-end (at the α interruption) as viral minus-strand DNA, and its 3′-terminus mapped at nucleotide 7437, or very close to the transcriptional start site of 35S RNA. Having ribonucleotides at the 5′-end (Guilley *et al.*, 1983) suggested an RNA priming event originated near or at the α interruption to initiate viral DNA synthesis. The small DNA of minus polarity is referred to as "strong stop" DNA or "sa-DNA" (Guilley *et al.*, 1983; Covey *et al.*, 1983; Turner and Covey, 1984). Events in viral DNA replication following the synthesis of this molecule are discussed later.

Three other caulimoviruses (CERV, FMV, SoyCMV) for which sequence information is available have base complementarity to the 3′-end of bean or wheat tRNA^{met}. In each case, the sequence occurs in the plus strand at or near the single interruption in the minus strand (Hull *et al.*, 1986; Richins *et al.*, 1987; Verver *et al.*, 1987).

Several lines of evidence indicate that the two strands of CaMV DNA are synthesized by separate events. The presence in infected plants of very small single-stranded DNA molecules of minus polarity (sa-DNA), but none of plus polarity, suggests that minus-strand synthesis is initiated independently. CaMV replicative complexes obtained by hypotonic extraction of plant organelle preparations utilize 35S RNA as a template for minus strand DNA synthesis (Pfeiffer *et al.*, 1984). Purified nuclei and inclusion bodies, largely freed of other cellular components (Marsh *et al.*, 1985), and CaMV replicative complexes isolated from synchronously infected turnip protoplasts (Thomas *et al.*, 1985) have been used *in vitro* to study CaMV replication.

In each case, viral minus strands and plus strands appeared to be synthesized on different templates. Minus-strand incorporation activity was resistant to actinomycin D whereas plus-strand synthesis was severely inhibited (Marsh *et al.*, 1985). Minus-strand DNA in replication complexes exists largely as single strands or DNA/RNA hybrids whereas plus strands are found only as DNA duplex molecules with regions of single-stranded minus polarity DNA (Marsh et al., 1985; Thomas et al., 1985).

The templates for CaMV DNA synthesis in replication complexes are partially nuclease resistant (Marsh et al., 1985; Thomas et al., 1985). This protection may be related to association with partially completed virions. DNA synthesis activity which sediments most rapidly in sucrose density gradients is largely unaffected by either deoxyribonuclease or ribonuclease. The protection of labeled products in fractions with sedimentation properties similar to CaMV particles suggests replication may be occurring in partially assembled virions (Marsh et al., 1985; Marsh and Guilfoyle, 1987; Thomas et al., 1985). Moreover, DNA polymerase activity is found in purified CaMV and may be encapsidated along with template (Menissier et al., 1984).

Figure 5 shows the model proposed by several laboratories for the replication of CaMV DNA (Hull and Covey, 1983a; Guilley et al., 1983; Pfeiffer and Hohn, 1983; Howell et al., 1983; Guilfoyle et al., 1983; Marco and Howell, 1984). The first step, labeled A in the figure, occurs in the nucleus. Other steps may occur in the cytoplasm in association with viral inclusion bodies.

The replication cycle starts with transcription of the viral minichromosome to produce the 35S RNA transcript in the nucleus. The latter is trans-

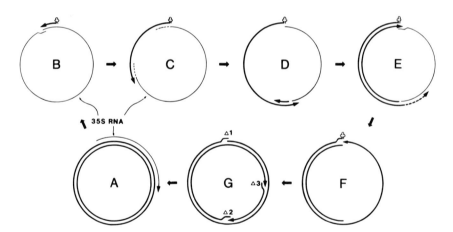

Fig. 5. Proposed steps in the replication of cauliflower mosaic virus DNA by reverse transcription. The replication cycle starts by production of a full-length redundant-end RNA (35S RNA) transcribed from minichromosomal DNA, as indicated in step A. This step occurs in the nucleus whereas all subsequent steps probably occur in the cytoplasm. In the second step (B), 35S RNA is the template for reverse transcription, initiating with a tRNA^{met} primer which binds to 35S RNA at a point about 600 nucleotides downstream from the 5' cap site. First-strand DNA synthesis continues past the overlap region where the DNA polymerase probably transfers from one end to the other of the RNA template (C). When first-strand synthesis proceeds past a point coinciding with the β_1 interruption, second-strand DNA synthesis is initiated (D). The completed genome has interruptions (overlaps) at three points as indicated by $\Delta 1$, $\Delta 2$ and $\Delta 3$ (step G). Figure redrawn from that of Marco and Howell (1984).

ported to the cytoplasm to initiate reverse transcription. The production, transport, and expression of the 19S RNA transcript of gene VI may be an antecedent for this event, since P62 is a major constituent of viral inclusion bodies and there is cytological evidence that the matrix material appears in infected cells before mature virus (Kitajima and Lauritis, 1969) (see Fig. 1). P62 and partially assembled virions may be required to inaugurate cytoplasmic events which follow transport of the 35S RNA out of the nucleus.

The second step (Fig. 5, step B) consists of the priming of first (minus)-strand synthesis by tRNA^{met}, which anneals to 35S RNA at a point about 590 nucleotides downstream of its 5'-end (Fig. 6). The identity of this transfer RNA primer as tRNA^{met} has been confirmed recently (Turner and Covey, 1984). Binding of the tRNA^{met} primer occurs at the α interruption with respect to the virion DNA map (Fig. 4). DNA polymerase (reverse transcriptase) initiates phosphodiester bond formation by adding deoxyribonucleotides to the 3'-OH of the tRNA molecule. As replication proceeds along the RNA template, the 5'-end of the template is soon reached whereupon DNA synthesis stops. At this point, the replication complex must switch to the 3'-end of 35S RNA in order for minus-strand synthesis to continue. Strand switching may occur after base pairing of the newly synthesized minus-strand DNA to complementary sequence on the 3'-end of 35S RNA in the 180-bp terminally redundant region of the molecule. The 5'-end of the RNA template has probably been removed at this point by RNase H activity of the reverse transcriptase.

In the third step (Fig. 5, step C), the replication complex has completed strand switching and has proceeded with minus-strand synthesis past the RNA overlap region. Synthesis continues until replication passes a primer site for the initiation of second (plus)-strand synthesis (step D). Plus strand is probably synthesized by the same DNA polymerase (viral reverse transcriptase). The first plus-strand primer site is at the β_1 interruption in plus-strand DNA (Fig. 4), and it probably consists of purine-rich sequences of RNA left

```
    ACCAUAGUCUCGGUCCAAA..................tRNA

         1           MetAsnArgPheLysAsn.....(region VII)
    ...TGGTATCAGAGCCATGAATCGGTTTAAAAACC.....CaMV DNA

         1           MetCysSerThrArgLys.....(region I)
    ...TGGTATCAAAGCCATGTGCTCAACTAGAAAAA.....FMV DNA

         1                            Met....(region I)
    ...TGGTATCAGAGCCATAGTGCTCAAAAATCATG.....CERV DNA
```

Fig. 6. Comparative sequences at the tRNA^{met} binding sites of three caulimoviruses. The 3'-terminus of bean cytoplasmic tRNA^{met} is shown along with the viral DNA sequences that show complementarity (underlined). The position indicated by "1" in each sequence is the zero point of each viral sequence. For viral abbreviations, see Table I.

in place (Pfeiffer and Hohn 1983; Covey, 1985) (see Fig. 7). Short remnants of ribonucleotides may be left at these sites as the result of incomplete RNase H activity of the gene V protein. Minor primer sites that initiate plus-strand synthesis at reduced frequency occur elsewhere (Maule and Thomas, 1985).

During minus-strand synthesis on the RNA template to produce an RNA–DNA hybrid, the RNA moiety is probably removed rapidly by RNase H activity associated with the reverse transcriptase (see Section II,G). This is characteristic of the reverse transcriptases of retroviruses (Verma, 1981) to which gene V of CaMV has a high degree of homology (Toh et al., 1985).

The mechanism by which the ends of completed minus strands are joined at the α interruption (Fig. 5, step F), so that DNA polymerase activity can copy the minus strand across this junction, is not known. Perhaps the complementary sequence of the tRNA^{met} primer has a role in this step. The specific interruptions in the completed CaMV genome (Fig. 5, step G) are a reflection of the manner in which the DNA is replicated. There are one to several ribonucleotides that remain *in situ* at the site of each priming event. The 5'-ends at these breaks in the DNA are at fixed positions, in contrast to the 3'-ends which are somewhat variable. The latter are probably the result of displacement of 5'-ends for various distances during DNA synthesis at these regions of the genome.

Perhaps the most persuasive evidence for reverse transcription of CaMV is the experimental demonstration of RNA splicing during virus multiplication. When a fragment of the leghemoglobin gene containing an intervening sequence was inserted into the cloned CaMV genome followed by inoculation of the modified DNA to plants, the intron disappeared from progeny virus DNA as if RNA splicing had occurred (Hohn et al., 1986). Such dele-

Gap	Virus	Sequence
B1 (G2)	CERV	TCTTTTT.AGAAGAGGGGGGGAA...TCATGTAGA
	CaMV	TTCTTTC....AGAGGGGAGGAGG..TTATCAGA
	FMV	TCTTTTT..AGAGAGGGGGAG.....TTAGGACATTT
	SoyCMV	TCATTTC..GAGGAGGGAGGAAG...TTACAAAAGGAA
B2 (G3)	CERV	AATCTTT..AAGGAGGGGGAGGAAATGGTGA
	CaMV	CATTTTAAGAGTAGGGGGG......TTGATTACTCGA
	FMV	CATTTTC..AGAGAGGGGGAAGGAAGTACTAA
	SoyCMV	AATTGTT..GAGGAGGGTGAGGA...TAAGGGAAAAA
B3 (G4)	FMV	ATTTTTT..AGAGAGGGGGAG.....TGATGGAAGAA
Consensus		YYTTYR$_{0-4}$RgAGGGggR$_{1-7}$...T

Fig. 7. Purine-rich regions of 35S RNA located near the positions of the single-stranded discontinuities of plus strands of different caulimoviruses. These tracts, left *in situ*, may serve as RNA primers for initiation of second-strand DNA synthesis. A consensus priming sequence is presented (R, any purine; Y, any pyrimidine).

tion at canonical intron–exon borders is explainable only by RNA splicing during DNA replication via an RNA intermediate.

Further evidence for the probable removal of intervening sequences by an RNA splicing mechanism during virus multiplication is provided by the nature of some deletion mutants of cauliviruses that accompany full-length viral genomes in natural infections. One of these, characterized by Hirochika *et al.* (1985), was about 800 bp shorter than the wild-type genome and comprised about 25% of the viral DNA present in infected cells. When sequences around the deletion were characterized, these corresponded to donor and acceptor consensus sequences for RNA splicing.

It is not yet clear which proteins specified by CaMV are required for viral DNA replication. Several observations suggest that the coat protein, perhaps in the form of a partially assembled capsid, probably participates in the process. Various subgenomic forms of viral DNA that are probably first-strand replicative intermediates have been found in association with virion-like particles. These consist of sa-DNA and various lengths, up to 8 kb, of minus-strand DNA which appear to be encapsidated. These particles sediment with CaMV virions on sucrose density gradients, band isopycnically like virus in CsCl (though slightly less dense), and precipitate with coat protein antibody, which indicates they exist primarily as virions. The behavior of these molecules during blotting experiments with and without prior RNase A digestion at low salt suggests they exist largely as RNA–DNA hybrids. The putative replicative intermediates react specifically with CaMV plus-strand probes (Marsh and Guilfoyle, 1987). Work with synchronously infected protoplasts also suggests that CaMV DNA replication is probably accompanied or preceded by encapsidation (Maule, 1985; Thomas *et al.*, 1985). In addition, small in-phase insertions (12 bp) in the coat protein gene cause loss of DNA infectivity (Daubert *et al.*, 1983), suggesting coat protein is required for DNA replication.

Highly purified preparations of CaMV have been found to contain forms which support limited *in vitro* synthesis of viral DNA. An 8-kb linear form of virus minus strand is the preferentially labeled product (Menissier *et al.*, 1984; Marsh and Guilfoyle, 1987). However, it has not been determined that incorporation is due to reverse transcription. Nevertheless, the slowly sedimenting, aphidicolin-resistant DNA replication complexes prepared by Laquel *et al.* (1986) from CaMV-infected leaves, which are probably similar to those of Marsh and Guilfoyle (1987), are inhibited by antibodies prepared to a synthetic peptide corresponding in amino acid sequence to the C-terminal of CaMV gene V.

A special feature of the coat protein gene of CaMV that may be relevant to the DNA replicating activity of partially assembled virions is the presence of a highly conserved array of cysteine residues, viz. the $CysX_2CysX_4HisCys$ motif (Covey, 1986) referred to herein as the "Cys-motif." This element is a highly conserved trait of the *gag* gene product of retroviruses that binds nucleic acids. It also occurs in the coat protein of other caulimoviruses such as FMV (Richins *et al.*, 1987) and CERV (Hull *et al.*, 1986). Covey (1986) has suggested that the Cys-motif is a universal characteristic of the coat proteins (or their precursors) of viruses that use tRNA as a primer for reverse transcription. Thus, the sequence could be active either in encapsidation or the initiation of reverse transcription in immature virions of caulimoviruses.

It is not known whether coat protein of CaMV will assemble into a virion-like structure in the absence of viral nucleic acid or the initial form of the CaMV genome that is encapsidated. Empty virions are never obtained in purified preparations of CaMV or other caulimoviruses. Hence, it seems unlikely that virus particles are assembled in the absence of viral nucleic acid.

One can speculate that 35S RNA may be the form of the viral genome that is encapsidated rather than viral DNA. Perhaps virion assembly is the mechanism by which 35S RNA is trapped in the inclusion body as a forerunner of reverse transcription.

G. The Reverse Transcriptase of CaMV

It now seems clear that the caulimoviruses code for their own DNA polymerase for replication of their genome. The enzyme is an RNA-dependent DNA polymerase (reverse transcriptase) that is probably endowed with multiple activities, including an acid protease and RNase H in addition to its polymerase activity.

The reverse transcriptase of Rous sarcoma virus, a retrovirus, is the most thoroughly characterized RNA-dependent DNA polymerase (Verma, 1981). The isolated enzyme consists of dimers of a single protein, in some cases with one subunit reduced in size by proteolysis. The purified enzyme utilizes both polyribonucleotides and polydeoxyribonucleotides as templates to direct the synthesis of complementary polydeoxyribonucleotides, with a preformed primer required to initiate DNA synthesis.

The reverse transcriptase of CaMV is encoded by gene V, the largest and most complex gene of the viral genome. It is also the most highly conserved gene among strains of this virus (Gardner *et al.*, 1981), and among different caulimoviruses. About 64% of the amino acid residues of gene V of CaMV, CERV, and FMV are identical (Table III) (Hull *et al.*, 1986; Richins *et al.*, 1987).

There is also a high level of homology between the DNA polymerase of the caulimoviruses and those of other reverse transcribing elements. The relatedness among these proteins has been pointed out by Toh *et al.* (1985) and Volovitch *et al.* (1984). CaMV gene V is most similar to a reverse transcriptase of a retrotransposon of *Drosophila,* the *copia*-like element designated "17.6" (Fig. 8).

The reverse transcriptases are large, complex proteins with several domains specifying different enzymatic activities. Those exemplified by retroviruses code for at least three different activities, viz. DNA polymerase and RNase H located on the N-terminal region and a DNA endonuclease activity located on the C-terminal region (Fig. 8). The N-terminal highly conserved region shows significant amino acid sequence homology to the polymerase genes of hepatitis B virus and CaMV, which both replicate their DNAs via an RNA intermediate. The extent of the relationships among these polymerases has been analyzed by Toh *et al.* (1985) on the basis of conserved clusters of amino acids, and five domains can be assigned (Fig. 8). The N-terminal domain, missing from some of the viral polymerases but present in CaMV and the 17.6 transposon, exhibits strong amino acid sequence homology with the *gag*-specific protease of Rous sarcoma virus and the *gag-pol* spacer of human adult T cell leukemia virus (HTLV). These sequences have a highly conserved tract of amino acids with close similarity to the active site Asp-Thr-Gly of the acid protease family (Toh *et al.*, 1985).

The central region of the proteins contains two blocks of sequence (II and III of Fig. 8) common to the 17.6 transposon and the retrovirus polymerases. The region has a highly conserved element Tyr-hydrophobic residues-Asp-Asp flanked by three hydrophobic residues, similar to a Tyr-Gly-Asp-Asp

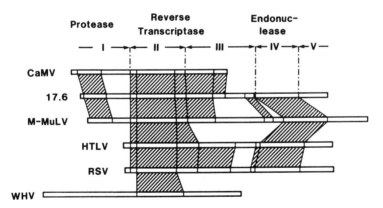

Fig. 8. Sequence homology profile of the polypeptide chain of various reverse transcriptases. Five regions divided on the basis of highly conserved sequences among the various retroid agents are indicated. Note that the cauliflower mosaic virus (CaMV) gene V sequence contains no homologous segment to tract IV, the endonuclease believed to bring about the integration of proviruses or retrotransposons into host DNA. 17.6, a *copia*-like transposon of *Drosophila;* RSV, Rous sarcoma virus; M-MuLV, Moloney murine leukemia virus; HTLV, human adult T cell leukemia virus; WHV, woodchuck hepatitis virus (a reverse transcribing virus related to human hepatitis B virus). Figure is redrawn from that of Toh *et al.* (1985).

flanked by hydrophobic residues that exists in polymerases of the picornaviruses and certain bacteriophages, suggesting that these residues play a pivotal role in polymerase function (Toh *et al.*, 1985). Kamer and Argos (1984) had previously pointed out a conserved 14-residue element in retrovirus reverse transcriptase, influenza virus, CaMV, and hepatitis B virus as a probable active site for these viral polymerases.

The first portion of the central conserved tract probably specifies RNase H activity with the DNA polymerase portion toward the C-terminal end. At this point the polymerase of CaMV ends, in contrast to the 17.6 retrotransposon and the retroviruses which have a run of amino acid sequences associated with DNA endonuclease activity (Toh *et al.*, 1985) (Fig. 8). The latter part of the protein is believed to be active in splicing the retroid elements into the chromosomes of the host organism, i.e., integration of provirus as the result of reverse transcription. CaMV, which appears to lack this activity, has not been found in an integrated state in plant cells.

When aligned, the amino acid sequences of the CaMV polymerase and the *Drosophila* 17.6 transposon bear striking resemblance. The homology values for the various regions diagramed in Fig. 8 are region I (109 amino acid residues) 19%, region II (325 amino acid residues) 35%, region III (187 amino acid residues) 33% (Toh *et al.*, 1985). These authors have suggested that CaMV may have descended from an ancestral *copia*-like element of insects.

The catalytic properties of CaMV DNA polymerase have not been well characterized, due in part to the template-bound nature of the enzyme and the lack of highly purified enzyme. Most investigators have studied subcellular fractions from infected plants which synthesize CaMV DNA on an endogenous template (Pfeiffer and Hohn, 1983; Guilfoyle *et al.*, 1983; Pfeiffer *et al.*, 1984; Modjtahedi *et al.*, 1984). Pfeiffer *et al.* (1984), for example, were unable to prepare template-dependent activity with fractions from CaMV-infected plants. In contrast, high levels of CaMV-specific endogenous activity were found.

Only two investigators have reported template-dependent activity with solubilized CaMV polymerase fractions (Volovitch *et al.*, 1984; Thomas *et al.*, 1985). In one case, a subcellular fraction specific to infected plants copied added poly(C) and several natural mRNAs. The poly(C) activity was isolated by phosphocellulose chromatography and eluted as a fraction specific to CaMV-infected plants, clearly distinct from a classical-type γ-like DNA polymerase present in both healthy and CaMV-infected plants (Volovitch *et al.*, 1984). Thomas *et al.* (1985) prepared a soluble template-dependent DNA polymerase by a freeze–thaw cycle of glycerol-containing fractions. The enzyme could copy added cowpea mosaic virus RNA to produce extensive tracts (greater than 2.5 kb) of cDNA. Pfeiffer *et al.* (1984) found that CaMV-specific DNA synthesizing activity was unaffected by aphidicolin, an inhibitor of plant DNA polymerase.

A single enzyme probably synthesizes both strands of CaMV DNA during the replication cycle, as indicated by a similar sensitivity of both plus- and minus-strand synthesis to a variety of inhibitors (Pfeiffer *et al.*, 1984). In addition, *in vitro*-labeled CaMV DNA produced by an active fraction from CaMV-infected *Brassica* protoplasts hybridized to several strand-specific clones, showing that DNA of both plus and minus polarities is synthesized on the endogenous template (Thomas *et al.*, 1985).

As mentioned previously, an 80-kDa protein has been detected in infected plants with antibody to a synthetic peptide corresponding to the 25-amino acid C-terminal sequence of CaMV gene V (Ziegler *et al.*, 1985). Laquel *et al.* (1986) have isolated CaMV replication complexes and assayed the associated activity using activated calf thymus DNA or poly(rCm) oligo(dG)$_{12-18}$ as template-primer. The polymerase was specifically inhibited by the immune serum prepared by Ziegler *et al.* (1985).

H. Insect Transmission of Caulimoviruses

The caulimoviruses have evolved an interesting biological relationship with insects (aphids) to bring about their dissemination and thus ensure their survival (for a review, see Shepherd, 1976). The virus during the act of transmission appears to be associated wholly with the mouthparts (stylets) of the aphid vector after it has acquired virus by feeding on infected plants. However, the insect does not serve as merely a mechanical device ("flying pin") for virus transmission by contaminated mouthparts.

Early observations showed that aphids could not acquire and transmit purified CaMV by feeding through a membrane (Pirone and Megehed, 1966). However, aphids allowed to probe first on infected plants could then acquire and transmit purified virus fed through a membrane. Moreover, experiments with aphid-nontransmissible strains of CaMV showed that these too could be acquired (from infected plants) by insects if they were allowed to first probe into plants infected with an aphid-transmissible strain (Lung and Pirone, 1973, 1974). These experiments demonstrated that a factor active in virus transmission, other than virus particles, is obtained from infected plants. Furthermore, aphid-nontransmissible strains must be defective in the production of this helper factor. This factor is now believed to be the product of CaMV gene II.

As mentioned previously, both insertions and deletions into gene II of cloned CaMV DNA inactivate aphid transmission without otherwise affecting the ability of the virus to infect and multiply in plants (Daubert *et al.*, 1983; Armour *et al.*, 1983; Woolston *et al.*, 1983).

Mapping experiments have also been done by exchanging restriction fragments between aphid-transmissible and aphid-nontransmissible isolates. The small *Bst*EII–*Xho*I fragment, which contains part of gene VII, the whole of gene I, and the 5′-half of gene II (see Fig. 4), contains the information controlling aphid transmission (Woolsten *et al.*, 1983; Daubert *et al.*, 1984).

Comparison of the DNA sequence for region II (Gardner et al., 1981) of aphid-nontransmissible strain CM1841 with that of two insect-transmissible strains (Frank et al., 1980; Balazs et al., 1982) revealed that only two amino acid changes occurred in the region between the ATG codon and the XhoI site at amino acid 99. One or both of these exchanges (Thr → Asn at amino acid 89 or a Gly → Arg at amino acid 94) could account for loss of the aphid transmission function of strain CM1841 (Daubert et al., 1983; Modjtahedi et al., 1985).

Aphid transmission tests of hybrid genomes of a second aphid-nontransmissible isolate of CaMV and a transmissible strain showed that the small BstEII–XhoI fragment of the latter conferred insect transmissibility on the defective strain (Campbell strain). DNA sequence comparisons of region II showed two changes in the relevant segment of gene II. One of these caused a Gly → Arg exchange at position 94 of the gene II product of the Campbell strain (Woolston et al., 1987), identical to one of the exchanges previously postulated to cause the defect in aphid transmission of strain CM1841 (Daubert et al., 1983). The exchange occurs in a region of the protein that has been reported to show a significant level of amino acid homology with the aphid helper component sequence of an RNA plant virus (Domier et al., 1987). This suggests that this region of the gene II protein plays a crucial role in insect transmission.

Of the three caulimoviruses that have been sequenced, gene II amino acid comparisons show a high degree of homology (37.4–40.9% identity). The extent of similarities, however, varies with different regions of the polypeptide, with regions toward the C-terminal and N-terminal parts showing higher degrees of homology (Hull et al., 1986; Richins et al., 1987). The functional relatedness of these genes is illustrated by the complementation for aphid transmission between defective strains of CaMV and two other caulimoviruses (Markham and Hull, 1985).

The 18-kDa protein found in plants infected with aphid-transmissible strains of CaMV (Woolston et al., 1983; Givord et al., 1984; Rodriguez et al., 1987), but not in plants infected by strains with extensive deletions in gene II (Givord et al., 1984; Modjtahedi et al., 1985; Harker et al., 1987), suggests that this polypeptide is probably the accessory factor that brings about insect transmission. The gene II product occurs in isolated inclusion bodies of CaMV (Harker et al., 1987; Rodriguez et al., 1987) and appears to influence structural characteristics of these bodies (Givord et al., 1984).

It is believed that the aphid accessory factors brings about virus acquisition and retention on aphid mouthparts, rather than playing a role in the infection process. Nevertheless, there are no data to distinguish between these two possible roles. Viruses carried on aphid mouthparts may enter plant cells by a different route than that introduced by other processes (e.g., during mechanical transmission). For example, viruses carried on aphid mouthparts may infect plants through plasmodesmata when aphids probe between cells rather than through cells. If so, perhaps an accessory factor is required for this process. It may be relevant that even viruses not normally transmissible by aphids, such as tobacco mosaic virus, can be transmitted by aphids if combined with poly-L-ornithine (Pirone and Shaw, 1973). This material is believed to stimulate pinocytosis. A similar role for the aphid accessory factor seems plausible.

III. GEMINIVIRUSES (SINGLE-STRANDED DNA VIRUSES)

A. Biology and Intrinsic Features of Geminiviruses

The geminiviruses are a more diverse group than the caulimoviruses. These viruses infect monocots as well as dicots and two types of insect vectors disseminate these viruses. Hence, this group may have some evolutionary advantages over the caulimoviruses. The geminiviruses are also more prevalent in nature and cause more serious economic losses, particularly in tropical and subtropical regions of the world. In spite of their importance, only recently have some of the basic intrinsic features of the group been defined. With the discovery of their single-stranded DNA genome a new group was created to accommodate these remarkable pathogens (Harrison *et al.,* 1977; Matthews, 1979).

Members of the geminivirus group are listed in Table IV. The unique morphology of their virions (Mumford, 1974; Bock, 1974) and their single-stranded DNA genome (Goodman, 1977a,b) stimulated interest in the group. While the properties of these viruses are now well defined, relatively little is known about their gene expression or replication.

Geminiviruses cause extreme damage in many important crop species, including several plants used in subsistence agriculture in underdeveloped areas. In several cases these viruses have destroyed industries built around cultivated crops, or completely prevented the culture of desirable crop species. For this reason, the epidemiology of geminivirus-incited diseases has been studied intensively. In North America the beet curly top virus (BCTV) essentially eliminated the sugar beet crop in the western United States early in the twentieth century. This disease and its transmission by the beet leafhopper (*Circulifer tenellus*), first reported in 1915 (Boncquet and Hartung, 1915), were studied extensively for many years. But only recently has information become available on the intrinsic nature of the virus. A virus of similar historical significance is maize streak virus (MSV), which was described and transmitted with leafhoppers in Africa almost 60 years ago (Storey, 1928).

Progress with the geminiviruses has been slow because most members of the group are not transmissible mechanically. Virus assays can be done only

TABLE IV

Geminiviruses with Their Principal Hosts and Insect Vectors

Virus[a]	Principal host	Insect vector[b]
Bean golden mosaic virus (BGMV)	*Phaseolus vulgaris*	*Bemisia tabaci*
Beet curly top virus (BCTV)	*Beta vulgaris*	*Circulifer tenellus*
Cassava latent virus (CLV)	*Manihot esculenta*	*Bemisia tabaci*
Chloris striate mosaic virus (CSMV)	*Chloris gayana, Triticum* sp., *Avena sativa, Hordeum vulgare*	*Nesoclutha pallida*
Digitaria streak virus (DSV)	*Digitaria sanguinalis*	*Nesoclutha declivata*
Euphorbia mosaic virus (EMV)	*Euphorbia prunifolia*	*Bemisia tabaci*
Maize streak virus (MSV)	*Zea mays,* many grasses	*Cicadulina mbila*
Squash leaf curl virus (SLCV)	*Cucurbita pepo, Cucurbita maxima*	*Bemisia tabaci*
Tomato golden mosaic virus (TGMV)	*Lycopersicon esculentum*	*Bemisia tabaci*
Tobacco leaf curl virus (TLCV)	*Lycopersicon esculentum*	*Bemisia tabaci*
Tobacco yellow dwarf virus (TYDV)	*Nicotiana tabacum, Datura stramonium, Phaseolus vulgaris*	*Orosius argentatus*
Wheat dwarf virus (WDV)	*Triticum aestivum*	*Psammotettix alienus*

[a] Goodman (1981a), Harrison (1985), and others have mentioned poorly characterized viruses which may be members of this group.

[b] *Bemisia tabaci* is a whitefly; all other insects named here are leafhoppers.

by injecting fractions into their insect vectors which must then be fed on plants for symptom development. Only after mechanically transmissible members of the groups were discovered, such as bean golden mosaic virus (BGMV), were the first geminiviruses characterized biochemically and found to contain single-stranded DNA. The basis of mechanical transmissibility is the ability to infect the superficial tissue of the plant. BGMV also readily infects mesophyll protoplasts in culture (Goodman, 1981b). Other members, e.g., *Chloris* striate mosaic virus (CSMV), are generally distributed throughout the plant except for the epidermis (Francki *et al.*, 1979). Such viruses are not mechanically transmissible.

Other geminiviruses may not infect all tissues of the host plant. BCTV, for example, is found only in association with the phloem (Esau and Hoeffert, 1973; Esau, 1977). These phloem-associated viruses are difficult or impossible to transmit manually by rubbing virus-containing extracts over leaves with an abrasive.

Cytological changes in geminivirus-infected plants are frequently associated with the phloem and frequently reflect the tissue distribution of the viruses. BCTV causes hypertrophy, hyperplasia, and necrosis of phloem parenchyma (Esau, 1933, 1934). The nuclear chromatin of such cells is depleted until virus particles nearly fill the nucleus (Esau and Hoeffert, 1973;

Esau, 1977). The virions of BGMV appear as aggregated masses or hexagonal arrays in nuclei of both phloem parenchyma and young sieve elements (Kim et al., 1978). Crystalline arrays of MSV (Bock et al., 1974) and tobacco leaf curl virus (TLCV) (Osaki and Inouye, 1978) occur in infected plants.

Geminiviruses are transmitted in nature either by leafhoppers (order Hemiptera, family Cicadellidae) or whiteflies (order Hemiptera, family Aleyrodidae). Both vectors acquire virus by feeding on infected plants and the viruses are absorbed through the gut into the hemolymph of the insect, where they persist for prolonged periods. Transmission is by secretion in the saliva of feeding insects. If acquired by prolonged feeding, the viruses may persist in the vector for the life of the insect (Bennett, 1971; Bock, 1974).

B. Properties of Virions and Viral DNA

The virions of geminiviruses are small asymmetric particles 18–20 nm in diameter by 30 nm in length and usually appear geminate. In outline, particles resemble short dumbells. Geminate particles have been reported for BCTV, BGMV, MSV, cassava latent virus (CLV), and TLVC. The paired particles appear angular, usually five-sided, and the contiguous side generally appears longer than the others. The center-to-center distance of intracellular packed arrays of BGMV is 18 nm (Kim et al., 1978).

The paired particles of BCTV in nuclei of phloem cells are found in extensive ribbon-like arrays (Esau, 1977). The particles of TLCV in nuclei of solanaceous plants appear as symmetrical paired arrays of virions in the form of rigid rodlike structures (Osaki and Inouye, 1978).

Infectivity trials indicate a relationship between the geminate character of virions and biological activity. Single particles of BCTV show little infectivity whereas paired and presumably intact particles are infectious (Larsen and Duffus, 1984). Electron micrographs of CSMV show that each virion (or paired particle) consists of two incomplete icosahedra with a $T = 1$ surface lattice (Hatta and Francki, 1979). Each structure consists of 22 capsomeres, each of about 8 nm diameter. Each paired particle contains 19% DNA which consists of a single molecule of single-stranded material (Francki et al., 1980; Goodman et al., 1980).

Limited analyses of the structural proteins of the geminivirus coat report only one protein. The molecular mass values for different viruses have been reported as: 28,000 kDa for MSV (Bock et al., 1977), 34,000 for CLV (Morris-Krsinich et al., 1985), 31,000 for BGMV (Reisman et al., 1979), 28,000 for tomato golden mosaic virus (TGMV) (Hamilton et al., 1983); 28,000 for CSMV (Francki et al., 1980), and 27,500 for tobacco yellow dwarf virus (TYDV) (Thomas and Bowyer, 1980).

Identification of the genomic material of a geminivirus (BGMV) as single-stranded DNA by Goodman (1977a,b) was based on a positive diphenylamine test, sensitivity to DNase and S_1 nuclease, insensitivity to RNase A or

0.3 N NaOH, reactivity with formaldehyde, and melting in a noncooperative manner (15% hyperchromicity) over a wide temperature range (20–70°C). Harrison et al. (1977) showed that the nucleic acid of MSV and CLV had similar properties and hence must be single-stranded DNA.

The single-stranded DNA of the geminiviruses is the smallest genome yet found among plant viruses that replicate independently. The molecular weight of BGMV DNA measured by sedimentation velocity was estimated to be only $6.5–7.8 \times 10^5$ (Goodman, 1977a,b). Contour lengths of MSV and CLV DNAs indicated a molecular weight of $0.7–0.8 \times 10^6$ (Harrison et al., 1977), while that of CSMV was estimated to be 0.71×10^6 (Francki et al., 1980). These sizes have since been confirmed by DNA sequencing.

Early investigators, observing that geminivirus DNAs separated as two components during gel electrophoresis, suggested that the viruses may have bipartite genomes, a common feature among RNA plant viruses. In gels the two forms appeared to be circular and linear molecules, with the latter arising by breakage of the circular form (Goodman et al., 1980). When DNA of BGVM was subjected to restriction analysis, however, the fragments sizes summed to about 5400 nucleotides, twice the value expected on the basis of contour length measurements (Haber et al., 1981). Moreover, when the DNA was used to infect bean mesophyll protoplasts, dilution kinetics were characteristic of a two-component system. Hamilton et al. (1982) and Bisaro et al. (1982) found that two types of molecules also occurred in the cloned DNAs of TGMV, again based upon analysis of DNA restriction fragments. Nevertheless, it was not proved that any geminivirus had a split genome until Stanley (1983) cloned the DNA components of CLV and found that two DNA molecules were required to initiate infections. Hamilton et al. (1983) found the same for the cloned DNA of TGMV.

In general, intact recombinant plasmids containing full-length copies of the bipartite genome are not infectious until viral DNA is cut free by restriction endonuclease cleavage at the cloning site. Stanley and Townsend (1986) found, however, that intact uncut plasmids of CLV could initiate infections if inoculations were made with both cloned components simultaneously. Restriction analyses suggested that this occurred by intermolecular recombination in vivo between DNAs 1 and 2. Ikegami et al. (1982) showed that native double-stranded-DNA of BGMV virus isolated from infected plants was infectious.

Restriction maps of native DNA of BGMV (Haber et al., 1983) and cloned DNA of TGMV (Bisaro et al., 1982) [see also Stanley (1983) and Hamilton et al. (1983)] showed that each portion of the bipartite genome is made up of about 2500 nucleotides, each with a different DNA sequence. Hybridization analysis showed large dissimilarities in nucleotide sequence except for a region of high homology that mapped to a unique location on each molecule of cloned DNA (Haber et al., 1983). This segment, later defined by DNA sequencing, is referred to as the "common region."

A bipartite genome is not common to all geminiviruses. Of four leafhopper-borne geminiviruses that have been cloned and sequenced (Howell, 1984; Mullineaux *et al.*, 1984; Stanley *et al.*, 1986; Donson *et al.*, 1987), only a single DNA molecule has been found to be required for infection, in contrast to the two DNA components characteristic of the whitefly-transmitted members of the group. For example, with BCTV, a leafhopper-borne member of the group, a single DNA molecule of 2993 nucleotides is fully competent biologically. Its cloned DNA is infectious to plants, and insects fed on these plants acquire virus and transmit it to other plants (Stanley *et al.*, 1986).

DNA 1 of CLV (Townsend *et al.*, 1986) and DNA A of TGMV (Rogers *et al.*, 1986), both whitefly-borne geminiviruses, replicate in plants without the second DNA component of the bipartite genome. This is analogous to the independent replication of the bottom RNA component of cowpea mosaic virus or the long particle RNA of tobacco rattle virus, i.e., split genome RNA viruses in which part of the genome is replication competent (see Chapter 14 in Volume 6 of this series). The experiments with CLV DNA were done with viral DNA transfected to protoplasts of *Nicotiana plumbaginifolia*. With DNA 1, supercoiled and progeny single-stranded DNAs were produced, while with DNA 2, neither type of DNA was made. In the experiments of Rogers *et al.* (1986), the virus genome with each DNA component as a tandem direct repeat was transformed to petunia plants with a Ti plasmid vector. In plants transformed with TGMV DNA A, free single-stranded and double-stranded viral DNAs were found. In transgenic plants transformed with TGMV component B DNA, neither free single-stranded nor double-stranded DNA was detected. Thus, component A must encode virus functions for both replication and encapsidation (Hamilton *et al.*, 1984; Rogers *et al.*, 1986). It will be interesting to learn what functions are encoded on the second DNA component.

Experiments with TGMV transgenic plants, in which component A transformed plants were crossed sexually with component B transformed plants, spontaneously produced virus-diseased progeny (Rogers *et al.*, 1986). Fully encapsidated and competent virus could be transferred from these plants to healthy indicator plants. Encapsidated, infectious virus was also present in component A transformed plants or component A infected protoplasts as shown by mechanical transfer of DNase-treated virus to component B transgenic plants (Sunter *et al.*, 1987). When healthy indicator plants were inoculated with virus from component A transformed plants, no disease was induced and no viral DNA could be detected. Hence component A alone is probably not able to spread cell to cell from the initially infected cells, suggesting that TGMV component B probably has functions which govern virus movement (Sunter *et al.*, 1987). In other experiments, however, an infectious coat protein mutant of CLV that did not produce mature virions moved systemically in infected plants as if packaging was not essential for efficient cell-to-cell movement (Stanley and Townsend, 1986).

C. Genetic Organization of Geminiviruses

The genetic organization of the geminiviruses has been defined by DNA sequencing. Seven viruses, three with whitefly vectors and four with leafhopper vectors, have been sequenced. The common regions and the uniqueness of the two DNA components of the whitefly-transmitted viruses were confirmed by these studies. In addition, each double-stranded counterpart of viral DNAs contained one or more open translational reading regions in each strand, suggesting that each may be transcribed bidirectionally.

The two viral DNA molecules of CLV, 2779 and 2724 nucleotides in length, are of unique sequence except for a region of about 200 nucleotides common to both DNAs (Stanley and Gay, 1983). This confirmed hybridization data (Haber *et al.*, 1983) for a region of similar sequence with BGMV DNAs. The common region of CLV has small repeated or near-repeated sequence elements with the potential to form a very stable hairpin structure with an unpaired loop of A,T residues (Stanley and Gay, 1983). This region may contain recognition signals for DNA replication as well as transcriptional promoters.

Common regions of about the same size and structure as that of CLV have been found for the bipartite genomes of TGMV and BGMV, also transmitted by the tropical whitefly (*Bemisia tabaci*). The common region in TGMV DNA has 200 bp and the potential to form a stable hairpin structure with a GC-rich stem of 11 bp and an AT-rich loop of 11–12 nucleotides (Hamilton *et al.*, 1984). The hairpin structure has a high level of homology with the stem-loop structure of CLV, but the remainder of the common regions of TGMV show only 29% homology with the common region of CLV (25% = random expectation). BGMV DNAs have a common region of 205 bp with a GC-rich 12-bp stem and an AT-rich 12-nucleotide loop (Howarth *et al.*, 1985). Sequences with the capacity to form highly stable stem-loop structures are also found in the common regions of DSV DNA (Donson *et al.*, 1987). MacDowell *et al.* (1985) and Stanley *et al.* (1986) pointed out that a 9-nucleotide sequence (TAATATTAC) that makes up part of the loop of the hairpin structure has been found in all geminiviruses that have been sequenced. Others have made analogies between the stable stem-loop structure of the geminiviruses and a similar structure in the origin of DNA replication of ϕX174, a single-stranded bacteriophage (Hamilton *et al.*, 1984; Donson *et al.*, 1987; Lazarowitz, 1987). This feature and its possible involvement in the replication of geminiviruses are discussed later.

When the DNAs of CLV DNA were screened for open reading regions with potential coding capacities greater than 10 kDa, both the viral DNA (plus) strand and its complement (minus strand) contained potential coding sequences lying outside the common region. DNA 1 of CLV, for example, coded for two overlapping proteins (Fig. 9). The complementary sequence of DNA 1 had the unusual feature of coding for seven proteins, several of

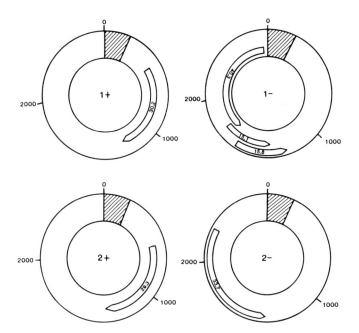

Fig. 9. Potential coding regions conserved among different whitefly-transmitted geminiviruses. The coding regions of cassava latent virus are used as an example [redrawn from Stanley (1985)]. Each region is identified by its potential coding capacity, given in kilodaltons. Virion sense DNA is indicated by the diagrams on the left, labeled 1+ and 2+ for the two components of the split genome. The complementary strands are shown on the right (1− and 2−). The indicated regions have corresponding open reading frames in bean golden mosaic virus and tomato golden mosaic virus. The cross-hatched area at the top of each diagram indicates the common region. The latter is a region of about 200 bases which is almost identical in each DNA component of the divided genome. Except for the common region, there is little sequence homology between the two DNA components.

which were overlapping (Table V). DNA 2 had sequence capable of coding for two proteins in each strand. Consequently, the two viral DNAs as a whole had the surprising capacity to code for a dozen different proteins greater than 10 kDa. The DNAs of the other geminiviruses are not as complex as that of CLV (Table V). Several of the nonconserved open reading regions of these viruses are probably fortuitous and most likely do not code for viral-specific proteins (Donson *et al.*, 1987). The predicted amino acid content of the putative 30.1-kDa protein of CLV DNA 1, as inferred from the DNA sequence, has a close approximation to the amino acid composition of the viral coat protein (Stanley and Gay, 1983). Identification of the long open reading region on DNA 1 (Fig. 9) as the coat protein gene was confirmed by Townsend *et al.* (1985) by S_1 nuclease mapping and hybrid-arrested translation.

TABLE V

Translational Reading Frames of the Genomes of Geminiviruses (from Their DNA Sequences)

Virus	Size of genome component in nucleotides	ORFs of coding capacity greater than 10 kDa	
		Plus strand (virion strand) (kDa)	Minus strand (kDa)
Whitefly-transmitted viruses			
Cassava latent virus (CLV)[a]	DNA 1: 2779	30.1,[b] 12.4	40.2, 27.0, 15.5, 15.6 15.1, 13.5, 10.6
	DNA 2: 2724	29.2, 13.5	33.6, 13.5
Bean golden mosaic virus (BGMV)[c]	DNA 1: 2646	27.7	40.2, 19.6, 15.6, 12.0
	DNA 2: 2587	29.7, 10.0	33.1
Tomato golden mosaic virus (TGMV)[d]	DNA A: 2588	28.7	40.3, 14.9
	DNA B: 2508	29.3	15.7, 21.1
Leafhopper transmitted viruses			
Beet curly top virus (BCTV)[e]	DNA 1: 2993	29.6, 12.1	40.9, 20.1, 19.4, 16.1
Digitaria streak virus (DSV)[f]	DNA 1: 2701	12.1, 26.6[b]	30.5, 16.8
Maize streak virus (MSV)[g]	DNA 1: 2687	10.9, 27.0,[b] 11.2	31.4, 17.8, 21.8, 13.0
Wheat dwarf virus (WDV)[h]	DNA 1: 2749	10.1, 29.4	14.6, 30.2, perhaps 17.8

[a] From Stanley and Gay (1983).
[b] Indicates probable coat proteins.
[c] From Howarth et al. (1985).
[d] From Hamilton et al. (1984).
[e] From Stanley et al. (1986).
[f] From Donson et al. (1987).
[g] From Howell (1984, 1985a); Mullineaux et al. (1984).
[h] From McDowell et al. (1985).

Sequence comparisons among the whitefly-borne, bipartite geminiviruses indicate that six open reading frames are conserved among the three viruses that have been sequenced (Hamilton et al., 1984; Howarth et al., 1985; Stanley, 1985). The size and distribution of these regions on the various strands of the two DNAs of CLV are shown in Fig. 9. Of these open reading frames, the long region on the virion sense strand probably codes for the coat protein of TCMV, as the analogous gene does for CLV (Hamilton et al., 1984). There is a high level of conservation of sequence at both the DNA and protein levels for the coat protein genes of the three whitefly-transmitted viruses that have been sequenced. Moreover, hydrophilicity plots for the coat protein amino acid sequences for TGMV and CLV are very similar, especially at the C-terminal end of the protein (Hamilton et al., 1984), perhaps accounting for the serological affinities among these viruses (Roberts et al., 1984). A more variable homology is shown between the tentative coat protein gene of MSV, one of the leafhopper-borne viruses, and CLV and TGMV (Mullineaux et al., 1985). There is no detectable serological relationship between MSV and these two bipartite geminiviruses (Roberts et al., 1984).

There are relatively high sequence homologies between the DNA components of the whitefly-transmitted viruses. For example, CLV DNA 1 has about 60% overall homology with TGMV DNA A; DNA 2 of CLV has 40% overall homology with TCMV DNA B (Hamilton et al., 1984). The homologies between the single DNAs of the leafhopper-borne and whitefly-borne viruses are less.

There are also amino acid sequence homologies between the proteins encoded by the two DNA components of CLV that suggest the two genomes arose from a common ancestral source. The virus sense strands of CLV DNA 1 and DNA 2 encode proteins of 29.2 and 30.1 kDa, respectively (Table III), with a 43% homology of the aligned sequences when substitutions are allowed for chemically similar amino acids (Kikuno et al., 1984). Similar sequence similarities were not found for the other open reading frames of the two DNA components. Amino acid sequence relationships between the leafhopper- and whitefly-borne viruses also suggest a common ancestry for these viruses (MacDowell et al., 1985; Mullineaux et al., 1985).

Of the DNAs of the four leafhopper-transmitted geminiviruses which have been cloned and sequenced, only a single DNA component has been found. With three of the viral DNAs, it has been possible to infect plants either directly or via Ti plasmid vectors with cloned material to show that the single cloned component is adequate to initiate infections in plants and to be transmissible by the leafhopper vector. As pointed out by Stanley et al. (1986), the biological competence of the 2993-nucleotide DNA of BCTV proves that the geminivirus genome comprises what is probably the smallest autonomous pathogen of eukaryotes.

Proof of the biological competence of cloned DNAs of two other leafhopper-borne geminiviruses has been accomplished by "agroinfection" (Grimsley et al., 1986b), the term being derived from *Agrobacterium*-mediated DNA transfer. The cloned DNAs of MSV and *Digitaria* streak virus (DSV) were inserted into Ti plasmid-derived shuttle vectors which were then conjugated into *Agrobacterium tumefaciens* containing a wild-type, virulent Ti plasmid helper (to supply *vir* functions for plasmid transfer to plants). The bacterium was injected into the stems of plants to initiate DNA transfer. Probably only a few cells in the vicinity of the injection became transformed by plasmid DNA, whereupon these few cells produced infectious virus which reproduced and spread throughout the plant.

In the case of MSV, a tandem linear insertion of the 2687-nucleotide genome into the Ti shuttle vector was used to initiate infection (Grimsley et al., 1987). The genome of DSV is 2701 nucleotides in length. A tandem dimeric insert of the cloned viral genome was made into the Ti vector (Donson et al., 1988). In neither case is it known how the replicating DNA originates from the tandem dimer. Either homologous recombination occurred between the repeated genomes to produce full-length circular viral DNA, or alternatively, replicative DNA arose directly from the T-DNA insert to produce infectious viral DNA (Grimsley et al., 1987). Regardless of the mechanism, the method is much more efficient than manual inoculation of viral DNA directly to plants. Most significantly, it permits infectivity assays with cloned viral DNA that cannot be transmitted in any other way.

None of the functions of the leafhopper-borne geminiviruses has been identified, although an open reading frame coding for a 27.0-kDa protein by MSV DNA sense strand has been found to be coat protein (Mullineaux et al., 1984). An open reading frame encoding a 29.4-kDa protein on the wheat dwarf virus (WDV) genome (MacDowell et al., 1985) and one encoding a 27.5-kDa protein on the DSV genome (Donson et al., 1987) may be the capsid protein genes. It is also possible to align the inferred amino acids of these genes with those of CLV and TGMV. Of 56 conserved amino acids, 20 are directly homologous in the four putative coat proteins (MacDowell et al., 1985). Four domains of familial homology can be distinguished between the putative coat proteins of MSV-Nigerian isolate and CLV and TGMV.

When comparing the DNA sequences of WDV and MSV DNAs, MacDowell et al. (1985) pointed out a highly conserved region that coincides with an open reading region of 17.8 kDa on minus strand MSV DNA (Table III), but no open reading region at the corresponding position on the WDV sequence. However, when the complementary strand of WDV was compared, an open reading region which encoded a 17.3-kDa protein was found. When the two sequences were aligned, they showed 70.8% homology. This degree of relatedness strongly suggests that the two open reading regions are functional in the two viruses. However, this 17.3-kDa protein in WDV can be accounted for only by the production and processing of a readthrough product from an open reading region upstream to the 17.3-kDa gene of WDV. A similar lack of an ATG triplet for a minus sense region encoding a putative protein of 16.9 kDa, also with a high level of homology to the 17.3-kDa protein of WDV, occurs in the DSV genome (Donson et al., 1987).

MacDowell et al. (1985) also pointed out other predicted amino acid homologies between the tentative genes of WDV and MSV and some of the whitefly-borne viruses if transcriptional splicing, frameshift, or termination suppression mechanisms are active in gene expression of these viruses. Donson et *al.* (1987) have analyzed the relatedness and conservation of inferred proteins of the four leafhopper-borne geminiviruses which have been sequenced (Fig. 10). Four conserved coding regions distributed over both strands (two occurring on the plus strand and two on the minus strand) of viral double-stranded DNA suggest that transcription is probably bidirectional.

The presence of possible 5'/3' expression signals and bidirectional transcription in relation to the open reading regions of geminiviruses have been noted often during the sequencing of these viruses (Hamilton *et al.*, 1984;

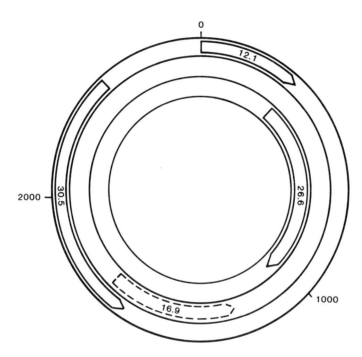

Fig. 10. Potential coding regions conserved among different leafhopper-transmitted geminiviruses that infect monocotyledonous plants. The coding regions of *Digitaria* streak virus are used as an example [from Donson *et al.* (1987)]. Each region is identified by its potential coding capacity, given in kilodaltons. Virion sense DNA is shown in a clockwise orientation. Considering both direct and familial relatedness in the amino acid sequences, the total homologies for the four conserved regions range from 49 to 86% for the three viruses (DSV, MSV, and WDV) that have been sequenced. The 26.6-kDa region of DSV, the probable coat protein gene, has 47% total homology with the 29.6-kDa region of beet curly top virus (BCTV), a leafhopper-borne geminivirus infecting dicotyledonous plants. Two of the coding regions of these viruses (the 30.5- and 16.9-kDa regions) possess homology to the 40.9-kDa region of BCTV (39 and 51% total homology, respectively). The regions for the 16.9-kDa protein of DSV (broken lined arrow) and wheat dwarf virus (WDV) do not have a 5' proximal ATG, although the comparable region of maize streak virus (MSV) does start with an ATG.

Mullineaux et al., 1984; MacDowell et al., 1985; Stanley et al., 1986; Donson et al., 1987). The work of Townsend et al. (1985) with CLV has provided definite evidence for this. Northern blots of poly(A) RNA isolated from CLV-infected plants showed transcripts of 1.7, 1.0, and 0.7 kb which mapped to DNA 1, and two major transcripts of 1.1 and 0.9 kb and a minor one of 1.35 kb which mapped to DNA 2. In S_1 nuclease mapping experiments, poly(A) RNA from infected plants protected two virion sense DNA 1 fragments of 0.6 and 1.5 kb in length. Complementary sense DNA 1 was protected by an abundant 1-kb transcript. Virion sense DNA 2 was protected by poly(A) RNA fragments of 1.0 and 1.3 kb. Its complementary sense DNA (component 2) was protected by a transcript of 0.85 kb (Townsend et al., 1985). There is also good evidence for bidirectional transcription of MSV (Morris-Krsinich et al., 1985).

The replication of geminiviruses seems to occur in the nucleus. Much speculation has centered on the involvement of the highly stable stem-loop structures in the common region of these DNAs as part of an origin of replication. These structures are similar in sequence to the DNA interaction and cleavage site for the ϕX174 gene A protein. The latter is involved, in addition to several host enzymes, in the production of a pool of double-stranded replicative forms of this single-stranded bacteriophage. The gene A protein specifically recognizes and cleaves at the sequence TGATATTAATAAC, which is remarkably similar to the conserved loop common to all geminiviruses (TAATATTAC).

Stanley (1985) has discussed the replication of geminiviruses and pointed out three plausible stages in the process. The first is probably the conversion of single-stranded viral DNA to a double-stranded replicative form (RF). Second, these molecules probably replicate to produce a pool of RF molecules. The last stage is probably the preferential production of plus strands for packaging into coat protein to form mature virions.

Single-stranded geminiviral DNA is infectious, suggesting that host factors alone may be adequate for synthesis of second-strand DNA on viral strand templates. However, the small 80-nucleotide DNA discovered in denatured MSV DNA preparations, reported and characterized by Donson et al. (1984) and Howell (1985a), is probably a primer for second-strand synthesis. The presence of covalently attached 5'-ribonucleotides suggests this role. The sequence of this tentative primer shows it to be complementary to nucleotides 1123–1206 of one of the two noncoding regions of the MSV genome. This molecule will prime second-strand synthesis on a single-stranded MSV template DNA catalyzed by the Klenow fragment of DNA polymerase (Donson et al., 1984). A similar molecule of 80 nucleotides has been found to be associated with DSV DNA (Fig. 11). It maps to approximately the same intergenic position near a conserved stem-loop structure, and has been found to prime second-strand synthesis of DSV (Donson et al., 1987).

MSU

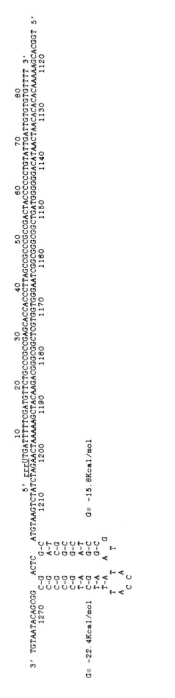

DSV

Fig. 11. Nucleotide sequence of the *in situ* primers for second-strand DNA synthesis of maize streak virus (top) and *Digitaria* streak virus (bottom). The sequence is shown in a base-paired alignment with viral plus strands. Unidentified ribonucleotides are indicated by r. The stop codon of the 16.9-kDa potential coding region of *Digitaria* streak (bottom) is indicated by (#) following the carboxy-terminal amino acids in single letter code. Information from Donson *et al.* (1984, 1987).

IV. PROSPECTS FOR USING DNA VIRUSES AS GENE VECTORS

DNA viruses may ultimately be useful vectors for introducing foreign DNA into plants and may offer certain advantages for this purpose. The genomes of these viruses are rapidly amplified in infected plants and thus may bring about gene expression in high copy number. In addition, they move rapidly from cell to cell to carry genetic information throughout intact growing plants. If developed as gene vectors, it might be possible to use the viruses to transform intact plants without needing to regenerate plants by tissue culture.

Currently, the only well-developed system for genetic transformation of plants is the Ti plasmid of *A. tumefaciens,* in which sequences inserted by the plasmid are integrated into plant chromosomes and transferred to succeeding generations through seed. With Ti plasmid vectors one or a few plant cells are transformed. These are detected by including drug-resistant genes in the T-DNA, selecting for drug resistance in tissue culture, and regenerating the selected cells to whole plants. The process may require several months. Using DNA viruses as transducing vectors on the whole-plant level might avoid the long delay in obtaining transformed plants. However, several technical obstacles must be overcome.

Molecular cloning of the viral genomes in bacteria is a prerequisite for gene vector construction since this is the only way to propagate and isolate large quantities of infectious viral DNA or noninfectious subgenomic pieces for modular construction of gene vectors to be introduced into plants. The advantages of cloning for restructuring caulimoviral DNA have been discussed elsewhere (Howell, 1984; Shepherd, 1985). Most of the work done so far in gene vector development with the DNA viruses has used CaMV since more is known about gene expression and replication of this virus.

1. Productive Infection-Type Vectors

This type of vector consists of a fully active virus that retains functions needed for virus replication and systemic development in the plant. Foreign DNA would be inserted into the viral genome such that it would not interfere with the expression of virus functions and its expression would be brought about by appropriate 5'/3' expression signals.

The first requirement in development of a vector of this type is to identify nonessential regions of the viral genome where insertions of foreign DNA can be made without disturbing virus functions.

Gronenborn *et al.* (1981) identified a nonessential region in the CaMV genome by making insertions into the unique *Xho*I site in the center of ORF II. A small deletion, including loss of the β_2-interruption (Fig. 4), had been observed in a mutant of the virus (Howell and Hull, 1978) and sequencing had shown 421 bp of ORF II to be missing from this strain (Howarth *et al.,*

1981). Of a variety of insertions placed in the *Xho*I site of an infectious clone of CaMV, one of 65 bp was infectious and stably propagated throughout the plant. Moreover, it was retained through several successive transfers of virus from systemically infected plants. Tolerance of a fragment of this size provides good evidence for a nonessential region in the CaMV genome.

Of larger inserts of 256, 531, and 1200 bp, only the first was stable in the gene II region of CaMV strain CM1841 (Gronenborn *et al.*, 1981). With the 531-bp insert, the DNA was infectious, but there was a delay in symptom development. When viral DNA was recovered from infected plants, most or all of the insert had been deleted. This suggests that, although CaMV will tolerate insertions in region II, the size must be no larger than about 250 bp if the recombinant virus is to be stably propagated.

Others (Gardner, 1982; Daubert *et al.*, 1983; Dixon *et al.*, 1983) have made insertions in the large intergenic region of the CaMV genome. This region extends from position 7336 to position 13 (695 bp) of the 8000-bp genome. Just to the right of the α discontinuity extending from nucleotide 13 to nucleotide 300 is a small open reading region, referred to as region VII (Fig. 4). This region is probably not functional since extensive insertions can be made in this region, or it can be completely deleted, without affecting biological activity of the virus (Dixon and Hohn, 1984).

Insertions into regions other than gene II or the large intergenic region destroy virus infectivity (Howell *et al.*, 1982; Dixon *et al.*, 1983; Daubert *et al.*, 1983). Insertions, even small ones which do not change the reading frame, generally inactivate the virus, except for small in-frame insertions into certain regions of gene VI.

2. Probable Loss of Cell-to-Cell Movement by Unencapsidated CaMV DNA

The behavior of larger insertions in region II or the large intergenic region suggests a more stringent limitation of the additional DNA on encapsidation than for replication. Inserts of 1200 bp into region II result in a systemic infection after a long delay (Gardner, 1982), i.e., 2 months as opposed to the usual 2 weeks. When the DNA of these plants was examined, some of the inserted DNA was still present in region II but most was deleted, including the flanking portions of region II. The inoculated DNA had obviously persisted in some manner, probably by replication in the initially infected cells. This may not, however, apply when larger DNA inserts (the size of plasmid cloning vectors) are spliced into CaMV region II. When such DNA has been inoculated to plants, it has not been infectious (Walden and Howell 1983a,b).

The unstable nature of large insertions in the CaMV genome may reflect a packaging limitation for DNA encapsidation. Reasonably small, circular DNAs may be a precondition for encapsidation in coat protein and perhaps only encapsidated DNAs move from cell to cell to establish systemic infections.

3. Gene II Replacements with Foreign Genes

The first successful use of one of these DNA viruses as a gene vector was that of Brisson *et al.* (1984). They substituted a bacterial methotrexate-resistant dihydrofolate reductase (*dhfr*) gene for region II of CaMV and mechanically inoculated the cloned DNA to plants. The resulting virus infection carried the *dhfr* gene systemically throughout the plant, and the plants became resistant to methotrexate sprays, which are very toxic to uninfected plants. The constructs used left as little nontranslated sequence as possible between the flanking genes. CaMV genes I–V are closely packed in native viral DNA with only 1–2 nucleotides between successive genes, and Brisson *et al.* (1984) reasoned that this feature may be necessary for efficient translation of the full-length transcript as a polycistronic messenger RNA. In practice, two plasmids with different length leader sequences before the *dhfr* gene were constructed and tested. In one, CaMV region II was almost completely removed except for five codons, plus a stop codon, before the coding region of the *dhfr* gene. In the second, the distance between the region I stop codon and the *dhfr* initiation codon was reduced even further, to 9 bp. With both plasmids only 1 bp existed between the *dhfr* stop codon and the start codon of region III.

When the DNAs were manually inoculated to plants, both caused CaMV infections which developed systemically. Both were stably propagated and maintained in infected plants through the first infection cycle. However, the *dhfr* gene in the DNA with the longer spacer sequence between coding regions was gradually lost during the second and third infection cycles in plants, while the DNA with the shorter spacer between genes I and II was stable during similar transfers. Other constructions with longer intergenic spacers at either the 5'- or 3'-ends of the *dhfr* coding region were considerably less stable, indicating that a close packing arrangement of genes may be crucial to stability and expression with a CaMV vector (Brisson *et al.*, 1984).

The *dhfr* gene used by Brisson *et al.* (1984) is a small plasmid-derived sequence (234 bp coding region) practically insensitive to methotrexate. Its small size probably allowed encapsidation of the restructured viral genome and cell-to-cell movement as a virion.

Whether larger genes can be transported and propagated as part of the CaMV virion is not known. CaMV strain CM4-184 is a mutant 600 bp smaller than the wild type (Howarth *et al.*, 1981). If its genome can be expanded to the maximum genome size tolerated by other strains, this would allow about 850 bp of foreign sequence to be inserted. If the sequences between the α-interruption and the initiation codon of region 1 are dispensible, as indicated by the observations of Dixon and Hohn (1984), a genome with about 1100 bp of foreign DNA should be biologically active. The Brisson *et al.* (1984) experiments clearly show that, with the proper construction of the untranslated spacer between regions, it is possible to obtain good expression of foreign genes with CaMV as the gene vector.

4. Complementation Experiments with Defective Viral Genomes

Another approach made to circumvent the low limit on the amount of DNA that can be accommodated by a caulimovirus vector is to construct artificial split genome systems in which functions from one component are complemented by a second component multiplying in the same cell. The missing functions in each component could then be replaced by foreign DNA. Such a system would be akin to the split genome of the whitefly-borne geminiviruses discussed earlier in this chapter, or to certain RNA viruses (see Chapter 14 of Vol 6 of this series). When inactive clones of CaMV with lethal modifications in different regions of the virus genome are inoculated to plants in pairs, infections frequently develop as if complementation had occurred (Howell *et al.*, 1982; Lebeurier *et al.*, 1982). However, the viral DNA recovered from such plants corresponds to that of wild-type virus, presumably a consequence of recombination between the mutant genomes. These results make it likely that, with a single viral genome, recombination will preclude the development of complementing systems (Howell *et al.*, 1982).

The recombining of mutant genomes of CaMV may occur by one of several mechanisms. Homologous intergenomic recombination probably accounts for most cases (Walden and Howell, 1983a,b). However, replicative recombination in which the reverse transcriptase makes intergenomic strand switches when copying the 35S RNA template is another possibility for which evidence exists (Dixon *et al.*, 1986). A third possibility that occurs when cloned genomes are cleaved free of vector with the same restriction endonuclease is head-to-tail dimerization of the released genomes. This provides a template for transcription of full-length 35S RNA, as well as 19S RNA, which together may be capable of setting up the infection process (Geldreich *et al.*, 1986).

Certain caulimovirus DNAs show little if any hybridization with CaMV DNA and may be different enough in sequence to preclude recombination. Hybridization tests have indicated there is relatively little homology between the DNAs of CaMV and dahlia mosaic virus, FMV, *Mirabilis* mosaic virus, or thistle mottle virus (Hull and Donson, 1982; Richins and Shepherd, 1983). However, most of these viruses do not have common hosts in which tests can be done. Yet FMV and certain CaMV strains have been observed to multiply together in *Nicotiana edwardsonni* for long periods without recombining.

Another approach worth investigation is to transfer virus functions to the host chromosome using the *Agrobacterium* Ti plasmid system. The removal of one or more essential virus genes to the host genome would allow the replacement of deleted regions of the virus with foreign DNA. A ploy of this sort would be analogous to the COS cell system developed by Gluzman (1981) for SV40 as a vector for mammalian cells. In this case, early virus functions (large T-antigen and permissivity factors) required for SV40 DNA replication were transformed into mammalian cells which subsequently expressed these functions.

Problems in gene stability may be encountered in the use of virus-based gene vector systems that have high error rates during replication (see van Vloten-Doting et al., 1985). The reverse transcriptase of retroviruses catalyzes the incorporation of an exceptionally large number of incorrectly paired bases when copying either ribonucleotide or deoxyribonucleotide templates (Battula and Loeb, 1974, 1975). The frequency of error may be as high as 1 in 6000. Thus, a high error rate can be expected for the CaMV polymerase, in contrast to the high fidelity of most DNA → DNA generating systems. Other DNA replicating systems with editing and repair mechanisms have a remarkably low error rate of about 1 nucleotide in 10^{10} (Reanney, 1984). The caulimoviruses have a DNA → RNA transcriptional step carried out in the nucleus by DNA-dependent RNA polymerase II and a subsequent high error RNA → DNA generating step during reverse transcription to produce the minus strand. Consequently, foreign DNA sequences in a caulimovirus vector may be subject to a high rate of spontaneous mutation during virus replication. There would be no selection pressure for maintenance of foreign DNA inserts as there might be for virus functions under the same conditions.

Overall, it remains to be seen how useful the caulimoviruses will be as recombinant DNA vectors. With a complementing system using partially deleted genomes, it seems likely that these viruses can be developed as useful gene vectors. They represent one of the few sources of DNA sequences that become highly amplified in plants. Vectors of this sort for gene transfer in plants are desirable and further efforts to explore the vector potential of these viruses seem justified.

The vector potential of the geminiviruses is largely untested at the moment. However, it seems likely that these viruses probably replicate by a more conventional DNA → DNA generating system, with better fidelity perhaps than that afforded by the caulimovirus system. Moreover, their DNA has been reported to move in plants in an unencapsidated state, which may relieve packaging limitations on the size of foreign DNA inserts that can be made.

REFERENCES

Al Ani, R., Pfeiffer, P., and Lebeurier, G. (1979). *Virology* **93**, 188–197.
Al Ani, R., Pfeiffer, P., Whitechurch, O., Lesot, A., Lebeurier, G., and Hirth L. (1980). *Am. Virol. (Inst. Pasteur)* **131E**, 33–44.
Armour, S. L., Melcher, U., Pirone, T. P., Lyttle, D. J., and Essenberg, R. C. (1983). *Virology* **129**, 25–30.
Balazs, E., Guilley, H., Jonard, G., and Richards, K. (1982). *Gene* **19**, 239–249.

Battula, N., and Loeb, L. A. (1974). *J. Biol. Chem.* **249,** 4086–4091.
Battula, N., and Loeb, L. A. (1975). *J. Biol. Chem.* **250,** 4405–4409.
Baughman, G. A., Jacobs, J. D., and Howell, S. H. (1988). *Proc. Natl. Acad. Sci.* **85,** 733–737.
Bennett, C. W. (1971). *Am. Phytopathol. Soc. Monogr.* **7,** 1–81.
Bisaro, O. M., Hamilton, W. D. O., Coutts, R. H. A., and Buck, K. W. (1982). *Nucleic Acids Res.* **10,** 4913–4922.
Bock, K. R. (1974). *Commonw. Mycol. Inst. Descript. Plant Viruses* **133,** 1–4.
Bock, K. R., Guthrie, E. J., and Woods, R. D. (1974). *Am. Appl. Biol.* **77,** 289–296.
Bock, K. R., Guthrie, E. J., Meredith, G., and Barker, H. (1977). *Am. Appl. Biol.* **85,** 305–308.
Boncquet, P. A., and Hartung, W. J. (1915). *Phytopathology* **5,** 348–349.
Bonneville, J. M., Volovitch, M., Modjtahedi, N., Demery, D., and Yot, P. (1984). *Adv. Exp. Med. Biol.* **179,** 113–119.
Breathnach, R., and Chambon, P. (1981). *Annu. Rev. Biochem.* **50,** 349–383.
Brisson, N., Paszkowski, J., Penswich, J. R., Gronenborn, B., Potrykus, I., and Hohn, T. (1984). *Nature (London)* **310,** 511–514.
Brunt, A. A., Barton, R. J., Tremaine, J. H., and Stace-Smith, R. (1975). *J. Gen. Virol.* **27,** 101–106.
Chauvin, C., Jacrot, B., Lebeurier, G., and Hirth, L. (1979). *Virology* **96,** 640–641.
Condit, C., and Meagher, R. B. (1983). *J. Mol. Appl. Genet.* **2,** 301–314.
Condit, C., Hagen, T. J., McKnight, T. D., and Meagher, R. B. (1983). *Gene* **25,** 101–108.
Covey, S. N. (1985). *In* "Molecular Biology of Plant Viruses" (J. Davies, ed.), Vol. 2, pp. 121–159. CRC Press, Boca Raton, Florida.
Covey, S. N. (1986). *Nucleic Acids Res.* **14,** 623–632.
Covey, S. N., and Hull, R. (1981). *Virology* **111,** 463–474.
Covey, S. N., Lomonossoff, G. P., and Hull, R. (1981). *Nucleic Acids Res.* **9,** 6735–6747.
Covey, S. N., Turner, D., and Mulder, G. (1983). *Nucleic Acids Res.* **11,** 251–264.
Daubert, S., Richins, R., and Shepherd, R. J. (1982). *Virology* **122,** 444–449.
Daubert, S., Shepherd, R. J., and Gardner, R. C. (1983). *Gene* **25,** 201–208.
Daubert, S. D., Schoelz, J., Li, D., and Shepherd, R. J. (1984). *J. Mol. Appl. Genet.* **2,** 537–547.
Dixon, L. K., and Hohn, T. (1984). *EMBO J.* **3,** 2731–2736.
Dixon, L. K., and Hohn, T. (1985). *In* "Recombinant DNA Research and Virus" (Y. Becker, ed.), pp. 247–275. Nijhoff, Boston.
Dixon, L. K., Koenig, I., and Hohn, T. (1983). *Gene* **25,** 189–199.
Dixon, L., Nyffenegger, T., Delley, G., Martinez-Izquierdo, J., and Hohn, T. (1986). *Virology* **150,** 463–468.
Domier, L. L., Shaw, J. G., and Rhoades, R. E. (1987). *Virology* **158,** 20–27.
Donson, J., Morris-Krsinich, B. A. M., Mullineaux, P. M., Boulton, M. I., and Davies, J. W. (1984). *EMBO J.* **3,** 3069–3073.
Donson, J., Accotto, G. P., Boulton, M. I., Mullineaux, P. M., and Davies, J. W. (1987). *Virology* **161,** 160–169.
Donson, J., Gunn, H. V., Woolston, C. J., Pinner, M. S., Boulton, M. I., Mullineau, P. M., and Davies, J. W. (1988). *Virology* **162,** 248–250.
Dudley, R. K., Odell, J. T., and Howell, S. H. (1982). *Virology* **117,** 19–28.
duPlessis, D. H., and Smith, P. (1981). *Virology* **109,** 403–406.
Esau, K. (1933). *Phytopathology* **23,** 679–712.
Esau, K. (1934). *Phytopathology* **24,** 303–305.
Esau, K. (1977). *J. Ultrastruct. Res.* **61,** 78–88.
Esau, K., and Hoeffert, L. L. (1973). *Virology* **56,** 454–464.
Favali, M. A., Bassi, M., and Conti, G. G. (1973). *Virology* **53,** 115–119.
Francki, R. I. B., Hatta, T., Grylls, N. E., and Grivell, C. J. (1979). *Am. Appl. Biol.* **91,** 51–59.
Francki, R. I. B., Halta, T., Baccardo, G., and Randales, J. W. (1980). *Virology* **101,** 233–241.
Frank, A., Guilley, H., Jonard, G., Richards, K., and Hirth, L. (1980). *Cell* **21,** 285–294.
Fujisawa, I., Rubio-Huertos, M., and Matsui, C. (1971). *Phytopathology* **61,** 681–684.

Fujisawa, I., Rubio-Huertos, M., and Matsui, C. (1972). *Phytopathology* **62**, 810-811.
Gardner, R. C. (1982). *In* "Genetic Engineering of Plants: An Agricultural Perspective" (T. Kosuge, C. P. Meredith, and A. Hollaender, eds.), pp. 121-142. Plenum, New York.
Gardner, R. C., Howarth, A. J., Hahn, P., Brown-Luedi, M., Shepherd, R. J., and Messing, J. (1981). *Nucleic Acids Res.* **9**, 2871-2888.
Gardner, R. C., Hiatt, W. R., Facciotti, D., and Shewmaker, C. K. (1984). *Plant Mol. Biol. Rep.* **2**, 3-8.
Geldreich, A., Lebeurier, G., and Hirth, L. (1986). *Gene* **48**, 277-286.
Giband, M., Mesnard, J. M., and Lebeurier, G. (1986). *EMBO J.* **5**, 2433-2438.
Givord, L., Xiong, C., Giband, M., Koenig, I., Hohn, T., Lebeurier, G., and Hirth, L. (1984). *EMBO J.* **3**, 1423-1427.
Gluzman, Y. (1981). *Cell* **23**, 175-182.
Goldberg, K. B., Young, M. J., Schoelz, J. E., Kiernan, J. M., and Shepherd, R. J. (1987). *Phytopathology,* **77**, 1704.
Goodman, R. M. (1977a). *Nature (London)* **266**, 54-55.
Goodman, R. M. (1977b). *Virology* **83**, 171-179.
Goodman, R. M. (1981a). *In* "Handbook of Plant Virus Infections and Comparative Diagnosis" (E. Kurstak, ed.). pp. 879-910. Elsevier, Amsterdam.
Goodman, R. M. (1981b). *J. Gen. Virol.* **54**, 9-21.
Goodman, R. M., Shock, T. L., Haber, S., Browning, K. S., and Bowers, G. R. (1980). *Virology* **106**, 168-172.
Grimsley, N., Hohn, B., Hohn, T., and Walden, R. (1986b). *Proc. Natl. Acad. Sci. U.S.A.* **83**, 3282-3286.
Grimsely, N., Hohn, T., Davies, J. W., and Hohn, B. (1987). *Nature (London)* **325**, 177-178.
Gronenborn, B., Gardner, R. C., Schaefer, S., and Shepherd, R. J. (1981). *Nature (London)* **294**, 773-776.
Guilfoyle, T. J. (1980). *Virology* **107**, 71-80.
Guilfoyle, T. J., Olszewski, N., Hagen, G., Kuzj, A., and McClure, B. (1983). *In* "Plant Molecular Biology" (R. Goldberg, ed.), pp. 117-136. Liss, New York.
Guilley, H., Dudley, R. K., Jonard, G., Balazs, E., and Richards, K. E. (1982). *Cell* **30**, 763-773.
Guilley, H., Richards, K. E., and Jonard, G. (1983). *EMBO J.* **2**, 277-282.
Haber, S., Ikegami, M., Bajet, N. B., and Goodman, R. M. (1981). *Nature (London)* **289**, 324-327.
Haber, S., Howarth, A. J., and Goodman, R. M. (1983). *Virology* **129**, 469-473.
Hahn, P., and Shepherd, R. J. (1980). *Virology* **107**, 295-297.
Hahn, P., and Shepherd, R. J. (1982). *Virology* **116**, 480-488.
Hamilton, W. N. O., Bisaro, D. M., and Buck, K. W. (1982). *Nucleic Acids Res.* **10**, 4901-4912.
Hamilton, W. D. O., Bisaro, D. M., Coutts, R. H. A., and Buck, K. W. (1983). *Nucleic Acids Res.* **11**, 7387-7396.
Hamilton, W. D. O., Stein, V. E., Coutts, R. H. A., and Buck, K. W. (1984). *EMBO J.* **3**, 2197-2205.
Harker, C. S., Woolston, C. J., Markham, P. G., and Maule, A. J. (1987). *Virology* **160**, 252-254.
Harrison, B. D. (1985). *Annu. Rev. Phytopathol.* **23**, 55-82.
Harrison, B. D., Finch, J. T., Gibbs, A. J., Hollings, M., Shepherd, R. J., Valenta, V., and Wetter, C. (1971). *Virology* **45**, 356-363.
Harrison, B. D., Barker, H., Bock, K. R., Guthrie, E. J., Meredith, G., and Atkinson, M. (1977). *Nature (London)* **270**, 760-762.
Hatta, T., and Francki, R. I. B. (1979). *Virology* **92**, 428-433.
Hirochika, H., Takatsuji, H., Ubasawa, A., and Ikeda, J.-E. (1985). *EMBO J.* **4**, 1673-1680.
Hohn, B., Balazs, E., Ruegg, D., and Hohn, T. (1986). *EMBO J.* **5**, 2759-2762.
Hohn, T., Richards, K., and Lebeurier, G. (1982). *Curr. Top. Microbiol. Immunol.* **96**, 193-236.

Hohn, T., Hohn, B., and Pfeiffer, P. (1985). *Trends Biochem. Sci.* **10,** 205–209.
Howarth, A. J. (1985). *In* "Genetic Engineering: Principles and Methods" (J. K. Setlow and A. Hollaender, eds.), pp. 85–99. Plenum, New York.
Howarth, A. J., Gardner, R. C., Messing, J., and Shepherd, R. J. (1981). *Virology* **112,** 678–685.
Howarth, A. J., Caton, J., Bossert, M., and Goodman, R. M. (1985). *Proc. Natl. Acad. Sci. U.S.A.* **82,** 3572–3576.
Howell, S. H. (1981). *Virology* **112,** 488–495.
Howell, S. H. (1982). *Annu. Rev. Plant Physiol.* **33,** 609–650.
Howell, S. H. (1984). *Nucleic Acids Res.* **12,** 7359–7375.
Howell, S. H. (1985a). *Nucleic Acids Res.* **13,** 3018–3019.
Howell, S. H. (1985b). *CRC Crit. Rev. Plant Sci.* **2,** 287–316.
Howell, S. H., and Hull, R. (1978). *Virology* **86,** 468–481.
Howell, S. H., Walker, L. L., and Dudley, R. K. (1980). *Science* **208,** 1265–1267.
Howell, S. H., Walker, L. L., and Walden, R. M. (1982). *Nature (London)* **293,** 483–486.
Howell, S. H., Walden, R. M., and Marco, Y. (1983). *In* "Plant Molecular Biology" (R. Goldberg, ed.), pp. 137–146. Liss, New York.
Hull, R. (1984). *Commonw. Mycol. Inst./Assoc. Appl. Biol., Descript. Plant Viruses* **295,** 4.
Hull, R., and Covey, S. N. (1983a). *Nucleic Acids Res.* **11,** 1881–1895.
Hull, R., and Covey, S. N. (1983b). *Trends Biochem. Sci.* **8,** 119–121.
Hull, R., and Donson, J. (1982). *J. Gen. Virol.* **60,** 125–134.
Hull, R., and Shepherd, R. J. (1976). *Virology* **70,** 217–220.
Hull, R., and Shepherd, R. J. (1977). *Virology* **79,** 216–230.
Hull, R., Shepherd, R. J., and Harvey, J. D. (1976). *J. Gen. Virol.* **31,** 93–100.
Hull, R., Covey, S. N., Stanley, J., and Davies, J. W. (1979). *Nucleic Acids Res.* **7,** 669–677.
Hull, R., Sadler, J., and Longstaff, M. (1986). *EMBO J.* **5,** 3083–3090.
Ikegami, M., Haber, S., and Goodman, R. M. (1982). *Proc. Natl. Acad. Sci. U.S.A.* **78,** 4102–4106.
Kamei, T., Rubio-Huertos, M., and Matsui, C. (1969). *Virology* **37,** 506–508.
Kamer, G., and Argos, P. (1984). *Nucleic Acids Res.* **12,** 7269–7282.
Kay, R., Chan, A., Doly, M., and McPherson, J. (1987). *Science* **236,** 1299–1302.
Kelley, D. C., Cooper, V., and Walkey, D. G. A. (1974). *Microbios* **10,** 239–246.
Kikuno, R., Toh, H., Hayashida, H., and Miyata, T. (1984). *Nature (London)* **308,** 562.
Kim, K. S., Shock, T. L., and Goodman, R. M. (1978). *Virology* **89,** 22–33.
Kitajima, E. W., and Lauritis, J. A. (1969). *Virology* **37,** 681–685.
Kornberg, A. (1979). "DNA Replication." Freeman, San Francisco, California.
Kozak, M. (1986a). *Adv. Virus Res.* **31,** 229–292.
Kozak, M. (1986b). *Cell* **44,** 283–292.
Kozak, M. (1986c). *Cell* **47,** 481–483.
Koziel, M. G., Adam, T. L., Hazlet, M. A., Damm, D., Miller, J., Dahlbeck, D., Jayne, S., and Staskawicz, B. (1984). *J. Mol. Appl. Genet.* **2,** 549–562.
Kruse, J., Timmins, P., and Witz, J. (1987). *Virology* **159,** 166–168.
Laquel, P., Ziegler, V., and Hirth, L. (1986). *J. Gen. Virol.* **67,** 197–201.
Larsen, R. C., and Duffus, J. E. (1984). *Phytopathology* **74,** 114–118.
Lazarowitz, S. G. (1987). *Plant Mol. Biol. Rep.* **4,** 177–192.
Lebeurier, G., Hirth, L., Hohn, T., and Hohn, B. (1980). *Gene* **12,** 139–146.
Lebeurier, G., Hirth, L., Hohn, B., and Hohn, T. (1982). *Proc. Natl. Acad. Sci. U.S.A.* **79,** 2932–2936.
Lung, M. C. Y., and Pirone, T. P. (1973). *Phytopathology* **63,** 910–914.
Lung, M. C. Y., and Pirone, T. P. (1974). *Virology* **60,** 260–264.
Lutcke, H. A., Chow, K. C., Mickel, F. S., Moss, K. A., Kern, H. F., and Scheele, G. A. (1987). *EMBO J.* **6,** 43–48.
MacDowell, S. W., Macdonald, H., Hamilton, W. D. O., Coutts, R. H. A., and Burk, K. W. (1985). *EMBO J.* **4,** 2173–2180.

Marco, Y., and Howell, S. H. (1984). *Nucleic Acids Res.* **12**, 1517-1528.
Markham, P. G., and Hull, R. (1985). *J. Gen. Virol.* **66**, 921-923.
Marsh, L., and Guilfoyle, T. (1987). *Virology* **161**, 129-137.
Marsh, L., Kuzj, A., and Guilfoyle, T. (1985). *Virology* **143**, 212-223.
Martinez-Izquierdo, J., and Hohn, T. (1987). *Proc. Natl. Acad. Sci. U.S.A.* **84**, 1824-1828.
Martinez-Izquierdo, J. A., Futterer, J., and Hohn, T. (1987). *Virology* **160**, 527-530.
Matthews, R. E. F. (1979). *Intervirology* **12**, 130-296.
Maule, A. J. (1985). *Plant Mol. Biol.* **5**, 25-34.
Maule, A. J., and Thomas, C. M. (1985). *Nucleic Acids Res.* **13**, 7359-7373.
Mazzolini, L., Bonneville, J. M., Volovitch, M., Magazin, M., and Lot, P. (1985). *Virology* **145**, 293-303.
Menissier, J., Laquel, P., Lebeurier, G., and Hirth, L. (1984). *Nucleic Acids Res.* **12**, 8769-8778.
Menissier, J., deMurcia, G., Lebeurier, G., and Hirth, L. (1983). *EMBO J.* **2**, 1067-1071.
Menissier, J., deMurcia, G., Geldreich, A., and Lebeurier, G. (1986). *J. Gen. Virol.* **67**, 1885-1891.
Meshi, T., Watanabe, Y., Saito, T., Sugimoto, A., Moeda, T., and Okada, Y. (1987). *EMBO J.* **6**, 2557-2563.
Modjtahedi, N., Volovitch, M., Sossountzov, L., Habricot, Y., Bonneville, J. M., and Yot, P. (1984). *Virology* **133**, 289-300.
Modjtahedi, N., Volovitch, M., Mazzolini, L., and Yot, P. (1985). *FEBS Lett.* **181**, 223-228.
Morris-Krsinich, B. A. M., Mullineaus, P. M., Donson, J., Boulton, M. I., Markham, P. G., Short, M. N., and Davies, J. W. (1985). *Nucleic Acids Res.* **13**, 7237-7256.
Mullineaux, P. M., Donson, J., Morris-Krsinich, B. A. M., Boulton, M. I., and Davies, J. W. (1984). *EMBO J.* **3**, 3063-3068.
Mullineaux, P. M., Donson, J., Stanley, J., Boulton, M. I., Morris-Krsinich, B. A. M., Markham, P. G., and Davies, J. W. (1985). *Plant Mol. Biol.* **5**, 125-131.
Mumford, D. L. (1974). *Phytopathology* **64**, 136-139.
Nishiguchi, M., Motoyoshi, F., and Oshima, N. (1978). *J. Gen. Virol.* **39**, 53-61.
Nishiguchi, M., Motoyoshi, F., and Oshima, N. (1980). *J. Gen. Virol.* **46**, 497-500.
Odell, J., and Howell, S. H. (1980). *Virology* **102**, 349-359.
Odell, J., Dudley, R. K., and Howell, S. H. (1981). *Virology* **111**, 377-385.
Odell, J. T., Nagy, F., and Chua, N.-H. (1985). *Nature (London)* **313**, 810-812.
Ohno, T., Tokamatsu, N., Meshi, T., Okada, Y., Nishigucki, M., and Kiho, Y. (1983). *Virology* **131**, 255-258.
Okada, K., Takebe, I., and Nagata, T. (1986). *Mol. Gen. Genet.* **205**, 398-403.
Olszewski, N. E., and Guilfoyle, T. J. (1983). *Nucleic Acids Res.* **11**, 8901-8914.
Olszewski, N., Hagen, G., and Guilfoyle, T. J. (1982). *Cell* **29**, 395-402.
Osaki, T., and Inouye, T. (1978). *Am. Phytopathol. Soc. Jpn.* **44**, 167-178.
Paszkowski, J., Shillito, R. D., Saul, M., Mandak, V., Hohn, T., Hohn, B., and Potrykus, I. (1984). *EMBO J.* **3**, 2717-2722.
Peabody, D. S., and Berg, P. (1986a). *Mol. Cell. Biol.* **6**, 2695-2703.
Peabody, D. S., and Berg, P. (1986b). *Mol. Cell. Biol.* **6**, 2704-2711.
Pfeiffer, P., and Hohn, T. (1983). *Cell* **33**, 781-789.
Pfeiffer, P., Laquel, P., and Hohn, T. (1984). *Plant Mol. Biol.* **3**, 261-270.
Phillips, G., and Gigot, C. (1977). *Nucleic Acids Res.* **4**, 3617-3625.
Pirone, T. P., and Megehed, E. (1966). *Virology* **30**, 631-637.
Pirone, T. P., and Shaw, J. G. (1973). *Virology* **52**, 274-276.
Plant, A. L., Covey, S. N., and Grierson, P. (1985). *Nucleic Acids Res.* **13**, 8305-8321.
Potrykus, I., Paszkowski, J., Saul, M. W., Petruska, J., and Shillito, R. D. (1985). *Mol. Gen. Genet.* **199**, 169-177.
Reanney, D. (1984). *Nature (London)* **307**, 318-319.
Reisman, D., Ricciardi, R. P., and Goodman, R. M. (1979). *Virology* **97**, 388-395.

Richards, K. E., Guilley, H., and Jonard, G. (1981). *FEBS Lett.* **134**, 67–70.
Richins, R., and Shepherd, R. J. (1983). *Virology* **124**, 208–214.
Richins, R. D., Scholthof, H. B., and Shepherd, R. J. (1987). *Nucleic Acids Res.* **15**, 8451–8466.
Roberts, I. M., Robinson, D. J., and Harrison, B. D. (1984). *J. Gen. Virol.* **65**, 1723–1730.
Rodriguez, D., Lopez-Abella, D., and Diaz-Ruiz, J. R. (1987). *J. Gen. Virol.* **68**, 2063–2068.
Rogers, S. G., Bisaro, D. M., Horsch, R. B., Fraley, R. T., Hoffman, N. L., Brand, L., Elmer, J. S., and Lloyd, A. M. (1986). *Cell* **45**, 593–600.
Saigo, K., Kugimiya, W., Matsuo, Y., Inouye, S., Yoshiaka, K., and Yuki, S. (1984). *Nature (London)* **312**, 659–661.
Sanders, P. R., Winter, J. A., Barnason, A. R., Rogers, S. G., and Fraley, R. T. (1987). *Nucleic Acids Res.* **15**, 1543–1558.
Schoelz, J. E., and Shepherd, R. J. (1988). *Virology* **162**, 30–37.
Schoelz, J. E., Shepherd, R. J., and Daubert, S. (1986). *Mol. Cell. Biol.* **6**, 2632–2637.
Schoelz, J. E., Shepherd, R. J., and Daubert, S. D. (1987). *In* "Molecular Strategies for Crop Protection" (C. J. Arntzen and C. A. Ryan, eds.), pp. 253–265. Liss, New York.
Shah, D. M., Horsch, R. B., Klee, H. J., Kishore, G. M., Winter, J. A., Turner, N. E., Hironaka, C. M., Sanders, P. R., Gasser, C. S., Aykent, S., Siegel, N. R., Rogers, S. G., and Fraley, R. T. (1986). *Science* **233**, 478–481.
Shepherd, R. J. (1976). *Adv. Virus Res.* **20**, 305–339.
Shepherd, R. J. (1979). *Annu. Rev. Plant Physiol.* **30**, 405–423.
Shepherd, R. J. (1985). *In* "Genetic Engineering: Principles and Methods" (J. K. Setlow and A. Hollaender, eds.), Vol. 8, pp. 241–276. Plenum, New York.
Shepherd, R. J., and Lawson, R. H. (1981). *In* "Handbook of Plant Virus Infections" (E. Kurstak, ed.), pp. 847–878. Elsevier, Amsterdam.
Shepherd, R. J., and Wakeman, R. J. (1971). *Phytopathology* **61**, 188–193.
Shepherd, R. J., Wakeman, R. J., and Romanko, R. R. (1968). *Virology* **36**, 150–152.
Shepherd, R. J., Bruening, G. E., and Wakeman, R. J. (1970). *Virology* **41**, 339–347.
Shepherd, R. J., Richins, R., and Shalla, T. A. (1980). *Virology* **102**, 389–400.
Shewmaker, C. K., Caton, J. R., Houck, C. M., and Gardner, R. C. (1985). *Virology* **140**, 281–288.
Shockey, M. W., Gardner, C. O., Melcher, U., and Essenberg, R. C. (1980). *Virology* **105**, 575–581.
Stanley, J. (1983). *Nature (London)* **305**, 643–645.
Stanley, J. (1985). *Adv. Virus Res.* **30**, 139–177.
Stanley, J., and Gay, M. R. (1983). *Nature (London)* **301**, 260–262.
Stanley, J., and Townsend, R. (1985). *Nucleic Acids Res.* **13**, 2189–2206.
Stanley, J., and Townsend, R. (1986). *Nucleic Acids Res.* **14**, 5981–5998.
Stanley, J., Markham, P. G., Callis, R. J., and Pinner, M. S. (1986). *EMBO J.* **5**, 1761–1767.
Storey, H. H. (1928). *Ann. Appl. Biol.* **15**, 1–19.
Summers, J., and Mason, W. S. (1982). *Cell* **29**, 199–201.
Sunter, G., Gardiner, W. E., Rushing, A. E., Rogers, S. G., and Bisaro, D. M. (1987). *Plant Mol. Biol.* **8**, 477–484.
Takatsuji, H., Hirockika, H., Fukushi, T., and Ikeda, J.-E. (1986). *Nature (London)* **319**, 240–243.
Thomas, J. E., and Bowyer, J. W. (1980). *Phytopathology* **70**, 214–217.
Thomas, C. M., Hull, R., Bryant, J. A., and Maule, A. J. (1985). *Nucleic Acids Res.* **13**, 4557–4576.
Toh, H., Kikuno, R., Hayashida, H., Miyata, T., Kugimiya, W., Inouye, S., Yuki, S., and Saigo, K. (1985). *EMBO J.* **4**, 1267–1272.
Townsend, R., Stanley, J., Curson, S. J., and Short, M. N. (1985). *EMBO J.* **4**, 33–37.
Townsend, R., Watts, J., and Stanley, J. (1986). *Nucleic Acids Res.* **14**, 1253–1265.
Turner, D. S., and Covey, S. N. (1984). *FEBS Lett.* **165**, 285–289.
van Vloten-Doting, L., Bol, J.-F., and Cornelissen, B. (1985). *Plant Mol. Biol.* **4**, 323–326.

Verma, I. M. (1981). *Enzymes* **14A,** 87–103.
Verver, J., Schijns, P., Hibi, T., and Goldbach, R. (1987). *J. Gen. Virol.* **68,** 159–167.
Volovitch, M., Dumas, J. P., Drugeon, G., and Yot, P. (1976). *In* "Acids Nucleiques et Synthese des Proteines Chez lex Vegetaux" (L. Bogorad and J. H. Weil, eds.). pp. 635–641. Centre National de la Recherche Scientifique, Paris.
Volovitch, M., Modjtahedi, N., Brun, G., Drugeon, G., and Yot, P. (1978). *Nucleic Acids Res.* **5,** 2913–2925.
Volovitch, M., Modjtahedi, N., Yot, P., and Brun, G. (1984). *EMBO J.* **3,** 309–314.
Walden, R. M., and Howell, S. H. (1983a). *Plant Mol. Biol.* **2,** 27–31.
Walden, R. M., and Howell, S. H. (1983b). *J. Mol. Appl. Genet.* **1,** 447–456.
Woolston, C., Covey, S., Penswick, J., and Davies, J. (1983). *Gene* **23,** 15–23.
Woolston, C. J., Czoplewski, L. G., Markham, P. G., Good, A. S., Hull, R., and Davies, J. W. (1987). *Virology* **160,** 246–251.
Xiong, C., Muller, S., Lebeurier, G., and Hirth, L. (1982). *EMBO J.* **8,** 971–976.
Xiong, C., Lebeurier, G., and Hirth, L. (1984). *Proc. Natl. Acad. Sci. U.S.A.* **81,** 6608–6612.
Young, M. J., Daubert, S., and Shepherd, R. J. (1987). *Virology* **158,** 444–446.
Ziegler, V., Laquel, P., Guilley, H., Richards, K., and Jonard, G. (1985). *Gene* **36,** 271–279.

Tumor Formation in Plants   16

A. POWELL
M. P. GORDON

I. Introduction
II. Crown Gall Tumors
    A. History
    B. General Background
    C. The Inducing Organism
    D. The *Agrobacterium tumefaciens* Tumor-Inducing Plasmid
    E. The *Agrobacterium rhizogenes* Tumor-Inducing Plasmid
    F. Host Range
    G. Agrobacteria Chromosomal Traits
    H. The Role of Opines in the Evolution of Plant Transformation by Agrobacteria
    I. Application to Genetic Engineering of Plants
III. Virus-Induced Tumors of Plants
    A. General Background
    B. Composition of Virions
IV. Habituated Plant Tissues and Genetic Tumors
    A. Habituated Plant Tissues
    B. Plant Genetic Tumors
V. Transfer of Genetic Information in the Biosphere
    References

I. INTRODUCTION

A plant tumor is recognized as tissue that has escaped the normal developmental restraints of the host plant. Plant tumors may be induced by bacterial transformation, by viruses, or by certain interspecific crosses. Knowledge of the transformation of plant cells by bacteria has advanced rapidly, in large part due to the techniques of molecular biology, thereby facilitating our understanding of plant growth, development, and differentiation. A burgeoning industry based on the "genetic engineering" of plants can trace its genesis to these approaches, providing promise in the application of these methods to increase the world's supply of food, fuel, and fiber.

This review emphasizes the developments occurring in the early 1980s. Helpful reviews of crown gall tumors include Chapter 13 in Volume 6 of this series (Gordon, 1977) and those by several other authors (Bevan and Chilton, 1982a; Zambryski et al., 1983a; Gelvin, 1984; Hooykaas and Schilperoort, 1984; Nester et al., 1984; Zambryski, 1988). Texts about plant tumors (Kahl and Schell, 1982) and the relationship of plants with microorganisms (Kosuge and Nester, 1984; Melchers and Hooykaas, 1987; White and Sinkar, 1988) have also been published.

II. CROWN GALL TUMORS

A. History

Crown gall disease is the undifferentiated growth of plant tissue resulting from the interaction between susceptible plants and virulent strains of *Agrobacterium*. The microorganism was discovered by Smith and Townsend in 1907, although the first accounts of this disease on grape vines in Europe were given by Fabre and Dunal in 1853 (quoted by Owens, 1928). It was quickly recognized that crown gall tumors could serve as a model for animal tumors since the tumorous transformation persists in the absence of inciting bacteria (Jensen, 1918).

The next major milestone was the observation by Braun that crown gall tumors could be grown *in vitro* in the absence of exogenously supplied auxins and cytokinins (Braun, 1956), phytohormones required by most plant tissues for growth *in vitro*. Braun introduced the concept of a "tumor-inducing principle," a material that passed from the tumor-inciting bacterium to the plant cell during the transformation process (Braun and Mandle, 1948). Subsequent major observations include the demonstration that oncogenicity of the inciting organism was associated with large bacterial plasmids, the tumor-inducing (Ti) plasmids (Zaenen et al., 1974; Watson et al., 1975), and the demonstration that a portion of the Ti plasmid called the T-DNA (transferred DNA) was incorporated, transcribed, and replicated in axenically cultured transformed plant cells (Chilton et al., 1977). Thus, T-DNA satisfies the definition of the "tumor-inducing principle."

B. General Background

The events of the formation of plant tumors incited by agrobacteria are depicted in Fig. 1. The transforming bacterium invades a plant at the site of a wound. Chemical messengers released by the wounded plant cells induce the transcription of the Ti plasmid-borne virulence genes (*vir*) in the bacte-

Tumor Formation in Plants

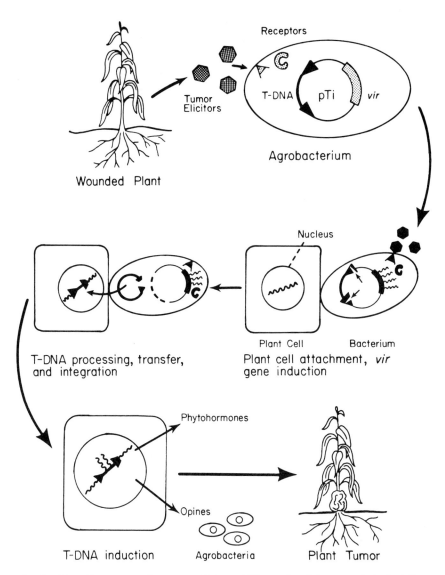

Fig. 1. Phenolic compounds, tumor elicitors (●), produced by wounded plant tissues react with the *virA* gene product (A) located in the membrane of the bacterial cell resulting in the modification of G. Modified A switches on transcriptional control element G, which in turn results in the transcription of the *vir* region. The products of the *vir* genes, especially *virD* and *virD₂*, cut the plasmid at two direct repeats (▶), forming the T-DNA which is transferred to and integrated into the nuclear DNA. The transcription and translation of the T-DNA in the plant results in the formation of opines that are catabolized by the inciting bacteria and of phytohormones that cause uncontrolled plant cell proliferation.

rium (Stachel *et al.*, 1985; Bolton *et al.*, 1986). The *vir* gene products effect the transfer of the T-DNA to the plant cell. The T-DNA, despite its prokaryotic origin, contains eukaryotic transcriptional and translational control sequences used by enzymes within the transformed plant cell so that the T-DNA is transcribed into messenger RNA which is translated into proteins. The T-DNA codes for enzymes for auxin and cytokinin biosynthesis; these phytohormones cause the transformed plant cells to grow rapidly, thereby forming a tumor. The T-DNA also codes for enzymes that synthesize unusual compounds termed "opines." Since the agrobacteria that carry the Ti plasmids can utilize these opines as carbon and nitrogen sources, the transformation process establishes an ecological niche for the bacteria (Petit *et al.*, 1983), a type of parasitic transformation termed "genetic colonization" (Schell *et al.*, 1979).

C. The Inducing Organism

The causative agent of crown gall disease, *Agrobacterium tumefaciens*, is a member of the family Rhizobiaceae which, according to "Bergey's Manual of Determinative Bacteriology, 9th Edition" may be divided as indicated in Fig. 2. DNA hybridization studies also indicate that rhizobia and agrobacteria are closely related (Kersters *et al.*, 1973).

The members of the genus *Rhizobium* fix nitrogen either in root nodules of leguminous plants or in suitable culture media (Evans *et al.*, 1985). Most species of the genus *Agrobacterium* induce tumors on dicotyledonous plants (Fig. 3). *Agrobacterium rubi* is found in tumors on blackberries, raspberries, and related plants, but the strain is not well characterized. Indeed, the designation of *A. rubi* as a separate species may not be justified. *Agrobacterium radiobacter* lacks a virulence plasmid, and produces no tumors on plants.

Strains of *A. tumefaciens* have been divided into three biotypes (I, II, and III) based upon various metabolic and physiological traits (Keane *et al.*,

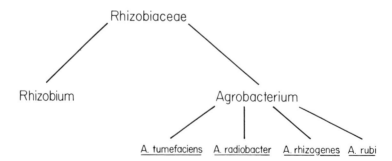

Fig. 2. The relationship of the members of the family Rhizobiaceae.

1970; Kerr and Panagopoulos, 1977). This classification has proved helpful in the isolation and characterization of strains of agrobacteria found in the field.

D. The *Agrobacterium tumefaciens* Tumor-Inducing Plasmid

There are several types of virulence plasmids in *A. tumefaciens*, classified according to the opines produced by the resultant crown gall and catabolized by the inciting bacteria. In this review, we emphasize octopine plasmids, recognizing that most of the functions encoded by the plasmid have analogs in other *Agrobacterium* virulence plasmids. The genetic domains of a prototype Ti plasmid (pTi) are shown in Fig. 4 and have been reviewed extensively (Nester *et al.*, 1984).

When the virulence plasmid is present in the bacteria, the strains are oncogenic (Van Larebeke *et al.*, 1974, 1975; Watson *et al.*, 1975), although the plasmid alone is not sufficient to cause transformation of plants. The agrobacteria chromosome contributes other functions necessary for *in planta* oncogenicity (Holsters *et al.*, 1978; Garfinkel and Nester, 1980; Douglas *et al.*, 1982, 1985). Most Ti plasmids are stably maintained in the bacteria despite their large size (200–900 kb). Two major regions of the Ti plasmid encode functions relevant to the tumorous transformation of plant cells. The products of the genes of the *vir* region initiate the transformation process in response to elicitors produced by wounded plant cells, but the *vir* region DNA has not been found in transformed plant cells (Chilton *et al.*, 1977). The *vir* region is necessary for plant transformation since strains with mutations in these genes are nononcogenic and proteins encoded by this

Fig. 3. (A) Tumors produced by *A. tumefaciens* on sunflower. (B) Tumors produced by *A. rhizogenes* on kalanchoe. Courtesy of F. White. (C) Tumors produced by *A. rubi* on blackberry cane. Courtesy of L. Moore.

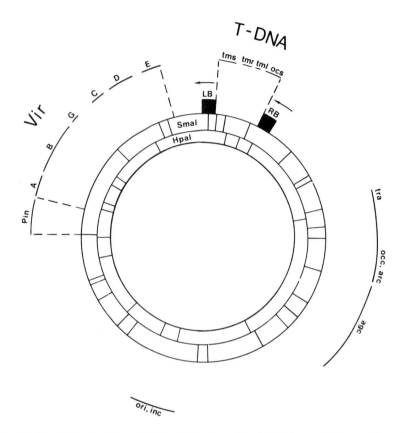

Fig. 4. Genetic loci of the pTiA6 octopine-type plasmid of *A. tumefaciens* with regions pertaining to virulence and tumor morphology shown. The restriction endonuclease fragments from *Sma*I and *Hpa*I cleavages are indicated. The T-DNA (transferred DNA) contains the genetic loci for octopine synthesis (*ocs*) and for the morphology mutants showing shooting (*tms*), rooting (*tmr*), and large (*tml*) phenotypes. The *vir* region (virulence) locus contains six virulence loci (*vir*A, *vir*B, *vir*C, *vir*D, *vir*E) and the nononcogenic, *pin*, region. The origin of replication (*ori*), the incompatibility (*inc*), agropine catabolism (*agc*), arginine catabolism (*arc*), octopine catabolism (*occ*), and transfer (*tra*) loci are also indicated.

region have functions used in the T-DNA transfer process (Garfinkel and Nester, 1980). The T-DNA (approximately 15 kb) contains genes responsible for the phenotype of the transformed plant cells, and is physically integrated into the transformed plant nuclear DNA (Chilton *et al.*, 1977, 1978, 1982; Thomashow *et al.*, 1980a; Willmitzer *et al.*, 1980; Ooms *et al.*, 1982b). Although not directly required for tumorigenicity, the *ori* region of the plasmid is required for the replication of the plasmid. The plasmids also have regions which include genes for opine and arginine catabolism. The *tra* region is required for the conjugal transfer of the plasmid between bacteria but is not involved with transfer of the T-DNA to plant cells.

1. The vir *Region*

Initially, the *vir* gene complex was defined as a region of the Ti plasmid which is necessary for the tumorigenicity of the plasmid but which is not transferred to the transformed cell (Garfinkel and Nester, 1980; Ooms *et al.,* 1980; DeGreve *et al.,* 1981; Klee *et al.,* 1982, 1983; Lundquist *et al.,* 1984, Stachel and Zambryski, 1986b). Subsequently, the *vir* region was shown to contain genes whose induction was elicited specifically by simple phenolic compounds, acetosyringone and its derivatives, produced by tobacco cells in culture and by wounded plants (Stachel *et al.,* 1985, Bolton *et al.,* 1986). Additional genes outside the *vir* region of the Ti plasmid also respond to the phenolic elicitors, but their role in oncogenicity is not clear (Okker *et al.,* 1984; Stachel *et al.,* 1985; Stachel and Nester, 1986; Stachel and Zambryski, 1986a). Postulated roles for the gene products of the *vir* region include T-DNA processing, transfer, and perhaps integration into the plant cells (Melchers and Hooykaas, 1987).

The *vir* region is approximately 40 kb and includes six complementation groups that map to the left of the T-DNA on the octopine Ti plasmid (Garfinkel and Nester, 1980; Ooms *et al.,* 1980; Klee *et al.,* 1983; Hooykaas *et al.,* 1984; Stachel and Nester, 1986). Complementation experiments have shown that the *vir* genes function in *trans* to the T-DNA (Hoekema *et al.,* 1983; de Framond *et al.,* 1983; Bevan, 1984; Hille *et al.,* 1982, 1984; An *et al.,* 1985; Klee *et al.,* 1985). This property has been useful for developing vectors for higher plant cell transformations (see Section I). The genetic and physical map of the *vir* region of the octopine plasmid (pTiA6) is shown in Fig. 5.

The nucleotide sequences of all of the octopine plasmid *vir* genes are known and transcripts from each of the loci have been identified or proposed (Das *et al.,* 1986; Stachel and Nester, 1986; Winans *et al.,* 1986; Yanofsky *et al.,* 1986; Ward *et al.,* 1988). The *vir*A and *vir*G genes are expressed constitutively in bacterial cells. Induction by the phenolic elicitors of the *vir* genes occurs in *Agrobacterium* but not in *Escherichia coli.* Expression of both the *vir*A and *vir*G genes is required for the induction of the other *vir* genes. The predicted amino acid sequence for the *vir*A region shows extensive homology to several regulatory proteins, such as the *env*Z protein of *E. coli* (Leroux *et al.,* 1987). These regulators are thought to act as environmental sensors and to signal the presence of an inducer molecule to a second protein which acts as a transcriptional activator. The *vir*A protein is localized exclusively in the inner bacterial membrane and is proposed to sense and signal the presence of the phenolic inducer molecules (Leroux *et al.,* 1987; Melchers *et al.,* 1987). *vir*A may participate in the switch of *vir*G to an induced form. Furthermore, in the induced condition, the enhanced transcription of *vir*G includes transcription from a second site upstream of the initiation site used in the constitutive expression of the gene (Das *et al.,* 1986; Stachel and Zambryski, 1986a; Winans *et al.,* 1988). *vir*G is most likely a transcriptional activator since it shares extensive amino acid sequence homology to other

	A INDUCIBILITY	SIZE(Kb)	B ORF'S	SIZE(kDa)	G MUTANT PHENOTYPE	C	D ROLE	E FUNCTION
B	+	9.5	11	–	avir.		Transfer	?
D	+	4.5	4	16 48 21 ? 72 or 76	avir.		Transfer	Endonuclease DNA binding protein
C	+	1.5	2	26 23	atten.		Accessory	?
E	+	2.2	2	7 60	very atten.		Transfer/ Accessory	SS DNA binding protein
A	–	2.8	1	92	avir.		Regulation	Phenolic Sensor
G	+	1.0	1	30	avir.		Regulation	Transcriptional Activator

Fig. 5. Map of the pTiA6 virulence region (*vir*) showing the genetic loci and their phenotype. The number of open reading frames (ORFs) observed in the nucleotide sequence, the direction of transcription, the size of the potential gene products, and the inducibility by cocultivation with plant cells are shown.

activator proteins such as *omp*R of *E. coli* (Winans *et al.*, 1986). The other genes in the *vir* region have been postulated to encode proteins involved in the formation of processed T-DNA intermediates utilized in the processing, transfer, and integration of the T-DNA into the plant cell DNA (Koukolíková-Nicola *et al.*, 1985; Gardner and Knauf, 1986; Horsch *et al.*, 1986; Stachel and Zambryski, 1986a,b; Albright *et al.*, 1987). Recently, the *vir*D operon was shown to encode a site-specific endonuclease which cleaves at unique locations within the 24-bp direct repeats flanking the T-DNA (Yanofsky *et al.*, 1986). The *vir*D operon also encodes a protein that binds to the 5' end of the T-DNA intermediate (Young and Nester, 1988). The *vir*E operon encodes a protein that binds to the specific single stranded DNA (Gietl *et al.*, 1987; Christie *et al.*, 1988; Citovsky *et al.*, 1988; Das, 1988). The *vir*B operon encodes eleven proteins, and their sequences suggest that they function in a membrane complex necessary for the transfer of the T-DNA intermediate from the bacteria to the plant cells (Engstrom *et al.*, 1987; Ward *et al.*, 1988).

The nopaline Ti plasmid *vir* genes also have been mapped into six complementation groups (Holsters *et al.*, 1980; Lundquist *et al.*, 1984; Hagiya *et al.*, 1985; Otten *et al.*, 1985). The *vir* genes from one type of plasmid can complement functions lacking in another type of virulence plasmid, including *A. rhizogenes* plasmids. This functional homology between the *vir* regions of various types of plasmids correlates with their sequence homology (Hoekema *et al.*, 1984; Hooykaas *et al.*, 1984). The nopaline plasmid has two novel genes whose induction is dependent on *vir*A, *vir*G, and the chemical inducer. The two genes, *tzs* encoding a cytokinin biosynthetic prenyltransferase and *iaa*P encoding an indole acetic acid (IAA) biosynthetic function,

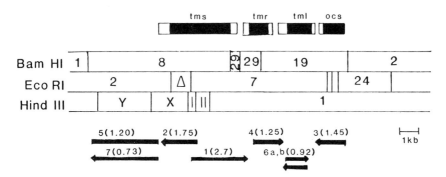

Fig. 6. Map of the pTiA6 transferred region (T-DNA) showing the genetic loci for mutant tumor morphology exhibiting shooting (*tms*), rooting (*tmr*), and large (*tml*) phenotypes and the locus for octopine synthesis (*ocs*). The fragments generated by cleavage with the restriction endonucleases *Bam*HI, *Eco*RI, and *Hin*dIII are shown and the identification number, size (in kilobases in parentheses), and direction of the transcripts are indicated.

may influence the plant cells at the site of infection during transformation (Kaiss-Chapman and Morris, 1977; Claeys *et al.*, 1978; Liu and Kado, 1979; Liu *et al.*, 1982; Regier and Morris, 1982; Akiyoshi *et al.*, 1985, 1987; John and Amasino, 1988; Powell *et al.*, 1988).

2. The T Region

The T region is the portion of the Ti plasmid found integrated stably in the nuclear DNA of crown gall tumors. Extensive knowledge now exists about the T-DNA regions from various examples of Ti plasmids, and the nucleotide sequences are known (Depicker *et al.*, 1982; Simpson *et al.*, 1982; Yadav *et al.*, 1982; Zambryski *et al.*, 1980, 1982; Barker *et al.*, 1983; De-Greve *et al.*, 1983; Dhaese *et al.*, 1983; Holsters *et al.*, 1983; Goldberg *et al.*, 1984; Klee *et al.*, 1984; Lichtenstein *et al.*, 1984). The products and functions of the T-DNA in the transformed cells are well described (Schröder *et al.*, 1981, 1983, 1984; Bevan and Chilton, 1982b; Gelvin *et al.*, 1982; Leemans *et al.*, 1982; Murai and Kemp, 1982; Schröder and Schröder, 1982; Willmitzer *et al.*, 1981b, 1982, 1983; Bevan *et al.*, 1983; Akiyoshi *et al.*, 1984; Inze *et al.*, 1984; Janssens *et al.*, 1984; Karcher *et al.*, 1984; Salomon *et al.*, 1984; Thomashow *et al.*, 1984, 1986; Winter *et al.*, 1984).

One major function encoded by the T-DNA in transformed cells is the synthesis of opines, a characteristic which identifies the categories of plasmids (Engler *et al.*, 1981; Bevan and Chilton, 1982a; Chilton, 1982; Nester *et al.*, 1984). Among the octopine and nopaline plasmids, portions of the T-DNA are conserved and contain the phytohormone biosynthetic genes. After integration into the DNA of the plant cell, the products of these genes produce the tumorous phenotype of the plant. The restriction endonuclease cleavage sites, genetic loci, and transcripts of the T_L-DNA (see Section II,D,4) of an octopine-type Ti plasmid are shown in Fig. 6.

3. T-DNA Genes Controlling Tumor Morphology

Plant cells transformed by agrobacteria are able to grow *in vitro* on defined media lacking auxin and cytokinin. Braun (1958) concluded that a key aspect of plant transformation is the production by the plant tumor cells of unusual levels of auxins and cytokinins.

Initially, functions were assigned to parts of the T-DNA based on the phenotype of tumors resulting from infection by bacterial strains containing mutated plasmid T-DNA. Mutagenesis of the octopine plasmid T-DNA (Ooms *et al.*, 1980, 1981; Garfinkel *et al.*, 1981; Leemans *et al.*, 1982; Ream *et al.*, 1983) and the nopaline plasmid T-DNA (Holsters *et al.*, 1980; Joos *et al.*, 1983) defined three loci in transformed plant tissue: *tmr* (*t*umor *m*orphology *r*oot), a *tms* (*t*umor *m*orphology *s*hoot), and a *tml* (*t*umor *m*orphology *l*arge) (Fig. 7). In cultures of normal plant tissues, an excess of auxins results in root formation and excess cytokinins results in shoot formation (Skoog and Miller, 1957). By analogy to growth of plant tissue with excess phytohormones, the tumors resulting from transformation by the *tmr* and *tms* mutants were postulated to have an excess of either auxins or cytokinins, thus suggesting a function for these loci (Ooms *et al.*, 1980, 1981, 1982a; Garfinkel *et al.*, 1981; Amasino and Miller, 1982).

Analysis of *E. coli* containing the cloned coding regions and determination of the enzymatic activity of agrobacteria strains containing these genes

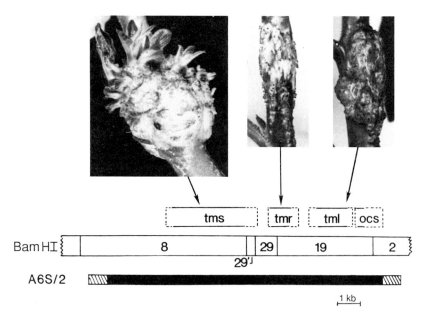

Fig. 7. Morphology of tumor mutants in the *tms*, *tmr*, and *tml* loci of the pTiA6. The restriction fragments after cleavage with *Bam*HI are shown above the map of the T-DNA found in the tumor, A6S/2.

showed that in *tms* transcript 1 from *iaa*M encodes a tryptophan monooxygenase activity (Thomashow *et al.*, 1984, 1986; Yamada *et al.*, 1985) and that transcript 2 from *iaa*H encodes an indole-3-acetamide hydrolase (Inze *et al.*, 1984; Schröder *et al.*, 1983; Thomashow *et al.*, 1984). These enzymes form one part of the biosynthetic pathway of IAA. In *tmr*, transcript 4 encodes a dimethylallyl-pyrophosphate: AMP dimethylallyltransferase (Akiyoshi *et al.*, 1984; Barry *et al.*, 1984; Buchmann *et al.*, 1985) and is part of the biosynthetic pathway for *trans*-zeatin. Transcript 3 from *ocs* encodes octopine synthase in octopine-type Ti plasmids (Murai and Kemp, 1982; Schröder *et al.*, 1983). The promoter elements for these genes as well as for other T-DNA genes are typical of those found in eukaryotic rather than prokaryotic organisms. The control sequences 5' to the transcription initiation sequences contain "TATA" and "CAAT" sequences, and polyadenylation signals are found at the 3' ends of the messenger RNAs (see Nester *et al.*, 1984). These genes apparently are transcribed by eukaryotic RNA polymerase II (Willmitzer *et al.*, 1981a).

Other T-DNA transcripts have functions which are less well characterized. Transcript 6a is required for the active secretion of the opines (Messens *et al.*, 1985). Transcripts 5, 6b, and 7 have functions which have not yet been defined, although *tml* mutations in some plants have been mapped to transcript 6b. The second part of the octopine T-DNA, the T_R DNA (see Section II,D,4) found in some tumors, has been shown to encode five transcripts, some of which are involved in mannopine or agropine biosynthesis in tumors (Karcher *et al.*, 1984; Salomon *et al.*, 1984; Winter *et al.*, 1984; Komro *et al.*, 1985).

The T-DNA in plant cells may not always be transcriptionally active. Methylation of T-DNA has been correlated with inactive T-DNA genes in some tumors (Hepburn *et al.*, 1983; Amasino *et al.*, 1984), while other tumors have no apparent methylation of the T-DNA and the T-DNA is transcribed (Gelvin *et al.*, 1983; Amasino *et al.*, 1984). Inactivation of the phytohormone biosynthetic T-DNA genes by methylation or deletion in transformed cells permits the emergence of revertant tissue (e.g., tissue having the normal plant phenotype).

4. T-DNA Sequences in Plant Genomic DNA

The borders of the T-DNA contain some conserved sequences necessary for processing, transfer, or integration of the T-DNA into the plant genome. The T-DNA borders in transformed cells are imperfect direct repeats of 24 bp at the ends of the plasmid T-DNA. The right border is required for oncogenicity since deletion of the right border produces avirulent plasmids. The DNA to be incorporated into the plant DNA is that which is to the left of the right border. Most tumor T-DNAs terminate near the left T-DNA border although this border is not required for transformation (Holsters *et al.*, 1980; Joos *et al.*, 1983; Shaw *et al.*, 1984; Wang *et al.*, 1984; Peralta and Ream, 1985a; Gardner and Knauf, 1986). In the octopine plasmid, an approximately

20-bp sequence to the right of the right border stimulates the incorporation of the T-DNA and has been called the "overdrive" sequence (Peralta and Ream, 1985a,b). Another "overdrive" sequence is found near the left T-DNA border as well (M. Yanofsky and H. Klee, personal communication).

Albright et al. (1987) observed that, when bacteria and plant cells are cultivated together or when bacteria are cultivated with acetosyringone, precise single-strand nicks are found within the right and left borders of the octopine plasmid T-DNA. This processing event occurs as a result of the induction of the *vir* genes and generates a predominantly single-stranded T-DNA intermediate as well as some double-stranded T-DNA intermediates (Koukolíková-Nicola et al., 1985). The first two open reading frames of the *vir*D operon appear to be responsible for the T-DNA nicking function (Yanofsky et al., 1986). The location within the 24-bp plasmid T-DNA borders of the single-strand nicks and the T-DNA/plant DNA borders from several tumors are shown in Fig. 8. How the T-DNA intermediates are transferred into the plant cell is not known but some aspects of the transfer appear to be similar to bacterial conjugation (Stachel and Zambryski, 1986b).

Plant protoplasts can incorporate and integrate into their genome fragments of DNA which have been introduced into the interior of the cell by means such as polyethylene glycol or electroporation. These *in vitro* transformations do not require either the 24-bp termini or the T-DNA (Potrykus et al., 1985). Thus, once DNA is delivered into the plant cell, integration into the plant DNA may proceed by mechanisms inherent in the recipient cells.

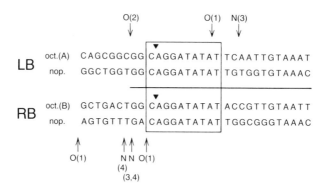

Fig. 8. Right (RB) and left (LB) border sequences of T-DNAs from octopine- and nopaline-type plasmids. The sequence is that of the "top" strand, i.e., the sequence reading from the left to right, 5' to 3'. The underlined sequence is the 24-bp direct repeat. The 10-base sequence enclosed in the box is identical in the borders from octopine and nopaline plasmids. O and N indicate the borders of the T-DNA in octopine and nopaline tumors, respectively. The number in parentheses indicates the reference identifying the sequence: (1) Holsters et al. (1983), (2) Simpson et al. (1982), (3) Yadav et al. (1982), (4) Zambryski et al. (1982). The ▼ indicates the site of the single-strand nick, made on the "bottom" strand during a putative T-DNA processing event (Albright et al., 1987).

The T-DNA region of the octopine plasmid has four copies of the 24-bp repeat. As a result, the T-DNA borders demarcate the 13-kb T_L region and the 8-kb T_R region separated by 1.8 kb of unincorporated plasmid DNA (Barker *et al.*, 1983). Apparently, the T_L and the T_R regions of the octopine T-DNA are transferred independently into the plant genome (Holsters *et al.*, 1983). The T_L region is always present at least once in the transformed plant cells (Simpson *et al.*, 1982; Holsters *et al.*, 1983) and the T_R region, if present, may or may not have more copies than the T_L DNA (Murai and Kemp, 1982; Ooms *et al.*, 1982a,b; Karcher *et al.*, 1984; Komro *et al.*, 1985).

The T-DNA found in tumors appears to be largely a faithful copy of the plasmid T-DNA, although some short scrambled plasmid sequences have been found outside the border defining the T-DNA (De Beuckeleer *et al.*, 1981; Simpson *et al.*, 1982; Kwok *et al.*, 1985). Generally, no major rearrangement of the T-DNA sequences within the borders has been detected in established tumors, nor would this be expected since the hyperplasia results from products of the RNA transcripts of the hormone biosynthetic genes. However, there may be some fluidity of copy number or sequence arrangement of the T-DNA in transformed plant cells, depending perhaps on the chromosome location, on mobile genetic element insertions, or on the selective pressure on the tissue in culture (Binns *et al.*, 1982; Waldon and Hepburn, 1983; Amasino *et al.*, 1984). There is no evidence for homologous recombination as a mechanism for the integration of T-DNA, nor is there evidence for preferred plant DNA sequences surrounding the T-DNA in tumors (Thomashow *et al.*, 1980b,c; Ursic *et al.*, 1983; Ursic, 1985). Work currently is in progress to investigate the possibility that plant DNA in a particular format, such as transcribed DNA, is more likely to be observed as a site of integration.

E. The *Agrobacterium rhizogenes* Tumor-Inducing Plasmid

Species of agrobacteria are classified as *A. rhizogenes* if they cause root proliferation on the host plant (see Fig. 3) (Kersters and De Ley, 1984). The basic mechanism of transformation by these strains is similar to that of *A. tumefaciens*. The virulence region of the Ri plasmid can substitute for the Ti plasmid *vir* region (Hoekema *et al.*, 1984; Hooykaas *et al.*, 1984; Huffman *et al.*, 1984); however, transformation with *A. rhizogenes* allows the regeneration of intact, fertile plants containing the *A. rhizogenes* T-DNA (Ackermann, 1977; Tepfer, 1984). Virulent strain A4a contains a large plasmid, with two T-DNA segments, T_R and T_L (Fig. 9) (Moore *et al.*, 1979; White and Nester, 1980a,b; Chilton *et al.*, 1982; White *et al.*, 1982, 1983; Huffman *et al.*, 1984).

The T_R segment of the Ri plasmid has a high degree of DNA homology to the IAA synthesis genes of the agrobacteria Ti plasmid (Huffman *et al.*, 1984). This region of the Ri plasmid has been shown to restore virulence to

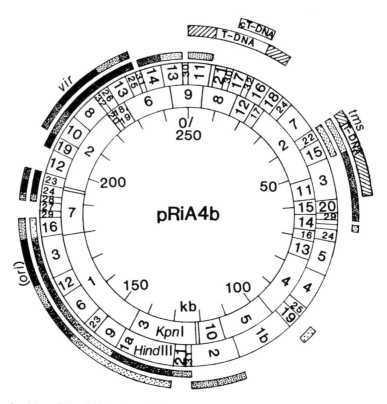

Fig. 9. Map of the pRiA4b plasmid of *A. rhizogenes* showing sites of cleavage by *Kpn*I and *Hin*dIII restriction nucleases. The two transferred (T-DNA) regions, the virulence region (*vir*), and the origin of replication (*ori*) are indicated. The cT-DNA is the region of homology between the untransformed *Nicotiana glauca* DNA and the plasmid. The intensity of stippling indicates the extent of homology between the *A. rhizogenes* Ri plasmid and nopaline (inner circle) and octopine (outer circle) Ti plasmids. Courtesy of G. Huffman.

mutants of the *tms* region of the Ti plasmid (Offringa *et al.*, 1986). The Ri plasmid T_R region is necessary for the formation of tumors on a variety of hosts. One exception is *Kalanchoe*, where weak root formation occurs after infection of stems with *A. rhizogenes* containing mutations in T_R (White *et al.*, 1985).

The T_L DNA of the Ri plasmid does not show homology with any other Ti plasmid. Biochemical functions of this region are yet to be identified. Virulence of *tmr* mutants of *A. tumefaciens* can be restored on *Kalanchoe* leaves by introduction of the Ri plasmid T_L DNA region, indicating that a part of the T_R DNA has *tmr*-like activity (White *et al.*, 1985). Mutants generated by transposon insertions onto the T_L DNA indicate the presence of four genetic loci in Ri plasmid T_L DNA which control the morphology of tumors on *Kalanchoe* leaves. *rol*A (root locus A) mutants exhibit a greater development of fine roots. Insertion into the *rol*B locus results in avirulence. Insertion into the *rol*C locus results in tumor formation with sparse root growth, while insertion into the *rol*D exhibits strong callus growth but no roots (White *et al.*, 1985). The entire nucleotide sequence (21 kb) of the Ri plasmid T_L DNA has been determined (Slightom *et al.*, 1986). The left and right borders contain a 24-bp sequence similar to that found in the *A. tumefaciens* T-DNA. Several copies of this border sequence were found in Ri plasmid T_L DNA which may account for the observation that it can be of variable length in different plant species. A total of 18 open reading frames are present, many of which have characteristics of eukaryotic genes.

Culture of roots from *A. rhizogenes* transformed tissues of most plant species leads to the formation of intact plants (Ackermann, 1977). The regenerated plants frequently show a bizarre phenotype characterized by highly crinkled leaves, loss of apical dominance, shortened internodes, and abnormal flowers and roots (Tepfer, 1984; Taylor *et al.*, 1985). The abnormal phenotype reverts to a normal phenotype and the normal phenotype is inherited as a dominant trait (Sinkar *et al.*, 1988). The reversion is correlated with the absence of one T_L DNA transcript but is not due to a major change in the T_L DNA (Durand-Tardif *et al.*, 1985; Sinkar *et al.*, 1988).

F. Host Range

1. General Considerations

The virulence of a given bacteria strain varies with the tissue, the age, and the nutritional state of the host plant as well as the temperature, pH, and other local factors. In addition, the age of the bacterial culture and the number of bacteria applied to a wound site can also affect the outcome. In some instances a large number of bacteria is needed. In other cases, application of a large number of bacteria produces a "hypersensitive" response with a resulting necrosis, but application of a small number of bacteria leads to tumor formation. Frequently, under the most favorable conditions only a small percentage of inoculated plants will develop tumors. Virulence, thus, is not an "all or none" phenomenon. *Agrobacterium* has an extensive host range and can infect almost all dicotyledonous plants. De Cleene and De Ley (1976) examined the susceptibility of 643 host plant species to transformation by one strain of *A. tumefaciens*. Anderson and Moore (1979) found extensive differences in the tumor-inciting activity of 176 strains of agrobacteria on a single host. The strength of the *vir* induction may reflect a response of the bacteria to a particularly favorable environment (Venade *et al.*, 1988).

Although monocots are commonly considered to be resistant to transformation by agrobacteria, both asparagus (Hernalsteens *et al.*, 1984) and members of the *Liliales* (Hooykaas-Van Slogteren *et al.*, 1984) have been transformed. The important cereal crops, rice, wheat, and corn have not, as yet, been transformed *in planta* by agrobacteria. Maize has been infected by the "agroinfection" system based on virulence functions provided by *A. tume-*

faciens (Grimsley *et al.*, 1987). Recently developed techniques of cell culture, DNA delivery, and plasmid construction may be useful with these plants.

2. Molecular Basis of Host Range of Agrobacteria

Although most species of *Agrobacterium* are tumorigenic on a wide variety of host plants, some strains show a very narrow host range and induce tumors on only one or a few plant species. Host range is primarily determined by the particular Ti plasmid harbored by the infecting strain, and chromosomal genes rarely contribute to host specificity (Loper and Kado, 1979; Thomashow *et al.*, 1980a).

The limited host range (LHR) strain Ag162 was isolated from grapevine tumors in the Soviet Union, and is tumorigenic only on grapevine and a few other plants. Determinants within both the T-DNA and the *vir* regions contribute to the narrow host range phenotype of Ag162. Within the T-DNA, a defective cytokinin biosynthetic gene limits host range (Buccholtz and Thomashow, 1984; Yanofsky *et al.*, 1985). However, other factors also contribute to host range since providing the WHR cytokinin gene *in trans* does not restore tumor formation on certain host plants (i.e., *Kalanchoe*, *Petunia*, etc.). Two genes from the WHR *vir* region, *vir*A and *vir*C, must also be introduced into the LHR strain in order to restore fully a WHR phenotype (Yanofsky *et al.*, 1985). Consistent with this observation, the LHR Ti plasmid (pTiAg162) shares little or no homology with the WHR (pTiA6) *vir*A and *vir*C loci in Southern blot hybridization assays. Interestingly, the LHR plasmid does contain a functional counterpart to the WHR *vir*A locus, since a cosmid clone from the LHR plasmid is able to complement (i.e., restore virulence to) WHR *vir*A mutant strains (Yanofsky *et al.*, 1986). However, this complementation only occurs on some host plants (i.e., grapevines, *Nicotiana glauca*, *Nicotiana rustica*) and not on others (i.e., *Kalanchoe*, *Petunia*), indicating that the LHR *vir*A locus is weak or inadequate for tumorigenesis on certain host plants. In addition, it was determined that the LHR *vir*A gene was insufficient for *vir* gene induction when compared to its WHR counterpart. The nucleotide sequences of the LHR and WHR *vir*A loci have been determined (Leroux *et al.*, 1987). They share only 45% homology at the amino acid level and, thus, represent widely divergent proteins.

The WHR strains are unable to induce tumors on certain grapevine cultivars (i.e., *Vitis lambrusco* cv. Steuben) on which the LHR strain induces tumors. This unusual twist to the host range story is correlated with a hypersensitive-like response which is induced in the host grapevine by WHR strains. Factors both within the T-DNA and *vir* regions contribute to hypersensitivity. Furthermore, strains with mutations in the WHR *vir*C locus no longer elicit the hypersensitive response and are tumorigenic (Yanofsky *et al.*, 1985). It is tempting to speculate that the LHR strain has sacrificed its ability to infect a wide range of host plants in order to outcompete WHR strains in the grapevine niche. Has this perhaps played a role in the loss of the *virC* locus from the LHR strain?

The host range of strains carrying Ti plasmid Bo542 is unusually broad, and the tumors produced *in planta* are large and rapidly growing. This plasmid has been called "super-virulent," and a partial basis for its action appears to be the rapid transformation of a large number of plant cells at a wound site (An *et al.*, 1985; Komari *et al.*, 1986). The locus controlling this phenotype is in the *vir* region of the plasmid (Komari *et al.*, 1986). Strains containing pTiBo542 can be useful in producing tumors on otherwise refractory plants such as alfalfa and soybean (Owens and Cress, 1985; Komari *et al.*, 1986).

G. Agrobacteria Chromosomal Traits

Douglas *et al.* (1985) determined two chromosomal sites, *chv*A and *chv*B, that are required for absorption to cells of some plants and for tumor formation. Other loci, such as *att* (Matthysee, 1987) and *psc*A (*exo*C) (Cangelosi *et al.*, 1987) may be important chromosomal determinants of virulence. The relatedness of *Agrobacterium* and *Rhizobium* is demonstrated by the finding that *chv*A and *nod*A are interchangeable in both species as are *chv*B and *nod*B (Dylan *et al.*, 1986; Cangelosi *et al.*, 1987). Furthermore, the Ti plasmid can be transferred to and is virulent in *Rhizobium* (Hooykaas *et al.*, 1977).

H. The Role of Opines in the Evolution of Plant Transformation by Agrobacteria

Opines were first discovered as unusual amino acid derivatives in crown gall tissues. These compounds are formed by the condensation of the α-amino group of L-amino acids with Krebs cycle α-keto acids, followed by the reduction of the Schiff's base. Alternatively, opines are formed by the reductive condensation of carbohydrates with amino acids. The agrocinopines are phosphorylated derivatives of unknown structure. A comprehensive discussion of the different opines is presented by Tempé and Goldman (1982).

Opines are synthesized in transformed plant tissues, often in large amounts, by enzymes coded by the T-DNA. The level of expression of T-DNA opine genes is very low in the inciting organism (Gelvin *et al.*, 1981). The opine synthesized in a given tumor is specified by the Ti or Ri plasmid which also contains prokaryotic genes for the catabolism of the specified opine so that the inciting organism can metabolize the opines. Thus, the bacteria have created an ecological niche in the host plant, giving them an advantage over organisms that cannot utilize the specific opine. Some opines also stimulate the transfer of the Ti plasmid to nonvirulent agrobacteria

(Genetello *et al.*, 1977; Kerr *et al.*, 1977). Therefore, under conditions where the catabolism of opines is advantageous, the spread of this ability is facilitated. This model is termed the "Opine Concept" (Tempé *et al.*, 1979; Petit *et al.*, 1983) and has been proposed as the driving force behind the evolution of crown gall transformation. The Ti and Ri plasmids are considered to be catabolic plasmids that have evolved the ability to provide their requisite substrates.

Other strains of agrobacteria have alternative schemes to ensure their survival in the plant ecosystem. A number of bacteria which do not induce tumors, particularly *A. radiobacter,* can utilize opines. In the case of *A. radiobacter* 84, the organism produces a bacteriocin, agrocin 84, which kills organisms that can utilize agrocinopines (e.g., some nopaline strains), thus eliminating competition (Kerr and Roberts, 1976). *A. radiobacter* 84 has been utilized as a means of controlling strains of *A. tumefaciens* that are prevalent in Australia. Similar bacteriocin-producing strains that would be effective in other geographic locales have not been found.

I. Application to Genetic Engineering of Plants

Increased understanding of the molecular basis of plant cell transformation has made it possible to incorporate exogenous genes, eukaryotic or prokaryotic, into plants. With prokaryotic genes, a suitable eukaryotic promoter sequence and a poly(A) addition site must be linked to the genes. Among the promoters now used are those derived from the nopaline synthetase or the octopine synthetase genes of the Ti plasmid (Herrera-Estrella *et al.*, 1983) and the 35S promoter of cauliflower mosaic virus (Odell *et al.*, 1985). The products of the transferred genes can be targeted to specific organelles (De Block *et al.*, 1985; Van den Broeck *et al.*, 1985), controlled by various environmental factors (Herrera-Estrella *et al.*, 1984; Lamppa *et al.*, 1985), or developmentally expressed in specific tissues (Sengupta-Gopalan *et al.*, 1985).

To utilize an *A. tumefaciens*-derived transformation system, the gene is incorporated between the 24-bp border sequences found in the T-DNA. Frequently, a disarmed vector is used in which the oncogenic functions within the T-DNA have been removed (de Framond *et al.*, 1983; Zambryski *et al.*, 1983b). The *vir* region is present to ensure the transfer and integration of the desired gene into the plant genome. Binary systems of compatible plasmids also can be used in which the *vir* region is carried on one plasmid and the transferred region between the 24-bp T-DNA borders is carried on another. The vector carrying the *vir* region can be a fully virulent Ti plasmid (An *et al.*, 1985) or it may be disarmed (de Framond *et al.*, 1983; Hoekema *et al.*, 1983; Zambryski *et al.*, 1983b).

When working with genes which have no selectable traits, it is necessary to incorporate a marker into the transferred DNA, permitting selection of those cells which are transformed. Kanamycin or G418 resistance derived

Fig. 10. Shoot (left) regenerated from a leaf disk; (right) exposed to an engineered strain of *A. tumefaciens*. (Courtesy of R. Horsch).

from the bacterial aminoglycoside phosphotransferase II gene has been found to be a useful trait to select transformed plant cells (Horsch *et al.*, 1985). Genes conferring resistance to hygromycin B (Waldron *et al.*, 1985) or to gentamicin (Hayford *et al.*, 1988) have been used to select for transformed plant cells.

Horsch *et al.* (1986) have developed a simple procedure for regenerating intact plants from transformed tissue (Fig. 10). Leaf disks are incubated with *A. tumefaciens* containing disarmed plasmids marked with kanamycin resistance. The disks are placed on a shoot-regenerating medium for 2 to 4 weeks in the presence of kanamycin, and transformed plants are selected by the ability to regenerate roots. The advantage of this method is that prolonged culture *in vitro* is not necessary, thus avoiding somaclonal variation frequently found in tissue culture (Shepard, 1982). Transformed shoots appear to form from single cells. The inheritance is usually Mendelian, although at times loss of expression or loss of genes has been reported (see references in Horsch *et al.*, 1985; Amasino *et al.*, 1984). The level of expression of the integrated genes varies between clones. Recent applications of this procedure include the production of herbicide-resistant and virus-resistant plants (Horsch *et al.*, 1985; Comai *et al.*, 1985; Abel *et al.*, 1986).

There are a number of other techniques that have been developed for obtaining plants with new properties. These techniques include selection of somaclonal variants (Larkin and Scowcroft, 1981; Shepard, 1982), fusion of protoplasts (Power et al., 1976; Melchers et al., 1978; Evans, 1983), and microinjection. The transformation of protoplasts with free DNA is facilitated by polyethylene glycol and high-voltage DC current (reviewed by Potrykus et al., 1985). The use of cauliflower mosaic virus, a DNA virus, as a vector has been described (Brisson et al., 1984).

III. VIRUS-INDUCED TUMORS OF PLANTS

A. General Background

The tumor-inciting virus that has been studied most intensely is the wound tumor virus (WTV). The name is derived from the observation that plants infected with this virus produce tumors at the site of wounding (Fig. 11). WTV is a member of a group termed phytoreoviruses, characterized by a multicomponent double-stranded RNA genome. Many, but not all, of these viruses produce tumors in host plants. WTV is carried by the agallian leafhopper, where it is able to multiply *in vivo* and on host tissues *in vitro*. The virus grows in a large variety of plants. In sweet clover, *Melilotus officinalis* clone c/o, large tumors are produced (Black, 1982). The virus can be titered in cell cultures of the insect host by fluorescent antibody labeling. Mechanical transmission of the virus to plants has not been accomplished. Details on the development of this field are available in Nuss (1984) and Black (1982).

B. Composition of Virions

The WTV virion is a polyhedral structure with at least seven structural proteins (Reddy and MacLeod, 1976). The two outer proteins can be removed by gentle proteolysis with no loss of infectivity in monolayers of vector cells. These proteins, however, play an important role in the vectorial transmission of the virus to plant hosts (Reddy and Black, 1977).

A number of enzymatic activities have been found associated with the intact virus. Black and Knight (1970) found a transcriptional activity that asymmetrically copied each double-stranded RNA segment. Other enzymatic activities included mRNA guanylyltransferase, guanine-7-methyltransferase, and 2'-O-methyltransferase (Rhoades et al., 1977; Nuss and Peterson, 1981). These latter enzymes modify mRNA molecules to form typical capped eukaryotic RNA messages. It is not known if these activities are encapsulated host enzymes or if they are coded by the virus.

The genome was shown to consist of 12 double-stranded segments of RNA by analysis of base composition (Black and Markham, 1963), melting profile (Gomatos and Tamm, 1963), and electron microscopy (Kleinschmidt et al., 1964). Nuss (1984) lists the estimated sizes of the 12 dsRNA segments, which range from 320 to 2900 kDa. There is one copy of each segment in a complete virion.

Fig. 11. Tumors produced by wound tumor virus on clover.

It has been reported that WTV-incited tumors can be grown *in vitro* in the absence of exogenous hormones (Nuss, 1984), even though the WTV content has decreased below detectable levels. It is not known whether the host phytohormone synthesis genes have been switched on by WTV, or whether the WTV has contributed genetic information for phytohormone synthesis (Black, 1957, 1972).

The mechanism for the tumorigenic transformation of the host plant and the association of this process with wounding are topics which can be investigated using recently developed techniques. The preliminary report of the cloning of all 12 segments of the WTV genome and the sequence determination of several of the segments represent important steps toward determination of the basis of WTV tumorigenesis (Nuss *et al.*, 1985).

IV. HABITUATED PLANT TISSUES AND GENETIC TUMORS

Habituated plant tissues and plant genetic tumors exhibit hormone autotrophic growth without the prior intervention of infectious agents. Habituated plant tissues have altered requirements for exogenous growth factors during propagation in culture, and plant genetic tumors appear in certain interspecific hybrids. In no case is a specific molecular mechanism known which accounts for these phenotypic changes.

A. Habituated Plant Tissues

Habituation is the loss by plant tissues of the requirement for auxin and cytokinin phytohormones for growth *in vitro*. Therefore, the tissues can be considered to be a form of a plant tumor. Some studies of habituated tobacco tissues have shown that tissues are capable of autotrophic growth as a result of the increase in the cellular growth factor(s) (Butcher, 1977). Whether this apparent increase in auxin and cytokinin levels is a result of increased production or decreased degradation by the tissue has not been resolved. Not all habituated tissue have elevated hormone contents, so altered sensitivity to the hormones may also explain the habituated phenotype.

In habituated plants in which all tissues including the leaf tissue have a meiotically inheritable cytokinin auxotrophic phenotype (Meins *et al.*, 1983), a plant gene can compensate for cytokinin function of a mutant *A. tumfaciens tmr* locus (Hansen and Meins, 1986), further supporting an analogy between crown gall tumors and habituated tobacco tissues.

A second type of phenotypic alteration of habituated tissue appears to be an epigenetic change (Meins, 1982). This has been deduced from experiments with tobacco pith cells in which habituation is induced by elevated temperature (Meins and Binns, 1978). The frequency of habituation is 100–

1000 times greater than that observed for somatic mutations (Carlson, 1974; Meins, 1975, 1982). Furthermore, tissues from plants regenerated from habituated tissues revert to the nonhabituated phenotype at frequencies much greater than that expected for somatic back mutations. No particular karyotopic alterations are associated with epigenetic habituation, and the proposal has been made that this habituation occurs by means of a positive feedback mechanism that may illustrate the developmental commitment of cells in the plant (Meins and Binns, 1978; Meins, 1982, 1988).

B. Plant Genetic Tumors

Tumors which occur on plants as a result of genetic crosses between species are reviewed elsewhere (Smith, 1972; Bayer, 1982). Such tumors arise in certain interspecific crosses of a variety of plants and also occur in the fungus, *Schizophyllum commune* (Leonard, 1975), although the physiological and genetic basis of the phenomenon has been studied most extensively in *Nicotiana*. The tumors occur in the absence of any discernible external agents and result in a change from the normal plant cell phenotype to a less differentiated phenotype with phytohormone autonomous growth characteristics (Fig. 12).

Tumor-prone *Nicotiana* hybrids produce tumors in all of the progeny and throughout the plant. The genus has been divided into two groups (Näf, 1958), with members of one group providing a genetic component for tumor initiation (I), and members of the second group containing other genetic features which control tumor expression (ee) (Ahuja, 1968). A member of one group may be capable of forming a tumorous hybrid when crossed with a member of the second group but usually does not form a tumor-prone hybrid with a member of the same group.

The genetic features which control the tumorous phenotype appear to be found on specific chromosomes (Ahuja, 1962; Hess *et al.*, 1976). It has been concluded that one *N. glauca* chromosome is necessary to form the tumorous hybrid with *N. langsdorffii* (Ahuja and Hagen, 1966; Smith, 1972; H. H. Smith, personal communication). In crosses between *N. longiflora* (initiator group, I) and *N. debneyi-tabacum* (ee) a single fragment of a *N. longiflora* chromosome has been shown to be sufficient for the tumorous phenotype (Ahuja, 1962, 1968).

Under certain growth conditions, tumorous hybrids grow as normal-appearing plants, but ionizing radiation, stress, phytohormone application, or specific culture conditions can influence the expression of the tumorous phenotype, suggesting that regulation of the expression of the preexisting genetic elements in the tumorous hybrid is responsible for the appearance of the tumors (Riegert, 1964; Braun, 1972, 1978; Smith, 1972; Bayer, 1982). Genetic tumors occur in specific developmental stages and in specific tissues of many species including tobacco (Smith, 1972). Often tumors develop in

Fig. 12. Tumor on a tobacco amphidiploid, containing two complements each of the *N. glauca* and *N. langsdorffii* chromosomes. (Courtesy of R. Amasino).

tissues with various degrees of polyploidy, suggesting a correlation between the presence or copy number of some genes and the expression of the tumorous phenotype.

Plant growth regulators such as auxins and cytokinins are some of the cellular phytohormones whose concentrations and presumably regulation are altered in genetic tumors. Some hybrids, such as *N. glauca* × *N. langsdorffii,* have more free auxin and more of the IAA precursor, tryptophan, than is found in normal plants (Kehr and Smith, 1953; Bayer, 1965). Tumorous hybrids do not require inclusion of auxin in culture conditions. Cytokinins added to tumor-prone hybrids may stimulate tumor formation (Ames, 1972).

Other species have been analyzed for genes responsible for the tumorous phenotype. A dominant gene (*Np*) on a single chromosome is responsible for the tumors that appear on pea pods of *Pisum sativum* Nuttal and Lyall, 1964; Dodds and Matthews, 1966). A single gene (*frs*) in the tomato hybrid *Lycopersicon esculentum* × *L. chilense* is responsible for the tumorous phenotype which is expressed as a result of environmental stress (Martin, 1966; Doering and Ahuja, 1967).

Genetic tumors appear to result from a mechanism distinct from that which causes habituation in plant tissues, although the behavior of the resultant tissues in both cases is similar in that they can be grown in culture without the addition of exogenous plant growth regulators. Furthermore, the tumor formation process in tumor-prone species can be reversed and the resulting normal plant tissue still retains the genetic capacity to produce tumors (Bayer, 1982). Similarly, when habituated tissue becomes phytohormone autonomous as a result of an epigenetic change, reversion to a normal phenotype can occur readily.

V. TRANSFER OF GENETIC INFORMATION IN THE BIOSPHERE

In the course of investigating the structure of *A. rhizogenes* T-DNA in *N. glauca,* it was found that normal uninfected plants contained sequences, termed cT-DNA or cellular T-DNA, which were homologous to the *rol*B, *rol*C, and *rol*D regions of pRi A4 T-DNA (White *et al.,* 1983). A similar observation of sequence homology between the Ri plasmid and genomes of morning glory and carrots has been made (Spano *et al.,* 1982; Tepfer, 1982). A number of varieties of *N. glauca* from various localities, including wild isolates from Peru, Bolivia, and Paraguay, were examined by Furner *et al.* (1986). All the plants were found to contain regions of homology to the T_L DNA of *A. rhizogenes,* A4a, although the restriction endonuclease cleavage patterns differed. A part of the cT-DNA from one variety of *N. glauca* was cloned, sequenced, and found to contain all of *rol*C, all but the carboxyl-terminus of *rol*B, and an undetermined amount of *rol*D. The plant cT-DNA sequences are found in the form of an inverted repeat. The reading frames are 83% identical in DNA sequence to their bacterial counterparts. Although the cT-DNA contains open reading frames, no transcription was detected or could be induced from these sequences. Other members of the genus *Nicotiana*, though not all, contain lesser amounts of cT-DNA.

The cT-DNA found in *Nicotiana* species may be the remnants of a previous infection by agrobacteria of the progenitor of the *Nicotiana* species followed by subsequent regeneration of plants. The variation in cT-DNA in different species may suggest that the cT-DNA is undergoing elimination from the genome. Alternatively, the variation may indicate that infection of individual species of *Nicotiana* occurred and that varying amounts of cT-DNA have been retained by the regenerated plants. This recovery phenomenon can be duplicated in the laboratory (Tepfer, 1984; Amasino *et al.*, 1984; Sinkar, *et al.*, 1988). The sequence divergence of the cT-DNA as compared to that of the Ri plasmid A4b T-DNA reflects evolutionary changes in plant sequences and the divergence of current bacterial plasmid sequences from the original inciting plasmid. The time of this putative transfer can be estimated to be several million years ago. The finding of cT-DNA implies that the bacterial and plant kingdoms have exchanged genetic material. There is, thus, a possibility that acquisition of bacterial DNA has played a role in the evolution of plants.

The possibility of the reverse flow of information, from plants to bacteria, is more enigmatic. The Ti plasmid T_L-DNA may be of plant origin. The open reading frames in T-DNA do have many of the characteristics of eukaryotic genes, and they are transcribed in the eukaryotic environment of the host plant cell. The genes for the synthesis of the phytohormones, IAA and isopentenyladenosine monophosphate, could be of plant origin, but there does not appear to be a high degree of identity between this region of the T-DNA and the genome of any plant thus far examined. In the absence of additional evidence we tentatively conclude that, despite their similarities to typical eukaryotic genes, the T-DNA genes in agrobacteria have evolved independently. Carlson and Chelm (1986) have suggested that the glutamine synthesis II gene has been transferred from the soybean to the closely associated bacterial symbiote. The absence of introns in the gene found in the bacterium was thought to be due to selective deletion or to the acquisition of a processed or psuedo plant gene (Brownlee, 1986).

The passage of genetic information from a eukaryotic organism to a bacterium appears to have occurred in the ponyfish, where a symbiotic bacterium, *Photobacterium leiognathi*, contains a CuZn superoxide dismutase similar to that of the host fish (Martin and Fridovich, 1981; Bannister and Parker, 1985). A similar situation may have occurred in microorganisms isolated from human urine, which produce a material immunologically related to human chorionic gonadotropin (Cohen and Strampp, 1972).

Although the details are not clear, there appears to be a whole spectrum of processes by which genetic information is passed between living organisms. In addition to the transfer of nucleic acids by processes such as transformation, transduction, and retroelement insertion, there are also situations known in nature where cellular organelles are transferred. The transfer of algal chloroplasts into mollusks has been observed (Hinde, 1984), as well as the transfer of nuclei from the alga *Chloreocolas* to its red algal host, *Polysiphonia* (Goff and Coleman, 1984).

The reversible flow of genetic information between kingdoms and between individual species is suggested by these considerations. The evolutionary significance of such phenomena, if experimentally documented, would be profound.

ACKNOWLEDGMENTS

The authors thank their colleagues, D. Akiyoshi, L. Albright, R. Amasino, G. Bolton, H. T. Bradshaw, R. Kanemoto, E. W. Nester, T. J. Parsons, J. S. Powell, M. Sheffer, V. Sinkar, and M. Yanofsky for suggestions and discussions concerning this chapter. P. Horbett produced the illustrations and K. Fowler and J. Zajone provided excellent secretarial assistance. This work has been supported by funds from the American Cancer Society, The National Science Foundation, and the United States Department of Agriculture.

REFERENCES

Abel, P. P., Nelson, R. S., De, B., Hoffman, N., Rogers, S. G., Fraley, R. T., and Beachy, R. N. (1986). *Science* **232**, 738–743.
Ackermann, C. (1977). *Plant Sci. Lett.* **8**, 23–30.
Ahuja, M. R. (1962). *Genetics* **47**, 863–880.
Ahuja, M. R. (1968). *Mol. Gen. Genet.* **103**, 176–184.
Ahuja, M. R., and Hagen, G. L. (1966). *Dev. Biol.* **13**, 408–423.
Akiyoshi, D. E., Klee, H., Amasino, R. M., Nester, E. W., and Gordon, M. P. (1984). *Proc. Natl. Acad. Sci. U.S.A.* **81**, 5994–5998.
Akiyoshi, D. E., Regier, D. A., Jen, G., and Gordon, M. P. (1985). *Nucleic Acids Res.* **13**, 2773–2788.
Akiyoshi, D. E., Regier, D. A., Gordon, M. P. (1988). *J. Bacteriol.* **169**, 4242–4248.
Albright, L., Yanofsky, M., Leroux, B., Ma, D., and Nester, E. W. (1987). *J. Bacteriol.* **169**, 1046–1055.
Amasino, R. M., and Miller, C. O. (1982). *Plant Physiol.* **69**, 389–392.
Amasino, R., Powell, A. L. T., and Gordon, M. P. (1984). *Mol. Gen. Genet.* **197**, 437–446.
Ames, I. H. (1972). *Can. J. Bot.* **50**, 2235–2238.
An, G., Watson, B. D., Stachel, S., Gordon, M. P., and Nester, E. W. (1985). *EMBO J.* **4**, 277–284.
Anderson, A. R., and Moore, L. (1979). *Phytopathology* **69**, 320–323.
Bannister, J. V., and Parker, M. W. (1985). *Proc. Natl. Acad. Sci. U.S.A.* **82**, 149–152.
Barker, R. F., Idler, K. B., Thompson, D. V., and Kemp, J. D. (1983). *Plant Mol. Biol.* **2**, 335–350.
Barry, G. F., Rogers, S. G., Fraley, R. T., and Brand, L. (1984). *Proc. Natl. Acad. Sci. U.S.A.* **81**, 4776–4780.

Bayer, M. H. (1965). *Am. J. Bot.* **52**, 883–890.
Bayer, M. H. (1982). *In* "Molecular Biology of Plant Tumors" (G. Kahl and J. S. Schell, eds.), pp. 33–67. Academic Press, New York.
Bevan, M. (1984). *Nucleic Acids Res.* **12**, 8711–8721.
Bevan, M. W., and Chilton, M.-D. (1982a). *Annu. Rev. Genet.* **16**, 357–384.
Bevan, M. W., and Chilton, M.-D. (1982b). *J. Mol. Appl. Genet.* **1**, 539–546.
Bevan, M. W., Barne, M., and Chilton, M.-D. (1983). *Nucleic Acids Res.* **11**, 369–385.
Binns, A. N., and Meins, F. J. (1973). *Proc. Natl. Acad. Sci. U.S.A.* **70**, 2660–2662.
Binns, A. N., Sciaky, D., and Wood, H. N. (1982). *Cell* **31**, 605–612.
Black, L. M. (1957). *J. Natl. Cancer Inst.* **19**, 663–685.
Black, L. M. (1972). *Prog. Exp. Tumor Res.* **15**, 110–137.
Black, L. M. (1982). *In* "Molecular Biology of Plant Tumors" (G. Kahl and J. S. Schell, eds.), pp. 69–105. Academic Press, New York.
Black, L. M., and Knight, C. A. (1970). *J. Virol.* **6**, 194–198.
Black, L. M., and Markham, R. (1963). *Neth. J. Plant Pathol.* **69**, 215.
Bolton, G. W., Nester, E. W., and Gordon, M. P. (1986). *Science* **232**, 983–985.
Braun, A. C. (1956). *Cancer Res.* **16**, 53–56.
Braun, A. C. (1958). *Proc. Natl. Acad. Sci. U.S.A.* **44**, 344–349.
Braun, A. C. (1972). "Plant Tumor Research." Karger, Basel, Switzerland.
Braun, A. C. (1978). *Biochim. Biophys. Acta* **516**, 167–191.
Braun, A. C., and Mandle, R. J. (1948). *Growth* **12**, 255–269.
Brisson, N. J., Paszkowski, J., Penswick, J. R., Gronenborn, B., Potrykus, I., and Hohn, T. (1984). *Nature (London)* **310**, 511–514.
Brownlee, A. G. (1986). *Nature (London)* **323**, 585.
Buccholtz, W. G., and Thomashow, M. (1984). *J. Bacteriol.* **160**, 327–332.
Buchmann, I., Marner, F. J., Schröder, G., Waffenschmidt, S., and Schröder, J. (1985). *EMBO J.* **4**, 853–859.
Butcher, D. N. (1977). *In* "Plant Tissue and Cell Culture" (H. E. Street, ed.), pp. 429–461. Blackwell, Oxford, England.
Cangelosi, G. A., Hung, L., Puvanesarajah, V., Stacey, G., Ozga, D. A., Leigh, J. A., and Nester, E. W. (1987). *J. Bacteriol.* **169**, 2086–2901.
Carlson, P. S. (1974). *Genet. Res.* **24**, 109–112.
Carlson, T. A., and Chelm, B. K. (1986). *Nature (London)* **332**, 568–570.
Chilton, M.-D. (1982). *In* "The Molecular Biology of Plant Tumors" (G. Kahl and J. S. Schell, eds.), pp. 299–320. Academic Press, New York.
Chilton, M.-D., Drummond, M. H., Merlo, D. J., Sciaky, D., Montoya, A. L., Gordon, M. P., and Nester, E. W. (1977). *Cell* **11**, 263–271.
Chilton, M.-D., Drummond, M. H., Merlo, D. J., and Sciaky, D. (1978). *Nature (London)* **275**, 147–149.
Chilton, M.-D., Tepfer, D. A., Petit, A., David, C., Casse-Delbart, F., and Tempé, J. (1982). *Nature (London)* **295**, 432–434.
Christie, P. J., Ward, J. E., Winans, S. C., and Nester, E. W. (1988). *J. Bacteriol.*, **170**, 2659–2667.
Citovsky, V., De Vos, G., and Zambryski, P. (1988). *Science* **240**, 501–504.
Claeys, M., Messens, E., Van Montagu, M., and Schell, J. (1978). *Quant. Mass Spectrom. Life Sci.* **2**, 409–418.
Cohen, H., and Strampp, A. (1972). *Proc. Soc. Exp. Biol. Med.* **152**, 408–410.
Comai, L., Facciotti, D., Hiatt, W. R., Thompson, G., Rose, R. E., and Stalker, D. M. (1985). *Nature (London)* **317**, 741–744.
Das, A., (1988). *Proc. Natl. Acad. Sciences U.S.A.* **85**, 2909–2913.
Das, A., Stachel, S., Ebert, P., Allenza, P., Montoya, A., and Nester, E. (1986). *Nucleic Acids Res.* **14**, 1355–1364.

De Beuckeleer, M., Lemmers, M., De Vos, G., Van Montagu, M., and Schell, J. (1981). *Mol. Gen. Genet.* **183,** 283–288.
De Block, H., Schell, J., and Van Montagu, M. (1985). *EMBO J.* **4,** 1367–1372.
De Cleene, M., and De Ley J. (1976). *Bot. Rev.* **42,** 389–466.
de Framond, A. J., Barton, K. A., and Chilton, M.-D. (1983). *Bio/Technology* **1,** 262–275.
DeGreve, H., Decraemer, H., Seurinck, J., Van Montagu, M., and Schell, J. (1981). *Plasmid* **6,** 235–248.
DeGreve, H., Dhaese, P., Seurinck, J., Temmers, H., Van Montagu, M., and Schell, J. (1983). *J. Mol. Appl. Genet.* **1,** 499–511.
Depicker, A., Stachel, S., Dhaese, P., Zambryski, P., and Goodman, H. M. (1982). *J. Mol. Appl. Genet.* **1,** 561–573.
Dhaese, P., DeGreve, H., Gielen, J. Seurinck, J., Van Montagu, M., and Schell, J. (1983). *EMBO J.* **2,** 419–426.
Dodds, K. S., and Matthews, P. (1966). *J. Hered.* **57,** 83–85.
Doering, G. R., and Ahuja, M. R. (1967). *Planta* **75,** 85–93.
Douglas, C. J., Halperin, W., and Nester, E. W. (1982). *J. Bacteriol.* **152,** 1265–1275.
Douglas, C. J., Staneloni, R. J., Rubin, R. A., and Nester, E. W. (1985). *J. Bacteriol.* **161,** 850–860.
Durand-Tardif, M., Broglie, R., Slightom, J., and Tepfer, D. (1985). *J. Mol. Biol.* **186,** 557–564.
Dylan, T., Ielpi, L., Stanfield, S., Kashyap, L., Douglas, C., Yanofsky, M., Nester, E., Helinski, D. R., and Ditta, G. (1986). *Proc. Natl. Acad. Sci. U.S.A.* **83,** 4403–4407.
Engler, G., Depicker, A., Maenhaut, R., Villarroel-Mandiola, R., Van Montagu, M., and Schell, J. (1981). *J. Mol. Biol.* **152,** 183–208.
Engstrom, P., Zambryski, P., Van Montagu, M., and Stachel, S. (1987). *J. Mol. Biol.* **197,** 635–645.
Evans, D. A. (1983). *Bio/Technology* **1,** 253–261.
Evans, H. J., Bottomley, P. J., and Newton, W. E. (eds.) (1985). "Nitrogen Fixation Research Progress." Nijhoff, Dordrecht, The Netherlands.
Furner, I. J., Huffman, G. A., Amasino, R. M., Garfinkel, D. J., Gordon, M. P., and Nester, E. W. (1986). *Nature (London)* **319,** 422–427.
Gardner, R. C., and Knauf, V. C. (1986). *Science* **231,** 725–727.
Garfinkel, D. J., and Nester, E. W. (1980). *J. Bacteriol.* **144,** 732–743.
Garfinkel, D. J., Simposn, R. B., Ream, L. W., White, F. F., Gordon, M. P., and Nester, E. W. (1981). *Cell* **27,** 143–153.
Gelvin, S. B. (1984). *In* "Plant–Microbe Interactions: Molecular and Genetic Perspectives" (T. Kosuge and E. W. Nester, eds.), Vol. 1, pp. 342–377. Macmillan, New York.
Gelvin, S. B., Gordon, M. P., Nester, E. W., and Aronson, A. I. (1981). *Plasmid* **6,** 17–29.
Gelvin, S. B., Thomashow, M. F., McPherson, J. C., Gordon, M. P., and Nester, E. W. (1982). *Proc. Natl. Acad. Sci. U.S.A.* **79,** 76–80.
Gelvin, S. B., Karcher, S. J., and DiRita, J. (1983). *Nucleic Acids Res.* **11,** 159–174.
Genetello, C., Van Larebeke, N., Holsters, M., Depicker, A., Van Montagu, M., and Schell, J. (1977). *Nature (London)* **265,** 561–563.
Gietl, C., Koulolikova-Nicola, Z., and Hohn, B. (1987). *Proc. Natl. Acad. Sci. U.S.A.* **84,** 9006–9010.
Goff, L. J., and Coleman, A. W. (1984). *Proc. Natl. Acad. Sci. U.S.A.* **81,** 5420–5424.
Goldberg, S. B., Flick, J. S., and Rogus, S. G. (1984). *Nucleic Acids Res.* **12,** 4665–4677.
Gomatos, P. J., and Tamm, I. (1963). *Proc. Natl. Acad. Sci. U.S.A.* **50,** 878–885.
Gordon, M. P. (1977). *In* "The Biochemistry of Plants" (A. Marcus, ed.), Vol. 6, pp. 531–570. Academic Press, New York.
Grimsley, N., Hohn, T., Davies, J. W., and Hohn, B. (1987). *Nature* **325,** 177–179.
Hagiya, M., Close, T. J., Tait, R. C., and Kado, C. I. (1985). *Proc. Natl. Acad. Sci. U.S.A.* **82,** 2664–2673.

Hansen, C. E., and Meins, F., Jr. (1986). *Proc. Natl. Acad. Sci. U.S.A.* **83**, 2492–2495.
Hayford, M. B., Medford, J. I., Hoffmann, N. L., Rogers, S. G., and Klee, H. J. (1988). *Plant Physiol.* **86**, 1216–1222.
Hepburn, A. G., Clarke, L. E., Pearson, L., and White, J. (1983). *J. Mol. Appl. Genet.* **2**, 315–329.
Hernalsteens, T. P., Thia-Toongx, L., Schell, J., and Van Montagu, M. (1984). *EMBO J.* **3**, 3039–3041.
Herrera-Estrella, L., Depicker, A., Van Montagu, M., and Schell, J. (1983). *Nature (London)* **303**, 209–213.
Herrera-Estrella, L., Van der Broeck, G., Maenhaut, R., Van Montagu, M., Schell, J., Timko, M., and Cashmore, A. (1984). *Nature (London)* **310**, 115–120.
Hess, D., Schneider, G., Lorz, H., and Blaich, G. (1976). *Z. Pflanzenphysiol.* **77**, 247–254.
Hille, J., Klasen, I., and Schilperoort, R. A. (1982). *Plasmid* **7**, 107–118.
Hille, J., van Kan, J., and Schilperoort, R. (1984). *J. Bacteriol.* **158**, 754–756.
Hinde, R. (1984). In "Algal Symbiosis" (L. J. Goff, ed.), pp. 97–107. Cambridge University, New York.
Hoekema, A., Hirsch, P. R., Hooykaas, P. J. J., and Schilperoort, R. A. (1983). *Nature (London)* **303**, 179–180.
Hoekema, A., Hooykaas, P. J. J., and Schilperoort, R. A. (1984). *J. Bacteriol.* **158**, 383–385.
Holsters, M., Silva, B., Van Vliet, F., Hernalsteens, J. P., Genetello, C., Van Montagu, M., and Schell, J. (1978). *Mol. Gen. Genet.* **163**, 335–338.
Holsters, M., Silva, B., Van Vliet, F., Genetello, C., De Block, M., Dhaese, P., Depicker, A., Inze, D., Engler, G., Villarroel, R., Van Montagu, M., and Schell, J. (1980). *Plasmid* **3**, 212–230.
Holsters, M., Villarroel, R., Gielen, J., Seurinck, J., DeGreve, H., Van Montagu, M., and Schell, J. (1983). *Mol. Gen. Genet.* **190**, 35–41.
Hooykaas, P. J. J., and Schilperoort, R. A. (1984). *Adv. Genet.* **22**, 209–283.
Hooykaas, P. J. J., Klapwijk, P. M., Nuti, M. P., Schilperoort, R. A., and Rorsch, A. (1977). *J. Gen. Microbiol.* **98**, 477–484.
Hooykaas, P. J. J., Hofker, M., Den Dulk-Ras, H., and Schilperoort, R. A. (1984). *Plasmid* **11**, 195–205.
Hooykaas-Van Slogteren, G. M. S., Hooykaas, P. J. J., and Schilperoort, R. A. (1984). *Nature (London)* **311**, 763–764.
Horsch, R. B., Fry, J. E, Hoffmann, N. L., Eichholtz, D., Rogers, S. G., and Fraley, R. T. (1985). *Science* **227**, 1229–1231.
Horsch, R. B., Klee, H. J., Stachel, S., Winans, S. C., Nester, E. W., Rogers, S. G., and Fraley, R. T. (1986). *Proc. Natl. Acad. Sci. U.S.A.* **83**, 2571–2575.
Huffman, G. A., White, F. F., Gordon, M. P., and Nester, E. W. (1984). *J. Bacteriol.* **157**, 269–276.
Inze, P., Follin, A., Van Lijsebettens, M., Simoens, C., Genetello, C., Van Montagu, M., and Schell, J. (1984). *Mol. Gen. Genet.* **194**, 265–274.
Janssens, A., Engler, G., Zambryski, P., and Van Montagu, M. (1984). *Mol. Gen. Genet.* **195**, 341–350.
Jensen, C. O. (1918). *Kgl. Vet. Landboh. Aarskr., Kopenhagen* pp. 91–143.
John, H. C., and Amasino, R. M. (1988). *J. Bacteriol.* **170**, 790–795.
Joos, H., Inze, D., Caplan, A., Soimann, M., Van Montagu, M., and Schell, J. (1983). *Cell* **32**, 1057–1067.
Kahl, G., and Schell, J. S. (eds.) (1982). "Molecular Biology of Plant Tumors." Academic Press, New York.
Kaiss-Chapman, R. W., and Morris, R. O. (1977). *Biochem. Biophys. Res. Commun.* **76**, 453–459.
Karcher, S. J., DiRita, V. J., and Gelvin, S. B. (1984). *Mol. Gen. Genet.* **194**, 159–165.
Keane, P. J., Kerr, A., and New, P. B. (1970). *Aust. J. Biol. Sci.* **23**, 585–595.

Kehr, A. E., and Smith, H. H. (1953). *Brookhaven Symp. Biol.* **6**, 55–96.
Kerr, A., and Panagopoulos, C. G. (1977). *Phytopathol. Z.* **90**, 172–179.
Kerr, A., and Roberts, W. P. (1976). *Physiol. Plant Pathology* **9**, 205–221.
Kerr, A., Manigault, P., and Tempé, J. (1977). *Nature (London)* **265**, 560–561.
Kersters, K., and De Ley, J. (1984). *In* "Bergey's Manual of Systematic Bacteriology," 9th Ed., pp. 244–254. Williams & Wilkins, Baltimore, Maryland.
Kersters, K., De Ley, J., Sneath, P. H. A., and Sackin, M. (1973). *J. Gen. Microbiol.* **78**, 227–239.
Klee, H. J., Gordon, M. P., and Nester, E. W. (1982). *J. Bacteriol.* **150**, 327–331.
Klee, H. J., White, F. F., Iyer, V. N., Gordon, M. P., and Nester, E. W. (1983). *J. Bacteriol.* **153**, 878–883.
Klee, H. J., Montoya, A., Horodyski, F., Lichtenstein, C., Garfinkel, D., Fuller, S., Flores, C., Peschon, J., Nester, E., and Gordon, M. (1984). *Proc. Natl. Acad. Sci. U.S.A.* **81**, 1728–1732.
Klee, H. J., Yanofsky, M. F., and Nester, E. W. (1985). *Bio/Technology* **3**, 637–642.
Kleinschmidt, A. K., Dunnebacke, T. H., Spendlove, R. S., Schaffer, F. L., and Whitcomb, R. F. (1964). *J. Mol. Biol.* **10**, 282–288.
Komari, T., Halperin, W., and Nester, E. W. (1986). *J. Bacteriol.* **166**, 88–94.
Komro, C. T., DiRita, V. J., Gelvin, S. B., and Kemp, J. D. (1985). *Plant Mol. Biol.* **4**, 253–263.
Kosuge, T., and Nester, E. W. (1984). "Plant–Microbe Interactions," Vol. 1. Macmillian, New York.
Koukolíková-Nicola, Z., Shillito, R. D., Hohn, B., Wang, K., Van Montagu, M., and Zambryski, P. (1985). *Nature (London)* **313**, 191–196.
Kwok, W. W., Nester, E. W., and Gordon, M. P. (1985). *Nucleic Acids Res.* **13**, 459–471.
Lamppa, G., Nagy, D., and Chua, N.-H. (1985). *Nature (London)* **316**, 750–752.
Larkin, P. J., and Scowcroft, W. R. (1981). *Theor. Appl. Genet.* **60**, 197–214.
Leemans, J., Deblaere, R., Willmitzer, L., DeGreve, H., Hernalsteens, J. P., Van Montagu, M., and Schell, J. (1982). *EMBO J.* **1**, 147–152.
Leonard, T. J. (1975). *Proc. Natl. Acad. Sci. U.S.A.* **72**, 4626–4630.
Leroux, T., Yanofsky, M. F., Winans, S. C., Ward, J. E., Ziegler, S. F., and Nester, E. W. (1987). *EMBO J.* **6**, 849–856.
Lichtenstein, C., Klee, H., Montoya, A., Garfinkel, D., Fuller, S., Flores, C., Nester, E., and Gordon, M. (1984). *J. Mol. Appl. Genet.* **2**, 354–362.
Liu, S. T., and Kado, C. I. (1979). *Biochem. Biophys. Res. Commun.* **90**, 171–178.
Liu, S. T., Perry, K. L., Schard, C. L., and Kado, C. I. (1982). *Proc. Natl. Acad. Sci. U.S.A.* **79**, 2812–2816.
Loper, J. E., and Kado, C. I. (1979). *J. Bacteriol.* **139**, 591–596.
Lundquist, R. C., Close, T. J., and Kado, C. I. (1984). *Mol. Gen. Genet.* **193**, 1–7.
Martin, F. W. (1966). *Ann. Bot.* **30**, 701–709.
Martin, J. P., Jr., and Fridovich, I. (1981). *J. Biol. Chem.* **256**, 6080–6089.
Matthysee, A. G. (1987). *J. Bacteriol.* **169**, 313–323.
Meins, F., Jr. (1975). *In* "Tissue Culture and Plant Science" (H. E. Street, ed.), pp. 233–264. Academic Press, New York.
Meins, F., Jr. (1982). *In* "Molecular Biology of Plant Tumors" (G. Kahl and J. S. Schell, eds.), pp. 1–31. Academic Press, New York.
Meins, F., Jr. (1988). *UCLA Symposium on the Molecular Basis of Plant Development*, in press.
Meins, F., Jr., and Binns, A. N. (1978). *Symp. Soc. Dev. Biol.* **36**, 185–201.
Meins, F., Jr., Foster, R., and Lutz, J. D. (1983). *Dev. Genet.* **4**, 129–141.
Melchers, G., Sacristan, M. D., and Holder, S. A. (1978). *Carlsberg Res. Commun.* **43**, 203–218.
Melchers, L. S., and Hooykaas, P. J. J. (1987). "Oxford Surveys of Plant Molecular and

Cellular Biology," Vol. 4 (B. J. Miflin and H. F. Miflin, eds.) pp. 167–220. Oxford Univ. Press.
Melchers, L. S., Thompson, D. V., Idler, K. B., Neuteboom, S. T. C., deMaagd, R. A., Schilperoort, R. A., and Hooykaas, P. J. J. (1987). *Plant Mol. Biol.* **9**, 635–645.
Messens, E., Lenaerts, A., Van Montagu, M., and Hedges, R. W. (1985). *Mol. Gen. Genet.* **199**, 344–348.
Moore, L., Warren, G., and Strobel, G. (1979). *Plasmid* **2**, 617–626.
Murai, N., and Kemp, J. D. (1982). *Proc. Natl. Acad. Sci. U.S.A.* **79**, 86–90.
Näf, U. (1958). *Growth* **22**, 167–180.
Nester, E. W., Gordon, M. P., Amasino, R. M., and Yanofsky, M. F. (1984). *Annu. Rev. Plant. Physiol.* **35**, 387–413.
Nuss, D. L. (1984). *Adv. Virus Res.* **29**, 57–93.
Nuss, D. L., and Peterson, A. J. (1981). *Virology* **39**, 954–957.
Nuss, D. L., Anzola, J. V., and Asamizu, T. (1985). *Abstr. Int. Congr. Plant Mol. Biol., 1st* **OR-13-08**, 31.
Nuttal, V. W., and Lyall, L. H. (1964). *J. Hered.* **55**, 184–186.
Odell, J. T., Nagy, R., and Chua, M. H. (1985). *Nature (London)* **313**, 810–812.
Offringa, I. A., Melchers, L. S., Regensburg-Tuink, A. J. G., Constantino, P., Schilperoort, R., and Hooykaas, P. J. J. (1986). *Proc. Natl. Acad. Sci. U.S.A.* **83**, 6935–6939.
Okker, R. J. H., Spainik, H., Hille, J., van Brussel, T. A. N., Lugtenbug, B., and Schilperoort, R. A. (1984). *Nature (London)* **312**, 564–566.
Ooms, G., Klapwijk, P. M., Poulis, J. A., and Schilperoort, R. A. (1980). *J. Bacteriol.* **144**, 82–91.
Ooms, G., Hooykaas, P. J. J., Moolenaar, G., and Schilperoort, R. A. (1981). *Gene* **14**, 133–150.
Ooms, G., Bakker, A., Molendijk, L., Wullems, G. J., Gordon, M. P., Nester, E. W., and Schilperoort, R. A. (1982a). *Cell* **30**, 589–597.
Ooms, G., Hooykaas, P. J. J., VanVeen, R. J. M., Van Beelin, P., Reginsburg-Tuink, T. J. G., and Schilperoort, R. A. (1982b). *Plasmid* **7**, 15–19.
Otten, L., Piotrowiak, G., Hooykaas, P., Dubois, M., Szegedi, E., and Schell, J. (1985). *Mol. Gen. Genet.* **199**, 189–193.
Owens, C. E. (1928). In "Principles of Plant Pathology," pp. 182–189. Wiley, New York.
Owens, L. D., and Cress, D. E. (1985). *Plant Physiol.* **77**, 87–94.
Peralta, E. G., and Ream, L. W. (1985a). *Proc. Natl. Acad. Sci. U.S.A.* **82**, 5112–5116.
Peralta, E. G., and Ream, L. W. (1985b). In "Advances in Molecular Genetics of the Bacteria–Plant Interaction" (A. A. Szaky and R. P. Legocki, eds.), pp. 124–130. Cornell University, Ithaca, New York.
Petit, A., Delhaye, S., Tempé, J., and Morel, G. (1970). *Physiol. Veg.* **8**, 205–213.
Petit, A., Chantal, D., Dahl, G. A., Ellis, J. G., Guyon, P., Casse-Delbart, F., and Tempé, J. (1983). *Mol. Gen. Genet.* **190**, 204–214.
Potrykus, I., Shillito, R. D., Saul, M. W., and Paszkowski, J. (1985). *Plant Mol. Biol. Rep.* **3**, 117–128.
Powell, G. K., Hommes, N. G., Kuo, J., Castle, L. A., and Morris, R. O. (1988). *Molec. Plant-Microbe Interactions* **1**, 235–242.
Power, J. B., Frearson, E. M., Hayward, C., George, D., Evans, P. K., Berry, S. F., and Cocking, E. C. (1976). *Nature (London)* **263**, 500–502.
Ream, L. W., Gordon, M. P., and Nester, E. W. (1983). *Proc. Natl. Acad. Sci. U.S.A.* **80**, 1660–1664.
Reddy, D. V. R., and Black, L. M. (1977). *Virology* **80**, 336–346.
Reddy, D. V. R., and MacLeod, R. (1976). *Virology* **70**, 274–282.
Regier, D. A., and Morris, R. O. (1982). *Biochem. Biophys. Res. Commun.* **104**, 1560–1566.
Rhoades, D. P., Reddy, D. V. R., MacLeod, R., Black, L. M., and Banerjee, A. K. (1977). *Virology* **76**, 554–559.

Riegert, A. (1964). *Experimentia* **20**, 518.
Salomon, F., Deblaere, R., Leemans, J., Hernalsteens, J.-P., Van Montagu, M., and Schell, J. (1984). *EMBO J.* **3**, 141–146.
Schell, J., Van Montagu, M., De Beuckeleer, M., De Block, M., Depicker, A., De Wilde, M., Engler, G., Genetello, C., Hernalsteens, J. P., Holsters, M., Seurinck, J., Silva, B., Van Vliet, F., and Villarroel, R. (1979). *Proc. R. Soc. London Ser. B* **204**, 251–266.
Schröder, G., and Schröder, J. (1982). *Mol. Gen. Genet.* **185**, 51–55.
Schröder, G., Kleipp, W., Hillebrand, A., Ehring, R., Koncz, C., and Schröder, J. (1983). *EMBO J.* **2**, 403–409.
Schröder, G., Waffenschmidt, S., Weiler, E. W., and Schröder, J. (1984). *Eur. J. Biochem.* **138**, 387–391.
Schröder, J., Schröder, G., Huisman, H., Schilperoort, R. A., and Schell, J. (1981). *FEBS Lett.* **129**, 166–168.
Sengupta-Gopalan, C., Reichert, N. A., Barker, R. F., Hall, T. C., and Kemp, J. D. (1985). *Proc. Natl. Acad. Sci. U.S.A.* **82**, 3320–3324.
Shaw, C. H., Watson, M. D., Carter, G. H., and Shaw, C. H. (1984). *Nucleic Acids Res.* **12**, 6031–6041.
Shepard, J. F. (1982). *Sci. Am.* **246**, 154–166.
Simpson, R. B., O'Hara, P. J., Kwok, W., Montoya, A. M., Lichtenstein, C., Gordon, M. P., and Nester, E. W. (1982). *Cell* **29**, 1005–1014.
Sinkar, V., White, F. F., Furner, I. J., Abramson, M., Pythoud, F., and Gordon, M. P. (1987). *Plant Physiol.* **86**, 584–590.
Skoog, F., and Miller, C. O. (1957). *Symp. Soc. Exp. Biol.* **11**, 118–131.
Slightom, J. L., Durnad-Tardif, M., Jouanin, L., and Tepfer, D. (1986). *J. Biol. Chem.* **261**, 108–121.
Smith, D. R., and Townsend, D. O. (1907). *Science* **25**, 671–673.
Smith, H. H. (1972). *Prog. Exp. Tumor Res.* **15**, 138–164.
Spano, L., Pomponi, M., Constantino, P., Van Slogtern, G. M. S., and Tempé, J. (1982). *Plant Mol. Biol.* **1**, 291–300.
Stachel, S. E., and Nester, E. W. (1986). *EMBO J.* **5**, 1445–1454.
Stachel, S. E., and Zambryski, P. C. (1986a). *Cell* **46**, 325–333.
Stachel, S. E., and Zambryski, P. C. (1986b). *Cell* **47**, 155–157.
Stachel, S. E., Messens, E., Van Montagu, M., and Zambryski, P. (1985). *Nature (London)* **318**, 624–629.
Stachel, S. E., Nester, E. W., and Zambryski, P. C. (1986). *Proc. Natl. Acad. Sci. U.S.A.* **83**, 379–383.
Taylor, B. H., Amasino, R. M., White, F. F., Nester, E. W., and Gordon, M. P. (1985). *Mol. Gen. Genet.* **201**, 554–557.
Tempé, J., and Goldmann, A. (1982). In "Molecular Biology of Plant Tumors (G. Kahl and J. Schell, eds.), pp. 427–449. Academic Press, New York.
Tempé, J., Guyon, P., Tepfer, D., and Petit, A. (1979). In "Plasmids of Medical, Environmental, and Commercial Importance" (K. N. Timmis and A. Pühler, eds.), pp. 353–363. Elsevier/North-Holland, Amsterdam.
Tepfer, D. (1982). *Colloq. Rech. Fruitieres–Bordeaux*, 2nd pp. 47–59.
Tepfer, D. (1984). *Cell* **37**, 959–967.
Thomashow, L. S., Reeves, S., and Thomashow, M. F. (1984). *Proc. Natl. Acad. Sci. U.S.A.* **81**, 5071–5075.
Thomashow, M. F., Panagopoulos, C. G., Gordon, M. P., and Nester, E. W. (1980a). *Nature (London)* **283**, 794–796.
Thomashow, M. F., Nutter, R. Postle, K., Chilton, M.-D., Blattner, F. R., Powell, A., Gordon, M. P., and Nester, E. W. (1980b). *Proc. Natl. Acad. Sci. U.S.A.* **77**, 6448–6452.
Thomashow, M. F., Nutter, R., Montoya, A. L., Gordon, M. P., and Nester, E. W. (1980c). *Cell* **19**, 729–739.

Thomashow, M. F., Hugly, S., Buchholz, W. G., and Thomashow, L. S. (1986). *Science* **231**, 616–618.
Ursic, D. (1985). *Biochem. Biophys. Res. Commun.* **131**, 152–159.
Ursic, D., Slightom, J. L., and Kemp, J. D. (1983). *Mol. Gen. Genet.* **190**, 494–503.
Van den Broeck, O., Timko, M. P., Kausch, A. P., Cashmore, A. R., Van Montagu, M., and Herrera-Estrella, L. (1985). *Nature (London)* **313**, 358–363.
Van Larebeke, N., Engler, G., Holsters, M., Van den Elsacker, S., Zaenen, I., Schilperoort, R. A., and Schell, J. (1974). *Nature (London)* **252**, 169–170.
Van Larebeke, N., Genetello, C., Schell, J., Schilperoort, R. A., Hermans, A. K., Hernalsteens, J. P., and Van Montagu, M. (1975). *Nature (London)* **255**, 742–743.
Vernade, D., Herrera-Estrella, A., Wang, K., and Van Montagu, M. (1988). *J. Bacteriol.* **170**, 5822–5829.
Waldron, C., and Hepburn, A. G. (1983). *Plasmid* **10**, 199–203.
Waldron, C., Murphy, E. B., Roberts, J. L., Gustafson, G. D., Armour, S. L., and Malcolm, S. K. (1985). *Plant Molec. Biol.* **5**, 103–108.
Wang, K., Herrera-Estrella, L., Van Montagu, M., and Schell, J. (1984). *Cell* **38**, 455–462.
Ward, J. E., Akiyoshi, D. E., Regier, D. A., Datta, A., Gordon, M. P., and Nester, E. W. (1988). *J. Biol. Chem.* **263**, 5804–5814.
Watson, B., Currier, T. C., Gordon, M. P., Chilton, M.-D., and Nester, E. W. (1975). *J. Bacteriol.* **123**, 255–264.
Weiler, E. W., and Spanier, K. (1981). *Planta* **153**, 326–337.
White, F. F., and Nester, E. W. (1980a). *J. Bacteriol.* **141**, 1134–1141.
White, F. F., and Nester, E. W. (1980b). *J. Bacteriol.* **144**, 710–720.
White, F. F., and Sinkar, V. P. (1988). In "Plant DNA Infectious Agents." (T. Hohn and J. Schell, eds.) pp. 149–177. Springer-Verlag, Wien, New York.
White, F. F., Ghidossi, G., Gordon, M. P., and Nester, E. W. (1982). *Proc. Natl. Acad. Sci. U.S.A.* **79**, 3193–3197.
White, F. F., Garfinkel, D., Huffman, G. A., Gordon, M. P., and Nester, E. W. (1983). *Nature (London)* **301**, 348–350.
White, F. F., Taylor, B. H., Huffman, G. A., Gordon, M. P., and Nester, E. W. (1985). *J. Bacteriol.* **164**, 33–44.
Willmitzer, L., De Beuckeleer, M., Lemmers, M., Van Montagu, M., and Schell, J. (1980). *Nature (London)* **287**, 359–361.
Willmitzer, L., Schmalenbach, W., and Schell, J. (1981a). *Nucleic Acids Res.* **9**, 4801–4812.
Willmitzer L., Otten, L., Simons, G., Schmalenbach, W., Schröder, J., Schröder, G., Van Montagu, M., DeVas, G., and Schell, J. (1981b). *Mol. Gen. Genet.* **182**, 255–262.
Willmitzer, L., Simons, G., and Schell, J. (1982). *EMBO J.* **1**, 139–146.
Willmitzer, L., Dhaese, P., Schreir, P. H., Schmalenbach, M., Van Montagu, M., and Schell, J. (1983). *Cell* **32**, 1045–1056.
Winans, S. C., Ebert, P. R., Stachel, S. E., Gordon, M. P., and Nester, E. W. (1986). *Natl. Proc. Acad. Sci. U.S.A.* **83**, 8278–8282.
Winans, S. C., Kerstetter, R. A., and Nester, E. W. (1988). *J. Bacteriol.* **170**, 4047–4054.
Winter, J. A., Wright, R. L., and Gurley, W. B. (1984). *Nucleic Acids Res.* **12**, 2391–2406.
Yadav, N. S., Vanderleyden, J., Bennett, D. R., Barnes, W. M., and Chilton, M.-D. (1982). *Proc. Natl. Acad. Sci. U.S.A.* **79**, 6322–6326.
Yamada, T., Palm, C. J., Brooks, B., and Kosuge, T. (1985). *Proc. Natl. Acad. Sci. U.S.A.* **82**, 6522–6526.
Yanofsky, M., Lowe, B., Montoya, A., Rubin, R., Krul, W., Gordon, M., and Nester, E. (1985). *Mol. Gen. Genet.* **201**, 237–246.
Yanofsky, M. F., Porter, S. G., Young, C., Albright, L. M., Gordon, M. P., and Nester, E. W. (1986). *Cell* **47**, 471–477.
Young, C., and Nester, E. W. (1988). *J. Bacteriol.* **170**, 3367–3374.

Zaenen, I., Van Larebeke, N., Teuchy, H., Van Montagu, M., and Schell, J. (1974). *J. Mol. Biol.* **86,** 109–127.
Zambryski, P. (1988). *Annu. Rev. Genet.* **22 A,** 1–30.
Zambryski, P., Holsters, M., Kruger, K., Depicker, A., Schell, J., Van Montagu, M., and Goodman, H. M. (1989). *Science* **209,** 1385–1391.
Zambryski, P., Depicer, A., Kruger, K., and Goodman, H. M. (1982). *J. Mol. Appl. Genet.* **1,** 361–370.
Zambryski, P., Goodman, H. M., Van Montagu, M., and Schell, J. (1983a). *In* "Mobile Genetic Elements" (J. A. Shapiro, ed.), pp. 505–535. Academic Press, New York.
Zambryski, P., Joos, H., Genetello, C., Leemans, J., Van Montagu, M., and Schell, J. (1983b). *EMBO J.* **2,** 2143–2150.

Genetic Manipulation of Plant Cells

17

ANTHONY J. CONNER
CAROLE P. MEREDITH

I. Introduction
II. Cell Selection
    A. Historical Perspective
    B. Screening for Variants
    C. Cell Selection Strategies
    D. Mutagenesis and Variant Frequency
    E. Characterization of Variants
    F. Nature of Genetic Variation in Cell Cultures
    G. Agricultural Perspective
III. Protoplast Fusion
    A. Historical Perspective
    B. Protoplast Isolation and Fusion
    C. Selection of Somatic Hybrids
    D. Cybrids and Partial Somatic Hybrids
    E. Organelle Segregation and Recombination
    F. Agricultural Perspective
IV. Transformation
    A. Historical Perspective
    B. *Agrobacterium*-Mediated Transformation
    C. Direct Gene Transfer
    D. Other Transformation Systems
    E. Comparison of Transformation Methods
    F. Agricultural Perspective
V. Concluding Discussion
    References

I. INTRODUCTION

Because of the significant advances in plant cell and tissue culture that have occurred since the 1930s, it is now possible to culture individual cells and regenerate complete plants from them in many species. These developments have served as the foundation for the genetic manipulation of somatic plant cells in culture. The ability to regenerate fertile plants from single modified cells makes it possible to study the expression and inheritance of the manipulated trait in complete plants. This approach thus offers a powerful means to improve the understanding of the molecular, biochemical, and physiological basis of plant traits and their regulation within the developmental complexity of whole plants. Furthermore, the opportunity exists to genetically manipulate crop plants to improve their agricultural merit, a subject that provides considerable motivation and justification for much of the current research.

In general terms, the genetic manipulation of somatic plant cells can be divided into three main areas: (1) The isolation of mutants via *cell selection*. (2) The synthesis of full or partial somatic hybrids via *protoplast fusion*. (3) The production of transgenic plants via *transformation*.

This chapter will provide a general overview with reference to more specific reviews where appropriate. Excellent sources of information are Chaleff (1981), Evans *et al.* (1983), Vasil (1984), Gleba and Sytnik (1984), and Green *et al.* (1987).

II. CELL SELECTION

A. Historical Perspective

The first variant phenotype to be commonly isolated from plant cell culture was hormone autotrophy—cell growth in the absence of exogenous plant growth regulators. This growth state, known as habituation, was initially observed by Gautheret (1946) in callus cultures of carrot (*Daucus carota*) and *Scorzonera*. Habituation is very stable and has since been reported to spontaneously arise in cell cultures of many plant species (Gautheret, 1955). Another early report of variant isolation was that of an orange-colored cell line of carrot containing high levels of carotene (Eichenberger, 1951).

The deliberate selection of cell lines with an altered phenotype was first reported by Melchers and Bergmann (1959). They exposed haploid cell suspensions of snapdragon (*Antirrhinum majus*) to both high and low temperature extremes and selected resistant cell lines. This study clearly outlined the potential of using the approaches of microbial genetics to isolate mutants from plant cell cultures. It was over a decade, however, before the feasibility of cell selection in plant cell cultures was generally recognized. Early reports included petunia (*Petunia hybrida*) cell lines resistant to streptomycin (Binding *et al.* 1970), and tobacco (*Nicotiana tabacum*) cell lines resistant to L-threonine (Heimer and Filner, 1970) and auxotrophic for certain amino acids (Carlson, 1970). It was subsequently shown that new phenotypes selected in cell culture can be transmitted to sexual progeny of regenerated plants (Carlson, 1973; Carlson et al., 1973; Maliga et al., 1975), indicating a genetic basis for the selected trait. These findings provided the impetus for the selection of numerous cell lines with many different phenotypes from cultures of a large number of plant species (see Chaleff, 1981; Flick, 1983; Berlin and Sasse, 1985). It soon became apparent from the characterization of selected cell lines that the new phenotypes could arise from nongenetic as well as genetic events. For this reason, selected cell lines are referred to as variants, with the term mutant being reserved for those situations in which a genetic basis has been firmly established.

B. Screening for Variants

Screening for phenotypic variants (as opposed to selection) simply involves making observations or measurements on a large population of cell lines or regenerated plants until rare individuals with a new phenotype are discovered. The observed variation has been commonly referred as "somaclonal variation." Examining large populations of regenerated plants is the most common approach to screening. Since this subject does not involve specific genetic manipulation at the cellular level, it is beyond the scope of this review, and the reader is referred to reviews by Larkin and Scowcroft (1981), Orton (1984), and Karp and Bright (1985) for further information.

Screening in cell culture is most often used for isolating variants with altered levels of specific biochemical compounds (see Conner, 1986). The principal requirement for an efficient cell screening protocol is a simple, sensitive, quantitative or semiquantitative assay for the desired phenotype that permits rapid testing of a large number of cell clones. In its simplest form it involves visual examination of cell colonies for colored constituents such as anthocyanins (e.g., Yamamoto et al., 1982) or naphthoquinones (e.g., Mizukami et al., 1978). Alternatively, a portion of a cell colony may be sacrificed for chemical analysis. This is the usual approach for variants producing elevated levels of biochemicals, for example, high nicotine-producing variants of tobacco (Ogino et al., 1978). A further method involves assessing cell growth over a range of media or incubation conditions. This has been used for the isolation of auxotrophic and temperature-sensitive variants (Gebhardt et al., 1981).

A more sophisticated approach to cell screening is cell sorting via flow cytometry. Many thousands of cells can be rapidly scanned based on their light diffraction, absorption, or fluorescence characteristics. By this means, estimates of cell size, pigmentation, and staining responses to specific dyes can be obtained as indicators of such parameters as viability, cell cycle phase, or biochemical accumulation (Brown, 1984). Furthermore, a large population of cells can be rapidly sorted into two subpopulations based on preselected criteria, thereby offering a convenient system for the isolation of variants.

C. Cell Selection Strategies

In contrast to cell screening, cell selection involves imposing a selection pressure on a population of cells so that only rare individuals with a specific phenotype are capable of survival or growth. When an effective selection strategy can be designed for a given phenotype, the efficiency of isolating a specific variant is greatly enhanced (Meredith, 1984).

Cell selection experiments can be performed using callus cultures (e.g., Chaleff and Ray, 1984), plated cells (e.g., Chaleff and Parsons, 1978a,b), or cell suspension cultures (e.g., Handa *et al.*, 1983). Of these approaches, selection via plated cells is generally considered the most desirable approach (Meredith, 1984; Conner, 1986). Compared to callus cultures, plated cells are more uniformly exposed to the selection agent and cross-feeding between variant cells and neighboring wild-type cells is minimized. Under these conditions, the frequency with which escaped wild-type cells and chimeras, consisting of a mixture of wild-type and variant cells, occur would be expected to be reduced. Selection via plated cells allows rare variants to be easily recognized as distinct, independent cell colonies. In contrast, cell growth in a suspension culture under selection may arise from more than one independent event and represent a mixture of variant types, thereby confusing subsequent analysis. Furthermore, selection in cell suspension cultures does not allow distinction between the rapid growth of rare variants or the slow, gradual adaptation of the whole cell population to the selection agent.

Cell selection can involve direct, rescue, or stepwise strategies and may be performed in a positive or negative manner. Direct selection involves a one-step, positive approach in which cells are cultured in the presence of a selection pressure and those capable of growth are isolated. This is the simplest form of cell selection and is commonly used for the isolation of resistance variants. For example, Chaleff and Parsons (1978a) isolated herbicide-resistance variants of tobacco by selecting for cell growth with normally toxic concentrations of picloram (500 μM) in the medium. Direct selection has also been used to isolate glycerol-utilizing variants of tobacco by selecting for growth on media with glycerol as the sole carbon source (Chaleff and Parsons, 1978b).

Rescue selection involves a two-step approach in which cells are first exposed to culture conditions that kill or permanently inhibit wild-type cells, then subsequently cultured under different conditions to allow the recovery of survivors. It may be performed in either a positive or negative manner. A positive approach allows the selection of variants resistant to the initial selection pressure. Examples include the recovery of chilling-resistant variants of pepper (*Capsicum annuum*) and *Nicotiana sylvestris* at 25°C after exposure to low temperatures (−3° to 5°C) for 21 days (Dix and Street, 1976), and the rescue of aluminum-resistant variants in *Nicotiana plumbaginifolia* by plating cells onto standard medium after exposure to aluminum-toxic medium in suspension culture (Conner and Meredith, 1985a). A negative approach to rescue selection is often used to recover auxotrophic or temperature-sensitive variants. The initial selection pressure is applied via counterselection agents such as bromodeoxyuridine or arsenate to kill or inhibit actively growing wild-type cells. The counterselection agents are then removed and the culture conditions changed to allow the recovery of formerly nongrowing "dormant" cells. Using this approach, Negrutiu *et al.* (1985) rescued auxotrophic variants of *N. plumbaginifolia* that could only grow in the presence of amino acid supplements. Likewise, Malmberg (1979) rescued temperature-sensitive tobacco variants that could grow at 26°C, but not at 33°C.

Stepwise selection is a positive approach involving the gradual increase in selection pressure over time. This strategy may favor variants resulting from mutations in organelles or from gene amplification. The increasing selection pressure promotes the growth of cells with a higher proportion of mutant organelles or higher copy number of an amplified gene. Gengenbach *et al.* (1977) gradually increased the concentration of *Helminthosporium maydis* Race T pathotoxin over five subcultures of maize (*Zea mays*) cell cultures. Maternal inheritance of the pathotoxin resistance in plants regenerated from the resistant variants, and biochemical experiments on isolated mitochondria suggested that resistance resulted from genetic events in the mitochondrial genome. Donn *et al.* (1984) selected a variant with increased resistance to the nonselective herbicide L-phosphinotricin in alfalfa (*Medicago sativa*) suspension cultures by gradually elevating the concentration of the herbicide. Resistance was associated with a 3- to 7-fold increase in the levels of glutamine synthetase, the target protein of L-phosphinotricin. Southern blot analysis, using as a probe a cDNA clone corresponding to glutamine synthetase, established a 4- to 11-fold amplification of a glutamine synthetase gene in the variant. Similarly, glyphosate-resistant petunia cells have been shown to possess an amplified 5-enolpyruvylshikimate-3-phosphate synthase gene (Shah *et al.*, 1986). Variants arising from stepwise selection are often unstable and lose their resistance when cultured in the absence of the selection agent, for example, methotrexate-resistant variants of petunia (Barg *et al.*, 1984). A further disadvantage of stepwise selection is the possible increased frequency of variants resulting from physiological adaptations rather than genetic events, due to a gradual adaptation of cells to the selection pressure.

The effectiveness of cell selection depends on the development of stringent selection conditions that enable rare variants to be recognized against a large background population of wild-type cells. The most appropriate strategy is clearly dependent on the phenotype being selected, and in some instances may also vary with the plant species. The aim is to optimize the selection conditions so that inhibited wild-type cells and growing variant cells can be readily distinguished. The selection pressure should be severe enough to reduce the frequency of false positives (escapes), but not excessive, otherwise the growth of potential variant cells may also be suppressed (false negatives). Factors that can improve the stringency of selection include the strict control of cell density, elimination of cell aggregates, and the isolation of variants not only on the basis of growth in the presence of a selection agent, but also on their ability to synthesize chlorophyll (Conner, 1986). The efficiency of cell selection may be improved in some instances by using feeder cells to assist in the recovery of rare variant cells. This is especially applicable to rescue selection strategies (Conner, 1986). After culturing *N. plumbaginifolia* cell suspensions in aluminum-toxic medium for 10 days, aluminum-resistant variants were rescued within 5 to 6 days after plating with feeder cells, compared to 14 days without feeder cells (Conner and Meredith, 1985a).

D. Mutagenesis and Variant Frequency

The treatment of cultured cells with mutagens prior to screening or selection of variants is generally considered to increase the frequency at which variants are isolated (Maliga, 1980). Commonly used mutagens include ethyl methanesulfonate, methyl methanesulfonate, *N*-methyl-*N'*-nitro-*N*-nitrosoguanidine, *N*-ethyl-*N*-nitrosourea, ultraviolet light, and X-rays. Optimum conditions for mutagenesis may depend on the mutagen, the culture system, and the plant species, and preliminary experiments are usually required to establish the most effective dosage. Treatments resulting in about 50% cell survival are commonly used (Maliga, 1980). After treatment, cells are cultivated for a few generations to allow phenotypic expression, then screened or selected for variants.

The frequency of spontaneous mutations is usually in the range of one in 10^6 to 10^8 cells (Maliga, 1980). The effective use of mutagenesis may increase variant frequencies up to 100-fold (Maliga, 1980). In some instances it has been possible to isolate variants only after mutagenesis. X-Irradiation of haploid protoplasts from *Datura innoxia* was necessary to isolate pigment-deficient variants via cell screening (Schieder, 1976). Likewise, UV mutagenesis was required to select variants with resistance to methylglyoxal bis(guanylhydrazone) in diploid tobacco cell cultures (Malmberg and McIndoo, 1984).

It is considered preferable to avoid mutagenesis when spontaneous variants can be readily isolated (Conner, 1986). Induced mutations are likely to increase the frequency of undesirable changes in other genes independent of the isolated phenotype. Such changes may reduce the vigor of variants, the morphogenic potential of the cell cultures, or the fertility of the regenerated plants, or may result in the segregation of recessive lethals in subsequent sexual generations.

E. Characterization of Variants

Phenotypic variants isolated from plant cell cultures are well known to arise from a variety of physiological or genetic events. An excellent discussion of these can be found in Chaleff (1981, 1983a), with general guidelines for the characterization of variants given in Conner (1986). Independently selected variants can be characterized into one of several categories: escapes, unstable variants, epigenetic variants, and mutants.

Although escapes do not strictly fall within the definition of variants, they are included here since they initially appear along with true variants, and are only discarded after further testing. Escapes (false positives) are identified by their failure to survive a further cycle of screening or selection and may represent a large proportion of the initial isolates (Conner, 1986). The development of stringent variant isolation strategies will reduce the frequency of escapes, resulting in more efficient variant isolation. For example, during the isolation of aluminum-resistant variants of *N. plumbaginifolia,* the escape rate was lower with rescue selection than direct selection (Conner and Meredith, 1985a).

Unstable variants lose their new phenotype after a period of growth in the absence of screening or selection. The majority of variants isolated via cell screening for increased biochemical production are unstable (Zenk, 1978). Likewise, large-scale cell selection experiments have reported a low proportion of stable variants (Conner, 1986). Unstable variants may arise from a physiological adaptation to the screening or selection pressure due to temporary changes in gene expression that rapidly disappear in the absence of repeated screening or sustained selection pressure. For example, transient cycloheximide resistance in tobacco variants was lost after a single subculture cycle in the absence of cycloheximide (Maliga *et al.,* 1976). The initially resistant phenotype appeared to result from the activation of an antibiotic metabolizing gene that was normally suppressed in cell culture. Variants arising from gene amplification may also be unstable (see Section II,C). This issue has not received much attention in plant cell genetics, but is well understood in animal cell cultures in which gene amplification is often observed as a resistance mechanism to cytotoxic chemicals and is commonly associated with the appearance of double minute chromosomes (Cowell, 1982). These supernumerary chromosomes lack a centromere and are only maintained in cells cultured under continual selection pressure.

Epigenetic variants are stable during mitotic cell divisions in cell culture, with the new phenotype being retained even after growth in the absence of continual screening or selection pressure. However, the new phenotype is lost either during plant differentiation from cell culture or during meiosis or zygotic embryogenesis. Epigenetic events are attributed to stable change in gene expression or biochemical activity (Chaleff, 1981; Meins, 1983). Hormone habituation (see Section II,A) is commonly attributed to epigenetic events. For example, cytokinin-habituated variants of tobacco were stable in primary callus cultures derived from cloned cells but lost this phenotype during differentiation/dedifferentiation, since secondary callus cultures (those derived from plants regenerated from variants) regained their cytokinin dependence (Binns and Meins, 1973).

The term mutant is reserved solely for situations in which the new phenotype is transmitted to the sexual progeny of plants regenerated from variants (Chaleff, 1981; Maliga, 1980; Meredith, 1984). In cases in which plant regeneration or fertility problems preclude such genetic analysis, molecular or biochemical evidence, such as altered nucleotide sequences or gene products, is also acceptable evidence (Meredith, 1984). Controlled pollination and examination of segregation patterns in progeny of regenerated plants allow the nature of inheritance to be determined. Occasionally, the new phenotype is expressed only in cell culture and not at the whole plant level, due to developmental regulation of gene expression (Chaleff, 1981, 1983a; Conner, 1986). Such instances can only be determined by testing tertiary cell cultures (those derived from sexual progeny of regenerated plants). There have been many instances in which abnormal segregation patterns have complicated the elucidation of inheritance patterns. Several reasons to account for the unusual segregation ratios in these mutants have been suggested (Conner, 1986).

F. Nature of Genetic Variation in Cell Cultures

The nature of the genetic variation originating in plant tissue cultures has been investigated primarily in plants regenerated from unselected cultures. The same genetic events are generally assumed to be also responsible for traits selected in plant cell cultures. General phenotypic features of the genetic variation observed among regenerated plants include homozygous as well as heterozygous changes, changes in both qualitative and quantitative traits, and high frequencies of occurrence. Hypotheses accounting for the origin of this variation have been reviewed by Larkin and Scowcroft (1981), Orton (1984), and Karp and Bright (1985).

Genetic changes most often reported in regenerated plants have a chromosomal basis and may involve complete chromosome sets (polyploidy), loss or gain of individual chromosome (aneuploidy), or structural rearrangements within individual chromosomes (deletions, duplications, inversions, and translocations) (D'Amato, 1985). Detailed cytological investigations on regenerated plants have revealed cryptic chromosomal rearrangements to be a common event (e.g., Orton, 1980; McCoy et al., 1982; Karp and Maddock,

1984; Criessen and Karp, 1985). Furthermore, Southern blot analysis of potato (*Solanum tuberosum*) regenerants using a probe encoding the 25S rRNA genes revealed plants with a 70% reduction in the copy number of the rRNA genes (Landsmann and Uhrig, 1985). A loss of rRNA intergenic spacer DNA has been observed in the progenies of both wheat and barley regenerants (Breiman *et al.*, 1987a,b). Restriction endonuclease analysis of mitochondrial DNA from regenerated plants has revealed DNA rearrangements in maize (Umbeck and Gengenbach, 1983; Kemble *et al.*, 1982) and potato (Kemble and Shepard, 1984). Such changes may provide the genetic basis for cell selection through changes in gene dosage, epistatic interactions between genes, and position effects influencing the expression of genes. The loss of genetic material may also result in localized regions of hemizygosity, thereby allowing the expression of recessive alleles in heterozygous diploid cell cultures. Mitotic recombination and gene conversion may also expose recessive alleles to selection. In most instances, chromosomal changes are likely to result in genetic events inherited in a recessive manner. Most cell selection experiments have been performed in diploid cell cultures and have recovered dominant mutations, with recessive mutations being selected only in monoploid cell cultures (Flick, 1983). However, there are many instances in which the selected phenotype is inherited in an irregular manner (Conner, 1986; see Section II,E). Since chromosomal rearrangements are also often inherited in an abnormal manner, it is possible that they may provide the underlying basis for some of the mutants selected from cell culture.

The action of transposable elements is frequently suggested to be a source of genetic variation among regenerated plants (Larkin and Scowcroft, 1981; Karp and Bright, 1985; Freeling, 1984). Groose and Bingham (1984) regenerated a population of alfalfa plants from a genotype heterozygous for purple flowers and recovered a white-flowered mutant with unstable recessive inheritance. This mutant reverted to a stable purple-flowered state at a low frequency *in planta* (≤1%), but at a much higher rate when recultured (22%) (Groose and Bingham, 1986). These results are best explained by the action of a transposable element which is especially active in tissue culture. The irregular inheritance patterns of some mutants selected from cell cultures (see Section II,E) may be a consequence of the gene stability characteristics of transposable elements.

Classical point mutations have also been established to occur *in vitro*. Among a population of 645 regenerated maize plants, Brettell *et al.* (1986) identified an individual with a slower electrophoretic mobility of alcohol dehydrogenase (ADH). Cloning and sequencing of the ADH gene from this mutant revealed a single nucleotide change responsible for the substitution of a valine residue in the place of a glutamic acid residue in the ADH polypeptide. The mutation frequency during selection from cell culture is often in the range characteristic of point mutations (see Section II,D). It is therefore probable that most of the mutants selected from cell culture with stable inheritance and typical Mendelian segregation patterns have originated as point mutations.

The molecular basis for the selection of variants in plant cell culture has been elucidated in only a few instances. The best understood phenomenon involves gene amplification (see Section II,C). Stepwise selection for resistance to herbicides inhibitory to specific enzymes resulted in the recovery of variants overproducing the sensitive enzymes (Donn et al., 1984; Steinrücken et al., 1986). In both of these instances, Southern blots established that overproduction was a consequence of amplification of the gene coding for the target enzyme. Selection of mutants with resistance to T-toxin in male sterile maize cell cultures was associated with mitochondria (Brettell et al., 1980; Gengenbach et al., 1977). Restriction endonuclease analysis of mitochondrial DNA revealed that resistant plants were missing a DNA fragment and had other DNA rearrangements (Gengenbach et al., 1981). Resistance was also accompanied by a reversion to male fertility and the disappearance of a mitochondrial protein (Dixon et al., 1982).

One of the distinguishing features of genetic variation among plants regenerated from tissue cultures is the high frequency with which it occurs. Recent advances in the molecular biology of plants have revealed considerable genetic fluidity and capacity for rapid genomic changes (Walbot and Cullis, 1985; Freeling, 1984). These changes are largely attributed to the movement of transposable elements and to gene amplification and loss, which are especially prevalent under stressful conditions. Plant tissue culture is generally perceived as a form of stress on the cultured cells (Freeling, 1984; Walbot and Cullis, 1985). One response to this stress may be the high incidence of genomic change.

G. Agricultural Perspective

Many variant cell lines with stable phenotypes of agricultural significance have been isolated from cell cultures. These include resistance to physical stress (e.g., freezing, chilling, heat, and drought), chemical stress (e.g., herbicides, fungicides, heavy metals, aluminum, salinity, and low phosphate), biotic stress (e.g., pathotoxins and viruses), as well as overproduction of nutritional components (e.g., amino acids) and secondary metabolites (e.g., alkaloids). Although such cell lines have been valuable for studying the biochemistry and physiology of stress resistance in plants, only in rare instances has the selected phenotype been expressed in regenerated plants and transmitted to subsequent generations in a genetically defined manner. Unequivocal evidence for the isolation of agriculturally important mutants from plant cell culture is limited to a few instances of herbicide resistance, pathotoxin resistance, amino acid overproduction, and aluminum resistance (Table I).

TABLE I
Agriculturally Important Mutants Selected from Plant Cell Culture

Trait	Species	Inheritance[a]	Reference
Herbicide resistance			
Picloram	Nicotiana tabacum	D	Chaleff (1980), Chaleff and Parsons (1978a)
Chlorsulfuron	N. tabacum	D,S	Chaleff and Ray (1984)
Sulfometuron	N. tabacum	D,S	Chaleff and Ray (1984)
Terbutryn	N. plumbaginifolia	M	Cseplo et al. (1985)
Pathotoxin resistance			
Pseudomonas tabaci	N. tabacum	S	Carlson (1973)
Drechslera maydis	Zea mays	M	Gengenbach et al. (1977), Brettell et al. (1980)
Fusarium oxysporum	Lycopersicon esculentum	D	Shahin and Spivey (1986)
Amino acid accumulation			
Methionine	N. tabacum	D	Carlson (1973)
Threonine	Z. mays	D	Hibberd and Green (1982)
Lysine	N. sylvestris	D	Negrutiu et al. (1984)
Tryptophan and phenylalanine	Oryza sativa	D	Wakasa and Widholm (1987)
Aluminum resistance	N. plumbaginifolia	D	Conner and Meredith (1985b)

[a] D, Single dominant mutation; S, single semidominant mutation; M, maternally inherited mutation.

A major limitation to the application of cell selection to crop improvement is that the underlying physiological basis of most agriculturally significant phenotypes is insufficiently understood to permit the design of effective selection strategies (Chaleff, 1983b; Meredith, 1984). Agriculturally important traits which are readily expressed in cell culture and, therefore, more amenable to cell selection include resistance to herbicides (Meredith and Carlson, 1982), mineral stress (Meredith, 1984; Conner and Meredith, 1987), and pathotoxins of plant pathogens (Meredith, 1984). A further major limitation is that most crop plants cannot presently be regenerated from cell cultures, especially from fast growing, fine cell suspensions or protoplasts. It is unfortunate that, when crop plants can be regenerated, it is usually only from slow-growing cultures with many lumpy, large aggregates of cells. Such cultures are less desirable for cell selection experiments (see end of Section II,C).

III. PROTOPLAST FUSION

A. Historical Perspective

It has been possible to isolate protoplasts from plant cells since the late 1800s (see Cocking, 1972). These early methods involved plasmolyzed cells that released protoplasts either spontaneously or following agitation after cell walls had been mechanically disrupted, usually via cutting. However, this approach was very inefficient. The technical breakthrough that allowed large-scale isolation of plant protoplasts was not made until Cocking (1960) used a cellulase from the fungus *Myrothecium verrucaria* to isolate protoplasts from tomato (*Lycopersicon esculentum*) root tips. Later, Nagata and Takebe (1970) succeeded in obtaining cell wall regeneration and cell division in tobacco protoplasts. This was soon followed by regeneration of complete plants, via callus, from tobacco protoplasts (Nagata and Takebe, 1971; Nitsch and Ohyama, 1971; Takebe *et al.*, 1971).

Although protoplast fusion had been observed with mechanically isolated protoplasts, high frequency fusion was not reported until Power *et al.* (1970) demonstrated interspecific fusion between maize and oat (*Avena sativa*) protoplasts following their treatment with $NaNO_3$. Carlson *et al.* (1972) used this approach to fuse protoplasts of *Nicotiana glauca* and *Nicotiana langsdorffii*, from which they regenerated the first reported somatic hybrid plant. Subsequent work found that treatments involving Ca^{2+} at high pH (Keller and Melchers, 1973) and polyethylene glycol (Kao and Michayluk, 1974; Wallin *et al.*, 1974) were generally more effective than $NaNO_3$ for the large-scale production of viable fusion products. More recently, electrical fusion of protoplasts has offered an entirely different approach (Vienken *et al.*, 1981; Watts and King, 1984).

B. Protoplast Isolation and Fusion

Routine, large-scale isolation of protoplasts is now possible from virtually any plant tissue of any plant species [see Evans and Bravo (1983) for a review]. Enzyme treatments usually involve mixtures of a cellulase, a hemicellulase, and a pectinase. Once the correct balance is known, yields of up to 10^6 protoplasts from 1 g of fresh plant tissue are not uncommon (e.g., Banks and Evans, 1976). In an intact cell, the cell wall provides mechanical support against the pressure of the protoplast. During protoplast isolation this is replaced by an osmotic pressure to prevent the protoplasts from bursting. Mannitol or sorbitol are the osmotica most often used, although other sugars or sugar alcohols will suffice.

The most common approach to protoplast fusion involves the addition of polyethylene glycol to a mixture of protoplasts. Immersion in 20–40% (w/v) polyethylene glycol (1500–6000 MW) causes protoplasts to aggregate within a few minutes. Actual fusion occurs upon dilution of the polyethylene glycol. A combination of polyethylene glycol treatment for protoplast agglutination, followed by elution with high Ca^{2+} at high pH to trigger fusion is now the preferred method (Evans, 1983; Pelletier and Chupeau, 1984).

Electric field-induced fusion offers an alternative to these chemical approaches. The initial approach (Vienken et al., 1981) involved bringing protoplasts into close membrane contact by an alternating, nonuniform electric field between two electrodes in a microchamber. This induced protoplasts to line up like a "string of pearls" along the electric field lines. Fusion was initiated via a single, rapid, high-intensity electric field pulse. More recently, Watts and King (1984) simplified the equipment required by designing an inexpensive, hand-held, transferable electrode that permits electrofusion of 1-ml volumes in standard cell culture dishes. Aggregating the protoplasts in an alternating radio-frequency field (10 V RMS, 0.5 MHz) for 10 sec overcame problems of protoplast migration and adherence to the electrodes. After a rapid 300 V DC pulse to trigger fusion, the electrode could be immediately transferred to the next culture.

C. Selection of Somatic Hybrids

The ideal protoplast fusion event involves one cell from each parent with the resulting fusion product, a heterokaryon, consisting of a mixture of cytoplasms containing chloroplasts, mitochondria, and nuclei from both parents. However, random protoplast aggregation often leads to multiple fusions involving only one parent or various combinations of the two parents. Furthermore, hybrid cells are often rare and become progressively diluted among the dividing parental cells, and may ultimately be lost. It is, therefore, important to identify and individually culture somatic hybrid cells at an early stage, or to develop a selection system for the preferential growth of somatic hybrids.

When fusion parents involve nongreen protoplasts (e.g., from albino mutants, dark-grown tissues, or cultured cells) and green protoplasts (e.g., from leaf tissue), somatic hybrids can be detected microscopically. Different colored, nontoxic, fluorescent dyes may also be used to tag each parent. Heterokaryons can then be identified by the presence of both dyes (Galbraith and Galbraith, 1979). After fusion, heterokaryons can be either micropipetted and individually cultured (e.g., Gleba and Hoffman, 1978; Menczel et al., 1978) or automatically separated via cell sorting in a flow cytometer (e.g., Redenbaugh et al., 1982). Centrifugation in density gradients has also allowed the recovery of fusion products at a level intermediate between the two parents (Harms and Potrykus, 1978). In some instances, somatic hybrids can be recognized on the basis of leaf morphology during plant regeneration (e.g., Ohgawara et al., 1985).

A method often used to screen for somatic hybrids involves complementation between different albino mutants. For example, Melchers and Labib (1974) fused haploid protoplasts from two chlorophyll-deficient, light-sensitive mutants of tobacco (*sublethal* and *virescent*). Somatic hybrids were selected on the basis of each mutant complementing the other's deficiency, resulting in dark-green calli that regenerated into plants with normal green leaves and resistance to high light intensity.

Selection systems that preferentially allow the growth of somatic hybrids are based on each parent being able to compensate for the inability of the other to grow on a specific culture medium. They may involve the formation of cell colonies from protoplasts and subsequent plant regeneration (Schieder, 1978, 1980; Ohgawara *et al.*, 1985) or the use of specific variant cell lines. Grimelius *et al.* (1978) fused protoplasts from two different tobacco mutants auxotrophic for nitrate reductase and unable to utilize nitrate as a nitrogen source. One mutant (*nia*-63) was unable to synthesize the nitrate reductase apoprotein, whereas the other (*cnx*-68) was deficient in the molybdenum cofactor. Since each parent can compensate for the other's deficiency, somatic hybrids could be selected on the basis of growth with nitrate as the sole nitrogen source. A similar approach using resistance mutants is also possible. White and Vasil (1979) fused protoplasts from two cell lines of *Nicotiana sylvestris* resistant to two different amino acid analogs (*S*-2-aminoethylcysteine and 5-methyltryptophan). Somatic hybrids were selected for their ability to grow in the presence of both inhibitors. Many other examples of genetic complementation can be found in Chaleff (1981), Evans (1983), and Pelletier and Chupeau (1984).

A genotype carrying both auxotrophic and resistance mutations can be used as a "universal hybridizer." For example, Hamill *et al.* (1983) produced a tobacco mutant with both nitrate reductase deficiency and streptomycin resistance. Protoplasts of this mutant can, therefore, be fused to any wild-type protoplast and somatic hybrids selected for growth on a medium containing nitrate as the sole nitrogen source plus streptomycin (Pental *et al.*, 1986).

Selection of somatic hybrids is not limited to the availability of biochemical mutants. Recovery of fusion products by metabolic complementation can also be based on the inactivation of the parent protoplasts by different metabolic inhibitors. For example, Nehls (1978) inactivated *Solanum nigrum* protoplasts with diethyl pyrocarbonate and petunia protoplasts with iodoacetate. Cell metabolism of both parents is irreversibly affected and neither is able to divide to form cell colonies. Somatic hybrids survive and grow as a result of complementation of the inhibited enzymes.

D. Cybrids and Partial Somatic Hybrids

Nuclear fusion does not always occur following protoplast fusion. The nuclei of a heterokaryon may subsequently segregate at cytokinesis into separate cells, or one of the parent nuclei may degenerate. In either situation, a cell is produced with mixed cytoplasms from both parents, but the nucleus from only one. Such cytoplasmic hybrids are commonly called *cybrids*. The production of cybrids may be enhanced by the heavy irradiation (X-rays, UV light, or γ-rays) of protoplasts from one parent to cause nuclear destruction. Such a nucleus does not usually contribute genetic material to a somatic hybrid; however, the organelle genomes (chloroplasts and mitochondria) are less affected. Using this approach Menczel et al. (1982) transferred chloroplasts from tobacco to *N. plumbaginifolia*. They irradiated protoplasts from a streptomycin-resistant tobacco mutant (chloroplast encoded) then fused them to *N. plumbaginifolia* protoplasts. Somatic cybrids were selected via streptomycin resistance and regenerated plants were morphologically and cytologically indistinguishable from the *N. plumbaginifolia* parent. Streptomycin resistance in these plants was maternally inherited and the *Eco*RI restriction pattern of their chloroplast DNA was identical to the original tobacco parent.

Cybrids can also be produced by direct simultaneous selection for "donor" organelles and "recipient" nuclei. Bourgin et al. (1986) fused protoplasts from two tobacco mutants, SR1 with streptomycin resistance that is maternally inherited as a chloroplast trait, and Val^r-2 with nuclear inherited valine resistance controlled by two recessive genes. The resulting cell colonies were selected for both streptomycin and valine resistance. Regenerated plants retained the typical phenotypic characteristics of the recipient Val^r-2 parent and had gained the streptomycin-resistant chloroplasts of the donor SR1 parent.

Another convenient approach to cybrid production involves the use of *cytoplasts*. High-speed centrifugation of protoplasts in a density gradient results in the separation of nucleated miniprotoplasts and enucleated cytoplasts. Maliga et al. (1982) prepared cytoplasts from a tobacco mutant with streptomycin-resistant chloroplasts. These were fused with *N. plumbaginifolia* protoplasts and cybrids selected via streptomycin resistance. Chloroplast transfer was confirmed by the tobacco-specific chloroplast DNA *Eco*RI restriction pattern in the regenerated *N. plumbaginifolia* plants, which also showed maternal inheritance of streptomycin resistance.

Even when nuclear fusion occurs in somatic hybrids, chromosomal instability during subsequent cell divisions often gives rise to cells with aneuploidy and chromosomal rearrangements. This is more common in interspecific hybrids. When the chromosomes of one parent are preferentially lost from a somatic hybrid, there exists the possibility that a few traits may have been transferred to the other parent via some form of recombination, integration, or addition event prior to genome elimination. An interesting example involves the fusion of protoplasts from a nuclear-inherited albino mutant of carrot (*Daucus carota*) ($2n = 18$) with protoplasts from *Aegopodium podagraria* ($2n = 42$) (Dudits et al., 1979). Plants regenerated from this somatic hybrid were morphologically and cytologically ($2n = 18$) similar to *D. carota*. However, these plants retained two traits from the *A. podagraria* parent, the ability to synthesize chlorophyll in their leaves and neurosporeum in their roots. Fusion of protoplasts from a nitrate reductase-deficient streptomycin-resistant mutant of tobacco (see end of Section III,C) with protoplasts from petunia allowed the recovery of nuclear somatic hybrids with the tobacco streptomycin-resistant chloroplasts (Pental *et al.*, 1986). However, after 6 to 12 months in cell culture, genomic incompatibility resulted in the loss of the majority of the tobacco nuclear genome. Regenerated plants resembled petunia, but had an extra peroxidase isozyme band not present in either parent, suggesting that a limited amount of the tobacco nuclear genome may have been retained in the cybrid.

Irradiation of the protoplasts of one parent prior to somatic hybridization may promote genome elimination, except for the possible transfer of one or a few traits. For example, Dudits *et al.* (1980) fused irradiated protoplasts of parsley (*Petroselinum hortense*) ($2n = 22$) with protoplasts of a nuclear albino mutant of carrot ($2n = 18$). Plants regenerated from the somatic hybrid were carrot-like in appearance ($2n = 19$) and expressed chlorophyll synthesis and various isozyme markers from the irradiated donor. Similarly, cytoplasmic male sterility was transferred from one rapeseed (*Brassica napus*) line to another. Donor protoplasts were X-irradiated and fused with protoplasts from a male fertile cultivar. A male sterile somatic hybrid plant was recovered that carried nonparental mitochondria and chloroplasts from the fertile parent. It did not have the sensitivity to low temperatures associated with the original male sterile phenotype (Menczel *et al.*, 1987).

When protoplasts of a nitrate reductase (NR)-deficient tobacco mutant were incubated with barley protoplasts, either irradiated or not, under conditions that would not support the growth of either protoplast type alone, a number of NR+ tobacco plants were obtained (Somers *et al.*, 1986). While some of these plants were shown to be revertants, others had NR subunits with the electrophoretic mobility and antigenic properties of the barley enzyme. This phenomenon was observed not only after fusion experiments, but also when irradiated barley protoplasts were incubated with the tobacco protoplasts without a fusion treatment.

E. Organelle Segregation and Recombination

The mixing of cytoplasms in somatic hybrids brings together organelle genomes (mitochondria and chloroplasts) from two parents. This is not normally possible via sexual hybridization. Independent segregation of mitochondria and chloroplast genomes during mitotic divisions in a somatic hybrid can lead to a variety of potential organelle combinations (Gleba and Evans, 1983; Galun and Aviv, 1983). Yarrow *et al.* (1986) and Barsby *et al.*

(1987) both fused protoplasts of two rapeseed genotypes, one with chloroplast-encoded resistance to triazine herbicides and the other with mitochondrial cytoplasmic male sterility, and obtained hybrids carrying both of these traits. Fusion, therefore, can permit the rapid and permanent exchange of organelles between two different cytoplasms, and offers a convenient approach to the study of cytoplasmic genetics in plants. Furthermore, the possibility of organelle recombination may allow genetic manipulation of organelle traits.

Chloroplast genomes quickly segregate after protoplast fusion, resulting in cells possessing only one parental type of chloroplast (see Pelletier and Chupeau, 1984). Only in rare instances are mixed populations of chloroplasts retained in regenerated plants (e.g., Chen et al., 1977). Many studies have failed to detect recombination between chloroplast genomes (Pelletier and Chupeau, 1984). This is presumably a consequence of rapid chloroplast segregation, since it is probably essential to maintain mixed chloroplast populations for several cell divisions to detect recombination events. The availability of stringent selectable markers on the chloroplast genome has permitted the recovery of an interspecific chloroplast recombination event prior to chloroplast segregation in a somatic hybrid between tobacco and *N. plumbaginifolia* (Medgyesy et al., 1985). These two parents differed with respect to three chloroplast genetic markers. The tobacco parent was streptomycin resistant, defective in chloroplast greening, and lincomycin sensitive. In contrast, the *N. plumbaginifolia* parent was streptomycin sensitive, normal green, and lincomycin resistant. Both antibiotic resistant traits are identified by the ability of callus to green on selective medium. However, in the tobacco parent, streptomycin resistance could not be expressed due to the defective chloroplasts. Calli from the population of fused protoplasts were screened for streptomycin resistance. Plants regenerated from one clone (pt 14) were classified as somatic hybrids based on leaf shape and the color, size, and morphology of the flowers. This somatic hybrid expressed a new combination of chloroplast genetic markers: streptomycin resistance, normal green coloration, and lincomycin sensitivity. Molecular analysis of the chloroplast genome of the pt 14 somatic hybrid confirmed that it arose via chloroplast recombination. The pt 14 chloroplast DNA contained three restriction enzyme sites specific to tobacco and four specific to *N. plumbaginifolia*.

In contrast to the rapid segregation and rare recombination in chloroplast genomes, mitochondrial genomes show considerable plasticity. Mixed populations of mitochondrial genomes are commonly observed in somatic hybrid cells. Often mitochondrial genomes will not segregate and will be maternally transmitted in a stable mixed pattern for several sexual generations (Pelletier and Chupeau, 1984). Furthermore, restriction enzyme digestion patterns of mitochondrial genomes from somatic hybrids characteristically show the appearance of new DNA fragments not present in either parent (Hanson, 1984; Menczel et al., 1987). The most common and simplest interpretation of these findings is that recombination between the mitochondrial DNA of the two parents has occurred.

F. Agricultural Perspective

It is clear that protoplasts from widely divergent species can be readily fused together and, in some instances, these somatic hybrid cells will divide to form cell colonies. Notable achievements in this respect include somatic hybrids between monocot and dicot plants. Kao et al. (1974) constructed somatic hybrids between soybean (*Glycine max*) and both maize and barley (*Hordeum vulgare*). In each instance, somatic hybrids formed cell colonies of up to 100 cells within 2 weeks. However, such wide somatic hybrids fail to regenerate plants.

Plants have been regenerated from intergeneric somatic hybrids in a number of instances, probably the best known being the tomato + potato somatic hybrid (Melchers et al., 1978).

As expected, somatic hybrids between plant species that cannot be crossed sexually are sterile or have exceedingly low fertility. They are, therefore, of limited agricultural value except as vegetatively propagated crops in which the harvested part is not a fruit or seed. With some crosses the somatic hybrid is easier to construct than the sexual hybrid, and plants regenerated from such combinations occasionally retain some fertility. An example is the tetraploid somatic hybrid between potato and *Solanum brevidens*, which retains both pollen and egg fertility (Austin et al., 1985). This somatic hybrid can now be used as a "bridging genotype" to transfer resistance to frost and potato leafroll virus from *S. brevidens* to potato using conventional breeding methods.

Protoplast fusion may have its greatest significance to agriculture in situations in which one parent (the recipient) is left intact, and only one or a few genes are contributed by the other parent (the donor). Such limited gene transfer may be achieved in several ways (see Section III,D):

1. Uniparental chromosome elimination with the possibility of a few recombination events occurring prior to this that result in gene transfer.

2. Radiation of the donor parent prior to fusion to fragment the chromosomes and promote the integration of chromosome segments into the recipient genome.

3. Construction of cybrids and the exchange of organelle genomes.

Characters associated with organelle genomes are the only traits of agricultural significance that have been successfully exchanged between plant species via protoplast fusion (Table II).

The current major limitation to the further exploitation of protoplast fusion for crop improvement is the inability to consistently regenerate major

TABLE II

Agriculturally Important Traits Transferred from One Species to Another via Protoplast Fusion

Trait[a]	Source species	Recipient	Reference
Cytoplasmic male sterility (m)	Nicotiana tabacum	N. sylvestris	Zelcer et al. (1978), Aviv et al. (1980)
	N. tabacum	N. plumbaginifolia	Menczel et al. (1983)
	Petunia axillaris	P. hybrida	Izhar and Power (1979)
	Raphanus sativus	Brassica napus	Pelletier et al. (1983)
Male fertility (m)	N. tabacum	N. sylvestris	Aviv and Galun (1980)
Triazine resistance (c)	Brassica campestris	B. napus	Pelletier et al. (1983)
	N. plumbaginifolia	N. tabacum	Menczel et al. (1986)
Tentoxin[b] resistance (c)	N. tabacum	N. sylvestris	Aviv and Galun (1980)

[a] (m), Mitochondrial-based trait; (c), chloroplast-based trait.
[b] Pathotoxin from *Alternaria alternata*.

crop plants from protoplasts and fusion products. Once this problem is overcome, limited gene transfer via protoplast fusion offers several advantages over conventional sexual backcrossing methods of gene transfer:

1. The germplasm base is extended beyond the limits of sexual hybridization.
2. Character transfer can be accomplished in a single asexual generation instead of seven to eight sexual generations.
3. The power to manipulate organelle genomes via substitution and recombination (Section III,E) greatly exceeds that possible via conventional alloplasmic substitutions.

IV. TRANSFORMATION

A. Historical Perspective

There were many attempts to transform plants with foreign DNA during the 1970s. Although several successes were claimed, these have generally remained unsubstantiated. The majority of experiments involved attempts to obtain expression of selectable markers from bacteria and intact heterologous eukaryotic genes from fungi and animals. The failure to obtain expression of such foreign genes in plants was attributed to differences between the gene expression "machinery" of plant versus prokaryotic or other eukaryotic cells.

A major advance occurred when chimeric genes were constructed in which the coding regions of foreign genes were inserted between the signals controlling gene expression in plants—upstream promoters and downstream polyadenylation sites (see Downy *et al.*, 1983). The natural gene transfer ability of *Agrobacterium tumefaciens* (see Section IV,B below) was exploited to insert into plant cells the bacterial Tn5 neomycin phosphotransferase coding sequence under the control of nopaline synthase expression signals from the T-DNA of *Agrobacterium*. This gene confers resistance to kanamycin and related antibiotics by detoxification through phosphorylation. The results (Bevan *et al.*, 1983; Fraley *et al.*, 1983; Herrera-Estrella *et al.*, 1983) convincingly established the integration and expression of the chimeric gene in plant cells. The evidence included the phenotypic expression of kanamycin resistance in plant cells, Southern blots to establish the presence of the chimeric DNA in transformed tissue, Northern blots to demonstrate the presence of an RNA transcript of the correct size, and activity of the neomycin phosphotransferase enzyme in transformed tissue. Subsequent studies established the sexual transmission of the foreign DNA to progeny of transgenic plants in segregation ratios typical of simply inherited Mendelian genes (De Block *et al.*, 1984; Horsch *et al.*, 1984, 1985). These initial experiments established the feasibility of plant transformation, and stimulated a rapid proliferation of studies on the construction of chimeric genes and their subsequent expression in transformed plants.

B. *Agrobacterium*-Mediated Transformation

Agrobacterium tumefaciens infects wounded plant tissue of many dicotyledonous plants (and a few monocots) and incites crown gall tumors via a natural genetic engineering event. The plant host plays a key role in this infection process in that exudates produced by wounded plant cells induce the expression of the *Agrobacterium* virulence genes that are critical to DNA transfer (Stachel *et al.*, 1985, 1986). During tumorigenesis, a specific region of the *A. tumefaciens* Ti (tumor-inducing) plasmid, the T-DNA, integrates into the nuclear DNA of plant cells. The plant cells express the T-DNA genes that code for enzymes responsible for phytohormone biosynthesis (thus accounting for the tumorous growth) and opine production. *A. tumefaciens* can be "disarmed" to produce nontumorigenic strains by deleting the phytohormone biosynthetic genes from the T-DNA. These genes and/or those responsible for opine synthesis can be replaced with other chimeric or intact plant genes. The genes responsible for the T-DNA transmission are not located within the T-DNA; consequently, disarmed *Agrobacterium* strains remain capable of transferring other incorporated DNA to plant cells. To a lesser extent, *Agrobacterium rhizogenes*, which harbors an Ri plasmid responsible for the hairy root disease in many dicotyledonous plants, has been manipulated for gene insertion into plants. Excellent reviews of *Agrobacterium*-mediated transformation of plants and the subsequent expression of transferred genes can be found in Fraley *et al.* (1986) and Klee *et al.* (1987). Since the use of the Ti plasmid for the genetic manipulation of plants is considered elsewhere in this volume (see Powell and Gordon, Chapter 16), only the selection and regeneration of transformed plant cells will be considered in this chapter.

Transformation can result from the *in vivo* inoculation of wounded plants or the *in vitro* cocultivation of plant protoplasts or tissue with *A. tumefaciens*. When tumorigenic strains of *A. tumefaciens* are used, transformed cells can be recognized by gall formation on intact plants or by hormone-autotrophic growth of plant cells cultured *in vitro*. In general, the expression of the phytohormone biosynthetic genes prevents the regeneration of normal plants from transformed cells. For this reason, disarmed *Agrobacterium* strains are used. In addition, selectable markers are incorporated into the T-DNA to allow transformed cells to be easily recognized by their *in vitro* growth on selection medium. The most commonly used selectable markers confer resistance to kanamycin, hygromycin, chloramphenicol, methotrexate, and glyphosate (Fraley *et al.*, 1986).

The initial studies on *Agrobacterium*-mediated transformation involved cocultivation of plant protoplasts with *A. tumefaciens* (Marton *et al.*, 1979; Wullems *et al.*, 1981; Herrera-Estrella *et al.*, 1983; Fraley *et al.*, 1983). However, there are a number of disadvantages with protoplast transformation (see Section IV,E) which could be overcome by a leaf disk transformation system (Horsch *et al.*, 1985). Leaf disks, cut from surface-sterilized leaves of tomato, tobacco, and petunia, were dipped into a suspension of *A. tumefaciens* harboring a chimeric gene for kanamycin resistance and a functional nopaline synthase gene in the T-DNA. After being blotted dry, the leaf disks were incubated for 2 days on regeneration medium with feeder cells to stimulate cell growth. They were then transferred to the same medium (without feeder cells) plus carbenicillin (to inhibit *Agrobacterium* growth) and kanamycin (to select for transformed cells). Transformation of regenerated plants was confirmed by nopaline analysis, Southern blots, and segregation of kanamycin resistance in seedling progeny in simple Mendelian ratios. General protocols for leaf disk transformation are outlined by Rogers *et al.* (1986) and can be adapted to any plant species susceptible to *Agrobacterium* infection from which adventitious plants can be regenerated from explant tissue.

Agrobacterium has also been used to transform seeds. Transformed plants of *Arabidopsis thaliana* have been obtained via cocultivation of imbibed seeds with *Agrobacterium* (Feldmann and Marks, 1987).

Cotransformation of two unlinked genes into one plant cell has been demonstrated. When tobacco leaf disks were incubated with *Agrobacterium* cultures containing two distinct plasmids, one carrying a kanamycin-resistance gene (selectable) and the other a nopaline synthase gene (unselectable), 3 of 11 kanamycin-resistant regenerated plants were also nopaline positive. Genetic analysis of the progeny of the three plants confirmed that all three resulted from cotransformation of both markers into the same cell (McKnight et al., 1987).

C. Direct Gene Transfer

Direct gene transfer involves the uptake of naked DNA through the plasma membrane of plant protoplasts and its integration into the plant genome. It is independent of biological vectors and only requires a gene under the control of plant expression signals and a protoplast culture system for the host plant. The permeability of the plasma membrane to DNA uptake is improved by the use of conditions similar to those employed for protoplast fusion (Section III,B).

Initial plant transformation experiments by direct gene transfer involved the uptake, integration, and expression of isolated Ti plasmid from *A. tumefaciens* into the protoplasts of petunia (Davey et al., 1980; Draper et al., 1982) and tobacco (Krens et al., 1982). However, these experiments failed to establish whether DNA uptake and integration could occur independently of the Ti plasmid functions. The definitive experiments that verified the feasibility of direct gene transfer were those of Paszkowski et al. (1984). They constructed a chimeric selectable marker gene with the bacterial Tn5 neomycin phosphotransferase coding sequence under the control of the 5' and 3' expression signals from the cauliflower mosaic virus gene IV (pABDI). Tobacco protoplasts were incubated for 30 min with pABDI, calf thymus carrier DNA, and polyethylene glycol, then cultured for 7 days without selection pressure. Microcolonies were then plated onto kanamycin selection medium and resistant cell colonies recovered. As expected, all controls failed to yield kanamycin-resistant cells. Transformation was confirmed in cell cultures and in all stages of development in regenerated plants by Southern blot hybridization and enzyme assay for neomycin phosphotransferase activity. Kanamycin resistance was inherited in seedling progeny as a single dominant trait (Paszkowski et al., 1984; Potrykus et al., 1985a).

More recent work has increased the frequency of transformation to 1 to 3% of the colonies formed after treatment (Shillito et al., 1985). Important improvements included the use of heat shock (45°C for 5 min), optimization of concentration and timing of polyethylene glycol treatment, and the use of electrical impulses (electroporation) to facilitate the reversible permeabilization of cell membranes and promote the transfer of DNA into plant cells. Protocols incorporating these modifications are described by Potrykus et al. (1985b). A comparably high transformation rate has been achieved with no electrical pulses by employing a $MgCl_2$ treatment just prior to transformation (Negrutiu et al., 1987). As with *Agrobacterium*-mediated transformation, direct gene transfer can effect the cotransformation of two unlinked genes into the same cell. When tobacco protoplasts were electroporated with both a kanamycin-resistance gene and a zein gene, on separate plasmids, up to 88% of the kanamycin-resistant cell clones also contained zein sequences (Schocher et al., 1986). The general applicability of direct gene transfer as a method for transforming plants is suggested from its successful use in a wide range of species, including tobacco (Paszkowski et al., 1984; Hain et al., 1985), *N. plumbaginifolia*, petunia, *Brassica rapa*, *Hyoscyamus muticus* (see Potrykus et al., 1985b), Italian ryegrass (*Lolium multifolium*) (Potrykus et al., 1985c), *Triticum monococcum* (Lörz et al., 1985), maize (Fromm et al., 1986), and soybean (Christou et al., 1987).

D. Other Transformation Systems

There are a few instances of plants being successfully transformed using a variety of additional approaches. Brisson et al. (1984) replaced the cauliflower mosaic virus open reading frame II with the coding sequence of a bacterial gene for dihydrofolate reductase that confers resistance to methotrexate. Turnip plants inoculated with this chimeric viral DNA became systemically infected and their leaves synthesized the bacterial dihydrofolate reductase enzyme. Furthermore, the transformed plants showed increased resistance to methotrexate sprays. The use of viruses as gene vectors for plant transformation is discussed in detail elsewhere in this volume (see Shepherd, Chapter 15).

Transformation has also resulted from the fusion of plant protoplasts with liposomes (phospholipid vesicles) and bacterial spheroplasts (bacterial equivalents to plant protoplasts). Using protoplast fusion techniques (see Section III,B), liposomes encapsulating a plasmid containing a gene conferring kanamycin resistance to plant cells were fused to tobacco protoplasts (Caboche and Deshayes, 1984; Deshayes et al., 1985). Enzyme assays and Southern blot hybridizations confirmed that plants regenerated from kanamycin-resistant cell colonies were transformed. Kanamycin resistance was also transferred to seedling progeny as a single dominant nuclear trait. In a similar manner to transformation by liposomes, tobacco protoplasts were fused to *Escherichia coli* and *A. tumefaciens* spheroplasts harboring wild-type or modified Ti plasmids with plant expression genes (Hasezawa et al., 1981; Hain et al., 1984). Selection for hormone autotrophic growth allowed the recovery of callus cultures that produced opines. Southern blot hybridization confirmed the presence of the T-DNA genes in the plant genomes.

Microinjection of DNA into plant protoplasts is yet another method that has been used to achieve transformation. Reich et al. (1986) injected the Ti plasmid from *A. tumefaciens* into protoplasts of alfalfa. Of 65 cell clones recovered, 17 were positive for opine synthesis.

An interesting variation on the microinjection theme is the use of a particle gun to fire microprojectiles into intact plant cells at high velocity. Tungsten particles (4 μm diameter) onto which plasmid DNA containing a chimeric chloramphenicol acetyltransferase (CAT) gene had been adsorbed were propelled into epidermal cells of onion (*Allium cepa*). Transient expression of the CAT gene was detected in the tissue at very high levels 3 days after the treatment (Klein *et al.*, 1987). Similar results have been obtained with maize (Weissinger *et al.*, 1987).

E. Comparison of Transformation Methods

Although all the methods that have been used to transform plants can be used with plant protoplasts, protoplasts are currently a requirement for liposome fusion, spheroplast fusion, and microinjection. The use of protoplasts is laborious and time-consuming, and the extended culture period required for plant regeneration often results in other unrelated genetic abnormalities (Larkin and Scowcroft, 1981; Orton, 1984; Karp and Bright, 1985). Furthermore, well-developed systems for regeneration of plants from protoplasts are only available in a limited number of species. Although direct gene transfer experiments have employed protoplasts, the requirement may not be absolute. Morikawa *et al.* (1986) demonstrated that tobacco mosaic virus RNA was taken up, expressed, and replicated in tobacco mesophyll cells subjected to an electric field pulse in the presence of viral RNA.

Transformation via liposome and spheroplast fusion is currently very inefficient, whereas microinjection requires considerable technical skill and only very small numbers of cells can be treated compared to other methods. Microprojectile bombardment has so far been used only to achieve transient expression of introduced DNA. Integration of a foreign gene via this method has not yet been reported.

Agrobacterium-mediated transformation offers considerable advantage over the other methods since it is not limited to protoplasts and transformed plants can be recovered from any plant tissue which regenerates adventitious plants. The main disadvantage of *Agrobacterium*-mediated transformation is that its natural tumorigenic host range is limited to dicots and a few monocot species (De Cleene and De Ley, 1976, 1981). It may be possible to devise protocols that allow *Agrobacterium* to transfer T-DNA into species outside its natural tumorigenic host range. For example, carefully defined inoculation conditions have recently allowed the expression of T-DNA-specific enzymes in maize seedlings (Graves and Goldman, 1986). Furthermore, maize seedlings become systemically infected with maize streak virus after infection with *Agrobacterium* carrying a tandemly repeated dimer of the virus genome in the T-DNA (Grimsley *et al.*, 1987). It may also be possible to artificially manipulate the host range of *Agrobacterium* through the modification of either the Ti plasmid-borne *vir* genes (Yanofsky *et al.*, 1985) or chromosomal genes affecting host range (Knauf *et al.*, 1982). The role of the host plant in inducing the *vir* genes has been exploited by Schäfer *et al.*

(1987) to transform a nonhost. Tissue of yam (*Dioscorea bulbifera*), a monocot, was incubated with *Agrobacterium* that had been treated with wound exudates of potato, a dicot *Agrobacterium* host. Hormone-independent tumors were produced in which nopaline could be detected. Southern analysis confirmed the integration of T-DNA into the nuclear DNA of the plant. T-DNA transfer did not occur in the absence of the potato exudates.

If *Agrobacterium* host range limitations cannot be overcome, transformation of the agriculturally important cereals and other grasses may necessitate the use of non-*Agrobacterium* approaches. Unfortunately, this taxonomic group is particularly recalcitrant to plant regeneration from protoplasts (Vasil, 1982), although recent success has been reported for rice (*Oryza sativa*) (Fujimura et al., 1985; Abdullah et al., 1986; Coulibaly and Demarly, 1986; Yamada et al., 1986) and maize (C. Rhodes, personal communication). Transgenic maize plants have recently been obtained via electroporation of protoplasts with DNA containing a chimeric kanamycin-resistance gene. Regenerated plants expressed the enzyme encoded by the gene, and Southern analysis confirmed the presence of the gene in the callus tissue from which these plants were regenerated (C. Rhodes, personal communication). Similar success with rice is imminent.

Approaches that might circumvent plant regeneration problems include microinjection of DNA directly into pollen or egg cells, thereby eliminating the need for a cell culture step (Flavell and Mathias, 1984). Direct DNA transformation may be adaptable to cultured plant tissues other than protoplasts, or even to pollen. One possibility is the use of laser beams to produce minute, self-healing holes in cell walls and membranes, thereby allowing foreign DNA in the medium to enter the cells. This method has been used successfully to transform animal cells (Tsukakoshi et al., 1984) and permitted the uptake of tobacco mosaic virus RNA and the proliferation of virus particles in tobacco pollen cells (Kurata and Imamura, 1984). Immature germ cells of rye (*Secale cereale*) were shown to be amendable to transformation when the cavity containing developing inflorescences was injected with plasmid DNA containing a chimeric kanamycin-resistance gene. A small proportion of the resulting seedling progeny (7 of 3023) was resistant to kanamycin, two of which were confirmed to carry and express the foreign gene (de la Peña et al., 1987).

Viral vectors offer several advantages over integrative vectors for transforming plants (Gardner, 1983). These include the ease of infecting whole plants rather than using cultured tissues, systemic spread throughout the plant, and multiple gene copies per cell. Disadvantages include narrow host range, limited space available for foreign DNA, and limited transfer via seed propagation. Nevertheless, viral vectors may be especially appropriate for vegetatively propagated plants and for woody perennial plants, in which a foreign gene could be added to an existing crop, thus obviating the need to replant with a genetically improved cultivar.

F. Agricultural Perspective

Genetic transformation is currently being attempted in virtually all crop plants worldwide. The most popular trait to monitor transformation is kanamycin resistance, an excellent selectable marker with which to establish efficient transformation protocols for specific crop plants. Future transformation of crop plants with agriculturally important genes will, in most instances, involve tight linkage between the desired gene and a convenient selectable marker such as kanamycin resistance. Transformants carrying the desired gene will therefore be identified via selection for the marker gene. Phenotypically normal transgenic plants that express foreign genes in a stable manner have already been obtained in a number of crop species (Table III). New crops are joining this group with such regularity that any attempt to produce a complete list would be futile.

The major limitation to the use of genetic transformation for crop improvement is the limited number of cloned genes offering immediate agricultural advantage if incorporated into crop plants. The initial emphasis is centering on resistance to herbicides, viruses, and insects.

An excellent example is resistance to the herbicide glyphosate. Glyphosate inhibits the enzyme 5-enolpyruvylshikimate-3-phosphate (EPSP) synthase, which is involved in the aromatic amino acid biosynthetic pathway in plants. Comai *et al.* (1983) selected a glyphosate-resistant mutant of *Salmonella typhimurium* that synthesized a resistant form of EPSP synthase. Resistance was due to a single base pair change, resulting in a single amino acid substitution (Stalker *et al.*, 1985). A chimeric gene containing the coding region of this resistant gene flanked by plant expression regulatory se-

TABLE III
Some Crop Species in Which Transgenic Plants Have Been Produced

Common name	Species	Reference
Alfalfa	*Medicago sativa*	Shahin *et al.* (1986)
Asparagus	*Asparagus officinalis*	Bytebier *et al.* (1987)
Bird's-foot trefoil	*Lotus corniculatus*	Jensen *et al.* (1986)
Cotton	*Gossypium hirsutum*	Umbeck *et al.* (1987)
Cucumber	*Cucumis sativus*	Trulson *et al.* (1986)
Flax	*Linum usitatissimum*	McHughen *et al.* (1986)
Lettuce	*Lactuca sativa*	Michelmore *et al.* (1987)
Maize	*Zea mays*	C. Rhodes (personal communication)
Poplar	*Populus* spp.	Fillatti *et al.* (1987)
Potato	*Solanum tuberosum*	An *et al.* (1986)
Rapeseed	*Brassica napus*	Pua *et al.* (1987)
Tomato	*Lycopersicon esculentum*	Horsch *et al.* (1985), McCormick *et al.* (1986)
White clover	*Trifolium repens*	White and Greenwood (1987)

quences was constructed and introduced into tobacco via *Agrobacterium*. Expression of this gene increased glyphosate resistance in the transformed plants (Comai et al., 1985). Glyphosate resistance can also be achieved via overproduction of the glyphosate-sensitive plant enzyme. Steinrücken et al. (1986) used stepwise selection in petunia cell suspension cultures to isolate a glyphosate-resistant variant that overproduced the herbicide-sensitive EPSP synthase protein by 15- to 20-fold. The EPSP synthase gene was cloned from this variant, and Southern blots established that enzyme overproduction was due to a 20-fold amplification of the EPSP synthase gene (Shah et al., 1986). A chimeric gene with the plant EPSP synthase coding region under the control of the strong cauliflower mosaic virus 35S promoter was then constructed and introduced into wild-type petunia. The high level of EPSP gene expression conferred by this promoter resulted in sufficient enzyme overproduction to confer glyphosate resistance on the transgenic plants (Shah et al., 1986).

Another active area of research is in the incorporation of viral genes, under the control of plant expression signals, into plants to achieve cross-protection to viral diseases. Transgenic tobacco and tomato plants carrying a chimeric tobacco mosaic virus coat protein gene showed delayed or even no symptoms when inoculated with tobacco mosaic virus particles (Abel et al., 1986; McCormick et al., 1986a). Similar results have been obtained with tobacco and tomato plants transformed with the coat protein gene from alfalfa mosaic virus (Tumer et al., 1987; Loesch-Fries et al., 1987). A different transformation strategy employs DNA complementary to viral satellite RNA. Tobacco plants transformed with DNA copies of satellite RNA from cucumber mosaic virus or tobacco ringspot virus exhibit markedly reduced disease symptoms when subsequently infected with the respective viruses (Harrison et al., 1987; Gerlach et al., 1987).

Genes from *Bacillus thuringiensis* strains that encode proteins toxic to certain insects are also being investigated as sources of resistance to some groups of insect pests. Tobacco hornworm larvae that feed on the leaves of plants transformed with chimeric genes for these toxins are killed (Adang et al., 1987; Fischhoff et al., 1987; Vaeck et al., 1987).

The use of transformation in crop plants is not limited to the addition of dominant genes to plant genomes. It might also become possible to reduce or even prevent the expression of existing genes in plants via the introduction of antisense DNA. Transcription of these "backward genes" produces an RNA complementary to an existing mRNA message. These bind together via duplex formation, thereby blocking translational expression of the mRNA. This approach has been well-documented in animal systems (e.g., Izant and Weintraub, 1985). Ecker and Davis (1986) have monitored transient CAT activity in carrot protoplasts 1–2 days after transformation via electroporation with various ratios of sense and antisense constructs of chimeric CAT genes. Ratios of 1 sense : 100 antisense DNA were required for greater than 95% reduction in CAT activity, although the amount of antisense DNA could be reduced by at least a factor of 2 when a nopaline synthase polyadenylation signal was included on the antisense construct. This suggests that a high expression level of antisense message is required to reduce the expression of existing genes in plants. Consequently, gene inactivation via antisense DNA is likely to be more appropriate for reducing the expression of plant genes normally expressed at low levels. Incorporation of antisense versions of viral genes into plant genomes may be an alternative approach to virus resistance in plants.

Regardless of the applications to crop improvement, transformation offers a powerful tool for studying the molecular biology of plant genes, especially the control of gene expression and the processing and translocation of synthesized proteins. Furthermore, the development and expression of selectable markers will facilitate multiple transformations of the same plant. The assembling of "quantitative traits" in this manner will permit experiments on gene dosage and gene interaction, and will allow the testing of genetic models of inheritance of quantitative traits.

V. CONCLUDING DISCUSSION

It is clear that plants can be genetically manipulated at the cellular level via cell selection, protoplast fusion, and transformation. These techniques offer opportunities to establish a series of independent mutants for a specific phenotype in the same genetic background. The construction of a series of isogenic lines in this manner will contribute valuable experimental material to further the understanding of the molecular, biochemical, and physiological basis of specific phenotypes in plants. The main advantage of using isogenic lines in this manner is that any difference in growth and metabolism between the mutant and wild-type genotype must be related to the trait in question. Some of the differences are likely to represent interesting pleiotropic effects and/or secondary metabolic consequences resulting from the expression of the genetically manipulated characters. However, examination of the more fundamental molecular or biochemical features is likely to reveal the primary nature of genetic traits in plants.

It is important that genetic manipulation at the cellular level be performed on a large scale. This will provide a wide range of potentially different mutants from which the most appropriate for the desired end use can be selected. In cell selection experiments only a small proportion of the initially selected variants may be mutants (see Section II,E), and the individual mutants may have arisen from independent events. Different mutants in the same or different genes may result in varying levels of gene expression with different tissue specificities. Furthermore, some mutants may have undesirable pleiotropic effects. With protoplast fusion, independent somatic hybrids have only one of many potentially different combinations of genetic material due to segregation and recombination of individual chromosomes and organelle genomes. In transformation experiments, the point of integration may disrupt the functioning of other important genes, resulting in undesirable pleiotropic effects, or may result in position effects that influence the level and tissue specificity of gene expression. Since the initial selection for transformation and mutation events usually involves a cell culture phase, it is possible that the selected events may only be expressed in cell culture and not in complete plants. This has been observed in some mutants selected from cell culture (Chaleff, 1983a; Conner, 1986) and in some mutants arising from *Agrobacterium*-mediated transformation (Horsch *et al.*, 1985). Isolation of a large number of independent events will circumvent many of these potential problems and allow the identification of individuals with the desired attribute.

Undesirable genetic changes, independent of the modification sought, may also occur during the cell culture phase associated with mutant isolation, protoplast fusion, or transformation. These may interfere with regeneration (even in highly regenerable genotypes), reduce fertility, and affect overall performance. A good example of these possible effects is evident among the large collection of aluminum-resistant variants selected from cell cultures of *N. plumbaginifolia* (Conner and Meredith, 1985b). Although *N. plumbaginifolia* is one of the most highly regenerable plant species, it was possible to regenerate plants from only 60% of the variants that exhibited stable aluminum resistance in cell culture. Plants regenerated from many of these variants had reduced fertility. Many also segregated in their seedling progeny for recessive lethal mutations in addition to aluminum resistance. Large-scale genetic manipulations at the cellular level will provide the opportunity for selecting, among many independent events, one mutant that retains all of its former attributes, but incorporates the single desired genetic change.

A principal requirement for the genetic manipulation of plant cells is an effective cell culture and regeneration protocol that allows efficient, large-scale, rapid regeneration from individual cells or protoplasts. While tissue culture procedures are well developed for various *Nicotiana* and other solanaceous species, elite genotypes of most crop plants can only be regenerated from single cells with great difficulty, if at all. Plant regeneration remains a major limitation to the application of genetic manipulation of plant cells for crop improvement. Genetic variation for tissue culture aptitude is well known in most crop plants (see Sharp *et al.*, 1984; Ammirato *et al.*, 1984) and there are several instances in which one to two simply inherited genes are known to control callus growth (e.g., Mok *et al.*, 1980) or regeneration (e.g., Reisch and Bingham, 1980). Attempts to breed for enhanced plant regeneration from tissue culture have proved highly successful, even after only one to two generations of selection (e.g., Bingham *et al.*, 1975; Beckert and Qing, 1985). Consequently, there appears to be plenty of potential for genetically manipulating the tissue culture aptitude of crop plants by conventional plant breeding approaches. This would overcome one of the major limitations to crop improvement via genetic manipulation of plant cells.

REFERENCES*

Abdullah, R., Cocking, E. C., and Thompson, J. A. (1986). *Bio/Technology* **4**, 1087–1090.
Abel, P. P., Nelson, R. S., De, B., Hoffman, N., Rogers, S. G., Fraley, R. T., and Beachy, R. N. (1986). *Science* **232**, 738–743.
Adang, M. J., Firoozabady, I., Klein, J., DeBoer, D., Sekar, V., Kemp, J. D., Murray, E., Rocheleau, T. A., Rashka, K., Staffeld, G., Stock, C., Sutton, D., and Merlo, D. J. (1987). *UCLA Symp. Mol. Cell. Biol., New Ser.* **48**, 345–354.
Ammirato, P. V., Evans, D. A., Sharp, W. R., and Yamada, Y. (eds.) (1984). "Handbook of Plant Cell Culture. Vol. 3: Crop Species." Macmillan, New York.
An, G., Watson, B. D., and Chiang, C. C. (1986). *Plant Physiol.* **81**, 301–305.
Austin, S., Baer, M. A., and Helgeson, J. P. (1985). *Plant Sci.* **39**, 75–81.
Aviv, D., and Galun, E. (1980). *Theor. Appl. Genet.* **58**, 121–127.
Aviv, D., Fluhr, R., Edelman, M., and Galun, E. (1980). *Theor. Appl. Genet.* **56**, 145–150.
Banks, M. S., and Evans, P. K. (1976). *Plant Sci. Lett.* **7**, 409–416.
Barg, R., Peleg, N., Perl, M., Beckmann, J. S. (1984). *Plant Mol. Biol.* **3**, 303–311.
Barsby, T. L., Chuong, P. V., Yarrow, S. A., Wu, S. C., Coumans, M., Kemble, R. J., Powell, A. D., Beversdorf, W. D., and Pauls, K. P. (1987). *Theor. Appl. Genet.* **73**, 809–814.
Beckert, M., and Qing, C. M. (1985). *Theor. Appl. Genet.* **68**, 247–251.
Berlin, J., and Sasse, F. (1985). *Adv. Biochem. Eng. Biotechnol.* **31**, 99–132.
Bevan, M. W., Flavell, R. B., and Chilton, M.-D. (1983). *Nature (London)* **304**, 184–187.
Binding, H., Binding, K., and Straub, J. (1970). *Naturwissenschaften* **57**, 138–139.
Bingham, E. T., Hurley, L. V., Kaatz, D. M., and Saunders, J. W. (1975). *Crop Sci.* **15**, 719–721.
Binns, A., and Meins, F. (1973). *Proc. Natl. Acad. Sci. U.S.A.* **70**, 2660–2662.
Bourgin, J. P., Missonier, C., and Goujaud, J. (1986). *Theor. Appl. Genet.* **72**, 11–14.
Breiman, A., Felsenburg, T., and Galun, E. (1987a). *Theor. Appl. Genet.* **73**, 827–831.
Brieman, A., Rotem-Abarbanell, D., Karp, A., and Shaskin, H. (1987b). *Theor. Appl. Genet.* **74**, 104–112.
Brettell, R. I. S., Thomas, E., and Ingram, D. S. (1980). *Theor. Appl. Genet.* **58**, 55–58.
Brettell, R. I. S., Dennis, E. S., Scowcroft, W. R., and Peacock, W. J. (1986). *Mol. Gen. Genet.* **202**, 235–239.
Brisson, N., Paszkowski, J., Penswick, J. R., Gronenborn, B., Potrykus, I., and Hohn, T. (1984). *Nature (London)* **310**, 511–514.
Brown, S. (1984). *Physiol. Veg.* **22**, 341–349.
Bytebier, B., Deboeck, F., De Greve, H., Van Montagu, M., and Hernalsteens, J. P. (1987). *Proc. Natl. Acad. Sci. U.S.A.* **84**, 5345–5349.
Caboche, M., and Deshayes, A. (1984). *C. R. Hebd. Seances Acad. Sci., Ser. C* **299**, 663–666.
Carlson, P. S. (1970). *Science* **168**, 487–489.
Carlson, P. S. (1973). *Science* **180**, 1366–1368.
Carlson, P. S., Smith, H. H., and Dearing, R. (1972). *Proc. Natl. Acad. Sci. U.S.A.* **69**, 2292–2294.
Carlson, P. S., Dearing, R. D., and Floyd, B. M. (1973). *In* "Genes, Enzymes and Populations" (A. M. Srb, ed.), pp. 99–107. Plenum, New York.

* The literature review for this chapter was completed in September 1987.

Chaleff, R. S. (1980). *Theor. Appl. Genet.* **58**, 91-95.
Chaleff, R. S. (1981). "Genetics of Higher Plants: Applications of Cell Culture." Cambridge University Press, Cambridge, England.
Chaleff, R. S. (1983a). *In* "Genetic Engineering of Plants: An Agricultural Perspective" (T. Kosuge, C. P. Meredith, and A. Hollander, eds.), pp. 257-270. Plenum, New York.
Chaleff, R. S. (1983b). *Science* **219**, 676-682.
Chaleff, R. S., and Parsons, M. F. (1978a). *Proc. Natl. Acad. Sci. U.S.A.* **75**, 5104-5107.
Chaleff, R. S., and Parsons, M. F. (1978b). *Genetics* **89**, 723-728.
Chaleff, R. S., and Ray, T. B. (1984). *Science* **223**, 1148-1151.
Chen, K., Wildman, S. G., and Smith, H. H. (1977). *Proc. Natl. Acad. Sci. U.S.A.* **74**, 5109-5112.
Christou, P., Murphy, J. E., and Swain, W. F. (1987). *Proc. Natl. Acad. Sci. U.S.A.* **84**, 3962-3966.
Cocking, E. C. (1960). *Nature (London)* **187**, 962-963.
Cocking, E. C. (1972). *Annu. Rev. Plant Physiol.* **23**, 29-50.
Comai, L., Sen, L. C., and Stalker, D. M. (1983). *Science* **221**, 370-371.
Comai, L., Facciotti, D., Hiatt, W. R., Thompson, G., Rose, R. E., and Stalker, D. M. (1985). *Nature (London)* **317**, 741-744.
Conner, A. J. (1986). *N.Z. J. Technol.* **2**, 83-94.
Conner, A. J., and Meredith, C. P. (1985a). *Planta* **166**, 466-473.
Conner, A. J., and Meredith, C. P. (1985b). *Theor. Appl. Genet.* **71**, 159-165.
Conner, A. J., and Meredith, C. P. (1987). *In* "Genetic Aspects of Plant Mineral Nutrition" (W. H. Gabelman and B. C. Loughman, eds.), pp. 69-77. Nijhoff/Junk, The Hague, The Netherlands.
Coulibaly, M. Y., and Demarly, Y. (1986). *Z. Pflanzenzuecht.* **96**, 79-81.
Cowell, J. K. (1982). *Annu. Rev. Genet.* **16**, 21-59.
Criessen, G. P., and Karp, A. (1985). *Plant Cell Tissue Organ Cult.* **4**, 171-182.
Cseplo, A., Medgyesy, P., Hideg, E., Demeter, S., Marton, L., and Maliga, P. (1985). *Mol. Gen. Genet.* **200**, 508-510.
D'Amato, F. (1985). *CRC Crit. Rev. Plant Sci.* **3**, 73-112.
Davey, M. R., Cocking, E. C., Freemann, J., Pearce, N., and Tudor, I. (1980). *Plant Sci. Lett.* **18**, 307-313.
De Block, M., Herrera-Estrella, L., Van Montagu, M., Schell, J., and Zambryski, P. (1984). *EMBO J.* **3**, 1681-1689.
De Cleene, M., and De Ley, J. (1976). *Bot. Rev.* **42**, 389-466.
De Cleene, M., and De Ley, J. (1981). *Bot. Rev.* **47**, 147-194.
de la Peña, A., Lörz, H., and Schell, J. (1987). *Nature (London)* **325**, 274-276.
Deshayes, A., Herrera-Estrella, L., and Caboche, M. (1985). *EMBO J.* **4**, 2731-2737.
Dix, P. J., and Street, H. E. (1976). *Ann. Bot.* **40**, 903-910.
Dixon, L. K., Leaver, C. J., Brettell, R. I. S., and Gengenbach, B. G. (1982). *Theor. Appl. Genet.* **63**, 75-80.
Donn, G., Tischer, E., Smith, J. A., and Goodman, H. M. (1984). *J. Mol. Appl. Genet.* **2**, 621-635.
Downey, K., Voellmy, R., Ahmad, F., and Schultz, J. (eds.) (1983). "Advances in Gene Technology: Molecular Genetics of Plants and Animals." Academic Press, New York.
Draper, J., Davey, M. R., Freeman, J. P., Cocking, E. C., and Cox, B. J. (1982). *Plant Cell Physiol.* **23**, 451-458.
Dudits, D., Hadlaczky, G., Bajszar, G. Y., Koncz, C., Lazar, G., and Horvath, G. (1979). *Plant Sci. Lett.* **15**, 101-112.
Dudits, D., Fejer, O., Hadlaczky, G., Koncz, C., Lazar, G. B., and Horvath, G. (1980). *Mol. Gen. Genet.* **179**, 283-288.
Ecker, J. R., and Davis, R. W. (1986). *Proc. Natl. Acad. Sci. U.S.A.* **83**, 5372-5376.
Eichenberger, M. E. (1951). *C. R. Seances Soc. Biol. Ses Fil.* **145**, 239-240.

Evans, D. A. (1983). *In* "Handbook of Plant Cell Culture. Vol. 1: Techniques for Propagation and Breeding" (D. A. Evans, W. R. Sharp, P. V. Ammirato, and Y. Yamada, eds.), pp. 291–321. Macmillan, New York.
Evans, D. A., and Bravo, J. E. (1983). *In* "Handbook of Plant Cell Culture. Vol. 1: Techniques for Propagation and Breeding" (D. A. Evans, W. R. Sharp, P. V. Ammirato, and Y. Yamada, eds.), pp. 124–176. Macmillan, New York.
Evans, D. A., Sharp, W. R., Ammirato, P. V., and Yamada, Y. (eds.) (1983). "Handbook of Plant Cell Culture. Vol. 1: Techniques for Propagation and Breeding." Macmillan, New York.
Feldmann, K. A., and Marks, M. D. (1987). *Mol. Gen. Genet.* **208**, 1–9.
Fillatti, J. J., Sellmer, J., McCown, B., Haissig, B., and Comai, L. (1987). *Mol. Gen. Genet.* **206**, 192–199.
Fischhoff, D. A., Bowdish, K. S., Perlak, F. J., Marrone, P. G., McCormick, S. M., Niedermeyer, J. G., Dean, D. A., Kusano-Kretzmer, K., Mayer, E. J., Rochester, D. E., Rogers, S. G., and Fraley, R. T. (1987). *Bio/Technology* **5**, 807–813.
Flavell, R. B., and Mathias, R. (1984). *Nature (London)* **307**, 108–109.
Flick, C. E. (1983). *In* "Handbook of Plant Cell Culture. Vol. 1: Techniques for Propagation and Breeding" (D. A. Evans, W. R. Sharp, P. V. Ammirato, and Y. Yamada, eds.), pp. 393–441. Macmillan, New York.
Fraley, R. T., Rogers, S. G., Horsch, R. B., Sanders, P. R., Flick, J. S., Adams, S. P., Bittner, M. L., Brand, L. A., Fink, C. L., Fry, J. S., Galluppi, G. R., Goldberg, S. B., Hoffmann, N. L., and Woo, S. C. (1983). *Proc. Natl. Acad. Sci. U.S.A.* **80**, 4803–4807.
Fraley, R. T., Rogers, S. G., and Horsch, R. B. (1986). *CRC Crit. Rev. Plant Sci.* **4**, 1–46.
Freeling, M. (1984). *Annu. Rev. Plant. Physiol.* **35**, 277–298.
Fromm, M. E., Taylor, L. P., and Walbot, V. (1986). *Nature (London)* **319**, 791–793.
Fujimura, T., Sakurai, M., Akagi, H., Negishi, T., and Hirose, A. (1985). *Plant Tissue Cult. Lett.* **2**, 74–75.
Galbraith, D. W., and Galbraith, J. E. C. (1979). *Z. Pflanzenphysiol.* **93**, 149–158.
Galun, E., and Aviv, D. (1983). *In* "Handbook of Plant Cell Culture. Vol. 1: Techniques for Propagation and Breeding" (D. A. Evans, W. R. Sharp, P. V. Ammirato, and Y. Yamada, eds.), pp. 358–392. Macmillan, New York.
Gardner, R. C. (1983). *In* "Genetic Engineering of Plants: An Agricultural Perspective" (T. Kosuge, C. P. Meredith, and A. Hollaender, eds.), pp. 121–142. Plenum, New York.
Gautheret, R. J. (1946). *C. R. Seances Soc. Biol. Ses Fil.* **140**, 169–171.
Gautheret, R. J. (1955). *Annu. Rev. Plant Physiol.* **6**, 433–484.
Gebhardt, C., Schnebli, V., and King, P. J. (1981). *Planta* **153**, 81–89.
Gengenbach, B. G., Green, C. E., and Donovan, C. M. (1977). *Proc. Natl. Acad. Sci. U.S.A.* **74**, 5113–5117.
Gengenbach, B. G., Connelly, J. A., Pring, D. R., and Conde, M. F. (1981). *Theor. Appl. Genet.* **59**, 161–167.
Gerlach, W. L., Llewellyn, D., and Haseloff, J. (1987). *Nature (London)* **328**, 802–805.
Gleba, Y. Y., and Evans, D. A. (1983). *In* "Handbook of Plant Cell Culture. Vol. 1: Techniques for Propagation and Breeding" (D. A. Evans, W. R. Sharp, P. V. Ammirato, and Y. Yamada, eds.), pp. 322–357. Macmillan, New York.
Gleba, Y. Y., and Hoffman, F. (1978). *Mol. Gen. Genet.* **165**, 257–264.
Gleba, Y. Y., and Sytnik, K. M. (1984). "Protoplast Fusion: Genetic Engineering in Higher Plants." Springer-Verlag, Berlin, Federal Republic of Germany.
Graves, A. C. F., and Goldman, S. L. (1986). *Plant Mol. Biol.* **7**, 43–50.
Green, E. C., Somers, D. A., Hackett, W. P., and Biesboer, D. D. (eds.) (1987). "Plant Tissue and Cell Culture." Liss, New York.
Grimelius, K., Eriksson, T., Grafe, R., and Müller, A. (1978). *Physiol. Plant.* **44**, 273–277.
Grimsley, N., Hohn, T., Davies, J. W., and Hohn, B. (1987). *Nature (London)* **325**, 177–179.
Groose, R. W., and Bingham. E. T. (1984). *Crop Sci.* **24**, 655–658.

Groose, R. W., and Bingham, E. T. (1986). *Plant Cell Rep.* **5,** 104–107.
Hain, R., Steinbiss, H. H., and Schell, J. (1984). *Plant Cell Rep.* **3,** 60–64.
Hain, R., Stabel, P., Czernilofsky, A. P., Steinbiss, H. H., Herrera-Estrella, L., and Schell, J. (1985). *Mol. Gen. Genet.* **199,** 161–168.
Hamill, J. D., Pental, D., Cocking, E. C., and Müller, A. J. (1983). *Heredity* **50,** 197–200.
Handa, A. K., Bressan, R. A., Handa, S., and Hasegawa, P. M. (1983). *Plant Physiol.* **72,** 645–653.
Hanson, M. R. (1984). *Oxford Surv. Plant Mol. Cell Biol.* **1,** 33–52.
Harms, C. T., and Potrykus, I. (1978). *Theor. Appl. Genet.* **53,** 49–55.
Harrison, B. D., Mayo, M. A., and Baulcombe, D. C. (1987). *Nature (London)* **328,** 799–802.
Hasezawa, S., Nagata, T., and Syono, K. (1981). *Mol. Gen. Genet.* **182,** 206–210.
Heimer, Y. M., and Filner, P. (1970). *Biochim. Biophys. Acta* **215,** 152–165.
Herrera-Estrella, L., De Block, M., Messens, E., Hernalsteens, J. P., Van Montagu, M., and Schell, J. (1983). *EMBO J.* **2,** 987–995.
Hibberd, K. A., and Green, C. E. (1982). *Proc. Natl. Acad. Sci. U.S.A.* **79,** 559–563.
Horsch, R. B., Fraley, R. T., Rogers, S. G., Sanders, P. R., Lloyd, A., and Hoffmann, N. (1984). *Science* **223,** 496–498.
Horsch, R. B., Fry, J. E., Hoffmann, N., Eichholtz, D., Rogers, S. G., and Fraley, R. T. (1985). *Science* **227,** 1229–1231.
Izant, J. G., and Weintraub, H. (1985). *Science* **229,** 345–352.
Izhar, S., and Power, J. B. (1979). *Plant Sci. Lett.* **14,** 49–55.
Jensen, J. S., Marcker, J. A., Otten, L., and Schell, J. (1986) *Nature (London)* **321,** 669–674.
Kao, K. N., and Michayluk, M. R. (1974). *Planta* **115,** 355–367.
Kao, K. N., Constabel, F., Michayluk, M. R., and Gamborg, O. L. (1974). *Planta* **120,** 215–227.
Karp, A., and Bright, S. W. J. (1985). *Oxford Surv. Plant Mol. Cell Biol.* **2,** 199–234.
Karp, A., and Maddock, S. E. (1984). *Theor. Appl. Genet.* **67,** 249–255.
Keller, W. A., and Melchers, G. (1973). *Z. Naturforsch.* **280,** 737–741.
Kemble, R. J., and Shepard, J. F. (1984). *Theor. Appl. Genet.* **69,** 211–216.
Kemble, R. J., Flavell, R. B., and Brettell, R. I. S. (1982). *Theor. Appl. Genet.* **62,** 213–217.
Klee, H., Horsch, R., and Rogers, S. (1987). *Annu. Rev. Plant Physiol.* **38,** 467–486.
Klein, T. M., Wolf, E. D., Wu, R., and Sanford, J. C. (1987). *Nature (London)* **327,** 70–73.
Knauf, V. C., Panagopoulos, C. G., and Nester, E. W. (1982). *Phytopathology* **72,** 1545–1549.
Krens, F. A., Molendijk, L., Wullems, G. J., and Schilperoort, R. A. (1982). *Nature (London)* **296,** 72–74.
Kurata, N., and Imamura, J. (1984). *Mitsubishi-Kasei Inst. Life Sci., Tokyo Ann. Rep.* **13,** 38.
Landsmann, J., and Uhrig, H. (1985). *Theor. Appl. Genet.* **71,** 500–505.
Larkin, P. J., and Scowcroft, W. R. (1981). *Theor. Appl. Genet.* **60,** 197–214.
Loesch-Fries, L. S., Merlo, D., Zinnen, T., Burhop, L., Hill, K., Krahn, K., Jarvis, N., Nelson, S., and Halk, E. (1987). *EMBO J.* **6,** 1845–1851.
Lörz, H., Baker, B., and Schell, J. (1985). *Mol. Gen. Genet.* **199,** 178–182.
McCormick, S., Nelson, R. S., Dube, P., Beachy, R. N., and Fraley, R. (1986a). "Abstracts: Tailoring Genes for Crop Improvement," p. 22. University of California, Davis, California.
McCormick, S., Niedermeyer, J., Fry, J., Barnason, A., Horsch, R., and Fraley, R. (1986b). *Plant Cell Rep.* **5,** 81–84.
McCoy, T. J., Phillips, R. L., and Rines, H. W. (1982). *Can. J. Genet. Cytol.* **24,** 37–50.
McHughen, A., Browne, R., Kneeshaw, D., and Jordan, M. (1986). *Int. Cong. Plant Tissue Cell Cult., 6th,* p. 127.
McKnight, T. D., Lillis, M. T., and Simpson, R. B. (1987). *Plant Mol. Biol.* **8,** 439–445.
Maliga, P. (1980). *Int. Rev. Cytol. Suppl.* **11A,** 225–250.
Maliga, P., Breznovits, A., Marton, L., and Joo, F. (1975). *Nature (London)* **255,** 401–402.
Maliga, P., Lazar, G., Svab, Z., and Nagy, F. (1976). *Mol. Gen. Genet.* **149,** 267–271.
Maliga, P., Lörz, H., Lazar, G., and Nagy, F. (1982). *Mol. Gen. Genet.* **185,** 211–215.

Malmberg, R. L. (1979). *Genetics* **92**, 215–221.
Malmberg, R. L., and McIndoo, J. (1984). *Mol. Gen. Genet.* **196**, 28–34.
Marton, L., Wullems, G. J., Molendijk, L., and Schilperoort, R. A. (1979). *Nature (London)* **277**, 129–130.
Medgyesy, P., Fejes, E., and Maliga, P. (1985). *Proc. Natl. Acad. Sci. U.S.A.* **82**, 6960–6964.
Meins, F. (1983). *Annu. Rev. Plant Physiol.* **34**, 327–346.
Melchers, G., and Bergmann, L. (1959). *Ber. Dtsch. Bot. Ges.* **71**, 459–473.
Melchers, G., and Labib, G. (1974). *Mol. Gen. Genet.* **135**, 277–294.
Melchers, G., Sacristan, M. D., and Holder, A. A. (1978). *Carlsberg Res. Commun.* **43**, 203–218.
Menczel, L., Lazar, G., and Maliga, P. (1978). *Planta* **143**, 29–32.
Menczel, L., Galiba, G., Nagy, F., and Maliga, P. (1982). *Genetics* **100**, 487–495.
Menczel, L., Nagy, F., Lazar, G., and Maliga, P. (1983). *Mol. Gen. Genet.* **189**, 365–369.
Menczel, L., Polsby, L. S., Steinback, K. E., and Maliga, P. (1986). *Mol. Gen. Genet.* **205**, 201–205.
Menczel, L., Morgan, A., Brown, S., and Maliga, P. (1987). *Plant Cell Rep.* **6**, 98–101.
Meredith, C. P. (1984). *In* "Gene Manipulation in Plant Improvement" (J. P. Gustafson, ed.), pp. 503–528. Plenum, New York.
Meredith, C. P., and Carlson, P. S. (1982). *In* "Herbicide Resistance in Plants" (H. M. LeBaron and J. Gressel, eds.), pp. 275–291. Wiley, New York.
Michelmore, R. W., Marsch, E., Seely, S., and Landry, B. (1987). *Plant Cell Rep.* **6**, 439–442.
Mizukami, H., Konoshima, M., and Tabata, M. (1978). *Phytochemistry* **17**, 95–97.
Mok, M. C., Mok, D. W., Armstrong, D. J., Rabakoarihauta, A., and Kim, S. G. (1980). *Genetics* **94**, 675–686.
Morikawa, H., Iida, A., Matsui, C., Ikegami, M., and Yamada, Y. (1986). *Gene* **41**, 121–124.
Nagata, T., and Takebe, I. (1970). *Planta* **91**, 301–308.
Nagata, T., and Takebe, I. (1971). *Planta* **99**, 12–20.
Negrutiu, I., Cattoir-Reynaerts, A., Verbruggen, I., and Jacobs, M. (1984). *Theor. Appl. Genet.* **68**, 11–20.
Negrutiu, I., DeBrouwer, D., Dirks, R., and Jacobs, M. (1985). *Mol. Gen. Genet.* **199**, 330–337.
Negrutiu, I., Shillito, R., Potrykus, I., Biasini, G., and Sala, F. (1987). *Plant Mol. Biol.* **8**, 363–373.
Nehls, R. (1978). *Mol. Gen. Genet.* **166**, 117–118.
Nitsch, J. P., and Ohyama, K. (1971). *C. R. Hebd. Seances Acad. Sci., Ser. D* **273**, 801–804.
Ogino, T., Hiraoka, N., and Tabata, M. (1978). *Phytochemistry* **17**, 1907–1910.
Ohgawara, T., Kobayashi, S., Ohgawara, E., Uchimiya, H., and Ishii, S. (1985). *Theor. Appl. Genet.* **71**, 1–4.
Orton, T. J. (1980). *Theor. Appl. Genet.* **56**, 101–112.
Orton, T. J. (1984). *Adv. Plant Pathol.* **2**, 153–189.
Paszkowski, J., Shillito, R. D., Saul, M., Mandak, V., Hohn, T., Hohn, B., and Potrykus, I. (1984). *EMBO J.* **3**, 2717–2722.
Pelletier, G., and Chupeau, Y. (1984). *Physiol. Veg.* **22**, 377–399.
Pelletier, G., Primard, C., Vedel, F., Chetit, P., Remy, R., Rousselle, P., and Renard, M. (1983). *Mol. Gen. Genet.* **191**, 244–250.
Pental, D., Hamill, J. D., Pirrie, A., and Cocking, E. C. (1986). *Mol. Gen. Genet.* **202**, 342–347.
Potrykus, I., Paszkowski, J., Saul, M. W., Petruska, J., and Shillito, R. D. (1985a). *Mol. Gen. Genet.* **199**, 169–177.
Potrykus, I., Shillito, R. D., Saul, M. W., and Paszkowski, J. (1985b). *Plant Mol. Biol. Rep.* **3**, 117–128.
Potrykus, I., Saul, M. W., Petruska, J., Paszkowski, J., and Shillito, R. D. (1985c). *Mol. Gen. Genet.* **199**, 183–188.
Power, J. B., Cummins, S. E., and Cocking, E. C. (1970). *Nature (London)* **225**, 1016–1018.
Pua, E. C., Mehra-Palta, A., Nagy, F., and Chua, N. H. (1987). *Bio/Technology* **5**, 815–817.

Redenbaugh, K., Ruzin, S., Batholomew, J., and Bassham, J. A. (1982). *Z. Pflanzenphysiol.* **107**, 65–80.
Reich, T. J., Iyer, V. N., and Miki, B. L. (1986). *Bio/Technology* **4**, 1001–1004.
Reisch, B., and Bingham, E. T. (1980). *Plant Sci. Lett.* **20**, 71–77.
Rogers, S. G., Horsch, R. B., and Fraley, R. T. (1986). *Methods Enzymol.* **118**, 627–640.
Schäfer, W., Görz, A., and Kahl, G. (1987). *Nature (London)* **327**, 529–532.
Schieder, O. (1976). *Mol. Gen. Genet.* **149**, 251–254.
Schieder, O. (1978). *Mol. Gen. Genet.* **162**, 113–119.
Schieder, O. (1980). *Z. Pflanzenphysiol.* **98**, 119–127.
Schocher, R. J., Shillito, R. D., Saul, M. W., Paszkowski, J., and Potrykus, I. (1986). *Bio/Technology* **4**, 1093–1096.
Shah, D. M., Horsch, R. B., Klee, H. J., Kishore, G. M., Winter, J. A., Tumer, N. E., Hironaka, C. M., Sanders, P. R., Gasser, C. S., Aykent, S., Siegel, N. R., Rogers, S. G., and Fraley, R. T. (1986). *Science* **233**, 478–481.
Shahin, E. A., and Spivey, R. (1986). *Theor. Appl. Genet.* **73**, 164–169.
Shahin, E. A., Spielmann, A., Sukhapinda, K., Simpson, R. B., and Yashar, M. (1986). *Crop Sci.* **26**, 1235–1239.
Sharp, W. R., Evans, D. A., Ammirato, P. V., and Yamada, Y. (eds.) (1984). "Handbook of Plant Cell Culture. Vol. 2: Crop Species." Macmillan, New York.
Shillito, R. D., Saul, M. W., Paszkowski, J., Müller, M., and Potrykus, I. (1985). *Bio/Technology* **3**, 1099–1103.
Somers, D. A., Narayanan, K. R., Kleinhofs, A., Cooper-Bland, S., and Cocking, E. C. (1986). *Mol. Gen. Genet.* **204**, 296–301.
Stachel, S. E., Messens, E., Van Montagu, M., and Zambryski, P. (1985). *Nature (London)* **318**, 624–629.
Stachel, S. E., Nester, E. W., and Zambryski, P. (1986). *Proc. Natl. Acad. Sci. U.S.A.* **83**, 379–383.
Stalker, D. M., Hiatt, W. R., and Comai, L. (1985). *J. Biol. Chem.* **260**, 4724–4728.
Steinrücken, H. C., Schulz, A., Amrhein, N., Porter, C. A., and Fraley, R. T. (1986). *Arch. Biochem. Biophys.* **244**, 169–178.
Takebe, I., Labib, G., and Melchers, G. (1971). *Naturwissenschaften* **58**, 318–320.
Trulson, A. J., Simpson, R. B., and Shahin, E. A. (1986). *Theor. Appl. Genet.* **73**, 11–15.
Tsukakoshi, M., Kurata, S., Nomiya, Y., Ikawa, Y., and Kasuya, T. (1984). *Appl. Phys. B* **35**, 135–140.
Tumer, N. E., O'Connell, K. M., Nelson, R. S., Sanders, P. R., Beachy, R. N., Fraley, R. T., and Shah, D. M. (1987). *EMBO J.* **6**, 1181–1188.
Umbeck, P. F., and Gengenbach, B. G. (1983). *Crop Sci.* **23**, 584–588.
Umbeck, P., Johnson, G., Barton, K., and Swain, W. (1987). *Bio/Technology* **5**, 263–266.
Vaeck, M., Reynaerts, A., Höfte, H., Jansens, S., DeBeuckeleer, M., Dean, C., Zabeau, M., Van Montagu, M., and Leemans, J. (1987). *Nature* **328**, 33–37.
Vasil, I. K. (1982). *In* "Plant Improvement and Somatic Cell Genetics" (I. K. Vasil, W. R. Scowcroft, and K. J. Frey, eds.), pp. 179–203. Academic Press, New York.
Vasil, I. K. (1984). "Cell Culture and Somatic Cell Genetics of Plants. Vol. 1: Laboratory Procedures and Their Applications." Academic Press, Orlando, Florida.
Vienken, J., Ganser, R., Hampp, R., and Zimmerman, U. (1981). *Physiol. Plant.* **53**, 64–70.
Wakasa, K., and Widholm, J. M. (1987). *Theor. Appl. Genet.* **74**, 49–54.
Walbot, V., and Cullis, C. A. (1985). *Annu. Rev. Plant Physiol.* **36**, 367–396.
Wallin, A., Grimelius, K., and Eriksson, T. (1974). *Z. Pflanzenphysiol.* **74**, 64–80.
Watts, J. W., and King, J. M. (1984). *Biosci. Rep.* **4**, 335–342.
Weissinger, A., Tomes, D., Sanford, J., Kline, T., and Fromm, M. (1987). *In Vitro Cell. Dev. Biol.* **23**, 75A.
White, D. W. R., and Greenwood, D. (1987). *Plant Mol. Biol.* **8**, 461–469.
White, D. W. R., and Vasil, I. K. (1979). *Theor. Appl. Genet.* **55**, 107–112.

Wullems, G. J., Molendijk, L., Ooms, G., and Schilperoort, R. A. (1981). *Proc. Natl. Acad. Sci. U.S.A.* **78,** 4344–4348.
Yamada, Y., Yang, Z. Q., and Tang, D. T. (1986). *Plant Cell Rep.* **5,** 85–88.
Yamamoto, Y., Mizuguchi, R., and Yamada, Y. (1982). *Theor. Appl. Genet.* **61,** 113–116.
Yanofsky, M., Lowe, B., Montoya, A., Rubin, R., Krul, W., Gordon, M., and Nester, E. (1985). *Mol. Gen. Genet.* **201,** 237–246.
Yarrow, S. A., Wu, S. C., Barsby, T. L., Kemble, R. J., and Shepard, J. F. (1986). *Plant Cell Rep.* **5,** 415–418.
Zelcer, A., Aviv, D., and Galun, E. (1978). *Z. Pflanzenphysiol.* **90,** 397–407.
Zenk, M. H. (1978). *In* "Frontiers of Plant Tissue Culture" (T. A. Thorpe, ed.), pp. 1–13. University of Calgary, Alberta, Canada.

Index

A

AC element, in transposon tagging, 126–128
N-Acetylglucosamine, 516
Actin, 435
  -binding proteins, 437–441
  G-Actin, 436
  properties, 436
  structure, 467
  types, 434
ADH, *see* Alcohol dehydrogenase
Agrobacteria
  chromosomal traits, 633
  genetic manipulation of cells, 672
  molecular basis of host range, 631
  role of opines, 633
Agrobacterium rhizogenes, tumor-inducing plasmid, 629
Agrobacterium tumefaciens
  causative agent of crown gall disease, 620
  expression of cloned seed storage protein genes, 314
  tumor-inducing plasmid, 621–622
Alcohol dehydrogenase, 346
Alga, hydroxyproline contents, 487
Amino acid
  composition of various AGPs, 509
  globulin storage proteins, 302–304
  sequence
    carrot AGPs, 509

carrot HRGPs, 492
hydroxyproline-rich glycoproteins, 486
intermediate filament proteins, 444–445
tomato HRGP, 494
tubulin monomers, 403–408
tRNA, 202
wheat prolamine sequences, 320
Aminoacyl-tRNA synthetases, 202
Aminoacylation, 173
  of tRNA, 199
Anaerobiosis, 362–367
Anthocyanin biosynthesis, 124, 125
Antibiotic resistance, 204, 207
Anti-calmodulin labeling, 461
Antirrhinum majus, see Snapdragon
Arabidopsis, actin genes, 435, 436
Arabidopsis thaliana, genome size, 10
Arabinoglactan proteins, 486
  biosynthesis, 512
  distribution, 506, 512
  function, 512
  structure, 507–511
Ascomycetes, small linear DNA replicons, 241

B

Bacteria
  insertion sequence elements, 87
  transposons elements, 87
  Tn3, 88–89

689

Barley
  anaerobic stress, 365
  prolamine storage proteins, 321–322, 330–332
Barley stripe mosaic virus, 104
Brassica hirta
  genome organization, 236
  small linear DNA replicons, 241
Brassica oleracea, see Cauliflower
Broad bean
  globulin storage proteins, 301
  plasmid forms, 240

C

Ca^{2+} ATPase, regulation of calcium levels, 476
Cadmium-induced proteins, 368–369
Calmodulin, 456
  -binding proteins, 464–466
  cell division, 472
  in chromosome movement, 461
  concentration in plants, 459, 463
  -dependent processes, 466–477
  -dependent proteins, 464
  function, 458
  kinetochore-to-pole MT depolymerization, 473
  localization, 460
  microtubule polymerization, 471
  NAD kinase, 469
  processes regulated by, 458–463
  structure, 456–458
Capping proteins, 439
Carrot
  amino acid sequence of AGPs, 509
  cell wall HRGPs, 486, 493
  tubulin synthesis during somatic embryogenesis, 433
Cauliflower, chloroplast sequences in mitochondrial genomes, 256
Caulimoviruses, 563–565
  biology and features, 566
  genetic organization, 573
  host ranges and, 566
  insect transmission, 591–593
  mapping of gene functions, 579–582
  properties of, 567–573
  replication, 582–588
  reverse transcriptase, 588–591
  transcription and translation, 573–579

cDNA hybridization analysis, 24
Cereals, prolamine storage proteins, 316
Chartins, 423
Chelation, 476
Chimeric genes, 270
Chlamydomonas reinhardtii
  cell wall HRGPs, 504
  chloroplast genome, 136
  α-tubulin isoform, 416
  chloroplast promoter regions, 164
  chloroplast RNA polymerase, 155
  gel electrophoresis of ribosomal proteins, 206
  open reading frames (ORF), 177
  rDNA, 140
  tubulin mutants resistant to colchicine, 420–421
  tubulin synthesis during flagellar outgrowth, 431
Chloroplasts
  codons, 212–217
  DNA, 134–139
  gene expression, 218–219
  genes, 138–139
  introns, 176–178
  mRNA, 174–176
    structure, 195–198
    trans-splicing, 178–179
  promoter function, 163
  promoter/protein interactions, 165
  promoter regions
    identification, 159
    of protein genes, 161–163
    rrn operon, 161
    tRNA, 160
  protein maturation, 217–218
  protein synthesis, 194
  ribosomes, 204, 206
    antibiotic resistance, 207
    elongation, 211
    initiation, 210–211
    in protein synthesis, 210–211
    proteins, 142, 206
    RNA, 205
    termination, 211
    translation factors, 207–210
  RNA polymerase, 143
    E. Coli, 155
    promoter regions, 159–165
    properties, 153–155
    purification, 152
    rRNA, 140, 169–171

Index

transcriptionally active chromosome, 153–154, 165
tRNA, 141, 171–174, 199–204
transcription, 152–169
  in vitro, 156
  in organello, 157
  run-on, 158
  soluble, 157
  termination, 165–169
    inverted repeat sequence, 165–167
    mRNA processing and, 167–169
Circular dichroism measurements, 320
Circular plasmid DNAs, 239
  oligomeric forms, 240
Cis-acting control elements
  regulation of gene expression, 4, 51
  of RbcS and LHCP genes, 55
  in seed protein, 59
  structural gene sets in eukaryotic organisms and, 62
Cis-splicing, 178
Cleavage
  in *E. coli,* 170
  myosin, 438
  tubulin monomers, 410
Cloning of plant regulatory gene loci, 67
Codons
  mitochondrial genes, 266
  recognition in chloroplasts, 213–216
  usage in chloroplasts, 212
  usage in HRGP cDNA, 495
Colchicine, binding in tubulins, 417–420
Complex reassociation kinetics, 6
Conglycinin, 298–300
Copper-induced proteins, 368–369
Cortical array, 397
Cotton, globulin storage proteins, 301
Crown gall tumors, 620–637
  Agrobacterium tumefaciens, 621, 622–623
  Agrobacterium, molecular basis of host range, 632
  genetic engineering application, 634
  history and background, 618–620
  host range, 632–633
  opines, 633
  T region, 625
  vir region, 623–624
Cytochrome, 145
Cytoplasmic mRNA complexity, 41
Cytoplasmic regulation of plant gene expression, 49

Cytoplasmic streaming, 467
Cytoskeleton, 394
  dynamics of, 396–398
  intermediate filaments, 396
  microfilaments, 396
  microtubules, 395
  proteins, 394

D

Datura stramonium, solanaceous lectins, 513
Desmin, intermediate filament protein, 442
Dicotyledonous plants, chloroplast protein genes, 161
DNA
  reassociation kinetics, 5–6
    and genome complexity, 5
    and genome size, 6, 7, 8
  viruses, 565
  use as gene vectors, 606–610
DNA-DNA hybridization studies, 258
Drosophila melanogaster
  heat-shock proteins, 348–349
  tubulin genes, 415
Drought, stress induced proteins, 359–362
Dynamic instability, 426–427
Dynein ATPase, 476

E

Elongation, protein synthesis in chloroplasts, 211
En/Spm element
  changes in phase, 117
  changes of state, 114
  differential effects of, 113–114
  in transposon tagging, 124–126
  transposition in plants, 118
Endosymbiont hypothesis, 252
Erythrocyte, β-tubulins expressed, 415
Escherichia coli
  codon usage, 212
  ribosomes, 206, 207
  thaumatin expression, 385
  transcription termination, 166
Euglena
  chloroplast genome, 137
  chloroplast RNA polymerase, 153
  introns, 177, 199
  rDNA, 140
  tRNA genes, 142

Eukaryotic organisms
  cis-elements and trans-factors of structural gene sets, 62
  regulation of gene expression, 40
  regulation of gene networks, 64
Evening primrose (oenothera berteriana) chloroplast sequences in mitochondrial genomes, 256
Extensins, 486

F

Fascin, 440
Fibrous actin, 436
Filamin, 440
Fimbrin, 440
Flowering plant, life cycle, 2
Fluphenazine, 468
Fodrin, 440
Fragmin actin-binding proteins, 440

G

Gametophyte, gene activity in, 37
Gelsolin-villin actin-binding proteins, 439
Geminiviruses, 563–565
  biology and features, 593
  genetic organization, 598–605
  properties, 595–597
Gene activity
  in the male gametophyte, 37
  in the sporophyte, 29–30
Gene activity measurement, 16
  RNA/DNA hybridization kinetics, 16–18
  RNA/single-copy DNA saturation, 24–26
Gene copy number, 13–14, 274
Gene expression
  alteration during heat stress, 348–358
  cis-acting elements, 51–63
  cold temperature induced, 358–359
  control elements, 51–57
  cytoplasmic regulation, 49
  in eukaryotic organisms, 62–63
  heat shock, 57–58
  hormonally regulated, 48
  housekeeping genes, 33
  light regulated, 51
  mRNA population, 11–12
  mRNA sequences in plant cells, 34
  number expressed during development, 38
  organ-specific mRNA sets, 33–34
  during plant ontogeny, 8
  posttranscriptional control, 43, 184, 219
  regulation of, 36, 39, 43–45
    seed protein genes, 58–62
    specific genes, 45–48
  regulation of globulin storage proteins, 313
  regulation of prolamines, 334–336
  RNA/DNA hybridization, 4, 16–19, 24–25
  trans-acting factors, 70
    identification, 63–70
    molecular analysis of, 70–72
  transcriptional control, 43, 184
  transposable elements, 112
Gene tagging, 123–128
  in Drosophila, 124
  in maize, 124
Genetic analysis, 66
Genetic manipulation, 653
  cell selection, 654
    mutagenesis, 658
    screening for variants, 655
    variant characterization, 659
    variation in cell cultures, 660
  direct gene transfer, 674
  protoplast fusion, 664
    cybrids, 666
  transformation, 671
    Agrobacterium-mediated, 672
    comparison of methods, 676
    direct gene transfer, 674
Genetic tumors, 639
Genome complexity, 5, 8–10
Genome sequences, 258
Genome size, 6, 8–10
  determination of size, 7
  plant and animal size ranges, 8
  repeated sequence copy number, 7
Genomes, 4–6, 12–13
  Arabidopsis thaliana, 10
  Drosophila melanogaster, 8
  Escherichia coli, 6
Glial cell intermediate filament protein, 443
Globulin storage proteins
  amino acid sequences, 302–304
  gene structure, 311
  glycosylation, 307
  Golgi apparatus in globulin transport, 309
  peptide insertions, 304
  proteolytic processing, 308

Index   693

regulation of gene expression, 313
synthesis, 304–310
Globular actin, 436
  polymerization, 436–437
Globulin storage proteins, 298
  soybean, 299–300
  structure, 302–304
  in various species, 301
Glycinin, 299
Glycosylation, 218
  in globulin storage proteins, 307
Golgi apparatus, in globulin transport, 305, 309
Gravitropism, 469
GTP hydrolysis, 425

H

Habituated plant tissues, 638
Heat shock gene expression, 57–58
  soybean, 58
Heat-shock proteins, 348–359
  and HRGPS, 503
Heavy meromyosin, 438
Heavy metals
  protein and peptide synthesis, 368–369
Hormonally regulated gene expression, 48
Housekeeping genes, 33
HRGP role in biological stress, 371
HRGP, see Hydroxyproline-rich glycoproteins
HSP, see Heat Shock Proteins
Hydroxyproline, 485
Hydroxyproline-rich glycoproteins, 348, 371, 486
  and heat-shock proteins, 503
  biosynthetic pathways, 498–500
  comparison to animal collagen, 500
  function in dicots, 500
  in nondicotyledonous species, 504–506
  response to stress and wounding, 501–504
  structure, 486–497

I

Integration sites
  analysis of, 120
Intermediate filament proteins, 441–446
  amino acid sequence, 444
  solubility, 443
Intermediate filaments, 396

Intracellular motility, 467
Introns, 198, 273
  in *Euglena*, 199
  in nuclear and cytoplasmic mRNA complexity, 41
  open reading frames, 177
  sequences, 176

K

Keratin, intermediate filament protein, 442
Kinesin, 423
Kluyveromyces lactis
  expression of thaumatin, 388

L

Lamins, 443
Lectins, *see* Solanaceous lectins
Life cycle of flowering plant, 2
Light-harvesting chlorophyll a/b/ binding protein, (LCHCP), 55
  regulation of gene expression, 56
Light-regulated genes, 48, 51
Liverwort, chloroplast genome, 230
Loci, cloning of plant regulatory gene, 67
Lyase activity, ubiquitin-protein, 529

M

Maize
  actin genes, 435, 436
  alcohol dehydrogenase (Adhl) gene, 102
  barley stripe mosaic virus (BSMV), 104
  chloroplast promoter regions, 164
  chloroplast mRNA, 174
  chloroplast RNA polymerase, 152
  chloroplast rRNA, 140
  chloroplast sequences in mitochondrial DNA, 253–255
  chloroplast transcription, 157
  cytoplasm, 246–248
  double- and single-stranded RNAs, 249
  episomes, 243–246
  gene fragments, 273
  gene tagging, 124
  genome reorganization, 285
  anaerobic stress, 362–367
  heat-shock proteins, 349, 356
  inactive genes, 272
  introns, 273
  master circles, 236

plasmid forms, 240, 248
prolamine storage proteins, 323, 332–334, 335
retroviral elements, 104
sequenced ribosomal RNA genes, 261
  chimeric genes, 270
shrunken (Shl) gene, 92
transposable elements, 83–86, 89, 90
  enhancer (En), 94
  molecular isolation of, 90
  mutator (Mu) system, 102–104
  waxy (Wx) locus, 90, 117
Marchantia polymorpha, see Liverwort
Master circles, 236
Meiosis, and genome size, 10
Messenger RNA abundance classes, 19
Microfilaments, 396
Microtubule proteins, 399–400
Microtubules, 395
Mitochondrial coding sequences, 260
  chimeric genes, 270–271
  introns, 273
  open reading frames, 269
  polypeptide genes, 266–269
  pseudogenes, 272
  rRNA, 261
  tRNA, 261–263
Mitochondrial DNA, 251–253
  base composition, 230
  chloroplast sequences, 253–258
  genetic complexity assessment, 258
  genomes
    composition, 233
    size, 232
  large circular species, 233–238
  linear replicons, 241–243
  small circular species, 239–241
Mitochondrial genomes
  conserved sequences, 258
  cytoplasm of maize, 246–248
  episomes of maize, 243–246
  gene copy number, 274–276
  plasmid of maize, 248
  protein synthesis, 259
  reorganizations, 283–286
  ribosome binding, 281–283
  RNA, 249
  transcription, 274, 277–280
Mitosis, and genome size, 10
Mitotic apparatus, localization of calmodulin, 460
Mitotic spindle, 398

Molecular analysis of *trans*-acting regulatory factors, 70
Monocots, HRGP evidence, 506
Monocotyledonous plants, chloroplast protein genes, 161
mRNA stability and turnover, 49
mRNA transcripts
  abundance classes, 19–23
  5' processing of termini, 174
  regulation in the nucleus, 45
mRNA/cDNA hybridization kinetics, 18–19
Mung Bean (*phaseolus aureaus*)
  chloroplast sequences in mitochondrial genomes, 256
Mustard, open reading frames, 177
Mutagenesis
  in genetic manipulation of cells, 658
  of infectious viroid cDNAs, 553–557
Mutator function, 112
Myosin, 438

N

N-terminal methionine, 217
NADPH-protochlorophyllide oxidoreductase, 524
Neuroblastoma cells, 432
Neurofilament proteins, 442
Nicotiana tabacum, see Tobacco
Nitella, presence of myosin, 441
Nitrate reductase, 522
Nivea recurrens, 105
Nonfunctional genes, 272
Nuclear mRNA complexity, 41
Nuclear transcripts in sporophytic organs, 44
Nucleotide sequence
  of carrot HRGP, 493
  viroids, 541–544

O

Oat, globulin storage proteins, 301
Oenothera berteriana
  gene fragments, 273
  sequenced ribosomal RNA genes, 261
  genome organization, 238
Oligomeric plasmid forms, 240
Onion root cells, actin binding proteins, 440–441

Index

Open reading frames, 177
  mitochondrial genomes, 269
Opines, role in plant transformation by agrobacteria, 633

P

Pea
  anaerobic stress, 365
  calmodulin concentration, 459
  chloroplast genome, 134
  chloroplast RNA polymerases, 153
  chloroplast transcription, 157
  globulin storage proteins, 301
    synthesis, 308
  globulin transport, 309
  heat-shock proteins, 349
  introns, 273
  mitochondrial genome size, 232
  5' processing of mRNA termini, 174
  rDNA, 140
  regulation of globulin gene expression, 313, 314
Peanut, globulin storage proteins, 301
Petunia hybrida, genome organization, 238
Phaseolin, 302
Phaseolus vulgaris, globulin storage proteins, 301
Phenothiazines, 465
Photosynthetic apparatus
  ATP synthase, 145
  cytochrome b/f complex, 145
  photosystem genes, 144–145
  RuBisCo, 143
Phragmoplast, 398
Phytoalexins, 371
Phytochrome, function in protein degradation, 522
Plant development, 2
  differential gene activity, 28
    sporophyte, 29–32
Plastids
  in higher plants, 179
  posttranscriptional regulation, 183–184
  regulation of gene expression, 179–185
  regulation of plastid RNA levels, 181–183
  run-on, 158, 182
  transcriptional regulation, 179–181
Polar tip growth, 468
Polycistrons, processing, 175
Polypeptide genes, 266
Posttranscriptional processes, 43–47
Posttranscriptional regulation of plant gene expression, 43, 46
Potato
  early studies of viroids, 539
  solanaceous lectins, 513
Preprophase band, 397
Prolamine genes
  barley, 330–332
  expression, 334–336
  maize, 332–334
  wheat, 330–332
Prolamine storage proteins
  barley, 321–322
  cereals, 316, 327–330
  maize, 323–326, 332–334
  organization and structure, 330–332
  regulation of gene expression, 334–336
  synthesis, 327
  wheat, 317–321
Promoter regions, 160
  protein interactions, 165
  rrn operon, 161
Protein phosphorylation, 466
Protein(s)
  actin binding, 437–441
  binding, 165
  calmodulin dependent, 456
  degradation, 522
    functions, 522–525
    measurement of turnover rate, 524
    mechanisms of degradation, 525–533
    pathways in organelles, 531–533
    senescence, 525
    ubiquitin, 527–531
  intermediate filaments, 441–446
  maturation of, 217–218
  microtubule-associated, 421–424
  ribosomal, 142
  stress-induced
    anaerobic, 362–367
    biological stress, 369–372
    cold temperatures, 358–359
    drought and salt, 359–362
    heat, 348–359
    heavy metals, 368–369
    ultraviolet light exposure, 367–368
  synthesis, 193, 210–211, 259
  translation factors, 208
Proteolysis
  function in protein degradation, 523
  processing of globulin storage proteins, 308

ubiqutin-dependent pathways in protein degradation, 527
vacuole proteolytic activities, 531–532
Protoplast fusion, 664
  isolation and fusion, 664
  organelle segregation and recombination, 668–670
  somatic hybrids, 665–668
Pseudogenes, 272

R

Regulation of gene expression, 3–4, 39–40
  cis-acting control elements, 4, 51
  cytoplasmic, 49
  eukaryotic cell, 40, 64
  gene networks, 64
  globulin storage proteins, 313
  LHCP, 55
  mRNA transcripts, 45
  plastids, 179–185
  posttranscriptional, 43
  prolamines, 334–336
  quantitative, 35
  rbcS, 55, 56
  trans-acting regulatory factors, 4
  transcriptional processes, 43–47
    hormonally regulated genes, 48
    in vitro analysis, 47
    LHCP, 56
    light-regulated genes, 48, 51
    RbcS, 56
    seed protein genes, 47
Repeated sequence copy number, 7
Replicons, 241
Restriction fragment analysis, 233
Ribosomal proteins, 142
Ribosomal RNA
  genes, 140
  processing, 169
  promoter regions, 161
Ribosome binding sites, 198
Ribulose bisphosphate carboxylase, 194, 197
Ribulose-1,5-bisphosphate carboxylase/oxygenase, 143
Ribulose-biphosphate carboxylase (rbcS), 55
  regulation of gene expression, 56
Rice
  introns, 273
  anaerobic stress, 365

RNA
  double- and single-stranded, 249
  nuclear vs. cytoplasmic complexity, 41–42
  processing, see Chloroplast, RNA processing
  RNA-DNA hybridization, 4, 16–19, 24–25, 258
  mRNA abundance classes, 19–23
  mRNA complexity, 25–28, 41
  sporophyte activity, 29–30
  tobacco leaf, 31–32

S

Saccharomyces cerevisiae
  tubulin forms, 415
Salt-stress induced proteins, 359–362
Seed protein genes, 47
  gene regulation, 58–62
  embryo-specific cis-control elements, 59
Seed storage proteins, 297
  globulin, 298–304
  prolamines, 316–330
Senescence, role in protein degradation, 525
Sequence complexity of mRNA population, 26
Serine proteinase inhibitors, 369–372
Severin actin-binding proteins, 440
Shine-Dalgarno sequences, 197, 220, 281
Snapdragon
  tam elements in, 105–107
  transposon tagging, 127
Solanaceous lectins
  biosynthesis and degradation, 515
  cellular and tissue distribution, 513
  distribution, 516
  function, 516
  structure, 513
Somaclomal variation, 284
Somatic embryogenesis, tubulin synthesis, 432
Sorghum
  anaerobic stress, 365
  small linear DNA replicons, 241
Soybean
  actin genes, 435, 436
  anaerobic stress, 363
  β-tubulin genes, 402
  globulin gene structure, 312
  heat-shock proteins, 349

Index

sequenced ribosomal RNA genes, 261
storage globulins, 299–316
storage proteins, 298
transposable elements, 108
variegation in, 108
Spinach (*Spinacea oleracea*)
   base modification, 174
   chloroplast promoter regions, 164
   chloroplast RNA polymerases, 153
   chloroplast sequences in mitochondrial genomes, 256
   gene copy number, 274
   5′ processing of mRNA termini, 174
   tRNA promoter regions, 160
Spm
   dependent alleles, 115–116
   presetting phenomena, 116
   suppressible alleles, 112–115
Sporophyte
   gene activity, 29
   nuclear transcripts, 44
Stem-loop structures, 279
Stoichiometric relationships, 233, 238
Storage globulins, 304–310
   gene expression, 313–316
   genes, 311–313
Streptomyces rochei, small linear DNA replicons, 241
Stress, ubiquitin-dependent proteolysis, 531
Structural genes, total number expressed during plant development, 38
Superabundant class transcripts, 22
Sucrose synthase gene, 92–93
Sugar beet
   gene copy number, 274
   master circles, 236
   plasmid forms, 240
Sunflower
   introns, 273
   oligomeric plasmid forms, 240
   storage globulins, 302

T

Tau factor, 422
Tektins, 443
Temperature
   cold stability of plant microtubules, 429
   cold temperature induced gene expression, 358–359
   heat stress, 348–358

Termination of protein synthesis in chloroplasts, 211
Thaumatins, 379
   biochemistry, 382–383
   expression in *E. coli,* 385
   expression in yeasts, 385–388
   isolation and characterization, 380–381
   molecular genetics, 383–384
   physiology, 382–383
   production of, 385–389
Thaumatococcus danielli, 389
Threonyl-tRNA, 201
Tissue cultures, hydroxyproline contents, 488
Tobacco
   chloroplast genome, 135, 230
   codon recognition, 213
   gene activity in sporophytic organ, 30
   gene expression in, 38
   heat-shock proteins, 349
   open reading frames, 177
   tRNA, 141, 213
Tomato
   actin genes, 436
   cell wall HRGP, 487, 492, 494
   heat-shock proteins, 349
   isolation of myosin, 441
   role of HRGP in wounded stem, 502
Tracheary element differentiation, 433
Trans-acting regulatory factors, 63
   regulation of gene expression, 4
   molecular analysis, 70
   and structural gene sets in eukaryotic organisms, 62
Transcription, 277, *see also* Chloroplast, transcription
   in organello, 157
   in vitro, 47, 156–159
   light-regulated genes, 48
   regulation of plant gene expression, 43–48
   termination, 165–169
Transfer RNA genes, 141
Transposable elements
   Ac/Ds system, 91–94
   effects on gene expression, 112–117
   En/Spm, 94
   gene tagging, 123–128
   in maize, 83–86, 89, 90, 94, 122
   in snapdragon, 105
   in soybean, 108
   molecular decription, 98–101

molecular isolation of, 90-92
mutator (Mu) system, 102-104
origin, 122
retroviral-like elements, 104
role in evolution, 120-121
suppressor-mutator (Spm), 94
Tam, 105-107
Tgm insertions, 108-111
Transposition
  excision and integration, 119
  models, 117
Transposon mutagenesis
  cloning of gene loci, 67
  tagging, 68
tRNA
  gene organization, 171
  nucleotidyltransferase activity, 173
  precursor molecules, 172
  processing, 171
Tropomyosin, 440
Troponin, 440
Tubulin
  amino acid sequence, 403-408
  assembly *in vitro*, 427
  autoregulation, 430
  colchicine, 417-419
  cold stability, 429
  dimer pool, 430
  domains, 410-414
  encoding of, 401-402
  flagellar outgrowth, 431
  formation in neuroblastoma cells, 432
  GTP hydrolysis, 425
  isoforms, 414
  isolation, 399
  isotypes, 414
  microtubule-associated proteins, 421-424
  monomers, 400-401
  mutants, 420-422
  pseudogenes, 401
  self-assembly, 424-429
  synthesis, 430-434
  vinblastine, 430
Tumors, 617
  crown gall, 618-636
  genetic, 638
  virus-induced, 636-638
  wound tumor virus, 636-638
Tyrosination/detyrosination, 416

U

Ubiquitin
  function in protein degradation, 527
  other functions, 531
Ultraviolet light exposure, 367-368

V

Variegation, 84, 108
Villin, 440
Vimentin, 442
Vir gene complex, 623
Viroids, 537-557
  biochemistry of, 538-539
  functions, 544-552
  infectivity of viroid cDNAs, 552-555
  in vivo structure, 545
  molecular structure, 539-540
  mutational analysis, 555
  nucleotide sequence, 540, 541-544
  pathogenicity, 557
  replication, 546-552
    enzymes involved, 547-548
  translation, 545-546
Virus-induced tumors of plants, 636-638

W

Wheat
  introns, 273
  prolamine storage proteins, 317-321, 330-332
Wound tumor virus, 638
  composition, 638
Wounding, role of HRGPs in plant defense, 501

Y

Yeast
  thaumatin expression, 385-388
  ubiquitin-dependent proteolysis, 527-528
  role in stress, 531

Z

Zein, prolamine storage proteins, 324, 332-334

Contents of Other Volumes

Volume 1—The Plant Cell

1. The General Cell
    Eldon H. Newcomb
2. Use of Plant Cell Cultures in Biochemistry
    Paul Ludden and Peter S. Carlson
3. The Primary Cell Walls of Flowering Plants
    Alan Darvill, Michael McNeil, Peter Albersheim, and Deborah P. Delmer
4. The Plasma Membrane
    Robert T. Leonard and Thomas K. Hodges
5. The Cytosol
    Grahame J. Kelly and Erwin Latzko
6. Development, Inheritance, and Evolution of Plastids and Mitochondria
    Jerome A. Schiff
7. Biochemistry of the Chloroplast
    Richard G. Jensen
8. Plant Mitochondria
    J. B. Hanson and D. A. Day
9. Microbodies—Peroxisomes and Glyoxysomes
    N. E. Tolbert
10. The Endoplasmic Reticulum
    Maarten J. Chrispeels
11. Ribosomes
    Eric Davies and Brian A. Larkins
12. The Golgi Apparatus
    Hilton H. Mollenhauer and D. James Morré
13. The Plant Nucleus
    E. G. Jordon, J. N. Timmis, and A. J. Trewavas
14. Protein Bodies
    John N. A. Lott

15. Plant Vacuoles
    Francis Marty, Daniel Branton, and Roger A. Leigh
16. Cyanobacteria (Blue-Green Algae)
    C. Peter Wolk

Index

Volume 2—Metabolism and Respiration

1. Assessment of the Contributions of Metabolic Pathways to Plant Respiration
   T. ap Rees
2. Enzyme Flexibility as a Molecular Basis for Metabolic Control
   Jacques Ricard
3. Direct Oxidases and Related Enzymes
   V. S. Butt
4. Electron Transport and Energy Coupling in Plant Mitochondria
   Bayard T. Storey
5. Nature and Control of Respiratory Pathways in Plants: The Interaction of Cyanide-Resistant Respiration with the Cyanide-Sensitive Pathway
   David A. Day, Geoffrey P. Arron, and George G. Laties
6. Control of the Krebs Cycle
   T. Wiskich
7. The Regulation of Glycolysis and the Pentose Phosphate Pathway
   John F. Turner and Donella H. Turner
8. Hydroxylases, Monooxygenases, and Cytochrome P-450
   Charles A. West
9. One-Carbon Metabolism
   Edwin A. Cossins
10. Respiration and Senescence of Plant Organs
    M. C. J. Rhodes
11. Respiration and Related Metabolic Activity in Wounded and Infected Tissues
    Ikuzo Uritani and Tadashi Asahi
12. Photorespiration
    N. E. Tolbert
13. Effects of Light on "Dark" Respiration
    Douglas Graham
14. Anaerobic Metabolism and the Production of Organic Acids
    David D. Davies
15. Effect of Low Temperature on Respiration
    John K. Raison
16. The Use of Tissue Cultures in Studies of Metabolism
    D. K. Dougall

Index

Volume 3—Carbohydrates: Structure and Function

1. Integration of Pathways of Synthesis and Degradation of Hexose Phosphates
   T. ap Rees
2. myo-Inositol: Biosynthesis and Metabolism
   Frank A. Loewus and Mary W. Loewus
3. L-Ascorbic Acid: Metabolism, Biosynthesis, Function
   Frank A. Loewus
4. Sugar Nucleotide Transformations in Plants
   David Sidney Feingold and Gad Avigad

Contents of Other Volumes

5. Branched-Chain Sugars: Occurrence and Biosynthesis
   Hans Grisebach
6. Biosynthesis and Metabolism of Sucrose
   Takashi Akazawa and Kazuo Okamoto
7. Occurrence, Metabolism, and Function of Oligosaccharides
   Otto Kandler and Herbert Hopf
8. Translocation of Sucrose and Oligosaccharides
   Robert T. Giaquinta
9. Structure and Chemistry of the Starch Granule
   W. Banks and D. D. Muir
10. Starch Biosynthesis and Degradation
    Jack Preiss and Carolyn Levi
11. Conformation and Behavior of Polysaccharides in Solution
    David A. Brant
12. Chemistry of Cell Wall Polysaccharides
    Gerald O. Aspinall
13. Structure and Function of Plant Glycoproteins
    Derek T. A. Lamport
14. The Biosynthesis of Cellulose
    J. Ross Colvin
15. Glycolipids
    Alan D. Elbein
16. Biosynthesis of Cell Wall Polysaccharides and Glycoproteins
    Mary C. Ericson and Alan D. Elbein

Index

Volume 4—Lipids: Structure and Function

1. Plant Acyl Lipids: Structure, Distribution, and Analysis
   J. L. Harwood
2. Membrane Lipids: Structure and Function
   John K. Raison
3. Degradation of Acyl Lipids: Hydrolytic and Oxidative Enzymes
   T. Galliard
4. The Role of the Glyoxylate Cycle
   Harry Beevers
5. Lipoxygenases
   T. Galliard and H. W.-S. Chan
6. Biosynthesis of Ethylene
   S. F. Yang and D. O. Adams
7. Biosynthesis of Saturated and Unsaturated Fatty Acids
   P. K. Stumpf
8. The Biosynthesis of Triacylglycerols
   M. I. Gurr
9. Phospholipid Biosynthesis
   J. B. Mudd
10. Phospholipid-Exchange Systems
    Paul Mazliak and J. C. Kader
11. Sulfolipids
    J. L. Harwood
12. Plant Galactolipids
    Roland Douce and Jacques Joyard

13. Biochemistry of Terpenoids
    W. David Loomis and Rodney Croteau
14. Carotenoids
    Sandra L. Spurgeon and John W. Porter
15. Biosynthesis of Sterols
    T. W. Goodwin
16. Sterol Interconversions
    J. B. Mudd
17. Biosynthesis of Acetate-Derived Phenols (Polyketides)
    N. M. Packter
18. Cutin, Suberin, and Waxes
    P. E. Kolattukudy
19. Biosynthesis of Cyclic Fatty Acids
    H. K. Mangold and F. Spener

Index

Volume 5—Amino Acids and Derivatives

1. Biochemistry of Nitrogen Fixation
    M. G. Yates
2. Ultrastructure and Metabolism of the Developing Legume Root Nodule
    J. G. Robertson and K. J. F. Farnden
3. Nitrate and Nitrite Reduction
    Leonard Beevers and Richard H. Hageman
4. Ammonia Assimilation
    B. J. Miflin and P. J. Lea
5. Assimilation of Inorganic Sulfate into Cysteine
    J. W. Anderson
6. Physical and Chemical Properties of Amino Acids
    Peder Olesen Larsen
7. Enzymes of Glutamate Formation: Glutamate Dehydrogenase, Glutamine Synthetase, and Glutamate Synthase
    G. R. Stewart, A. F. Mann, and P. A. Fentem
8. Aminotransferases in Higher Plants
    Curtis V. Givan
9. Synthesis and Interconversion of Glycine and Serine
    A. J. Keys
10. Arginine Synthesis, Proline Synthesis, and Related Processes
    John F. Thompson
11. Synthesis of the Aspartate Family and Branched-Chain Amino Acids
    J. K. Bryan
12. Sulfur Amino Acids in Plants
    John Giovanelli, S. Harvey Mudd, and Anne H. Datko
13. Aromatic Amino Acid Biosynthesis and Its Regulation
    D. G. Gilchrist and T. Kosuge
14. Histidine Biosynthesis
    B. J. Miflin
15. Amino Acid Catabolism
    Mendel Mazelis

Contents of Other Volumes

16. Transport and Metabolism of Asparagine and Other Nitrogen Compounds within the Plant
    P. J. Lea and B. J. Miflin
17. Accumulation of Amino Acids and Related Compounds in Relation to Environmental Stress
    G. R. Stewart and F. Larher

Index

Volume 6—Proteins and Nucleic Acids

1. The Nuclear Genome: Structure and Function
   William F. Thompson and Michael G. Murray
2. Enzymatic Cleavage of DNA: Biological Role and Application to Sequence Analysis
   S. M. Flashman and C. S. Levings III
3. RNA: Structure and Metabolism
   T. A. Dyer and C. J. Leaver
4. Biosynthesis of Nucleotides
   Cleon W. Ross
5. DNA and RNA Polymerases
   Tom J. Guilfoyle
6. Nucleic Acids of Chloroplasts and Mitochondria
   Marvin Edelman
7. Proteins of the Chloroplast
   Katherine E. Steinback
8. Plant Proteinases
   C. A. Ryan and M. Walker-Simmons
9. Proteinase Inhibitors
   C. A. Ryan
10. Lectins in Higher Plants
    Halina Lis and Nathan Sharon
11. Seed Storage Proteins: Characterization and Biosynthesis
    Brian A. Larkins
12. Protein Biosynthesis: Mechanisms and Regulation
    Donald P. Weeks
13. Tumor Formation in Plants
    M. P. Gordon
14. Biochemistry of Plant Viruses
    George Bruening

Index

Volume 7—Secondary Plant Products

1. The Physiological Role(s) of Secondary (Natural) Products
   E. A. Bell
2. Tissue Culture and the Study of Secondary (Natural) Products
   Donald K. Dougall
3. Turnover and Degradation of Secondary (Natural) Products
   Wolfgang Barz and Johannes Köster
4. Secondary Plant Products and Cell and Tissue Differentiation
   Rolf Wiermann
5. Compartmentation in Natural Product Biosynthesis by Multienzyme Complexes
   Helen A. Stafford

6. Secondary Metabolites and Plant Systematics
   David S. Seigler
7. Stereochemical Aspects of Natural Products Biosynthesis
   Heinz G. Floss
8. Nonprotein Amino Acids
   L. Fowden
9. Amines
   T. A. Smith
10. Coumarins
    Stewart A. Brown
11. Phenolic Acids
    G. G. Gross
12. Enzymology of Alkaloid Metabolism in Plants and Microorganisms
    George R. Waller and Otis C. Dermer
13. Biosynthesis of Plant Quinones
    E. Leistner
14. Flavonoids
    Klaus Hahlbrock
15. Lignins
    Hans Grisebach
16. Cyanogenic Glycosides
    E. E. Conn
17. Glucosinolates
    Peder Olesen Larsen
18. Vegetable Tannins
    Edwin Haslam
19. The Betalains: Structure, Biosynthesis, and Chemical Taxonomy
    Mario Piattelli
20. Phenylalanine Ammonia-Lyase
    Kenneth R. Hanson and Evelyn A. Havir
21. Oxygenases and the Metabolism of Plant Products
    V. S. Butt and C. J. Lamb
22. Transmethylation and Demethylation Reactions in the Metabolism of Secondary Plant Products
    Jonathan E. Poulton
23. Glycosylation and Glycosidases
    Wolfgang Hösel

Index

Volume 8—Photosynthesis

1. Thylakoid Membrane and Pigment Organization
   R. G. Hiller and D. J. Goodchild
2. Photosynthetic Accessory Proteins with Bilin Prosthetic Groups
   A. N. Glazer
3. Primary Processes of Photosynthesis
   P. Mathis and G. Paillotin
4. Photosynthetic Electron Transport and Photophosphorylation
   Mordhay Avron
5. Photosynthetic Carbon Reduction Cycle
   S. P. Robinson and D. A. Walker
6. The C_4 Pathway
   G. E. Edwards and S. C. Huber

Contents of Other Volumes

7. Crassulacean Acid Metabolism
    C. B. Osmond and J. A. M. Holtum
8. The C_2 Chemo- and Photorespiratory Carbon Oxidation Cycle
    George H. Lorimer and T. John Andrews
9. Chlorophyll Biosynthesis
    Paul A. Castelfranco and Samuel I. Beale
10. Development of Photosynthetic Function during Chloroplast Biogenesis
    J. W. Bradbeer
11. Light-Energy-Dependent Processes Other than CO_2 Assimilation
    J. W. Anderson

Index

Volume 9—Lipids: Structure and Function

1. Analysis and Structure Determination of Acyl Lipids
    Michael R. Pollard
2. β-Oxidation of Fatty Acids by Specific Organelles
    Helmut Kindl
3. Oxidative Systems for Modification of Fatty Acids: The Lipoxygenase Pathway
    Brady A. Vick and Don C. Zimmerman
4. Lipases
    Anthony H. C. Huang
5. The Biosynthesis of Saturated Fatty Acids
    P. K. Stumpf
6. Biochemistry of Plant Acyl Carrier Proteins
    John B. Ohlrogge
7. Biosynthesis of Monoenoic and Polyenoic Fatty Acids
    Jan G. Jaworski
8. Triacylglycerol Biosynthesis
    Sten Stymne and Allan Keith Stobart
9. Galactolipid Synthesis
    Jacques Joyard and Roland Douce
10. Sulfolipids
    J. Brian Mudd and Kathryn F. Kleppinger-Sparace
11. Lipid-Derived Defensive Polymers and Waxes and Their Role in Plant–Microbe Interaction
    P. E. Kolattukudy
12. Lipids of Blue-Green Algae (Cyanobacteria)
    N. Murata and I. Nishida

Index

Volume 10—Photosynthesis

1. The Molecular Basis of Chloroplast Development
    J. Kenneth Hoober
2. Composition, Organization, and Dynamics of the Thylakoid Membrane in Relation to Its Function
    J. Barber
3. Rubisco: Structure, Mechanisms, and Prospects for Improvement
    T. John Andrews and George H. Lorimer

4. The CO_2-Concentrating Mechanism in Aquatic Phototrophs
   Murray R. Badger
5. Biochemistry of C_3–C_4 Intermediates
   Gerald E. Edwards and Maurice S. B. Ku
6. Control of Photosynthetic Sucrose Formation
   Mark Stitt, Steve Huber, and Phil Kerr

Index

Volume 11—Biochemistry of Metabolism

1. Introduction: A History of the Biochemistry of Plant Respiration
   David D. Davies
2. Control of Metabolism
   H. Kacser
3. Enzyme Regulation
   Jacques Ricard
4. The Regulation of Glycolysis and the Pentose Phosphate Pathway
   Les Copeland and John F. Turner
5. Control Involving Adenine and Pyridine Nucleotides
   Philippe Raymond, Xavier Gidrol, Christophe Salon, and Alain Pradet
6. Electron Transfer and Oxidative Phosphorylation in Plant Mitochondria
   Roland Douce, Renaud Brouquisse, and Etienne-Pascal Journet
7. Regulation of Mitochondrial Respiration
   Ian B. Dry, James H. Bryce, and Joseph T. Wiskich
8. Metabolism of Activated Oxygen Species
   Erich F. Elstner
9. Folate Biochemistry and the Metabolism of One-Carbon Units
   Edwin A. Cossins

Index

Volume 12—Physiology of Metabolism

I. Cellular Organization
1. Comparative Biochemistry of Plant and Animal Tubulins
   Peter J. Dawson and Clive W. Lloyd
2. Subcellular Transport of Metabolites in Plant Cells
   Hans Walter Heldt and Ulf Ingo Flügge
3. Compartmentation of Plant Metabolism
   T. ap Rees
4. The Role of Calcium in Metabolic Control
   E. F. Allan and A. J. Trewavas

II. The Metabolism of Organs and Tissues
5. Temperature and Metabolism
   Brian D. Patterson and Douglas Graham
6. Metabolic Responses to Stress
   David Rhodes

7. Plant Responses to Wounding
   Eric Davies
8. Fruit Ripening
   G. A. Tucker and D. Grierson

Index

Volume 13—Methodology

1. Immunochemistry for Enzymology
   Nicholas J. Brewin, David D. Davies, and Richard J. Robins
2. The Use of Mutants for the Study of Plant Metabolism
   P. McCourt and C. R. Somerville
3. The Use of Plant Cell Cultures in Studies of Metabolism
   M. C. J. Rhodes and Richard J. Robbins
4. The Application of Mass Spectrometry to Biochemical and Physiological Studies
   John A. Raven
5. NMR in Plant Biochemistry
   Justin K. M. Roberts
6. Electron Spin Resonance
   R. Cammack

Index

Volume 14—Carbohydrates

1. Hexose Phosphate Metabolism by Nonphotosynthetic Tissues of Higher Plants
   T. ap Rees
2. Recent Advances in Sugar Transport
   W. J. Lucas and M. A. Madore
3. Ascorbic Acid and Its Metabolic Products
   Frank A. Loewus
4. Fructans
   C. J. Pollock and N. J. Chatterton
5. Structure and Chemistry of the Starch Granule
   Keiji Kainuma
6. Biosynthesis of Starch and Its Regulation
   Jack Preiss
7. Starch Degradation
   Martin Steup
8. Structure and Function of Plant Cell Walls
   A. Bacic, P. J. Harris, and B. A. Stone
9. Biosynthesis of Plant Cell Walls
   D. P. Delmer and B. A. Stone
10. Structure and Biosynthesis of Plant N-Linked Glycoproteins
    G. P. Kaushal, T. Szumilo, and A. D. Elbein
11. Recent Progress in α-Amylase Biosynthesis
    T. Akazawa, T. Mitsui, and M. Hayashi

Index